A Dictionary of

Geography

FOURTH EDITION

SUSAN MAYHEW

OXFORD
UNIVERSITY PRESS

OXFORD
UNIVERSITY PRESS

Great Clarendon Street, Oxford OX2 6DP

Oxford University Press is a department of the University of Oxford.
It furthers the University's objective of excellence in research, scholarship,
and education by publishing worldwide in

Oxford New York

Auckland Cape Town Dar es Salaam Hong Kong Karachi
Kuala Lumpur Madrid Melbourne Mexico City Nairobi
New Delhi Shanghai Taipei Toronto

With offices in

Argentina Austria Brazil Chile Czech Republic France Greece
Guatemala Hungary Italy Japan Poland Portugal Singapore
South Korea Switzerland Thailand Turkey Ukraine Vietnam

Oxford is a registered trade mark of Oxford University Press
in the UK and in certain other countries

Published in the United States
by Oxford University Press Inc., New York

First published 1992 as *The Concise Oxford Dictionary of Geography*
First edition (with Anne Penny) 1992
Second edition 1997
Third edition 2004
Fourth edition 2009

British Library Cataloguing in Publication Data
Data available

Library of Congress Cataloging in Publication Data
Data available

Typeset by SPI Publisher Services, Pondicherry, India
Printed in Great Britain
on acid-free paper by
Clays Ltd., St Ives plc

ISBN 978-0-19-923180-5 (pbk)
     978-0-19-923181-2 (hbk)

How to use this book

This dictionary provides coverage in one volume of the terms used in both human and physical geography. There are over 6300 definitions across the following fields: cartography, surveying, remote sensing, statistics, meteorology, climatology, biogeography, ecology, simple geology, soils, geomorphology, population, migration, settlement, agriculture, industry, transport, development, and diffusion.

Headwords are printed in bold type and appear in alphabetical order. However, some entries contain further definitions. These have been included under the headword to avoid unnecessary repetition and to indicate some of the wider applications of the headword. Many entries have an asterisk. This points the reader to cross-references. In a very few cases the cross-reference as indicated does not have the exact wording as in the entry, but is close enough to make further reference possible.

References to geographical authors include surname, initials, and date. The use of these details in a card or electronic library catalogue will yield the book referred to.

Most journals are easily identifiable by their abbreviations, less simple abbreviations include:

AAAG - Annals of the Association of American Geographers
ESPL - Earth Science & Planetary Letters
TIBG - Transactions of the Institute of British Geographer

aa *See* BLOCK LAVA.

abiotic Not living, non-biological, usually describing factors in an ecosystem: atmospheric gases, humidity, salinity, soil mineral particles, water, and so on. An **abiotic environment** is one without any life. M. Gray (2003) argues that abiotic diversity is valuable, but threatened by human agency, which can destroy, damage, or pollute physical systems.

ablation Loss of snow and ice from a glacier by *sublimation, melting, and evaporation; and from the *calving of icebergs, and avalanches. In temperate and subpolar regions melting is the major form of ablation; in the Antarctic, it is calving. The rate of loss varies with air temperature, relative humidity, wind speed, insolation, aspect, and the nature of the surface. In snowfields, ablation includes snow removed by the wind, and is affected by aspect, depth of snow, and the nature of the underlying surface.

The **ablation sub-system**, where annual ablation exceeds annual accumulation, lies between the *firn line and the glacier snout. Where ablation occurs at the edges of glaciers, debris accumulates to form **ablation moraine/ablation till**. An **ablation valley** is a subsidiary valley formed beneath the crest of a lateral moraine and the valley side. Iturrizaga (2001) *Geogr. J.* **54** notes that 'true ablation valleys' can result from differences in *insolation, and may act in the development of lateroglacial moraine valleys.

aborigine A member of an indigenous people existing in a land before invasion or colonization from outside. For Canadian aboriginal peoples, see *Atlas of Urban Original Peoples*; for Australian, the Aboriginal Environments Research Centre, U. Adelaide; for Bolivian, Denevan (1966) *Nat. Tech. Info. Service*.

abrasion Also known as corrasion, this is the grinding away of bedrock by fragments of rock which may be incorporated in ice (*glacial abrasion), water (marine abrasion, *fluvial abrasion), or wind (*aeolian abrasion). In fluvial environments, the main agent of abrasion is the *bed load. The mass of solid material removed varies with the size, density, and velocity of the particles, and the density of the vector bearing these particles. See Sklar and Dietrich (2004) *Water Resour. Res.* **40** on a model for bedrock abrasion.

((⊕)) SEE WEB LINKS

• Sklar on the efficiency of particle abrasion in rivers.

abrasion platform *See* SHORE PLATFORM.

abrasion terrace A former *shore platform, now above sea level, either through *tectonic uplift of the mainland, or *eustatic lowering of sea level—and thus indicative of an emergent coastline. See Hanß in A. Alsharhan et al. (1998) for a discussion of coastal terraces.

absolute drought In the UK, a period of fifteen days on none of which more than 0.25 mm of rain falls. National definitions vary with climate; in Libya, droughts are recognized only after two years without rain. These arbitrary definitions give no indication of the impact of drought.

(⊕) SEE WEB LINKS

• Provides more definitions of drought.

absolute humidity The density of the water vapour present in a mixture of air and water vapour. Cold air cannot contain as much water vapour as warm air, so has a lower absolute humidity than warm air. Low absolute humidity results in more evaporation than in air with high absolute humidity. *See also* RELATIVE HUMIDITY.

absolute plate motion The movement of a crustal plate in relation to a fixed point, such as a hot spot. See Gripp and Gordon (1990) *Geophys. Res. Letts* **17** for calculations of absolute plate motion.

(⊕) SEE WEB LINKS

• Shows an absolute plate motion calculator.

abstraction The selection and conceptualization of a phenomenon, or some aspect of it; a way of viewing a real world object, usually a simplification—such as a *model. **Data abstraction** captures the essential information needed to describe a spatial phenomenon and is fundamental in *geographic information science.

See Smaalen in M. J. Kraak and M. Molenaar, eds (1996) for a hierarchical rule for geographic information abstraction.

abundance The total number of individuals of a certain species present in an area, generally estimated by sampling methods (such as capture-recapture), and which may vary according to competition, predation, and resources.

An **index of abundance** is a relative measure of the size of a population or sub-unit of the population, such as a year class. The abundance, for example, of fish may be measured as number/weight of fish caught per standard unit of fishing effort, in a given area, and for a given time span, and then compared to earlier data. See Pearce and Ferrier (2001) *Biol. Conserv.* **98**, 1 on modelling the relative abundance of species for regional conservation planning.

abyssal, abysso- At depths of over 3000 m below sea level, hence **abyssopelagic zone**—that part of deep lakes, oceans, or seas characterized by specific forms of plankton and nekton which inhabit open water—and

abyssobenthic zone—the bottom of a deep lake, ocean, or sea. An **abyssal plain** is the deep sea-floor, formed of abyssal deposits, with a gradient of less than 1 in 10000. The Canada Abyssal Plain, lying between Canada and the Alpha ridge, is the largest of the Arctic sub-basins, with an average depth of 3658 m. **Abyssal hills**, which interrupt an abyssal plain, are 50–250 m high. See Kitchell et al. (1978) *Palaeobiol.* **4**, 2 on **abyssal** traces and megafauna.

accelerator A factor which increases the momentum of a boom or slump in an economy, so that a small change in demand, for example, may lead to a greater industrial growth (or decline). Mehta and Theodore's study (in L. Simmons, ed. 2004) of 'temping' agencies as 'accelerators of churning in low-wage labour markets' is an interesting example.

access A. Sen (1999) measures **access to advantage** by one's access to basic needs: satisfying goods—like food; freedoms—as in a labour market; and capabilities. S. W. Allard (2004), in an examination of access to social services in three American cities, finds that poor populations in urban centres generally have greater spatial access to social services than poor populations in the suburbs, and that the potential demand for services is much greater in central city areas than in suburban areas.

accessibility The ease of approach to one location from other locations: in terms of the distance travelled, the cost of travel, or the time taken. Accessibility relies on location, where the relativity of space is estimated in relation to transport infrastructures (J.-P. Rodrigue 2006) and distance, derived from the *connectivity between locations. It is a key element in *transport geography.

Topological accessibility is measured in relation to a system of nodes and paths (transportation network), assuming that accessibility is significant only to specific elements of a transport system, such as airports or ports. **Contiguous accessibility** is a measurable attribute of every location, as space is considered in a contiguous manner. See Miller and Wu (2000) *GeoInformatica* **4**, 2) on *GIS software for **space-time accessibility** measures.

Physical accessibility is the spatial separation of people from the supply of goods and services; see, for example, Orcao and Diez-Cornago (2007) *Area* **39**, 3 on physical access to health services in Spain, and Gage and Calixte (2006) *Pop. Studs.* **60**, 3 on maternal health services in Haiti. **Social accessibility** is the ability of an individual to reach a resource or location, as affected by class structures, income, age, educational background, gender, or race. See Parks (2004) *Econ. Geog.* **80**, 2 on social accessibility in Los Angeles, and the UK Transport Studies Group's *Social Accessibility Mapping Project*.

accordant Complying with; thus, **accordant drainage** has evolved in conformity with the underlying geological structure: domes show a radial pattern, for example. The **law of accordant junctions** (Playfair's law) states that tributaries join a stream or river at the same elevation as that of the larger watercourse; there is no sudden 'drop' in the level of the tributary, which is therefore *graded to the level of the junction. See Niemann et al. (2001) *ESPL* **26**, 12 for a quantitative evaluation of Playfair's law. **Accordant summits** are hill or mountain tops of approximately the same elevation, whose presence has been seen as confirmation of the cycle of erosion theory (Ojany (1978) *Geo. J.* **2**, 5).

accreting margin *See* CONSTRUCTIVE MARGIN.

accretion **1.** The growth of land by the offshore deposition of sediment, forming *spits and *tombolos. Accretion is most active in *estuaries, particularly within the tropics. See Robin et al. (2007) *U. Caen Coastal Morphodynamics Gp.* on *swash bar and hook spit formation.
  2. The increase in size of a continent by the addition of **accretion terranes**. A **tectonostratigraphic terrane** is a fault-bounded geologic entity characterized by a distinctive geologic history that differs markedly from that of adjacent terranes (Hamilton (1990) *Phil. Trans. R. Soc. London A* **331**).
  3. The growth of a landform by the addition of deposits; *seif dunes grow by accretion. See H. S. Edgell (2006).
  4. The increase in size of particles by additions to the exterior, as in the formation of *hailstones. See Zheng and List (1995) *Procs Conf. Cloud Phys.*

acculturation The adaptation to, and adoption of, a new culture. This may occur simultaneously as two cultures meet, but occurs more often as an immigrant group takes to the behaviour patterns and standards of the receiving group. A major example is acculturation among African Americans; see H. Landrine and E. Klonoff (1996), and Sclemper (2007) *J. Hist. Geog.* **33**, 2 on acculturation at the regional scale.

accumulated temperature From a specific date, the length of time for which mean daily temperatures have been above, or below, a stated temperature; the total time for which temperatures varied from that standard. Bartholomew and Williams (2005) *Crop Sci.* **45** illustrate the response of cool-season grass to accumulated temperature.

accumulation **1.** The input of ice to a glacier. Weaver (1975) *Arct. & Alpine Res.* **7**, 3 suggests that it is greatest in shaded uplands. The **accumulation zone** (**accumulation sub-system**) is between the source and the *firn line, where the input of snow, firn, and ice exceeds *ablation. For snowfall in an accumulation zone to emerge as ice at the glacier snout takes about 100 years (Tuffen et al. (2002) *Sed. Geol.* **149**).
  2. The reinvestment of surplus value, in the form of capital, in order to increase that capital. Accumulation is a key feature of *capitalism because, in order to remain in business, capitalists have not only to preserve the value of their capital, but to add to it. 'The accumulation of [a] firm's net worth determines the growth rate of capital and the growth rate of the economy' (Chatelain (2004) *Econ. Letts.* **85**). D. Harvey (1982) argues that the obligatory

accumulation of capitalism has fostered *uneven development. Somel (2004) *ERC W. Paper* 0411 argues that the terms of trade between North and South help maintain a gap in capital accumulation. See also Hart-Landsberg and Burkett (2006) *Hist. Materialism* **14**, 3 on China and the dynamics of transnational accumulation.

Smith's views on accumulation and phases of capitalist development (2005, *Hist. Materialism* **13**, 4) are contested by O'Brien (2007) *Hist. Materialism* **15**, 1.

acidification A soil-forming process whereby organic acids (from humus) increase hydrogen ion concentration, as in the transformation of *brown-earth soils into acid brown earth, in humid temperate forest regions. For the Model of Acidification of Groundwaters in Catchments (MAGIC), see Cosby (1995) *Hydrol. & Earth Sys. Scis* **2**, 4. See also Koptsik et al. (2001) *Water Air & Soil Poll.* **139** on soil acidification in boreal forests.

acid rain When *fossil fuels are burned, dioxides of sulphur and nitrogen are released into the air; these dissolve in atmospheric water to form acid rain. In addition, nitrogen oxides combine with volatile organic compounds to form ground-level smog. These pollutants lead to the acidification of lakes and streams (making some of them incapable of supporting aquatic life), impair visibility, weaken forests, and degrade buildings.

Industrial development has discharged increasing quantities of such atmospheric pollutants; in the USA, the electric power industry accounts for around 70% of total annual SO_2 emissions and slightly over 20% of total annual Nitrogen Oxide emissions (US EPA Acid Rain Prog. 2006 Prog. Rept). See Xu and Sudo (1997) *Fuel & Energy Abstr.* **38**, 4 on acid rain in China. Oxides of sulphur and nitrogen may also be 'exported'; more than half of Taiwanese acid rain is from mainland China (Govt. Info. Office, Rep. China (Taiwan)). However, the UK Environment Agency and the US Environmental Protection Agency both record sharp falls in atmospheric sulphur dioxide. See also K. Satake (2000).

acid soil Soil with a pH under 7, such as *podzols and *brown earths. Acidity in a soil may be due to the *leaching out of cations when *precipitation exceeds *evapotranspiration. Other factors include the nature of the vegetation (and thus the *humus) and of the parent rock. Adams and Evans (1989) *Eur. J. Soil Sci.* **40**, 3 propose a model of soil acidification.

acre A unit of area, defined in British law as 4840 square yards (about 0.4 hectares).

actant A person, creature, or object playing an active role. Smith (2007) *PHG* **31**, 4 claims that actant networks 'fold the spaces and times of cities in ways that question the privileging of geometrical space . . . and linear time . . . in explanations of global and world cities'.

actinometer A device that may be used to record *insolation at the earth's surface.

action space The area in which an individual moves and makes decisions about her or his life. 'The living city is the principal locus of action space and enabler of social change, as well as the source of fundamental concepts in economic theory' (Ikeda (2007) *Rev. Austrian Econ.* **20**, 4). See also Susilo and Kitamura (2005) *Transport. Res. Record* 1902/2005.

active layer A highly mobile layer, periodically thawing, located above the *permafrost in *tundra regions, and ranging in depth from a few centimetres to 3 m. Its thickness depends on slope angle and aspect, drainage, rock and/or soil type, depth of snow cover, and ground-moisture conditions; clearing vegetation will increase the depth and mobility of the active layer. See Zhang et al. (2006) *Nat. Snow & Ice Data Center* on active layers in the Russian Arctic, and M. Lecki (2006). The mobility of the active layer is due to the tundra vegetation, which scarcely binds it together, so that it moves on slopes as gentle as 2°. Thawing may occur daily or only in summer; on refreezing, the active layer may expand, especially if *silt-sized particles predominate.

  *Periglacial processes in the active layer include *frost heaving, *frost thrusting, *ice wedging, *gelifluction, and the formation of *patterned ground. *See* THERMOKARST.

active margin Also called Pacific margin, this is a type of destructive *plate margin characterized by *ocean trenches, *earthquakes, *andesitic *volcanic chains, and young *fold mountains (Schellart and Lister (2005) *Earth & Planet. Sci. Lett.* **231**).

active volcano A volcano known to have erupted in recent times, or which is likely to erupt. Examples include Mt Etna, Sicily; see Behncke et al. (2004) *Geomorph.* **63**, 3–4.

activity segregation The *spatial separation of the sexes during the working day, as in a mining village, where the mines were an exclusively male preserve and the kitchen an exclusively female one. 'The gender division of labour helps to reinforce activity segregation where men and women . . . use space differently and unequally . . . this has meant that men are free to inhabit the public sphere and public spaces, while women's lives have been restricted to the private spaces of the home' (M. Longan (no date) U. Valparaiso).

activity space The space we live in from day to day; that part of *action space with which an individual interacts daily. There seems to be a hierarchy of activity spaces for most people, increasing in spatial extent: from family space, to *neighbourhood, economic space, and then urban space. Newsome et al. (1998) *Transport.* **25**, 4 present a model of urban activity space.

actor–network theory (ANT), insists that researchers should refute all pre-given distinction between classes of possible actors, rejecting artificial divides such as local/global, agency/structure, natural/social, and so on, and focusing on network building and network consolidation. Humans and non-humans alike are treated as possible actors; each giving meaning to any situation. The basis of every action and decision depends on the actor's subjective interpretations, which are made and mutually adjusted in interaction with others.

Smith (2003) *PHG* **27**, 5 argues that neither global networks, nor space and time, are static and fixed, but are made, remade, and unmade. He describes the way, with ANT, global cities can be viewed and researched as intermediaries in networks. For a clear demonstration of the application of ANT to a geographical topic, see Johannessen (2005) *Tourist Studs.* **5**, 2.

adaptive environmental management A management style that sees technocentric resource extraction as based on irrational, biophysically oversimplified, and short-term tactics. It recognizes non-linear feedback, and dynamic and unpredictable effects, and stresses a pragmatic and heuristic approach. See Noble in B. Mitchell, ed. (2001).

adaptive radiation 'The evolution of diversity within a rapidly multiplying lineage. It can cause a single ancestral species to differentiate into an impressively vast array of species inhabiting a variety of environments' (D. Schluter 2000). Kassen et al. (2004) *Nature* **431**, 7011 argue that the ecological gradient may limit the size of adaptive radiations.

adiabat A line plotted on a thermodynamic diagram, usually on a *tephigram, showing as a continuous sequence the temperature and pressure states of a parcel of air with changing height. An **adiabatic change** is a change in temperature, pressure, or volume, involving no transfer of energy to or from another material or system. In an adiabatic process, compression is accompanied by warming, and expansion by cooling. An **adiabatic temperature change** thus results from a pressure change. The speed at which the temperature of rising air falls with altitude is the **adiabatic lapse rate**. Dry, rising air expands with height. The energy needed for this expansion comes from the air itself in the form of heat.

The resulting change in temperature is expressed in the equation:

$$\frac{Dt}{dz} = \frac{-g}{Cp}$$

where Dt is the temperature change, g is the acceleration due to gravity, dz is the height change, and Cp is the specific heat of the air parcel. This change is the **dry adiabatic lapse rate** (DALR): 9.84 °C/1000 m. The temperature change sustained by any parcel of dry air is calculated using Poisson's equation.

If the rising air becomes saturated to *dew point, condensation of vapour will begin. This condensation is accompanied by the release of *latent heat, which partly offsets the cooling with height, so that the rate of cooling of moist air—the **saturated adiabatic lapse rate** (SALR)—is lower than the DALR. In the lower *troposphere, the vapour content of air is high so the latent heat of condensation is high; SALRs may be as low as 5 °C/1000 m. In the cold, dry, high troposphere, though, there is little vapour ready for condensation, so the SALR may be close to the DALR. Quantitative expressions of the SALR are therefore quite complex. See B. Haurwitz (2007).

administrative principle W. Christaller (1933; trans. C. W. Baskin 1966) proposes that, in a region with a highly developed system of central administration, settlement is so arranged that one major centre administers six centres of lesser rank, each of which, in turn, oversees a further six centres. The number of settlements, from highest rank downwards thus follows the sequence 1, 7, 49, 343 . . . This hierarchy is known as k = 7.

k = 7

Administrative principle

adret The sunny slopes of a hill or valley side; warmer because they receive more *insolation. In the Northern Hemisphere adret slopes face south; in the Southern Hemisphere, north. Beckford and Barker (2007) *Geog. J.* **173**, 2 note that the Jamaican landscape categories 'front ridge' and 'back ridge' are a local tropical variant of adret and *ubac slopes.

adsorption In *soil science, the addition of ions or molecules to the electrically charged surface of a particle of clay or humus. In this way, minerals become bonded to soil particles (M. Ashman and G. Puri 2002). An **adsorption complex** is an 'organic and inorganic substance in soil that can absorb ions or molecules' (K. Peverill et al. 1999).

advanced capitalism A key feature of advanced capitalism is the possession of capital by fewer and fewer owners. M. Castells (2002) stresses the fundamental importance of collective consumption in advanced capitalism. See R. Hernandez (2002) on the mobility of workers under advanced capitalism.

advanced economy (economically advanced economy) 'A term used by the International Monetary FUND (IMF) for the top group in its hierarchy:

advanced economies, countries in transition, and developing countries. It includes: Australia, Austria, Belgium, Canada, Denmark, Finland, France, Germany, Greece, Hong Kong, Iceland, Ireland, Israel, Italy, Japan, South Korea, Luxembourg, Netherlands, NZ, Norway, Portugal, Singapore, Spain, Sweden, Switzerland, Taiwan, UK, and USA' (CIA (2007) *World Factbook, appendix B*). The IMF also includes Cyprus and Slovenia (IMF (2007) *WEO Groups & Aggregates Info.*).

advection The 'horizontal transport of energy and mass by parcels of air in convective flow systems' (Sivertsen (2007) *EMS7/ECAM8 Abstracts* 4). In meteorology, the term usually means the horizontal transfer of heat. **Advective transfers** occur in air streams, ocean currents, and surface *run-off, partly redressing the *insolation imbalance between the tropics and the poles. For **advection fog**, *see* FOG.

aeolian Of the wind: aeolian processes include erosion, transport, and deposition, and work best with sparse, or absent vegetation. Baas (2007) *Geomorph.* **91**, 3–4 describes aeolian sand transport as 'a classic dissipative process with non-linear dynamics'. **Aeolian landforms** include desert *zeugen and *yardangs. See I. Livingstone, ed. (1998).

aeration zone The zone between the soil-moisture zone and the capillary zone, just above the water-table, which acts as a dispenser and buffer in the infiltration of groundwater flow (Diankov and Nitcheva (2005) *Bulgarian. Acad. Scis*). See also Sobotovich et al. (1991) *Atom. Energy* **70**, 6 on the reduction of groundwater contamination since Chernobyl, 1986.

aerobic Describing any living organism which depends on atmospheric oxygen to release energy from foodstuffs during respiration. *Compare with* ANAEROBIC.

aerological diagram A chart plotting the factors which determine the movement of air. Variations of temperature, pressure, dry and saturated *adiabatic lapse rates, and *saturated mixing ratio lines are plotted against height in a *tephigram. An adiabatic chart may be used to predict the *convective condensation level.

aerosol A suspension of droplets or particles in a gas; more precisely, of particles with a maximum diameter of 1 μm (*fog and *mist are thus aerosols). In meteorology, the term is often used to describe the particles suspended within the air, such as minute fragments of sea-salt, dust, organic matter, and smoke. These enter the atmosphere by natural processes such as vulcanicity, and by human agency such as burning *fossil fuels. Aerosols absorb heat and may act as *condensation nuclei. NASA posts a daily global aerosol map on the web.

affect A non-conscious, but intense, experience. Shouse (2005) *MC J.* **8**, 6 explains that 'feelings are *personal* and *biographical*, emotions are *social*, and affects are *prepersonal*'—in other words, affect is our unconscious response which precedes our conscious feelings and decisions. 'The importance of affect', Shouse explains, is that 'in many cases the message consciously received may be of less import to the receiver of that message than his or her non-conscious affective resonance with the source of the message'. See Thrift (2004) *Geografiska B* **86**, 1, Ash and Thrift (2005) *Antipode* **37**, Tolia-Kelly (2006) *Area* **38**, 2, and Thien (2005) *Area* **37** for consideration of affect in geography.

afforestation Planting trees. Tree-planting can stabilize soils by increasing *interception and reducing *run-off, reduce flooding through the reduction of silting, improve soil fertility, and provide timber and firewood. See Zhang et al. (2007) *Indus. Rept.* 1/07, eWater CRC.

aftershock A smaller tremor or series of tremors occurring after an *earthquake. Aftershocks result from static stress changes and 'dynamic' stresses (transient, oscillatory stress changes), that can weaken faults. Such dynamically weakened faults may fail after the seismic waves have passed, and might even cause earthquakes that would not otherwise have occurred (Kilb et al. (2003) *Nature* **408**). See also Johnson et al. (2008) *Nature* **451**.

age dependency The dependency of people too young or too old to be employed full-time on the contributions of those in full employment. 'The actual dependency burden is determined not just by the sizes of the various functional age groups, but also by a host of social, cultural, and institutional factors' (Yap et al. (2005) *J. Cross-Cult. Gerontol.* **20**, 4).

ageing, geographies of Spatial variations in demographic ageing, and the distribution of the elderly at local, regional, and national scales, are key issues in geographies of ageing: see Ferry et al. (2006) *EPRC* on variations within the EU, and van Steen and Pellenbarg (2006) *Tijdschrift* **97**, 1 & 2 on the Netherlands.

Mulder (2007) *Pop. Space & Place* **13**, 4 notes that when elderly people change residence they more often move closer to their children than further away. 'However, adult children changing residence move towards or away from their parents equally frequently.' See also Williams et al. (2000) *Tourism Geog.* **2**, 1 on international retirement migration.

Sunley (2000) *TIBG* **25**, 4 reports a North–South divide in income and benefit take-up among pensioners in the UK, with lower private pension coverage in low-income areas. 'The social status of people throughout life is reflected in the way their deaths are perceived and in the practices concerned with their dying' (Lloyd (2004) *Ageing & Soc.* **24**, 2).

Agenda 21 A set of proposals, made at the 1992 UN Conference on Environment and Development, Rio de Janeiro, to promote *sustainable development. It sets out environmental strategies for managing coasts, oceans, and water, monitoring and reducing chemical waste, eradicating radioactive waste, and conserving natural vegetation and soils (through sustainable farming). Other proposals include measures to improve health care, reduce poverty, and develop fair and environmentally friendly trade policies. The aim is to promote global sustainable development at a more fundamental level than traditional aid programmes, based on common needs and interests. The requirement is for local governments to develop their own 'Local Agenda 21' in order to promote and implement sustainable development.

(((⊕))) **SEE WEB LINKS**

• Agenda 21 online.

agent An entity which can take a role; a function, service, or identity. Müller (2007) *Area* **39**, 2 calls the geographer 'an active agent who moulds the production of meaning'. Bondi (2005) *Antipode* **37**, 3 holds that neoliberalism envisages people as autonomous, individualized, self-directing, decision-making agents. 'Interdependencies between agents are constrained by social and spatial structure, but, over time, structure and agency are mutually constituted' (Plummer and Sheppard (2006) *J. Econ. Geog.* **6**).

age–sex pyramid A set of two histograms set back to back on a vertical axis, depicting the numbers of the two sexes by age group. In many less developed countries the pyramid will have a very wide base; these are termed **progressive pyramids** because they suggest future population growth. **Regressive pyramids** characteristic of advanced economies, where there are fewer children and more old people, are more cylindrical; population is likely to fall.

age structure The composition of a nation by age groups. The age structure of the UK population has become older, the median age rising from 34.1 years in 1971 to 38.4 in 2003 and projected to rise to 43.3 in 2031 (*UK National Statistics*). Europe, with more than 15% of the population over 60, suffers from *age dependency, while South Asia, with 45–55% of its population under 16, is concerned to limit population growth. *See* AGE–SEX PYRAMID.

agglomerate A coarse-grained volcanic rock, made of sharp/sub-angular lava fragments, set in a fine matrix. Some are *pyroclastic; others are deposits from *mud flows or *lahars.

agglomeration 1. Within a geographical area, a concentration of economic activity brought about by, among others, *external economies (such as a pool of skilled labour), economies of scale, *cumulative causation, local authority planning, and chance. 'The agglomeration of economic activities is a phenomenon as old as cities themselves' (E. Soja 2000). Phelps and Ozawa (2003) *PHG* **27**, 5 distinguish between proto-industrial, industrial, and post-industrial agglomerations, with no common, formative processes underlying each. Pflüger and Südekum (2008) *J. Econ. Geog.* **8**, 1 believe that agglomeration processes depend on specific technical properties; Roos (2005) *J. Econ. Geog.* **5**, 5 claims that the influence of geography on agglomerations is

agglomeration economies

12

Age–sex pyramid

small: 'at most, ~36% of the total variation can be explained by the effects of geography'. Fan and Scott (2003) *Econ. Geog.* **79**, 3 find a strong positive relationship between spatial agglomeration and productivity in China. See Feldman in G. Clark et al., eds. (2000).

2. In meteorology, the process by which cloud droplets grow by assimilating other droplets.

agglomeration economies For an industry, the benefits of locating in a densely peopled and highly industrialized situation, where the market is large, but spatially concentrated. Transport costs are thus low, and specialized industries can evolve, since local demand is high. Further benefits arise because of *functional linkages and *external economies. See Park (2002) *AI & Soc.* **16**, 3 on a Swedish Science Park which has intensified knowledge generation by facilitating technical innovation, education, and research.

Roos (2005) *J. Econ. Geog.* **5**, 5 claims that the effect of agglomeration economies is over five times larger than the effects of geography, while Viladecans-Marsal (2004) *J. Econ. Geog.* **4**, 5 shows that, in Spain, agglomeration economies influence the location of manufacturing activity. 'Empirical analyses often find that factor endowments are more powerful in accounting for trade and production patterns than are agglomeration economies, but that both are important' (Hanink and Cromley (2005) *AAAG* **95**, 3).

aggradation The deposition of unconsolidated deposits by *aeolian, marine, or *fluvial processes, when the quantity or calibre of the load is greater than the competence of the transporting medium to carry it, or as a result of *mass movement. See Daniels (2003) *Geomorph.* **56**, 3–4 on aggradational *pedogenesis on floodplains. For **aggradation surface**, see Wang and Grapes (2008) *Geomorph.* **96**, 1–2.

aggregate A group of soil particles held together by electrostatic forces, *polysaccharide gums, and *cementation by carbonates and iron oxides.

aggressivity In *karst geomorphology, the ability of water to dissolve calcium carbonate. See Staford et al. (2008) *Int. J. Speleol.* **37**, 2.

agrarian Describing an agricultural system which combines horticulture and animals.

agribusiness An integrated farming system, linking farm operations with upstream and downstream manufacturing and distribution: a total production system from inputs (such as seed and fertilizers) through cultivation on the farm to agro-processing and marketing. (Sutton (2001) *Sing. J. Trop. Geog.* **22**, 1). See Kristiansen (2002) *Sing. J. Trop. Geog.* **23**, 1 on small-scale agribusiness entrepreneurs in Indonesia, and Wilson and Thompson (2003) *Rev. Ag. Econ.* **25**, 1 on timing in agribusiness.

agricultural geography The study of spatial patterns in agricultural activity. 'Agriculture is an integral part of a wider food and fibre production system, encompassing input supply, farming, food processing, wholesaling and retailing' (G. Robinson 2004) Rural land-use patterns, cultural landscapes, and encroachment of urban sprawl continue to inform work in rural geography; see, for example, Hart (2003) *Geogr. Rev.* **93**.

   Antle and Stoorvogel in N. Heerink et al., eds (2001) discuss models of an agricultural system; see also Antle and Stoorvogel (2006) *Env. & Dev. Econ.* **11**. Other agricultural geographers are concerned with the relationship of agricultural systems and economic development; see Tsakok and Gardner (2007) *Am. J. Ag. Econ.* **89**, 5, but see also Evans et al. (2002) *PHG* **26**, 3.

agricultural revolution A period of rapid change in agriculture, usually associated with increases in output; for example, the English Agricultural Revolution, from 1750 onwards, when the use of rotations, nitrogen-fixing crops, and mixed farming with manuring increased output. See M. Overton (1996). *See also* GREEN REVOLUTION.

agricultural system Any farming type is an agricultural system. 'Inputs' include seeds, water, pesticides, herbicides, fertilizer, and livestock, and the 'plant' is the buildings, machinery, and land. See Antle and Stoorvogel (2006) *Env. & Dev. Econ.* **11**.

agroforestry Any agricultural system which includes planting, or encouraging, trees. Tree roots can bind soil and limit soil erosion, deep-rooted trees can tap new nutrient sources, leguminous trees can fix atmospheric nitrogen and improve soil fertility, leaf litter can add organic matter, and tree cover can moderate temperatures; trees may also provide food, fodder, firewood, and timber. See Dove (2004) *Agroforest. Sys.* **61**.

Agulhas current A warm *ocean current off south and south-eastern Africa. Its movements result from large-scale wind stress curl (Matano et al. (1999) *J. Phys. Oceanog.* **29**, 2).

aid Transferring resources from developed to less developed countries. **Bilateral aid** is from one donor to a recipient country, while **multilateral aid** comes from a group of countries. **Emergency aid** is short-term, generally given in response to disasters: 'windfalls in relief and reconstruction aid can also increase the risk of (perceived) fraud, corruption, mismanagement and dispossession by the government and its cronies, aggravating the plight of the most vulnerable' (LeBillon and Waizenegger (2007) *TIBG* **32**, 3). **Structural aid** is given for long-term *development; see Seguiti (2003) *Public Budget. & Finance* **23**, 2 on EU structural aid. A donor of **tied aid** can transfer resources more cheaply as its aid is tied to its own exports (Kemp (2005) *Pacif. Econ. Rev.* **10**, 3).
  Bebbington (2004) *PHG* **28**, 6 criticizes NGOs: 'their geographies of intervention do not reflect the geographies of poverty and livelihood in the Andes; and their strategies of intervention do not respond to the economic and spatial dynamics of poor people's livelihoods . . . they do less and less to address the deeper processes of . . . development that produce poverty and inequality of opportunity.' See Swain (2006) *TIBG* **31**, 2 on aid and market principles 'designed by the government of the United States and the international financial institutions . . . and implemented by, them in various ways, for the more vulnerable societies, often as stringent structural adjustment programs' (N. Chomsky 1999).

aiguille French for a *pyramidal peak, sharply pointed due to *nivation and *frost wedging.

air mass An area of the *atmosphere extending for hundreds of kilometres that, horizontally, has generally uniform properties, especially of temperature and *humidity, and with similar vertical variations of temperature and pressure throughout. Air masses obtain these attributes from their areas of origin, known as source regions, and become less uniform with movement into different areas, and with the effects of *wind shear.

Air masses are classified by the source region: **continental air mass**, **c**, or **maritime air mass**, **m**; and by the latitude of the source area: the arctic, **A**, the poles, **P**, and the tropics, **T**. These two combine to distinguish most air masses, such as **cA—continental arctic**, **mP—maritime polar**, **mT—maritime tropical**, and so on. **Equatorial**, **E**, and **monsoon**, **M**, tags also exist. Browell et al. (2003) *J. Geophys. Res.* **108**, D20 identify nine air mass types, quantifying their frequency as a function of altitude, and Chen et al. (2007) *J. Geophys. Res.* **112**, D10S05 classify air mass types along coastal New England. See Peterson et al. (1981) *J. Appl. Met.* **20**, 3 on turbidity and synoptic air masses. *See also* SECONDARY AIR MASSES.

air parcel An abstract volume of air. A simple forecast of the behaviour of an air parcel is made by assigning to it the relevant qualities, such as humidity or temperature, and then predicting its movements according to the laws of physics.

air pollution The presence of substances in the air in quantities which can affect animal or plant life, human health and welfare, or can unreasonably interfere with life or property. Common pollutants include carbon dioxide, carbon monoxide, lead, nitrogen oxides, ozone, smoke, and sulphur dioxide. Gulliver and Briggs (2004) *Atmos. Env.* **38**, 1 suggest that reducing car use by encouraging walking or cycling may increase exposure to air pollution: the longer journey times, compared to car use, would lead to raised overall exposures, unless cuts in the volume of traffic bring compensatory falls in pollution levels. See Gulliver and Briggs (2005) *Env. Res.* **97** on modelling traffic management plans, emission technologies, fuel pricing strategies, and transport policies.

(⊕) SEE WEB LINKS

• Description of types of air pollution.
• EU emission standards.

air pressure *See* ATMOSPHERIC PRESSURE.

air–sea interaction *See* OCEAN–ATMOSPHERE INTERACTION.

Aitken nuclei *See* NUCLEI.

alas In a *periglacial landscape, a large, steep-sided, flat-bottomed depression which can be several kilometres across. Lakes often occur in alases, which may then develop into *pingos.

Alaska current A warm *ocean current, an offshoot of the North Pacific current, washing the southern coasts of Alaska. See Royer (1981) *J. Mar. Res.* **39**.

albedo That proportion of solar radiation reflected from a surface, such as clouds (low from thin stratus cloud; up to 80% from thick strato-cumulus), or bare rock. Lighter, whiter bodies have higher albedos than darker, blacker bodies. Earth's total albedo is about 35%.

Land surface albedo is an important variable in General Circulation Models. Lofgren (1995) *J. Clim.* **8**, 10 finds that thermally induced circulations between oceans and land at low latitudes can be affected by the land surface albedo. See also Liang et al. (1999) *J. Appl. Met.* **38**, 6 on the retrieval of land surface albedo from satellite observations.

alcohol, geographies of Alcohol is at once a social problem, a leisure activity, a pleasure, an accelerator of violence, and central to identity formation (Jayne et al. (2008) *PHG* **32**, 2). In geographical studies of alcohol, drunkenness features heavily. It's good, therefore, to read that Eldridge and Roberts (2008) *Area* **40**, 3 challenge the urban savage/social drinker binary, 'which wrongly presumes consumers cannot be both'. In Europe, a broad division was made between a northern dry area, where beer is the leading beverage, consumed at weekends and outside mealtimes; and a southern wet area, where wine is the main beverage, usually drunk at meals. Subsequently, European alcohol choices have started to converge (Allamani et al. (2000) *Substance Abuse* **21**, 4).

P. Chatterton and R. Hollands (2003) see alcohol-related establishments as the driving force in recent city-centre regeneration; Roberts (2006) *Cities* **23** reports that cities at night now attract vast numbers of people. See also D. Hobbs et al. (2003). However, Roberts et al. (2006) *Urb. Studs* **43**, 7 conclude that British free market attitudes to extended licensing hours will undermine the government's professed aspirations for an 'urban renaissance' of cultural inclusion and animation. Smith (2003) *Tijdschrift* **94**, 4 makes the expected link between alcohol and deprivation in the city—where alcohol use is more visible. Working-class cities are deemed to be places for excessive partying and wild stag or hen nights (Chatterton and Hollands (2001) *U. Newcastle on Tyne*), and a drinking culture remains firmly tied to the North-East industrial heritage (Nayak (2003) *Env. & Plan. D* **21**, 1). Shaw, (2000) *Australian Geogr. Studs* **38**, 3 observes that when Aboriginality was conflated with alcoholism, it tended to be exoticized and romanticized in its traditional 'bush' setting, but demonized in its urban setting. Kokorev (1997, cited in Round (2006) *Geografiska B* **88**, 1) finds that Magadan Oblast, in Russia's far north-east, has the country's highest alcohol consumption rates; Magadan is the location of a notorious gulag. Henry (2006) *Paper IGU Conf.* argues that differences between a city Central Business District in Hobart, Tasmania, where alcohol is banned, and the Wine and Food Festival, where it is not, result from the 'different ontological claims made about their spaces'.

S. Winlow and S. Hall (2006) discuss alcohol's central connection to interpersonal violence and disorder; see also Silverman and Della-Giustina (2001) *Urban Studies* **38**. Tatlow et al. (2000) *J. Community Health* **25** demonstrate a clear link between the density of shops selling alcohol and alcohol-related hospital admissions in San Diego county. Alavaikko and Osterberg (2000) *Addiction*, 95 (suppl. 4) consider the way the increasingly globalized alcohol supply and international supply-side initiatives impact on strategies to reduce alcohol-related harm, and how international trade treaties affect alcohol regulation and policy.

alcove A semicircular, steep-sided cavity on a rock outcrop, caused by water erosion, especially spring-sapping. See Reneau (2000) *Geomorph.* **32**, 1 for photographs.

Aleutian current A cold ocean current of the North Pacific. The **Aleutian trench** lies at the boundary of the North American and Pacific plates; try Choi (1990) *Pac. Geol.* **5**.

alfisol A soil order of the *US soil classification. Alfisols are young, acid soils, with a clay-rich B horizon, commonly occurring beneath deciduous forests. *See* BROWN EARTH.

algae A diverse group of simple plants that contain chlorophyll and can photosynthesize. An **algal bloom** is a dense spread of algae caused by changes in water chemistry and/or temperature; Shukla et al. (2008) *Appl. Maths & Comput.* **196**, 2 find a positive correlation between increasing nutrients in water and algal density—leading to *eutrophication. Moran et al. (2001) *CSIRO Land & Water Tech. Rept.* 20/01 question the proposed link between phosphorus entering channel systems and algal blooms.

alienation The estrangement or separation of individuals from one another, said to include isolation, powerlessness, and meaninglessness. K. Marx (1844, 1975) saw this as a result of separating the workers from the benefits of industrialization as enjoyed by the owners of the means of production. L. Wirth (1938) saw alienation as a product of urban life. Others cite political causes; see *Human Rights Watch*, 20 December 2000, on Israeli Arab students' feelings of alienation from their education; McEwan (2003) *Geoforum* **34**, 4 on cultures of alienation amongst black women in South Africa, and Davies and Niemann (2002) *New Pol. Sci.* **24**, 4 on alienation—especially as it relates to globalization—in everyday life. A major contemporary theme is the alienation of some Muslims from 'Western' lifestyles; 'young British Muslims are increasingly found to be in the precarious position of having to choose their loyalties, being impacted by radical Islamism on the one hand, and . . . British multicultural citizenship on the other' (Abbas (2007) *J. Intercult. Stud.* **28**). See also Polat (2004) *Midwest Polit. Sci. Assoc.* on Muslim alienation and modernity.

alkaline soil Any soil, such as a *rendzina, which has a pH above 7. Alkalinity usually reflects a high concentration of carbonates, notably of sodium and calcium.

Alleghanian orogeny An *orogeny from the Early Carboniferous to the late Permian, caused by a collision between North America and Africa, affecting an area from Alabama to Newfoundland, and forming the Appalachians. See Fain (1987) *GSA Centennial Field Guide.*

Allerød A north-west European *interstadial, from *c.*12000 to 10800 years BP; in Britain not usually distinct from the *Bølling. Pollen records point to a cool temperate flora, with birch. See Willard et al. (2007) *Pal. Pal. & Pal.* **251**.

allochthonous Rock formed of materials brought from outside, for example from sediments transported by ice; *till is an example. *Compare with* AUTOCHTHONOUS.

allogenic Caused externally. Thus, an **allogenic stream** is one—for example, the Nile—fed from outside the local area, where precipitation and run-off generate the flow. **Allogenic succession** is an alteration in a plant *succession brought about by some external factor.

allometry The study of the relative rates of change of two variables of a system. See Bull, (2003) *GSA Bull.* on allometric change of landforms, Hood (2002) *Restoration Ecol.* **10**, 2 on landscape allometry and the restoration of tidal channels, and Evans and McClean (1995) *Zeitschrift. Suppl.-Band*, **101** on *cirque scale and allometry.

alluvial Referring to *fluvial contexts, processes, and products. The form of an **alluvial channel** (cut in alluvium) reflects the *load and discharge of the river rather than the *bedrock. Dade (2000) *Geomorph.* **35**, 1 relates sediment grain size and transport mode to the channel pattern of a river. An **alluvial fan** is a triangular landform; its shape results from the river swinging back and forth during *aggradation while the apex is fixed at the point where the river emerges from the uplands. Sanderson et al. (2000) *GSA Bull.* **112**, 12 suggest that fan aggradation results from anthropogenic and climatic forcing. See also Pope et al. (2008) *Geomorph.* **94**, 1–2. An **alluvial cone**, similar in origin, tends to have a steeper slope.

 Alluvial filling is *sedimentation from fluvial channels. During low water, **alluvial flood plains** are mostly dry, but when most of the *flood plain is swamped, coarse material is transported short distances, temporarily deposited, and picked up during the next flood. Major controls on alluvial flood plains include sediment, flow regime, and differing base levels (O'Connor et al. (2003) *Geomorph.* **51**, 1–3). **Alluvium** is a general term for deposits laid down by present-day rivers, characterized by sorting (coarser alluvium is found in the upper course of rivers; finer in the lower courses), and *stratification (coarse material is overlain by finer). **Alluvial landforms** include *deltas, river beds, and flood plains.

alluviation The accumulation of sediment. See DeLong et al. (2007) *Geomorph.* **90**, 1–2. See Blum and Törnqvist (2000) *Sedimentol.* **47**, Suppl. on change in sediment yield and discharge, and alluviation during relative sea-level fall. **Cut and fill alluviation** results in a shifting channel eroding old surfaces and creating new surfaces (Hauer et al. (2007) *J. Am. Water Resources Ass.* **43**, 1).

Alonso model William Alonso's 1964 explanation of urban land use and land values is grounded on the concept of *bid rents. If the level of goods and services is constant, land prices should decrease with distance from the centre. The poor will live near the city centre, where high housing densities more than offset high land costs; the better-off will live at lower densities at the edge of the city (on more, but cheaper, land). Each household represents a balance between land, goods, and *accessibility to the workplace. However, commuting costs rise with distance from the centre, reducing disposable income, and higher-income groups may prefer the accessibility to the *CBD offered by the inner city to the space, quiet, and cheaper land of the suburbs, thus causing *gentrification. Hua (2001) *Int. Reg. Sci. Rev.* **24**, 3 reviews and re-presents Alonso's model.

alpha index (α index) A measure of connectivity using the ratio of the number of circuits in a network to the maximum possible in that network. The higher the alpha index, the more a network is connected. A value of 1 indicates a completely connected network, but 'it is very rare that a network will have an alpha value of 1, because this would imply very serious redundancies' (J.-P. Rodrigue 2006). The alpha index is given as:

$$\alpha = \frac{u}{2v - 5}$$

where u is the maximum number of independent cycles in a graph, estimated through the number of nodes (v), links (e), and of sub-graphs (p); $u = e - v + p$.

Alps A chain of *fold mountains running from eastern France and northern Italy through Switzerland to Austria. An **alpine glacier** is a valley glacier, and **alpine topography** describes features of upland glaciation: *cirques, *glacial troughs, *hanging valleys, and *truncated spurs. The **alpine orogeny** occurred in the *tertiary, creating fold mountains as far apart as the Andes, Japan, and East Indian Ocean *island arcs. See Rosenbaum (2002) *Tectonophys.* **359** on the movements of Africa and Europe during the Alpine orogeny.

alto- Prefix referring to *clouds between 3000 and 6000 m high.

ambient Surrounding, of the surroundings. **Ambient temperature** is the temperature surrounding a given point. The **ambient air standard** is a quality standard; see J. Stedman et al. (2005), DEFRA, on ambient air quality assessment in the UK.

ambivalence In *post-colonial theory, 'the mixture of charm and revulsion in the relationship between the colonizer and the colonized' (H. Bhabha 2004). It is inherent in all colonial relationships; colonizers don't care to be copied exactly, and ambivalence can lead to *hybridization in the colonizers ('going native'). Jorgenson and Tylecote (2007) *Landsc. Res.* **32**, 4 use the concept in exploring urban responses to 'areas in which the spontaneous growth of vegetation through natural succession suggests that nature is in control'.

amenity Pleasantness; those aspects of an area such as housing, space, and recreational and leisure activities which make it attractive to live in. **Amenity migration** is the purchase of first or second homes in rural areas prized for their aesthetic and/or recreational values (McCarthy (2008) *PHG* **32**, 1). It has been aided by rising incomes, the easing of limits on foreign property ownership in many countries, developments in transport and communications technologies, and increasing exposure to images of prized rural landscapes. 'Amenity migration and much associated tourism is driven at least in part by temporarily deterritorialized and significantly globalized visions of the rural' (P. Cloke et al. 2006).

amphidromic point The point around which *tides oscillate; while there are tides along the coasts of East Anglia and the Netherlands, there is a point in the sea between the two where there is no change in the height of the water. High water rotates around the amphidromic point; anticlockwise in the Northern Hemisphere, clockwise in the Southern.

ana- Rising. An **anabatic wind** is an upslope wind driven by heating—usually daytime insolation—at the slope surface under fair-weather conditions (AMS (1959) *Glossary of Meteorology*; which see). These winds tend to be less frequent, and weaker, than their corresponding, descending, *katabatic winds.

anabranching channel A *distributary channel which leaves the main channel, sometimes running parallel to it for several kilometres, and then rejoins it; a channel 'separated by vegetated semi-permanent alluvial islands, excised from an existing floodplain, or formed by within-channel or deltaic accretion' (Nanson and Knighton (1998) *ESPL* **21**, 3). Jansen and Nanson (2004) *Water Resources Res.* **40**, W04503 show that anabranching can sustain a stable river pattern where a single channel could not transport sediment. Anabranching channels differ from *anastomosing channels in that they are undivided.

anaerobic Describing any organism or process which can or must exist without free oxygen from the air, such as the anaerobic bacteria which are responsible for the process of *gleying.

ana-front A front at which the warm air is ascending. An ana-front is normally the initial state of a cold front, developing into a kata-front as the depression becomes more occluded. See Sansom (1951) *Q. J. Royal Met. Soc.* 77.

anastomosis The division of a river into a stable multi-channel system with levées, backswamps, and large, stable islands. An **anastomosing channel**, unlike an *anabranching channel, has distributaries of its own. Valley incision starts at outflow points; up-valley incision begins where the flow rejoins the main channel (Schumm et al. (1996) *GSA Bull.* **108**). In time, the channel segments join, making a new anabranch of the anastomosing-channel system. Many anastomosed systems cannot be categorized; one common factor seems to be low specific stream powers, thus explaining the river's inability to maintain channel capacity via sediment transport and erosion; see Makaske (1998) *Earth Sci. Rev.* **53**.

anchor tenant A pioneer tenant in a new development. A state may promote the 'anchor' with preferential treatments in terms of land prices, tax exemptions, infrastructure support, and administrative services, in order to persuade the downstream and upstream firms along its supply chain to follow suit (Lee and Saxenian (2008) *J. Econ. Geog.* **8**, 2).

andesite A fine-grained volcanic rock. The **andesite line** is the boundary between the rocks of the oceanic *crust and islands and the rocks of the continental crust in a circum-Pacific belt; the boundary between oceanic *sima and continental *sial.

angle of repose The angle at which granular material comes to rest, varying inversely with fragment size in perfectly sorted materials, but directly in those imperfectly sorted; inversely with fragment density; directly with fragment angularity, roughness, and compaction; inversely with height of fall of material on free cones; and directly with moisture content up to saturation point, but inversely thereafter (van Burkalow (1945) *GSA Bull.* **56**, 6).

animal geographies Animals occupy human social life and space in many ways, yet, it is argued, humans have turned them into commodities (J. Wolch and J. Emel 1998), or have degraded them by using them as pets (J. Berger 1980). Major themes in animal geographies include: the relationships and interactions between animals and humans; the spaces and places occupied by animals in human culture; the environmental, social, and cultural implications of the representation and use of animals; the meanings and identities ascribed to animals and animal issues, and questions of animal welfare and rights. For a summary of animal geographies, see Emel et al. (2002) *Soc. & Animals* **10**, 4.

   Geographers using *actor–network theory argue that the divisions between people and animals are subject to change and negotiation (S. Whatmore and L. Thorne 1998; Hobson (2007) *Pol. Geog.* **26**, 3) while ecofeminists see the 'human chauvinism' towards animals as part of a hierarchy topped by European-American males (G. Gaard 1993). Wolch and Emel (op. cit.) argue that Western representations of animals such as wolves underpin the racism and sexism which label 'others' as savage. (Interestingly, in Russian representations, the wolf is the benevolent 'caretaker' of the forest, cleaning up the carrion).

annular drainage *See* DRAINAGE PATTERNS.

anomie The lack of traditional social patterns within a group; a lack of norms leading to conflict and confusion. In a study of second-generation British-Barbadians moving to Barbados, Potter and Phillips (2006) *AAAG* **96**, 3 note that they are often thought by the locals to be mad. L. Wirth (1938) sees anomie as the result of urbanism; Entriken (2002) *Geografiska B* **84**, 1 thinks anomie and multiculturalism are 'two related, yet often competing, visions of modern society' (but gives no more supporting evidence than Wirth).

Antarctic Denoting regions south of the Antarctic circle, 66° 32′ S. Within the **Antarctic circle**, the sun does not rise on 21 June (winter solstice in the Southern Hemisphere) or set on 22 December (summer solstice in that hemisphere).

Antarctic meteorology Winters are severe, with characteristic double temperature minima, i.e. two separate occasions of minimum temperature, due to the absence of *insolation for several winter months, and to the frequent exchange of air with that of lower latitudes. *Blizzards are common. Temperatures rise in late summer as the long waves of the *westerlies bring incursions of warmer air. Nevertheless, precipitation is still almost always in the form of snow, since maximum temperatures, occurring at the summer (December) solstice, rarely exceed 0 °C. See J. King and J. Turner (1997).

Antarctic plate Antarctica seems to be part of a single tectonic plate, but there is clear palaeomagnetic evidence that, in the past, there has been a large relative rotation between different parts of the continent (Gordon (2000) *Nature* **404**). Whittaker et al. (2007) *Science* **318**, 83 detail major Australian and Antarctic plate reorganizations.

antecedent Prior to, before, as in **antecedent drainage** patterns: the Wind River, Wyoming, for example, probably pre-dates the dome it now cuts through, the river's erosion keeping pace with the uplift of the dome. **Antecedent moisture** is the amount of moisture already present in the soil before a specified rainstorm.

anthrogeomorphology The study of human effects on the physical landscape.

anthropocene era The geological era, beginning in the 18th century, in which humans are a significant, sometimes dominating, environmental force (Crutzen and Stoermer (2000) *Glob. Change Newsl.* **41**). During this era humankind has: increased its population tenfold; exhausted 40% of the known oil reserves; contributed to a 30% increase in atmospheric CO_2; transformed nearly 50% of the land surface (with significant consequences for biodiversity, nutrient cycling, soil structure, soil biology, and climate); dramatically altered coastal and marine habitats (50% of mangroves have been removed and wetlands have shrunk by one-half); and increased extinction rates (Sanderson et al. (2002) *Bioscience* **52**, 10).

anthropogenic Brought about by human agency.

anticline *See* FOLD.

anticlinorium *See* FOLD.

anticyclone A region of relatively high *atmospheric pressure, frequently thousands of kilometres across, often formed as response to *convergence in the upper *atmosphere, and also known as a high. As air near ground level flows into an anticyclone, its absolute *vorticity decreases, causing *divergence, and the descent of air (Tan and Curry (1993) *Monthly Weather Rev.* **121**). **Anticyclonic circulation** is clockwise in the Northern Hemisphere, and anticlockwise in the Southern Hemisphere.

Cold anticyclones (*continental highs*) form as continental interiors, such as Siberia, lose heat in winter through *terrestrial radiation, and cool the air above. They are marked by subsidence, which inhibits cloud formation and maximizes radiative cooling, making them self-sustaining. Keegan (1958) *J. Meteorol.* **15**.

Subtropical anticyclones are warm, and form due to subsidence below convergence resulting from the westerly sub-polar *jet stream at the northern limit of the *Hadley cell. The semi-permanent subtropical anticyclone over the North Atlantic (the *Azores high*) strongly influences weather and climate of much of North America, western Europe, and north-west Africa (Davis et al. (1997) *J. Climatol.* **10**, 4). Subtropical anticyclones bring stable atmospheric conditions, and fine, hot, dry weather. In *mid-latitudes, anticyclones often locate beneath the leading edge of ridges in the upper-air *westerlies, where they may be associated with *blocking weather patterns; see Wiedmann et al. (2002) *J. Climate* **15**, 23.

anticyclonic gloom In winter, in north-west Europe, fog, or poor visibility caused by broad, persistent sheets of strato-cumulus cloud at the base of an *inversion trapping polluted air below. The inversion results from the arrival of a cold *anticyclone.

antidune A ripple on the bed of a *fluvial channel, travelling upstream. Antidunes are sediment bedforms, formed beneath *standing waves by fast, shallow water flows. A critical *Froude number of 0·84 may be taken as the discriminator between dunes and antidunes (Carling and Shvidchenko (2002) *Sedimentol.* **49**, 6).

anti-globalization A colloquial term for the stance of those who, internationally, coordinate, organize, and mobilize opposition to *neoliberalism and the political power of *transnational corporations. Most anti-globalization activists use other titles: 'the Global Justice Movement'; the 'Movement of Movements' (a critical alliance of disparate movements espousing anti-capitalist politics, anti-globalization, anti-war, and radical ecological issues); and the 'alter-globalist movement' (promoting humanist values and 'economic justice'). A major concern is that past North–South trade agreements have been unfair to the South.

J. Agnew et al., eds (2003) think the claim that anti-globalization protests are global is a bold one: 'in the absence of a locatable centre of authority, anti-globalization protesters mirror the "non-place" of power by choosing targets that symbolize global imperial sovereignty.' De Filippis (*On-Line Conf. Commun. Organ. & Dev.*) thinks *local* organizers should fight neoliberal globalization, while M. Neocleous (2003) argues that 'a non-territorial democratization of power...a geography of insurrection as opposed too the geography of order' is needed to fight globalization. Wissenburg (2004) *Org. & Env.* **17**, 4 calls for a 'radical articulation of environmental interests' by anti-globalization activists.

anti-natalist Limiting population growth. The People's Republic of China has pursued anti-natalist policies, notably the 'one-child' strategy (although J. Dreze and A. Sen (1989) claim that economic security for China's rural population hastened the fall in birth rates). Barber (2003) *Cross-Cultural Res.* **37** believes that countries with large populations are more likely to pursue anti-natalist policies in general, and to discourage single parenthood in particular, to justify controlling population. However, Coe (2004) *Reproductive Health Matters* **12**, 24 argues that pro- and anti-natalism are 'the greatest obstacles to making real and long-lasting improvements to sexual and reproductive health and rights'.

antitrades Westerly winds in the upper atmosphere above, and in contrast to, the easterly *trade winds at ground level.

apartheid The system of racial segregation first promulgated by the largely Afrikaner National Party of South Africa in 1948. **Petty apartheid** meant the separation of facilities such as lavatories, transport, parks, and theatres into white and non-white. On a much larger scale was the allocation of 12% of the land area into independent republics ('homelands, or bantustans') for the African population, which comprised 69% of the population when the policy began, in 1954. These 'homelands' were to be governed and developed separately from white South Africa, while allowing African workers strictly limited rights to live in the white areas, as and when their labour was required. J. Robinson (1996) suggests that the organization of urban space into racially segregated living areas was central to the persistence of the racial state. 'Without a gathering of the racially defined African population into spatially contained areas and the evolution of specific methods of administration and governance in these areas, the implementation of various racial policies would have been held hostage to the racial and physical "chaos" of the early twentieth-century city.'

With the election of South Africa's first democratic government in 1994, the last vestiges of apartheid were officially removed, but the policy will leave a mark on the South African landscape and its society for many years to come; see Benjaminsen et al. (2006) *AAAG* **96**, 3 and Ramutsindela (2007) *Tijdschrift* **98**, 4. 'What investigations into, say, apartheid in South Africa...reveal is the way in which these racist geographies were embedded in a wider field of utter indifference' (Jacobs (2000) *TIBG* **25**, 4).

aphelion The point furthest from the sun of any body orbiting the sun; the earth is at aphelion 152 000 000 km from the sun in July. *See* MILANKOVITCH CYCLES.

aphotic zone In any watery environment, the deeper zone, which is not penetrated by light.

applied climatology The systematic study of climatology for an operational purpose, such as agriculture.

applied geomorphology 'As geomorphology is the study of the surface of the Earth, its form, evolution and the processes that sculpture it, so applied geomorphology can be defined as the application of this knowledge in the resolution of engineering, planning and environmental/resource management problems' (Jones (1983) *Quart. J. Engineering Geol. & Hydrogeol.* **16**, 3). 'The social relevance of geomorphology' (anon., U. Victoria, Canada). See P. Fookes et al., eds (2005).

applied meteorology The use of climate information in decision-making, impact assessments, seasonal climate forecast applications and verification, climate risk and vulnerability, development of climate monitoring tools, urban and local climates, and climate as it relates to the environment and society. It includes weather modification, satellite meteorology, radar meteorology, boundary layer processes, air pollution meteorology, agricultural and forest meteorology, and applied meteorological numerical models.

appropriate technology A technology which evolves or is developed in response to a particular set of needs, and in accordance with prevailing circumstances (Intermediate Techn. Dev. Group); 'a development strategy that would be appropriate to the material, technical, and organizational resources of nations poor but proud, and that at the same time would enhance local and national autonomy' (Brown (1979) *Dev. Econs.* **17**, 1).

aquaculture The use of waters, other than the sea, for agricultural production, usually fish.

aquiclude A rock, such as London clay, which does not allow the passage of water through it; an impermeable rock. Such a rock will act as a boundary to an *aquifer.

aquifer A rock, such as chalk, which will hold water and let it through. Water runs into aquifers where the rock is exposed to the surface or lies below the *water-table. See Water Science and Technology Board (2008), *Prospects for Managed Underground Storage of Recoverable Water*. A **confined aquifer** is sandwiched between two impermeable rocks, and the water-table marks the upper limit of an **unconfined aquifer**. See A. Klimchouck, ed. (2000).

Arabian plate A minor lithospheric *plate, currently colliding with the Iran plate; see Al-Damegh et al. (2005) *Earth & Planet. Sci. Lett.* **231**.

arch Along a coast, an arch is made when caves on either side of a headland meet. Arches are temporary; roof falls cut off the seaward end of the arch, which is then left as a *stack.

Arctic Denoting regions within the Arctic Circle, i.e. north of 66° 32′ N (often taken as 66½° N). Within these regions the sun does not set on 21 June (the summer solstice in the Northern Hemisphere) nor rise on 22 December (the winter solstice in the Northern Hemisphere). In climatology the Arctic is defined in terms of the treeless zone of *tundra and of the regions of *permafrost in the Northern Hemisphere. **Arctic air masses** are exceedingly cold, with the Arctic Ocean as their source region. During winter, they are very dry, and little different from a continental *polar (cP) air. In summer, the air is more *maritime. A **maritime Arctic (mA)** air mass is generally similar to maritime polar air. Initially it has continental characteristics, but when it crosses warmer water, it is warmed from below, and in consequence becomes unstable.

Arctic meteorology The Arctic regions experience an annual cycle of winter 'night' and summer 'day'. Most weather results from the intensely cold ground air which is chilled by contact with land, which loses heat from strong *terrestrial radiation—winter clouds are scarce. Only infrequently do depressions penetrate the *inversions so formed. Winter temperatures are close to −40 °C. Snowfall is slight, but winds cause frequent blizzards and drifting. In spring, days are longer and sunny, but temperatures remain low because of the high *albedo of the snow surface. In summer, some depressions bring thicker cloud and light rain, and snow- and ice-melt in June and July keep air temperatures low.

Since the early 1970s, sea-level pressure over the Arctic has decreased; the stratospheric polar vortex has become colder and has been persisting longer into spring; and water mass characteristics of the Arctic Ocean have changed. Ostermeier and Wallace (2003) *J. Climate* **16**, 2 link these changes to a shift in the *North Atlantic Oscillation.

Arctic sea smoke A form of *steam fog. As very cold air passes over a warmer sea, it is rapidly heated. Convection lifts moisture up, which quickly recondenses to form fog.

areal differentiation How areas of the earth's surface differ; a study of the way human and physical phenomena vary over the earth's surface. As an approach to geography, areal differentiation has come in and out of favour (R. Hartshorn 1939; D. Cosgrove 1998; D. Massey 2005). D. Livingstone (2003) uses 'place', 'region', and 'circulation' to understand the areal differentiation and spatial flows of scientific knowledge.

areal scouring Landscapes of areal glacial *scouring have been described as comprising irregular depressions with intervening bosses scraped by ice and labelled 'knock and lochan' topography, but Johansson et al. (2001)

Geografiska A **83** suggest that an etched bedrock surface is a prerequisite for this type of landscape to develop.

Area of Outstanding Natural Beauty (AONB) In the UK, an area in which development is very carefully considered, so that the beauty of the landscape is not diminished. The Countryside and Rights of Way Act 2000 empowers local authorities to establish a Conservation Board for their own AONB, in order to conserve it, enhance it, and increase public understanding and enjoyment of its qualities. See McKenna et al. (2005) *Area* **37**, 3.

areic Without surface drainage, that is, without streams or rivers. Areas of permeable rocks, such as limestone, often lack surface drainage.

arenaceous Sandy in texture, or applied to rocks composed of cemented, usually quartz, sand. Arenaceous rocks include quartzites and greywackes.

areography The analysis of the range of species (R. Hengeveld 1992).

arête A steep knife-edge ridge between *corries or *glacial troughs in a glacially eroded, mountainous region. Arêtes are possibly formed by the backward extension of corries into a mountain mass, or moulded by *nivation and *frost wedging.

argillaceous Clay-like in composition and texture, referring to rocks containing *clay minerals and clay-sized particles, for example, shale.

aridisol A soil order of the *US soil classification, found in arid environments. It is a *desert soil, predominantly composed of minerals, and often high in accumulations of water-soluble salts.

aridity Dryness; aridity is the most common criterion for the definition and classification of drylands, occurring when *evapotranspiration exceeds precipitation (Adamo and Crews-Meyer (2006) *Appl. Geog.* **26**, 1). An **aridity index** is the ratio of annual potential evaporation to precipitation (M. Budyko 1974). See Arora (2002) *J. Hydrol.* **265**.

arkose An *arenaceous rock composed of quartz and more than 25% feldspar.

armouring The coarse surface material which protects the finer material below (Gibbins (2007) *Freshwater Biol.* **52**, 12). The armouring of a river bed is caused by selective entrainment of smaller sediments, and reduces further entrainment of smaller sediments. An **armour layer** is generally well sorted, and one or two grains in thickness. Migratory during floods, it will re-form when normal flow resumes (Vericat et al. (2006) *Geomorph* **76**, 1–2).

arroyo A *gully, rectangular in cross-section, usually with a sandy floor, made of fine-grained cohesive sediments in arid environments. It results from the entrenchment of stream channels when run-off increases. See Bierman (2005) *ESPL* **30**.

Areas of Outstanding Natural Beauty

artesian basin A *syncline of permeable rocks, outcropping at the crest. Water from rain or streams seeps into this *aquifer; the rock becomes saturated, and the water is under pressure. If a hole is sunk to tap the water, an **artesian well** forms—water will initially flow upwards without pumping. Try B. Radke et al. (2000) on the Great Australian Artesian Basin.

artificial neural network (ANN) A computing methodology which
simulates the human brain as it processes spatial date problems. Thurston has
written an easily accessible summary, noting that 'in almost every instance
where *GIS is being used, [artificial intelligence] applications could potentially
be developed for the purpose of enhancing decision-making capabilities'.

(((())) SEE WEB LINKS

• Thurston's summary of artificial neural network.

ash *See* PYROCLAST.

ash cone *See* SCORIA.

ash fall The precipitation of volcanic ash (tephra) during and after a volcanic
eruption; see Kimura et al. (2005) *Island Arc* **14**, 4. Ash may sterilize soils, bury
crops, and cause roofs to collapse under its weight (Cronin and Neil (2000)
Bull. Volcanol. **62**, 3).

ash flow Watenabe et al. (1999) *J. Volc. & Geotherm. Res.* **89**, 1 suggest that
the ash flows from Mt Unzen, 1991, were formed by the fracturing of the
cooling lava blocks, and were thereafter entrained into the plumes at the tops
of the moving flows. *See* PYROCLAST.

aspect The direction in which a valley side or slope faces. In deeply cut east–
west oriented valleys, the slopes facing the equator receive more sun and are
more attractive to settlement than the shaded sides of the valley.

Aspect may be an important factor in the formation of landforms, since
slopes facing away from the equator may be 6 °C colder than their opposites.
Beaty (1961) *J. Geol.* **51** estimates gradational processes to be two to three
times as active on northward-facing slopes in the Northern Hemisphere; in the
Jotunheim of Norway, 70% of corries lie on the north side of the massif (Evans
(1969) *Geografiska A* **59**). Dikau in J. Raper, ed. (1989) uses aspect as a criterion
in the description of landforms, as do Dymond et al. (1995) *ESPL* **20**, 2.

asphalt A naturally occurring, semi-solid, bituminous rock that can be made
from crude oil.

assarting In medieval Britain, the expansion of agriculture into the forested
areas (Beaver in S. Wooldridge and S. H. Beaver 1950).

assimilation Also known as *acculturation, or integration, this is the
integration of an immigrant, outsider, or subordinate group into the dominant,
host community.

The concept of assimilation implies that the minority group eventually take
on the values of the host, or *charter group; Akinwumi (2006) *Geografiska B* **88**,
2 asserts that the USA requires immigrants to put aside previous national
identities and heritages; 'for immigrants to the United States, "correct"
consumption became a symbol of assimilation' (Domosh (2004) *TIBG* **29**, 4).
'In today's debates about immigrant assimilation, the degree to which

immigrants remain spatially concentrated is treated as a measure of their assimilation, or willingness to assimilate' (Ellis (2006) *Tijdschrift* **97**, 1). Sometimes the indigenous population is required to assimilate with a non-indigenous group; Lester (2006) *Geogr. Res.* 44, 3 argues that Australian aborigines were to retain their lands as sites upon which missionaries could prepare them for assimilation as black, Christian Britons.

Wei Li (2006) challenges traditional notions of the urban-to-suburban transition in assimilation, revealing emerging ethnic spatial patterns and friction within the host community. R. Smith (2005) argues for transnationalism as a crucial component of assimilation; 'assimilation and pluralism are not mutually exclusive' (Burnley (2005) *Geogr. Res.* **43**, 4).

assisted areas Those areas of Great Britain where regional aid may be granted under European Community law (UK National Statistics).

associated number (König number) The associated number of a *node is the number of *edges from that node to the furthest node from it. A low number indicates a high degree of *connectivity; for an example of its use, see Wood (1975) *Geografiska B* **57**, 109–19.

association (plant association) A *plant community unit; a floral assemblage with a characteristic dominant and persistent species.

asthenosphere That zone of the earth's *mantle which lies beneath the relatively rigid *lithosphere, between 50 and 300 km below the surface, and representing a mechanical boundary between more rigid regions above and below. It is approximately commensurate with that zone of the mantle which transmits seismic waves at low velocity. The asthenosphere represents the location in the mantle where the melting point is most closely approached, and the transition between a rigid lithosphere and a viscous asthenosphere is gradational (Stüwe (1999) *Surveys Geophys.* **20**). The asthenosphere is composed of hot, semi-molten, and therefore deformable, rock, and may be plate- or density-driven, or both (Hansen et al. (2006) *Geology* **34**, 10).

asylum migration The international movement of *refugees and persons who, while having suffered generalized repression, violence, and poverty, do not qualify as refugees under the strict requirements of the 1976 UN protocol. 'Asylum seeking is embedded in complex ways in the linkages and structures of transnational entrepreneurism, informal economic activity and the shadow labour markets of the host economy' (B. Jordan and F. Düvell 2002).

Economic hardship and economic discrimination against ethnic minorities lead to higher inflows of asylum seekers; political oppression, human rights abuse, violent conflict, and state failure are also important determinants, as are migration networks and geographical proximity (Neumayer (2005) *Int. Studs Qly.* **49**, 3). Attempts by the EU to reduce access to asylum systems or to curtail the rights of asylum seekers have generated an increase in illegal migration (Boswell (2003) *Int. Affairs* **79**, 3). 'If Western European countries want to tackle the root

causes of asylum migration, then they need to undertake policy measures that
promote economic development, democracy, respect for human rights and
peaceful conflict resolution in countries of origin' (Neumayer, op. cit.).

asymmetry Lacking symmetry. An **asymmetrical valley**, common in
present or past *periglacial environments, has one side steeper than the other,
since *aspect can be critical in determining where frost forms; see Meikeljohn
(1992) *S. African Geogr. J.* **74**.

Atlantic Period In north-west Europe, the period from around 7500 to
5000 years BP of wet, oceanic climate, when temperatures were warmer than
at present. See Uścinowicz et al. (2007) *GSA Special Paper* **426** on sea-level
oscillations in the Atlantic period.

Atlantic-type coast A coastline where the trend of ridges and valleys runs
transverse to the coast. If the coastal lowlands are inundated by the sea, a *ria
or *fiord coastline may result. The coast of south-west Eire is an example.
Compare with PACIFIC COAST.

atmosphere A gaseous envelope gravitationally bound to a celestial body;
in the case of the Earth, with an average composition, by volume, of 79%
nitrogen, 20% oxygen, 0.03% carbon dioxide, and traces of rare gases. This
surprisingly uniform composition is achieved by *convection in the
*turbosphere and by *diffusion above it, especially above 100 km, where
diffusion is rapid in the thin atmosphere, and stirring is weak. Also present are
atmospheric moisture, ammonia, ozone, and salts and solid particles. The
atmosphere is commonly divided into the *troposphere, the *stratosphere, and
the *ionosphere. Since the troposphere contains the majority of the
atmospheric mass, and virtually all of the atmospheric water vapour, most
weather events occur within it.

atmospheric boundary layer *See* PLANETARY BOUNDARY LAYER.

atmospheric cells Air may move, with a circular motion, northwards or
southwards in a vertical cell; see Trump et al. (1982) *Boundary-Layer Met.* **24**
for a model of a closed mesoscale atmospheric cell. Atmospheric cells are
major components in the transfer of heat and momentum in the atmosphere
from the equator to the poles. Horizontal cells (winds) also fulfil this role; see
Abitol and Ben-Yosef (1994) *J. Opt. Soc. Am. A* **11**, 7.

atmospheric correction The retrieval of surface characteristics from
remotely sensed imagery by removing the atmospheric effects. See Lane et al.
(1999) *J. Appl. Met.* **38**, 12.

atmospheric heat engine The system of energy which drives and
controls the nature of the pressure, winds, and climatic belts of the Earth's
surface, with *solar radiation as the heat source, and *terrestrial radiation as the
sink. Mechanical energy is expended by the 'engine' in the form of atmospheric

a

ITCZ = Inter-Tropical Convergence Zone
PF = polar front
➡ warm air
⇨ cold air

polar tropopause

Polar cell

PF

mid-latitude tropopause

60° N

Ferrel cell

30° N

tropical tropopause

Hadley cell

ITCZ

EQUATOR

not to scale

Atmospheric cells

processes. Haine and Marshall (1998) *J. Phys. Oceanog.* **28**, 4 characterize the poleward heat transport as a diffusion by baroclinic eddies. Barry et al. (2002) *Nature* **415** regard the baroclinic zone as a heat engine which generates eddy kinetic energy by transporting heat from a warmer to a colder region.

atmospheric pressure The pressure exerted by the *atmosphere as a result of gravitational attraction exerted on the column of air lying above a particular point. Atmospheric pressure, measured in millibars, decreases logarithmically with height.

atoll A *coral reef, ring- or horseshoe-shaped, enclosing a tropical *lagoon. Most of the world's atolls are found in the Indian and Pacific oceans, and are said to be sensitive to fluctuations in relative sea level—although Woodruffe et al. (1999) *Marine Geol.* **160**, 1, in a study of West Island, Cocos, dispute this assumption. Single colony, semicircular corals (**microatolls**) grow to a level constrained by exposure at low spring tides (Spencer and Viles (2002) *Geomorph.* **48**, 1-3).

Atterberg limits Limits used to classify soils. The **liquid limit** is the minimum moisture content at which the soil can flow under its own weight; the **plastic limit** is the minimum moisture content at which the soil can be rolled into a 3 mm diameter thread without breaking; the **shrinkage limit** is the moisture content at which further water loss will not cause further shrinkage. See Fener et al. (2005) *Cana. Geotech. J.* on the Atterberg limits of cohesive soils.

attrition The wearing away or fragmentation of particles of debris by contact with other such particles, as with river pebbles.

aulacogen A long-lived, sediment-filled graben oriented at a high angle to either a neighbouring modern ocean or a neighbouring orogenic belt; an intracontinental graben (Umeji (1988) *Int. J. Earth Scis.* **77**, 2). An aulacogen can be seen as the failed arm of a triple-limbed spreading centre (Jansen (1975) *Trans. Geol. Soc. S. Afr.* **78**).

aureole *See* METAMORPHIC AUREOLE.

autecology The ecology of particular species and individual organisms, particularly the relationships between species and their environments and within species, and especially the way organisms act within communities. See Sundberg et al. (2006) *J. Biogeog.* **33**, 8.

authority constraint An area (or 'domain') that is controlled by certain people or institutions that set limits on its access to particular individuals or groups (Raubal in S. Winter, ed. 2002).

autochthonous Referring to features and processes occurring within, rather than outside, an environment. An **autochthonous rock** has been formed *in situ*; coal is an example.

autogenic succession A succession which begins as a result of disturbance by an external physical factor, but whose sequence in community composition results primarily from interactions among the species. Sometimes the early species actually enhance the success of the later ones, perhaps by increasing the nutrient status or water-holding capacity of the soil, as in a psammosere (C. Townsend et al. 2003).

automobility The study of the economic, technological, social, and political consequences of the pervasive and accelerating culture of the car. See Featherstone et al. (2004) *Theory Cult. Soc.* **21**.

autotrophe An organism which uses light energy to synthesize sugars and proteins from inorganic substances. Green plants are by far the most common autotrophes.

avalanche A rapidly descending mass, usually of snow, down a mountainside. **Powder avalanches** consist of a moving amorphous mass of snow. **Slab avalanches** occur when a large block of snow moves down a slope and can cut a swathe through the soil and sometimes erode the bedrock if the snow is wet. See A. V. Briuchanov (1967).

Avalanches of other substances are forms of *mass movement, and are distinguished by the type of material involved. **Debris avalanches** are typically triangular in shape, and are associated with morphological discontinuities, such as scarps and road cuts; see Guadagno et al. (2005) *Geomorph.* **66**, 1–4 on man-made cuts and debris slides. **Rock avalanches** occur when a jointed rock loses internal cohesion, until some sections are held together only by the friction between blocks. If this frictional force is lessened through water seepage, or if lateral support is removed, failure will occur. See Mitchell et al. (2007) *Landslides* **4**, 3.

avalanche wind The rush of air formed in front of an *avalanche. Its most destructive form, the avalanche blast, occurs when an avalanche stops abruptly.

avulsion The abandonment of all or part of a channel belt in favour of a new course (Stouthaven and Berensen (2007) *Sed. Geol.* **198**). Makaske (2001) *Earth-Science Revs.* **53** argues that bed aggradation reduces channel carrying capacity, increases overbank flooding and sedimentation, and eventually results in avulsion. See Jerolmack and Paola (2007) *Geomorph.* **91**, 3–4 for a cellular model of river avulsion.

awareness space Any locations known of by an individual before a decision about such places is made. The development of awareness space is highly dependent upon the perceived distance, type of street, mode, and ease with which one can travel between any given nodes, but especially anchor points (Malm et al. (2008) *Security J.* **21**).

azonal Referring to a soil without *soil horizons, such as a young soil developing on a bare rock surface. Alluvium and *sand dunes are examples of azonal soils.

back-arc basin A zone of tectonic enlargement, and substantial, and strongly asymmetric, sedimentation. A back-arc basin forms when an *island arc is split along the line of its magmatic axis, forming a remnant arc, which then glides away. Well-developed back-arc basins widen by **back-arc spreading**, which is thought to resemble normal oceanic crust formation, but in conjunction with a *subducting plate. See Ohara (2006) *Island Arc* **15**, 1 for a general view of the mantle process beneath a back-arc basin spreading ridge.

backreef A sea area landward of a reef, usually including a lagoon. Dahlgren and Marr (2004) *Bull. Marine Sci.* **75**, 2 argue that backreefs are of critical ecological value to a coral reef ecosystem. See K. Sullivan-Sealey et al. (2002).

backshore The part of a beach between the ordinary limit of high tides and the point reached by the very highest tides. On cliffed coastlines, the backshore is the section of cliff foot and *shore platform affected only by storm tides; on low, shingle coasts it can take the form of a *berm. Udo and Yamawaki (2007) *Geomorph.* **60**, 1–2 find that waves have larger effects than winds on backshore sedimentation.

backswamp On a *flood plain, a marshy area where floodwater may be confined between the river *levées and higher ground. See Takehiko et al. (2006) *Geog. Res.* **44**, 1 on mangroves and backswamps, and Patrick and Ellery (2007) *Afr. J. Ecol.* **45**, 3.

backwall The curved, steep head wall of a landslip scar of *tafone (Matsukura and Tanaka (2000) *Geografiska A* **82**, 1) of *nivation hollows (Thorn and Hall (2002) *PPG* **26**, 4), or of *cirques (Hughes et al. (2007) *Geomorph.* **88**, 3–4).

backwash The return flow of water downslope to the sea after the *swash has moved upshore. Steep waves and long waves are linked with stronger backwash, and flatter beach gradients; see the great C. King (1959), and Hardisty et al. (1984) *Marine Geol.* **61**, 1. When the path taken by the backwash differs from the path of the swash, *longshore drift occurs. See Bird et al. (2007) *Geogr. J.* **173**, 2, on backwash erosion.

backwash effect G. Myrdal (1970) states that a 'growing point' established by the location of a factory, or any other expansional move, will attract other businesses, skilled labour, and capital. However, it will also have backwash effects that keep down, or even impoverish, out-regions; see Allen et al. (2005)

Asia Pac. Viewpt **46**, 2. Barkley et al. (1995) *Papers Reg. Sci.* **74** identify different types of spread-backwash processes—but Malecki (2003) *AAAG* **93**, 4 writes that 'old ideas, such as cumulative causation and spread and backwash, are translated into the disequilibrium of growth poles and into evolutionary economics and the new economic geography'.

backwearing The erosion of a slope at a constant angle as it retreats; *parallel slope retreat. See Röhmer (2007) *Geomorph.* **86**, 3–4 on backwearing in *inselberg formation; Fjellanger and Sørbel (2007) *Norw. J. Geol.* **87** on backwearing and palaeic landforms, and MacGregor et al. (2001) American Geophysical Union, Fall Meeting Abstract #HD42D0382M, on headwall backwearing in cirques.

badlands Arid lands, generally bare of vegetation, dissected to form ravines and sharp-crested hills. This dissection is aided by the lack of vegetation, high *run-off, and heavy sedimentation, which increases *abrasion. Non-fluvial processes, such as mass wasting, *piping, and tunnel erosion, are also important. Badland topography develops best in weakly consolidated sediments. See B. Rorke and A. Yair (1989).

bajada (bahada) A series of *alluvial fans, typically laid down by *ephemeral streams, which has coalesced along the foot of the mountains to form a gently sloping plain of unconsolidated sediments. The *weathering of the mountain front may also supply debris. Alluvium on the lower part of a *pediment may also make a bajada. See Bull's classic paper (1977), *PPG* **1**; Leeder and Mack (2001) *J. Geol. Soc.* **158**) for a clear review, plus a great title; and Wainwright et al. (2002) *J. Arid Envs.* **51**.

balanced growth In macroeconomics, the growth of output and capital stock at the same rate.

balanced neighbourhood A neighbourhood which contains groups from all levels of society. Such a neighbourhood does not usually occur spontaneously and has to be planned, usually in order to counteract the processes of social *segregation; Uttermark (2003) *Urb. Studs.* **40**, 3 concludes that planning balanced neighbourhoods in the Netherlands mitigates the problematic integration of ethnic minorities, but Tunstall (2003) *Housing Theory & Soc.* **20**, 3 thinks that the UK policy of 'tenure mix' 'appears to have been used as a euphemism, initially for privatisation and latterly, for 'social mix'.

balance of payments The ratio between the payments made by one country to other nations, and the revenue it receives from them. If receipts exceed outgoings, the balance is positive. Bristow (2005) *J. Econ. Geog.* **5**, 3 thinks that modern socio-economic systems have not only to achieve a sustainable balance of payments, but social objectives, such as income redistribution, and basic health care. Ramos (2006) *ersa* **06** concludes that regional differences in balances of payments are not significant.

bank erosion The erosion of material from the side of a river channel. Hooke (1979) *J. Hydrology* **42** identifies corrasion and slumping as the major processes, and these appear to be associated with river flow levels and antecedent precipitation conditions, respectively. Rates of erosion vary with speed of flow (rates are highest on the outer bank of meander bends or where bars in the channel have diverted the *thalweg), bank vegetation and moisture content, and bank composition (Parker et al. (2008) *Geomorph.* in press).

bankfull discharge The *discharge of a river which is just contained within the banks. This is the state of maximum velocity in the channel, and of maximum *competence (V. P. Singh, ed. 1987). Timár (2003) *Quat. Sci. Revs* **22**, 10 shows that sinuosity is a function of the channel slope and bankfull discharge; Wu et al. (2008) *Geomorph.* **100**, 3–4 show that bankfull channel dimensions result from the accumulative effect of several consecutive years' discharge and sediment load conditions.

bank retreat The advance or retreat of a channel bank line (Rüther and Olsen (2007) *Geomorph.* **89**, 3–4). See Amiri-Tokaldany et al. (2003) *J. Am. Water Resources Ass.* **39**, 4.

bank storage Water held in the river bank which may contribute to stream flow, but, in arid conditions, may cause a decrease in discharge as water percolates from the river to the banks; see Burt et al. (2002) *J. Hydrol.* **262**, 1. Chen et al. (2006) *J. Hydrol.* **327**, 3–4 present a technical paper on bank storage and baseflow separation.

banner cloud 'Clouds resembling a banner, or flag, which form on the leeward side of ridges or peaked mountains' (Schween (2007) *Geophys. Res. Abstr.* **9**).

bar **1.** On a gently sloping coastline, a submarine accumulation of marine sediment, which may be exposed at low tide. Most bars form where steep, *destructive waves break; such bars can be called **break-point bars**, with crests generally running parallel to the coast. **Bay bars** extend across an *estuary or a bay: 'it seems likely that [Chesil Beach, UK] began as a bay bar . . . when the sea level was lower' (I. West 2007). Some bay bars entirely enclose the inlet, and a *lagoon may then form on the landward side; see Roberts and Slater (2007) *Holocene* **17**. **Offshore bars**, located further out to sea, are thought to result from the breaking of larger waves, which erode the sea bed and throw up material ahead of them to form ridges (Hoefel and Elgar (2003) *Science* **29**).

  2. In a *glacial trough, such as the Nant Ffrancon of North Wales, a transverse rocky barrier (Embleton (1961) *Transactions and Papers, IBG*, **29**).

  3. Within a river, a deposit of alluvium, dropped where velocity and turbulence are low, and which may form temporary islands. **Alternating bars** develop as patches of alluvium, often regularly spaced, along opposite sides of a straight channel. **Braid bars**, roughly diamond shaped, are generally aligned along the channel course; Best et al. (2003) *J. Sed. Res.* **73**, 4 provide 'a *unique* [authors' italics] insight into the process-product relationship of braid-bar'. **Point bars** form on the inner curves of a meandering river where *discharge is low; see Pyrce and Ashmore (2005) *Sedimentol.* **52**, 4.

barchans *See* SAND DUNE.

baroclinic A term applied to sections of the *atmosphere where trends in pressure (pressure surfaces) are at an angle to trends in temperature. (The precise definition refers to the intersection of isobars and isopycnals—levels, or surfaces, of equal density—but, for most purposes, isotherms can replace isopycnals). The number of intersecting isobars and isopycnals is a measure of **baroclinicity** (Nakamura (1992) *J. Atmos. Sci.* **17**). Moving polewards, temperatures fall very rapidly, while pressure remains constant in that direction, but falls with height (Garreaud (1999) *Monthly Weather Rev.* **127**). At some point, the slope for isotherms intersects with the slope for isobars, so that the two intersect in a **baroclinic zone**; most often in *mid-latitudes in winter. Barry et al. (2002) *Nature* **415** see the baroclinic zone as a transport mechanism.

In the mid-latitude baroclinic zones spontaneous generation of weather systems such as depressions and thunderstorms is common. (Sanders and Hoffman (2002) *Weather & Forecast.* **17**, 4). These are **baroclinic disturbances**, characteristically, on *synoptic charts, with strong meridional pressure gradients in the constant-pressure surfaces and vertical *wind shear. When the temperature gradient along the *meridians is very steep, *atmospheric cells break down into cyclonic and anticyclonic *eddies. This failure is known as **baroclinic instability** (Lesiur et al. (2000) *J. Turbulence* **1**, 1), and is characterized by the ascent of warmer, and the descent of colder, air.

barometric gradient *See* PRESSURE GRADIENT.

barotropic Describing *atmospheric conditions where trends in pressure align with trends in temperature, as in the ideal *air mass; the reverse of *baroclinic.

barrage A structure across a river or estuary, built in order to restrain or use water. Brichieri-Columbi and Bradnock (2003) *Geog. J.* **169**, 1 discuss the development of a barrage for the Ganges–Brahmaputra–Meghna waters: which 'would cause negligible population displacement, and making maximum use of existing river channels would minimize the environmental impacts associated with large canal or dam construction'.

barrier In *geomorphology, a general term for any offshore depositional form, usually running parallel to the coast or across an *estuary, and above water at normal tides. Barrier islands offshore of Mozambique appear to have been initiated as spits, with subsequent sedimentation, and anchoring and protection by beachrock and aeolianite formation (Armitage et al. (2006) *Geomorph.* **82**, 3–4).

barrier beach A long, narrow beach, running parallel to the coastline, and unsubmerged by high tides. See Chadwick et al. (2005) *Procs Inst. Civil Engineer. Maritime Engineer.* **158**, 4 on the barrier beach at Slapton Ley.

b

(SEE WEB LINKS)

• West (2008) shows a barrier beach.

barrier island A long, narrow, offshore island, usually having beaches and *dunes on the seaward side, often with *lagoons on the landward side, such as the islands enclosing Pamlico Sound, off the coast of North Carolina. The formation of barrier islands has been explained as the drowning of beach ridges (Stutz and Pilky (2002) *J. Coastal Res.* **SI 36**) or the *progradation of *spits (van Mauren (2005) *Marine Geol.* **224**).

barrier reef A *coral reef, stretching along a line parallel with the coastline but separated from it by a wide, deep *lagoon, the most famous of which is the Great Barrier Reef off north-west Australia (see D. Hopley et al. 2007).

barrow A communal burial mound built from the Stone Age until Saxon times. Long barrows, up to 100 m long and 20 m wide, were the earlier form, while round barrows were introduced during the Bronze Age; see P. Ashbee (1970 and 1960, respectively).

basal complex (basement complex) The ancient *igneous and metamorphic rocks which lie beneath Precambrian rocks and constitute the *shield area of the earth's continents. See Gutiérrez et al. (2006) *GSA Bull.* **118** on a basal complex in the Canary Islands.

basal ice Many glaciers display basal layers of debris-rich ice. Lawson et al. (1998) *J. Glaciol.* **44** argue that supercooling is the principal mechanism responsible for creating debris-laden basal ice at the Matanuska Glacier. Hall and Glasser (2003) *Boreas* **32**, 1 model basal ice temperatures.

basal slipping (basal sliding) The advance of a glacier by comparatively rapid creep close to its bed, occurring where the ice at the base is at its *pressure melting point. The major types are enhanced basal creep, where ice, deformed by stresses created by jagged bedrock, bends round obstructions; regelation (pressure melting); and slippage over a layer of water at the bed. Basal sliding, occurring in sporadic movements, accounts for the majority of flow in warm glaciers, especially where the gradient is steep, the ice thin, or where *meltwater is present. Glasser and Bennett (2004) *PPG* **28**, 1 believe that patterns of basal sliding strongly influence patterns of glacial erosion. See also Cohen et al. (2000) *J. Glaciol.* **46**, 155, and Cooke et al. (2006) *PPG* **30**, 5.

base flow The usual, reliable, background level of a river, maintained by seepage from groundwater and by *throughflow, which means that the river can maintain the base flow during dry periods. The **base flow index** is the contribution of base flow to a stream. Total flow to stream systems consists of surface run-off, interflow, and base flow; streamflow partitioning methods are used for finding the relative contributions of each (try Morgan (1972) *TIBG* **56**). With prolonged drought, base flow itself will diminish, the rate of flow falling in a depletion curve (Gebert et al. (2007) *J. Am. Water Resources Ass.* **43**, 1).

Base flow

base level The theoretically lowest level to which the course of a river can cut down. This level may be sea level, the junction between a tributary and the main river, or the level of a waterfall or lake, but streams rarely erode as far as base level. Base level may alter due to *eustatic or *isostatic change, and be termed **positive changes of base level** if the land sinks relative to the sea, or **negative changes of base level** if it rises. (Try Schumm (1992) *J. Geol.* **101** on river response to base level change; *see also* REJUVENATION.) **Marine base level** is the lowest point at which marine erosion occurs, perhaps as low as 180 m below the surface.

basement Intensely folded metamorphic or plutonic rocks, often *Precambrian, covered by relatively undistorted sedimentary rocks. See Valdiya (2002) *PPG* **26**, 3.

base saturation The degree of saturation of the *adsorption complex of the soil; that is, its occupation by exchangeable basic cations, or cations other than aluminium and hydrogen, expressed as a percentage of the total cation-exchange capacity. Halvorson (1980) *EM2894* explains exchangeable cations, cation exchange capacity and base saturation, and the relationship to soil fertility.

base surge A type of pyroclastic surge that forms at the base of volcano eruption columns and travels outward during some hydroclastic eruptions. As base surges move laterally away from the eruptive vent, they lose heat, sediment load, and velocity, which directly affects the resulting deposits. The sedimentary structures observed within surge deposits are also directly related to the amount of liquid water involved (Brand (2004) *Rocky Mt. & and Cordill. Meeting GSA Abstract* 37-5). See also R. Fisher and H.-U. Schmincke (1984).

basic activity (basic workers) Activities which contribute directly to the wealth of a city or region because they bring in money from outside; that is, export activities. See McGregor et al. (2000) *J. Reg. Sci.* **40**, 1. 'Activities can be basic, if paid by money external to the region under study, or nonbasic, if they are targeted to the local demand' (Dentinho and Silveira (2007) *Port. Econ. J.*).

basin A major relief *depression, guided by structure, or formed by *erosion. **Basin and range** describes a landscape where ridges made of asymmetric *fault blocks alternate with lowland basins. In the USA, the basin and range country lies between the Sierra Nevada and the Wasatch Mountains; see W. S. Baldridge (2004).

bastide A planned, fortified strong point and centre of economic development created in the Middle Ages, mostly in France. The rectilinear street pattern contrasts strongly with the irregular, cramped layout of most medieval towns. This morphology was used by French colonizers in Canada (Louder (1979) *Cahiers de géo Québec* **23**).

batholith A massive, frequently *discordant, intrusion of coarsely textured plutonic rocks, at least 100 km^2 in area and extending 20–30 km down into the layer of *magma, which may be composed of several plutons.

(⊕) SEE WEB LINKS

• The Idaho Batholith.

bathymetry Measuring water depth, mainly of seas and oceans but sometimes of deep lakes, such as the Aral Sea (Shibuo et al. (2006) *Water Resources Res.* **42**).

baydzharakh (baydjarakh) A dome-shaped polygon, some 3–4 m high and up to 20 m wide, found in a *periglacial landscape, and formed as the *ice wedges begin to thaw.

beach An accumulation of *sediment deposited by waves and *longshore drift along the coast. The upper limit is roughly the limit of high tides; the lower, of low tides. The beach classifications **dissipative** and **reflective** seem to have fallen out of favour, but see Sherman and Bauer (1993) *PPG* **17** if necessary. *See also* DRIFT-ALIGNED BEACH.

beach budget (beach sediment budget) The balance between the material deposited on, and eroded from, a beach; see Davidson-Arnott and Van Heyningen (2003) *Sedimentol.* **50**, 6. During winter, waves tend to be destructive, and gradients steepen, but, with summer conditions, waves become constructive, and gradients slacken; a sequence detailed in Engels and

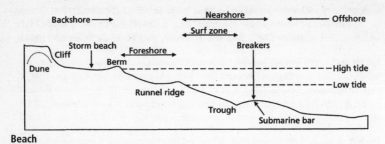

Beach

Roberts (2005) *J. Sed. Res.* **75**, 6. See Martinez and Psuty (2004) *Ecol. Studs* **171** on *foredunes and beach budgets.

beach ridge A wave-deposited ridge, made of sand and reworked underlying beach material, running parallel to a shoreline, beyond the present limit of ordinary tides. Changes in beach ridge morphology and orientation reflect environmental factors such as changes in wave climate and wind regime (Bristow (2006) *Sedimentol.* **53**, 4).

beach rocks Hard coastal sedimentary formations, consisting of various beach sediments, lithified through the precipitation of carbonate elements (Vousdoukas et al. (2007) *Earth Sci. Revs* **85**, 1–2).

bedding plane The surface separating distinct rock *strata.

bedforms (sedimentary bedforms) **1.** In hydrology, accumulations of sediment on the river bed, formed by variations in flow competence, and ranging in size from ripples in the sand, a few centimetres apart, to 'dunes' tens of metres in length.

River bed form is related to energy level; as flow energy and sediment transport competence increase, bedforms change from ripples to dunes, which alters the ratio of inertial to gravitational forces in the flow field. This form-flow feedback is a reflection of river energy reorganizing the sediment mass in the most effective configuration for that flow condition.

2. Migrating megaripples are bedforms that appear in the surf zone of sandy coasts; see Gallagher et al. (1998) *Nature*, **394**.

((())) SEE WEB LINKS

• The excellent USGS bedform sedimentology site.

bed load Material moved along a river bed by *traction and *saltation. Bed load usually comprises sands and pebbles, but when the water level is high and the current strong, large material boulders may be moved (Laronne (2001) *Water Resources Res.* **37**, 5).

bed profile *See* LONG PROFILE.

bedrock The unweathered rock which underlies the soil and *regolith or which may be exposed at the land surface. **Bedrock gouges and cracks** are small cracks, gouges, and indentations created on bedrock surfaces. Both *chattermarks* and *crescentic gouges* are shallow crescentic furrows: in crescentic gouges the convexity is turned forward in the direction of ice flow, in chattermarks it is turned backwards (Glasser and Bennett (2004) *PPG* **28**, 1). *Crescentic cracks* are vertical fractures of the rock without the removal of bedrock fragments, with a forward-turned concavity.

bed roughness (bed-floor roughness) The frictional force of a river bed (also called *channel roughness*). A rough bed of boulders, pebbles, and potholes exerts more friction than a smooth, silky channel. Lamarre and Roy (2005) *Geomorph.* **68**, 1–2 examine the effects of roughness on mean flow velocity. *See* MANNING'S ROUGHNESS COEFFICIENT. The term is also applied to a glacier bed; see Rippin et al. (2004) *Ann. Glaciol.* **39**.

behavioural environment The behavioural environment model emphasizes the importance of perception in human geography, the significance of subjective experience, and the potential of man as an active agent in the environment. See F. Boal and D. Livingstone (1989).

behavioural geography Behavioural geography set out to humanize *homo economicus*, to recognize that people do not have complete information, are not always distance minimizing, are embedded in networks of social relations, and therefore may base decisions on factors other than sheer economic rationality. It specifically recognized the role of cognition and the importance of social and cultural values and constraints, plus all the institutional, economic, and physical factors that characterize the public, 'objective' environment (Couclelis and Golledge (1983) *AAAG* **73**, 3). Behavioural geography was criticized as inherently positivist, severing individuals from their social and cultural contexts. Nonetheless, behavioural approaches have been gaining ground in the field of economics; see Strauss (2008) *J. Econ. Geog.* **8**, 2, who calls for a renewed behavioural economic geography, 'without adopting, wholesale, its ways of conceptualising economic man (sic)'.

behaviourism The view that the actions of an individual occur as responses to stimuli. Through constant repetition, the individual learns to make the same response to a given stimulus. Mental maps have especially been associated with the behavioural approach; see Soini (2001) *Landsc. & Urb. Plan.* **57**, 3–4. Norton (2003) *Area* **35**, 4 thinks that cultural geography might benefit from consideration of radical behaviourism.

belonging The concept of belonging is intimately tied to place, as any understanding of community and affinity with specific landscapes (place-making) simultaneously constructs a sense of socially recognized membership (Trudeau (2006) *Cult. Geogs* **13**). Belonging captures 'the desire for some sort of attachment, be it to other people, places, or modes of being, and ways in which individuals and groups are caught within wanting to belong, wanting to become' (Probyn 1996, quoted in Nelson and Hiemstra (2008) *Soc. & Cult. Geog.* **9**, 3).

belonging, geographies of The geographical study of the practices, meanings, and processes of belonging: from the body, the house or home, the neighbourhood, the region, the nation, and the global. See Till (1999) *Ecumene* **6**; Zilberg (2004) *Landsc. Res.* **29**, 3 on diasporic geographies of belonging and being; and Veronis (2006) *Env. Plan A* **38**, 10 on the way immigrant, ethnic/racial, visible minority groups build identity and community, and Holston and Appadurai (in J. Holston, ed. 1999) on belonging, everyday experiences, and tussles over urban space.

Amster and Lindquist (2005) *Asia Pac. J. Anthrop.* **6**, 1 consider transnational relationships in South-East Asia; Ikeotuonye (2007) *SGIR Turin Conf.* examines borders, frontiers, and the state's own image of itself, and M. Joseph (1999) illustrates the emergence of nomadic identity. See Sparke (2006) *Pol. Geog.* **25**, 2 on 'seemingly unbounded global visions of belonging'.

Benguela current The eastern boundary current of the South Atlantic subtropical gyre, which begins as a northward flow off the Cape of Good Hope, then skirting the western African coast equatorwards until around 24–30 °S (Gyory et al. (no date) CIMAS).

Benioff zone An inclined zone of earthquakes, plunging below the earth's surface at an angle between 30° and 80°, but commonly at around 45°, and extending to a depth of 300–400 km. Benioff zones are associated with the downward movement of a lithospheric *plate at a destructive plate margin. See Scalera (2007) *Geofisica Int.* **46**, 1.

benthic Occurring at the base of bodies of water: lakes, oceans, and seas. **Benthos** is the life attached to the bottom, or moving in the bottom mud.

Bergeron–Findeisen theory A mechanism by which ice crystals grow from the vapour supplied by cloud droplets. In mixed clouds, where ice crystals and supercooled droplets coexist, the water vapour diffuses from supercooled droplets to ice crystals because the vapour pressure over ice is lower than over liquid water at the same temperature (Castellano et al. (2004) *Atmos. Envir.* **38**, 39). See Korolev (2007) *J. Atmos. Scis* **64**, 9 on the limitations of the Wegener–Bergeron–Findeisen mechanism in the evolution of mixed-phase clouds.

bergschrund A deep, tensional *crevasse formed around the head of a *cirque glacier. The crevasse forms as ice falls away downslope. Often, a sequence of bergschrunds forms. Johnson's bergschrund hypothesis (1904, *J. Geol.* **12**) explained headwall retreat by vigorous frost weathering in a narrow zone; Lewis (1940) *Geog. Rev.* **30** suggested that meltwater forced its way down far below the bergschrund, explaining the great height of some cirque headwalls. For the limitations of bergschrund theory, see D. R. Stoddart (1997).

berm A low embankment or ridge on a sand beach, constructed by *swash or breaking waves. Weir et al. (2006) *Geomorph.* **82**, 3–4 examine two models of berm development: vertical growth at spring tides or 'following significant beach cut due to substantial swash overtopping', and horizontal progradation at neap tides through the formation of a proto-berm, located lower and further seaward of the principal berm.

beta index A simple measure of connectivity which can be derived from the formula β arcs/nodes, where nodes are road junctions and arcs are connections between the nodes as straight lines. Beta index ranges from 0.0 for network, which consists just of nodes without any arcs, through 1.0 and greater where networks are well connected.

betterment migration Migration in the hope of social and economic advancement (Clark in P. Clark and P. Slack 1972). In a study of migration to early 19th-century London, Humphries and Leunig (2007) *LSE W. Papers* **101**/07 come to the surprising conclusion that an individual's height worked better than literacy and age as a predictor of betterment migration.

bevelled cliff A sea cliff with an upper, gentle slope, but a steep, lower slope.

bid-rent theory W. Alonso (1964) notes that when a purchaser acquires land, he acquires two goods (land and location) in one transaction, and a single payment is made for the combination. Thus it is possible to trade off a quantity of land against location. The **residential bid price curve** is 'the set of

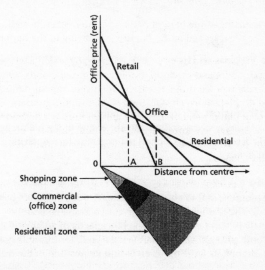

Bid-rent

prices for land the individual could pay at various distances while deriving a constant level of satisfaction'. As individuals consider residential locations at increasing distances from the city centre, they assess the price of land that would allow them to buy enough land (and other goods) to provide as much utility (satisfaction) as a given price, and amount of land, at the city centre. The **bid price function** for the urban firm describes the prices which the firm is willing to pay at different distances from the city centre in order to achieve a certain level of profits. Egan and Nield (2000) *Urb. Studs* **37** develop a model of the Alonso type which casts light upon the intra-urban location decisions of hotels.

bifurcation ratio In a drainage basin, the ratio of the number of streams of a given order to the number of streams of the next, higher order. 'In general, small streams are tributaries of the next bigger stream in such a way that flow architecture develops from the lowest scale to the highest scale' (Reis (2006) *Geomorph.* **78**, 3–4).

Stream ordering after Strahler

Bifurcation ratio

binary distribution A *city-size distribution in which a number of settlements of similar size dominate the upper end of the hierarchy, said to be characteristic of nations with a federal political structure, such as Australia. *See also* RANK-SIZE RULE.

biocapacity The capacity of a given area to generate an enduring supply of renewable resources and to absorb its spillover wastes. Earth's biocapacity is determined by the biologically productive area available and the bioproductivity (yield) per hectare. The biocapacity of an area of land is affected by both physical and human factors: Iraq, for example has a low biocapacity, despite its large oil reserves and its position in the once 'fertile crescent'.

The European Union's *Ecological Footprint and Biocapacity* (2006) outlines the world's ability to regenerate resources and absorb waste in a limited time period.

b

bioclimatology The scientific study of the relationship between organisms and climate. See Di Filippo et al. (2007) *J. Biogeog.* **34**, 11. Global interactions between climate and vegetation may be studied by using *remote sensing (Goward (1989) *J. Climate.* **2**, 7). Energy demands can be reduced by the use of bioclimatic principles in building design; see Krstic (1998) *Renewable Energy* **15**.

biodiversity The number and variety of living organisms, from individual parts of communities to ecosystems, regions, and the entire biosphere, including: the genetic diversity of an individual species; the subpopulations of an individual species; the total number of species in a region; the number of endemic species in an area; and the distribution of different ecosystems. Greater plant diversity leads to greater primary productivity, because there is a greater chance that a more productive species would be present at higher diversity, and from the better 'coverage' of habitat heterogeneity caused by the broader range of species traits in a more diverse community (Timan (1999) *Ecology* **80**, 5). The effects of global warming on biodiversity are, as yet, unclear; Botkin et al. (2007) *Bioscience* **57**, 3 explain why.

The **biodiversity gradient** describes the greater biodiversity of living organisms at the tropics than at the poles in the biomass, the number of individuals, and, in many taxonomic groups, the number of species (Hawkins (2001) *Trends Ecol. & Evo.* **16**). Diversity gradients described for the northern hemisphere seem to be invalid for the southern hemisphere (Platnick (1991) *Natural History* **25**) and 'despite this recognition of the generality of latitudinal diversity gradients, our knowledge is biased towards some taxonomic groups, regions and ecosystems' (Boyero (2006) *Ecology Info.* **32**).

biodiversity crisis M. Novacek (2001) reports that Earth is suffering the fastest mass extinction in its history, which is clearly a direct result of human activity, 'yet the public remains unaware of the crisis in sustaining biodiversity'. E. O. Wilson (1992) calculates that if the annual extinction rate is 0.27%, and tropical forests contain 50% of all species, then about 0.14% of all species are committed to extinction each year. Unless immediate action is taken to restrain the rate of tropical habitat loss, the twenty-first century will see another major extinction catastrophe. See Pimm et al. (1995) *Science* **269**, 5222, and Gaston and Fuller (2007) *PPG* **31**, 2.

biodiversity hotspot Myers et al. (2000) *Nature* **403** identify 25 global hotspots, which together comprise only 1.4% of the earth's terrestrial surface, but which, among them, contain 44% of all species of vascular plants, and 35% of all species of four vertebrate groups. Examples include the Mediterranean basin, Madagascar, the Western Ghats, and Sri Lanka.

biogenic Produced or brought about by living organisms.

biogeochemical cycle The cyclical movement of energy and materials within *ecosystems. See W. Schlesinger (1997).

biogeographic realm A continent- or subcontinent-sized area, separated from other biogeographical realms by physical barriers to plant and animal migration, such as major mountain ranges, and oceans: for example, the Afrotropical Biogeographic Realm, which contains one of the highest levels of biodiversity on Earth (see Udvardy (1975) *IUCN Occ. Paper* **18**). Each biogeographic realm may include a number of different biomes. The use of *ecozone* for biogeographic realm is a fairly recent development; see J. Schultz (2005).

biogeography The spatial analysis, and the search for patterns, in any biological feature: demographic, ecological, genetic, morphological, or physiological; see B. Cox and P. Moore (1999). 'Biogeography transcends classical subject areas ... **ecological biogeography** is concerned with ecological processes occurring over short temporal and short spatial scales, whereas ... **historical biogeography** is concerned with evolutionary processes over millions of years on a large, sometimes global, scale' (Crisci (2001) *J. Biogeog.* **28**, 2). Brown and Lomolino (2000) *Glob. Ecol. & Biogeog.* **9** write that 'the discipline of biogeography is again, as in the 1960s and 1970s, in transition and ferment'. Mackey et al. (2008) *J. Biogeog.* **35**, 2 observe two basic methodologies in the study of **biogeographical regionalization**: the integrated survey method and the parametric approach: 'to help reconcile these different approaches, we propose a simple, four-step, flexible and generic framework.'

biogeomorphology Any study uniting biota with geomorphic form and process (Osterkamp and Friedman (1997) *Procs USGS Sed. Workshop*). Many biogeomorphological studies concentrate on the processes that link biotic and geomorphic systems, such as bioweathering, bioerosion, bioconstruction, biotransformation, biostabilization, and bioprotection (Naylor et al. (2002) *Geomorph.* **47**, 1).

Many 'soft engineering' schemes for environmental management use biogeomorphological systems; see Goodson et al. (2002) *Geomorph.* **47** on *river restoration.

biohistory The interplay between human society and the *biosphere (Medley et al. (2003) *AAAG* **93**, 1). S. Boyden (1987) gives an account of the stages of civilization, from the primeval and early farming stages to the current high-energy phase, discussing economics, health and disease, and ecology.

biological control The attempt to reduce numbers of pests by the use of predators, either from within the community, or by introduction from outside, rather than by chemicals. See Strobel (1992) *Sci. Amer.* **265** on the biological control of weeds; see also Mannion (1993) *GeoJournal* **31**, 4. Determination of the origins of species is important in the search for biological control agents in the source region (Wardill et al. (2005) *J. Biogeog.* **32**, 12). Clough et al.

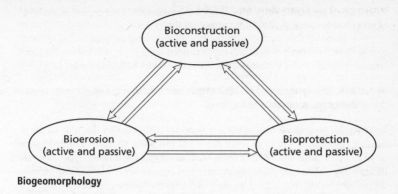

Biogeomorphology

(2005) *J. Biogeog.* **32**, 11 observe that natural biological control may be more effective in certain regions than in others.

biological crust A soil crust dominated by algae, lichens, or mosses. Biological soil crusts stabilize soils against wind erosion, and contribute to some of the organic dusts (McTainsh and Strong (2007) *Geomorph.* **89**, 1-2).

biological invasion The establishment of species in ecosystems to which they are not native. Invasive species cause significant ecological harm: they can alter ecosystem processes, act as vectors of disease, and reduce biodiversity. Worldwide, out of 256 vertebrate extinctions with an identifiable cause, 109 are known to be due to biological invaders (Olsen and Santanou (2002) *Am. J. Agric. Econ.* **84**, 5). Simberloff and Alexander (in P. Calow, ed. 1998) estimate that around a quarter of the value of US agricultural output is lost to non-indigenous plant pests or the costs of controlling them.

Case studies abound, among them the UK Environment Agency's material on American signal crayfish and Japanese knotweed, which is also of interest for its use of anthropomorphic phrasing: 'because [Japanese knotweed] does not originate in the UK, it does not compete *fairly* with our native species' (author's italics). Olson and Roy (*Univ. Maryland, Ag. & Res Econ., W. Paper* 03–06) develop an economic model of a biological invasion, and Peterson (2003) *Q. Rev. of Biol.* **78** claims that the geographic course that invasions are able to take can be predicted.

biological magnification The build-up of a toxin as it moves up the *food chain. For example, pesticides wash into waterways, are eaten by aquatic organisms which are in turn eaten by large birds, animals, or humans, and become concentrated in tissues or internal organs. Since aquatic food chains have more links, biological magnification is greater; see, for example, Sutherland et al. (2000) *Arch. Env. Contam. & Toxic.* **38**, 4.

biological oxygen demand (BOD) An indicator of the polluting capacity of an effluent where pollution is caused by the take-up of dissolved oxygen by micro-organisms that decompose the organic material present in the effluent. Excess biological oxygen demand loads caused large-scale fish kills in the Cape Fear and Neuse River estuaries (Mallin et al. (1999) *Ecol. Appl.* **9**).

biomass The total mass of all the organisms inhabiting a given area, or of a particular population or *trophic level.

biome An ecological zone whose uniformity is defined by the type of plant life that dominates within it, since plant life will often give a strong indication of other ecological features of a zone, such as animal life and soil type. Because biomes are defined by plant life rather than region, some biomes can stretch around the world. For example, the Boreal Forest biome (defined by the predominance of conifers) covers much of northern Europe, Russia, and Canada (Woodward, no date, *Env. Literacy Council*). Other major biomes are: *tundra, temperate (deciduous) forest, *tropical rain forest (selva), tropical grassland (*savanna), temperate grassland (steppe), and *hot deserts. A biome is an idealized type; local variations within a biome are sometimes more significant than variations between biomes.

biophysical diversity The diversity of the natural environment, which controls the intrinsic quality of the natural resource base which is utilized for production. It includes soil characteristics and their productivity, the biodiversity of natural (or spontaneous) plant life, and of the soil biota. It takes account of physical and chemical aspects of the soil, surface and near-surface physical and biological processes, hydrology, microclimate, and also variability and variation in all these elements (Brookfield and Stocking (1999) *Glob. Env. Change* **9**, 2). Geertsima and Pojar (2007) *Geomorph.* **89**, 1–2 recognize several types of biophysical/ecological diversity: site diversity, soil diversity, and derivative habitat or ecosystem (including aquatic ecosystems) diversity.

biopiracy A description of the purchase, by actors in *more economically developed countries, of biological materials at exploitatively low prices. V. Shiva (1997) maintains that this system of exploitation, continuing under the auspices of the World Trade Organization, now treats biopiracy as a 'natural right of Western corporations, necessary for the development of Third World communities'. Shiva writes that Western capital is now seeking out new colonies, new properties—the interior spaces of women, plants, and animals—to invade and exploit.

bioprospecting The methodical search of wildlife for designs, genes, natural compounds, and whole organisms which have a potential for product development. Bioprospecting can be seen as: a conservation process that offers financial benefits, knowledge transfer, and training to biologically rich, but economically poor, nations, as *biopiracy, or as 'one more troubling example of "post-modern ecological capital" in action, representing the further commodification of nature for profit purposes' (Castree (2003) *TIBG* **28**, 1).

biosphere The zone where life is found; the outer portion of the geosphere and the inner portion of the *atmosphere. This extends from 3 m below the ground to some 30 m above it. The biosphere also comprises that region of waters, some 200 m deep, where most marine and freshwater life is found. *Gaia theory postulates biosphere-scale evolution; Free and Barton (2007) *Trends in Ecol. & Evol.* **22**, 11 ask whether and why the biosphere might tend towards stability and self-regulation (but don't look for a clear-cut answer). See Phillips (2008) *PPG* **32**, 1 on goal functions in ecosystem and biosphere evolution.

biotic Living or having sustained life. **Biotic interaction** is a mutual or reciprocal interaction between living organisms, in which the distribution of one species is influenced by the distribution of other species. Biotic interaction can be within the same group, for example, competition, facilitation, and parasitism in plants, or between groups, as best exemplified by relationships in food webs: herbivory, predation, and symbiosis (Guisan et al. (2006) *J. Appl. Ecol.* **43**). See Gaines in D. Sax et al. 2005 on what species invasions reveal about biotic interaction.

   Biotic potential is the maximum population that an area can support. As the local ecological system approaches the biotic potential of a site, local stability rises, but global sustainability falls; as the system develops feedback loops and complex interactions, it becomes better adapted to local conditions and more stable to local impacts. In the long run, conditions keep changing, so that the system becomes sub-optimal to the changed conditions (O'Neill (2001) *Ecology* **82**, 12). **Biotic homogenization** describes the process by which species invasions and extinctions increase genetic, taxonomic, or functional similarity; 'evidence is growing that changes in land cover are likely to be accompanied by the ongoing homogenization of biotas, and processes in which native species are replaced by a relatively small set of alien species (Gaston and Fuller (2007) *PPG* **31**, 2). See Cassey et al. (2008) *J. Biogeog.* **35**, 5.

bioturbation The stirring or mixing of sediment or soil by organisms, especially by burrowing or boring. Statzner and Sagnes (2008) *Geomorph.* **93**, 3–4 assess the joint effects of bioturbaters with different mechanistic abilities.

birth rate The crude birth rate is the number of live births per year, per 1000 population. This does not take the age structure of a population into account, and hence the numbers of women who are capable of giving birth in any year, making accurate comparisons difficult. Because of this, many demographers prefer to use a **standardized birth rate** which indicates what the crude birth rate would have been for a population if the age and sex composition of that population were the same as in a population selected as standard, allowing comparisons to be made between geographical areas, or across social groups within a society (see Fargues (1997) *Pop. & Dev. Rev.* **23**, 1 for a demonstration of standardization).

black box The view that the workings of the human mind cannot be analysed; all that can be observed is the input and output. See Fox (1995) *J. Trans. Geog.* **3**, 2.

black earth *See* CHERNOZEM.

black economy The part of the job market which is not reported to the tax authorities and does not appear in official statistics. 'Black' and 'grey' market activities encompass a very diverse set of practices that range from individual bribes and favours to get things done (C. Humphrey 2002), to protection rackets, and to the illegal appropriation of Western aid payments. The trafficking of women and children is a particularly sinister new form of black economy (Smith and Stenson (2006) *PHG* **30**, 2). See also Kligman and Limonelli (2004) *Eur. Studs. Conf. Chicago*.

blanket bog A continuous covering of bog, needing high humidity and rainfall, mostly comprising peat; only steep slopes and rocky outcrops are dry. Huang (2002) *J. Biogeog.* **29**, 2 finds that human impact was the dynamic force responsible for the destruction of the woodland and the formation of blanket bog in upland Ireland. In the UK Peak District blanket bog has suffered a decline in species diversity as a result of 'air pollution, overgrazing, inappropriate or accidental burning, peat extraction and past drainage' (*Blanket Bog Action Plan* 1, section 6.4).

blind valley A steep-sided karst valley, abruptly terminating in a *streamsink.

blizzard A wind, over 50 kph, which whips up particles of ice and dry, powdery snow, reducing visibility to less than 200 m. See Poulos et al. (2002) *Weather & Fcst* **17**, 5.

bloc An alliance of governments, groups, or parties with a common purpose. Trading blocs, such as the *North America Free Trade Agreement are usually geographically contiguous. Poon et al. (2000) *TIBG* **25**, 4 dismiss the notion that international trade is increasingly organized around, three economic blocs centred around the United States, European Union, and Japan—the 'triad'—and show that investment intensity patterns do not currently conform to any bloc-like formation, but exhibit instead globally diffused network regions. Badinger and Breuss (2006) *Rev. Int. Econ.* **14**, 4 investigate the gains for each member from enlarging a trade bloc, finding that small countries increase in competitiveness by attaining easier access to a larger market, while large countries may have advantages over small countries via relatively more group ties, high market power, and related terms-of-trade effects, advantages in competitiveness due to economies of scale and larger absolute endowments (particularly human capital), larger product varieties (reflecting the greater scope for specialization), and technological advantages.

blockbusting A technique, used by estate agents in the USA, of inclining residents of a white neighbourhood to move out because they fear that the district is to be taken over by black families. See Hanlon (2008) *Annual Meeting AAG* Boston.

block field (block stream) A sheet of angular rock fragments in spreads or lines in a *periglacial landscape. Blockfields (*felsenmeer*) develop on slopes under 5° (A. L. Washburn 1979), or under 10° (Caine in White (1976) *Quat. Res.* **6**). See van Steijn et al. (2002) *PPG* **26**, 4. Thompson and Syverson (2006) *GSA Abstracts with Progs*, **38**, 4 maintain that this debris is formed *in situ* by *freeze–thaw. Others suggest that the blocks have ridden down on the top of saturated debris during *gelifluction (Creemeens et al. (2005) *Geomorph.* **70**, 1–2).

blocking A meteorological condition when a pressure system remains stationary, for perhaps weeks. Blocking anticyclones develop when the *Rossby pattern changes from zonal to strongly *meridional, often forming one or two high-level, closed anticyclonic circulations; see Schopf et al. (1981) *Dyn. Atmos. Oceans* **5**. The upper air flow then guides depressions around the edge of the anticyclone(s). Renwick and Revell (1999) *Month. Weath. Rev.* **127**, 10 suggest that *Rossby wave propagation links anomalous convection in the tropics and blocking over the south-east Pacific Ocean.

block lava (aa) This lava has a thick skin broken into jagged blocks. See Navarro-Ochoa et al. (2002) *J. Volcan. & Geotherm. Res.* **117**, 1–2.

block mountain An area of upland identified with an uplifted area bounded by *faults. The Sierra Nevada is a massive block mountain range that broke free along a bounding fault line and has been uplifted and tilted (Huber (1987) *US Geol. Surv. Bull.* **1595**).

Blockschollen flow A glacier flow in glaciers with steady velocity almost to the edges, then falling sharply, creating a highly erosive, strong shear force near the valley walls. The lower threshold of Blockschollen flow is a mean velocity to width ratio of 0.16 (Finsterwalder (1950) *J. Glaciol.* **1**).

blowhole A crack in the top of a cliff through which air and sea water blow. The blowhole is fed from the seaward end via joints and tunnels.

blow-out (blowout) A saucer-, cup-, or trough-shaped depression or hollow formed by *deflation on a pre-existing sand deposit. The adjoining accumulation of sand, the depositional lobe, derived from the depression, and possibly other sources, is normally considered part of the blow-out (Hesp (2002) *Geomorph.* **48**, 1–3). Blow-outs usually result from the removal of vegetation; see Dech et al. (2005) *Catena* **60**, 2.

bluff A steep, almost vertical, cliffed section of a river bank.

body The body may be seen as: the smallest unit of geography upon which may be inscribed power and resistance (Camp (2002) *J. Southern Hist.* **68**); a map of meaning and power (Simonsen (2000) *TIBG* **25**); or a cultural representation of masculinity or femininity (R. Ainly 1998). It is also a form of reference by which supposedly 'disembodied' dominant cultures designate certain groups—the elderly, ethnic minorities, females, the obese, the disabled, and so on—as *other (N. Duncan 1996; A. Blunt and G. Rose 1994; and S. Pile and N. Thrift 1995). The figure of the body can be a metaphor for understanding socio-spatial relations in contemporary culture (Dyck and O'Brien (2003) *Canad. Geogr./Géog. canad.* **47**, 4).

    Sexed bodies can create spaces, as in a gay pride parade (Johnston (2001) *Annal. Tour. Res.* **28**, 1). The notion of the **raced body** tied to the land 'is not novel, having been used to discredit urban Aboriginals, claiming that once they leave the land, they cease to be Aboriginal' (Wazana (2004) *Refuge* 01–03–2004). Every researcher—including you—is situated in her or his own body; *embodied*.

bohemian Someone with creative or literary interests, who is socially unconventional. Florida (2002) *J. Econ. Geog.* **2** finds strong relationships between high concentrations of bohemians and concentrations of high human capital, with a particularly strong relationship between the concentrations of bohemians and concentrations of high-technology industry. See Wojan et al. (2007) *J. Econ. Geog.* **7**, 6.

bolide A meteoric fireball, or asteroid; see Dawson and Stewart (2007) *PPG* **31**, 6.

Bølling The mild Bølling–Allerød interstadials began *c.*13000, and ended between 12800 and 12300 years BP (Brauer (2002) *Quat. Int.* **88**). In Britain the Bølling is not usually distinguished from the *Allerød.

Bora (fall-wind) A cold winter wind blowing down from the mountains onto the eastern Adriatic coast. See Pullen et al. (2007) *J. Geophys. Res.* **112**.

border A boundary line established by a state, or a region, to define its spatial extent—'central multi-scalar nodes where power, place and identity intersect' (Kaiser and Nikiforova (2006) *Ethnic & Racial Studs.* **29**, 5). Borderlands are fundamental to the way that citizenship, identity, and the nation are imagined, policed, and performed. While in the European Union borders have become increasingly porous, struggles over the demarcation of borders have caused major conflicts in, for example, the former Yugoslavia; see Corson and Turregano (2002) *GeoJournal* **57**, 4 and Klemenčić and Gosar (2000) *GeoJournal* **52**, 2, who also examine 'the unintended consequences of creating an uncontrolled political and security space in the midst of a zone of conflict'.

K. Ohmae (1999) is not alone in identifying a move to a borderless world, brought about by *globalization, a conclusion contested by Yeung (1998) *TIBG* **23**, 3 and Sparke et al. (2004) *TIBG* **29**, 4. See also W. Maley et al. (2000) on refugees and the 'myth of the borderless world'.

b

(⊕)) SEE WEB LINKS

• The e-journal *borderlands*, published by the University of Adelaide.

bore A positive surge that may form at flood tide in shallow estuaries with a large tidal range (Wolanski et al. (2004) *Estuarine, Coast. & Shelf Sci.* **60**, 4).

Boreal (Period) A division of the *Flandrian, *c.*10 000–8 000 years BP.

bornhardt A dome-shaped rock outcrop more than 30 m high, and sometimes several hundred metres in width, commonly rising above erosional plains in the tropics, but also found in unglaciated uplands in high latitudes. The origin of these features is problematic: faulting, lithological control, cross folding, and contrasts in fracture density have been cited. C. Twidale and J. R. Vidal Romaní (2005) give an excellent summary.

borral A soil order of the *US soil classification. *See* CHERNOZEM.

Böserup model E. Böserup's (1965) view that increases in population size stimulate agricultural change. 'Pre-industrial farmers, according to the Boserup model, were almost always technically capable of increasing the productivity of their land by applying methods (soil tillage, additional weeding, application of organic fertilisers and irrigation) which allowed it to be cropped more frequently and fallowed for shorter periods. But because such methods typically demanded higher labour inputs per unit of production, they tended to be adopted only when population pressure and the resulting land scarcity forced farmers to work harder in order to maintain existing levels of subsistence' (Henley (2005) *Asia Pac. Viewpt* **46**, 2). Brookfield (2001) *Asia Pac. Viewpt* **42** has doubts, observing that agricultural intensification has more often been inspired by price incentives to surplus production, rather than though population pressure.

boulder clay A now outmoded term for glacial *till.

boulder field As distinct from a *block field, this is an area of boulders which is the result of *spheroidal weathering. See Sevon (1987) *N-E Section GSA* **5**.

boundary detection Mostly in biogeography and plant ecology, the identification of the zones of greatest change in vegetation and associated abiotic/biotic factors; see Jacquez et al. in T. Poiker and N. Chrisman, eds (1998). Kent et al. (2006) *PPG* **30**, 2 are a model of clarity in discussing boundary detection. The extent to which geographically isolated regions support local diversification should predict the placement of biogeographical boundaries (Patten and Smith-Patten (2008) *J. Biogeog.* **35**, 3).

boundary layer Any layer of the *atmosphere significantly affected by the earth's surface. The **laminar boundary layer** is the few millimetres above the surface; the **turbulent boundary layer** denotes the conspicuously turbulent part of the atmosphere.

bounded rationality The idea that individuals strive to be rational, after first greatly simplifying the choices available. In choosing between a small number, the decision-maker acts as a *satisficer. Conlisk (1996) *J. Econ. Lit.* **34** argues for incorporating bounded rationality in economic models.

Bowen ratio The ratio of sensible heat to *latent heat. The Bowen ratio–energy balance (BREB) is a micrometeorological method often used to estimate latent heat flux because of its simplicity, robustness, and cost (Todd et al. (2000) *Ag. & Forest Met.* **103**, 4).

braid bars *See* BAR.

braided channel A river channel in which *bars and islands have been deposited, differing from *anastomosis since the islands are less permanent. Braiding occurs with frequent variations in discharge: when the river cannot carry its load; where the river is wide and shallow; where banks are easily eroded; and where there is a copious bed load. Richardson and Thorne (2001) *Geomorph.* **38**, 3–4 show that the division of the flow into multiple threads is a prerequisite for bifurcation. See Bledoe and Watson (2001) *Geomorph.* **38**, 3–4 for an excellent summary.

braunerde *See* BROWN EARTH.

Brazil current A weak boundary current, flowing in the upper 600 m of the water, carrying warm subtropical water along the coast of Brazil—from about 9° S to about 38° S, where it joins the Malvinas Current. The two currents create the strong Brazil-Malvinas Confluence thermohaline frontal region (Bischof et al., no date, *CIMAS*).

break point (breaking point) The point at which the field of influence of one settlement ends, and that of another begins. See Haberkern (2002) *Issues in Shopping Centre Trade Areas* for W. Reilly's (1931) original equation, and a number of modifications to it.

break point bar *See* BAR.

Bretton Woods A shorthand for the system of international finance created in 1944 with the establishment of the *International Monetary Fund, the International Bank for Reconstruction and Development (*World Bank), and the creation of a system of fixed exchange rates. This system lasted until 1971, when it gave way to floating exchange rates.

In 2007, the World Economic Forum and the Reinventing Bretton Woods Committee (RBWC) made specific proposals for improvements in the exchange rate system that could help to prevent large, persistent economic imbalances of the kind that currently threaten world financial stability.

(((()))) SEE WEB LINKS

• The International Monetary System, the IMF, and the G-20.

brown earth (brown forest soil) A free-draining *zonal soil associated with deciduous woodland. Generally, brown earths have a thick litter layer and a *humus-rich A *horizon, containing iron and aluminium sesquioxides in small, crumb-like *peds. The weakly developed, lighter, B horizon contains blocky peds; there is little *leaching. See Storrier and Muir (1962) *Eur. J. Soil Sci.* **13**, 2.

brownfield site Any previously used land and/or premises that is not currently fully in use, and may also be vacant, derelict, or contaminated. In an outstanding paper, dealing with both principles and practice, Raco and Henderson (2006) *Local Env.* **11** assess the role of brownfield development in urban and regional policy in the UK.

buffer state A generally neutral state, lying between potentially belligerent neighbours; 'protective devices made obsolete by modern systems of communication' (Spykman (1947) *Geog. Rev.* **32**). See V. Pholsena and R. Banomyong (2007) on Laos.

buoyancy The quality of an air parcel allowing it to rise through, and remain suspended within, the *atmosphere, due its relative warmth, and therefore lighter density. Should its water evaporate, the parcel will suffer an *adiabatic heat loss, and lose buoyancy. See Dupilka and Reuter (2006) *Weather & Fcst* **21**, 3 on forecasting tornado potential.

bush encroachment The change in vegetation from open *savanna, or mixed grass and woodland, to scrubland. Occurring also in Australia and South Africa, bush encroachment affects the agricultural productivity and biodiversity of 10–20 million ha of South Africa (Ward (2005) *Afr. J. Range & Forage Sci.* **22**, 2). Its probable causes are: increased grazing levels; reduced use of burning to create grassland; changes in rainfall regimes; and the interactions between these factors. However, writing in 2005, Ward (op. cit.) writes, 'many people believe that we understand the causes of bush encroachment. We do not.'

bush fallowing A farming system whereby the fallow period is longer than the farming period. Hance (1969, in Omofonmwan and Kadiri (2007) *J. Hum. Ecol.* **22**, 2) argues that bush fallowing is a major hindrance to food production in Nigeria, but Chokor (1993) *Env. Manage.* **17**, 1 argues that 'modern' methods, which shorten the fallow period, disrupt 'the conservation principles inherent in the traditional man–nature relationship'.

business cycle Economies seem to rise and fall cyclically. Saks and Wozniac (2007) *IZA Discuss. Paper* **2766**, link long-distance migration with business cycle fluctuations. Wall (2006) *FRB St Louis, W. Paper* 2006–035A finds regional variations, and an increase in the time period between successive expansions in Japan, 1976–2005.

bustee (basti) In India, a *shanty town; reading S. Mehta (2005) should be mandatory. N. Islam (1996) estimates that almost 50% of Dhaka's population live in bustees, and a major portion of this population is female. To combat negative views of bustees, see Mahmud (2003) *Cities* **20**, 5.

butte A small, flat-topped, unvegetated, and very steep-sided hill of layered strata. It may be a volcanic neck or a cone that has resisted erosion, or the residue of a *mesa.

C3 and C4 plants C3 plants make up over 95% of earth's plant species. They flourish in cool, wet, and cloudy climates, where light levels may be low, because the metabolic pathway is more efficient there. With enough water, the stomata can stay open and let in more carbon dioxide. However, carbon losses through photorespiration are high. C4 plants are able to raise the intercellular carbon dioxide concentration at the site of fixation, thus reducing, and sometimes eliminating, carbon losses by photorespiration. C4 plants inhabit hot, dry environments, have very high water-use efficiency, and can double the C3 photosynthesis, but C4 metabolism is inefficient in shady/cool environments. Less than 1% of Earth's plant species can be classified as C4, but they account for around 20% of global gross primary productivity (Wand et al. (1999) *Glob. Change Biol.* **5**, 6).

Large shifts in carbon isotopes can indicate a vegetative change from C4 to C3 plants (see L. Huenneke, no date, Jordana Basin LTER).

cadastre 1. A record of the area, boundaries, location, value, and ownership of land, achieved by a **cadastral survey**.

2. The term is also used in *geographic information systems (GIS). CARIS (Computer Aided Resource Information System) is a GIS system for hydrography (see Mioc, Anton, and Nickerson, no date, for a web-based GIS model for predicting flood events) and cadastral systems (see Steudler (2004) *GIS@development* for a worldwide comparison of cadastral systems.)

caesium analysis Caesium-137 was deposited as fallout in the 1950s and 1960s, and, in most environments, was absorbed into the soil. If this fallout was uniform, negative deviations in the distribution of caesium-137 would represent soil erosion, while positive deviations would indicate deposition. See Hardy et al. (2000) *J. Hydrol.* **229**, 3–4.

Cainozoic *See* CENOZOIC.

calcrete A *duricrust made mostly of calcium carbonate, forming in arid climates via capillary action and prolonged evaporation. See Stokes et al. (2007) *Geomorph.* **85** on groundwater, pedogenic processes, and fan dynamics in calcrete development.

caldera A sunken crater at the centre of a volcano, formed when the magma founders, so that the centre of the volcano collapses. Kennedy and Stix (2007) *GSA Bull.* **119** examine magmatic processes and caldera collapse in New

Hampshire; and Furuya (2003) *Earth Planets Space* **55** constructs an optimum source model of caldera formation.

calibration 'The process by which the controlling parameters of a model are adjusted to optimize the model's performance, that is the degree to which the model's output resembles the reality that the model is designed to simulate . . . While validation may actually be impossible in some cases (see Oreskes et al., 1994 *Science* **263**, 641–646), calibration is obviously among the most essential obligation of modelling' (Clarke in P. Atkinson et al., eds. 2005). Lajoie and Hagen-Zanker (2007) *Cybergeo: systs., modél., géostat.* **405** create and calibrate a robot land-use model.

California current A cold *ocean current off the California coast, given to anomalous conditions, that can be termed a Californian *El Niño (Mullin (1998) *Glob. Change Biol.* **4**, 1).

calving The breaking away of a mass of ice from an iceberg, ice front, or glacier, caused by: wind and/or wave action; over-extension of an ice shelf; or a collision with an older iceberg. Calving represents the major form of *ablation from a glacial system. See Benn et al. (2006) *PPG* **30**, 3 and Skvarca et al. (2002) *Quat. Sci. Revs.* **21**.

cambering In *periglacial landscapes, the rounding at the edge of a cap rock underlain by clays. The clays are squeezed out as they thaw, and the overlying cap rock sag. See Horswill et al. (1976) *Phil. Trans. Royal Soc. A* **283**; and Kimber et al. (1998) *TIBG* **23**, 3.

Cambrian The oldest period of *Palaeozoic time, approximately 570–500 million years BP; the time when most of the major groups of animals first appear in the fossil record.

(((●))) SEE WEB LINKS

• Provides an overview of the Cambrian period.

canyon A valley with very steep sides and no floor, such as the Grand Canyon of the Colorado River. A canyon differs from a *gorge in that the sides are stepped, reflecting alternating rock resistances. See Spamer (2006) *Annal. Imposs. Res.* **12**, 2, who sounds off about the name 'Grand Canyon', but is quite sane in his book of 1984. **Submarine canyons** are deep troughs in the sea bed, sometimes as prolongations of river valleys on land; see Morais et al. (2007) *Marine Ecol.* **28**, and suppl. 1. Canyons may also form via earth flows, turbidity currents, submarine springs, or the slipping of sediments.

capability constraint In *time–space geography, a limit to one's actions, due to biological needs, like sleep, and to restricted facilities, like access to transport. See Golob and Regan (2001) *Transp. Res C* **9**, 87–91 and Axhausen in D. Ettema and H. Timmermans, eds. (1997).

capacity The maximum amount of debris that a stream can move as *bed load. Capacity depends on *discharge, and on the nature of the load; for the same volume, a stream may be able to carry more gravel than boulders. See Prosser and Rutoni (2000) *PPG* **24**.

capillary In a soil, the fine spaces between soil particles. **Capillary action** is the movement of a liquid in a very narrow space, such as a soil pore, due to surface tension, cohesion, and adhesion. See N. Brady and R. Wei (1998).

capital One of the four *factors of production, capital includes all the items designed by society to further the creation of wealth. Plant, machinery, and buildings are **fixed capital**—they earn profit without circulating further; see Melachroinos and Spence (2001) *PHG* **25**, 3. D. Harvey (1982) considers the coercion that fixed capital exerts upon capitalists. **Circulating capital (floating capital)** includes raw materials, fuels, components, and labour inputs which are then sold—in the form of the product—at a profit; see Henderson (1998) *Antipode* **30**, 2. **Financial capital** is the money necessary for production; although G. Duménil and D. Lévy (2004) argue that financial capital is unproductive.

Commercial capital mediates in the circulation of commodities for a fee, but has no long-term control over it; 'for Marx, a specialized commercial capital and financial capital drives down the costs of circulation and finance—reducing the deductions they make from the profits of industrial capital and speeding up capital turnover and reinvestment' (Dunford and Pickvance (2007) *PPG* **31**, 4).

Knowledge capital encompasses 'talents, skills, know-how, know-what, and relationships' (T. Stewart 2001). Amin and Thrift (2007) *PHG* **31**, 2 link urbanism with **fast capitalism**—rapid information and communication technologies together with their impact on self, society, and culture. See P. Taylor (2004).

Capital is seen as a social relation, symbolizing the economic relationships between individuals and/or groups; 'the result of social labour achieved in the creation of goods and services' (Tickell and Peck (1992) *PHG* **16**; see also Pasinetti in M. Baranzini and R. Scazzieri, eds. 1986).

Geographers focus on capital as a source of *uneven development; Walker (1978) *Rev. Radical Polit. Econ.* **10** is still worth quoting: '[the] mobility of capital means that capital may use location as a strategy against labour, local development becomes more dependent on outside capital, and development comes and goes over time . . . and capital generates a permanent reserve of stagnant places; a lumpengeography of capital.' *See* SPATIO-TEMPORAL FIX. Grettler (2001) *Env. Hist.* **6** argues that capitalism has been the greatest force for environmental change in America; see also M. W. Lewis (1992).

capital intensive With a high ratio of *capital to labour. Barthelmie (2007) *Geog. Compass* **1**, 3 notes that, because wind farms are highly capital-intensive,

European wind farms are generally owned by utilities, which have access to more attractive financing packages.

capitalism An economic system based on private property and private enterprise, with a major proportion of economic activity carried on by private profit-seeking individuals or organizations, and other material *means of production largely privately owned. Capitalism never stands still. Its central imperative—the search for profit and wealth creation—'drives a perpetual process of economic flux' (Boschma and Martin (2007) *J. Econ. Geog.* **7**, 5). This flux is powered by innovation: 'the production, acquisition, absorption, reproduction, and dissemination of knowledge is seen by many as the fundamental characteristic of contemporary competitive dynamics' (Gertler (2003) *J. Econ. Geog.* **3**).

'Innovation and growth are at the heart of the capitalist system . . . As new large-scale technologies develop; new forms of firm and factory organisations result in increasingly global value chains producing for global market' (Kaplinsky (2008) *Geog. Compass* **2**, 1). See also Metcalfe et al. (2006) *Camb. J. Econ.* **30**, 7–32. *Accumulation is at the heart of capitalism—Buck (2007) *Antipode* **39**, 4 notes the periodic crises of over-accumulation. A feature of **advanced capitalism** is the possession of capital by fewer and fewer owners.

Capitalism is not uniform: 'nation-states have different trajectories of capitalist development, in which there is considerable variation in the role of markets' (J. Hollingsworth and R. Boyer 1997). Peck and Theodore (2007) *PHG* **31**, 6 note 'the persistence of . . . real institutional differences—and the instituted space-economies with which they are associated' and Soskice (1997) *Indust. & Innov.* **4** argues that different national institutional frameworks support different forms of capitalism. See H. Wai-Chung Yeung (2006) and Buck (op. cit.) on Chinese capitalism; E. M. Wood (2003) on English capitalism, M. Wiener (1985) on British capitalism, 'enervated by gentry values'; and Clarke (2007) *TIBG* **32**, 1 on British capitalism 'that conceals its destructive effects under the cloak of oldness'. Colas (2005) *Contemp. Polit.* **11**, 2–3 explains that capital constantly searches for fresh markets, leading inevitably to 'spatial expansion, as new lands, peoples and resources are exposed to the "law of value"'. 'Spatial differentiation is the fundamental building block of capitalism' (Clark (1985) *Econ. Geog.* **61**, 3). **Vagabond capitalism** is an increasingly global capitalist production that can discard many of its commitments to place: 'at worst, this disengagement hurls certain people into forms of vagabondage; at best, it leaves people in all parts of the world struggling to secure the material goods and social practices associated with social reproduction' (Katz (2001) *Antipode* **33**, 4).

D. Harvey (1985) holds that the inner contradictions of capitalism are expressed through the formation and re-formation of geographical landscapes. Cox, Ohio State U., picks up on this: 'instead of towns organized around cathedrals we have towns organized around "central business districts". Capitalist societies are characterized by extraordinarily differentiated geographic divisions of production.' Similarly, Clark (op. cit.) claims that 'the *longue durée o*f a capitalist development trajectory diverted Britain away from the path bequeathing continental European cities their particular flavour in favour of a voracious, London-centred colonial capitalism'. Brown (2007) *PHG* **32,** 4 sees enclosure—the appropriation and privatization of previously common, public or open access resources—as promoted by global capitalism, and commodification and capital.

'Capital reproduces the peaceful civility of the private contract amongst seemingly free and equal individuals so long as the exploited party accedes to its own subordination—once there is significant resistance to the terms and conditions of the contract, the hidden hand of the market becomes the visible fist of the state' (Colas, op. cit.). I. Boal et al. (2005) go further: 'war is virtually *synonymous* with the capitalist state.' 'Urban transformation processes are largely the outcome of space wars, in that deliberate and systematic creative destruction is part of the logic behind current capitalist space economy. These spatial struggles have often been staged in city streets' (Hansen (2008) *Liminalities* **4**, 1).

Buck (2007) *Variant* **28** doesn't believe that human-induced global warming will finally destroy capitalism: 'while capitalism may survive, this is not to say that we can safely embrace rosy visions of utopian futures.' See Peck and Theodor (op. cit.) and Faulconbridge (2008) *J. Econ. Geog.* **8**, 4 on varieties of capitalism.

capitalist state The state provision of *infrastructure, *external economies, and a regulatory framework facilitates *accumulation. The state can limit or increase the freedom of market actors, raise or lower taxes, intervene in major financial crises, and defend and secure its territory—none of which can be accomplished by *transnational corporations. The **theory of the capitalist state** argues that such functions legitimize the state's existence. B. Jessop (2002) focuses upon the structural coupling and co-evolution of accumulation regimes and political regimes.

It is argued that the state's capacity to manoeuvre the accumulation regime has been weakened by international finance, trading *blocs, and so on, leading to a loss of national sovereignty; see Zaky (2005) *Res. Pol. Econ.* **22**.

capital–labour ratio The amount of capital per unit of work. The capital–labour ratio is at the heart of the 'one-sector regional growth model' (McCombie (1998) *Urb. Studs* **25**): as people move from lower real waged cities-regions to higher ones, the capital–labour ratio in stronger regions decreases, and increases in weaker regions. The process stops as soon as the capital–labour ratio is the same in all regions. P. McCann (2001) shows how, in time, this development improves the welfare of society as a whole.

capture (river capture) When a river is extending its channel upstream by *headward erosion, it may come into contact with the headwaters of a less vigorous river. The headwaters from the minor river may be diverted into the more rapidly eroding channel. There is often a sudden change of stream direction at the point of capture; this is the **elbow of capture**. See Mather (2000) *Geomorph.* **34**, 3 for two examples of river capture; see also Bishop (1995) *PPG* **19**.

River capture

carbon cycle Carbon is supplied to the *biosphere as carbon dioxide during volcanic eruptions. Most of this is dissolved in the sea or built into calcareous sediments, which then form limestones and dolomites. As these rocks are folded and raised above sea level, they are subjected to solution by weak carbonic acid and form sediments once more. This is the largest and slowest of the carbon cycles. See Yuan in D. Yuan and Z. Cheng, eds (2002). The shortest cycle involves respiration by plants and animals whereby carbon dioxide is expired, and photosynthesis by plants, changing carbon dioxide and water into organic compounds. Plant species affect ecosystem carbon uptake via biomass production, and carbon release via decomposition. Dorrepaal (2007) *J. Ecol.* **95**, 6 holds that differences in the way plants respond to climate change may feed back to the atmospheric carbon balance, and the climate at a global scale. Uncertainty in understanding the carbon cycle has significant cost implications for a climate stabilization policy, 'with cost differences denominated in trillions of dollars' (Smith and Edmons (2006) *Tellus B* **58**, 5).

Since 1957, only about half of the CO_2 emissions from fossil-fuel combustion have remained in the atmosphere; the rest being taken up by land and oceans. In the face of increasing fossil-fuel emissions, the rate of carbon absorption by the land and ocean has accelerated over time (Canadell et al. (2007) *Proc. Nat. Acad. Sci. USA* **104**). There is no guarantee that the 50% discount will continue; if it disappears we will feel the full climatic brunt of CO_2 emissions from fossil fuels. Climate models that include descriptions of the carbon cycle predict that terrestrial uptake of carbon will decrease in the next century as the climate warms (Miller (2008) *Nature* **451**).

carbonation A form of solution where carbonic acid, formed by the solution of atmospheric CO_2 in water, dissolves minerals. Carbonation is best seen in the solution of calcium-rich rocks, such as limestone, but carbonic acid will also dissolve silicates.

Carboniferous A period of *Palaeozoic time, stretching approximately from 345 to 280 million years BP, and subdivided into the Mississippian and the Pennsylvanian periods.

(((()) SEE WEB LINKS

• An overview of the Carboniferous period.

carbon management The monitoring of carbon emissions and their removal through **carbon sequestration**; that is, the storage of carbon dioxide. Smith et al. (2000) *Glob. Change Biol.* **6** believe that energy crops have the greatest potential for carbon sequestration in Europe, and Bourque et al. (2007) *Mitigation & Adapt. Stats. Glob. Change* **12**, 7 provide a methodology for forestry and carbon sequestration.

CO$_2$ may also be stored underground (see Holt et al. (1995) *Energy Cons. & Manage.* **36**) or below the ocean thermocline. Hoffert et al. (2000) *Science* **298** discuss CO$_2$ sequestration in deep seas, and Herzog et al. (2000) *Sci. Am.* **282** believe that ocean injections can substantially decrease peak atmospheric CO$_2$ (although some CO$_2$ would return to the atmosphere (Kheshgi et al. (1994) *Energy* **19**, 967)). See Harvey (2003) *Geophys. Res. Lett.* **30**, 5 on the impact of deep-ocean carbon sequestration on atmospheric CO$_2$. The pH impacts of ocean CO$_2$ disposal could be diminished by accelerating carbonate mineral weathering (Caldeira and Rau (2000) *Geophys. Res. Lett.* **27**). Harvey (2004) *Climatic Change* **63**, 4 warns that the impact of carbon sequestering on atmospheric CO$_2$ rapidly decreases over time, as there is a reduction in the rate of absorption of atmospheric CO$_2$ by the terrestrial biosphere and the oceans.

carbon sink A phenomenon, such as a forest or ocean, which can absorb atmospheric carbon dioxide. Palmroth et al. (2006) *Procs Nat. Acad. Sci. US* **103** report that the above-ground sink strength in forests controls the allocation of carbon below ground.

carnivore Any animal which eats only animal flesh. At the top of a *pyramid of numbers, **top carnivores** are usually the rarest, largest, and most complex animals. With their large body size, low abundance, and large home range requirements, top carnivores are particularly vulnerable to changes in habitat (Raffaelli (2004) *Science* **306**).

carrying capacity The maximum potential number of inhabitants which can be supported in a given area. The upper limit is set at the point where the environment deteriorates. Carrying capacities are far from being universal constants, but alter with value judgements and objectives. 'In human society, institutional arrangements are likely to alter the carrying capacities and desired levels of populations, and carrying capacities in the shorter term may differ greatly from those in the longer term' (Seidl and Tisdell (1999) *Ecol. Econ.* **31**, 3). See Daily and Ehrlich (1996) *Ecol. Applics* **6**, 4 on increasing equality of opportunity and increasing Earth's carrying capacity.

Seidl and Tisdell (op. cit.) maintain that carrying capacity is a political concept 'generally highlighting that exponential growth, and thus environmental pressures, have to be curbed'. Clarke (2002) *Pop. & Env.* **23**, 4 recounts the difficulties in trying to convert the 'vague carrying capacity concept into a functional, qualitative method' and Benjaminsen et al. (2006) *AAAG* **96**, 3 decry 'the continued dogged insistence on the application of rigid carrying capacity rules and concepts'.

cartel A system whereby producers divide up the market between themselves, avoiding direct competition and not encroaching on each other's share of the market. See Walker in R. Friman and P. Andreas, eds (1999) on drug cartels.

cartography The production and study of maps and charts. Cartography is a system of information which is used to communicate something of the real world to other people; the map is a *model, to be decoded by the map reader; see Lloyd (1989) *AAAG* **79**, 1. All maps are approximations; their clean lines and colours don't reflect the muddled nature of the reality they represent, and they can easily be used to support a point of view; 'a whole arsenal of devices may be used for [persuasive cartography] . . . to influence opinion rather than to inform the reader' (N. Thrower 1996). So that the reader is not distracted by 'noise'—anything which stands in the way of understanding—the map has to be encoded using easily understandable signs, symbols, lettering, and lines; see Board and Taylor (1977) *TIBG* **2**, 1. F. Farinelli (1989) observed that reason 'is nothing more than the protocol of . . . the cartographic image . . . the history of geography is the history of the confusion between the model and the reality'. See J. Pickles (2004) and C. Jacob (2006).

cascade 1. A sequence of small waterfalls.

2. A configuration in which the output of one component is the input of another. Thus, the catchment hydrological cascade is: bedrock, cascade, step-pool, plane-bed, riffle-pool, and dune-ripple (Branefirun and Roulet (2002) *Hydrol. & Earth. Sys. Sci.* **6**, 4). See Couthard and Van Der Wiel (2007) *Geomorph.* **91**, 3–4. 'River systems can viewed at distinctive hierarchical levels that represent a cascade of geomorphological interrelationships'.

(⊕) SEE WEB LINKS

● AUSRIVAS, Parsons et al. (2002).

cascading systems *See* SYSTEMS.

case hardening The production on a rock outcrop of a resistant *weathering rind mostly made of iron and magnesium hydroxides; sometimes of amorphous silica. Case hardening in limestone occurs when the calcium carbonate dissolved at the surface is re-precipitated internally, most commonly close to the surface from which it was just dissolved (Donovan (2002) *Geol. Today* **18**, 4). Case hardening may be related to the production of visors over some *tafoni entrances. See R. Dorn (1998).

cash cropping Growing a crop to produce goods for sale or for barter, rather than for the subsistence of the farmer and family. Tungittiplakorn and Dearden (2002) *Biodiv. & Conserv.* **11**, 11, in a study of increased cash cropping in Thailand, find that, with reduced need for hunting, cash cropping may aid biodiversity conservation. Lim and Douglas (1998) *Asia Pac. Viewpt* **39**, 3 describe the results of pressure from the Malaysian government to switch from shifting cultivation to cash cropping: land shortages, increased use of agricultural chemicals, and the failure of sustainable agriculture.

caste A position in society inherited at birth, and from which there is no transfer throughout life. The system is at its strongest in the Indian subcontinent; see Müller-Böker (1988) *Mt. Res. & Dev.* **11**, 2 on caste and spatial organization in Nepal, and Ostwald and Baral (2000) *Geograf. Annal. B* **82**, 3 on caste, gender, and forest protection in Orissa, India.

cataract A step-like succession of waterfalls. Cataracts, such as those on the Nile, are often associated with the 'rungs' formed by the erosion of horizontally bedded rocks.

catastrophic In geomorphology, any process which is rare, and acts briefly with great force and suddenness. See, for example, Clayton and Knox (2008) *Geomorph.* **93**, 3–4 on the catastrophic drainage of Lake Wisconsin, and Guthrie and Evans (2007) *Geomorph.* **88**, 3–4.

catch crop A fast-growing plant which is intercropped between the rows of the main crop, often used as *green manure. Dressel (1992) *Proc. 2nd Gumpensteiner Lysimeter Day* (in German), shows that the use of a catch crop instead of fertilizer can reduce nitrate discharge by up to 50%.

Catchment Abstraction Management Strategies (CAMS) A strategy by the *Environment Agency to license water abstraction in the UK.

( ) SEE WEB LINKS
• Sets out CAMS policy.

catchment area **1.** The area served by a city, or some function such as retailers, airports, schools, etc. *See* URBAN FIELD.
   2. An alternative term for a *drainage basin/watershed.

catena A sequence of soil types arising from the same parent rock, but distinct from each other because of the variations—such as *leaching and *mass movement—arising from topography. De Alba et al. (2004) *Catena* **58**, 1 present a model of catena evolution.

cave A large, natural, underground hollow, usually with a horizontal opening. **Karst caves** result from solution and *corrosion; see Miller (2006) *GSA Special Paper* **404**.

Catena

cavitation The caving-in of bubbles in a liquid, close to a solid surface, causing shock waves. In a river, cavitation usually occurs downstream of an obstruction. It may aid the fluting and potholing of massive, unjointed rocks (Whipple et al. (2000) *GSA Bull.* **112**, 3).

CBD *See* CENTRAL BUSINESS DISTRICT.

celerity In a river, the square root of the product of the acceleration due to gravity and the mean depth of the river flow. See Richardson and Carling (2006) *Geomorph.* **82**, 1–2.

cellular automata A set of identical cells, each of which occupies a node of a regular, discrete, infinite spatial network (Bandini et al. (2001) *Parallel Comput.* **27**, 5). See M. Batty (2005) on understanding cities with cellular automata, Benenson et al. (2002) *Env. & Plan. B* **29**, and Miller in J. Wilson and A. Fotheringham, eds. (2005).

cellular clouds Roughly hexagonal patches of clouds. The open cell patterns typically found behind cold fronts can be determined by a large-scale sinking motion of a convectively unstable layer (Krishnamurti (1975) *J. Atmos. Scis* **32**).

cellular convection *Convection occurring in semi-regular cells. See Hsu and Sun (1991) *Boundary Layer Met.* **57**, 1–2 on mesoscale cellular convection over the East China Sea.

cellular modelling Cellular models in geomorphology can be defined as representing the modelled landscape with a grid of cells, over which the development of the landscape is determined by the interactions between cells (for example, fluxes of water and sediment) using rules based on simplifications of the governing physics. In fluvial geomorphology, cellular models use simplified or 'relaxed' versions of the complex flow equations (Coulthard et al. (2007) *Geomorph.* **90**, 3–4; see also Nicholas (2005) *ESPL* 30, 5).

cementation The binding together of particles by adhesive materials, as in sedimentary rocks. See Nash and Smith (2003) *Geomorph.* **50**, 1–3 on calcretes.

Cenozoic (Cainozoic) The most recent *era of earth's history stretching approximately from 65 million years BP to the present day.

(⊕) SEE WEB LINKS
• An overview of the Cenozoic period.

census An investigation, usually into the size and nature of a population, but occasionally into other things, such as traffic. A **census tract**, known in the UK as an *enumeration district, is a small unit of area used in collecting, recording, and reporting census data. These units may change over time; **census geography design** is the delineation of appropriate geographical base units from which to create, and use, geography; see Vickers and Rees (2007) *JR Statist. Soc. A* **170**, 2 and Rees and Martin in P. Rees et al., eds (2002).

Center for Spatially Integrated Social Science (CSISS) A centre focusing on spatial analysis as an integrating force for the social sciences. Their website will lead you to their really useful *GIS cookbook: 'a collection of simple descriptions and illustrations of GIS methods, written with minimal GIS jargon.'

(⊕) SEE WEB LINKS
• The CSISS website.

central business district (CBD) The heart of an urban area, usually located at the meeting point of the city's transport systems, containing a high percentage of shops and offices. High accessibility leads to high land values, and therefore intensive land use. Consquently, development is often upwards. Within the CBD, specialist areas, such as a jewellery quarter, benefit from *external economies. Vertical land-use zoning is also common, so that retail outlets may be on the ground floor, with commercial users above them and residential users higher up. See Chang (2007) *Urb. Plan. & Dev.* **133**, 2 on measuring and assessing the economic activities of CBDs. The CBD is under threat from traffic restrictions, and out-of-town developments, such as superstores. 'Arguments on the future [of the CBD] range from a speculated centre-less form to one of centrality' (Wong (2004) *Land Use Policy* **21**).

Methods of delimiting the CBD include the *central business height index, recording the percentage of floor space given over to CBD functions, charting high level pedestrian flows, and surveying pavement chewing gum.

central business height index (CBHI) A measure of the intensity of land use within the *CBD. The total floor area of all storeys of a building is compared with the ground-floor area of the building under consideration. The higher the index, the more intensive the urban land use. The index can also be used to delimit the CBD. Nahm (1999) *GeoJournal* **49**, 3 notes that the CBHI increased fourfold in Seoul, 1953–98.

central flow theory An interlocking network model developed at GaWC. 'In central flow theory, flows generate a network; a space of flows. Central flow theory produces a spatial network of urban settlements' (Taylor (2007) *GaWC Res. Bull.* **238**). City networks are constituted by the interlocking of cities by commercial agents in the everyday course of their business practice; central flow theory entails network mutuality, and this suggests that urban policy makers should focus on cooperation with other cities.

centrally planned economy A synonym for a communist economy, also known as a **command economy**. The *factors of production are owned by the state, and their deployment is planned from the administrative capital. True central planning requires an enormous quantity of information, difficult to gather at the centre. See Zhou (2005) *Env. & Plan. A* **37**, 6 on 'an innovative region from a centrally planned economy in Beijing'.

central place A settlement or nodal point which, by its functions, serves an area round about it for goods and services. **Central place theory** is an explanation, advanced by W. Christaller (1933, trans. 1966), of the spatial arrangement, size, and number of settlements. Christaller modelled settlement locations in a hexagonal pattern to examine and define settlement structure and the size of the hinterland. His initial assumptions were of an *isotropic plain proportional to distance; and *economic man. Economic factors include demand density, economies of scope, localization economies, and agglomeration economies.

While Christaller's theory has waned in popularity, a number of studies use central place theory; see Daniels (2006) *Annal. Tourism Res.* **34**, 2 on central place theory and sports tourism; Dennis et al. (2002) *J. Retailing & Consumer Services* **9**, 4 on central place theory and UK retail hierarchy; Clarkson et al. (1996) *Int. J. Retail & Distrib. Manage.* **24**, 6 on UK supermarket location; and Dean (2003) *China Qly* **174** on hierarchical space in China. See also R. Schiller

k = 4 network, explanation

- higher-order centre
- lower-order centre
- ---→ direction and proportion of custom from lower-order centres to higher-order centres

(a) k = 7 network, explanation; (b) k = 7 network (an alternative orientation)

(a)

(b)

- higher-order centre
- lower-order centre
- ---→ direction of custom from lower-order centre

higher order in orientation (a)   higher order in this orientation

The marketing principle, k = 3, the G-system. (Top right sector shown in full detail.)

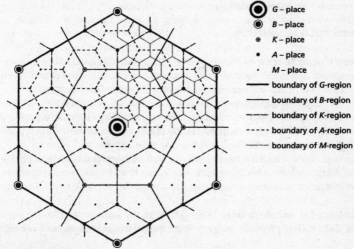

- ◉ G – place
- ◎ B – place
- ⊙ K – place
- • A – place
- · M – place
- —— boundary of G-region
- —— boundary of B-region
- —·—· boundary of K-region
- ----- boundary of A-region
- ······· boundary of M-region

Central place theory

city-poor sector
city-rich sector

1.
2.

■ city-rich sector
□ city-poor sector

3.

Central place theory (continued)

(2001). Taylor (2007) *GaWC Res. Bull.* **238** takes exception to 'this simple equating of cities with hierarchy and competition'.

central vent eruption A volcanic eruption from a single vent or a cluster of centrally placed vents, fed by a single pipe-like supply channel. Lava from a central vent eruption accumulates to form a conical volcano; see Takada (1997) *Bull. Volcanol.* **58**, 7.

centrifugal forces In *human geography, those forces which encourage a movement of people, business, and industry away from central urban areas; these include: intra-city location costs (Anas (2004) *J. Econ. Geog.* **4**), improvements in transport systems (Alonso-Vilar, *ECINEQ WP* 2006–57), the needs of spatially dispersed clients (Grote (2007) *J. Econ. Geog.* **8**, 2); and knowledge spillovers (Baldwin and Forslid (2000) *Economica* **67**). Koch (2007) *UN Univ. Res. Paper* 2007/**45**, lists the dispersion of demand, reputation (a new organization might takes risks that, if successful, would raise their profile), competition effects (where organizations locate away from direct competitors), and high transport costs (when a business might locate near the end-users).

centripetal acceleration In meteorology, the force acting into the centre of a high- or low-pressure system, which makes winds blow along a curved path. It is crucial in *tropical cyclone and *tornado formation; see Wurman and Gill (2000) *Monthly Weather Rev.* **128**, 7.

centripetal drainage *See* DRAINAGE PATTERNS.

centripetal forces Those forces which move people, business, and industry towards a centre—accessibility, *functional linkages, *agglomeration economies, and *external economies—thus encouraging the growth of large central places. Porteous in R. Martin, ed. (1999) cites labour market externalities, intermediate services, technological and informational spillovers, and socio-institutional and cultural factors as centripetal forces. Grote (2008) *J. Econ. Geog.* **8**, 2 holds that, with ever-falling transaction costs, centripetal forces decline over time.

chain migration A process which depends on a small number of pioneer migrants, who settle in a new place. A feedback process is set up, from the destination to the origin, generating higher than expected levels of migration to these pioneers' neighbourhoods. A metropolis then becomes both segmented and segregated (Skop et al. (2006) *Urb. Geog.* **27**, 5). See also Johnston et al. (2007) *AAAG* **97**, 4.

chalk A pure form of limestone composed of the shells of minute marine organisms, plus particles of calcium carbonate.

channel A watercourse. The nature of any river channel is a result of the interaction of: the material properties of the bed and banks; the flow hydraulics; and sediment transport within the river. Any alteration in any one of them affects the other two. Narrow, stable channel stretches are linked to finer bed material and high levels of river vegetation; wider, unstable channels are associated with larger bed material and less riparian vegetation (Rowntree and Dollar (1996) *S. Afric. Geog. J.* **78**, 1).

The **channel capacity** of a river is its cross-sectional area in square metres, but the borders of a channel are not easy to establish; see Rajaguru et al. (1995) *ESPL* **20**. **Channel resistance** slows or impedes the flow; friction with the bed is the major cause. See Righetti and Armanini (2002) *J. Hydrol.* **269**, 1 and Wohl (2007) *Phys. Geog.* **28**, 3.

channel classification Howard et al. (1994) *J. Geophys. Res.* **99**, primarily on the basis of bed morphology, define five channel types: live bed sand alluvial; live bed gravel alluvial; threshold gravel alluvial; mixed bedrock-alluvial; and bedrock. In a study of mountain drainage basins, Montgomery and Buffington (1997) *GSA Bull.* **109**, 5 distinguish seven types: colluvial and bedrock, plus five alluvial channel types (cascade, step-pool, plane-bed, pool-riffle, and dune-ripple). Heritage et al. (2001) *J. Geol.* **109** use cluster and discriminant analyses to classify: bedrock anastomosed, mixed anastomosed, pool-rapid, braided, and alluvial single-thread. These channel types fit on a continuum from bedrock-dominated channels to fully alluvial systems.

channel flow Run-off within the confines of a channel. See L. B. Leopold et al. (1994). **Compound channel flow** occurs in *meandering channels, where the magnitude and directions of the main channel and the upper layer differ, causing sizeable internal *shear stress, and a consequent loss of energy. For meandering compound channels the important parameters affecting the boundary shear distribution are sinuosity (Sr), amplitude (a), relative depth (β), and the width ratio (α) and the aspect ratio (δ); see Khatua and Patra (2007) *Procs 5th Australian Stream Manag. Conf.* Compound channel flow also occurs during floods, when velocity differs between the main channel and the flooded area. See Helmio (2002) *J. Hydrol.* **269** and Fukuoka et al. (1999) *J. Hydrosci. & Eng.* **17**, 2.

channel management Any proposals for changing a fluvial channel should: establish past and present morphology, stability, and responsiveness to change; determine the impact of the proposed change on the sedimentary regime; estimate the effects of change in the sedimentary regime on channel morphology, and on bank pore water pressure, and therefore on bank stability. See P. Downs and K. Gregory (2004). Social and environmental conflicts of interest also arise: damming a river will strip out the finer silt, and increase the downstream concentration of coarser silt, leading to increased erosion, for example. Larsen et al. (2007) *Landscape & Urb. Plan.* **79**, 3–4 demonstrate the need for taking a landscape-level, long-term view of managing a river channel.

channel order *See* STREAM ORDER.

channel roughness *See* BED ROUGHNESS.

chaos theory 'Chaos Theory deals with simple, deterministic, non-linear, dynamic, closed systems . . . [that] are extremely sensitive to initial conditions, resulting in an unpredictable chaotic response to any minute initial difference or perturbation' (Reitsma (2003) *Geoforum* **34**, 1). Chaotic behaviour is not random—current conditions are still linked to future conditions—and chaotic systems do not repeat themselves exactly, but may behave in a loosely recurrent fashion. See Phillips (2003) *PPG* **27**, 1.

character displacement The accentuation of differences among similar species whose distributions overlap (or their minimization in regions where they do not). It results from natural selection; intensifying the aesthetic traits useful for species' discrimination, or for using different parts of a *niche, thus avoiding direct competition. See Montoya and Burns; Russo et al.; and Meiri et al.; all in (2007) *J. Biogeog.* **34**, 12.

charter group A group representing the most typical culture of a host community. The charter group creates a system that segregates it from other 'ethno-groups', who may be gradually assimilated, but in such a way that the charter group maintains its dominance. See O. Yiftachel (1999) and Johnston et al. (2007) *AAAG* **97**, 4.

chatter mark *See* PERIGLACIAL.

check dam A small, often temporary, dam constructed across a watercourse to reduce flow velocity and/or counter soil erosion. See Mustafa (2005) *AAAG* **95**, 3 and Castillo et al. (2007) *Catena* **70**, 3 on the effectiveness of check dams for soil erosion control.

chelate A chemical substance formed from the bonding of compounds to metallic ions. Intense podzolization in humic-iron podzols is linked to mineral chelation (Dearing et al. (2001) *Geophys. J. Int.* **144**). See Siebner-Freibach et al. (2004) *Soil Sci. Soc. Am. J.* **68** on the interaction of iron chelating agents with clay minerals. The *leaching out of chelates is **cheluviation**. **Chelation**, the formation of chelates, is a process whereby relatively insoluble materials may become soluble, and be released into the soil, increasing nutrient availability to plants.

Chelford interstadial An interstadial during the Devensian glaciation, between 65000 and 60000 years BP. *See* PLEISTOCENE.

chemical weathering The breaking down *in situ* of rocks by *carbonation, *hydrolysis, *oxidation, *solution, and attack by organic acids. Bouchard and Jolicoeur (2000) *Geomorph.* 32, 3–4 suggest that hydrology could be the main factor controlling chemical weathering. See Muhs et al. (2001) *Soil Sci. Soc. Amer. J.* 65 on the impact of climate and parent material on chemical weathering; S. Colman and D. Derthier (1986); and Kump et al. (2000) *Ann. Rev. Earth & Planet. Scis* 28 on chemical weathering and climate stability.

chernozem A *zonal soil (US soil classification mollisol, sub-order boroll) with a deep A *horizon, rich in *humus from decomposed grass, and dark in colour. The B horizon is lighter brown, but often absent. The lower horizons are often rich in calcium compounds.

Chernozem development has been associated with temperate continental climates with marked wet and dry seasons, providing enough moisture to permit the decay of the grass litter into humus, but not enough for *leaching to be significant. However, Eckmeier et al. (2007) *Geoderma* **139**, 3–4 claim that there is no consensus on the factors controlling the formation, conservation, and degradation of central European chernozems.

chestnut soil A *zonal soil (US soil classification mollisol, sub-order xeroll), found in more arid grasslands. The *xerophytic nature of much of the grassland under which chestnut soils develop retards the development of humus, and there is an accumulation of calcium carbonate in the B *horizon. Liu et al. (2007) *J. Arid Envs.* **68**, 4, in a study of soils in the semi-arid steppes of China, find that chestnut soils are the most prone to soil erosion. See also Babayan and Protopopov (1997) *Eurasian Soil Sci.* **30**, 10.

chevron A V-shaped, triangular, erosional microform, characteristically developed on the shallow flanks of *cuestas in arid zones. See Baas (2007)

Geomorph. **91**, 3–4 on the effect of different spatial resolutions on the 'visibility' of chevrons.

choropleth A map showing the distribution of a phenomenon by graded shading to indicate the density per unit area of that phenomenon; the greater the density of shading, the greater the density in reality. Choropleths give a clear, but generalized, picture of distribution, but may mask finer details. The choice of values for the classes, and the areal units for which data are available, will affect the visual picture; *see* SCALE. See Goldsberry et al. (2004) *Procs Annual Meeting NACIS* on classification choice and animated choropleths; see also Murray in P. Forer et al., eds (2007) on geographic space and choropleth display.

chronosequence A soil chronosequence is a group of soils whose properties vary primarily as a function of age variability (V. Holliday 2004). See Sauer et al. (2007) *Catena* **71**, 3.

c.i.f. pricing (cost, insurance, freight pricing) A form of uniform delivered pricing where prices are quoted with reference to a port; see Fan et al. (2000) *J. Reg. Sci.* **40**, 4.

cinder cone A cone formed by fragments of solidified lava thrown out during a volcanic explosion. See Cascadden et al. (1997) *New Mexico Bureau of Mines, Bull.* **156**.

circuit of capital Money *capital is needed to pay for *factors of production, such as raw materials, plant, energy, and labour. Commodities may then be produced which are more valuable than the cost of all the inputs; surplus value has thus been added. The commodities are sold (turned back into money) so that surplus value leads to reinvestment. *Accumulation occurs with each circuit of capital, and the spatial integration of different productive nodes within society gives concrete substance to the different moments of the money circuit of capital (De Angelis (1998) *Conf. Paper, Eur. Ass. Evol. Pol. Econ.*). As the circuit of capital expands into effective markets, the circuit is diversified (Lee (2006) *TIGB* **31**, 4). If the circuit of capital is broken, or if there is over-accumulation, capitalists might 'switch circuits'; see T. Hall (2001). D. Harvey (1996) discusses the different temporalities of short-term financial, medium-term industrial, and long-term infrastructural capital, and the pressures and distortions that these can create within the overall circuit of capital.

circulation 1. The movement of capital, labour, goods, and services through an economy. See Hughes on alternative forms of economic organization (2005, *PHG* **29**, 4); on transnational trade and governance (2006, *PHG* **30**, 5); and on the circulation of knowledge (2007, *PHG* **31**, 4).

2. In population geography, short-term, repetitive movements of individuals, where there is no intention to change residence permanently.

Adepojou in P. Kok (2006) observes that, in southern Africa, poverty has drawn more people into circulation or temporary migration, and notes that 'brain drain' is being transformed into 'brain circulation'. See also Curry and Kobarski (1998) *Asia Pac. Viewpt* **39** on circulation in Papua New Guinea.

circumpolar vortex A vigorous, westerly flow of air in the upper and middle *troposphere over mid-latitudes. Frauenfeld and Davis (2003) *J. Geophys. Res.* **108** think the Northern Hemisphere circumpolar vortex could be a good indicator of climate change. 'During the summer–fall season, the trend towards stronger circumpolar flow [in the Southern Hemisphere] has contributed substantially to the observed warming over the Antarctic Peninsula and Patagonia, and to the cooling over eastern Antarctica and the Antarctic plateau' (Thompson and Solomon (2002) *Science* **296**).

cirque Also known as a **corrie** or **cwm**, this is a circular, armchair-shaped, hollow cut into bedrock during glaciation, up to 2 km across. The steep side and back walls are subject to intense physical weathering (Matsuoka and Sakai (1999) *Geomorph.* **74**, 1–4); the front opens out downslope.

The formation of cirques remains unclear. Haynes (1968) *Geografiska A* **50**, 221–34 concludes that bedrock jointing determines the overall cirque morphology, and Hooke ((1991) *GSA Bull.* **103**, 1104–80) finds that periglacial freeze–thaw weathering and glacial quarrying facilitate erosion of the headwall. Turnbull and Davies (2006) *ESPL* **31**, 9 show that cirques can be initiated by large, deep-seated failures of rock slope. Fluvial erosion and large-scale slope failures can be important for the creation of hollows, which are then deepened during glaciations; see Wilson and Smith (2006) *Geograf. Annal. A* **88**, 3.

During glaciation, ice is thought to undergo *rotational slipping, perhaps *overdeepening the cirque floor (Vilborg (1977) *Geografiska A*, **59**, 3–4), although Hooke ((1991) *GSA Bull.* **103**) suggests that glaciohydraulic supercooling is central to overdeepening. (See also the classic paper by Lewis (1938) *Geol. Mag.* **75** on a meltwater hypothesis of cirque formation.) Cirques seem to grow by headward extension, biting back into the mountain mass until only *arêtes or *pyramidal peaks remain; see Oskin and Burbank (2005) *Geology* **33**, 12 on cirque retreat.

Aspect plays an important role in the orientation of cirques: in the Fuegian Andes, for example, the cirques which are oriented towards the south-east are the most abundant. Glacial ice is preserved in these due to the limited exposure to sunlight, or the effects of humid air masses (Planas et al. (2002) *Geomorph.* **87**, 1). Evans (2006) *Geomorph.* **80**, 3–4 considers the effects of geology, relief, and region on the *allometric development of cirques in Wales. Gordon (1977) *Geograf. Annal. A* **59** shows that, in general, the larger a cirque, the longer the period of glacial occupancy.

citizenship 'The process of political engagement between diverse groups and individuals . . . urban citizenship involves a politics of connection within, and beyond, particular cities' (Painter (2005) *ICRRDS, U. Durham*). The liberal/individualist approach to citizenship stresses the importance of individual rights; the communitarian approach emphasizes the duties of a citizen towards fellow members of the community they belong to (H. van Gunsteren 1998). Radical pluralists view the community critically; hence, they uphold policies that maintain differences; social differentiation without exclusion, variety, and publicity (Ganapati (2008) *J. Plan. Ed. & Res.* **27**). Mansvelt (2008) *PHG* **32**, 1 has it that citizenship is produced through consumption practice 'with specific commodities and the social practices surrounding them'. See Desforges et al. (2005) *Citizenship Studs* **9**, 5 for a wide-ranging review of geographies of citizenship. *See* RIGHT TO THE CITY .

city A large urban centre, functioning as a central place, and providing specialized goods and services. There is no worldwide, or even European, agreement over limiting figures of population size or areal extent for a city. Geographies of the city include: the politics of public space (D'Arcus (2004) *Space & Polity* **8**, 3), and how the city is experienced by different urban dwellers: 'urban space equals difference, not merely coexisting differences and different perspectives, but contesting differences and unequal power relations' (Cupers (2005) *Int. J. Urb. & Reg. Res.* **29**, 4). I. Sagan and M. Czepczyński, (2004) discuss the quality of urban life; T. Hall and M. Miles (2003) write on urban futures; see Jones and Evans (2006) *Urb. Studs* **43**, 9 and Bell and Jayne (2003) *Local Econ.* **18**, 2 on urban regeneration (Atkinson (2003) *Urban Studs* **40** complains of the blandness of regenerated cities). Glaeser et al. (2001) *J. Econ. Geog.* **1**, 1, argue that the critical urban amenities are: a rich variety of services and consumer goods; aesthetics and physical setting; weather; and ease of movement within the city. See Bell (2007) *PHG* **31**, 1 on the **hospitable city**, and Hannigan (2007) *Geog. Compass* **1**, 4 on **casino cities**, which 'elevate consumption and spectatorship at the expense of economic and social equality'.

A city may drive capital accumulation (Martin and Ottaviano, *CEPII W. Papers* **1996–14**); Glückler (2007) *Reg. Studs* **14**, 1 sees cities as business development nodes. A city may be a creative field: R. Florida (2005) argues that urban fortunes increasingly turn on the capacity to 'attract, retain and even pamper a mobile and finicky class of "creatives" '. A city may be a multilingual and multi-ethnic community expressed through its infrastructure, public spaces, and festivals; Podmore (2006) *Soc. Cult. Geog.* **7**, 4, observes that gay men produce clearly visible territorial enclaves in inner-city areas. Jouve (2007) *Env. & Plan. C* **25**, 3 perceives towns and cities as political spaces, 'in which relations are redefined through a play of confrontation and between social groups'. Amin (2006) *Urb. Studs.* **43**, 5–6 sees cities as endless inhabited sprawls: 'the contemporary city bears little resemblance to imaginings of the times when urbanism stood for citizenship, the ideal republic, good government, civic behaviour and the ideal public sphere.'

Painter (2005 *ICRRDS Background Paper*) sees access to public space as 'fundamental to the *right to the city'; the core elements of which are 'the promotion of equal access to the potential benefits of the city for all urban dwellers, democratic participation of all inhabitants in decision-making processes and [the] realization of inhabitants' fundamental rights and liberties' (Colin 2005 UNESCO). See Reynolds (2004) *Econ. Geo.* **80**, 4 on social justice and the right to the city, and Amin (2006) *Urb. Studs* **43**, 5–6 on the 'Good City'.

city network Taylor (2007) *TIBG* **32**, 2 argues that cities, in promoting commerce, efficiency, and innovation, create 'spaces of flows, economic chains, and networks', which are central to economic expansion.

city of refuge San Francisco's City of Refuge ordinance prevents local police from collecting information about a person's immigration status if they have not been charged with any crime (Ridgley (2007) *Urb. Geog.* **29**, 1).

city region The area around a city which serves, and is served by, the city (also called *umland or urban field). See Knight (2006) *Gov. Office for NE*, on developing the Tyne and Wear city region; for the theoretical background, see Harrison (2007) *J. Econ. Geog.* **7**, 3.

city-size distribution The frequency with which the settlements of a country or region occur in certain, arbitrarily defined, population size groups. Generally, large settlements are less frequent than small ones, but when size is graphed against frequency, several patterns may be recognized. Iyer (2003) *Eurasian Geog. & Econ.* **44**, 5 documents the change from a relatively uniform city-size distribution 'in accordance to Marxist ideals' to increasing evenness in post-Soviet Russia. See Overman and Ioannides (2001) *J. Urb. Econ.* **49**, 3.

City-size distribution

city/state relations J. Jacobs (1992) identifies two opposing 'moral syndromes' which give rise to different social spaces, spaces of flows, and spaces of places: cities are seen as commercial, honest, cooperative, enterprising, and conscientious; states are seen as guardians, supporting loyalty, tradition, ostentation, exclusion, and, when necessary, vengeance. Taylor (2007) *TIBG* **32**, 2 concludes that the **city-state** is not appropriate in a globalizing world, since the guardian functions of the city-state impinge too greatly on economic expansion.

class Within a society, a set of people who are of the same economic position, and who may share the same tastes and social status. In capitalist societies, class is defined by socio-economic status, but *post-Fordism has created a new division: specialist, skilled workers, in highly paid positions with a large degree of self-determination, and low-paid workers with few skills, working in poor conditions with little or no security. This pattern is common in service employment, but is exacerbated among women (McDowell (2006) *Antipode* **38**, 4; Ward et al. (2007) *Geoforum* **38**, 2).

Bondi and Christie (2000) *Geoforum* **31**, 3 observe that 'the spatial mobility of middle-class lifestyles are taken for granted in ways that sharply remind those struggling to participate in British society of their precarious hold on, if not their exclusion from, the mainstream . . . This self-consciousness about social position operates in relation to gender as much as class.' Lekhi (2000) *New Pol. Sci.* **22**, 3 maintains that class is 'a process that is, by necessity, only ever evolving and never fully constituted'. 'Working classness is seen as a dynamic and relational category which is simultaneously economic and cultural. It rests not only on the material and labour market position of individuals, households and communities, but also on symbolic value and cultural practice, intertwining a number of interpretations of class position and class subjectivity' (Stenning (2008) *Antipode* **40**, 1).

classification Geography uses a range of classificatory systems: **intrinsic classification** depends on natural differences or 'breaks' in the features studied; **extrinsic classification** uses arbitrarily defined class limits; **monothetic classification** uses one criterion; **polythetic classification** uses a number of criteria; **attribute-based classification** is based on 'present' or 'absent' evidence (a climate may or may not have a dry season); **variable-based classification** forms classes on the basis of the degree to which a variable is present.

clast A rock fragment, eroded from a rock mass, and deposited in a new location. See Evans et al. (2007) *Geograf. Annal.* **89**, 2 on clast microfabrics; Bourke et al. (2007) *Lunar & Plan. Sci. Conf.* **38** on fluvial clasts; and Clark (1995) *Geology* **19**, 2 on **clast pavements**.

clay Mineral particles < 0.002 mm. When dry, clay is hard; when wet it swells and becomes pliable and sticky. **Clay colloids** are finely divided clays, dispersed in water, with a negative surface charge that attracts positively charged ions; they are among the most reactive constituents of a soil (W. van Olphen 1991). **Clay micelles** are individual particles, platey in form, < 2 μm in diameter, having a negative charge, and therefore able to attract cations within a soil; try Murrmann and Koutz in S. Reed, ed. (1972).

clay minerals A group of hydrous aluminium silicates, such as kaolinite, created by intense rock weathering; see Egli et al. (2002) *Clay Minerals* **37**, 2. Clay minerals affect the structure of soils, as they expand when wet (Barshad (1952) *Clays & Clay Min.* **1**).

cleavage 1. A division in society due to political or partisan allegiance; see Kitschelt (1992) *Pol. & Soc.* **20**, 1 on cleavage development in post-communist eastern Europe, and Secor (2001) *Polit. Geog.* **20**, 5 on socio-political cleavages in Turkey.
   2. The ability of a rock to split along a **cleavage plane**.

cliff A steep rock face, usually facing the sea; see Lee et al. (2001) *Geomorph.* **40**, 3–4 on probabilistic models for different cliffs. While an **active cliff** is still subject to the forces of marine erosion, an **abandoned cliff** is protected from wave attack by a *wave-cut platform or by a *barrier beach. As the abandoned cliff is exposed to *subaerial denudation it becomes less steep, and its upper edge more indented. See Hutchinson (1984) *DTIC* ADA144040.

climate A summary of mean weather conditions over a time period, usually based on thirty years of records. Climates are largely determined by location with respect to land- and sea-masses, to large-scale patterns in the *general circulation of the atmosphere, *latitude, altitude, and to local geographical features. Safford provides a summary of the role of radiation in climate (1999, *J. Biogeog.* **26**, 4), and of the regional atmospheric circulation (*J. Biogeog.* **26**, 3). Feedbacks between land surface and atmosphere are also important determinants of climate (Knorr and Schnitzler (2006) *Clim. Dynam.* **26**; G. Bonan 2002). Gallup et al. (1999) *Int. Reg. Sci. Rev.* **22**, 2, argue that climate has large effects on income levels and income growth.

climate sensitivity The response of climate to a change in a variable; see Tomassini et al. (2007) *J. Climate* **20**, 7 and Cash (2007) *Tellus A* **59**, 2 on the origin of climate sensitivity differences, defined by Denby et al. (2002) *Tellus A* **54**, 5 as the change in climate brought about by increasing radiative forcing (radiation). See Voigt et al. (2007) *Glob. Change Biol.* **13**, 8 on trophic climate sensitivity.

climatic change During the last 55 million years the earth has been cooling; during the last million years there have been alternating *glacials and *interglacials. *See* LITTLE ICE AGE.
   External causes of climatic change include: changes in solar output (Haigh (2001) *Science* **294**, 5549); changes in the number of sunspots, which seem to have an eleven-year cycle (Foukal et al. (2004) *Science* **306**, 5693), changes in the ellipticity of the earth's orbit, which follow a 100000-year cycle (Goldsmith (March 2007) *Natural History*), and changes in the earth's axis

of rotation, which alters the season of perihelion, and which follow roughly a 100000-year cycle (Zachos et al. (2001) *Science* **292**, 5517). *See* MILANKOVITCH CYCLES.

Internal causes include changes in the distribution of land and sea, *continental drift, and changes in the atmosphere–surface–ocean system. 'Although vulcanicity is seen as a possible mechanism of climatic change, under present rates of eruptions it is difficult to envisage volcanic activity acting independently of other factors in causing significant and lasting global climate changes' (Wyrwoll and McConchie (1986) *Climate Change* **8**, 3).

Nyberg et al. (2007) *Nature* **447** report that the average frequency of major hurricanes decreased from the 1760s until the early 1990s, and that the phase of enhanced hurricane activity since 1995 appears to be a recovery to normal hurricane activity, rather than a direct response to increasing sea surface temperatures. Nicholls and Alexander (2007) *PPG* **31**, 1 find a clear pattern of increasing warm extremes and decreasing cold extremes in global climates since the early 1990s. A. E. Dressler and E. A Parson (2006) summarize the key scientific literature on climate change. See Comrie (2007) *Geog. Compass* **1**, 3 on climate change and human health. *See also* GREENHOUSE EFFECT.

climatic climax community *See* CLIMAX COMMUNITY.

climatic geomorphology (climatomorphology) The association of landform types with different climates, such as the evolution of *periglacial landforms in *tundra climates. However, the geomorphic response to climatic changes, fluctuations, and episodes varies spatially and temporally. Some recent landforms result from the long-term and essentially continuous activity of particular processes. Others are surface manifestations of processes that hitherto were active but cryptic and yet others result from the impact of storm or catastrophic events, frequently, though not in every instance, on surfaces previously rendered vulnerable. 'Few, if any, minor landforms can be accepted without question as indicators of climatic change or fluctuation' (Twidale (1997) *Geomorph.* **19**). See J. Büdel (1982), M. Gutierrez Elorza (2005), and Gunnell (2000) *Geomorph.* **35**, 1–2.

climatology The study of the origins and impacts of climates.

climax community An *ecosystem which experiences a turnover of species, but with an overall steady state. This steady-state ecosystem is characterized by a complex, highly integrated community structure with high species diversity, but relatively low individual population densities not subject to serious fluctuation, and can be seen as the final stage of a *succession. An example is a mature oak woodland. 'Equilibrium or climax communities are rarely affected by intense natural disturbances and are characterised by a predominance of larger, long-lived and often deepburrowing species. Such species can physically modify their surroundings' (Kröger et al. (2006)

Estuarine, Coastal & Shelf Sci. **67**). See Ren et al. (2007) *Acta Ecologica Sinica*
27, 9 on the natural restoration of a degraded rangeland.

climograph A line graph of monthly average temperature against average
humidity. The shape and location of line indicates the nature of climate. See
R. G. Bailey (1998).

closed system A system marked by clear boundaries that prohibit the
movement of energy across them. *Entropy in a closed system can never
decrease. See R. Chorley (1964).

closure The delimitation of, or setting boundaries to, an enquiry. Massey
(1999) *TIBG* **24** lists the methods of closure commonly used by geographers:
oversimplification; bounding a region by process, rather than space; and
limiting an enquiry to a specified time period. 'The process of closing research
must address the motives behind research topics and questions; the reasons
for preferring a particular methodology in addressing a given set of research
questions, including the theoretical and practical issues raised by closure; and
the importance of place in shaping our understandings of reality' (Brown
(2004) *TIBG* **29**, 3). The choice of time–space model is also significant in
closure; see Lane (2001) *TIBG* **26**, 2.

cloud A visible, dense mass of water droplets and/or ice crystals, suspended
in the air, and generally forming when air is forced to rise: at a *front, over
mountains, or because of *convection. Clouds mirror atmospheric processes;
the approach and passage of a warm front, for example, often follows the
sequence: cirrus, cirro-stratus, alto-stratus, nimbo-stratus. At active
*ana-fronts these clouds may take on a more cumulus form. Atmospheric
*convection currents are generally indicated by the presence of cumulus or
cumulo-nimbus clouds. A cumulus cloud will often form over a heated surface
and then shift with the wind, so that further cumulus is formed over the same
spot; see *cloud street. *Turbulence is a common cause of stratus cloud, which
is often trapped beneath an *inversion; it also creates a nearly continuous
sheet of strato-cumulus cloud. See C. D. Ahrens (2000).

cloud classification Clouds may be classified by form and by height.
   1. Low-level clouds (0–2 km above sea level): stratus, S—extensive, shallow
cloud sheet, often yielding *drizzle/light rain; strato-cumulus, Sc—shallow
cloud sheet, in roughly recurring *cumuliform masses, often yielding drizzle/
snow; cumulus, Cu—separate, hill-shaped clouds with flat, and often level,
bases, often at the same height; cumulo-nimbus, Cb—large, high *cumulus,
with tops often formed of ice crystals, dark bases, and often showery.
   2. Medium-level clouds (2–4 km above mean sea level): alto-cumulus,
Ac—shallow cloud sheet broken into roughly regular, rounded clouds; alto-
stratus, As—featureless, thin, translucent cloud sheet; nimbus, Ns—extensive,
very dark cloud sheet, usually yielding precipitation.

3. High clouds (tropical regions 6–18 km high, temperate regions 5–14 km high, polar regions 3–8 km high): cirrus, Ci—separate, white, feather-like clouds; cirro-cumulus, Cc—shallow, more or less regular patches or ripples of cloud; cirro-stratus, Cs—shallow sheet of largely translucent cloud.

Suffixes may be added: *capillatus*—like a feather, or thread; *congest*—growing rapidly, in cauliflower form; *fractus*—broken or ragged; *humilis*—shallow; *lenticularis*—like a lens, especially of alto-cumulus, cirro-cumulus, and strato-cumulus; *radiates*—banded. See Li et al. (2007) *Remote Sens. Env.* **108** on cloud classification.

cloud droplet *See* PRECIPITATION.

cloud seeding Any technique of inducing rain from a cloud, usually by dropping on it crystals of dry ice (frozen CO_2), or silver nitrate, as *condensation nuclei. Ćurić et al. (2007) *Met. & Atmos. Phys.* **95**, 3–4 find that accumulated hail and rain precipitation at the surface are significantly reduced if seeding is at an early stage of cloud development.

cloud street Parallel lines of small cumulus: irregular over land, downwind from a sunny slope, and regular over warm seas. The cloud-free areas come from the mixing of saturated moist cloud air with overlying warmer, drier air: 'an individual cloud is capable of drying out a large area of the surrounding layer through cloud-top entrainment and thus is able to produce significant gaps between the clouds' (Chlond (1992) *Boundary-Layer Met.* **58**, 1–2).

Club of Rome A society that aims to understand the workings of the world as a finite system, and to suggest alternative options for meeting critical needs. Concerns have included inequality on global and regional scales, pollution, unemployment, and inflation. Its *Limits to Growth* (1972) argued the necessity of slowing down population growth in order to ward off *Malthusian checks.

(⊕) SEE WEB LINKS

• Club of Rome website.

cluster A geographic concentration of interconnected companies, specialized suppliers, service providers, associated institutions, and firms in related industries (Porter (1998) *Harvard Bus. Rev.* **76**, 6). Ketels in (2004) *Innovative City and Business Regions*, **3** identifies the critical characteristics of clusters as: proximity—firms need to be sufficiently close in space to allow spillovers and the sharing of common resources; linkages; some level of active interaction; and critical mass—there need to be enough participants for the interactions to have a meaningful impact. B. Asheim (2005) adds localization economies and specialization to this list, and Storper and Venables (2004) *J. Econ. Geog.* **4**, 4 stress the importance of face-to-face contact. Florida (2003) *City & Community* **2**, 1 sees 'concentrations of talented people who power innovation and economic growth' as vital. Glaeser, in G. Clark (2003) argues that places with higher levels of human capital are more innovative and grow more rapidly and robustly over time; see also Cushing (2001) *W. Paper, U. Texas at Austin*. Giuliani (2007) *J. Econ. Geog.* **7**, 2 stresses the importance of embeddedness in local business networks. Bathelt (2005) *Reg. Studs* **39** argues that clusters can only create new knowledge if the cluster firms have linkages with external markets and employ a mix of local and non-local transactions. Clustering and globalization are closely intertwined; see M. Fujita et al. (2001).

Maskell and Malmberg (2007) *J. Econ. Geog.* **7**, 5 conceive clusters in evolutionary terms: the growing cluster is made up of firms with similar competencies, reducing expenditure on locational search costs, and enjoying *agglomeration economies. Further growth results from *cumulative causation, the presence of support institutions, and the power of the name of the locality as an attraction to knowledge workers: the benefits of being physically located within, for example, Silicon Valley are perceived to be worth the costs of relocation. See also Chapman, MacKinnon, and Cumbers (2004) *TIBG* **29**, 3.

M. Porter (1990) sees clusters as the key to economic growth and poverty reduction in backward regions that are not too remote from existing economic centres. European policy-makers in particular have turned to cluster policy because of a shift in priorities from macro- to microeconomic issues (Ketels, op. cit.). Sadler (2004) *Reg. Studs* **38** argues that an effective regional cluster policy needs to identify clusters on a number of variables, including knowledge flows. The use of the cluster model as a paradigm of economic development has been criticized. Bathelt and Taylor (2002) *Geografiska B* **84**, 2 argue that clusters 'depend on specific circumstances in "time–space" and, because of their very transience and specificity, those conditions might be very difficult if not impossible to create through the blunt instruments of policy'. See Rutherford and Holmes (2008) *J. Econ. Geog.* **8**, 4 on the state's role in the reshaping of *actor networks in clusters.

cluster analysis A type of multivariate analysis. Kent (2006) *PPG* **30**, 3 reviews recent changing patterns in the use of cluster analysis. For similarity analysis/minimum variance, see Ward (1963) *Am. Stat. Ass. J.* **5**; for unweighted pairs group average, see Williams et al. (1966) *J. Ecol.* **54**. For ordination techniques/detrenched correspondence analysis, see M. O. Hill (1979); for non-metric multidimensional scaling, see T. Cox and M. Cox (2000). CANOCO software does most multivariate techniques.

coalescence theory *Bergeron–Findeisen's theory cannot explain the formation of all tropical rainfall, since ice crystals are often absent in tropical clouds. Langmuir's coalescence theory (1974) suggests that the small droplets in clouds grow larger by coalescence until they are heavy enough to fall. As they fall, they collide with other droplets, growing more. (Not every collision results in coalescence.) The deeper the cloud, the more the drops grow; up to about 5 mm in diameter. See Xue and Wang (2007) *J. Atmos. Scis.* **31**, 15.

coastal dune A ridge or hill which forms when marine deposits of sand are blown to the back of the beach. 'Dune initiation is often site-specific and dependent on the mechanisms by which sand can be delivered from the shoreface to points where it will be reworked into available accommodation space' (Aagaard et al. (2007) *Geomorph.* **83**, 1–2).

coastal flooding The inundation of coastal areas may be caused by *storm surges, *tsunami (Dawson and Stewart (2007) *PPG* **31**, 6), or linked to *El Niño (Tingt et al. (2007) *Geol. Today* **23**, 6). D. Sobien and C. Paxton (1998) SR/SSD 98-2 provide a primer on coastal flooding. See Carlson et al. (2008) *Ground Water* **46**, 1 on the poisoning of aquifers by the inundation of salt water from storm surges. Nunn (2007) *Geog. Res.* **45**, 2 points out that coastal flooding is often destructive at the time, but, 'as many delta dwellers know', not always so thereafter.

coastal geomorphology Stephenson (2006) *Prog. Phys. Geog.* **30**, 1 provides an overview of recent coastal research, and M. Schwartz ed. (2005) is recommended reading.

cobble A stone, 60–200 mm, partly or wholly rounded by wave action or running water.

cockpit karst (kegelkarst) A landscape of star-shaped hollows surrounded by steep, rounded hills, found in tropical *karst. The cockpits, up to 100 m deep, usually containing a *streamsink, are the hollows (*dolines) formed by the solution of limestone, and now floored with alluvium. See Fleurant et al. (2006) *J. Geol. Soc.* special issue.

coefficient of localization A measurement of the degree of concentration of a given phenomenon, such as industry, over a set of regions (also called the **index of concentration**). Values of the coefficient, L, lie between 0 (even distribution) and 1 (extreme concentration). Lafourcade and Mion (2007) *Reg. Sci. & Urb. Econ.* **37**, 1 use the coefficient to study the geographic distribution of manufacturing, and the size of plants.

cognitive dissonance A mismatch between what is perceived and what is (between cognition and reality), so that an individual may seem to act irrationally. 'Cognitive dissonance theory argues that individuals strive to minimize cognitive discomfort through fostering behavioural consistency' (Hobson (2006) *Area* **38**, 3 on recycling). See Lundholm and Malmberg (2006) *Geografiska B* **88**, 1.

cognitive mapping The acquisition, coding, storage, manipulation, and recall of spatial information within the mind (also called mental mapping). Cognitive mapping simplifies the complexity of the landscape, and a mental map influences behaviour. See R. Downs and D. Stea, eds. (2005). **Cognitive trails** represent 'the creation of knowledge by movement through the natural and intellectual environment' (Turnbull (2007) *Geog. Res.* **45**, 2).

cohesion Adhesion; the force holding materials together. Cohesion is a measure of strength; a higher cohesion value results from low friction angles (Thermann et al. (2006) *IAEG* **484**).

cohort A group of people who experience a significant event, such as birth or leaving school, during the same period of time. **Cohort analysis** traces the subsequent vital history of cohorts; see Frejka and Callow (2001) *Pop. & Dev. Rev.* **27**, 1 on cohort reproductive patterns in low-fertility countries. A **cohort effect** is any effect associated with being a member of a group born at roughly the same time and bonded by common life experiences; Cohen and Iams (2007) *Int. Mig. Rev.* **41**, 3, for example, see a US **cohort effect** in the fact that foreign-born people have lower retirement incomes than do US-born people. See Franklin and Plane (2004) *Geograf. Anal. B* **60**, 1 on the contribution of the cohort effect to fertility decline in Italy. The major problem of cohort analysis is to distinguish between the effects on the cohort of getting older (age effects), of common experiences like National Health orange juice (cohort effects), and particular historical events, like a war (period effects).

   Cohort fertility is the total of live births born to a particular birth or marriage group; see Lutz and Skirbekk (2005) *Pop. & Dev. Rev.* **31**, 4.

col 1. In the landscape, a pass between two peaks or ridges. Landscape cols may have been formed by: the *headward erosion of a *cirque, *river capture, the beheading of *dip-slope valleys by *scarp retreat, or the localized *differential erosion of a ridge.

   2. In meteorology, a narrow belt of relatively low pressure, but not a depression, between two *anticyclones. Here, isobars are few and therefore winds are slack.

cold glacier (cold-based glacier, polar glacier) A glacier with its base well below 0 °C, unlubricated by *meltwater, and therefore frozen to the bedrock. *Accumulation is slow; snow may take 150 years to turn to ice. Cold glaciers move very slowly; rates of 1–2 m per year are not uncommon (Fitzsimmons et al. (2000) *Geol. Soc. London*, special pub. **176**), therefore they cause very little erosion through *abrasion, although *plucking may be reinforced. Where subglacial sediments are frozen to a cold glacier base, shear failure will occur at the bottom of the frozen zone, as high pore water pressures develop there (Owen and Derbyshire (1993) *Zeitschrift* **76**). See Glasser et al. (1998) *J. Glaciol.* **44** on Kongsvegen, Svalbard, and its role in landform genesis; see also Vaughn and White (2000) *Geology* **28** on entrainment at cold glacier beds.

cold low (cold pool, polar low) An area of low pressure and temperature in the middle *troposphere. These lows don't appear on *synoptic charts, but are important in *Arctic and *Antarctic meteorology (Serreze et al. (1989) *Meteor. & Atmos. Phys.* **51**, 3–4). **Cut-off low** systems are synoptic-scale low pressure systems, formed as a result of meridional shifts of jet streams in the upper troposphere, and playing a role in exchanges between troposphere and stratosphere (Gimeno et al. (2007) *Met. Atmos. Phys.* **96**).

collective farm A farming organization identified with socialist regimes. The farm is owned by the state, but leased to the members of the collective, and the workers are shareholders, not state employees. The collective is, in theory, self-governing, although the state may set production targets. See Rausing (2007) *J. Royal Anthr. Inst.* **13**, 2 on the end of a collective farm in post-Soviet Estonia.

collectivism A school of thought which maintains that the *factors of production and the means of distribution should be owned by all and not by individuals who might pursue their self-interest at a cost to the state. Huang (2006) *Urb. Geog.* **27**, 6 argues that gating and neighbourhood enclosure help to define a sense of collectivism in urban China.

collision margin The boundary between two continental plates. These are hybrid margins since the two slabs of continental crust that eventually collide were initially separated by oceanic crust, with passive or destructive margins; when the oceanic crust is consumed, the two continental crusts collide. The lower continental plate forms a double layer of crust together with the upper plate, as in the Himalayas; see D. S. N. Raju et al. (2005).

collision theory This states that raindrops grow by colliding and coalescing with each other, especially in tropical maritime *air masses. *See also* BERGERON–FINDEISEN THEORY and COALESCENCE THEORY.

colluvium The mixture of soil and unconsolidated rock fragments deposited on, or at the foot of, a slope. See Coe et al. (2008) *Geomorph.* **96**, 3–4 on colluvium and debris flows. Colluvial boulder deposits are relatively stable colluvium, consisting of linear fields of well-varnished hillslope boulders, ubiquitous on desert hillslopes (Liu and Broecker (2008) *Geomorph.* **93**, 3–4).

colonial discourse theory European colonizers tended to construct the identities of colonized peoples and lands as *other: undeveloped, primitive, and immature; as homogeneous objects, rather than sources of knowledge; see Anand (2007) *New Polit. Sci.* **29**, 1 on Western colonial representations of the (non-Western) other. The colonizer was represented as having a duty which entailed both financial and emotional cost (and for many these costs were very real).

   Colonial discourse analysis critically examines the role played by these *representations in colonialism and imperialism, 'rather than focusing on texts, systems of signification, and procedures of knowledge generation . . . a fuller understanding of colonial powers is achieved by explaining colonialism's basic geographical dispossessions of the colonized' (Harris (2004) *AAAG* **96**, 1). 'The city-state of Singapore is first represented in terms of a narrative of masculine heroism' (S. Mathur 1997).

a

Continent

Marginal sediments

Ocean trench

Continent

b

Mountain range

c

Mountain range

Fracture zone

——— Sea level	- - - Marginal sediments
▬▬▬ Oceanic crust	
■■■ Continental crust	〰〰 Marginal sediments which have been faulted and metamorphosed
⠂⠄⠂ upper mantle	

Collision margin (after J. F. Deaxy and J. M. Bird 1970)

colonialism The acquisition and colonization by a nation of other territories and their peoples. This term took on a more specific meaning in the late 19th century when colonists saw it as the extension of 'civilization' from Europe to the 'inferior' peoples of 'backward' societies: 'the institution of British legal regimes . . . was underpinned by a certain type of violence that served to facilitate colonialism . . . These legal landscapes were founded . . . upon a mythological hierarchy wherein white peoples were accorded the label of civilized in contrasting them from savage, non-white (read: non-European) peoples' (Rossiter (2007) *J. Hist. Geog.* **33**, 4). Colonialism also entailed colonial philanthropy: 'a new doctrine of responsibility toward the unprivileged, a doctrine which received its chief impulse from the Evangelical emphasis on the value of the human soul, and hence, of the individual' (E. Howse 1953). Colonialism is also seen as a search for raw materials, new markets, and new fields of investment; see Giordano and Matzke, and Redclift (both in (2001) *PHG* **25**, 4). Characteristics of colonialism include inequality between rulers and subjects; political and legal domination by the imperial power; and the exploitation of the subject people. Many commentators link colonialism with *uneven development; see, among others, Dutt (1992) *Am. Econ. Dev.* **82**, 2 on India, and K. Cox (2002).

Sometimes, but not always, colonialism was accompanied by **colonization**; that is, the physical settling of people from the imperial country. Lester (1998) *TIBG*, **23**, 4 argues that 'in their new colonial environment, settlers, who were initially divided by prevailing British social boundaries of class, gender, religion and ethnicity, soon faced an imperative for solidarity'. Although independence from former colonization has been achieved almost everywhere, most accept that it has been replaced by *neo-colonialism; see Dutt (1992) *World Dev.* **20**, 8.

command economy A communist economic system where the state controls macroeconomic policy and entrepreneurial activity, but allows some local freedom for decisions on employment and consumption (also called a *centrally planned economy*). See D. Shaw (1999) on the Soviet command economy and its transition to capitalism.

commercial agriculture The production of agricultural goods for sale. See Keys (2004) *Land Degrad. & Dev.* **15** on the transition to commercial farming in southern Yucatán.

commercial spaces (commercial hospitality spaces) Commercial hospitality is becoming increasingly important for the branding and promoting of cities. Latham (2003) *Urb. Studs* **40**), detects a new form of public culture, based around cafés, bars, and restaurants, and Laurier and Philo (2004) *Env. & Plan. A* **36**, 3 represent cafés as 'sites where economic, political and cultural matters come into contact and are mutually transformed'. These joyful judgements are shared by Anderson (2004) *Annals Am. Acad. Polit. Soc. Sci.* **595** and Esperdy (2002) *Architectural Design* **72**.

D. Hobbs (2003) takes a more cynical view, seeing the creation of hospitality as a means of generating customer loyalty, but Bell (2007) *PHG* **31**, 1 argues that it shouldn't be seen as merely calculative. Jayne et al. (2006) *PHG* **30** stress the downside of pub culture, and Hobbs (op. cit.) illustrates the affects and effects of 'the alcohol-fuelled night-time economy'—from boozy matinées to random drunken brawls.

comminution Changing rock debris to fine powder via *abrasion and *attrition. See Hiemstra et al. (2007) *Boreas* **36**, 4 on comminution in subglacial tills. *See also* ROCK FLOUR.

commodity An object of economic value, intended for exchange in a *capitalist system, such as labour and money. **Commodity chains** (**commodity networks**) are the connections between the production, circulation, and consumption of goods; for example, the chains which bring cut flowers from Kenya to the UK. See A. Hughes and S. Reimer (2004) and J.-P. Rodrigue et al. (2006).

Ioris (2007) *Capital Nature Socialism* **18**, 1 shows how the **commodification** of water in Brazil has created social and environmental inequities; Naidoo (2007) *Rev. Afr. Polit. Econ.* **34**, 111 describes the negative impacts of the commodification of service delivery in South Africa; and Thornes and Randalls (2007) *Geografiska A* **86**, 9 describe the commodification of the atmosphere. R. Sack (1992) sees commodification as a feature of *modernity. See Marks and Bezzoli (2001) *Urb. Forum* **12**, 1 on **commodified culture**.

Common Agricultural Policy (CAP) Since the early 1960s, the European Union's CAP has been highly protective, encouraging production through high price support, underpinned by import levies, state intervention and trading, and export subsidies. This helped to transform the European Union from a net importer of major temperate agricultural products to a major net exporter. EU exports needed heavy subsidization to be saleable internationally, and consequently depressed world prices. The EU has also maintained preferential access to regulated amounts of imports from members' former colonies and traditional trading partners.

'The Common Agricultural Policy (CAP) is at a crossroads. There are many who say that it should disappear altogether and others who point out that, without it, the great majority of farmers in the European Union (EU) would go out of business . . . Some of the most insistent demands from new EU Members will be for help with their antiquated systems of minute rural holdings; while farmers in the EU-15 countries are having to deal with the complication of switching from production subsidies to single farm payments which require compliance with environmental objectives' (*The Future Financing of the Common Agricultural Policy*, 2006–6). See Dwyer et al. (2007) *Reg. Studs* **41**, 7 on sustainable rural development and the CAP.

communication geography 'The geography of messages and messengers' (Thrift in R. J. Johnston 2000). The study of information and communication technologies (ICT) by geographers has evolved over the past third of a century from a concentration on friction of distance and spatial organization toward a set of four interrelated social approaches: 'ICT as a set of contested terrains, ICT as a means of perception, ICT as a form of embodiment, and ICT as virtual places or spaces—distanciated social contexts' (Adams and Ghose (2003) *PHG* **27**, 4). 'Media enthusiasm over the Internet often tends to obscure the much larger, and more important, telecommunications networks that comprise the nervous system of the global and most regional economies' (Costello (2000) *PHG* **24**).

communism Historically, the principle of communal ownership of all property; basic economic resources are held in common. See Shaw and Oldfield (2008) *Pol. Geog.* **27**, 1.

community 1. In ecology, a naturally occurring, non-random, collection of plant and animal life within a specified environment. Badano and Cavieres (2006) *J. Biogeog.* **33**, 2 discuss the view that increases in habitat complexity due to the presence of ecosystem engineers lead to higher **community diversity**.

2. In human geography, an interacting group of people living in the same territory: town, village, suburb, or *neighbourhood; see Lee and Saxenian (2008) *J. Econ. Geog.* **8**, 2 on the Taiwanese transnational technical community in Silicon Valley. Ryan-Nicholls and Racher (2004) *Rural & Remote Health* **4**, 244 see community as a subject with its own construction of reality, unique needs, values, and assets, while Lacy (2000) *Rural Soc.* **65** sees community as the container for the development of social cohesion, human endeavour, empowerment, and place formation. 'In taking up . . . the classical distinction between Gemeinschaft and Gesellschaft, networking practices in the transient word of projects might be differentiated into network communality, network sociality and network connectivity' (Grabner (2004) *PHG* **24**, 4). A. Amin (with N. Thrift, 2004), uses the concept of **communities of practice**: the process of social learning that occurs when people who have a common interest in some subject or problem collaborate over an extended period to share ideas, find solutions, and build innovations. Communities of practice act as performative spaces for learning by homogeneous groups of employees, not only via spatial proximity, but also at a distance (J. Peck and H. Yeung, eds. 2003).

community-supported agriculture (community-shared agriculture) (CSA) A CSA is organized around a contract between a farmer, commonly an organic producer, and a set of local residents, who share the risks of the farming enterprise by contributing money up front for a 'share' of the harvest prior to the farming season; see O'Hara and Stagl (2001) *Pop. & Env.* **22**. Wells et al. (1999) *Community Develop. J.* **34** hold CSA brings a growing circle of people into a closer relationship with place—'farming, nature, each other'.

commuting The movement from suburban or rural locations to the place of work and back. In a Montreal-based study, Shearmur (2006) *Urb. Geog.* **27**, 4 concludes that differences in commuting length between different places of work are, by and large, independent of possible explanatory factors (such as residential location, economic sector, occupation, and income). Despite their shorter average overall commutes, women travel farther than men to reach jobs in the *CBD (Shearmur op. cit.). Murata and Thisse (2005) *CEPR Disc. Paper* **4936** find that low transportation costs lead to the dispersion of economic activities. Elias et al. (2003) *Reg. Studs* **37**, 8 find that accessibility to job openings in surrounding regions significantly increases the likelihood of commuting, but that labour mobility decreases with access to job opportunities in neighbouring regions.

comparative advantage The advantage of some nations or regions to produce goods better and more cheaply than less favoured nations or regions; an important concept in understanding regional specialization. Llang (2004) *RURDS* **16**, 3 finds that comparative advantages from improved technical efficiency, scale economies, and growing economic openness contribute positively and significantly to regional specialization in China.

comparative cost analysis An analysis of costs in order to establish a least-cost location; see R. Hayter (2004). Guyatt et al. (2002) *Health Policy & Plan.* **17**, 2 apply the method to combating malaria in Kenya.

comparative urbanism The knowledge, understanding, and generalization of what is true of all cities and what is true of one city at a given point in time. 'It should not surprise us that each and every place is different or even unique in some ways—this is the idiosyncratic nature of place. What begs our attention is why separate places can be very similar in certain respects' (Nijman (2007) *Urb. Geog.* **28**).

compatibility 1. Being able to accommodate more than one form of activity; for example, 'to achieve compatible physical and economic coexistence of rural residences, agriculture and cities, through preservation of resources, clustering of development contiguous to cities, buffering uses, and adequate infrastructure' (*Leavenworth County Plan*).
   2. To harmonize existing land-use datasets in order to make comparisons within and between countries (Jensen (2006) *J. Land Use Sci.* **1**, 2–4).

competence The largest size of particle that a river can carry. Competence varies with water depth, but is reduced with channel widening (Cheetham (1979) *GSA Bull.* **90**, 9). The sixth power law suggests that a doubling, for example, of river velocity would raise competence by 2^6, i.e. by a factor of 64. Baker and Ritter (1975) *GSA Bull.* **86**, 7 demonstrate the analysis of shear stress as an estimate of competence. *See also* HJULSTRÖM DIAGRAM.

competition 1. Rivalry between suppliers providing goods or services for a market (Jonathan Law, ed. 2006). Essletzbichler and Rigby (2007) *J. Econ. Geog.* **7**, 5 seek to show how competition may produce distinct economic spaces: 'as competition unfolds over time, new spaces of economic activity are created.' See H. Garretsen and S. Brakman (2005) on competition between decision-makers.

2. In biogeography, competition occurs when a necessary resource is sought by a number of organisms. 'Vegetation dynamics, in a modelling context, comprise the processes of competition for resources among individuals or plant functional types' (Smith et al. (2001) *Glob. Ecol. Biogeog.* **10**, 6).

complementarity A mutual dependency based on an ability to produce goods in one area which are needed in another, as with the import of Japanese manufactured goods to Australia, and the export of Australian agricultural goods to Japan (D. Rumley 1999). Nazara et al. (2006) *J. Geog. Sys.* **8**, 3 explore the degree to which complementarity and competitive interaction at one level in the hierarchy persist at lower or higher levels.

complexity science 'Complexity theory . . . takes on new meanings as it circulates in and through a number of actor-networks, and, specifically global science, global business and global New Age' (Thrift (1999) *Theory Cult. & Soc.* **16**, 3; bizarrely, this is the only use of the term in this paper). At the heart of all definitions, however, is the realization that, in general, the reductionist methods used in science cannot be used in studying complex systems: 'the interaction among constituents of the system, and the interaction between the system and its environment, are such that the system as a whole cannot be fully understood simply by analysing its components' (P. Cilliers 1998). 'Each entity has different relations to others, and, *where* an entity is in the system has significance for the unfolding behaviour of entities individually, and of the system collectively' (O'Sullivan (2004) *TIBG* **29**, 1). This means that scaling up from local to global behaviour is not straightforward. As P. W. Anderson (1972) *Science* **177**, 393 shows, 'more is different'; the whole is very much more than the sum of the parts. See Trudghill (in S. Trudghill and A. Roy, eds. 2003) for a wonderfully user-friendly discussion of models, narratives, and constructs in complexity theory. 'The resulting geography . . . is expected to reach the public sooner or later, even if the effect—according to complexity theory itself—cannot be foreseen in detail' (O'Sullivan (2004) *TIBG* **29**, 3).

At the time of writing, the last comment must come from Martin and Sunley (2007) *J. Econ. Geog.* **7**, 5, who, in a model of exegesis, write that 'to refer to complexity "theory" is perhaps to exaggerate the degree of conceptual coherence and explanatory power associated with the notion'. For **complexity in geomorphology** see (2007) *Geomorph.* **91**, 3–4, special issue.

compressive flow (compressional flow) Where a glacier is slowing up, the flow is compressive; in zones of compression, the ice may erode the valley floor. Compressional flow on the *stoss side of hills causes till sheets to shear, and debris bands to stack (Gore et al. (2003) *Antarc. Sci.* **15**, 2). See also Swift et al. (2006) *Q. Sci. Revs* **25**, 13–14.

compulsory purchase order A directive to enforce the sale of land from a private landowner to national or local governments (Gallent and Shaw (2007) *J. Env. Plan. & Manage.* **50**, 5).

concave slope A slope which declines in steepness with movement downslope, also known as a *waning slope. Claessens et al. (2007) *Geomorph.* **74**, 1–4 find that steep concave slopes are one of the main factors responsible for landslides in Uganda. Most often, the concave element of a slope occurs as a basal concavity. Where the scarp is not being undercut, the base and the crest make straight-line intersections with the mid-section to form a sharp basal concavity and crestal convexity (Nash and Beaujon (2006) *Geomorph.* **75**, 3–4). The basal concavity is poorly developed in humid temperate environments (J. Gerrard 1988).

concealed unemployment A situation whereby individuals know that there are no jobs available and don't register as unemployed, even though they would like a job; official figures thus under-represent the true situation. P. Alcock (2003) uncovers extensive concealed unemployment in the UK among officially 'economically inactive' men.

concentration and centralization The tendency of economic activity to congregate in a restricted number of central places, encouraged by *functional linkages, and *external and *agglomeration economies. See Hanson in G. L. Clark et al., eds (2000). Falcioğlu and Akgüngör (2008) *Eur. Plan. Studs* **16**, 2, in a study of centralization in Turkey, confirm the Krugman hypothesis (1991, *J. Pol. Econ.* **99**) that industries become more concentrated with economic integration. Rodríguez-Pose and Zademach (2003) *Urb. Studs* **40**, 10 find that proximity appears to play a distinctive role in the geography of mergers and acquisitions in Germany.

concentration dynamics A change in the concentrations of suspended sediment in a river. Concentrations usually rise with increased river velocity.

concentric zone theory R. Park and E. Burgess (1925) suggested that the struggle for scarce urban resources, especially land, led to competition for land and resources, which ultimately led to the spatial differentiation of urban space into zones. They predicted that cities would take the form of five concentric rings with areas of social and physical deterioration concentrated near the city centre and more prosperous areas nearer the periphery. In the post-war period, Park and Burgess's model fell out of favour as critics suggested that the models were overly simplistic. However, M. Davis (1992) used the concentric rings model to describe Los Angeles as a city with an inner core of 'urban decay metastasizing in the heart of suburbia' (Brown, no date, *USBC Spatial Rev.*).

Concentric zone theory

concordant Complying with. In geomorphology, relief and drainage may be concordant with geological structure. A **concordant coast** is parallel to the ridges and valleys of the country.

condensation The change of a vapour or gas into liquid form, accompanied by the release of *latent heat, altering the *adiabatic temperature change in rising air. See Wang and Wang (1999) *J. Atmos. Scis* **56**, 18 on the classic turbulence condensation theory. Condensation in meteorology can be caused by the cooling of a constant volume of air to *dew point; the expansion of a parcel of air without heating; the evaporation of extra moisture into the air; the fall in the moisture-holding capacity of the air due to changes in volume and/or temperature; and contact with a colder material or *air mass. The likelihood of water vapour condensing will depend on the *saturation vapour pressures of water and ice at a given temperature, and/or the presence of *condensation nuclei, since water vapour can cool well below 0 °C before condensation occurs. *See* CONDENSATION NUCLEI; BERGERON–FINDEISEN THEORY. The **condensation level** is the height at which rising air will cool, condense, and form clouds.

condensation nuclei Microscopic *atmospheric particles that attract water droplets, that may then *coalesce to form a raindrop. Condensation nuclei (*hygroscopic nuclei*) come from pollen, salt from sea spray, dust from volcanic eruptions and soil erosion, and particulate air pollution (C. Park 2001). Many small condensation nuclei cause many small droplets to remain liquid well below normal freezing point, and small ice nuclei may increase the rainfall rate from single clouds, which is the principle of *cloud seeding (Graf (2004) *Science* **303**, 5662). *See* HYGROSCOPIC NUCLEI.

conditional instability *See* INSTABILITY.

cone volcano A volcanic peak with a roughly circular base, tapering to a point. Cones may be built solely of *lava or of *scoria, or of both. Parasitic cones form round smaller vents on the flanks of the volcano. See Frazzetta et al. (1983) *J. Volc. & Geotherm. Res.* **17**.

congelifluction (congelifluxion) *See* GELIFLUCTION.

congelifraction *See* FREEZE–THAW.

congeliturbation *See* GELITURBATION.

congestion 'The nation's principal means of allocating scarce road space among competing users during periods when too many people want to use that limited space at the same time' (A. Downs 2004). 'Combinations of public transport/cycling/traffic management measures (traffic calming, park and ride), are not enough to reduce traffic in city centres; restraint measures, such as traffic calming, parking controls or road pricing are needed as well. To achieve public/political acceptability for such policies, improved public transport, providing alternative means of travel, needs to be in place before the restraints are introduced. The danger is that once the situation improves, the cars will immediately return.

A reduction in congestion will have the potential to induce extra trips and traffic, which would not otherwise have occurred; all assessments of measures to reduce congestion should take account of the potential for induced traffic' (*House of Commons, Res. Paper* **98/16**). This term can apply to an excessive demand for any facility.

conglomerate. A *sedimentary rock composed of rounded, water-borne pebbles which have been naturally cemented together.

coniferous forest (boreal forest) Natural forest, found between 55° and 66°N, where winters are long and very cold, summers short and warm, and annual precipitation is around 600 mm. Pure stands are common and species relatively few. Trees are evergreen, and the needle-shaped leaves restrict surface area, thus preventing loss of water via transpiration. Undergrowth is sparse. Animal species are dominated by insects and seed-eating rodents;

larger animals include deer, bear, and wolves. See Väisänen (1995) *IUFRO XX World Congr.*

connectivity The degree to which the *nodes of a network are directly connected with each other. A graph is connected if every node can be reached from some other node. The higher the ratio of the *edges to the nodes in a network, the greater the connectivity. See Urban and Keitt (2001) *Ecology* **82** on graph theory as applied to connectivity in heterogeneous landscapes. Connectivity in a network is said to increase as economic development proceeds; see G. Vigar (2002).

Taylor (2002) *Urb. Studs* **39** measures a city's connectivity by summing the products of every firm's service value in the city with their service values in all other cities. US cities are generally less globally connected than their European Union and Pacific Asian counterparts—New York is the only US city with more non-US cities than US cities in its top ten of global connections (C. Taylor 2005, Brookings Institution).

consequent stream Any stream whose course is controlled by the initial slope of a land surface (A. Strahler and A. Strahler 1979).

conservation The protection of natural or man-made resources and landscapes for later use. A distinction is made between conservation and preservation; while a **conservationist** recognizes that people will use some of the fish in a lake, a preservationist would ban fishing entirely. 'If the twentieth century question was how to stop extinction, then perhaps the twenty-first-century challenge is how to avoid the total collapse of the biosphere, our life support system' (W. Adams 2004).

Conservation is a highly political enterprise: McKenna et al. (2005) *Area* **37** argue that a failure to conserve coastal dunes in Ireland results 'more from management deficiencies than from shortcomings in scientific understanding'. Duffy (2006) *Pol. Geog.* **25**, 1 casts light on the struggles encountered in environmental politics over access to key natural resources; and Sundberg (2003) *Pol. Geog.* **22**, 7 observes that the existence of democratic regimes and formal institutions does not guarantee that environmental projects will be implemented through demographic processes. C. Hambler (2004) favours establishment of an environmental fund into which polluters pay and from which societies can draw, such that those whose development opportunities are curtailed for global conservation objectives can access compensatory development funds. See Vogiatzakis et al. (2006) *PPG* **30** on using *GIS for conservation.

conservation biology A sub-discipline of biology concerned with the impacts of people on the environment and the conservation of biological diversity (see Murray et al. (2002) *Austral. Ecol.* **27**; see also Klinkenberg (2001) *Canad. Geogr./Géog. canad.* **45**, 3). Van Kleunen and Richardson (2007) *PPG*

31, 4 argue for more integration between conservation biology and invasion biology.

conservation of angular momentum The momentum of a body taking a curved path is *angular momentum*. A body's angular momentum will remain constant unless changed by some force. This is important in understanding *jet streams; a body of air at the equator will move at the same speed as the earth. When this body of air moves polewards, it must increase its velocity to maintain its angular momentum, since earth's radius decreases polewards. At 30°, for example, the air moves very much faster than the earth below (which is much slower than at equatorial speed). This very fast-moving stream of air is a jet stream.

conservative margin A plate margin where the movement of the plates is parallel to the margin, such as the San Andreas Fault. See Chéry (2008) *Earth & Pl. Sci. Lett.* **269**, 3–4.

consolidation The reform and reorganization of land ownership in order to redress *fragmentation. Crecente et al. (2002) *Land Use Policy* **19**, 2 review land consolidation in Galicia, Spain. Ramutsindela (2007) *Tijdschrift* **98**, 4 argues that the South African government's approach to land restitution accounts in part for the former bantustans. See Gonzalez et al. (2004) *Ag. Systs.* **82**, 1 on the practicalities of land consolidation.

constant slope A straight, sloping element of a hillslope, located either in the middle of a slope profile, or at the base of the *free face. Burt (2003) *PPG* **27**, 4 finds that a constant slope is characteristic of valley incision, but that it cannot survive at its original angle once downcutting stops. Calvache et al. (1997) *Geomorph.* **21**, 1 note that particle size influences the longitudinal constant slope profiles of alluvial fans.

constructive margin A boundary where two *plates are spreading apart from one another and new oceanic crust is being created and added to the outer shell of the planet. See Jackson and Gunnarsson (1990) *Tectonophys.* **172** and Edmonds et al. (2003) *Nature* **421**, 6920, on the slow-spreading Gakkel Ridge.

constructive wave A low-frequency (6–8 per minute) spilling wave, with a long wavelength and a low crest, running gently up the beach. *Swash greatly exceeds *backwash (which is reduced by *percolation), leading to deposition. Spilling waves usually occur on gently sloping beaches, and may form beach ridges and *berms (Hughes (2004) *Coastal Eng.* **51**, 11–12; Coelho and Veloso-Gomes (2004) *J. Coast. Res.* **SI-39**). See Cooper et al. (2000) *Water & Env. J.* **14**, 2 on beach profiling.

consumer **1.** Organisms in all the *trophic levels, (herbivores, *carnivores, omnivores, and parasites) except for *producers. **Primary consumers** feed only on plant material; **secondary consumers** feed on primary consumers, and so on. Sullivan and Rohde (2002) *J. Biogeog.* **29**, 12 see tight consumer-resource relations as 'pulling a system towards equilibrium'.

2. A buyer of goods and services for her/his/others' personal satisfaction, as opposed to income generation. Jackson (2002) *PHG* **26**, 1 sees garment labels you're looking for an intelligent gift such as: "you're an 'alternative' person"; "you'd like to help the Indian economy"; as rounds of *commodification in an all-encompassing **consumer culture**. Slocum (2004) *Env. & Plan. A* **36** is more upbeat: 'citizens . . . cannot be wholly embodied by constructions such as the consumer . . . [and] can position themselves as reasoning *publics'.

consumption The utilization of consumer goods and services for the satisfaction of the present needs of individuals, organizations, and governments. Paolucci (2001) *Antipode* **33**, 4 illustrates the way that the contemporary city is 'a concrete manifestation of the centrality of consumption'. Jackson (2004) *TIBG* **29**, 2 examines the way that producers have 'customized' their products for different markets, and demonstrates the resilience of local consumption cultures in the face of *globalization.
'Alternative' consumption is a practice in the North that is underpinned by concern about social injustice and unfair labour practices in the South, and the practices adopted as part of a 'radical' lifestyle (Bryant and Goodman (2001) *TIBG* **29**, 3). Barnett et al. (2005) *Antipode* **3** see ethical consumption as a significant moment in consumer activism; 'one in which large numbers of people are explicitly mobilised in support of various political causes through a shared identity as consumers, but where the spatiality of this mobilisation exceeds the scale of the nation-state'.

consumption, geographies of Mansvelt (2008) *PHG* **32**, 1 identifies the themes that draw attention to shaping of consumption in place: consumption as an arena in which governance, regulation, and citizenship are produced; relationships between consumption and forms of urban space; and the way in which practice is embedded in social and spatial contexts which extend beyond the act of purchase. See Guano (2006) *Gender, Place & Cult.* **13** on sites for the consumption of culture in Genoa. McNeill (2008) *PHG* **32**, 3 argues that the hotel as hybrid of public and private space breaks down the artificial analytical distinction between 'public' (the high street, the mall, the department store) and 'private' consumption spaces (the home, the garden). 'To see private urban malls as iconic of contemporary retailing is to simplify the complexity of consumption spaces and practices' (Crewe (2000) *TIBG* **24**, 2). See Goss (1999) *AAAG* **89**, 1 on temporal-spatial contexts of consumption.

contested landscape A site of competing and often contrasting discourses (Rofe and Oakley (2006) *Geog. Res.* **44**, 3). T. Marsden et al. (2003) see the productivist rural landscape as a contested landscape 'where decision making reflects local economic priorities and development projects are likely to cause less contention'.

contextual effect The effect of their local community on the decisions that people make, so that they may be influenced by the views of others in the community; 'converted', in other words, by their neighbours. Mondak in D. C. Mutz et al., eds. (1996) reports that some contextual effects emerge when citizens use their local environment as a default source of information in the absence of national information. Cutler (2007) *Pol. Geog.* **26**, 5 thinks that local conditions will be more influential on issues that are: clearly related to a particular political judgement; easy to observe; with localized policy impacts; and where national-level information is scant or open to multiple interpretations. Pearce et al. (2008) *J. Epidemiol. & Comm. Health* **62** study the contextual effect on diet. *See also* EMBEDDEDNESS.

contiguous zone A zone of the sea beyond the *territorial seas of a nation, over which it claims exclusive rights.

(⊕) SEE WEB LINKS

• UN discussion of the Treaty on the law of the sea.

continental climate A climatic type associated with the interior of large land masses in mid-latitudes. Without the moderating influence of the sea, summer and winter temperatures are extreme. Precipitation is low, as the region is distant from moisture-bearing winds. The European Union PESETA project predicts that the greatest temperature rises related to global warming will be in continental climates; Klingbjer et al. (2005) *Geografiska A* **87**, 1 consider the effect of global warming on continental climate and glacier dynamics. Bergh et al. (2003) *For. Ecol. & Manage.* **183**, 1–3 compare net primary productivity in continental and maritime climates.

continental crust The outer, rigid surface of the earth which forms the continents. Wilde et al. (2001) *Nature* **409**, 175–8 find evidence for the existence of continental crust on the Earth 4.4 gigayears ago. Continental crust is rich in feldspars and granitic rocks and about 35 km thick (Gao et al. (1998) *Geochimica et Cosmochimica Acta* **62**, 11)—thicker, but less dense, on average, than *oceanic crust. Collins (2002) *Geology* **30**, 6 explains the role of cycles of tectonic switching in the production of continental crust. Crust-mantle interactions occur in the lower continental crust (Yang et al. (2007) *Prog. Nat. Sci.* **17**, 2).

continental drift The theory that continents which are now separate were united in a supercontinent, suggested by Alfred A. Wegener (1916). Wegener's ideas were vindicated in the mid-1960s when the development of plate-tectonic theory provided a new framework for planetary-scale tectonics, explicable only in terms of continental drift (Harper (2001) *Geol. Today* **17**, 4): Hess (1962), Buddington memorial volume *Geol. Soc. Am.* 599–620, identified sea-floor spreading as a geologic mechanism to account for Wegener's moving continents. See Jackson (1993) *AAAG* **83**, 2 on the role played by palaeomagnetic studies in continental drift research.

Dobson (1992) *AAAG* **82**, 187–206 suggests that convection drives circular plate motion while gravity drives lateral motion. Convection currents well up at high-pressure centres, spiral outward, transfer to low-pressure cells, spiral inward, and descend at low-pressure centres. In most instances, upwelling and descent of the asthenosphere occur at opposing plate centres rather than at plate margins. Cells migrate laterally in a global pattern driven by gravity. Sea-floor spreading and subduction occur because of differential rates of lateral plate motion. Bokelmann (2001) *Geology* **30**, 11 believes that movements of the mantle play an important role in driving the plates.

In 2006, NASA scientists released the first direct measurements of continental drift, showing that the Atlantic is gradually widening, and that Australia is receding from South America and heading for Hawaii (Anonymous (2006) *New Scientist* **192**, 2578; p. S24). Silver and Behn (Carnegie Institution) suggest that plate tectonics may have halted at least once, and may do so again. See Pratt (2000) *J. Sci. Exploration* **14**, 4 on the challenges confronting plate tectonics.

continental shelf The gently sloping submarine fringe of a continent, ended by a steep continental slope which occurs at around 150 m below sea level. Mossop (2007) *Ocean Dev. & Int. Law* **38**, 3 identifies potential threats to marine biodiversity on the continental shelf, and considers the legal rules that coastal states may use to protect marine biodiversity.

contingency An event or state of affairs dependent on another, uncertain event or occurrence. **Contingence** is the influence of the history of a system in determining its response to an input (Lane (2001) *TIBG* **26**). In the context of a fluvial system, for example, the response of a length of bank to a hydrological event will be influenced by the bank morphology resulting from a previous event. Hysteresis—the dependence of the equilibrium position of a system on what happens during the process of dynamic adjustment—seems to be the human geographer's term for contingence (J. Rogers Hollingsworth and R. Boyer, eds. (1997); Vandenberghe and Woo (2002) *PPG* **26**, 4; Jordaan (2005) *PHG* **29**).

contour A line on a map joining places of equal heights, and sometimes equal depths, above and below sea level. The **contour interval** is the vertical change between consecutive contours. **Contour ploughing** is a method of ploughing parallel to the contours rather than up or down a slope. It is used to check soil erosion and the formation of *gullies.

conurbation A group of towns forming a continuous built-up area as a result of *urban sprawl (also called *metropolitan area*). See the classification outlined by Raybould et al. (2000) *Update* **25**, 2–4S; see also S. Brunn (2003).

convection The process whereby heat is transferred from one part of a liquid or gas to another, by movement of the fluid itself. Convection carries excess heat from the earth's surface and distributes it through the *troposphere

(Fanaki (1971) *Boundary-Layer Met.* **1**, 3; Takayabu et al. (2006) *J. Met. Soc. Japan* 84A).

Thermal convection (*free convection*) is propelled by buoyancy (R. Barry and R. Chorley 2003). **Mechanical convection** (*forced convection*) is the upward movement of an air parcel over mountains, at fronts, or because of *turbulence (Sorbjan (2006) *Boundary-Layer Met.* **119**, 1). Kijazi and Reason (2005) *Theoret. & Appl. Climatol.* **82**, 3–4 observe that wet conditions during *El Niño years are associated with **enhanced convection** (convection with a higher Bowen ratio). See Ludlam and Scorer (2006) *Qly. J. Royal Met. Soc.* **79**, 341.

convection rain When upward *convection occurs in a parcel of moist air, the rising air will cool. Further cooling will cause condensation of the water vapour in the air, and rain may result. If the air is very moist, the cooling results in condensation, and hence the release of *latent heat. This causes the rising air to accelerate, and very tall *cumulo-nimbus clouds form; see Sui et al. (1997) *J. Atmos. Scis* **54**, 5.

convective condensation level The point on a thermodynamic diagram where a sounding curve (representing the vertical distribution of temperature in an atmospheric column) intersects the saturation mixing ratio line corresponding to the average mixing ratio in the surface layer (approximately the lowest 500 m). The dry adiabat through this point determines, approximately, the lowest temperature to which the surface air must be heated before a parcel can rise dry-adiabatically to its lifting condensation level without ever being colder than the environment. This temperature, the convective temperature, is a useful parameter in forecasting the onset of convection (AMS online).

convective instability Instability that is caused by the rising of very dry air over warm, moist air below. C.-G. Rossby (1932) argued that it exists when one of these conditions is met over a layer of atmosphere: the lapse rate of wet-bulb temperature exceeds the moist-adiabatic lapse rate; the equivalent potential temperature decreases with height; the wet-bulb potential temperature decreases with height. Convective instability of the terrestrial atmosphere is largely determined by water vapour. See Gutzler (1996) *J. Atmos Scis* **53**, 19.

convenience distance The ease, or otherwise, of travel. A town 50 km away may be well served by transport and thus 'nearer' than one 20 km away which is less accessible. **Convenience goods** are *low-order goods like milk, and occasional groceries, which are frequently bought locally, with little consideration of the prices.

convergence **1.** The meeting of *tectonic plates. See Rosenbaum (2002) *Tectonophys.* **359**, 1–2 on the convergence of Africa with Europe, and Gelabert et al. (2004) *Geologica Acta* **2**, 3 on the convergence of the Indoaustralian and Eurasian plates.

2. In meteorology, air streams flowing to meet each other. An area of **convergence** is an area of **rising air**. **Convergence** can occur aloft over dense, cold air and is not necessarily confined to a layer bounded by the surface. Byers and Rodebush (1948) *J. Meteor.* **5** were the first to show a correlation between convective rainfall and convergence of low-level winds. These areas of low-level convergence were often restricted to well-defined lines of convergence with widths of the order of 1–2 km (Crook and Klemp (2000) *J. Atmos. Scis* **57**, 6). In the upper troposphere, convergence causes air to subside, creating *anticyclonic conditions at ground level (Hastenrath (2007) *Dynam. Atmos. & Oceans* **43**, 1–2).

3. Within human geography, the use of the same production methods or practices by firms operating in different national-institutional spaces. Gertler (2001) *J. Econ. Geog.* **1**, 1 describes the convergence between Anglo-American, German, and Japanese models of economic growth, and Domosh (2004) *TIBG* **24**, 4 explores the convergence of economic and cultural approaches to *imperialism. See also A. Leyshon and J. Pollard (2000).

convergence space Spaces for actors, movements, and struggles at particular moments in time; 'a heterogeneous affinity between various social formations, such as social movements' (Routledge (2003) *TIBG* **28**, 3); 'a world made of many worlds' (Marcos in K. Abramsky, ed. 2001). 'Convergence spaces are comprised of place-based, but not necessarily, place-bound movements . . . relational achievements involving a practical relational politics of solidarity. Such solidarity takes place in the form of changing and overlapping circuits of relations that are enacted both virtually through the internet and materially in particular forums where connections are grounded in place- and face-to-face based moments of articulation' (Cumbers et al. (2008) *PHG* **32**, 2). These places of protest enter the collective memory of the participants of convergence spaces, thereby reinforcing the moral commitment of activists to their struggles (Routledge, op. cit.).

converging margin *See* DESTRUCTIVE MARGIN.

convex slope A slope, or slope element, that gets progressively steeper downhill. It may be determined by structure (*exfoliation domes), by lithology (Allison and Higgitt (1998) *Catena* **33**, 1), or by landslides (Di Luzio et al. (2004) *Geomorph.* **60**, 3). Convex slopes may result from weathering and debris transport—see J. Wilson and J. Gallant (2000). See Lin and Oguchi (2004) *Geomorph.* **63**, 3 on drainage density and convex slopes.

conveyor belt **1.** An expansive upwards, polewards flow of air. As a mid-latitude depression develops, a **warm conveyor belt** rises at *c*.20 cm s^{-1}, ahead of the cold front, and a reciprocal **cold conveyor belt** descends from medium/upper levels, below the warm conveyor, well ahead of the surface warm front; see Pomroy and Thorpe (2000) *Monthly Weather Rev.* **128**, for a useful diagram. Schultz (2001) *Monthly Weather Rev.* **129** suggests that cyclones with well-defined warm fronts show a sharp break between the cyclonic and anticyclonic paths of the cold conveyor belt, while cyclones with weaker warm fronts have a broad transition zone between the two.

2. In the oceans of the world, a global system of currents; *see* THERMOHALINE CIRCULATION.

cooperative A type of business organization owned by its employees or customers. **Agricultural cooperatives** are collective projects initiated by local producers which aim to organize the implementation, collection, and transformation of the members' production, and to ensure their reputation through collective actions. This means that they have to reconcile economic performance (marketing) with their social responsibilities towards their members (Filippi and Torre (2003) *Int. J. Technol. Manage.* **26**, 2–4).

Agricultural cooperatives have been encouraged in the developing countries, where several farms pool resources to jointly purchase and use agricultural machinery; see anon. (2005) *Rev. Int. Cooperation* **98**, 2 on cooperatives in Colombia. Rice and Lavoie (2005) *Canad. Geogr.* **49**, 4 evaluate **business cooperatives** in Canadian economic growth.

coral reef An offshore ridge, mainly of calcium carbonate, formed by the secretions of small marine animals. Corals flourish in shallow waters over 21 °C and need abundant sunlight, so the water must be mud free, and shallow. Fringing reefs lie close to the shore, while barrier reefs are found further from the shore, in deeper water. Reef health is mapped by: live coral cover (Mumby et al. (2004) *Marine Pollut. Bull.* **48**), diversity (Hodgson et al. 2004), and remote sensing (Kutser et al. (2006) *Limnol. & Oceanog.* **48**).

Coral reefs are the most biodiverse marine ecosystems on the planet; M. L. Reaka-Kudla (1997) estimates that they harbour nearly one million species globally. However, the health of coral reefs is declining (see C. Wilkinson 2004), due to overfishing, nutrient enrichment, and coral diseases, at the local scale, and ocean warming, acidification, and sea-level rise at the global scale; see also Hoegh-Guldberg (1999) *Marine & Freshw. Res.* **50**, Hempel and Morozova (2001) *Bull. Marine Sci.* **69**, 2 on coral reef management from the benthos up, and the US NOAA Coral Reef Conservation Programs.

A **coral atoll** is a horseshoe-shaped ring of coral which almost encircles a calm lagoon. Many coral reefs are hundreds of metres deep; Charles Darwin's theory (1842) is that deep reefs formed during a long period of subsidence. Thus, coral forms in shallow waters and then sinks. Woodfine et al. (2008) *Sedimentol.* **52** seem to support the subsidence theory. See Braithwaite et al. (2000) *Int. J. Earth Scis.* **89**, 2 on revisiting Darwin's model. Woodruffe et al. (1999) *Marine Geol.* **160**, 1 dispute Darwin's theory. There is also the possibility of a positive feedback loop: sea level rises, providing more habitat for coral reefs, the reefs flourish and produce more CO_2, the additional CO_2 causes additional temperature increase, and sea levels rise even more; see G. Camoin and P. J. Davies (1998).

core. 1. The central part of the Earth, the core is the point within the Earth where S-waves cannot penetrate, marked by an abrupt increase in pressure. It is believed to be composed primarily of a nickel-iron alloy (along with abundant platinum-group elements), with a liquid outer zone, 2900–5000 km deep, and a solid inner zone, 5000–6370 km deep.

2. *See* CORE REGION.

core–frame concept A model of the central area of the city that shows a core of intensive land use indicated by high-rise buildings (the *CBD), beyond which is the frame, with less intensive land use: warehousing, wholesaling, car parking, medical facilities, and so on.

core–periphery The core—a central region in an economy, with good communications and high population density, which conduce to its prosperity—is contrasted with the periphery—outlying regions with poor communications and sparse population (for examples, *see* UNEMPLOYMENT). 'Either defined in geographical or sociological terms, the center represents the locus of power and dominance and importantly, the source of prestige, while the periphery is sub-ordinate. Simply put, a center–periphery relationship is about hierarchy' (Azaryahu (2008) *Soc. & Cult. Geog.* **9**, 4).

Cores are associated with high wages, high technology, and high profit inputs and outcomes. Geographically, these processes have tended to concentrate and segregate—this produces places where core processes dominate and places where peripheral processes dominate. 'For short-hand purposes these may be designated as "core" and "periphery" but they must never be seen as purely one or the other: so-called "core countries" encompass numerous, if minority, peripheral processes; and the opposite is so for "peripheral countries"' (Brown et al. (2002) *GaWC Res. Bull.* **236**). 'When transport costs fall below a critical value, a core–periphery spontaneously forms, and nations that find themselves in the periphery suffer a decline in real income' (Krugman and Venables (1995) *Qly J. Econ.* **110**, 4). Lanaspa and Sanz (2001) *Papers Reg. Sci.* **80** add congestion costs and infrastructure to Krugman's model, and Baldwin and Forslid (2000) *Economica* **67**, 267 introduce Romerian product innovation growth into the model. M. Fujita and J.-F. Thisse (2002) develop a core–periphery model in which the agglomeration effects from concentrating R&D activity in the core, combined with relatively low transportation costs, generate sufficient value added to more than compensate the periphery for the loss of R&D activity.

The model has been criticized—it has been argued that *uneven development is not the inevitable consequence of development, but of the particular *mode of production used to bring about that development (Harris in L. Blume and S. Durlauf 2006). Copus (2001) *Eur. Plan. Studs* **9**, 4 finds the validity of conventional (spatial) models of peripherality 'increasingly questionable'.

core region In a nation, a centre of power where innovation, technology, and employment are at a high level, such as Madrid or south-east England. See Harrison (2007) *J. Econ. Geog.* **7**, 3 on England's Core Cities Group, and Pain, no date, *GaWC Res. Bull.* **220** on the London core and the global 'mega-city region'.

(((∰))) SEE WEB LINKS

• Brand and Bhatti (2006) on the core region of Madrid.

corestone *See* BOULDER FIELD.

CORINE The Coordination of Information on the Environment is an ecological commission of the European Communities located in Brussels, Belgium. Its main goals are to coordinate information and action within the EU, to define and protect biotopes, to combat air pollution, and to preserve the ecology of the Mediterranean region.

Coriolis force An apparent, rather than real, force which causes the deflection of moving objects, especially of air streams, through the rotation of the earth on its axis. It shows up, for example, in the movement of an air stream, relative to the rotating earth beneath it. The Coriolis force is proportional to the wind speed $U(z)$, and the Coriolis parameter, $f = 2 \, \Omega \sin\Phi$—where Φ is the latitude and Ω is the Earth's rate of rotation—and acts perpendicular to the wind direction (Orr et al. (2005) *Weather* **60**, 10).

corporate governance The system by which companies are directed and controlled; it is concerned with systems, processes, controls, accountabilities, and decision-making at the heart of and at the highest level of an organization. Issues of corporate governance are central to corporate social responsibility (Gouldson (2006) *Area* **38**, 4). See also A. Bongenaar (2001) and G. Clark and D. Wójcik (2007). There is a growing European market for corporate governance; large continental companies appear to be adopting shareholder-friendly practices more consistent with global financial market imperatives than their national traditions; see Bauer et al. (2008) *J. Econ. Geog.* **8**, 4.

corrasion The erosive action of particles carried by ice, water, or wind (although the term is most often used in a fluvial context). Corrasion is another term for *abrasion.

corridor A limb of one state's territory cutting through another state, usually for access. The most famous is the Polish corridor (Howkins (2006) *J. Transp. Geog.* **4**, 4). A **corridor of commerce** extends along a line of communication, such as the Mekong River (Furlong (2006) *Pol. Geog.* **25**, 4), or the UK 'M4 corridor' (Coe and Townsend (1998) *TIBG* **23**, 3).

corrie *See* CIRQUE.

cosmopolitan distribution When 'everything is everywhere' (Smith (2007) *J. Biogeog.* **34**, 10).

cost–benefit analysis The attempt to compare the total social costs and benefits of implementing a decision, usually expressed in money terms; 'the decision-making tool of choice in contested planning decisions' (J. Adams 1995). Costs and benefits include money, externalities—for example pollution, noise, and disturbance to wildlife—and external benefits, such as reductions in travelling time or traffic accidents. Rouwendal and van der Straaten (2008) *CPB Disc. Papers* **98** note that 'when costs and benefits of actual projects are listed and have to be traded off against each other as to determine the net result, consensus often disappears'. They discuss a hedonic method as an attempt to compute the monetary value of the benefits objectively. Hanley et al. (2003) *J. Env. Manag.* **68**, 3 observe that decisions over the size of the population affected are crucial in the calculation of aggregate benefits and costs. Castree (2003) *PHG* **27**, 3 argues that cost–benefit analysis is one mechanism of 'achieving proxy commodification and can serve as [a precursor] to the creation of real markets'. See A. Boardman et al. (2006).

cost–space convergence Especially in the last 50 years, innovations in air, rail, road, and sea transport technology, communications, and handling technology have caused cost/space (and time/space) to shrink, but differentially. These differential spatial effects largely stem from unequal investment in modal capacity, routes, and terminals at international, national, and local scales, and tend to enhance the importance of the largest demand centres in developed countries (Knowles (2006) *J. Trans. Geog.* **14**, 6). *See* CORE–PERIPHERY.

cost surface A three-dimensional 'contour model' representing the variation in costs over an area. See Martin et al. (2002) *Health & Place* **8** for a *GIS-based example.

Cost surface

cottage industry The production of finished goods by a worker at home, sometimes together with her or his family. See Kristiansen (2002) *Sing. J. Trop. Geog.* **23** on Indonesia.

counter-radiation The long-wave radiation emitted from the earth to the *atmosphere after it has absorbed the shorter-wave radiation of the sun.

countertrade A reciprocal trading system between two countries, useful for nations wishing to import high technology but lacking foreign exchange, or exporting goods for which world demand is low. Choi et al. (1999) *J. Int. Bus. Studs* **30**, 1 argue that international countertrade is an organizational remedy for uncertainty and institutional deficiency.

counter-urbanization The movement of population and economic activity away from *urban areas—see Mitchell (2004) *J. Rur. Studs* **20**, 1 for an exhaustive dissertation on the meaning of this term, and Escribano (2007) *Tijdschrift* **98**, 1 for a more nuanced view.
     Increased mobility and improved communications, higher incomes, changing household composition, changing technology in manufacturing, and growth of the service sector have led to increased commuting, counter-urbanization, and the urban–rural employment shift (Hodge and Monk (2004) *J. Rur. Studs* **20**, 3). Between 1970 and 2000, the shift of employment and population from the urban core to the suburban fringe carried with it huge inequalities in the distribution of income among Detroit area municipalities. The lowest-income communities, located in the areas suffering the greatest population and employment decline, concentrate in and around the central city; the most affluent are in Detroit's outlying north-west suburbs (Jacobs (2003) *Urb. Studs* **40**).

country rock A pre-existing rock which has suffered later igneous *intrusion.

Countryside Stewardship Scheme (CSS) Introduced in the UK in 1991, the CSS encourages land managers, in practice mostly farmers, to restore and maintain traditional features of the countryside for wildlife conservation and the enjoyment of the public. See Wilson and Hart (2001) *Sociologia Ruralis* **41**, 2 for an evaluation.

coupling constraint In *time–space geography a limit to an individual's actions because of the necessity of being in the same space and time as other individuals. Fixed activities dictate strict coupling constraints while flexible activities allow more fluid coupling in space and time (Miller (2005) *Geog. Analys.* **37**).

cover crop A fast-growing crop planted in the rows between the main crop to protect the soil from erosion caused by heavy rainfall (Tengberg et al. (1997) *Geograf. Annal A.* **79A**, 1–2).

crag and tail A large rock outcrop, with a rampart of *till on its lee side, formed by erosion on the up-glacier side of the outcrop and deposition of debris in the low-pressure lee area. Glasser and Bennett (2004) *Prog. Phys. Geog.* **28**, 1 define **micro crag and tails** as small tails of rock that are preferentially protected from glacial abrasion in the lee of resistant grains or mineral crystals on the surface of a rock.

crater A circular depression around the vent of a volcano. Craters form the summit of most volcanoes. They occur where lava overflows and hardens or where the walls collapse as the *magma sinks down the vent after an eruption. **Funnel-shaped craters** are typical of *stratovolcanoes, while **kettle-shaped craters** are characteristic of *shield volcanoes.

craton A core of stable continental crust within a continent and composed wholly or largely of *Precambrian rocks with complex structures. Two types are recognized: platforms, which are parts of cratons on which largely undeformed sedimentary rocks lie, and *shields. See Zheng et al. (2004) *Geology* **32**, 3 on the North China Craton.

creative cluster Florida (2004) *Next American City* **5** identifies as crucial to creativity: creative human capital; innovative activity associated with high-tech agglomerations; and the quality of a place as 'liberal, tolerant, and capable of attracting skilled people able to generate new ideas'. Storper and Venables (2004) *J. Econ. Geog.* **4**, 4 argue that face-to-face contacts are the most fundamental aspect of proximity, particularly important in environments where information is imperfect, rapidly changing, and not easily codified.

creative industries Those industries that focus on creating and exploiting intellectual property; for example, fashion designs, computer games, or providing creative services for business, such as advertising (also called the *creative economy*).
   In 1998, the cultural industries employed 5% of the total UK workforce at the time, making up 4% of UK GDP—a larger share than for any traditional manufacturing industry (UK Department of Media, Culture and Sport, *Creative Industries Mapping Document*). Peck (2005) *Int. J. Urb. & Reg. Res.* **29**, 4 argues that 'an unobserved creative milieu, that attracts artists, increases local economic dynamism', while Wojan et al. (2007) *J. Econ. Geog.* **7**, 6 feel that '**creative milieu**' remains an amorphous construct.

creep The slow, gradual movement downslope of soil, *scree, or glacier ice. Most creep involves a deformation of the material, i.e. *plastic flow.

creolization The hybridization of a culture, as it absorbs and transforms forces from outside; the production of new local forms in response to *globalization; 'a term now widely used to refer to processes of cultural change in the Caribbean and elsewhere, that contributed to the development of culturally hybrid social forms' (Lambert (2004) *PHG* **28**). Carter (2005) *Area*

37, 1 thinks that creolization overcomes 'the fixed and essentialized assumptions regarding both identity and territory'. Adams and Ghose (2003) *PHG* **27**, 4 discuss **cultural creolization**—tandoori pizza, anyone?

Cretaceous The youngest period of *Mesozoic time from *c.*136 to 65 million years BP.

(⊕) SEE WEB LINKS

• An overview of the Cretaceous period.

crevasse A vertical or wedge-shaped crack in a glacier, varying in width, from centimetres to tens of metres. Its maximum depth is about 40 m, as, at that depth, ice becomes plastic and any cracks merge together. **Transverse crevasses** occur when the ice extends down a steep slope. **Longitudinal crevasses** form parallel with the direction of flow as the ice extends laterally. **Marginal crevasses** occur across the sides of a glacier as friction occurs between the ice and the valley walls. **Radial crevasses** fan out when the ice spreads out into a lobe.

crime, geography of The analysis of crime—its effects, the offences, and the offenders—to understand the interactions between crime, society, and space. In the UK, Australia, New Zealand, and North America, the *inner city is a crime hot spot; Canada's Research and Statistics Division notes that 'several indicators were found to have a significant effect on crime levels [in Saskatoon] including higher proportions of single people and youth not attending school as well as lower average household incomes'. The residential pattern of the offenders resembles the pattern of offences. See Monmonier (2006) *PHG* **30**, specifically 375–8, on mapping crime.

Feminist geographers stress that, while most girls are brought up to fear violence by strangers in public places (thus hugely limiting female spatial mobility), most violence against women is located within the home. It's difficult for many women to acknowledge that their homes are not a place of safety; see Meth (2003) *Geoforum* **34**, 3. Brunel (2005) *Gender Place & Cult.* **12**, 3 shows that organizations devoted to reducing domestic violence are still rare in urban environments and virtually non-existent in rural ones. On the perception of crime, Stenning (2005) *TIBG* **30**, 1 addresses the 'wider tendencies to characterize and stigmatize working-class communities'. Pain (2000) *PHG* **24** reviews work on fear of crime, social identity, and exclusion, and 'local ethnographies of fear'.

Herbert and Brown (2006) *Antipode* **38**, 4 note that the onset of *neoliberalism in the United States coincided with 'an unprecedented expansion of punishment practices that intensify social divisions rooted in class and race'. Pallot (2005) *TIBG* **30**, 1 argues that the use of the peripheries as a place of exile and detention in the USSR was driven by the pursuit of national economic goals by a powerful central state. Post-1992, the continued expulsion of people to the peripheries may be seen as Moscow cleaning itself of 'unwelcome elements' at the expense of provincial Russia. S. Herbert

(2006) finds that the community is 'unbearably light' as a basis for political action or a foil against violent crime—'only rarely can people rally around a sense of common place'—but Martin (2003) *AAAG* **93**, 3 is more sanguine about the role of place in motivating activism. See also Martin (2000) *Geography* **21**, 5.

critical geography (critical human geography) The umbrella term for a group of geographical concepts and procedures that are centred on opposition to repressive and inequitable power relations in: *capitalism, *class, *colonialism, disability, ethnicity, gender, race, and sexuality. Critical geographers stress the role of dominance and confrontation in the production and reproduction of *landscape, *place, and *space: see, for example, Thompson-Fawcett and Bond (2003) *Prog. Plan.* **60**, 2 on the urbanist discourse.

Although Blomley (2007) *PHG* **31**, 1 judges that critical geography has become deeply entrenched within the academy, many share Cloke's frustration with 'our apparent inability to retain a critical political edge in human geography' (2002, *PHG* **26**). For Baeten (2002, *Geografiskal B* 3–4), urban geography 'fails to crystallize in a convincing political project that would provide a credible alternative for the poverty-generating capitalist shaping of today's city'. However, Peck and Wills (2000) *Antipode* **32**, 1–3 note that being radical in the late 1960s 'involved a very different cluster of beliefs, ideas and affiliations than might be expected today'. Mitchell (2000) *2nd Int. Crit. Geog. Conf.* **2** argues that critical geography should be 'for people, not just ourselves'. Oberhauser (2007) *AAAG* **97**, 2 judges C. Katz (2004) as 'a model for research in critical geography'. See Simonsen (2004) *Geoforum* **35**, 5 and Blomley (2008) *PHG* **32**, 2 for reviews of this sub-discipline; see also Oswin (2008) *PHG* **32**, 1. Unwin's words (2000, *TIBG* **25**, 1) are inspirational: 'A critical geography needs to engage with the everyday practices of all of us who live in the places that we do; it needs to focus on the needs and interests of the poor and the underprivileged; it remains a very modern enterprise, retaining a belief that it is possible to make the world a "better" place.'

critical isodapane The cost of a good increases with transport costs; the critical isodapane is the *isoline where this increase is exactly offset by the savings from cheaper labour.

critical race theory A theoretical approach used to deconstruct racial ideologies, and analyse the politics of 'race' and the links between nationalism—or localism—and racism (Dunn (2006) *Geog. Res.* **4**, 4). The interdisciplinary field of critical race studies pursues the claim that race has no necessary epistemological status in itself, but depends on the context and organization of its production for its political effects (Anderson (2008) *Cult. Geogs* **15**, 2).

crop rotation The practice of planting a succession of crops in a field over a period of years. Rotations can maintain field fertility since different crops use different soil nutrients, so excessive demands are not made of one nutrient. In certain rotations, plants like legumes (peas and beans) are grown to restore fertility. See D. Stone (2005) on medieval agriculture, and Sainju et al. (2006) *J. Env. Qual.* **35** on contemporary practice.

cross-bedding (current bedding) In a sedimentary rock, the arrangement of beds at an angle to the main bedding plane.

cross-correlation analysis A way of identifying any changes in land use that have occurred in a previously mapped area. See Butson and Fraser (2005) *3rd Int. Wkshp. Analysis Multi-Temp. Rem. Sens.* for an evaluation of the technique.

crude rate A *vital rate which is not adjusted for the age and sex structure of a population.

crumb A spheroidal cluster of soil particles, i.e. a type of *ped.

crust The outer shell of the earth including the continents and the ocean floor; the *lithosphere. *Sial overlies *sima in the continental crust; the oceanic crust is mostly sima.

cryofront *See* CRYOTIC.

cryogenic (cryergic) Processes carried out by ground ice. Needle ice and frost heaving are connected with the development of cryogenic phenomena (Grab et al. (2004) *Geografiska. B* **86**, 2). *See* FROST HEAVING; FROST WEDGING; CRYOPLANATION.

cryopediment In *periglacial environments, a low-angle, concave, piedmont footslope developed by slope retreat, brought about through frost *weathering and the sapping of slopes, and by *nivation surfaces. Cryopediments may coalesce to form a **cryopediplain** (Tzudek (1993) *Permaf. & Perig. Procs* **4**, 1—with useful photographs).

cryoplanation The lowering and smoothing of a landscape by *cryogenic processes. 'To deny the fairly widespread existence of features commonly called cryoplanation benches, terraces, and pediments would be foolish: to claim that there is anything approaching an adequate explanation of their origin(s) would be even more foolish . . . there is little to say about cryoplanation other than the concept needs to be challenged directly' (Thorn and Hall (2002) *PPG* **26**, 4).

cryosphere The ice at or below the earth's surface, including *glaciers, *ice-caps and *ice sheets, *pack ice and *permafrost. Climatic cooling increases the cryosphere (Lewis et al. (2007) *GSA Bull.* **119**, 11). The mass balance of the cryosphere is most likely contributing to sea-level rise (J. Bamber and A. Payne 2004).

cryostatic pressure The pressure exerted on rocks and soil when freezing occurs. As the *freezing front advances, the pressure of the trapped soil moisture increases (Mackay and Mackay (1976) *Can. J. Earth Scis* **13**). This pressure can separate individual grains of soil, forming a mass of fluid mud, which may dome up the ground, or form mud blisters.

cryotic Having temperatures below 0 °C. Van Everdingen (1976) *Can. J. Earth Scis.* **13** proposes that 'cryotic' be used to describe permafrost on the basis of temperature, and that 'frozen' be restricted to ground in which part or all of the pore water is ice.

cryoturbation A general term describing all frost-based movements of the *regolith, including *frost heaving and *gelifluction. Grab (2005) *PPG* **29**, 2 analyses cryogenic mounds.

cuesta A ridge with a *dip slope and a *scarp slope. *See* ESCARPMENT. For **cuesta topography**, see Pánek et al. (2008) *Geomorph.* **95**, 3–4.

cultural capital The symbols, ideas, tastes, and preferences that can be strategically used as resources in social action, and serve as markers of collective identity and social difference. Cultural capital includes cultural goods: pictures, books, dictionaries, instruments, machines, and so on. See Hubbard (2002) *Env. & Plan. A* **34**. The commodification of cultural capital is carried out by the **cultural industries** which convert cultural capital to economic capital. R. Florida (2002) argues that urban fortunes increasingly turn on the capacity to 'attract, retain and even pamper a mobile and finicky class of "creatives"'.

It's possible to be high in cultural capital, but low in economic capital: S. Fernandes (2006) for example, notes that Cuban arts and popular culture have been commercialized to entice foreign investment. Ottaviano and Peri (2006) *J. Econ. Geog.* **6**, 1 find that US-born citizens living in cities of increasing **cultural diversity** earned significantly higher wages. Khawaja and Mowafi (2007) *Scandi. J. Pub. Health* **35**, 5 find that cultural capital has significant associations with general and mental health status in Lebanese women.

cultural distance A gap between the culture of two different groups, as between Vietnamese immigrants and the Czech majority in the Czech Republic (Drbohlav and Dzúrová (2007) *Int. Mig.* **45**, 2).

cultural ecology An investigation that focuses on the dynamic interactions between human societies and their environments. Culture is seen as the primary adaptive mechanism used by human societies to deal with, understand, give meaning to, and generally cope with their environment (T. Darvill 2002). 'In a large part, cultural ecology in geography constituted a response to the appeal by Brookfield (1964, *Econ. Geog.* **40**, 283–303), that the landscape could not be understood without attention to the behaviour of land managers, who, after all, are the critical agents modifying or transforming the landscape' (Chowdhury and Turner (2006) *AAAG* **96**, 2).

cultural-economy An approach to economics centred on the premiss that the pursuit of prosperity has always been a cultural performance; Amin and Thrift (2007) *PHG* **31**, 2 argue that 'economic life is so shot through with cultural inputs and practices at all levels that "culture" and "economy" cannot be seen as separate entities'.

cultural geography The study of the impact of human culture on the landscape; 'the ways in which place and identity are embedded in a range of cultural landscapes, and the ways in which those social and material landscapes have reflected and influenced various experiences and notions of movement' (Mains (2004) *J. Cult. Geog.* **22**, 1). As things travel, their meanings and material nature can change; 'things themselves therefore have complex cultural geographies' (Crang in P. Cloke et al., eds 2006). 'The new cultural geography constitutes a powerful expressive form, giving voice to the effects of dislocation, disembodiment, and localization that constitute contemporary social orders' (Blair (1998) *Am. Lit. Hist.* **10**, 3).

   Mitchell (2002) *Antipode* **34**, 2 gives a breezy account of cultural geography, later writing that 'inequality, domination, oppression, exclusion, and power (and still the possibility for good lives despite all this)—at all scales from the household to the globe—are the true object of cultural geographic study' (2004, *J. Cult. Geog.* **22**, 1). Lorimer (2005) *PHG* **29**, 1 writes, somewhat diffusely, on the tendency for cultural analyses to cleave towards a conservative, categorical politics of identity and textual meaning, which can, he argues, be overcome 'by allowing in much more of the excessive and transient aspects of living'. See Lorimer (2007) *PHG* **31**, 1 for more; see also K. Anderson et al., eds. (2004).

cultural hearth The location in which a particular culture has evolved. Voigt-Graf (2004) *Glob. Networks* **4**, 1 uses the term as 'the country, region or place of origin of migrants and their descendants'.

cultural landscape The impact of cultural groups in shaping and changing the natural landscape. 'Landscape itself is a cultural image; a way of symbolizing, representing, and structuring our surroundings' (P. Cloke et al., eds 2006). Meinig 1979 (in D. Meinig and J. Jackson, eds 1979) sees that individuals interpret the landscape in many ways, 'all of them dependent on the viewer's mental ideas evoked by . . . their previous experiences'. Röhring (2006) *IASCP* argues for the cultural landscape as a regional common good.

cultural turn The increased focus, from the 1990s onwards, on culture as a geographical agent and product: 'a reorientation of human geography's interdisciplinary concerns toward the wide field of cultural studies' (Barnett (1998) *Antipode* **30**, 4). 'It is because all human realities are expressed through cultures that a cultural turn was needed in geography' (Claval (2007) *Tijdschrift*

98, 2). 'With the cultural turn in economic geography, the emphasis on economic necessity . . . has been contested by an agency-orientated cultural reading' that views engagement in such spaces as about the search for fun, sociality, distinction, discernment, the spectacular, and so forth, and more recently by a geographically sensitive approach that ascribes agency to affluent populations and economic rationales to deprived populations (Williams and Paddock (2003) *Geografiska B* **85**, 3).

culture 'The stuff of culture [is] meaning, symbols, signification' (Barnes (2003) *Urb. Geog.* **24**, 6). In a study of MTV, Rosati (2007) *TIBG* **32**, 4 argues that 'cultural forms—and the authority of their production—are always implicated in a wider field of struggle over how the practice and meaning of daily life will be organised and in whose interests'. 'Popular culture . . . is one of the ways in which people come to understand their position both within a larger collective identity and within an even broader geopolitical narrative' (Dittmer (2005) *AAAG* **95**, 3). Brons (2006) *Tijdschrift* **97**, 5 sees culture as a 'map for behaviour'. See Hedberg and Kepsu (2003) *Geografiska B* **85**, 2, on migration to Sweden as an expression of culture.

cultures of work The attitudes, values, beliefs, modes of perception, and habits of thought and activity in the working environment. Cowen (2005) *Antipode* **37**, 4 documents changing cultures of work in Canada, associated with the shift towards higher education: increasing mobility and individualism in career choices. Faulconbridge (2008) *J. Econ. Geog.* **8**, 4 shows how managers in TNCs act as **cultural entrepreneurs**, driving change in institutionalized cultures of work through strategies that alter the cognitive frames of workers. See Jarvis (2003) *Area* **34**, 4.

cumec A measurement of *discharge. One cumec is one *cu*bic *me*tre of water per second.

cumulative causation The unfolding of events connected with a change in the economy, as a consequence of the *multiplier effect. Cumulative causation can be set in motion where the expansion of the cluster, through attraction of firms with complementary competencies, adds to the initial attractiveness of the cluster (Maskers and Malmberg (2007) *J. Econ. Geog.* **7**, 5). Kaldor (1981) *Econ. appliquée* **34**, 4 argues that any initial differences in productivity between regions are sufficient to produce progressive, self-reinforcing divergence—but this demonstrates how the nature of 'historicity' in the new economic geography models is very circumscribed (Martin and Sunley (2006) *J. Econ. Geog.* **6**, 4). Improvements made through cumulative causation are made at a cost to some other part of the economy; although underdeveloped regions offer the advantage of low-wage labour, these benefits tend to be offset by the agglomeration economies in the industrialized regions. 'Therefore in order to overcome the [negative] effect of cumulative causation intervention on the part of government is necessary' (Cibulskiené and Butkus (2007) *Jahrbuch für Regionalwissenschaft* **27**). The process is not limitless and can be reversed through negative externalities, such as congestion costs, or reductions in trade costs. Simulation models tend to show that large changes in trade costs may be required to cause deconcentration and that at intermediate levels of trade costs concentration remains high (Weiss (2005) *ADB Institute Discuss. Paper* 2005/04).

Cumulative causation

cumuliform In the shape of a *cumulus cloud. **Cumuliform convection** is vertical convection.

cumulo-nimbus A low-based, rain-bearing *cumulus cloud, dark grey at the base and white at the crown, which spreads into an anvil shape, as it is levelled by strong upper-air winds.

cumulus An immense, heaped cloud with a rounded, white crown and a low, flat, horizontal base, extending as high as 5000 m. Updraughts within the cloud are strong—up to 10 m s^{-1}, and cloud growth is often rapid, through *entrainment and the saturation of surrounding air. **Cumulus congestus** is a swelling, small cumulus which often becomes *cumulo-nimbus.

current A horizontal movement of water, as along a stream or through an ocean. The rate of flow of a river current varies with depth because friction operates along the bed and sides (R. Chorley et al. 1985). **Tidal currents** are associated with the rise and fall of the sea, and the velocities of ebb and flow vary with the morphology of the coast and any outflow of fresh water; see FitzGerald et al. (2002) *Geomorph.* **48**, 1–2. **Rip currents** are generally strong, shore-normal, jet-like flows that originate within the surf zone and are directed seaward through the breakers. They form in the nearshore zone and balance the inflow of seashore currents, influence the morphology of the shoreline, and may be important for transporting fine sediments offshore (MacMahan et al. (2005) *Marine Geol.* **214**, 1–4). *Ocean currents are driven by the planetary winds; see Hidaka (1952) *J. Oceanograph. Soc. Japan* (available online, captivating English, many equations).

current bedding In a sedimentary rock, bedding which is oblique to the 'lie' of the formation as a whole. The structure is not due to tilting or folding, but develops when sandbanks are built up in shallow water, or where sand dunes accumulate from wind-blown sands, where the pattern of bedding reproduces all or part of the outline of the dunes.

curvature In meteorology, the wind speed divided by the radius of curvature of the bending air stream. Conventionally, *cyclonic curvatures have positive, and anticyclonic curvatures negative, radii of curvature in the Northern Hemisphere. See W. Saucier (2003).

cusp A small hollow in a beach, U-shaped in plan; the arms pointing seawards. Beaches tend to have a series of cusps, formed when outgoing *rip currents and incoming waves set up nearly circular water movements. They form by self-organization, wherein positive feedback between swash flow and the developing morphology initiates the development of the pattern, and negative feedback—from the circulation of flow within the beach cusps—causes pattern stabilization (Coco et al. (2003) *J. Geophys. Res.* **108**, C3). Ciriano et al. (2005) *J. Geophys. Res. Oceans* **110**, C02018 suggest that beach cusp evolution might control low-mode edge wave dynamics.

cuspate foreland An accretion of sand and shingle, shaped by *longshore drift and constructive waves from two directions. See Roberts and Plater (2007) *Holocene* **17**, 4.

customs union A common market encompassing two or more states within whose boundaries there is free trade; with no tariffs or barriers to the movement of goods. In 1996, Turkey established a customs union with the EU; Falcioğlu and Akgüngör (2008) *Eur. Plan. Studs* **16**, 2 find that Turkey's economic integration with the EU is a significant factor in explaining the change in industry concentration.

cut-off low *See* COLD LOW.

cwm *See* CIRQUE.

cyberspace The information 'space' created by information technologies, notably the internet, the worldwide web, and virtual reality. Kitchen (1998) *PHG* **22**, 3 argues that spatiality is central to understanding cyberspace. 'Cyberspace is firmly embedded in the social spaces seen in the "real" world, and this relationship is reflected in how cyberspace is created, conceptualized and studied' (M. Dodge and and R. Kitchen 2001). Valentine and Holloway (2002) *AAAG* **92**, 2, showing how on-line spaces are used, encountered, and interpreted within young people's off-line, everyday lives, demonstrate the way the real and the virtual are mutually constituted. M.-P. Kwan (2001) proposes a behavioural model of cyber-accessibility and examines the way notions underlying conventional accessibility measures such as impedance and opportunity set can be extended for measuring individual cyber-accessibility, and suggests that behavioural theories and models may provide a theoretical foundation for cybergeography. N. Lin (2001) contends that social capital has been on the ascent in the past decade—in the form of networks in cyberspace. 'The tales of total cyberspace are contradicted even by its own, very material, necessities: The world of "friction-free" money-flows is organised from places like the City of London and Canary Wharf, where the physicality of location, the implacable materiality of those pompous buildings, even the symbolism of the precise address, is of fundamental importance' (D. Massey, Open2Net).

cycle of poverty A vicious spiral of poverty and deprivation passing from one generation to the next. Poverty leads very often to inadequate schooling and then to poorly paid employment. As a result, the affordable housing is substandard—low housing costs in poor neighbourhoods attract migrants from rural areas—leads to overcrowding, overuse of facilities and services, and can also contribute to the perpetuation of the cycle of poverty (Okoko (2004) *Habitat Int.* **28**, 3; *US Dept. Housing & Urb. Dev.* (*HUD*) 1995). Children growing up in such areas start off at a disadvantage, and so the cycle continues; see Chilton et al. (2007) *Indian J. Med. Res.* **126**. J. Lin and C. Mele (2005) argue the need for class-specific policies in America, designed to raise educational levels, improve the quality of public schools, create employment, reduce crime, and strengthen the family: 'only a simultaneous attack along all fronts has any hope of breaking the cycle of poverty.'

cyclogenesis The formation of *cyclones, especially mid-latitude depressions. Cyclogenesis occurs in specific areas, such as the western North Atlantic, western North Pacific, and the Mediterranean Sea, and is favoured where thermal contrasts between air masses are greatest. Cyclogenesis occurs when *divergence in the upper *troposphere removes air more quickly than it can be replaced by convergence at ground level. The net result is low pressure. The significance of upper-air movements in cyclogenesis is also indicated by the link with *Rossby waves; see Molinari et al. (forthcoming) *J. Atmos. Sci.* Surface depressions develop below the downstream, or eastern, limbs of Rossby waves, where the airflow is divergent. The routes of mid-latitude cyclones (*depression tracks) closely parallel the movements of the upper-air *jet stream.

cyclomatic number In *network analysis, the number of circuits in the network. It is given by:

$$\mu = e - v + p$$

where e = number of edges, v = number of vertices (nodes), p = number of graphs or sub-graphs. A high value of the cyclomatic number indicates a highly connected network. Gorman and Maleki (2002) *Telecomm. Policy* **2** note the relationship between the level of economic development of a region and the cyclomatic number of its major transport networks.

cyclone A *synoptic-scale area of low *atmospheric pressure with winds spiralling about a central *low. As air near ground level flows into a cyclone, its absolute *vorticity increases, and it is therefore subject to horizontal *convergence, causing air to ascend. This rising causes cooling, which often leads to *condensation, so that *precipitation is associated with cyclones. See Kristjánsson and McInnes (1999) *Q'ly J. Royal Met. Soc.* **25**, 560 on Greenland and North Atlantic cyclones; see McTaggart-Cowan et al. (2007) *Month. Weather Rev.* **135**, 12 on Hurricane Katrina. See also Löptien et al. (2008) *Clim. Dynamic.*, online.

cyclone wave The wave-like distortion of flow in the middle and upper troposphere associated with a mid-latitude depression. See Orlanski and Katzfey (1991) *J. Atmos. Scis* **48**, 17.

cyclostrophic Referring to the balance of forces in a horizontal, tightly circular flow of air. **Cyclostrophic flow** is a form of gradient flow parallel to the isobars where the centripetal acceleration exactly offsets the horizontal *pressure gradient. The **cyclostrophic wind** is the horizontal wind velocity producing such a centripetal acceleration; see J. Dutton (2002). It equals the real wind only where the *Coriolis force is small, or where wind speed and curvature, and hence centripetal acceleration, are great.

daily urban system The commuting area around a city. Bretagnolle and Pumain (2001) *Cybergeo* **335** show that this area varies according to the definition of 'city'.

dambo A gently sloping, shallow *savanna wetland, with no permanent river channel, often wider upstream than downstream (perhaps through chemical sapping), and often infilled with sands and clays. Core boulders indicate *etchplanation; see von der Heyden (2004) *PPG* **28**, 4.

Dansgaard–Oeschger (D–O) cycles Very rapid climate changes—up to 7 °C in some fifty years—during the *Quaternary period, especially during the most recent glacial cycle. The cause of D–O cycles has yet to be comprehensively established. See Genty et al. (2003) *Nature* **421** on the precise dating of Dansgaard–Oeschger climate oscillations in western Europe.

Darcy's law When the *Reynolds number is very low, the velocity of flow of a fluid through a saturated porous medium is directly proportional to the difference in water pressure at the two sites:

$$V = \frac{h}{P_l}$$

where h is the height difference between the highest point of the water-table and the point at which flow is being calculated, V the velocity of flow, P the coefficient of permeability for the medium in question, and l the length of flow. Darcy's law does not hold good for well-jointed limestone which has numerous channels and fissures. **Darcy's equation** can be written in several ways: see GroundwaterSoftwater.com (March 2004). See also Buchan and Cameron in S W. Trimble et al., eds (2003).

dead cliff A sea cliff no longer subject to wave attack, with an emerged strandline, often overlain with *shingle at its foot. Dead cliffs result either through a fall in sea level or because they are protected by a broad beach. See A. Guilcher (1958).

dead ice (stagnant ice) Static glacier or ice-sheet material. Ice may stagnate after a surging extension is abandoned, or at the end of a glacial period. As stagnant ice melts, the remnants are covered in *ablation moraine; see Kjær et al. (2001) *Sedimentol.* **48**, 5 and Ebert and Kleman (2004) *Geomorph.* **62**, 3. Dead ice in valleys may generate *kame terraces.

death rate The number of deaths in a year per 1000 of the population, as measured at mid-year. This is a crude rate—no allowance is made for different distributions of age and sex. For example, in 2003, Sri Lanka had a crude death rate of 5.9%, compared with 10.2% in the UK, 2004. The low rate for Sri Lanka is a reflection of the youth of its population—24.3% are under 14, as opposed to 17.2% in the UK. A **standardized death rate** compares the rate with a real or assumed population which is chosen as standard.

debris Material such as scree, gravel, sand, or clay formed by *weathering. Through the air, debris is transported by *saltation and *deflation; in water, debris moves by rolling, by saltation, in solution, and in suspension. Debris can be carried on a glacier—supraglacially—within a glacier—englacially—or subglacially. **Debris entrainment** is the process whereby ice picks up material: by the freezing of basal ice to the bed as the ice flows forward (Cuffy et al. (2000) *Geology* **28**, 4); through the drag between ice and bedrock particles; via the closure of debris-filled basal cavities in the ice; by the refreezing of meltwater; and on the glacier surface. **Debris flow** is the very rapid downslope movement of saturated material, guided by stream channels, and both less deep seated and rarer than a landslide. See Blikra and Nemec (1998) *Sedimentol.* **45** on debris flow, deposits, and controls on debris-flow activity, Tunusluogu et al. (2007) *Nat. Haz. Earth Sys. Sci.* **7** on ANN and debris flows, and Hartshorn and Lewkowicz (2000) *Zeitschrift* **44**.

decalcification The *leaching of calcium carbonate from a soil *horizon by the downward movement of soil water. See van den Berg and Loch (2000) *Eur. J. Soil Sci.* **51**, 1.

decentralization (deconcentration) The devolution of decision-making powers to the lowest levels of government authority. Bryld (2001) *Democratization* **8**, 3 questions the use of decentralization as a development strategy. However, Faguet (2003) *Crisis States, W. Paper* **29** argues that decentralization in Bolivia 'put real power over public resources in the hands of ordinary citizens'. National governments may use decentralization to regenerate declining regions; see K. Altes (2002) *Urban Studies* **39**, 8 on decentralization in the Netherlands. In 2003, the government of the Republic of Ireland began the transfer of complete government departments to provincial locations.

(())) SEE WEB LINKS

• Decentralization in the Republic of Ireland.

deciduous forest In the cold season of temperate latitudes, a tree's water supply is restricted when the temperature falls below 0 °C. To lessen water loss, deciduous trees shed their leaves until the spring brings more available moisture. See Bradshaw et al. (2005) *Ecography* **28**, 2.

decision-making Some concern themselves with the *political nature* of the decision-making unit. Thus Archibugi (2000) *New Left Rev.* **4** proposes replacing the state as decision-maker with 'new flexible frameworks based on the rights of the global citizen, freed from territorial restrictions'. Meanwhile, Hinchcliffe (2001) *TIBG* **26**, 2 observes that precautionary and *participatory forms of decision-making predominate in environmental geography. Pickerill and Chatterton (2006) *PHG* **30**, 6 value the **consensus-based decision-making** characteristic of alter-globalization (globalization which puts democracy, environmental justice, and environmental protection before economic considerations) activists; Herbert (2007) *Pol. Geog.* **26**, 5 writes that, in contrast to 'the anti-democratic tendencies within the WTO', these activists hoped that their consensus decision-making would illustrate the merits of their cause. See Dengler (2007) *Pol. Geog.* **26**, 4 on the formation and evolution of ad hoc organizational spaces of decision-making.

As to *who* makes the decisions, Trauger (2007) *Tijdschrift* **98**, 1 reports on giving decision-making to the workers, which 'enhances the employees' job satisfaction and makes it more likely that opportunities will be spotted' (Stam (2007) *Economic Geog.* **83**, 1). Aitken in W. Craig et al., eds. (2002) explores the potential of Public Participation *GIS to enable local issues to 'jump'. See Viehe et al. (2006) *Geografiska B* **88**, 2, on the DaisyGIS decision-making model.

It is argued that 'satisficing' (as opposed to 'maximizing') is a much more faithful account of the way people make choices: 'people pick the alternative that satisfies a given level of aspirations without looking necessarily for the best possible option' (Landemore (2004) *J. Moral Philos.* **1**, 177). Medda and Nijkamp (2003) *Integr. Assess.* **4**, 3 observe a tendency to favour simpler, model-free modes of planning based on satisficing or compromise. Moreover, 'decision support models tend to be complex and difficult to understand, often resulting in their neglect by policy practitioners' (Lootsma (1999) IIASA/DAS workshop); thus, Strauss (2008) *J. Econ. Geog.* **8** advocates economic decision-making that 'combines insights from behavioural economics and cognitive science with both quantitative and qualitative methods, and the theorisation of context, embeddedness and the role of institutions'.

declining region A region suffering the economic decline associated with factory closures, outmoded industry, and high unemployment. For government action in declining areas, see Charles et al. (2005) *OECD Territorial Review* (Newcastle and the North-East). See also Klosterman et al. (2006) *Int. J. Env. Tec. & Manage.* **6**, 1–2, on the What if?™, *GIS system, as used in evaluating remedial strategies in declining regions.

décollement The movement of underlying strata that is moved during folding, as it adheres to the upper layers. See Osborn et al. (2006) *Geomorph.* **75**, 3–4.

decomposer *See* DETRITIVORE.

deconcentration *See* DECENTRALIZATION.

deconstruction Bringing alternative layers of significance in a *discourse to light, pointing out contradictions and unspoken assumptions (J. Derrida 1989). Doel (2005) *Antipode* **37**, 2 provides an entertaining, instructive, account. Colling (2006) *TIBG* **31**, 2 and Jones et al. (2007) *TIBG* **32**, 2 make very heavy weather of deconstruction in geography. See Dixon and Jones in J. S. Duncan et al., eds (2004).

deep ecology A view that long-term human self-interest is served best by serving the needs of nature. See Haigh (2002) *Area* **34**, 3 and A. Naess et al. (2002).

deepening A decrease in the central *atmospheric pressure, usually of a low pressure system. See Killer and Petty (1998) *Monthly Weather Rev.* **126**, 9.

deep weathering The *weathering of rocks, up to hundreds of metres in depth, by air and water, forming *saprolite—see Smith et al. (2002) *Catena* **49**, 1–2 on the premisses of deep weathering. Migon and Lidmar-Bergstrom (2001) *Earth-Sci. Revs* **56**, 1 hold that deep weathering has played a major part in the geomorphological evolution of Europe.

defensible space A community managed and controlled by its inhabitants, as an anti-crime strategy (O. Newman 1972). Cozens et al. (2001) *Property Manage.* **19**, 3 argue that creating defensible space should be a recognized crime prevention strategy.

deflation The dendation and lowering of a surface by wind action. Washington et al. (2006) *Geophys. Res. Lett.* **33**, L09401 show that contemporary deflation from the Bodélé Depression, Chad, is controlled by topography and wind stress. See French and Demitroff (2001) *Permaf. Perig. Procs* **12** on deflation hollows in periglacial environments.

deflocculation *See* FLOCCULATION.

deforestation The complete clearance of forests by cutting and/or burning. Between 1995 and 2005, the world's top deforesters were Brazil, Indonesia, and Burma. Grainger (2008) *Procs. Nat. Acad. Scis* **105**, 2 shows that it is difficult to demonstrate tropical deforestation convincingly, and that a 'forest return' effect may be operating, in which forest regeneration in some areas offsets deforestation (but not biodiversity loss) elsewhere.

deglaciation The process by which glaciers thin and recede—through climatic change (Turner et al. (2005) *Geografiska A* **87**, 363), or by increased *calving when sea levels rise. Deglaciation is generally accompanied by the release of *meltwater (Tarasov and Peltier (2006) *Q. Sci. Revs* **25**, 7–8), and the formation of recessional *moraines (Barrows et al. (2002) *Q. Sci. Revs* **21**, 3).

deglomeration The movement of activity, usually industry, away from *agglomerations, perhaps when congestion makes further agglomeration in a region difficult and expensive. See Werker and Athreye (2004) *J. Evol. Econ.* **14** on deglomeration, agglomeration, and evolutionary economic geography. **Deglomeration economies**, such as lower transport costs, or longer commuting times and congestion externalities (Y. Kanemoto et al. 2005), lead to *decentralization.

degradation The lowering and flattening of a surface through erosion.

degree-day Generally, a measure of the departure of the mean daily temperature from a given standard; one degree-day for each °C or °F departure above (or below) the standard during one day. See Roeber (1998) *Weather & Fcst.* **13** 3 on degree-day and forecasting.

deindustrialization The decreasing significance of the manufacturing sector, both in terms of employment, and national production. Bowen (2006) *Prof. Geogr.* **58**, 3 distinguishes between **positive deindustrialization**—a normal feature of economic development, since, as manufacturing increases, workers are increasingly employed in services—and **negative deindustrialization**—'characterized by a vicious cycle linking falling incomes, rising unemployment, the erosion of competitiveness', and one result of the 'credit crunch'. In a study of South Africa, Barchiesi and Kenny (2002) *Int. Inst. Sociale Geschiedenis* **47** observe that it is also marked by low-wage economies, the casualization of employment, and increasing *social exclusion.

delta A low-lying area at the mouth of a river, formed from *alluvial deposition, that occurs as the river's silt-carrying capacity is checked when it slows in the more tranquil waters of a lake or sea. For the Yellow River delta, where marine influences are weak, Wu (2006) *Geografisk A* **88**, 1 finds that the factors controlling the delta are, in rank order: precipitation, sediment trapping by *check dams, terracing, tree- and grass-planting, and water diversion. See also Hori et al.'s careful, detailed study of the Yangtse delta (2002, *Sed. Geol.* **146**, 3–4). Gao (2007) *Geomorph.* **85**, 3–4 thinks that river deltas may be limited when sediment retention approaches zero; when the delta front advances into deep waters; or by the combination of sea-level rise and subsidence. **Inland deltas** form in hot, arid inland drainage areas, where water is lost through evaporation; see McIntosh (1983) *Geog. J.* **149**.

demand cone An image of the continuous reduction of demand for a good with distance from the market, reflecting transport costs. See Nien (2006) Department of Economics, FuJen Catholic U.

demographic transition (DTM) An account, but not a complete explanation, of changing rates of fertility, mortality, and *natural increase. Four stages may be recognized: in the High Stationary Stage, birth and death rates are high and the death rate fluctuates from year to year; the Early Expanding Stage is characterized by high birth rates, falling death rates, and increases in population; in the Late Expanding Stage death rates are low and fertility is declining, but population is still increasing; and, lastly, in the Low Stationary Stage birth and death rates are low and the birth rate fluctuates. There seems to be a fifth stage where birth rates fall below death rates so population levels fall.

Death rates fall because of improved conditions and health care. The reasons for falling fertility are less clear, but include: anti-child labour laws; pensions (which remove the need for children as a support in old age); a higher proportion of women in work; and increasing costs of bringing up children, as standards of living rise. Everywhere, mortality decline appears to have played a central role for fertility decline. 'Female literacy and aggregate income do not seem to matter as much' (Conley et al. (2007) *NBER W. Paper* **2892**). Reher (2004) *Pop. Space Place* **10**, 19 notes that, with falling child death rates, fertility falls rapidly in the developed world, and more slowly in more recent transitions. In an analysis of developing countries, Zlidar et al. (2003) *Pop. Reports*, M **17** note that fertility has fallen by an average of 1% per year, due to increased contraception use, better child survival rates, and better maternal health care. Because of their late start to the DTM, developing countries will not catch up demographically with more developed regions for over 200 years: 'in fact, less developed regions never seem fully to catch up' (UN Dept. Econ. & Soc. Affairs, Pop. Division). See McNicoll (2006) *Pop. & Dev. Rev.* **32**, 1 on politics and the East Asian demographic transition. 'The end of the transition is also the end of the explanatory paradigm, and no other paradigm is, at present, capable of enlightening us on the future' (Vallin (2002) *Pop. & Dev. Rev.* **28**, 1).

The **second demographic transition** is marked by: a later age at marriage; an increase in the proportion of adults living alone or cohabiting; slower rates of remarriage; increased fertility outside of marriage; and delaying or forgoing childbearing. 'With respect to *timing, process,* and *explanation*, the second demographic transition is characterized by no less diversity than the historical first transition' (Lesthaeghe and Moors (2000) *Rev. Pop. & Soc. Policy* **9**).

demography The observed, statistical, and mathematical study of human populations, concerned with the size, distribution, and composition of such populations.

dendritic drainage *See* DRAINAGE PATTERNS.

dendrochronology An absolute dating technique, using the growth rings of trees. See Gartner et al. (2002) *Dendrochronologia* **20**, 1–2.

dendrogeomorphology The *dendrochronological interpretation of geomorphic processes (Alestalo (1971) *Fennia* **105**). Dendrogeomorphology concentrates mostly on the stem and only to a lesser extent on roots; root analyses concentrate mostly on measuring the length of an exposed part of the root and indicating its age, which allows the estimation of erosion volume. See Vandekerckhove et al. (2001) *Catena* **45** on dendrochronological root analysis; see also Malik (2008) *Geomorph.* **93**, 3–4.

density dependent factors The checks to population growth, such as *competition, which result from overcrowding; factors which connect the dynamics of human-dominated landscapes, and 'natural' ecological processes (Fagan et al. (2001) *Landsc. Ecol.* **16**).

density gradient The rate at which the intensity of land use or the density of population falls with distance from a central point. See Tannier and Pumain (2005) *Cybergeo* (*systèmes, modélisation, géostatistiques*) **307**, online 20 April 2005.

denudation A general name for the processes of weathering, transport, and erosion.

dependency ratio The ratio between the number of people in a population between the ages of 15 and 64 and the **dependent population**: children (0–14) and elderly people (65 and over). It is used as a rough way of quantifying the ratio between the *economically active population and those they must support, but the age limits are somewhat arbitrary as, in the UK, for example, 30% of over-16-year-olds go on to higher education, and are therefore still dependent, either on the state or, increasingly, on their parents or bank managers.

dependency theory The view that the development of the advanced economies has caused the underdevelopment of less economically developed nations; see A. G. Frank (1967). 'In effect, dependency theory, as with previous theories of development and underdevelopment, did not say "never," but only "not yet"' (Weinstein (2008) *122nd Ann. Meet. Am. Hist. Ass.*). D. Slater (2004) finds that while dependency theory possesses a continuing relevance, owing to its concern with the marginalized and oppressed, a post-colonial version would overcome a 'hyper' focus on class conditions by introducing concerns with agency, discourse, and knowledge-making. Shepherd (2005) *TIBG* **30**, 2 thinks dependency theory has been ignored because 'it fails to conform to the neoclassical hard core'.

depopulation The decline, in absolute terms, of the total population of an area, more often brought about by out-migration than by a fall in fertility or excessive mortality. Depopulation typically occurs in more remote areas; it is initiated by loss of employment, effected through out-migration, and exacerbated by the 'vicious cycle' of decline depicted in the Drudy–Gilg model (P. J. Drudy 1978; A. W. Gilg 1983; Millward (2005) *Canad. Geogr./Géog. canad.* **49**, 2). See Zaslavsky (2007) *J. Reg. Sci.* **40**, 4 on depopulation and Russia's demographic crisis.

depressed area An area, usually within a developed nation, where capital is scarce, and labour, plant, and *infrastructure are underemployed.

depression An area of low pressure (roughly, below 1000 mb); *see* MID-LATITUDE DEPRESSION. **Depression tracks** are influenced by the courses of *jet streams, energy sources—such as warm seas—and mountain barriers.

deprivation Loss; lacking in desired objects or aims. Within the less developed countries deprivation can be acute: water, housing, or food may be lacking. Within the developed world basic provisions may be supplied but, in comparison with the better-off, the poor and the old may well feel a sense of deprivation. This is **relative deprivation**, which entails comparison, and is usually defined subjectively. In the UK, inequalities in health seem to be more marked in deprived areas than in more affluent ones (Norman et al. (2005) *Soc. Sci. & Medicine* **60**), and in more affluent Western societies relative deprivation seems to be more important than absolute deprivation in determining population health (R. G. Wilkinson 2005). 'The impetus for social change comes not so much from changes in absolute deprivation, but from relative deprivation and it is this which perhaps above all threatens the sustainability of globalisation' (Kaplinsky (2008) *Geog. Compass* **2**, 2).

The **cycle of deprivation** is the transmission of deprivation from one generation to the next: lack of knowledge and skills all contribute to a cycle of deprivation, leaving the poor with few resources and many barriers to overcome.

(⊕) SEE WEB LINKS
- Describes circle of deprivation.
- UK indices of deprivation.

deregulation Rolling back state 'interference' in social and environmental life; 'state regulation has a "light touch", and more and more actors become self-governing within centrally prescribed frameworks and rules' (Castree (2008) *Env. & Plan. A* **40**).

desakota One or more urban cores in densely settled rural areas of intensive agriculture. See Sui and Zeng (2001) *Landscape & Urb. Plan.* **53**, 1 on urbanization in Asian desakota.

desalination (desalinization, desalting) The removal of dissolved salts and minerals from saline water to produce fresh, drinkable water. Commercial desalination processes include: reverse osmosis, electrodialysis, multieffect distillation, and vapour compression. The key issues associated with desalination are cost, energy use, and the environmental impacts of brine disposal and feedwater intake (Trimble et al. (2005) *Encyc. Water Sci.*).

desert An area of very low precipitation, sometimes specified as below 250 mm/yr^{-1} or where *evapotranspiration exceeds precipitation. A **desert pavement** is a continuous mantle of flat-lying, densely packed, partially overlapping pebbles, typically overlying a soft, silty layer (Quaid (2001) *Geology* **29**, 9). Most **desert soils** lack clear soil *horizons, as the climate is too dry for chemical *weathering or humus formation. *Leaching occurs only after occasional rain, and is soon reversed by evaporation. **Desert varnish** is a film of iron oxide or quartz on desert rocks. Silica, slowly dissolved from other minerals, gels together to form a glaze on the rock face. Detrital grains, organic compounds, and pollutants from local environments become entombed in the coating as it forms (Perry (2006) *Geology* **34**, 7).

deserted village A derelict village site. Most English deserted villages were abandoned in the late Middle Ages, and traces—building lines, lumps of masonry, clumps of nettles, and isolated churches—may still be seen in the landscape. See Dyer (1982) *Ec. Hist. Rev.* **35**, 1.

desertification A process of land degradation in drylands, induced by climatic factors and human use (Squires (2006) *Encycl. Soil Sci.*). Desertification undermines the capacity of a community to survive, and is especially acute in the *Sahel. The discussion of the causes and consequences of desertification is complex and little understood. 'Occasional droughts, due to the seasonal factors or inter-annual variations of rains, and severe droughts of long periods can be caused or aggravated by human influence on the environment, reduction of vegetation covering, change of the effect of albedo, local climate changes, greenhouse effect, etc.' (UNCCD 2004). However, J. Letera et al. (2006) challenge negative perceptions of pastoralism. 'Herd mobility is now recognised as a rational strategy in unstable environments, and some countries have taken steps to facilitate movement of herds' (see Touré (2004) *IIED issue paper* **126**). Rasmussen et al. (2001) *Glob. Env. Change* **11**, 4 conclude that 'broad generalisations on land degradation processes, based on local-scale studies, are risky. Significant variations exist at the landscape level, and trends in ecosystem dynamics are sometimes totally reversed even within small regions . . . conclusions concerning "irreversible" degradation, based on few years or decades of observations may be premature.'

desilication (desilification) The removal of silica from a *soil profile by intense weathering and *leaching. It occurs when water that is undersaturated with silica percolates through a soil at the correct pH, and is facilitated by freely draining conditions in a hot, humid environment (R. Schaetzl and S. Anderson 2005).

deskilling Breaking down jobs into smaller units, each to be tackled separately, so that low levels of skill—and therefore cheaper labour—are required for restricted tasks. 'When combined with the separation of residential ownership from direct control, deskilling meant industrial capital could draw on a much larger pool of labour over much broader geographic space' (Moore (2002) *Antipode* **34**, 2). Man (2004) *Women's Studs & Int. Forum* **27**, 2 suggests that deskilling is also an inherent part of Canadian immigration policy. Although L. Grossman (1998) questions the assumption that contract farming necessarily brings about deskilling, Hendrickson and Heffernan (2002)

Sociologia Ruralis **42** claim that the deskilling of farmers is well advanced in parts of the USA. Epp in R. Epp and D. Whitson, eds. (2001) refers to the **political deskilling** of rural communities through neoliberal political reforms.

destatization *See* DETERRITORIALIZATION.

destructive margin The zone where two *plates meet, and oceanic crust is subducted beneath either oceanic or continental lithosphere, changing to the more ductile, fluid-like, denser rocks of the aesthenosphere via increases in temperature and pressure. Subduction creates strong compression at the plate margins, piling up unconsolidated segments and pieces of rock slabs into a wedge (R. Krebs 2003).

destructive wave A plunging wave, with a short wavelength, a high frequency (13–15 per minute), and a high crest. *Backwash greatly exceeds *swash. Destructive waves comb beach material seawards (Joliffe (1978) *PPG* **2**).

determinism The view that human actions are stimulated and governed by some outside agency. **Environmental determinism** assumes that the physical environment is the primary determinant of cultural forms. During the late 19th and early 20th century, some geographers claimed a link between imperialism and environmental determinism (see Smith in J. N. Entrikin et al., eds 1989)—a view that has long been rejected; 'if political ecology has taught us anything it is that we can do better than environmental determinism' (P. Robbins 2004). Barnes (2001) *PHG* **25**, 4 argues that proponents of economic geography tended towards technological, rather than environmental, determinism.

deterritorialization The decline in the role of the state. 'Certain *transnational corporations have more financial power, within a state, than the state itself' (Kapferer (2005) *Anthrop. Theory* **5**, 3). 'The nation-state may be in decline but it is giving way to a relatively original state order or political/ economic formation with multiple state-like effects that is able to act in ways systemic with deterritorializing global processes' (Hudson (2002) *PHG* **26**, 3). Globalization has been explicitly seen as deterritorialization by J. A. Scholte (2000), but Eldon (2005, *TIBG* **30**, 1) is reluctant to accept Scholte's straightforward understanding of territory: 'work proposing an idea of deterritorialization requires an explicit theorization of what territory is.' Akinwumi (2006) *Pol. Geog.* **25**, 7 seems to see the euro as a deterritorialization strategy. For tax havens, 'the legal advantages of non-sovereignty are significant' (R. T. Naylor 2002).

Deterritorialization makes it increasingly difficult to maintain a convincing nationalistic discourse that involves 'cultural purity' (Gjerde and Onishi (2000) *Hum. Dev.* **43**). Connell and Gibson (2004) *PHG* **28**, 3 claim that the expansion of world music exemplifies the deterritorialization of cultures.

detritivore An animal which feeds on fragments of dead and decaying plant and animal material. Detritivores have a vital part to play in *food webs and nutrient recycling.

detritus Fragments of weathered rock which have been transported from the place of origin. See Doyle (2006) *Geomorph.* **77**, 3–4 on detritus loss rates; try Valdiya (2002) *PHG* **26**, 3.

development The use of resources to relieve poverty and raise living standards; the means by which a traditional, low-technology society is changed into a modern, high-technology society, with a corresponding increase in incomes. This can be done through mechanization, improvements in infrastructure and financial systems, and the intensification of agriculture. This definition is based on the more obvious distinctions in living standards between developed and less developed countries. Narrowly economic definitions of development have been criticized, and many believe that true development includes social justice; 'a focus on "global social justice" could provide a framework to think interesting and possibly dangerous thoughts about how to take the international development agenda beyond the relatively instrumental approach of the Millenium Development Goals' (Maxwell (2008) *ODI Opinion*).

Smith (2002) *Area* **34** argues against development as 'some neutral concept of progress'; Potter (2001) *Area* **33** argues that development is a non-linear process, and 'one should not mimic the experience of "developed" countries'. 'For many local communities, the maintenance of social and cultural practices is central to **participatory development** and just as important as income gains and poverty reduction' (Connell (2007) *Sing. J. Trop. Geog.* **28**, 2). 'As indigenous movement platforms and concepts are increasingly institutionalised . . . their demands for culturally appropriate government and development are in practice implemented as governmentally and developmentally appropriate culture' (Andolina et al. (2005) *Pol. Geog.* **24**, 6). O'Reilly (2007) *AAAG* **97**, 3 observes that women's participation in development is a dialogic process that includes negotiating how women fieldworkers will participate within their own organizations; see also Sharpe et al. (2003) *TIBG* **28**, 3. **Development indicators** are used by the World Bank. These indices of development are simply concerned with statistics and do not indicate social structures and patterns of behaviour; there is no definitive definition of what development should be for each society, and no blueprint for how to achieve it.

Ethnodevelopment interprets culture and indigenous identity in ways that make it compatible with neoliberal social policy, but potentially restrict indigenous agency (Laurie et al. (2005) *Antipode* **37**, 3).

(((())) SEE WEB LINKS

• World Bank development indicators.

development area A depressed area in need of investment in industry and *infrastructure.

development control A measure by central government to regulate the location of new industry in the UK. See German et al. (2008) *Planning* **1752**.

Devensian The last full glacial series in Britain, lasting from about 70000 to about 10000 years BP. During this time, sea levels were 130–160 m below current levels; Ireland was joined to Britain, and Britain to northern Europe and Scandinavia by a broad dry plain. The later part of the Devensian embraces pollen zones 1–111 as defined by Godwin (1940).

devolution The transfer of powers from central government to regional and local governments. UK devolution has been based on the principle that decision-making powers be devolved to the appropriate level of *governance, creating new assemblies in Wales and Scotland for the purpose. Wood et al. (2005) *Pol. Geog.* **24**, 3 find that UK businesses have been reconfigured to match these new institutions more closely. 'Devolution of state responsibility only occurs with localization if the state is not also decentralized' (Hayes-Conroy (2008) *Pol. Geog.* **27**, 1). In response to *nationalism, devolution may be seen as one way to avoid ethnic unrest, but neither the devolution of powers to the Basque Autonomous Community in Spain nor more than a decade of power-sharing within this region has led to a peaceful settlement of the Basque conflict (Mansvelt-Beck (2008) *Nations & Nationalism* **14**, 1). **Devolution** *discourses draw on nationalist identities to create new systems of governance: see Raco (2003) *TIBG* **28**.

Devonian A period of *Palaeozoic time stretching approximately from 395 to 345 million years BP.

(⊕) SEE WEB LINKS

• An overview of the Devonian period.

dew A type of condensation where water droplets form on the ground, or on objects close to it. Dew forms when strong night-time *terrestrial radiation cools the ground to chill the overlying air to *dew point. See Gidron (1999) *Agric. & Met.* **96**, 1–3.

dew point The temperature that a body must be chilled to for it to become *saturated with respect to water, so that condensation can begin; also seen as the temperature of a chilled surface just low enough to attract *dew from the *ambient air. The dew point of an *air mass varies with its initial temperature and humidity; see Dodd (1965) *Monthly Weather Rev.* **93**, 2.

dialect A variant of a language in terms of vocabulary grammar, syntax, and pronunciation. Dialect differences arise and persist because of geographical barriers (the Ohio River, for instance, helps to define the division between the dialect areas of the North and the South USA), political boundaries, settlement patterns, migration and immigration routes, territorial conquest, and language contact (Rickford (2002) *Discover*). See also Langevelde and Pellenbar (2001) *Tijdschrift* **92**, 3. **Dialectology** is the study of social and linguistic variations within a language; **dialect geography** is the study of local differentiations in a speech area: see J. K. Chambers and P. Trudgill (1998). **Dialectometric** techniques analyse linguistic variation quantitatively, to map geographic patterns of individual linguistic variants, such as which word is used for a particular concept in a language area, or which sounds are used in particular words (Kretzschmar (2006) *Literary & Ling. Comp.* **21**, 4).

diapir An upward-directed, dome-like intrusion of a lighter rock mass into a denser cover. See Petford and Clemens (2001) *Geology Today* **16**, 1.

diaspora The dispersion of people from their homelands. 'Those in diaspora may be seen as subject to power relations, tensions, disconnections and . . . notions of belonging, identity and community' (Mavroudi (2007) *Geog. Compass* **1/3**). 'Diaspora involves feeling "at home", in the area of settlement while retaining significant identification outside it' (B. Walter 2001). 'Citizenship is inherently geographic in that it is used to determine who can participate in (trans)national affairs and, more broadly, who is seen as "belonging". The bifurcation of this legal and cultural sense of belonging and representation is even more intensified for diaspora populations' (Mains (2007) *Focus on Geog.* **50**, 1). Drozdzewski (2007) *Soc. & Cult. Geog.* **8**, 6 discusses the way that different groups of Polish people contextualize their diasporic identity, and attribute meaning to place. See also Blunt (2007) *PHG* **31**, 5. Among others, the New Zealand government has a **diaspora strategy**— engaging expatriates to extend international marketing opportunities, without requiring them to return home (Larner (2007) *TIBG* **32**, 3).

diatreme Cone-shaped, sub-surface pipes or tubes, mostly extending into igneous dykes at depth (Lorenz (1986) *Bull. Volcanol.* **48**).

difference Difference exists only with reference to another; in the case of the social sciences, this may be patriarchy, 'whiteness', 'colour', *feminism, *colonialism, or *post-colonialism, all of which may be defined by what they are *not*. 'The differentiation of an owning (capitalist) class from . . . the working class is foundational for capitalism in ways that gender, racial or sexual differences, however violent and life-destroying, are not' (Malden (2008) *PHG* **31**, 1, paraphrasing N. Castree and D. Gregory 2006). 'Civil wars in Sri Lanka and sectarian fighting in Iraq are violent expressions of the difficulties of incorporating difference in the nation state' (Staeheli (2008) *PHG* **32**, 4). 'Walls to limit the movement of people between two territories—whether between Israel and the Palestinian territories or between Mexico and the United States—suggest that difference may no longer be accommodated within the liberal state' (Staeheli, op. cit.). I. Rivera-Bonilla (2000) analyses the way gating residential areas recreates spatial markers of class and racial differentiation. 'What seems common in all these **geographies of difference** is a combination

of a distinct geographical and ethnic make-up, cultural and linguistic
assertions resulting in aspiration for varying degrees of political autonomy'
(Parajuli in J. Grim, ed. 2001). See also Permezel and Duffy (2007) *Geog. Res.*
45, 4).

differential erosion The selective erosion of surfaces, so that softer rocks,
such as clays and shales, or lines of weakness, such as joints and faults, are
eroded more rapidly than resistant, competent, and unjointed materials.

diffusion The widespread dispersal of an innovation from a centre or
centres. T. Hägerstrand's model of diffusion (1968) implies the existence of a
*mean information field which regulates the flows of information around a
regional system; flows that are moderated by barriers which can obstruct the
evolution of information into innovation. 'All the basic ideas that have come to
be associated with diffusion research can be found in Lawrence Brown, 1981'
(Lutz (2006) *PHG* **30**, 4).

   'To the extent that a high proportion of personal contacts are local, the
diffusion of information must also be spatially constrained: the further people
are from early adopters the later they are likely to be to adopt the
innovation. To the extent that people's non-local contacts are hierarchical
(more likely to be with people in big cities than in small), the diffusion of
information must also be hierarchically constrained: the lower down the
urban hierarchy people live, the later they are likely to be to adopt the
innovation . . . the diffusion of innovations reflects supplier behaviour as
well as adopter behaviour' (Webber (2006) *PHG* **30**, 4). See Belhedi (2007)
Cybergeo **372**.

digital divide The differential access to computers, information, the
internet, and telecommunications, globally, regionally, nationally, and locally,
that also comprises unequal access to job opportunities, resources, and
training. 'To characterise the digital divide solely as a technological divide is a
(common) misconception . . . To some extent it is true that the digital divide is
a new name for an old phenomenon: uneven spatial distribution of resources'
(Crampton (2001) *Pres. to Ass. Internet Researchers*). 'Advanced technologies
have a pro-rich bias: they are essentially designed for developed countries'
(J. James 2002). For less economically developed countries, strategies for
closing the divide aim to leapfrog the earlier stages of economic development,
bringing advanced communication technologies to areas of deprivation, in a
way suited to local socio-economic conditions; for example, the 'phoneshop'
system (James (2000) *Development and Change* **31**, 4). Dasgupta et al. (2001 *W.
Bank Working Paper; WPS 2567*) highlight the importance of government
policy: 'simulations based on the econometric results, suggest that feasible
reforms could sharply narrow the digital divide during the next decade for
many countries in Africa, Asia, and Latin America.' See the World Bank's
annual *Little Data Book* for data on technology and infrastructure from over
200 countries.

In England and Wales, the areas with the highest proportion of households with broadband tend to be in dense, middle-class suburbs. For UK government initiatives, see the Dept. Trade and Industry *Policy Action Team 15 Report*.

digital elevation model (DEM) An array of regularly spaced elevation values referenced horizontally either to a Universal Transverse Mercator (UTM) projection, or to a geographic coordinate system. See USGS (2000) *Fact Sheet 040–00* for a thorough and helpful summary.

digital terrain analysis (geomorphological analysis, landform parametrization, land surface analysis) The quanitification of terrain. A **digital terrain model** (*digital elevation model*) is a digital relief map. See Deng (2007) *PPG* **31**, 4; Lacroix et al. (2002) *Env. Model. & Software* **17**, 2; and Thwaites (2002) *New Forests* **24**, 2.

dike *See* DYKE.

dilatancy theory Emphasizes the role played by variations of stress in *till found at the base of moving ice masses. A local reduction in stress causes the till to settle and become more compact. See Crozier (1975) *Canad. Geogr.* **19**. Dilatancy theory has also been used in *earthquake prediction—see F. Naeim (2001). **Dilatation** is *pressure release.

diminishing returns, law of The principle that further inputs into a system produce ever lower increases in outputs. Radelet, Clemens, and Bhavnani (2005) *Finance and Dev.* **42**, 3 find that *aid has a positive relationship with growth, but with diminishing returns.

dip The angle of inclination of a rock down its steepest slope, that is to say, the direction at right angles to the *strike. Dip is the angle between the maximum slope and the horizontal. A **dip slope** occurs where the slope of land mirrors the slope of the underlying *strata.

direct cell *See* ATMOSPHERIC CELLS.

dirt cone A cone of ice, up to 2 m high, covered with a thin layer of debris, found in the *ablation zone of a glacier.

disability; geographies of disability, welfare, and social exclusion Hunt (1966) *Union of the Physically Impaired against Segregation* distinguishes between *impairment*, which relates to individually based bio-physical conditions, and *disability*, 'which is about the exclusion of disabled people from "normal" or mainstream society'. 'The dominant societal stereotype of disability as a "pitiful" state reinforces the view that people with disabilities are somehow "less than human"' (R. Imrie 1996). 'The biggest obstacle to disabled people's meaningful inclusion into mainstream community life is negative public attitudes' (Barnes in L. Barton and M. Oliver, eds 1997). Wilton (2004) *TIBG* **29**, 4 outlines the multiple strategies of control of the disabled in the workplace; notably the denial of access to accommodation. B. Gleeson (1998) sees space as 'a social artefact that can be used to disable, rather than enable people with physical impairments'. I. Kawachi and L. F. Berkman, eds (2003) explore the way physical and cognitive disability affect and are affected by a neighbourhood's physical and social environment.

(⊕) SEE WEB LINKS

• Disability awareness in action.

discharge The quantity of water flowing through a cross-section of a stream or river in unit time. Discharge (Q) is usually measured in *cumecs, and calculated as $A \times V$ where A is the cross-sectional area of the channel and V is the mean velocity.

discordant Cutting across the geological grain, as in a stream cutting across an *anticline.

discourse A specific assembly of categorizations, concepts, and ideas that is produced, reproduced, *performed*, and transformed in a particular set of practices. Schott (2007) *Tijdschrift* **98**, 2 defines discourse as 'a social process which can be called "inclination of sense" or, in a more sophisticated manner, "genesis of meaning" . . . It is appealing to use the word discourse in Foucault's sense of a system of exclusion.'

Discourses create their own 'regimes of truth'—the acceptable formulation of problems, and solutions to those problems (M. Foucault 1980). 'Although many of the urban studies that undertake discourse analysis are interested in urban social justice . . . some critics charge that talking about social justice is not enough, and as critical urban researchers we need to create it' (Lees (2004) *PHG* **28**, 1). See Nijman (2007) *Tijdschrift* **98**, 2 on 'the prevailing political and ideological discourse [that] seeks to maintain the Haitian identity'.

discrete choice modelling Using a group of statistical techniques to model the way people choose between alternatives; see Duncombe et al. (2001) *Int. J. Pop. Geog.* **7**.

discrimination The exclusion of specific groups from certain sectors of society. 'This type of exclusion is to the disadvantage of the immigrant, but to differentiate it from xenophobic discrimination or racism . . . the concept *systemic discrimination* would cover the situation where exclusion is connected to the ethnic background of individuals, but is not a result primarily of hostility towards "the other"' (Bäcklund (2003) *Geografiska B* **85**, 1). Hyndman (2007) *AAAG* **97**, 2 shows how, in post-tsunami Sri Lanka, the implementation of no-build setback areas along the coastline stirred feelings of discrimination. See also C. Flint, ed. (2004).

discrimination model A risk-based model used in the insurance industry; see Rubel (2006) *J. Specialists Grp. Wk*, **31** and Pierret (2001) *Quart. J. Econ.* **116**, for examples.

diseconomies Financial drawbacks. **Diseconomies of scale** may occur when an enterprise becomes too large, sites become constricted, the flow of goods is congested, or the workforce is alienated. See Barcena-Ruiz and Begona Garzon (2005) *Eur. J. Pol. Econ.* **21**, 1 on economic integration and privatization under diseconomies of scale. *See* ECONOMIES OF SCALE.

dispersed settlement A settlement pattern of scattered, isolated dwellings. See Cruickshank (2006) *Norw. J. Geog.* **60**, 3 on the fight in rural Norway for a dispersed settlement pattern.

dissection The incision of valleys by river erosion. A **dissected plateau** is a level surface which has been deeply cut into by rivers. See Stanford et al. (2000) *ESPL* **25**, 9.

dissimilarity index A measurement of the overall difference between two percentage distributions. Dissimilarity indices are often used in studies of occupational, or racial, *segregation. Jones (2008) *Urb. Geog.* **29**, 3 develops a typology to explain the different pathways by which the dissimilarity index may increase or decrease in a metropolis.

dissipative Describing a *non-linear system where energy is lost at positions far from equilibrium. A **dissipative beach** is gently sloping, running from the seaward edge of the forezone to the surf zone; see Ruggiero et al. (2004) *J. Geophys. Res.* **109**, C06025.

dissolved load Material carried in solution by a river; capable of passing through a 0.45-μm membrane filter (S. Trimble et al., eds (2006) *Encycl. Water Sci.*).

dissolved oxygen Oxygen from the atmosphere and from photosynthesis, dissolved in the upper levels of bodies of water. Dissolved oxygen decreases with depth, rising temperatures, and the oxidation of organic matter. See Kung and Ying (1991) *Geog. J.* **157**, 1.

distance Absolute distance is expressed in physical units such as kilometres and is unchangeable. 'The geography of the world as we experience it is being twisted and contorted so as to bear very little relation to the physical distances that are involved. The human aspect of this is sometimes called **social distance** or **cultural distance**—those gulfs that can exist in the understanding and in the history of experiences that we bring to the world. Physical proximity is not necessarily a good measure of social and cultural distance' (D. Massey, BBC/OU Open2.net—Free Thinking).

    Relative distance includes any other kind of distance such as **time-distance**, measured in hours and minutes, and changing with varying technology. 'It's the relative distance that counts' (Chauvin and Gaulier (2002) *TIPS Annual Forum*). Breschi and Lissoni (2006) *CESPRI W. Papers* **184** show that spatial distance and technological distance are both proxies of **social distance**. Carrere and Schiff (2004) *World Bank W. Paper Series* **3206** find that 'distance has become more important over time for a majority of countries'.

distance decay The lessening in force of a phenomenon or interaction with increasing distance from the location of maximum intensity; the **inverse distance effect**. Seidl et al. (2006) *BMC Health Serv. Rev.* **6**, 13 find an inverse relationship between the geographical distance from a patient's home to a clinic, and the likelihood of the patient actually turning up there. Gimpel et al. (2008) *Pol. Geog.* **27**, 2 find that technology has lessened, but not eradicated, the **friction of distance**. See Farhan and Murray (2006) *Annals Reg. Sci.* **40**, 2.

distributary A branch of a river or glacier which flows away from the mainstream and does not return to it.

disturbance regime A regime marked by short, but recurring, episodes of high magnitude, or at critical sites. **Disturbance-related landforms** tend to be polygenetic and heterogeneous, and are long-term/permanent adjustments to fires, volcanoes, etc. Church (2002) *Freshw. Biol.* **47**, 4 argues for a disturbance-related interpretation of mountain rivers.

divergence The spreading out of an air mass into paths of different directions, linked with the vertical shrinking of the atmosphere (see R. Barry and R. Chorley (1992) fig. 3.6). Upper-air divergence is associated with ground level depressions (Sutcliffe and Forsdyke (1950) *Quart. J. RMS* **76**, 328), and closely related to vertical *vorticity; see Chen and Bosart (1979) *J. Atmos. Scis* **36**, 2.

diverging margin *See* CONSTRUCTIVE MARGIN.

diversification A spread of the activities of a firm or a country between different types of products or different markets, in order to reduce over-dependence; as in the Arab Gulf (Malecki and Ewers (2007) *PHG* **31**, 4). **Horizontal diversification** is diversification between same-type investments; see Achilladelis and Antonakis (2001) *Res. Policy* **30**, 4. **Vertical diversification** is investing between different types of investment. Diversification for a small open economy is inevitably protracted, involving a change in economic structure. Supportive conditions include low inflation, stable exchange rates, and a labour-market environment with few worker–employer conflicts. Diversification is not an inescapable route to output stability (Chan (2006) *Monetary Authority of Macao*). Herzer et al. (2006) *Appl. Econ.* **38**, 15 find that export diversification plays an important role in economic growth.

diversity The abundance of species within an ecosystem. **Alpha diversity** is the diversity of species within a particular area or ecosystem, expressed by its species richness (Moreno et al. (2006) *INCI* **31**, 1). **Beta diversity** is the difference in diversity of species between two or more ecosystems, expressed as the total number of species unique to each of the ecosystems compared; see Vellend (2001) *J. Veg. Sci.* **12**, 4 on beta diversity and species turnover.

divestment (divestiture) In business, the sale of an asset; the opposite of an investment. Geographical analyses of divestment focus on: the conduits through which surplus or excess effects are discarded; the practices of divestment; disinvestment as practice; and the connections between divestment practice. Gregson et al. (2007) *TIBG* **32**, 2 show how surplus/excess things are routinely divested through specific conduits; throwing something *out* is actually throwing it *in* to another location—a bin, a skip, a charity shop.

division of labour The partitioning of a production process into separate elements. **Gender divisions of labour** separate tasks into male and female; see Padmanabhan (2008) *J. Int. Dev.* **20**. A **social division of labour** divides workers by product: Mitsui (2003) *30th ISBC, Singapore* sees social divisions of labour as crucial to small and medium enterprises in manufacturing, unable to survive if they simply rely on own limited resources and specialized skills. Meijers (2007) *Reg. Studs* **41**, 7 detects **spatial divisions of labour** at the intra-urban level in the Randstad, the Flemish Diamond, and the Rhein–Ruhr Area, 'whereas at the inter-urban level this division of labour is diminishing'. Within the 'new' international division of labour, capital and control functions are concentrated in the advanced economies or global core, while labour has been relocated to developing economies on the global periphery (Tonkiss (2008) *Sociol. Compass* **2**, 2). May et al. (2007, *TIBG* **32**, 2) explore the emergence of a new **migrant division of labour** in London

doldrums Regions of light, variable winds, low pressure, and high temperature and humidity, occurring over the East Pacific, the East Atlantic, and from the Indian Ocean to the West Pacific; see Graham et al. (1988) *J. Atmos. Scis* **46**.

doline (dolina) A closed, steep-sided, and flat-floored depression in *karst country. The sides are 2–10 m deep; the floor 10–1000 m wide. **Solution dolines** form when solution enlarges a point of weakness in the rock into a hollow. **Subsidence dolines** form when limestone caves develop below insoluble deposits. The majority of dolines are corrosion forms.

(⊕) SEE WEB LINKS
• A description of dolines.

dome An uplifted section of rocks, highest at the centre, from which the rocks dip all around. **Volcanic domes** may be formed from slow-moving, *viscous lava, rounded as the result of pressure from lava below (Jousset and Okada (1999) *J. Volc. & Geotherm. Res.* **89**, 1). A **plug dome** is a small, irregular dome within a crater, possibly with projecting, spiny extrusions.

domestic space The home as a distinct, private, space. It may be produced through the inhabitants' possessions (see Rose (2003) *TIBG* **28**, 1 on postcards, and Brace (2007) *AAAG* **97**, 1 on the 'temperance household'), or through *capitalist and patriarchal reproduction. Lau (2006) *Mod. Asian Studs* **40** argues that the division of domestic space within South Asian households reflects the social status of women. Halford (2006) *Gender, Work, Organiz.* **13**, 4 argues that, in the modern Western world, the construction of fatherhood has been underpinned by the spatial separations of work/home, of public/ private. Jepson (2005) *Antipode* **37**, 4 chronicles how Mexican-American women in South Texas changed a domesticated space into one of empowerment. 'The apartheid system dichotomized physical space into masculine and feminine categories, marginalizing the feminine. Urban and public space is masculine while rural and domestic space is feminine. Only men who were economically productive were permitted in urban work spaces, and the associated housing system. Women were relegated to the inferior physical and social space of the homelands where they were expected to farm, raise children, and care for the sick and elderly' (G. Elder 2003).

dominant discharge Usually, *bankfull discharge, but dominant discharge has also been equated to *effective discharge; see Simon et al. (2004) *Geomorph.* **58**, 1–4.

donga In South Africa, a steep-sided gully resulting from severe *soil erosion.

dormant volcano An inactive, but not extinct, volcano.

dormitory town A settlement made up largely of daily commuters, and with relatively few retail outlets, as the commuters will use city-centre services, or out-of-town shopping centres.

doughnut city A city with empty buildings, crime, and vandalism in the centre, and wealth, happiness, and family life in the suburbs. See Priemus (2004) *Housing Theory Soc.* **21**, 1.

downward transition region In John Friedmann's *core–periphery model, a peripheral region marked by depleted resources, low agricultural productivity, or outdated industry.

downwearing, declining slope retreat A model of hillslope retreat where the slope angle decreases over time due to a combination of soil *creep, rain splash, and *sheet wash that causes slope convexities and concavities at the expense of straight hillslope segments.

drag The braking effect of friction, imparted by a fluid on bodies passing through it.

drainage The naturally occurring channelled flow formed by streams and rivers. A **drainage basin** is the area of land drained by a river and its tributaries (the term is synonymous with river basin, and termed *watershed* by North

Americans) Stream channel length, and equilibrium conditions within the system are revealed by the statistical relationships between the various parameters; see Gardiner (1978) *TIBG* **3**, 4.

drainage basin geometry Drainage basin geometry postulates that, as the *stream order increases, (a) the number of streams decreases, (b) the average stream length increases, and (c) the average slope decreases. See Strahler (1952) *GSA Bull.* **63** and Sólyom and Tucker (2004) *J. Geophys. Res.* **109**, F 03012.

drainage density The total length of streams per unit area. Any attempt to calculate drainage density is impeded by the difficulty of calculating total stream length, as the exact point at which a stream starts is problematical. See Lin and Oguchi (2004) *Geomorph.* **63**, 3–4.

drainage network evolution The drainage basin can increase its area and extend its channels by landslides at the edge of the network; by *headward erosion; by the extension upslope of underground *pipes; and by rill formation. See Coulthard and Van De Wiel (2007) *Geomorph.* **91**, 3–4 on *self-organization in drainage basin evolution, and Sagri et al. (2007) *Geomorph.* **94**, 1–2 on tectonics and drainage basins.

drainage patterns The pattern of a drainage network is strongly influenced by geological structure. **Anastomotic drainage,** developing on nearly horizontal, coarse sediments, is the division of a river into several channels, and **annular**, or **radial drainage**—where the major rivers radiate from a centre—develops on domes, particularly where belts of resistant rock are separated by weaker belts. **Centripetal drainage** moves into a centre created by a crater or depression. **Dendritic drainage**—a branching, tree-like network—is most common on horizontally bedded or crystalline rocks, with uniform geology. **Rectangular drainage** has tributaries running at right angles to the major river and occurs on rocks with intersecting, rectangular joints and faults; **contorted drainage** is rectangular drainage on complex metamorphosed rocks. **Trellised drainage** resembles a trained fruit tree, usually found on dipping or folded sedimentary or weakly metamorphosed rocks. See Riedel et al. (2007) *Geomorph.* **91**, 1–2 on structural controls on drainage patterns.

dreikanter A stone with three clearly cut faces, like a Brazil nut, formed by sand-blasting in desert environments.

drift-aligned beach A beach which has developed parallel to the line of *longshore drift. If the wave direction becomes more right-angular, the beach realigns itself.

dripstone Secondary calcite deposits, precipitated as *vadose water drips from/flows over cave roofs, walls, and floors (Tabarosi and Stafford (2003) *Int. J. Speleol.* **32**, 1–4).

drizzle Rain droplets, up to 0.2 mm across, and with a fall speed of around 0.8 m^{s-1}.

drought A long, continuous period of dry weather. Drought over Africa has increased, especially over West Africa. Northern and East Asia show positive trends and central Asia and the Tibetan Plateau show decreasing trends. In South Asia all trends are insignificant. Drought extent over Australia has decreased. Over the Americas, trends are uniformly negative and mostly significant (Sheffield and Wood (2008) *J. Climate* **21**, 3).

drumlin A long hummock or hill, egg-shaped in plan, and deposited and shaped under an ice sheet or very broad glacier, while the ice was still moving. The *stoss is blunt, while the lee is shallow, with its point indicating the direction of ice flow. Drumlins have been formed via a variety of mechanisms; see Clarke et al. (2005) *Q. Sci. Revs* **24**. **Drumlin swarms**, or **drumlin fields** are not uncommon. Briner (2007) *Boreas* **36**, 2 argues that the bedforms in the New York drumlin field indicate fast ice flow. **Rock drumlins** are more commonly known as roches moutonnées.

dry adiabatic lapse rate (DALR) *See* ADIABAT.

Dryas Part of the characteristic threefold late-glacial sequence of climatic change and associated deposits following the last, Devensian, ice advance and prior to the current, Flandrian, interglacial. The colder Dryas phases mark times of cold, *tundra conditions throughout what is now temperate Europe (Britain did not experience the Older Dryas). The **Oldest Dryas** (Dryas I/Pollen Zone Ia) lasted from about 16000 to 15000 years BP, with tundra-grassland characterized by *Dryas octopetala* (mountain avens); the **Older Dryas** (Dryas II/Pollen Zone Ic) from 12300 to 11800 years BP; and the **Younger Dryas** (Dryas III/Pollen Zone III) from about 11000 to 10000 years BP. *See* PLEISTOCENE.

dry farming Farming without irrigation, using mulching, frequent fallowing, working the soil to a fine tilth, and frequent weeding to conserve water. See R. Shetto and M. Owenya (2007).

dry valley A valley, usually in chalk, or *karst, with no permanent watercourse. These valleys may have been cut during *periglacial phases, when the *permafrost would stop *meltwater from soaking through chalk or limestone, and thus permit *dissection.

dual economy A. Lewis (1954) identified two sectors in an economy: the capitalist sector—'that part of the economy which uses *reproducible capital* and pays capitalists for the use thereof'—and the subsistence sector—'all that part of the economy which is not using reproducible capital'. Per capita output is lower in the subsistence sector than in the capitalist sector, through a lack of capital. With additional capital, more workers can move from the subsistence sector to the capitalist sector. The process of economic development then depends on the transfer of surplus labour from the subsistence to the capitalist sector. The nature of risk in a peasant-dominated subsistence sector tends to hinder the creation of employment and the mobilization of surplus labour in the capitalist sector. 'For this reason the path of industrialization is not smooth and trouble-free, or susceptible to analysis using the tools of equilibrium economics' (Ghosh (2007) *J. Econ. Methodol.* **14**, 1).

dumping The off-loading of goods at below cost, usually as exports.

dune *See* SAND DUNE; COASTAL DUNE.

duricrust A hard surface layer on tropical upland soils. See Clarke (2006) *Geomorph.* **73**, 1.

dust (atmospheric dust) The effect of dust aerosols on climate depends on both dust distribution in the atmosphere and the physical and chemical properties of the dust particles themselves (Tegen (2003) *Quat. Sci. Revs* **22**). Changes in atmospheric dust, derived mostly from deserts, can trigger large-scale climate responses (Zhang et al. (2002) *Earth Planet. Sc. Lett.* **202**). Yang et al. (2008) *Atmos. Chem. Phys.* **8** demonstrate the modulation of the monsoon climate by dust. Dust particles contribute to the greenhouse effect by absorbing and emitting radiation; see K. Kohfeld and I. Tegen (2007).

Dust Bowl Parts of Colorado, Kansas, New Mexico, Texas, and Oklahoma, which were severely afflicted by *drought and dust storms in the 1930s. One preconception is that the Dust Bowl was created by farmers ploughing the prairies, but G. Cunfer (2005) thinks it was probably induced over many years by unusually low rainfall and unusually high air temperatures: 'the Dust Bowl would have happened even without farmers ever being there.'

dust storm Washington et al. (2003) *AAAG* **93**, 2 highlight the primacy of the Sahara as a source of dust, and Qian et al. (2002) *J. Climate* **15**, 10 find that winter temperatures and spring cyclones trigger dust storms.

duyoda In a *periglacial landscape, a steep-sided, shallow, and often circular depression formed as *baydzharakhs collapse. A duyoda is smaller than, but can develop into, an *alas.

dyke (dike) A discordant vertical or semi-vertical wall-like igneous intrusion; see McDonnell et al. (2004) *Mineralogical Mag.* **68**, 5 on the Slieve Gullion ring dyke, Ireland.

dynamic equilibrium In a landform, a state of balance, in spite of changes taking place within it. For example, dynamic equilibrium in an alluvial stream would be 'delicately adjusted to provide, with available discharge and with prevailing channel characteristics, just the velocity required for the transportation of the load supplied from the drainage basin' (Biedenharn et al. (2000) *Geomorph.* **34**, 3–4).

dynamic meteorology The study of those motions of the atmosphere that are associated with weather and climate. For all such motions, the discrete molecular nature of the atmosphere can be ignored, and the atmosphere can be regarded as a continuous fluid medium (J. Holton 1992). See Weber and Névir (2008) *Tellus A* **60**, 1 on DSI (Dynamic State Index) analysis in dynamic meteorology.

earth flow A form of *mass movement, where water-saturated, weak-slope material flows under the action of gravity at speeds between 10 cm s^{-1} to 10 cm day^{-1}. A rapid earth flow generally quickly follows the initial fall-slide phase (Di Crescenzo and Santo (1999) *Geogr. Fis. Din. Quat.* **22**). Di Crescenzo and Santo (2005) *Geomorph.* **66**, 1–4 find relationships between: the height of the landslide crown zone and the slope relief energy; the slope relief energy and the sliding zone area; and the landslide area and the basin area.

earth pillar An upstanding column of soil that has been sheltered from erosion by a natural capstone on top, common where boulder-rich moraines have been subject to gully erosion.

earthquake A sudden and violent movement, or fracture, within the earth, followed by a resultant series of shocks. Earthquakes occur in narrow, continuous belts of activity which correspond with *plate junctions; see Witter et al. (2003) *GSA Bull.* **115**, 10. The scale of the shock of an earthquake is known as the **magnitude**; the most commonly used scale is the *Richter scale, while the **intensity** of an earthquake is measured by the *Mercalli scale.

Earthquake waves are of three basic types: P, primary, push waves travel from the focus by the displacement of surrounding particles and are transmitted though solids, liquids, and gases; S, secondary or shake waves travel through solids; and L, long or surface waves travel on the earth's surface. Computers have been used in earthquake wave analysis since the 1980s; see Tan et al. (2008) *Turkish J. Earth Sci.* **17**. For earthquake prediction, see Brace et al. (1966) *J. Geophys. Res.* **71**, 16 on *dilatancy theory; Kuo et al. (2006) *Ground Water* **44**, 5 on radon-222, and Chen et al. (2002) *Geology* **30** on older fault scarps.

easterlies Winds from the east. Easterlies blow in low latitudes in both high and low *tropospheres, and in high latitudes in the lower troposphere (Ichigawa and Yasunari (2006) *J. Climate* **19**, 7). The downward propagation of easterly winds occurs throughout the equatorial zone; see Baldwin et al. (2001) *Rev. Geophys.* **39**. **Easterly waves** are weak troughs of low pressure in *tropical areas. See Fuller and Stensrud (2000) *Monthly Weath. Rev.* **128**, 8 on easterly waves and the North American monsoon, and Ross and Krishnamurti (2007) *Monthly Weath. Rev.* **135**, 12 on easterly waves and tropical cyclones.

ecodevelopment 'The ideal of ecodevelopment action [places] its national emphasis upon economic equity, social harmony and environmental balance in the local pursuit of individual fulfilment, household self-sufficiency and community self-reliance' (R. Riddell 1981). Mahanty (2002) *World Dev.* **30**, 8 thinks the current emphasis on *what* conservation programmes do should be balanced with seeing *how* conservation programmes are undertaken.

ecological balance The equilibrium between, and amicable coexistence of, organisms and their environment. Changes in air temperature, precipitation, and carbon dioxide levels, plus lesser ecological changes, may alter the ecological balance between communities: 'slight increases in individual mortality of one community, coupled with increased success rates of another, could eventually cause changes in the geographical distribution of forests, grasslands, and deserts' (Masek (2001) *J. Biogeog.* **28**, 8). See Ding et al. (1998) *PPG* **22** on the concept of ecological balance and the definition of *desertification.

ecological climatology (eclimatology) An interdisciplinary approach to the role of terrestrial ecosystems in the climate system. The central theme is that terrestrial ecosystems, through their cycling of energy, water, chemical elements, and trace gases, are important determinants of climate (G. Bonan 2002; but see Beerling (2003) *Trends Ecol. Evol.* **18**, 2).

ecological crisis A state of human-induced ecological disorder that could lead to the destruction of this planet's ecosystem to the extent that human life will at least be seriously impaired for generations, if not destroyed. Evidence to suggest that this crisis has already been reached includes *deforestation, increasing levels of atmospheric carbon dioxide, and current rates of energy use; see C. Merchant (1989). Prudham (2003) *AAAG* **93**, 3 suggests that the destruction of particular natures under capitalism results not in absolute or final limits, but in ongoing political struggle over the social (re)production of new natures, for renewed accumulation and broader societal goals.

ecological efficiency The ability of the organisms at one *trophic level to convert to their own use the potential energy supplied by their foodstuff at the trophic level directly beneath them; the energy content of prey consumed by a predator population, divided by the energy content of the food consumed by the prey population (R. Wootton 1998). Ecological efficiency tends to be inversely correlated with primary production (P. Herring 2002).

ecological energetics The study of the flow of energy from the sun through and up the *trophic levels, expressed in calories. With movement up each trophic level, there is a very great loss of the energy available as food. 'A more appropriate name for ecological energetics could refer to limitations by a single component of the resource' (Brooks et al. (1996) *Ecoscience* **3**, 3). See also Kooijman et al. (1999) *J. Theor. Biol.* **197**.

ecological explosion The enormous increase in numbers of a living organism: a flu virus, the prickly pear, the grey squirrel; 'the bursting out from control of forces that were previously held in restraint by other forces' (C. Elton 1938, 2000).

ecological fallacy A wrong assumption about an individual based on aggregate data for a group; the presence of a relationship between two variables at an aggregated level that is due simply to aggregation, rather than to any real association (Openshaw (1984) *Env. & Plan. A* **16**). See Dark and Bram (2007) *PPG* **31**, 5 on the *modifiable areal unit problem and the ecological fallacy; see G. King (1997) on a solution.

ecological footprint The demand of a society on biocapacity (Rees in J. Dewulf and H. Van Langenhove, eds. 2006); the area of land functionally required to support a human society. In more economically developed countries, this will include land beyond the territory inhabited by that society; that is, the land from which its imports are sourced. Chen and Lin (2008) *Energy Policy* **36**, 5 identify the factors that affect the annual growth of China's ecological footprint. Wackernagel et al. (2004) *Land Use Policy* **21**, 3 calculate national and global ecological footprints, and Bagliani et al. (2008) *Ecol. Econ.* **65**, 3 disaggregate an ecological footprint into its energy and non-energy components.

ecological overshoot The shortfall in Earth's biological capacity to meet the demands of its human population. Wackernagel et al. (2002) *Procs Nat. Acad. Sci.* **99**, 14 argue that human demand may well have exceeded the biosphere's regenerative capacity; in 1961 humanity's load was 70% of the capacity of the global biosphere, rising to 120% in 1999.

ecological politics Advocacy for, or work towards, protecting the natural environment from degradation or pollution. See Philip (2001) *Capitalism, Nature, Socialism* **12**, 2.

ecological restoration The re-creation of entire *communities of organisms, closely modelled on those occurring naturally, together with long-term maintenance and management to ensure integrity, and stability. See Abella et al. (2007) *Natural Areas J.* **227**, 1.

ecology The study of the interrelationships between organisms and their surrounding, outer world; the study of animals and plants in relation to each other and to their habitats. **Production ecology** (*community ecology*) is the study of *communities in terms of the throughput of energy and chemical compounds; key concepts are *primary production, *trophic levels, and *nutrient cycles. See Chesson and Huntly (1997) *Am. Naturalist* **150** and Shea and Chesson (2002) *Trends Ecol. & Evol.* **17**, 4.

economically active population The total population between the ages of 15 and 65 in any country, used in the calculation of dependent population.

economic base theory The view that economic activity in a city can be broken down into two components: activity which meets local, internal demand, and activity which meets non-local demand. The latter is basic and city forming because, it's argued, it is demand from beyond the city which causes the city to grow. See Hill and Brennan (2000) *Econ. Dev. Q'ly* **14**, 1. Taylor (2007) *TIBG* **32**, 2 finds economic base theory limited and simplistic.

economic determinism The thesis, as advanced by Marx and Engels, that economic factors underlie all of society's decisions. The social relations specific to a particular *mode of production are said to structure social relations between classes, underpinning the legal and political systems. This implies that all political, cultural, and social life can be predicted from the prevailing mode of production. This extreme view has been severely criticized as it denies the existence of free will and individual independence. A more moderate view sees the relations of production as a constraint on the ways in which individuals and superstructures can develop, 'but once the chains of causality anchoring consciousness and value to collective production of material goods are broken, as they are in "humanist" Marxism, the theory loses much of its effective explanatory edge' (Duncan (2005) *PHG* **29**, 479).

economic development Taylor (2007) *GaWC Res. Bull.* **238** distinguishes between a growth in the economy, but with the same division of labour—*economic growth*—and the production of new commodities—*economic expansion*—where the division of labour is altered, creating a more complex economy. Taylor calls the former 'old work': 'State economic planning has produced such through increasing old work. But this has its limitations.' The second, he associates with cities 'where cycles of new work generate vibrant, cosmopolitan city economies'. Frenken and Boschma (2007) *J. Econ. Geog.* **7**, 5 view the process of economic development as an evolutionary branching process of product innovations. Essletzbichler in K. Franken, ed. (2007) finds significant positive relationships between diversity and stability on the one hand, and growth and instability on the other. Park (2003) *Pol. Geog.* **22**, 8 argues that the politics of local economic development in South Korea engendered regionalism and uneven development.

economic distance The distance a commodity may travel before transport costs exceed the value of the freight. Economic distance can be reduced by better physical infrastructure, improvements in the institutional framework, better cost recovery and maintenance, simpler transit procedures, better information, and the reduction of unofficial payments/corruption (Crochet et al. (2004) *IBRD*, World Bank).

economic dynamism Vigorous activity, progress, and strength in an economy. The Ewing Marion Kauffman Foundation measures economic dynamism by: the share of jobs in fast-growing firms; the degree of job turnover (a product of new business start-ups and existing business failures);

and the value of companies' initial public offering (the event of a firm's first sale of stock shares). J. Jacobs (1984) sees the economic dynamism of cities as the outcome of the closely related processes of innovation and import replacement. Wojan et al. (2007) *J. Econ. Geog.* **7**, 6 find that a relative abundance of *bohemians positively influences local economic dynamism.

SEE WEB LINKS

• Website of the Ewing Marion Kauffman Foundation.

economic geography The analysis of the spatial distribution of the transportation and consumption of resources, goods, and services, and their effects on the landscape; 'taking seriously the relations between economic and other social and bio-physical processes, rather than analyzing the economic as either separable from or foundational to such other processes' (Sheppard in S. Bagchi-Sen and H. Lawton Smith 2006).

There are two distinct new economic geographies in the Anglo-American literature. The first uses sophisticated spatial modelling to explain uneven development and the emergence of industrial clusters—by considering centripetal and centrifugal forces, especially economies of scale and transport costs. See Krugman (1998) *Oxford Review of Economic Policy* **14**, 2; for a critical review, see Martin (1999) *Camb. J. Econ.* **23**.

The second approach also attempts to explain the emergence of industrial clusters, but emphasizes relational, social, and contextual aspects of economic behaviour, particularly the importance of knowledge and learning—which takes place most effectively through personal contacts at the local–regional level (Amin (1999) *International Journal of Urban and Regional Research* **23**). This economic geography emphasizes aspects of economic behaviour that are considered intangible by the first; see Perrons (2004) *Econ. Geog.* **80**, 1. See Gertler (2003) *J. Econ. Geog.* **3** on tacit knowledge and the economic geography of context.

Bathelt and Glückler (2003) *J. Econ. Geog.* **3** argue for a relational economic geography, wherein economic actors are situated in contexts of social and institutional relations, and economic processes are contingent in these agents' strategies. Their key words are: organization, evolution, innovation, and interaction. Crevoisier (2004) *Econ. Geog.* **80**, 4 in an innovative milieux approach to economic geography, holds that space—or, more precisely, territory—is the matrix of economic development, and that economic mechanisms transform space. N. Coe et al. (2007) is very favourably reviewed. For a review of the 'institutional turn' in economic geography see Peck (2005) *Econ. Geog.* **81**.

Evolutionary economic geography seeks to integrate growth and innovation theories, and looks at endogenous reasons for regional economic development. 'Evolutionary Economic Geography explains the spatial evolution of firms, industries, networks, cities and regions from elementary processes of the entry, growth, exit and (re-)location of firms' (Frenken and Boschma (2007) *J. Econ. Geog.* **7**, 5). 'The "evolutionary turn" in economic geography has led to increasing emphasis on coevolution among technologies, organizations and territories. 'The weakness of this approach, however, is a focus on broad coevolutionary pictures that pays little attention to coordination processes that guide interdependent actions on the ground' (Lee and Saxenian (2008) *J. Econ. Geog.* **8**, 2).

economic growth The growth in wealth of a nation, as measured by an increase in *gross national product, or in national income. Kang and Meernik (2005) *J. Politics* **67**, 1 and Capasso (2004) *J. Econ. Surveys* **18**, 3 suggest that government fiscal and monetary policies are critical in improving economic growth; Gupta and Mitra (2004) *Dev. Policy Rev.* **22**, 2 stress the crucial role of health care. 'The critical driving force of economic growth is not the super-normal profits that technological change generates but rather the continuous creation of opportunities for further technological development' (Carlaw and Lipsey (2003) *J. Econ. Surveys*, **17** 3).

Regulation theorists explore the way 'economic and non-economic procedures can be articulated to produce a relatively stable, coherent and dynamic framework, which can in turn secure the expanded reproduction of capitalism' (Jessop (1990) *Econ. & Soc.* **19**, 2). R. Boyer (1990) sees the variability of economic and social dynamics, in both time and space, as 'the central question' for regulation theory; see B. Jessop and N.-L. Sum (2006).

economic man A theoretical being who has perfect knowledge of an economy and the ability to act in his or her interests to maximize profits; the concept of the *satisficer is more realistic.

economic rent Economic rent is not synonymous with profit, since opportunity cost is built into the concept: the real cost of choosing one alternative good or service in terms of the sacrifice of the next best alternative. Kaplinsky (1998) *IDS Discuss. Paper* **365** uses the term 'value' to refer to various forms of economic rent; see Coe et al. (2004) *TIBG* **29**, 4.

economic system Economic systems mainly deal with the relationships between production/supply and consumption/demand: manufacturing, regulation, circulation, and distribution. P. Hall and D. Soskice (2001) argue that national economic systems often experience external shocks, but adjust, and restore equilibrium. W. Eucken (1952) holds that 19th-century laissez-faire shows that 'the economic system cannot be left to organize itself', a view confirmed by the economic crises of recent years, and the calls for better regulation. N. Brenner and R. Kiel (2006) highlight the interplay between the global and the local within the economic system.

economies of scale The benefits of producing on a large scale: as the volume of production increases, cost per unit article decreases. After a certain volume, this fall in cost will be halted as diseconomies arise, but only at very high levels of production, if at all. Pinch and Henry (1999) *Reg. Studs* **33** argue that regional specialization is sustained by economies of scale that 'lock' regions into particular development paths. Parr (2002) *Env. & Plan. A* **34** has a classification of internal and external economies of scale.

economism The belief that economic values are all that is real or important; human beings are motivated mainly by economic drives. Thornton (2007) *Antipode* **39**, 5 thinks that American-style economism 'poses a grave threat to the core values of Islam and the West alike'. 'Economism is the claim that decision makers and theorists have overestimated the contribution that the economic realm can make to policy making' (Wolff and Haubrich in M. Moran et al., eds. 2006).

ecosystem A community of plants and animals within a particular physical environment that is linked by a flow of materials through the non-living (abiotic) as well as the living (biotic) sections of the system. In practice, the hierarchy is: cell, organism, population, community, ecosystem, landscape, biome, and biosphere, although T. Allen and T. Hoekstra (1992) point out that this hierarchy incorporates conceptionally different, non-homogeneous levels of organization, and should therefore be regarded with suspicion. Seastedt et al. (2008) *ESA Online J.*, DOI: 10.1890/070046, suggest that the focus in **ecosystem management** should be less on restoring ecosystems to their original state and more on sustaining new, healthy ecosystems that are resilient to further environmental change.

ecotone A region of rapidly changing species between two *ecosystems. Within the ecotone, local factors, such as soil and groundwater conditions, determine species (Walker et al. (2003) *J. Veg. Sci.* **14**. See DeDeckker and Forester in C. DeDeckker and J. P. Peypouquet (1988) on the **aquatic ecotone**.

ecotope A defined *niche or niche space within a habitat.

ecotourism The development and management of tourism such that the environment is preserved. The income from tourism adds to the investment into landscape conservation. Lake and Hanson (2000) *Urb. Geog.* **21**, 1–4 argue that most of the key concepts in ecotourism are chaotic; 'their uses imply or signify a great range of things, to the point where nearly anything can be deemed sustainable ecotourism' (D. Fennell 2003). Klack (2007) *Geog. Compass* **1**, 5 thinks that **sustainable ecotourism** should give equal weighting to ecological integrity, economic viability, and social justice.

ecozone *See* BIOGEOGRAPHIC REALM.

ecumene (oecumene) The inhabited areas of the world, as opposed to the non-ecumene which is sparsely or not at all inhabited. The ecumene of a nation is its more densely inhabited core. These very simplified classifications pose difficulties of delimitation.

edaphic Of the soil; produced or influenced by the soil.

eddy A roughly circular movement within a current of air or water. An eddy may be a *vortex (Thompson and Young (2006) *J. Phys. Oceanog.* **36**, 4), show *helicoidal flow, or the cylindrical motion of rollers, or appear as a surge phenomenon (a short-lived outbreak of greater velocity in a flow; see Mass and Allbright (1989) *Monthly Weather Rev.* **117**, 11).

edge In *network analysis, another term for the *link between two *nodes.

edge city A self-contained employment, shopping, and entertainment *node, located at the periphery of a pre-existing city, that allows its inhabitants to live, work, and consume in the same place. J. Garreau (1991) lists the criteria for an edge city: at least 465000 m^2 office space; at least 56000 m^2 retail space; more jobs than bedrooms; identification as a distinct 'place'; and a total dissimilarity to a 'city' of thirty years ago. Edge cities have developed with mass car ownership and a 'whole system of supportive infrastructures, from highways to service stations, to drive-through fast food centres, out-of-town malls and auto-access leisure and retail complexes . . . which thereby renders it at least *functionally* a city' (MacLeod (2003) ICRRDS *Rept Office Dep. PM*).

effective discharge In a channel, the discharge that, on average, transports the largest proportion of the annual sediment load, calculated by multiplying the frequency of different discharges by the corresponding sediment transport rate. See Doyle et al. (2005) *Water Resour. Res.* **41**, W11411 on effective discharge analysis.

efficiency The ratio of the work done to the effort used; usually the relation of output to input. **Spatial efficiency**—the organization of space in order to minimize costs—is central to *location theory; see Borrás and Pastor (1998) *Location Sci.* **6**, 1–4.

effluent Human, agricultural, or liquid, industrial waste, introduced into water.

EFTA Current members of the **European Free Trade Association**, established in 1960, are Iceland, Liechtenstein, Norway, and Switzerland.

(((⊕))) SEE WEB LINKS
• The EFTA website.

Ekman layer In the *atmosphere, the transition stratum between the surface *boundary layer and the free atmosphere. **Ekman layer rectification** describes the way the Ekman layer velocity profile differs with variations in surface winds (McWilliams and Huckle (2006) *J. Phys. Oceanog.* **36**, 8). The **Ekman spiral** models changes in wind speed with height in the *planetary boundary layer (Wang and Huang (2004) *J. Phys. Oceanog.* **34**, 5).

electoral geography The geographical analysis of elections; the study of the spatial patterns of voting and power. Johnston et al. (2004) *Pol. Geog.* **23**, 4 find that 'similar people vote differently in different places' (see Pattie and

Johnston (1998) *Area* **30** on the *contextual effect). Other concerns include the influence of voting decisions upon the environment, and the drawing of constituency boundaries; in India, for example, constituency boundaries had to be redrawn frequently to take into account the latest census figures, reorganization of states, graduation of some of the union territories into states, and bifurcation or trifurcation of some states, to fulfil the demands and aspirations of the people within a new federal structure (Singh (2000) *Pol. Geog.* **19**, 4).

El Niño A disruption of the equatorial Pacific, east–west *Walker circulatory cell. The gradient between the low pressure of the warm, western tropical Pacific and the high pressure of the cold, eastern tropical Pacific decreases, weakening the easterly trade winds, and allowing warm surface water to move eastwards. The rising branch of the circulation cell, and therefore the *precipitation associated with it, moves with the water. An El Niño is only one element of a dual-phase oscillating ocean–atmosphere system; when the system reverts to its 'normal' phase, Pacific waters cool off the coast of Ecuador and Peru, and become warm again in the Western Pacific. Sometimes the Eastern Pacific becomes unusually cool; this is a *La Niña event. This entire ocean–atmosphere system is the El Niño, and the 'normal' phase (sometimes developing into La Niña) is the **El Niño–Southern Oscillation**. Dawson and O'Hare (2000) *Geog.* **85**, 3 give an exceptionally clear explanation. See Moy et al. (2002) *Nature* **420** on the variability of the El Niño–Southern Oscillation.

El Niño events have been related to abnormally heavy rain in the southern USA (*Science Daily*, 2008/01/080128113104) and western South America, a decrease in tropical cyclones in the Atlantic, and to droughts in Indonesia and Australia. See Nicholls (1988) *Bull. Am. Met. Soc.* **69**, 2 on El Niño–Southern Oscillation Impact Prediction.

eluviation The movement of material in true solution or colloidal suspension from one place to another within the soil; soil horizons that have lost material through eluviation are **eluvial**.

embeddedness The state of being located or secured within a larger entity or *context. The economic life of a firm or market is territorially embedded in its peculiar social and cultural relations; in place-specific characteristics, infrastructure, operating environments; and conditions of production. *Transnational corporations in the USA are 'highly constrained by dynamic and deep capital . . . [while] their Japanese counterparts are effectively bound by complex but reliable networks of domestic relationships' (Yeung (2000) *TIBG* **23**, 3).

Hess (2004, *PHG* **28**, 1) argues that there are interconnected societal, network, and territorial dimensions of embeddedness: **territorial embeddedness**—the extent to which an 'actor' is anchored in particular territories or places; **societal embeddedness**—the cultural, political, institutional, and regulatory framework the actor is located in; and **network embeddedness**—the structure of

relationships among a set of individuals/organizations. Granovetter in
N. Nohria and R. Eccles, eds (1992) defines **structural embeddedness** as 'the
connectedness of not only two parties, but the extent of interconnection among
third parties or mutual contacts of dyadic partners'. Orderud (2007) *Geografiska
B* **89**, 4 refers to an embeddedness 'where trust acts as a reinforcement . . . as a
capacity restraint and a socially constructed need for face-to-face meetings'. See
Feagan (2007) *PHG* **31**, 1 on the 'quality turn' and embeddedness in food supply;
see also Jones (2008) *PHG* **32**, 1.

embodiment A perspective that roots social existence and experience
through the body and the cognition of the self. The body is symbolically
constructed as one incorporates socially constructed labels and assumptions
relating to the body; 'these take on the weight of unquestionable fact once
applied to self and others' (Adams and Ghose (2003) *PHG* **27**, 4). 'Intellectual
production is always materialized through human bodies, and nonhuman
objects . . . knowledge never arrives from pure brainpower, from only sparking
synapses. It is the outcome of embodied practice' (Barnes (2004) *PHG* **28**, 5).

Perception is seen as a practical bodily involvement; an active process
relating to our ongoing projects and practices. 'This means that the human
body is unique in playing a dual role both as the vehicle of perception and the
object perceived' (Simonsen (2005) *Geografiska B* **87**, 1). H. Nast and S. Pile
(1998) consider the relationship between space and embodiment: bodies are
relational and territorialized—they are woven together with space in 'intricate
webs of social and spatial relations that are made by, and make, embodied
subjects'. 'Our understanding of spaces and places must incorporate the ways
in which bodies are constituted and contested, and acknowledge their
challenge to conventional boundaries' (Little and Leyshon (2003) *PHG* **27**, 3).
S. Franklin and S. McKinnon (2001) list the 'vastly different scales of
embodiment', from the gene, to the body, to the family or species, to the
nation, to the commodity form and cyberspace.

emergence The creation of new phenomena, requiring new laws and
principles, at each level of organization of a complex, and often non-linear,
system. This concept from physics has been applied to geomorphology; see
Harrison (2001) *TIBG* **26**, 3.

emigration The movement of people from one place to another, usually
from one country to another. Green et al. (2008) *NZ Geogr.* **64**, 1 observe that
the 'push–pull' model incompletely describes the motivations of those who
migrate, and who are not necessarily either pushed or pulled, but motivated by
things like a desire for a change, or for more adventure. Hugo (2006), *Geogr.
Res.* **44**, 2 shows how increasing gradients of difference between nations: in the
pattern of growth in the workforce, in income and poverty levels, and in
governance have been important drivers of emigration. In a study of
emigration from rural Albania, Vullnetari and King (2008) *Glob. Netwks* **8**, 1
remind us that migration statistics are about real people.

emissions credits Through photosynthesis, green plants remove carbon dioxide from the atmosphere. Mindful of rising levels of atmospheric carbon dioxide and *global warming, the *Kyoto Protocol proposed that, if nations increase their forested areas, they should be granted emissions credits to count in their CO_2 **emission standards** (their permissible, and legally binding, pollution maxima); see Depledge and Lamb (2005) *UNFCCC*. A further proposal, **emissions trading**, would allow a nation with *greenhouse gas emissions below its agreed target to trade its 'credit' to over-target nations. The impact of emissions trading will depend on the structure of the system (Voon (2000) *J. Int. Law & Policy* **10**, 1).

emotional geographies The study of the location of emotions in both bodies and places; the emotional relations between people and environments; and the task of representing emotional geographies; see L. L. Bondi et al. (2007). Emotional geography has 'a common concern with the spatiality and temporality of emotions, with the way they coalesce around and within certain places' (J. Davidson et al. 2007). One of its central themes is the examination of how feelings mediate both our conscious and unconscious behaviours in the places where we live out, or 'enact', our lives (Hallman et al. (2007) *Soc. & Cult. Geog.* **8**, 6). See Thien (2005) *Area* **37** for consideration of emotion in geography. See also Anderson and Smith (2000) *TIBG* **26**, 1.

enclave A small area within one country administered by another country. Enclaves often appear to become the regions of international tensions and wars: East Bengal, West Berlin, Cabinda, Northern Ireland, and the Palestinian territories (E. Vinokurov 2007).

enclosure The conversion of common or open fields into private, exclusive parcels on which sole proprietorship gave rents and/or management decisions to individual owners (Williamson in J. Eatwell, M. Milgate, and P. Newman, eds 1987).

encounter, geographies of 'Social behaviours draw upon previous encounters' (Hitchings (2007) *TIBG* **32**, 3). Valentine (2008) *PHG* **32**, 3 observes that 'encounters in public space . . . always carry with them a set of contextual expectations about appropriate ways of behaving which regulate our coexistence. These serve as an implicit regulatory framework for our performances and practices.' R. Fincher and K. Iveson (2008) argue that libraries are spaces of encounter that have a redistributive function, offering free and equal access and a safe space for individuals and groups. See also Laurier and Philo (2006) *Area* **38**, 4. Tolia-Kelly (2006) *Area* **38**, 2.

endemic Occurring within a specified locality; not introduced. **Endemism** describes the restriction of a species (or other taxonomic group) to a particular geographic region, and is due to factors such as isolation or response to soil or climatic conditions. Giokas et al. (2008) *J. Biogeog.* **35**, 1 outline a technique to identify areas of endemism.

endogenetic (endogenic) 'From within'. In geomorphology, this means those forces operating below the crust which are involved in the formation of surface features. In human geography, it is those forces acting from within, for example, a society. An **endogenous variable** in economics is a variable explained within the theory under consideration.

endogenous growth theory An economic theory claiming that economic growth is generated from within a system as a direct result of internal processes, underlining the way some regions create an internal mechanism for promoting or perpetuating their growth. Firms operating in places with pools of skilled professionals, for example, are more innovative, leading to faster technological and productivity growth (Ó Huallacháin and Leslie (2007) *J. Econ. Geog.* **7**, 6). Frenken and Boschma (2007) *J. Econ. Geog.* **7**, 5 observe that the more product varieties are already present in a firm or city, the higher the probability that new product varieties can be generated through recombination of old routines—'this feedback relationship is non-linear in that the potential for new ideas rises more than proportional with the stock of existing ideas'. See also D. Felsenstein et al., eds (2001).

energy The physical capacity for doing work. Nearly all our energy derives from the sun, and technical progress has reflected more and more sophisticated uses of energy, from wind and water, through *fossil fuels, to *nuclear power.

In the early stages of industrialization, energy consumption is closely related to levels of economic development, and per capita *GNP—see Zeng et al. (2008) *Science* **319**, 5864, on China. Mature economies tend to be more energy efficient, perhaps because technology improves, and the emphasis shifts to service industries (M. Carr 1997). Even so, the *advanced economies still account for most of the world's energy consumption.

World demand for energy has increased so much that an **energy crisis** (a potential shortage of energy) has been identified; see the Roosevelt Institution's 25 ways of solving the energy crisis. This crisis, together with the adverse environmental effects associated with the burning of fossil fuels (*greenhouse effect, *acid rain) has led to increased emphasis on energy conservation. **Energy intensity** is energy consumption per unit GDP.

((⊕)) SEE WEB LINKS

● Energy Administration Information.

englacial Within a glacier.

ENGO Environmental NGO.

ENSO *See* EL NIÑO–SOUTHERN OSCILLATION.

enterprise zones 'Place-oriented policies that are a tool of regional development, and refer to geographically targeted areas chosen for development that are designated on the basis of unemployment, poverty, population, age of housing stock, and other criteria. Firms that locate in the area and create jobs are given tax credits, abatements, and exemptions. The underlying assumption is that firms and employees in the zone area benefit

N

Orkney Islands

Shetland Islands

Invergordon

Tayside

Clydebank

Tyneside
Sunderland
Hartlepool

Londonderry

Workington
(Allerdale)
Belfast

Middlesborough

Wakefield

Glanford
Scunthorpe

North East Lancs

Speke

Delyn

Rotherham

Telford

Corby

Dudley

Wellingborough

Milford Haven
waterway
Swansea

Isle of Dogs
North West Kent

- Enterprise Zones (26)

Development Areas

Intermediate areas

Special status for
Northern Ireland

UK Enterprise zones

because of a reduction in the price of capital and/or labour, and there is expanded investment and employment generation through deregulation' (S. Sridhar (2004) *Nat. Inst. Public Finance & Prop. N. Delhi W. Paper* **19**). In a study of US enterprise zones, A. Peters and P. Fisher (2002) find that while early econometric studies of the effects of state enterprise zones usually found positive results, more recent results have been much less favourable.

entisol In *US soil classification, young soils, high in mineral content and without developed soil *horizons. *See* RANKER.

entitlement A guarantee of access to benefits through rights, or through legal agreement. By entitlements, A. Sen (1981) means one's ability to achieve the commodities necessary for basic needs (including food and shelter), either through income-generating work or from direct production of these goods. The size of one's entitlement determines the amount of food (and other commodities) one can command. Thus Smith (2001) *PHG* **25**, 140 writes that access to food in urban areas is affected by a wide range of entitlements and blockages most of which revolve around the fact that unlike rural areas where self-sufficiency or exchange is possible, cities operate on the basis of a cash economy. Sen uses the **entitlement approach** to explain famines. Watts (2000) *Zeitschrift* **88** comments that 'one of the great strengths of Sen's approach to hunger is that its entitlements are part of a larger architecture of thinking about development as a state of well-being and choice or freedom' See also G. Robinson (2004).

entrainment 1. The picking up and setting into motion of particles, either by wind, water, or *ice. The main entrainment forces are provided by impact, *lift force, and *turbulence.

   2. In *meteorology, the incorporation of buoyant air into a cloud; see Moeng (2000) *J. Atmos. Scis* **57**, 21.

entrepôt A transhipment point where goods are held without incurring customs duties. See Dobbs (2002) *Sing. J. Trop. Geog.* **23**, 3 on Singapore as an entrepôt.

entrepreneur An organizer, singly or in partnership, who takes risks in creating, investing in, and developing a *firm from its inception through to hoped-for profitability as goods and services are marketed. In recent decades, most industrialized receiving countries have experienced a rise of immigrant and ethnic entrepreneurship: see R. Kloosterman and J. Rath (2003). Jessop and Sum (2000) *Urb. Studs* **12** hold that an entrepreneurial city pursues innovative strategies intended to maintain or enhance its economic competitiveness *vis-à-vis* other cities and economic spaces. A. Cronin and K. Hetherington (2008) see an **entrepreneurial city** as characterized by business-led urban development, technological innovation, social capital featuring skilled labour, and social and *environmental *sustainability, and supported by local authorities, educational and training institutions, and its own heritage and cultural industries.

entropy A measure of disorder in a system; the higher the entropy, the greater the disorder. Entropy is a widely used measure of the compositional diversity of a spatial unit in *urban geography. Its values reflect the range of possible mixing—large values reflecting the greatest demographic diversity (Ellis et al. (2007) *Urb. Geog.* **28**, 6).

enumeration district A unit of census survey; in the UK, about 150 households.

environment The surroundings. The **natural environment** includes the nature of the living space (sea or land, soil or water), the chemical constituents and physical properties of the living space, the climate, and the assortment of other organisms present. The **phenomenal environment** includes changes and modifications of the natural environment made by man. The effect of the environment on man is modified, in part, by the way the environment is perceived, and human geographers distinguish this—the **subjective environment**—from the **objective environment**—the real world as it is. Eden (2001) *PHG* **25**, 1 shows that nature and the environment are not only differently perceived and classified by different groups but also used differently in association with policies and actions which can then materially influence what is being perceived. Simmons (2000) *Geog.* **85** describes the 'many layers' of environment as 'something more like a double helix of mind and matter, whose gyre seems to be ever-widening, spinning unpredictable combinations of society and economy into the main space, but also throwing off minor gyres, which might be short-lived but might equally be the germs of the main arms of the future'.

Environment Agency An independent public body, created to protect and improve the environment in England and Wales. Its remit includes tackling flooding and pollution incidents, reducing industry's impacts on the environment, cleaning up rivers, coastal waters, and contaminated land, and improving wildlife habitats.

(⊕) SEE WEB LINKS
• The Environment Agency website.

environmental accounting The inclusion of economic and environmental information in a common framework to measure the contribution of the environment to the economy and the impact of the economy on the environment. Environmental accounts enable governments to set priorities, monitor economic policies more precisely, enact more effective environmental regulations and resource management strategies, and design more efficient market instruments for environmental policies (Lange et al. (2003) *Nat. Resources Forum* **27**, 1). Pulselli et al. (2006) *Ecol. Econ.* **60**, 1 use environmental accounting in calculating an index of sustainable economic welfare. See C. Barrow (2005).

An **environmental audit** is the practice of assessing, checking, testing, and verifying a process of environmental management, and/or the product(s) arising from it; see Alvarez-Larrauri and Fogel (2008) *J. Cleaner Prod.* **16**, 1 on environmental audits in Mexico.

environmental determinism The view that human activities are governed by the environment, primarily the physical environment. J. M. Blaut (2005) challenges the central thesis of environmental determinism, while J. Sachs (2005) reconsiders the role of the environment in shaping human history. Judkins et al. (2008) *Geog. J.* **174**, 1 think that Jared Diamond's books are pure environmental determinism: 'this resurgence threatens negative consequence if uncritically adopted by policymakers.'

environmental economics 'Understanding why environmental issues exist, identifying if it is worth remedying the situation, and developing solutions' (Rolfe (2008) *Australian J. Agric. & Resources Econ.* **52**, 1).

environmental gradient A gradual and continuous change in communities and environmental condition. The gradients can be related to environmental factors such as altitude, temperature, and moisture supply. See Damgaard (2003) *Ecol. Modelling* **170** on modelling plant competition along an environmental gradient.

environmental impact A change in the make-up, working, or appearance of the environment. An **Environmental Impact Assessment** considers the probable consequences of human intervention in the environment, seeking to restrict environmental damage. In most developed countries, the methodology includes provision for the responsible authority to produce a draft, release this for public comment, and gather responses from a variety of perspectives: other agencies and levels of government, environmental and community groups, interested corporations, resource users, and ordinary citizens. This generates information that might otherwise be excluded from administrative decision making, and although this information may not directly influence the decision, it makes environmental and democratic values more visible and legitimate than before (Turnbull (2004) *Geog. J.* **170**, 1). See S. Atapattu (2006).

environmentalism A concern for the environment, and especially with the bond between humans and the environment; not solely in terms of technology but also in ethical terms. In 2004, M. Shellenberger and E. Nordhaus savaged modern environmentalism as 'no longer capable of dealing with the world's most serious ecological crisis'. 'If we are to understand modern conservation of nature, we need to see environmentalism not as a thing, but as a process' (L. Vivanco 2006). Wade-Benzoni et al. (2007) *Presidential Studs Qly* **37**, 4 find that self-perceptions of environmentalism are changed by subtle manipulations of context and, in turn, affect environmental behaviour.

Market environmentalism is defined by T. Anderson and D. Leal (2001) as a mode of resource regulation that promises both economic and environmental health via market means, offering a 'virtuous fusion of economic growth, efficiency, and environmental conservation'. See Bailey (2007) *AAAG* **97**, 3 on territorializing market environmentalism, and Davis (2006) *Geog. J.* **172**, 2.

environmental justice 'The fair treatment and meaningful involvement of all people regardless of race, color, national origin, culture, education, or income with respect to the development, implementation, and enforcement of environmental laws, regulations, and policies' (US Environmental Protection Agency). See Ranco and Suagee (2007) *Antipode* **39**, 4. Remedies for environmental deterioration recognize the intrinsic value of 'nature'; seeing humans as part of, rather than owners of, the biotic community (*Swarthmore Coll. Bull.* 2005). See Buckingham (2004) *Geog. J.* **170**, 2 on environmental justice and ecofeminism.

environmental lapse rate The fall in temperature of stationary air with height, averaging 6 °C per 1 000 m. This fall of temperature with height in stable air is due to a fall in the density of the air, varying greatly with the time of day, and with the nature of the *air mass concerned. See Schultz et al. (2000) Monthly Weather Rev. **128**, 4143-8, and Houston and Niyogi (2007) Monthly Weather Rev. **135**, 3013-32.

Environmentally Sensitive Area (ESA) A fragile *ecosystem area where the conservation or preservation of the natural environment is sustained by state controls and/or grants, such as the eastern South Downs, UK (Boardman (2003) *TIBG* **28**, 2).

environmental perception The way in which an individual perceives the environment; the process of evaluating and storing information received about the environment. Decision-makers cannot but base their judgements on the environment as they perceive it. See J. Hobbs and C. Salter (2006) and Y.-F. Tuan (1974).

environmental policy The statement by a supranational, national, or regional government of its approach to environmental protection. Pezzey (2004) *Scand. J. Econ.* **106**, 2 contrasts environmental policy, 'which internalises externalised environmental values', with sustainability policy, 'which achieves some form of intergenerational equity'. See C. Barrow (2005); try Tews et al. (2003) *Eur. J. Pol. Res.* **42**, 4.

environmental realism The notion that the environment is a 'real thing'; environmental idealism posits a set of underlying, stable, and consistent values that relate to the character, sense, and quality of nature. See P. Macnaghten and J. Urry (1998).

environmental science The study of environments; traditionally the physical environment, or, more widely, social and cultural environments.

Major themes include: environmental accounting; environmental performance evaluation; reporting; participative processes in environmental policy-making; sustainable management; and the effectiveness of Local Agenda 21. See the special 2007 issue of *Environment and Urbanization* **19**, 1. O'Riordan (2004) *TIBG* **29**, 2 notes that environmental science 'has become highly political . . . the antagonistic pressures of established power and economic hegemony are never far away'.

Environment Directorate An agency of the OECD, the environment directorate provides governments with the analytical basis to develop policies that are effective and economically efficient, including country performance reviews, data collection, policy analysis, projections and modelling, and the development of common approaches.

(⊕) SEE WEB LINKS

• The Environment Directorate on the OECD website.

eolian *See* AEOLIAN.

epeirogeny Broad, and generally large-scale, limited, or slow vertical movements of the Earth's crust (Chardon et al. (2006) *Geomorph.* **82**, 3–4) which do not involve much alteration in the structure of the rock; hence **epeirogenic**—caused by the relatively gentle raising or lowering of the earth's crust.

ephemeral Short-lived. R-strategist plants are ephemeral in that they grow and reproduce rapidly when conditions are favourable, dying within a short space of time; see Boulos (2003) *J. Biogeog.* **30**, 12. **Ephemeral streams** flow only during and after intense rain, and are typical of arid and semi-arid areas; see Billi (2007) *Geomorph.* **85**, 1–2.

epicentre The point of the earth's surface which is directly above the *focus of an *earthquake, usually 0–50 km below it. See Bondár et al. (2004) *Geophys. J. Int.* **156**, 3 on epicentre accuracy based on seismic network criteria.

epilimnion The upper layer of a body of water, penetrated by light, thus enabling photosynthesis. This zone is warmer, and contains more oxygen, than the layers below.

epiphyte A plant growing on another plant but using it only for support and not for food, most commonly found in *tropical rain forests. See Zotz et al. (1999) *J. Biogeog.* **26**, 4 on the community composition, structure, and dynamics of epiphyte vegetation.

epistemology The philosophical theory of knowledge which considers how we know what we know, and establishes just what ought to be defined as knowledge. In *geography, the term is used to indicate the examination of geographical knowledge—how it is gained, sent, changed, and absorbed.

O'Sullivan (2004) *TIBG* **29**, 3 distinguishes between epistemology (1) 'experimenting on theories' and epistemology (2) learning from models.

epochs An interval of geological time; several epochs form a period, several periods an era. An epoch is ranked as a third-order time unit (M. Allaby 2006).

equal area map A map so drawn that a square kilometre in one portion of the map is equal in size to a square kilometre in any other portion. Equal area maps of the whole globe tend to be elliptical in shape, and severely distort the shapes of regions far from the equator.

Peters' projection

Interrupted equal area

equatorial rain forest *See* TROPICAL RAIN FOREST.

equatorial trough A narrow zone of low pressure, between the two belts of *trade winds, arising from high *insolation, especially in the centre of continents in summer; see Quinn and Burt (1967) *J. Appl. Met.* **6**, 6. Also known as the *inter-tropical convergence zone, the equatorial trough is not constant in position, breadth, or intensity; see Chao (2000) *J. Atmos. Scis* **57**, 641–651.

equifinality The development of different initial states to similar end states; similar landforms, for example, might arise as a result of quite different sets of processes and histories. Thus from the landform alone, without additional evidence, it might be difficult to identify the particular set of causes or to differentiate different feasible causes. See Beven in B. Rhoads and C. Thorn, eds (1996)

equilibrium In physical geography, the state of an open system where the inputs, throughputs, and outputs of energy or matter are in balance. R. Whittaker (1998) identifies four extremes in a continuum: **dynamic equilibrium** (fluctuation about a moving average); **dynamic non-equilibrium**; **static equilibrium** (no perceived change); and **static non-equilibrium.** See Whittaker (2000) *Glob. Ecol. Biogeog.* **9**, 1. **Spatial equilibrium** is regional equality, brought about by the interaction of labour moving to high-wage areas and capital moving to regions where wage rates are low, and usually expressed in terms of wages: 'workers and firms are indifferent among alternative locations as they have eliminated any systematic difference in indirect utility and profits through migration' (Ottoviano and Perri (2005) *J. Econ. Geog.* **6**, 1).

equilibrium line In glaciology, the point at which expansion of the glacier by accumulation is balanced by ice losses through *ablation. See Carrivick and Brewer (2004) *Geografiska A* **86**, 1.

equilibrium species Species that show characteristics consonant with a stable *niche. Dispersal is less important, perseverance is more significant than recovery from adverse conditions, and the survival of the young is more important than high fecundity.

equinox Equinoxes, those times when day and night are of equal length, occur twice a year. The **spring**, or **vernal**, **equinox** is on 21 March and the **autumn equinox** is on 22 September. On these dates, the sun is directly overhead at the equator. The changes in day length result from the changes in the tilt of the earth with respect to the sun.

era The largest unit of geological time. The approximate datings of the eras are:

ERA	DURATION IN MILLIONS OF YEARS BEFORE PRESENT
Precambrian	4600–570
Palaeozoic	570–225
Mesozoic	225–65
Cenozoic	65–0

erg Arid, sandy desert, particularly within the Sahara. The Rub Al Khali aeolian system of the Arabian Peninsula is the world's largest erg (Bray and Stokes (2004) *Geomorph.* **59**, 1–4).

erodibility The vulnerability of a material to erosion. This will vary according to the nature of the material: degree of consolidation, organic content, particle size, hardness, and strength, among others; on the extent and type of ground vegetation; thus on *land use. See Darboux and Le Bissonnais (2007) *Eur. J. Soil Sci.* **58**, 5 on measuring soil erodibility. Quantifying soil erodibility is crucial for forecasting dust events (Zender et al. (2003) *J. Geophys. Res.-Atmos.* **108**, 4543).

erosion The removal of part of the land surface by wind, water, gravity, or ice. These agents can only transport matter if the material has first been broken up by weathering. Some writers use a very narrow interpretation of the word, claiming that erosion refers only to the transport of debris; see Stallard in P. Hancock and B. J. Skinner, eds (2000). If so, *denudation would be the correct term for the weathering as well as the transport of rocks.

erosion surface A relatively level surface produced by erosion. Much of Africa is composed of extensive plains, often cutting *discordantly across varied geological structures, and these have therefore been identified as erosion surfaces as such (Lister (1965) *Rec. Geol. Surv. Malawi* **7**). See also Ojany (1978) *GeoJournal* **2**, 5. The whole concept is bound up with the theory of the cycle of erosion.

erosivity The ability of rainfall to cause erosion, which depends on rainfall intensity and drop size, as well as the material it lands on. Rainfall intensity is used as the primary erosivity indicator at shorter times and smaller areal scales. Larger areas and greater time scales generally require use of run-off as the measure of erosivity. See Cerro et al. (1998) *Soil Sci. Soc. Am. J.* **62**.

erratic A large boulder of rock which has been transported by a glacier so that it has come to rest on country rock of different *lithology; all erratic blocks in northern Germany, for example, originated in Scandinavia (Bernhardi (1832) *Jahrbuch für Mineralogie, Geognosie, und Petrefaktenkunde*, **3**). Agassiz (1840–1) *Procs. Geo. Soc.* **3**, 2 used the presence of erratics as evidence of glaciation in the British Isles.

ESA *See* ENVIRONMENTALLY SENSITIVE AREA.

escarpment A more or less continuous line of steep slopes, facing in the same direction, and caused by the erosion of folded rock (A. Bowen and J. Pallister 2001).

esker A long ridge of material deposited from *meltwater streams running *subglacially, roughly parallel to the direction of ice flow, and ranging from tens of metres to several hundred kilometres, as in Finland. Eskers wind up and down hill because subglacial streams are under great hydrostatic pressure, and can flow uphill. Brennan (2002) *Geomorph.* **32**, 3–4 identifies and investigates five types of esker. Thomas and Montague (1997) *Quat. Sci. Rev.* **16** argue that glacial lake systems are crucial for the development of esker systems.

estuary That area of a river mouth which is affected by sea tides. A **wave-dominated estuary** has waves as the dominant shaping force; in a **tide-dominated estuary**, tidal currents dominate geomorphologically (Komatsu (1999) *Cretaceous Res.* **20**, 3).

eta index (η) An expression of the relationship between a network as a whole and its *edges. For a worked example, see Alberto and Manly (2006) *J. Biogeog.* **33**, 4.

etchplain A tropical *planation surface where deep weathering has etched into the bedrock. Fluvial denudation may lay the etchplain bare. See Stengel and Busche in X. Yang, ed. (2002) on a Late Proterozoic etchplain in Namibia.

ethical geography Ethics are the principles of right and wrong conduct; the basis for doing what is right, and the discernment of what is right (W. R. F. Browning 1997); morals are the accepted norms and standards of conduct of a society, community, or nation (J. Last 2007). Smith (2001) *PHG* **25**, 2 seeks to clarify the difference between these two concepts: 'moral philosophers would doubtless agree that writing about certain things being right or wrong counts as ethics, but would wish to consider how such normative claims are made.' Cloke (2002) *PHG* **26** asks geographers to live ethically and act politically, but Evanoff (2007) *Eth. Place. Env.* **10**, 2 argues that a single environmental ethic that can be universally applied in all geographic settings and across cultures 'cannot be formulated'. Madge (2007) *PHG* **31**, 5 provides a wide-ranging review of research ethics in geography.

ethical trade Trade based on social and environmental responsibility in existing large-scale businesses. Freidberg (2003) *Soc. & Cult. Geog.* **4**, 1 notes that supermarket supply networks are characterized by highly unequal power relations and fetishized standards: 'cleaning up down South comes cheap.' Hughes (2001) *J. Econ. Soc.* **1** argues that for ethical trading strategy to make a more significant difference, attention needs to be paid to worker welfare at sites of production, and the terms of trade between retailers and suppliers.

ethnic group (ethnicity) 'The only working general definition of ethnicity is that it involves the common consciousness of shared origins and traditions' (Allison in I. McLean and A. McMillan 2003). Ethnicity in a group may become pronounced as a result of migration: 'in Trinidad and Guyana, Hindus were at the bottom of the social structure . . . When they migrated to Britain, they were still in an ethnic quandary, with the white British population thinking of them derogatorily as "Paki" (subcontinental Indian)' (Kong (2001) *PHG* **25**, 2).

However, Peach (2000) *PHG* **24**, 4 observes that the minorities in the 1960s to the 1980s which were sometimes compressed into a single 'black race' category are now being teased out into their constituent ethnicities, classes, and genders. Whites are also being unpacked from their homogenized categorization: see Shaw (2006) *Antipode* **38**, 4. 'Transnationalism is transforming the ethnic enclave from being a feature of deprivation in the inner city into affluent closure in the suburbs; ethnic villages of the audible rather than the visible minorities are appearing' (Simpson et al. (2008) *Urb. Studs* **45**, 1).

ethnic segregation The evolution of distinct neighbourhoods recognizable by their characteristic ethnic identity. Massey and Denton (1988) *Social Forces* **67** distinguish five dimensions of residential segregation: evenness, exposure, concentration, centralization, and clustering. Phillips et al. (2007) *TIBG* **32**, 2 discuss multiple readings of ethnic segregation.

External causes (imposed by the *charter group) of ethnic segregation include discrimination, low incomes—which direct them towards inner-city locations—and the need for minorities to locate near the *CBD since much of their employment is located there. Internal causes (springing from the ethnic group) include a desire to locate near facilities serving the group, such as specialized shops and places of worship, desire for proximity to kin, and protection against attack; see Fossett (2006) *J. Math. Soc.* **30**, 3–4.

Clark and Morrison (2008) *Pop. Res. & Policy Rev.* **27**, 3 find that an upgrading of housing stock may promote ethnically homogeneous apartment-building 'neighbourhoods' within an otherwise unsegregated ethnically diverse area, and Fossett (2006) *J. Math. Soc.* **30**, 3–4 reports that reductions in housing discrimination may not necessarily lead to large declines in ethnic segregation in the short run. Ellis et al. (2004) *AAAG* **940** reveal that segregation by work tract is considerably lower than by residential tract. Lobo et al. (2007) *Urb. Geog.* **28**, 7 argue that ethnicity itself is wielded as a potent force as a group moves into a neighbourhood and eventually dominates it. Phillips (2007) *Geog. Compass* **1**, 5 explores the politics of data collection, categorization, and representation in ethnic segregation research.

ethnocentricity Making assumptions about other societies, based on the norms of one's own. This may lead to the development of global theories based, for example, on Western conditions. 'Ethnocentric accounts of "hate" and "love" are continually disingenuous to the flows and rhythms of the development and experience of affective capacities of individual bodies, and thus are negligent in their theoretical project' (Tolia-Kelly (2006) *Area* **38**, 2).

ethnography The study of the customs, habits, and behaviour of specific groups of people (usually people in non-literate societies). Herbert (2000) *PHG* **24**, 4 believes that ethnography is an underused methodology in geography. 'This neglect is especially injurious to the discipline because ethnography provides unreplicable insight into the processes and meanings that sustain and motivate social groups. These processes and meanings vary across space, and are central to the construction and transformation of landscapes; they are both place-bound and place-making. Ethnography's potential contribution to geography is thus profound.' Hart (2006) *Antipode* **38**, 5 holds that using critical conceptions of spatiality can extend and enrich global ethnography. Hörschelman and Stenning (2008) *PHG* **32**, 2 argue that ethnographic research is most valuable when it is employed 'within an explicitly spatial imaginary, placed within a multi-sited context and critically intervening in broader theoretical debates'. Would it not then be geography?

euphotic zone The upper layer of a body of water receiving light and thus where photosynthesis is possible. Deep-sea benthic communities are largely controlled by the organic carbon flux from the euphotic zone (McClain et al. (2006) *J. Biogeog.* **33**, 9).

European Economic Area A single, free-trade grouping, comprising the member states of the *European Union and *EFTA.

((())) SEE WEB LINKS
• The European Economic Area defined on the European Commission website.

European Union (EU) A *free trade area comprising Austria, Belgium, Bulgaria, Cyprus, the Czech Republic, Denmark, Estonia, Finland, Germany, Greece, Hungary, Ireland, Italy, Latvia, Lithuania, Luxembourg, Malta, the Netherlands, Poland, Portugal, Romania, Slovakia, Slovenia, Spain, Sweden, and the UK. Designed initially as an economic unit, the European Union now attempts uniformity in social policies.

((())) SEE WEB LINKS
• The European Union website.

European Water Framework This directive's key aims are: expanding the scope of water protection to all waters, surface waters, and groundwater; combining emission limit values and quality standards; getting the citizen involved more closely; and streamlining legislation. See Carter (2007) *Geog. J.* **173**, 4 and Kay et al. (2007) *PPG* 31, 1.

((())) SEE WEB LINKS
• The EU Water Framework Directive website.

eustasy A worldwide change of sea level, which may be caused by the growth and decay of ice sheets, by the deposition of sediment, or by a change in the volume of the oceanic basins. 'It is commonly assumed that melting of polar ice sheets induced, for example, by climate warming yields a nearly eustatic trend in sea level, although it has long been recognized that self-gravitation and loading effects will lead to significant departures from eustasy' (Mitrovica et al. (2001) *Geophys. J. Int.* **147**, 3).

eutrophication The process by which *ecosystems, usually lakes, become more fertile environments as detergents, sewage, and agricultural fertilizers flow in; the enrichment of nutrients (De Anda et al. (2001) *Lakes & Reservoirs: Res. & Manage.* **6**, 4). For eutrophication in the UK, see National Expert Group on Transboundary Air Pollution (NEGTAP) 2001.

evaporite A deposit formed when mineral-rich water evaporates; the most common are of gypsum (hydrated calcium sulphate) and halite (sodium chloride). See Ayora et al. (2001) *Geology* **29**, 3 on Mesozoic and Tertiary **evaporite basins**.

evapotranspiration The release of water vapour from the Earth's surface by evaporation and transpiration. Since evapotranspiration is so variable, physical geographers prefer to use the concept of **potential evapotranspiration** (PE). This is the greatest amount of water vapour which could be diffused into the atmosphere given unlimited supplies of water.

evolutionary economic geography The evolutionary economic geography approach aims to understand actions of economic actors and paths of change in the context of time and space. It explains how behaviour of agents is situated and conditioned, but not determined, by structures accumulated at the level of organizations and the environment (Boschma (2004) *Reg. Studs* **38**, 9). However, as Martin and Sunley, in a vital summary of the sub-discipline (2007 *J. Econ. Geog.* **7**), point out, just as 'there is as yet no single, coherent, or widely agreed body of theory or methodology that defines evolutionary economics', so there is a wide variety of approaches to evolutionary economic geography. Key approaches include using the evolutionary principles of variety, selection, and retention (Essletzbichler and Rigby (2007) *J. Econ. Geog.* **7**). See also Maskell and Malmberg (2007) *J. Econ. Geog.* **7**.

exclave A portion of a nation which lies beyond national boundaries. The Kaliningrad Oblast has become a Russian exclave within the enlarged EU (Prozorov (2007) *Pol. Geog.* **26**, 3). Newman (2006) *PHG* **30**, 2 argues that transition zones are exclaves demarcated by the hard lines separating them from the external world around them.

exclusion *See* SOCIAL EXCLUSION.

exfoliation The *sheeting of rocks and their disintegration, thought to be due to *thermal expansion, at least on small structures. Waragai (1998) *Arct. Alp. Res.* **30** finds that high rates of change in rock temperature may produce exfoliation sheets in *periglacial environments. Le Pera and Sorriso-Valvo (2000) *Geomorph.* **34**, 3–4 think exfoliation boulders, now found at the ground surface, may have been formed at depth.

exhumation The removal of young deposits to reveal the underlying structure of older rocks.

exogenetic 1. Applying to processes which occur at or near the Earth's surface.

2. In human geography, as a result of outside, environmental influences.

expatriates, migrants, diaspora strategies New Zealand is a nation of 4 million people, plus 1 million expatriates: 'whereas early efforts involved encouraging highly skilled migrants to return to their home countries, today a range of new techniques [is] being used to engage expatriates in activities in their countries of origin without requiring them to return' (Larner (2007) *TIBG* **32**, 3).

extended family A family unit of relatives by blood and by marriage as well as two parents and their children. Kovacheva in C. Wallace, ed (2002) concludes that extended family and household are 'an indispensable means for survival in the difficult economic situation'. Smith and Stenning (2006) *PHG* **30**, 2 find that remittance economies from migrants to their extended family back home link households with wider circuits of value.

extending flow The extension and thinning of a glacier, often marked by an *ice fall. Extending flow occurs near the *equilibrium line, and where the velocity of the glacier increases, for example, down a rock *step. Extending flow can transmit material from the surface of a glacier to its base, thus increasing its powers of *abrasion. It is also responsible for *crevasses and is typical of the zone of *accumulation of ice. King (1970) *Geografiska A* **52**, 3–4 notes that, in zones of extending flow, ice reaches the bed tangentially, diminishing the ice's erosive capacity, and thus diminishing erosion.

extensive agriculture Farming with low inputs of capital and labour, generally with low yields per acre, and associated with regions of cheap available land where high revenues are unimportant.

external economies The cost-saving benefits of locating near factors which are external to a firm, such as locally available skilled labour, training, and research and development facilities. Moran et al. (2002) *PHG* **26**, 4 writes that regional specialization is sustained by external economies of scale that impart certain advantages and 'lock' regions into particular development paths. See also Crescenzi et al. (2007) *J. Econ. Geog.* **7**, 6.

externality A side effect on others following from the actions of an individual or group. This effect is not brought by those affected and may be unwished for. Two types of externality are recognized: **public behaviour externalities** covering property, maintenance, crime, and public behaviour, and **status externalities** resulting from the social and ethnic standing of the household. McCarthy (2004) *Geoforum* **35** argues that capitalists today can profit from resource use irrespective of the externalities this use generates. See also Helmsing (2001) *Dev. & Change* **32** on externalities, learning, and governance.

extinction The end of the existence of a species or group of taxa, or the end of their ability to reproduce. Gaston (2008) *PPG* **32**, 1 observes that rare species are disproportionately vulnerable to extinction in the short term. Loss of a species to chance extinction leaves an area of resource space unoccupied; extinctions could trigger behavioural or evolutionary adaptations (Phillips (2008) *PPG* **32**, 1). Van Kleunen and Richardson (2007) *PPG* **31**, 4 consider the links between species traits and extinction risk and invasiveness. See Kiesling and Aberhan (2007) *J. Biogeog.* **34**, 9 on geographical distribution and extinction risk. Losos and Schluter (2000) *Nature* **408** explain the relationship as the outcome of the effect of area on immigration and extinction rates. 'If the twentieth-century question was how to stop extinction, then perhaps the twenty-first-century challenge is how to avoid the total collapse of the biosphere, our life support system' (W. Adams 2004). See Sanderson et al. (2002) *Bioscience* **52**, 10 on human activity and extinction rates. *See also* BIODIVERSITY CRISIS.

extra-tropical cyclone A cyclonic disturbance outside the tropics; for example, a *mid-latitude depression.

extrusion A formation of rock made of *magma which has erupted onto the earth's surface as lava and has then solidified. The crystals in extrusive rocks are small, since the lava solidifies rapidly, giving little time for crystal growth. Extrusions emerge from *fissure eruptions and *volcanoes. See Redfield et al. (2007) *Geology* **35**, 11.

exurb American for *dormitory settlement. In many cases, amenity migration and the 'urbanization of the rural' it produces can be understood as the first signs of exurban development (McCarthy (2008) *PHG* **32**, 1).

eye The calm area at the centre of a *tropical cyclone.

F

fabric The physical make-up of a rock or sediment. **Fabric analysis** can determine the *dip and orientation of particles in a sediment: see Carr and Goddard (2007) *Boreas* on fabric analysis and glacier dynamics; Cifelli et al. (2006) *Eos Trans. AGU* **87** on magnetic/mineral fabric analysis and regional deformation; and Brown and Bell (2007) *Bull. Volcan.* **69**, 8 on fabric analysis and mass wasting. Millar (2005) *Geomorph.* **72**, 1–4 seriously doubts the time/cost effectiveness of fabric analysis to identify periglacial landforms.

facet A flat surface on a rock or pebble, produced by *abrasion. **Facet mapping** is the quantification of boulder morphology; try Heslop et al. (2004) *Lunar & Planet.* **35**, 1445.

facies The characteristics of a rock, such as fossil content, or chemical composition, which distinguish it from other formations and give some indication of its formation. Torsvik et al. (2004) *J. Geol. Soc.* **161**, 4 present a facies review of earth geography from 400 to 250 Ma.

factor analysis A multivariate statistical technique, used to simplify analysis; see R. Rummel (1970). For an example of its use in geography, see Brown and Raymond (2007) *Appl. Geog.* **27**, 2.

factorial ecology The investigation of urban spatial structure by *factor analysis. The classic study is Murdie's (1969) *U. Chicago Res. Paper* **116**. For the problems associated with factorial ecology, see Martinez-Martin (2005) *ITC Diss. Series* **127**, 44–5.

factors of production The resources required to produce economic goods. They are land (including all natural resources), labour (including all human work and skill), capital (including all money, assets, machinery, raw materials, etc.), and entrepreneurial ability (including organizational and management skills, inventiveness, and the willingness to take risks). For each of these factors there is a price, i.e. rent for land, wages for labour, interest for capital, and profit for the entrepreneur. Desmet and Fafchamps (2005) *J. Econ. Geog.* **5**, 3 state that perfect mobility of capital and labour tends to bring about an even distribution of economic activity across space, since the use of land leads to decreasing returns to the mobile factors of production—but Puga (2002) *J. Econ. Geog.* **2**, 4 disagrees. Jiménez (2003) *Growth & Change* **34**, 2 observes that public capital is believed to increase the productivity of the private factors of production whereas human capital is thought to contribute to the production process as an additional input. Ettinger (2008) *PHG* **32**, 1 refers to **intangible factors of production** such as trust, cooperation, and the sharing of knowledge.

Maskell and Malmberg (1999) *Eur. Urb. Reg. Studs* **6**, 1 point out that, historically, while some factors of production were ubiquitous, the cost of certain others varied between locations. They go on to discuss the process whereby some previously important locational factors are actively converted into ubiquities, which they label, with apologies, 'ubiquitification': 'ubiquitification tends to undermine the competitiveness of firms in the high-cost areas of the world. When international markets are opened up and when knowledge of the latest production technologies and organizational designs become globally available, firms in low-cost areas become more competitive.' Usher (2003) *Canad. Geogr./Géog. canad.* **47**, 4 shows how an aboriginal household allocates the factors of production to optimize income flows from both the market and subsistence economies.

factory farming A system of livestock farming in which animals—most commonly, pigs, laying hens, broiler chickens, and veal calves—are kept indoors, with very restricted mobility, for most of their lives. Factory farming has created its own set of environmental damage: raised output of greenhouse gases (farm animals account for 15–20% of methane emissions), and 'manure surpluses' (J. Wolch and J. Emel, eds. 1998).

factory system A concentration of the processes of manufacturing—fixed capital, raw material, and labour—under one roof, in order to provide the mass production of a standardized product or products. Ngai and Smith (2007) *Work Employment Soc.* **21** comment on China's 'unique factory system—a workforce of permanently young workers'.

fair trade A *commodity chain set up to promote worker welfare, minimize environmental damage, and allow producers to increase their living standards, and control within the supply chain, through *development projects. See Hughes in D. Burch and G. Lawrence, eds (2007) on supermarkets and ethical/fair trade, and Hughes (2006) *Geoforum* **37**, 6 on UK food and clothing retailers, and ethical trading.

fall A form of *mass movement in which fractured rock and soil separates into blocks and falls away from the parent slope. A fall starts with the detachment of soil or rock from a steep slope along a surface on which little or no shear displacement takes place. The material then descends mainly through the air by falling, bouncing, or rolling (Varnes in R. Schuster and R. Krizek, eds 1996).

fallow Agricultural land which is not cropped, but left unused, to restore its natural fertility. In response to rising grain costs, in September 2007 the EU abolished, for one year, its ruling that farmers must leave 10% of their land fallow.

family In geolinguistics, a genetic group of languages with numerous cognates and regular correspondences, arising as a language diffuses and diverges over time. See Millar (2003) *J. Linguistics* **7**, 1 and Sharma (2005) *J. Linguistics* **9**, 2 on new varieties of English.

famine A relatively sudden flare-up of mass death by starvation, usually relatively localized, and usually associated with a sharp rise in food prices, the sale of household goods, begging, the consumption of wild foods, and out-migration. P. Walker (1989) describes famine as a 'socio-economic process which causes the accelerated destitution of the most vulnerable, marginal and least-powerful groups in a community, to a point where they can no longer, as a group, maintain a sustainable livelihood'. See Woo-Cumings (2002) *ADB Institute Research Paper* **31** on famine in North Korea.

A. Sen (1982) sees famine as distinguishable from chronic hunger and deprivation, in that speedy intervention can prevent it. Sen developed the concept of *entitlement, identifying declining wages, unemployment, rising food prices, and poor food-distribution systems, as causes of starvation. Lin and Dennis (2000) *Econ. J.* **110**, 1 endorse Sen's approach, adding a decline in food availability as a further factor. Pluemper and Neumayer (2007) http://ssrn. com/abstract=920852 suggest that it can be politically rational for a government to remain inactive in the face of severe famine threat. Howe and Devereux (2004) *Disasters* **28**, 4 note that definitions of famine (including this one) tend to be vague, and propose new famine scales based on magnitude and/or intensity.

(⊕) SEE WEB LINKS

• The famine early warning systems network.

farm fragmentation The division of a farmer's land into a collection of scattered lots. Fragmentation usually results from inheritance but may also reflect *bush fallowing. See Pan et al. (2005) *IGU Commission Land Use* on farm fragmentation in Ecuador.

fault A fractured surface in the earth's crust along which rocks have travelled relative to each other. Usually, faults occur together in large numbers, parallel to each other or crossing each other at different angles; these are then described as a **fault system**. The slope of the fault is known as the *dip. Where rocks have moved down the dip there is a **normal fault**; where rocks have moved up the dip, there is a **reverse fault**. A **thrust fault** is a reverse fault where the angle of dip is very shallow and an **overthrust fault** has a nearly horizontal dip. A **fault plane** is the surface against which the movement takes place. A **tear fault** is where movement along the fault plane is lateral. This latter type of fault may be termed a **strike-slip fault**. Regions, such as the Harz of Germany, that are split by faults into upland *horsts or depressed *rift valleys are said to be **block faulted**.

(⊕) SEE WEB LINKS

• The USGS Visual Glossary provides definitions and outstanding photographs.

Fault

fault block A section of country rock demarcated by faults, which has usually been affected by *tectonic movement. See Young et al. (2002) *Basin Res.* 4, 1 on the Hammam Faraun Fault Block, Egypt.

fault breccia A zone of angular rock fragments, located along a fault-line, and formed by the grinding action associated with movement either side of the fault. Killick (2003) *S. African J. Geol.* **106** outlines a system of fault rock classification.

fault scarp A steep slope resulting from the movement of rock strata down the *dip of a normal fault; Locke et al. (1992) *Bull. Seismological Soc. America* **82** describe the Teton Range fault scarps. A **fault-line scarp** is formed when faulting juxtaposes stronger and weaker rocks, and the latter are eroded.

faunal realms The simplest groupings of the world's animals. The *holarctic* realm covers the *nearctic* (most of North America, plus Greenland) and the *palearctic* (extra-tropical Asia, Europe, and North Africa). The *neotropical* realm covers Central and South America; the *Ethiopian* Africa south of the Sahara and Arabia. The *oriental* realm is tropical Asia, with an ill-defined boundary between it and the *Australian* realm (New Zealand, Australia, Oceania, and some of South-East Asia). See Smith (2004) *Complexity* **10**, 2 for an evolutionary understanding of faunal realms, albeit in very ornate language.

fecundity The potential of a woman/women to bear live children. Fecundity in a population is, of course, closely linked to the proportion of women of childbearing age. See Bhalotra and van Soest (2006) *IZA Dis. Paper* **2163** on fertility and neonatal mortality in India.

federalism A two-tier system of government. The higher, central government is usually concerned with matters which affect the whole nation such as defence and foreign policy, while a lower, regional authority generally takes responsibility for local concerns such as education, housing, and planning, although, of course, the division of responsibilities varies from case to case. Cutler (2004) *J. Federalism* **34**, 2 argues that federalism may reduce voters' ability to hold their governments accountable.

Federations were designed to preserve regional characteristics within a united nation, or to constrain *nationalist elements. Thus, the Nigerian federation was substantially amended after the 1967 Biafran war; see Philips (2005) *African Studs Rev.* **48**, 2.

feedback The response within a *system to an action or process. **Negative feedback** causes the situation to revert to the original; in a river system, erosion is a negative feedback mechanism that restores stream stability by lowering channel gradient and increasing bed material size (Nunally (1985) *Env. Manage.* **9**, 5). **Positive feedback** causes further change: Perry and Enright (2002) *J. Biogeog.* **29** note that positive feedback loops between *fire and vegetation have degraded tropical and humid forest ecosystems.

felsenmeer A surface of broken rock fragments found in *periglacial environments. See Ebert and Kleman (2004) *Geomorph.* **62**, 3–4 on circular moraines on felsenmeer.

feminist geography Feminist geographies are concerned with the relationships between gender relations and space—how space impacts upon gender relations, and how gender relations express themselves spatially. The defining characteristic is a concern with *patriarchy—feminist geographies question the patriarchal and hierarchical assumptions on which geography is based, and emphasize the oppression of women and the gender inequality between men and women. G.-W. Falah and C. Nagel (2005) examine gender relations as they are mediated by Islamic practices: 'Muslim women are not only subject to patriarchal forces within their own communities and societies, but are also subjected to patriarchal structures of the west.' Gender can be understood as an organizing principle of societies; a social relation that shapes the forms, functions, structures, and governance of cities (Bondi (2006) *estudos feministas/études feministes*). Women's subjectivities, relationships, and symbolic productions occur within processes and practices of development, generating multifaceted experiences and diverse challenges to exclusion (Radcliffe (2006) *PHG* **30**, 4). M. Domosh and J. Seager (2001) discuss the gendering of space, and the ways that geographies restrict women's access to, and movements within, space. J. K. Gibson-Graham (2006) and Pratt and Yeoh (2003) *Gender, Place & Cult.* **10** provide accounts of the entanglement of gender in the global workings of the economy; Nagar et al. (2002) *Econ. Geog.* **78** examine the 'double marginalization' of subjects and space characteristic of dominant accounts of globalization: 'Women are sidelined, as is gender analysis more broadly, and southern countries are positioned as the feminized other to advanced economies.' Hyndman (2003) *ACME* **2**, 1 argues that feminist geopolitics 'provides more accountable, embodied ways of seeing and understanding the intersection of power and space'; that it 'refers to an analytic that is contingent on context, place, and time, rather than a new theory of geopolitics or a new ordering of space' (Hyndman (2007) *Prof. Geogr.* **59**, 1). See also Sharp (2007) *PHG* **31**, 3.

L. Staeheli and E. Kofman, eds (2004) argue that a **feminist political geography** recognizes a different outline of the political from other forms of *political geography: 'the political is not just about differences—either between people or between perspectives; it is also about the webs of power and social relationships that are the basis of connections.'

Dias and Blecha (2007) *Prof. Geogr.* **59**, 1 believe that feminist geographies should be seen not solely as a separate sub-discipline in Geography, 'but as a critical perspective useful to all subdisciplines'; Longhurst (2001) *PHG* **25**, 4 claims that feminist geography can be seen as a 'strong protagonist in a complex and mature politics of geographical knowledge'. L. Nelson and J. Seager (2004) offer a useful, concise, and densely referenced overview of the history of feminist geography.

feminization The social process which has resulted in certain occupations, such as nursing, being regarded as women's work. Watson and Stratford (2008) *Soc. & Cult. Geog.* **9**, 4 suggest that the feminization of HIV/AIDS invokes an ethic of care for women and their dependants in situations where risk means both hazard and vulnerability.

This term is also used to describe the increasing presence and influence of women in previously male-dominated occupations or institutions (see D. Perrons 2004). Gallo (2006) *Glob. Netwks* **6**, 4 explores the relationship between a 'feminization of migration' and the construction of masculine identities among migrants from Kerala.

ferrallitization The effect on a soil of strong, tropical *leaching and intense weathering. Organic matter is rapidly destroyed (so there is little *humus), and bases and silica are leached. Sesquioxides in the B2 *horizon form a hardpan that may be revealed by the erosion of the upper horizons. In *US soil classification, ferrallitic soils are oxisols.

Ferrel cell An *atmospheric cell lying between the Polar cell and the Hadley cell. It transfers warm air to high latitudes and shifts cold air back to the subtropics, where it is warmed. The Ferrel cell is a consequence of the stationary and transient eddy circulations in the mid-latitudes. See Salustri and Stone (1983) *J. Atmos. Scis* **40**, 5.

ferricrete A soil *horizon made from the *cementation of iron oxides at or near the land surface. See Phillips (2000) *Geografiska A.* **82**, 1 on groundwater and ferricrete formation.

ferruginous Of or containing iron or iron rust. A **ferruginous soil** is a very deep, *zonal soil found in warm temperate climates without a dry season, or in tropical *savannas/bushlands. The A *horizon is dark red-brown with a weak crumbling structure; the B horizon is stained red by ferruginous gravel. In *US soil classification, a ferruginous soil is an ultisol. See B. Warkentin (2006).

fertility The level of childbearing; in an individual, but more often in a society or nation. Crude *birth rate is the simplest measure of fertility, but does not relate the number of births to the number of women of childbearing age, while the **general fertility rate**, or **fertility ratio**, shows the number of births in a year per 1000 women of reproductive age (generally 15–45, sometimes 15–49). **Cohort fertility rates** show the number of births to women grouped according to either their year of birth or their year of marriage. The **total fertility rate** is the average number of children that would be born per woman, if women experienced the age-specific fertility rates of the year in question throughout their childbearing lifespan. Globally, fertility rates vary widely; in 2006, the rate for the UK was 1.84; for Malawi, 5.7. Abernethy (1999) *Pop. & Env.* **21**, 2 argues for the **fertility opportunity hypothesis**—that humans are alert to environmental signs that indicate whether conditions for childbearing and nurture are more or less optimal.

There are very strong global correlations between fertility rates and per capita *GNP, fertility rates and women's education, and fertility and infant mortality rates (Dasgupta in K.-G. Mäler and J. Vincent, eds 2001). Abernethy (2004) *AAAS Annual Meeting* proposes a link between energy and fertility. Fertility has been declining in industrial societies since the late nineteenth century; a decline which preceded easily available artificial contraception, and while western Europe witnessed a fractional increase in fertility in the late 1970s and early 1980s, fertility rates have declined consistently since then. The United Nations (*World Population Prospects: The 2006 Revision*) make the following predictions:

REGION	AVERAGE NUMBER OF CHILDREN PER WOMAN, 2005–10
World	2.83
More developed regions	1.60
Less developed regions	2.75
Least developed regions	4.63
Africa	4.67
Asia	2.34
Latin America and Caribbean	2.37
Europe	1.45
North America	2.00

fetch The distance that a sea wave has travelled, from its initiation to the coast where it breaks. The fetch controls the energy and height of a wave, and the longest fetch, and hence the dominant wave direction, will affect beach orientation and the direction of *longshore drift. See Stephenson and Brander (2004) *PPG* **28** on the effect of fetch on a beach, and Davidson-Arnott et al. (2005) *Geomorph.* **68**, 1-2 on sediment transport and fetch.

feudalism A system, common in Europe in the Middle Ages, where access to farm land was gained by service to the owner: the **feudal lord**. Initially, no money was involved in transactions between the serf and the lord, although the payment of cash in lieu of service became common in the later Middle Ages; see T. Aston, ed (1987). In 2006, the British island of Sark, part of the territory of Guernsey, voted to abandon feudalism.

fiard (fjärd) An inlet of the sea with low banks on either side, common along the Gulf of Finland, and formed by the post-glacial drowning of the Fenno-Scandian shield.

field capacity The volume of water which is the maximum that a soil can hold in its *pores after excess water has been drained away. Field capacity generally increases with finer soil textures, but Gabrielle and Bories (1999) *Transport in Porous Media* **35**, 1 note that field capacity hardly applies for sandy soil, because of its high hydraulic conductivity.

field system The layout and use of fields. The extent and use of fields varies with the natural environment, the nature of the crops and livestock produced, and aspects of the culture of the farming community such as inheritance rights and available technology. See M. Aston (1983) for medieval field systems in Britain; M. Widgren and J. Sutton, eds (1999) on the Engaruka field system, Tanzania; Xiong (1999) *Toung Pao* **85**, 4-5 on the highly political equal-field system in China and Japan; and McCoy (2005) *J. Polynesian Sociol.* **116**, 4.

filière *See* GLOBAL COMMODITY CHAIN.

filtering down The movement of progressively poorer people into housing stock. It is suggested that when the rich move away from the city to newly built houses, the next social and occupational class moves in. While there are reservations to this theory (not only the rich get new homes, many higher-status housing areas withstand infiltration, and the well-off *gentrify run-down areas), Harris and Wahba (2002) *Int. J. Urb. & Reg. Res.* **26**, 1 illustrate filtering in low-income settlements in cities in the developing world.

financial exclusion The disbarring of the disadvantaged and the poor from financial services. Financially excluded people typically: lack a bank account; rely on alternative forms of credit (such as doorstep lenders and pawnbrokers); and lack key financial products such as savings, insurance, and pensions. An estimated 8% of UK households (around 2.8 million adults) lacked a bank account of any kind during 2002-3 (HM Treasury 2004).

The socio-spatial characteristics commonly associated with financial exclusion include low-income households, single- and lone-parent households, social housing tenants, and residence within a deprived urban area (E. Kempson and C. Whyley 1999). 'The concept of exclusion is problematic, not only because of its predominant association with economic (labour market) participation but also because of its highly moral undertones' (Midgley (2005) *Area* **37**, 3). Financial exclusion often reinforces other aspects of social exclusion: see Chakravarty (2006) *Reg. Studs* **40**. Pollard and Samers (2007) *TIBG* **32**, 3 observe that the 'excluded' may also comprise a whole range of individuals who prefer, on religious, moral, or other grounds, not to engage with Western financial institutions. See Li et al. (2001) *Env. & Plan. A* **33** on ethnic minorities, banks, and overcoming financial exclusion in Los Angeles.

financial services, geography of Mergers, acquisitions, and takeovers, online and phone banking databases of customers' credit rating, and the central processing of cheques have radically altered the spatial distribution of providers of financial services. The 1990s saw a 28% fall in the number of UK bank branches (French et al. (2008) *Antipode* **40**, 1). The spatial implications of these changes have been: disproportionately more bank closures in less affluent areas; a shift in *central business district land use from financial services to restaurants and bars; and the development of call centres. Bristow et al. (2000) *Env. & Plan. A* **32**, 3 note the propensity to site call centres close to existing concentrations of allied activity, with preferences for densely populated areas mediated by needs to maintain employee access and avoid staff turnover problems: 'this has important implications for the spatial division of labour, with call centre growth likely to reinforce existing spatial unevenness in employment in key service activities.' See Sokol (2007) *Growth and Change* **38**, 2 on 'spaces of flows' and 'flows of value' in the geography of financial services.

Interestingly, Argent (2003) *Bureau Rural Sci., Canberra*, on financial services in rural and remote areas of Australia, reports that a diverse range of service modes, from face-to-face branches to virtual banking, has helped to reduce potential *financial exclusion. Wojcik (2007) *Soc. Sci. Res. Network* confirms the leading position of the UK in international finance. London, for example, has four distinct clusters of financial services: a very cohesive City of London cluster featuring banks, insurance, auxiliary finance, law, and recruitment firms; a less cohesive West End cluster, with distinctive cluster zones, such as banks near Mayfair, and advertising in Soho; an incipient general cluster north of the City of London featuring architecture, and business support; and the law cluster that straddles the City of London and the West End (Corporation of London 2003). See Faulconbridge (2004) *Area* **36**, 3 on Frankfurt and London.

fines Small stone particles, whose poor thermal conductivity may aid the development of *patterned ground in *periglacial environments; see Haugland

(2006) *Arct. Antarct. & Alp. Res.* **38**, 1. For fines in desert landforms, see A.
Goudie and N. Middleton (2006).

finger lake A long, narrow lake occupying an over-deepened basin in a
glacial trough.

finger plan The development of new towns or suburbs along routes, road or
rail, radiating from the city centre, since planners see commuting as a fact of
life which should be made as efficient as possible. Copenhagen, for example,
has a finger plan: leisure areas are close to the urbanized areas, which are
located along public transport axes (Østergård (2007) *Ministry Env. + Plan09*).

fiord (fjord) A long, narrow arm of the sea which is the result of the
'drowning' of a glaciated valley; the classic text is J. P. Syvitski et al. (1987).
    Fiords are distinctive because of their great depth, and the overdeepening of
their middle sections which are deeper than the water at the mouth.
Søgnefjord, for example, is 1 200 m deep, but its mouth is only 150 m below sea
level. The shallow bar at the seaward end of the fiord is thought to represent
the spreading and thinning of ice as it was released from its narrow valley and
spread out over the lowland; see Nesje et al. (1992) *Geomorph.* **5**, 6 on
Søgnefjord. Norway's 2004 nomination for its fjords to be added to the
UNESCO World Heritage List makes interesting reading: the geomorphology is
outlined, together with issues of management, developmental pressures, and
environmental pressures. Glaciation may not be the only fiord-forming factor;
see Nesje (op. cit.) on fracture systems and Søgnefjord.

fire ecology The study of the impacts of fire on ecosystems, fire ecology can
be considered a sub-discipline of landscape ecology as both span the
temporal, spatial, and social dimensions of landscapes (Bowman and Franklin
(2005) *PPG* **29**, 2). The primary focus has been to understand the direct and
indirect effects of fire disturbance on plants and, to a lesser extent, animals
(Bowman and Boggs (2005) *PPG* **30**, 2). See also Bowman (2007) *PPG* **32**, 2;
and Roberts (2001) *PPG* **25**, 2 on tropical fire ecology. See Mistry (2003)
J. Biogeog. **30**, 7 for a collection of papers on fire ecology. Kalabokidis et al.
(2007) *Area* **39**, 3 find that natural and/or agricultural vegetation, grazing, and
topography are the major factors in landscape wildfire dynamics.

firm In *neoclassical economics, an ordered, autonomous, rational
independent unit which utilizes the factors of production to produce goods
and services. Revenue is kept high enough to cover costs and to generate
profit. In Marxist analysis, a firm is a complex organization embodied in a logic
of *accumulation, driven to increase profits by avoiding the costs of
community or environmental degradation, exploiting labour, and
manipulating interactions with governments and unions (see Susman and
Schutz (1983) *Econ. Geog.* **59**, 2).
    Economic geographers are interested in the nature of a firm: its dimensions
of gender, sexuality, corporate culture, and cultural embeddedness, and its

locating factors, including *institutional thickness. See Taylor and Asheim (2001) *Econ. Geog.* **77**, 4 on the concept of the firm in economic geography, and Pollard (2003) *Econ. Geog.* **3**, 4 on firm finance.

firn Ice formed when falls of snow fail to melt from one season to another. As further snow accumulates, its weight presses on earlier snow, compacting and melting it to a mass of globular particles of ice with interconnecting air spaces. Where temperatures are around 0 °C, snow can turn to firn within five years, but the process takes much longer in very cold conditions. See Parry et al. (2007) *Annals Glaciol.* **46** on firn within the Greenland ice sheet. The **firn line** is the line, close to the equilibrium line, at which firn forms, and varies with *aspect (Menzies (1951) *J. Glaciol.* **1**, 9). See König et al. (2000) *Procs. EARSeL-SIG-Workshop* on equilibrium and firn-line detection.

First World A misleading term for western Europe, Japan, Australia, New Zealand, and North America; the first areas to *industrialize. Widely used synonyms include 'the developed world', 'the North', 'the more economically developed countries' (MEDCs), and 'the advanced economies'.

fish farming The rearing of fish in pools or tanks. In China, manuring has traditionally taken place by raising ducks and allowing their droppings to fall into the water. In a somewhat left-field approach to the topic, Tejeda and Townsend (2006) *Gender, Tech. & Dev.* **10**, 1 discuss the different meanings of fish farming for different family members as a key issue in understanding relative successes and failures in fish farming; look, too, for Ho's atmospheric account of the Chinese dyke-pond system (2006, *Institute of Science in Society*).

fissure eruption A volcanic eruption where *lava—usually very fluid and basic—wells up through fissures in the earth's crust and spreads over a large area. See Walker (1995) *Geol. Soc. Memoirs* **16** on the Antrim plateau, Northern Ireland; see also Mege (2004) *J. Volcan. & Geothermal Res.* **131**, 1 on fissure eruptions and dyke fracture length.

Flandrian The present *interglacial, during which there has been a global rise in sea level—the **Flandrian transgression**, caused by the melting of *ice sheets and *glaciers (Smith et al. (2006) *ESPL* **8**, 5). The Flandrian is divided into the *Pre-Boreal, *Boreal, and *Atlantic.

flashy In hydrology, applied to a natural watercourse which responds rapidly to a storm event, hence **flash flood**; a very sudden, brief, and dramatic flood event—rising and falling limbs are steep, and the period of peak flow is short. Flash floods are major hazards in deserts, for, although rainfall is rare, the hard-baked ground may be impermeable, and there is little or no vegetation to intercept the rainfall or slow the floodwater. For the prediction of flash floods, see Mukhopadhyay et al. (2003) *Hydrol. Procs* **17**, 4 and Foody (2004) *J. Hydrol.* **292**, 1–4. Flash flooding is the major process in the formation of *wadis—which, incidentally, should therefore be avoided as camp sites. See Chhopel

(2006) *Procs Int. Workshop Flash Flood Forecasting* on flash floods and glacial lake outbursts, and Chan (1997) *Disaster Prevention & Management* **6**, 2 on urbanization and flash flooding in Kuala Lumpur; see also Gupta in O. Hassan et al., eds (2000).

(((●))) SEE WEB LINKS

• USGS description and depiction of effects of flash flooding precipitation on desert geomorphology during El Niño years.

flexible accumulation The use of innovative industrial technologies, adaptable inter-firm relations, variable organizational structures, and flexible consumption, in response to *competition from *newly industrializing and *less developed countries, and to the saturation and fragmentation of markets within *more economically developed countries. Wong (2004) *Land Use Policy* **21**, 1 argues that digitization has created new spatial dynamics underlain by flexible specialization and flexible accumulation. 'Flexible accumulation under neoliberal globalisation produces a highly labile space-time most fully achieved through the spatially and temporally unmoored labouring body of the unauthorised immigrant' (Smith and Winders (2008) *TIBG* **33**, 1).

flexible spaces Space which admits of, or is revealed as having, multiple uses and meanings, which overlap in time and space. Crang (2000) *City* **4**, 3) outlines the case for seeing space as fluid, folded in complex dimensions, and eventful. Wilson (*AAG Annual Meeting* 2007) contends that flexible spaces are thoroughly integrated into global circuits of capital, while Hsu and Lin (*AAG Annual Meeting* 2007) demonstrate the contradictory and conflictual nature of the socio-spatial relations which underlie the production of flexible space, questioning the possibility of successfully creating such spaces. ICT and the developing knowledge-based economy are driving forces for a more flexible space. Thus, cyberspace creates and sustains an increasingly ambient virtual space 'that is still flexible enough for individuals and communities to appropriate their share—especially as the opportunities to acquire technologies and the skills to develop and use them broaden' (R. Murthy (2006) *First Monday, Spec. Issue* **4**).

flexible specialization A strategy at company level of permanent innovation, and accommodation to, rather than control of, ceaseless change. This strategy is based on innovation, multi-use equipment, flexible, skilled workers, or on subcontracting to major firms in sweatshops and homeworking (S. Sassen 2006). Locke and Romis (2007) *MIT Sloan Manag. Rev.* **48**, 2 show that suppliers that adopt flexible specialization, through multi-skilled workers operating in groups, working with incentive bonuses and empowered to stop the production line to ensure quality, had better outcomes for workers—in terms of wages and working conditions—and for firms—in terms of productivity gains, efficiency, and compliance. See also Essletzbichler (2003) *Reg. Studs* **37**, 8. M. Dunford and L. Greco (2006) criticize theorists who see flexible specialization as a replicable alternative to the mass production system and a way out of the crisis of Fordism. Press (2008) *J. Econ. Geog.* **8**, 4 finds that instability for firms is substantially higher under flexible specialization.

flocculation The process whereby very small soil particles, usually clays, aggregate to form *crumbs. Flocculation plays an important role in the transport and behaviour of fine sediment in aquatic systems—see Dyer and Manning (1999) *J. Sea Res.* **41** and Curran et al. (2004) *Marine Geol.* **203**, 1–2. In some subsoils of arid areas, downward translocation of soluble salts leads to the breakdown of these crumbs in the process of **deflocculation**.

flood Floods occur when peak discharge exceeds channel capacity, and this may be brought about naturally by intense *precipitation (Wilson and Rashid (2005) *Canad. Geogr./Géog. canad.* **49**, 1), snow- and ice-melt (Kaczmarek (2003) *Risk Analys.* **23**, 3), the rifting of barriers, such as ice dams; the failure of man-made structures (Cioffi and Gallerana (2003) *Rivers Res. & Applications* **19**, 1); deforestation and urbanization, which reduces *infiltration and *interception (Boardman (2003) *TIBG* **28**); and by land drainage and the straightening and embankment of rivers (Gilvear and Black (1999) *Hydrol. Scis J.* **44**, 3). Macklin and Rumsby (2007, *TIBG* **32**, 2) show that the incidence and size of **extreme floods** have markedly decreased over the last 50 years. Extreme upland flooding appears to be associated with negative *North Atlantic Oscillation index values.

Flood prevention and **flood control measures** include afforestation, the construction of relief channels and reservoirs, water meadow areas in which to divert flood water, and a ban on building in flood-prone environments, such as *flood plains. These measures may increase with per capita income, individual preparedness, and/or experience with flooding, but may decrease with distance from a river, acceptability of flood risk, and provision of environmental information (Zhai et al. (2006) *J. Am. Water Resources Ass.* **42**, 4). In *more economically developed countries, there is a move for flood costs to be borne by the private citizen (Salthouse (2002) *Insurance Res. & Practices* **17**, 1; Penning-Rowsell and Wilson (2006) *TIBG* **31**). See Ono (2002) *Australian Geogr. Studs* **40**, 2 on problems and conflicts arising from flood control in Hokkaido. Chen and Hou (2004) *J. Am. Water Resources Ass.* **40**, 1 develop a multi-criterion, fuzzy recognition model for flood control.

flood frequency analysis (FFA) The calculation of the statistical probability that a flood of a certain magnitude for a given river will occur in a certain period of time. Each flood of the river is recorded and ranked in order of magnitude with the highest rank being assigned to the largest flood. The *return period here is the likely time interval between floods of a given magnitude and can be calculated as:

$$\frac{\text{number of years of river record} + 1}{\text{rank of a given flood}}$$

All methods of FFA are methods of extrapolation, which requires the fitting of a model, and this, in turn, needs an assumption about the underlying distribution generating flood events: 'not only is this not known for extreme hydrological events beyond the observed record, but it is untestable within human timescales' (Kidson and Richards (2005) *PPG* **29**, 3).

flood plain The relatively flat land stretching from either side of a river to the bottom of the valley walls. Flood plains are periodically inundated by the river water; hence the name. Flood plains are often ill drained and marshy, and characteristic *fluvial features include meanders, levées, and ox-bow lakes.

The **flood-plain hydrology system** is dependent on interactions among dynamic, non-linear physical and biological processes linking water, heat, and materials (biota, sediment, plant-growth nutrients), flux, and retention to fluvial landscape change. The key processes driving biogeochemical patterns and cycles include flood-caused scour and sedimentation (cut and fill alluviation), routing of river water and nutrients above and below ground, channel movement (avulsion). Groundwater routing through the flood plain and upwelling back to the surface involves penetration of river water into zones of high hydraulic conductivity (subsurface palaeochannels) within the bed sediments that are created by channel scour and subsequent filling with sorted gravel and cobbles. Strong interactions between short-duration, high stream-power floods, channel and sediment movement, increased roughness due to presence of vegetation and dead wood, and upwelling of groundwater creates a complex, dynamic distribution of resource patches, referred to as the shifting habitat mosaic (*University of Montana Floodplain biocomplexity site*).

See Poole et al. (2002) *Freshwater Biol.* **48**, 10 on geomorphic controls on flood-plain hydrology and connectivity. For flood-plain management, see Evers and Urban (2003) *COFIS abstracts*).

floral (floristic) realms Areas characterized by the indigenous plant species and not by the *biome. Moreira-Muñoz (2007) *J. Biogeog.* **34**, 10 presents a detailed revision of 19th- and 20th-century floralistic realm classifications.

flow 1. A river of rock, earth, and other debris saturated with water. Flows are classified by the size of the particles: debris flow refers to coarse material; earth flow to soil; and mud flow to clay. See Quinta-Ferreira (2007) online *Bull. Engineer. Geol. & Env.* **66**, 1 on the natural and man-made causes of an earth flow in Portugal, 2000; and Dowdeswell et al. (1987) *ESPL* **13**, 8 reconstruct an active debris flow from mudlines. See also Stoffel and Bollschweiler (2007) *Geophys. Res. Abs.* **9** on the dendrogeomorphological reconstruction of a past debris flow.

  2. The movement of goods, people, services, and information along a *network. P. Dicken (2003) observes that flows of goods, people, and information are rising, based on changes in scale, and driven by innovations in

information and communication technologies. Janelle in Hanson et al., eds (2004), illustrates newly emerging systems of flows and nodes, and Hughes (2007) *PHG* **31**, 4 reviews the geographical literature on the flows and networks of knowledgeable capitalism. Jensen (2006) *Mobilities* **1**, 2 offers a 'new' perspective on flows in the contemporary city, and Carr et al. (2005) *J. World Business* **40**, 4 write about **talent flow**—whereby economically valuable individuals migrate between countries. **Spaces of flows** are the networks that bind a world system of cities (M. Castells 1989).

flowage The movement of solids such as ice or rock without fracturing. Butler (2001) *PPG* **25**, 2 argues that debris **flowages**, producing distinct landscape units, should be considered as disturbance corridors. The term is also used in tectonics (Dickinson (2004) *PPP* **213**, 3–4).

fluid stressing The erosion of weak, cohesive rocks by the force of water in a river. The effect of this force depends, among other factors, on the strength of the bed, the percentage of clay in the bed, and the velocity and *turbulence of the water. **Fluid-stressing** is effective for weakly cohesive muds, 'where the current may initiate its own defects' (Gibling and Tandon (1997) *Sedimentol.* **44**). Evorsion is an alternative term.

flume In hydrology, an apparatus placed across a watercourse to measure *discharge.

fluvial Of, or referring to, a river, including the organisms within a river or the landforms produced by river action. **Fluvial deposition** dumps material worked or deposited by rivers; 'at current sea levels, the locus of fluvial deposition is not necessarily the ocean, estuary, or delta, but floodplains in and upstream of the fluvial-estuarine transition zone' (Phillips and Slattery (2006) *PPG* **30**, 4). **Fluvial erosion** is the destruction of bedrock on the sides and bottom of the river; the erosion of channel banks; and the breaking down of rock fragments into smaller fragments; see Powell et al. (2001) *Water Resources Res.* **37** and Wijdenes et al. (2001) *ESPL* **26**. **Fluvial processes** include erosion, flow processes, and sediment and solute transport in rivers; see Pitty, ed. (1979) *Geoabstracts* **127–47** and Reid et al. in A. Brown and T. Quine, eds (1999).

fluvial geomorphology The study of the processes and pressures operating on river systems. Changes in the independent variables of *discharge, *sediment *load supplied to reach, and valley slope, give rise to adjustments in the dependent variables of sediment load and particle size, hydraulic characteristics, and morphologies, all of which interact with each other. Dollar (2004) *PPG* **8**, 3, provides a progress report.

fluvio-glacial Of, or concerned with, watercourses from melting glaciers. See Price in K. Clayton, ed. (1973) and Gretener and Strömquist (1981) *Geografiska A* **63**.

flysch A soft, easily erodable, series of alternating clays, shales, and sandstones. The Carpathian flysch have been intensively studied: see Margielewski (2006) *Geomorph.* **77**, 1–2.

focal area In geolinguistics, an area of relative uniformity, as indicated by sets of shared linguistic features, which acts as the central area of a *dialect: for example, the central area of industrial Scotland. See J. Chambers et al. (2008). Britain (2005) *Linguistics* **43**, 5 on Estuary English looks interesting.

focus The point of origin of an earthquake. Akira (2000) *J. Geog.* **109**, 6 explains how multi-beam echo sounding and seafloor imaging provides information on an earthquake focus.

fog A cloud of water droplets suspended in the air, limiting visibility to less than 1000 m. Fog forms when a layer of air close to a surface becomes slightly supersaturated and produces a layer of cloud, that is, when vapour-laden air is cooled below *dew point. In **advection fog**, this cooling is brought about as warm, moist air passes over cold sea currents. See Nakanishi and Niino (2006) *Boundary-Layer Met.* **119**, 2, in a highly technical paper, on predicting advection fog.

   Radiation fog forms during cloudless autumn nights when strong *terrestrial radiation causes ground temperatures to fall. Moist air is chilled by contact with the ground surface. See Sachweh and Koepke (1995) *Geophys. Res. Lett.* **22**, 9 on radiation fog and urban climate, and Underwood et al. (2004) *J. Appl. Met.* **43**, 2 on radiation fog in California. Where cold air streams across warm waters, **steam fog** forms. This is common when relatively warm surface air over lakes in *frost hollows convects into the cold *katabatic airflow above it, and is also the mechanism behind *Arctic sea smoke; see Walker (2003) *Weather* **58**, 5 on radiation and steam fog. **Frontal fog** forms when fine rain falling at a warm front is chilled to dew point as it falls through cold air at ground level. See also Roach (1995) *Weather* **50** on land fog.

föhn When moist air rises over a mountain barrier, it cools at the slow saturated adiabatic lapse rate; precipitation is common. Once past the mountains, the air, now much drier, descends, warming at the dry adiabatic lapse rate, higher than the saturated rate by some 3 °C/1000 m. A dry, warm, gusty wind, which can reach gale force, results. In summer, desiccation brings a serious risk of bush fires; in winter, snow melt can be rapid. See T. McKnight and D. Hess (2000) on föhn/Chinook winds, and Chan (2005) *Croatian Met. J.* **40** on föhn winds in Hong Kong.

fold A buckled, bent, or contorted rock. Folds result from complex processes including fracture, sliding, *shearing, and *flowage. Delcaillau et al. (2006) *Geomorph.* **76**, 3–4 report on recent fold growth in India. An arch-like upfold is an **anticline**, a downfold is a **syncline**. A complex anticline is an **anticlinorium** (see J. Conley 1973 on the Blue Ridge anticlinorium) and a complex syncline is a **synclinorium** (see Röhlich (2007) *Bull. Geosci.* **82**, 2 on the Prague

Anticline   Syncline   Anticlinorium   Overturned fold   Synclinorium

Recumbent fold

Fold

synclinorium). In an **overturned fold** the upper limb of the syncline and the lower limb of the anticline dip in the same direction. In **recumbent folds** the beds in the lower limb of the anticline and the upper limb of the syncline are upside down. G. Bennison and K. Moseley (2003) have excellent block diagrams of various fold types.

fold mountain An upland area, such as the Alps or Andes, formed by the buckling of the earth's crust. Many fold mountains are associated with destructive or collision margins of *plates. **Young fold mountains**, such as the Caucasus and Alps, were formed by the Alpine orogeny of 65 million years BP, and reach elevations of 10000 m. **Old fold mountains**, such as the Grampian mountains of Scotland, were created by earth movements pre-dating the Alpine orogeny but have been extensively eroded. Some old fold mountains have been uplifted and re-eroded.

food chain Plants (primary *producers) and *consumers at various *trophic levels are interconnected in their diet and in their role as sources of food. A food chain is a linear sequence representing the nutrition of various species from the simplest plant through to top *carnivores, as in: rose→greenfly→ladybird→sparrow→sparrowhawk. (This direct pathway is too simplified, and plants and animals are usually linked together in a *food web.) Schoener (1989) *Ecology* **70** shows that **food-chain length** is governed by the total energy available to a given trophic level. See Post and Takimoto (2007) *Oikos* **116** on variations in food-chain length, and Steinberg et al. (1995) *Procs Nat. Acad. Sci. USA* **92** on the evolutionary consequences of food-chain length.

food conversion ratio The ratio of the number of calories of a prey required to produce one calorie for a predator. Many strategies for sustainable agricultural development strive to improve the food conversion ratio while minimizing negative environmental impacts; see Phillips (2005) *Pacific States Marine Fisheries Commission*.

food production, industrialization of The increasing intensification and capitalization of the production, transport, storage, and retailing of foods, especially of 'Westernized' products, associated with *agribusiness and *transnational corporations. 'Very powerful corporations dominate many sectors. Primary producers are locked into tight specifications and contracts. Consumers may benefit from cheaper food but there are quality implications and health externalities. As consumer confidence has been shaken, new quality agencies have been created. Tensions have emerged about the state's role as facilitator of industrial efficiencies. Food policy is thus torn between the pursuit of productivity and reduced prices and the demand for higher quality, with implications for both producers and consumers in the developing world' (Lang (2003) *Dev. Policy Rev.* **21**, 5/6).

food security The status of people who live without hunger, fear of hunger, or starvation. 'Food security depends on robust food systems that encompass issues of availability, access and utilization—not merely production alone' (Gregory et al. (2005) *Phil. Trans. R. Soc. B* **360**, 21). 'A key component of these newer definitions of food security is attention to building local capacity to produce and distribute food and control food supplies . . . [and] to keep decision-making power within the community rather than losing it through dependence on external sources of food' (Anderson and Cook in J. Harris 2000). In southern Africa, for example, climate is among the most frequently cited drivers of **food insecurity**, while in the Indo-Gangetic Plain of India, labour and the availability and quality of ground water for irrigation rank higher than the direct effects of climate (Gregory et al., loc. cit.)

Korf and Bauer (2002) *IIED Gatekeeper Series* **106** explain the need to address food availability, access, and utilization simultaneously. Slocum (2006) *Antipode* **38**, 2, in a study of a community food coalition in New York, reveals that 'people of color disproportionately experience food insecurity, lose their farms and face the dangerous work of food processing and agricultural labor'. See Hyman et al. (2005) *Food Policy* **30** on poverty and food security mapping; and P. Porter (2006).

foodshed The flow of food from the area where it is grown into the place where it is consumed; recently, a way of looking at and thinking about local, sustainable food systems: 'a socio-geographic space: human activity embedded in the natural integument of a particular place' (Kloppenburg et al. (1996) *Agric. & Hum. Values* **13**). Starr et al. (2003) *Agric. & Hum. Values* **20** think foodsheds are embedded in the moral economy of a particular place, 'just as watersheds reattach water systems to a natural ecology'.

food web A series of interconnected and overlapping *food chains in an *ecosystem. A small change in the number of species in a food web can have consequences both for community structure and ecosystem processes (Emmerson et al. (2005) *Glob. Change Biol.* **11**, 3).

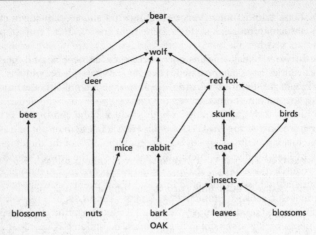

Food web

footloose industry An industry whose location is not influenced strongly by access either to materials or markets, and which can therefore operate within a very wide range of locations. Any form of 'direct line' business, operated almost entirely through telephone and fax lines, would be an example; see T. Friedman (2005) and J. R. Bryson and P. W. Daniels (2004). Such industries are also liberated from locational constraints by **footloose capital**.

Stichele and de Haan (2007) *SOMO*, look at the consequences of investment in garment making—a classic footloose industry. The negative impacts they detail include the problems of companies leaving the African countries where they temporarily had a presence. 'This creates ever more desperate attempts by countries to keep the investors, in an industry that has already cost countries too much, by offering better incentives. For instance, a company . . . was able to use the desperation of . . . Uganda for foreign investment to get the government to provide and invest in buildings and infrastructure, secure loans and credit facilities. The company left the country without repaying any of its debts, leaving behind a destitute workforce that did not even have enough money left to pay the bus fare home. And this happened after the company had already abandoned factories in Tanzania and Kenya.' This paper demands to be read.

Martin and Rogers (1995) *J. Int. Econ.* **39** developed the **footloose capital model**, with the basic assumptions that capital is mobile across regions, but moves without owners; the demand linkage is broken, as there is no market increase; the supply linkage is broken as the mobile factor responds to changes in nominal, not real, incomes; and that labour is mobile between sectors, but not across regions. A small selection of the many papers on this model includes Tafenau (2006) *25th SCORUS Conference*; Dupont and Martin (2006)

J. Econ. Geog. **6**; and Toulemonde (2006) *J. Urb. Econ.* **59**. R. Baldwin et al. (2003) evaluate this model.

foraminifera Usually marine micro-organisms of plankton and *benthic animals with calcite skeletons, found over much of earth's ocean beds. Foraminifera are very sensitive to temperature, and their fossils may be used to reconstruct past environments.

forcing (radiative forcing) The net flux of *radiation in or out of a system. The five major greenhouse gases account for about 97% of the direct radiative forcing by long-lived gases. Interannual variations in the growth rate of radiative forcing due to CO_2 are large, and probably related to natural phenomena such as volcanic eruptions and ENSO events, as well as to *anthropogenic activity (Hoffman et al. (2006) *Tellus B* **58**, 5). Precisely the same forcing can produce hugely divergent evolutions of the climate system given infinitesimally small differences in initial conditions (Washington (2000) *PPG* **24**, 4; E. N. Lorenz 1963).

A **forcing factor** would therefore be a factor that affects forcing (for example, Cuffy and Vimeux (2001) *Nature* **412**), but Tooth (2008) *PPG* **32**, 1 lists climate, tectonics, and human activity as forcing factors in an arid geomorphology, and Burningham (2008) *Geomorph.* **97**, 3–4 writes of sea-level rise as a forcing factor on estuaries. (The literature is littered with references to forcing factors—clearly the buzzwords—but the writers seem to mean major/dominating factors.) Coco and Murray (2007) *Geomorph.* **91**, 3–4 go one further, and refer to a **forcing template** of spatial structures and geological constraints in the evolution of sand patterns.

förde An elongated bay in an area of glacial deposition, formed when glacial tongue basins are drowned by rising water levels. The type locations are northern Germany and Denmark.

Fordism Henry Ford (1863–1947) paid workers high wages in return for intensive work. The term Fordism was coined by Antonio Gramsci (1971) to describe a form of production characterized by an assembly line (conveyor belt) and standardized outputs, linked with the stimulation of demand brought about by low prices, advertising, and credit. 'Under Fordism, it was presumed that growing social equity . . . was the sine qua non of economic growth' (Peck (2002) *J. Econ. Geog.* **2**). Fordism is associated with the spatial separation of the development of the product, at the centre of research and development (usually in a more developed country), and the actual sites of the production of a standardized product (often in a developing country). See Pietrykowski (1995) *Econ. Geog.* **71**, 4. McDowell (2003) *TIBG* **28**, 1 writes that in Britain, the old gender order of industrial Fordism is collapsing and the traditional moral certainties of that period, perhaps most dominant in Britain in the 1950s, are being renegotiated.

foredune A ridge of irregular sand dunes, typically found adjacent to beaches on low-lying coasts, and partially covered with vegetation. Hesp (2002) *Geomorph.* **48**, 1 reviews the initiation, dynamics, geomorphology, and evolution of foredunes.

foreign direct invesment (FDI) The purchase of real assets abroad; through acquiring land, buying existing foreign businesses, or constructing buildings, or mines. **Inward foreign direct investment** is the acquisition of real assets within a country by non-residents.

T. Moran et al. (2005) argue that FDI almost inevitably delivers economic development benefits. 'For host economies, especially developing ones, FDI from other developing countries can add to inflows of other external financial resources, including FDI from developed countries, commercial bank lending, portfolio investment and ODA. For poorer developing countries, it can be significant, accounting for over half of total FDI inflows into several LDCs' (UNCTAD 2006).

However, J. Dunning (1997) characterizes FDI as a bearer of scarce capital and technology that lacks both local linkages and technology transfer. 'Increasing locational "tournaments" to attract FDI . . . may have reduced the benefits to the host countries, as have the increasing skill of the managers of MNEs in making their investments more "footloose"' (Buckley and Ghauri (2004) *J. Int. Bus. Studs* **35**). Phelps (2008) *Reg. Studs* **42**, 4 argues that FDI is currently skewed towards the interests of TNCs, rather than the interests of local and national communities and governments.

foreign exchange The foreign capital earned by a country's exports. Since the currency of many less developed countries is not accepted by international markets, it often becomes necessary to earn foreign exchange in order to buy imports. See Venables (2004) *World Trade Rev.* **3** on the problems of nations that are small, remote, and poor.

forensic geoscience Criminal and civil investigations, utilizing geology, geomorphology, botany, biology, and statistics. See Morgan and Bull (2007) *PPG* **31**, 1.

formal game An exercise in *game theory. Formal gaming has become less popular in geography, but see Padon (1999) *J. Urb. Affairs* **21**, 2 on a game theoretic model of local public good provision.

formal region A region marked by relative uniformity of characteristics, such as the Scottish Highlands. This is a term falling out of use; Symanski and Newman's 1973 paper (*Prof. Geogr.* **25**, 4) probably contributed to its downfall.

formal sector All those types of employment which offer regular wages and hours, which carry with them employment rights, and on which income tax is paid—compare with the *informal sector. However, these two sectors are by no means entirely separate; see Arimah (2001) *Afr. Dev. Rev.* **13**, 1 on informal and formal sector linkages in Nigeria.

fossil fuel Any fuel found underground, buried within sedimentary rock: coal, oil, and natural gas. Doose (2004) *Geochem. Soc. Pub.* **9** argues that world resources of oil and natural gas, including future discoveries, will only be able to meet demand until roughly 2020. Coal will then be the dominant fuel source, projected to meet demand until roughly the year 2070.

fractals A descriptive aspect of geomorphology. Two important types of fractal statistics exist: self-similar fractals and self-affine fractals (Turcotte (2007) *Geomorph.* **91**, 3–4). See also Pelletier (2007) *Geomorph.* **91**, 3–4.

fractus *See* CLOUD CLASSIFICATION.

FRAGSTATS A computer software program designed to compute a wide variety of landscape metrics for categorical map patterns. The original FRAGSTATS future patch updates, and releases can be downloaded from the internet.

(⊕) SEE WEB LINKS

• The Fragstats download page.

Framework Convention on Climatic Change (FCCC) A convention established by the United Nations in 1992 in order to research climatic change, and to implement strategies to slow down, and adapt to, climatic change, replaced by the *Kyoto Protocol.

(⊕) SEE WEB LINKS

• The UN Framework Convention on Climatic Change.

free atmosphere That part of the atmosphere, above about 500 m, which is generally free of the influence of the Earth's surface.

free face (fall face) An outcrop of rock which is too steep for the accumulation of soil and rock debris.

free trade Trade between countries which takes place completely free of restrictions; there are no *tariffs and *quotas. This allows specialization in member states of free trade areas, and should lower costs because both competition and markets are increased. Beine and Colombe (2006) *J. Econ. Geog.* **7**, 1 find that free trade between Canada and the United States clearly leads to more export diversification in the long run, and Keeling (2008) *PHG* **32**, 2 finds new regional mobilities driven by the expansion of trade blocs such as the European Union or the North American Free Trade Area. However, Killick (2001) *Dev. Policy Rev.* **19** discredits the neoliberal argument that free trade and economic liberalization will generate prosperity in the developing world; see also Panagariya (2005) *Foreign Affairs Special edn.* 'Global trade is by no means free trade . . . actions on the part of the USA through the US Trade Representative's Office and in relation to trade-related intellectual property rights continually threaten to undermine the legitimacy of trade regimes' (Glasmeier and Conroy in J. Peck and H. Wai-chung Yeung, eds 2003) 'The free trade doctrine is now performed routinely under the auspices of the World Trade Organization. Nevertheless, it remains a local epistemology, whose truth-like status is kept insulated from rigorous challenge by alternative epistemologies and practices' (Sheppard (2005) *TIBG* **30**).

free trade zone A designated area, often within a *less economically developed country, where normal tariffs and quotas do not apply. It is common for the conditions of employment there to be more repressive; such zones are often mandatorily union free, and working conditions can be harsh (Traub-Werner (2006) *Antipode* **38**, 1). See Wei and Leung (2005) *Growth & Change* **36**, 1 on the Waigaoqiao free trade zone, Shanghai, and Mullings (2004) *J. Econ. Geog.* **4** on Antigua's free trade zone. See Shen (2004) *Prof. Geogr.* **56**, 4 on a shift from cross-border protectionism to cross-border free trade zone in Hong Kong: 'a rearticulation of the state cutting through the geoeconomic and geopolitical time-space envelopes.'

freeze–thaw The weathering of rock said to occur when water, which has penetrated joints and cracks, freezes. In a bright and breezy paper, Murton (2007, *Planet Earth* **4**) finds that the long-held explanation that freezing, expanding water shatters rocks is 'probably not very significant in nature because it requires some pretty unusual conditions. The rock must essentially be water-saturated and frozen from all sides, to prevent the piston-like effect of freezing water driving the remaining liquid water into empty spaces or out of the rock through an unfrozen side or crack. So we need to look for another explanation.'

freezing front The edge of frozen or partially frozen ground. In areas of seasonal freezing with no *permafrost, the freezing front moves downward through the earth. In areas of permafrost, the front can also move upwards: 'during frost heaving of soil when the soil is freezing from the top down, water in the soil pores flows upward to the freezing front because of a gradient in the soil moisture pressure (or tension). This occurs even when the soil pores are not saturated. Ice lenses form and grow at or slightly above the freezing front and cause great uplifting forces' (Henry (2000) *US Army Corps of Engineers*).

freezing nucleus A *nucleus on which a water droplet will freeze to form an ice crystal. Ice nuclei thus formed are much less common than *condensation nuclei, but their effectiveness rises as the temperature falls below 0 °C. They grow by *sublimation if the *ambient air is saturated with respect to water. See Auer (1971) *J. Atmos. Scis* **28**, 2.

freight rate This cost of transporting goods reflects a number of factors besides basic transport costs, such as the nature of the commodity. Non-breakable, non-perishable items, like coal, are carried most cheaply as they can be carried in bulk on open wagons. The more careful the handling required, the more expensive is the freight rate. Sophisticated manufactured goods can bear high freight rates because of their greater value. Raw materials are carried for less so that they can be moved over greater distances. Hummels (1999) U. Chicago, for example, finds for the United States that the ad valorem freight rate is 7.6% for food and live animals, but only 2.25% for machinery. Falling air-freight rates mean that around 30% of US imports now go by air. However, studies based on this data indicate an implicit willingness to pay for time saved at the rate of around 0.5% of the value of goods shipped for each day saved, indicating the massive premium on proximity (Hummels (2001) mimeo, Purdue U.).

Distance is an important factor. Many freight rates are tapered; that is, the rate per tonne-mile or tonne-kilometre drops as the distance increases, but this change in rates is expressed in a series of distance 'bands' so that, on a graph, the relationship between cost over unit distance and distance would appear as a series of downward steps rather than a smooth diagonal line. Plummer and Sheppard (2006) *J. Econ. Geog.* **6**, 5 take to task S. Brackman et al. (2001) for setting transport costs, and thus distance, as an **external quasi-freight rate**.

friction The force which resists the movement of one surface over another. Friction between the surfaces of two mineral grains is related to the hardness of the mineral, the roughness of the surface, and the number and area of the points of contact between the grains. It is of major significance in any study of the movement of sediment since the forces moving the sediment must be greater than the resistance provided by friction. In meteorology, the **frictional force** is the roughness and irregularity of the Earth's surface that reduces wind speeds. The **friction layer**, where this effect is strongest, roughly comprises the lowest 100 m of the atmosphere. See Arya (1985) *J. Appl. Met.* **24**, 9.

friction coefficient The ratio of the force that maintains contact between an object and a surface and the frictional force that resists the motion of the object; the ratio of two forces acting, respectively, perpendicular and parallel to an interface, between two bodies under relative motion (Blau (2001) *Tribology Int.* **34**, 9). Under the same friction coefficient different normal stresses cause different friction levels (Kame and Yamashita (2003) *Geophys. J. Int.* **155**, 3). In a rockfall, the friction coefficient is the strongest factor in determining velocity (Dorren (2003) *PPG* **27**, 1). See also Smith et al. (2007) *PPG* **31**, 4.

friction of distance As the distance from a point increases, the interactions with that point decrease, usually because the time and costs involved increase with distance. Distance need not be reckoned solely in spatial terms; the frictional effect of distance 'on the ground' is far less in a lowland area with good communications than in an upland area of difficult terrain, but has slackened with improvements in transport and communications. 'Human activities tend to organize with respect to geographic location due to the friction of distance and the consequent competition for advantageous location . . . Influencing the relationships between people, place, and activity are the technologies to mitigate the friction of distance' (Miller (2007) *Geog. Compass* **1**, 3). Ellegård and Vilhelmson (2004) *Geografiska A* **86**, 4 argue that geographical immobility and proximity in everyday life indicate 'the continuing and often neglected importance of the friction of distance'.

fringe belt At the edge of a town or built-up area, a zone of varied land use: Victorian hospitals and cemeteries, located beyond the city for reasons of public health; recreation facilities, such as playing fields, riding stables, and golf courses; and utilities, such as water and sewage works. Many of the functions of the fringe belt have been squeezed out from the town centre due to congestion, high land prices, the need for a special site, or disturbances in the central area. Sometimes further urban expansion leapfrogs the fringe belt. Whitehand and Gu (2006) *PHG* **30**, 3 argue that, in relation to the fringe-belt concept, the need for intra-urban centrality is culturally conditioned.

front The border zone between two *air masses which contrast, usually in temperature. A **warm front** marks the leading edge of a sector of warm air; a **cold front** denotes the influx of cold air. Fronts are intensely *baroclinic zones, about 2000 km long and 2000 km wide, moving at around 14 km per day. In *mid-latitude depressions, fronts develop as part of a horizontal wave of warm air enclosed on two sides by cold air. These **frontal wave forms** move from west to east in groups known as **frontal wave families**.

The basic classification into warm fronts, with a slope of 1 in 100, and steeper cold fronts is further divided by the type of air movement at the front. In **ana-fronts**, the warm sector air is rising, and a succession of cloud types and precipitation results. In **kata-fronts**, the warm sector air is descending, clouds are few, and precipitation is reduced to a drizzle.

frontier That part of a country which lies on the limit of the settled area; the term indicates outward expansion into an area previously unsettled by a particular state. 'The frontier arrives and passes, in a historical sense, replacing one set of social relations with another' (Simmons et al. (2007) *AAAG* **97**, 3). Some frontiers have occurred where two nations advance from different directions, leading to boundary disputes. A **settlement frontier** marks the furthest advance of settlement within a state while the **political frontier** is where the limit of the state coincides with the limit of settlement.

Jepson (2006) *Econ. Geog.* **82**, 3 links the **agricultural frontier** with land rents, transportation costs, and the institution of private property; the fringe of market-oriented agriculture and ranching, which advances on subsistence farming or uncultivated wilderness, as the case may be. See also Rindfuss et al. (2007) *AAAG* **97**, 4 on land change in agricultural frontiers. Pallot (2005) *TIBG* **30**, 1 sees the Soviet use of the peripheries as a place of exile as driven by a powerful state intent on expanding the **resource frontier**; see also Hanlon and Halseth (2005) *Canad. Geogr./Géogr. canad.* **49**, 1 on the greying of resource

1.

2.

(a) ana-fronts

(b) kata-fronts

Fronts

communities in northern British Columbia. Davidson (2007) *TIBG* **32**, 4 and
Butler and Lees (2006) *TIBG* **31**, 4 write on **gentrification frontiers**.

frontogenesis The development of *fronts and frontal wave forms.
Frontogenesis occurs in well-defined areas; for example, off the east coast of
North America (see Bell and Bossart, (1989) *QJ Royal Met. Soc.* **115**, 498).

Fronts are less common in the tropics, where contrasts between *air masses are less marked, but see Chen et al. (2007) *Monthly. Weather Rev.* **136**, 1.

frontolysis The change and decay of *fronts. See Parker (1998) *QJ Royal Met. Soc.* **124**, 549.

frost Frozen dew or fog forming at or near ground level. **Black frost** is a thin sheet of frost without the white colour usually associated with frost, occurring when few or no ice crystals are formed, because air in the lower atmosphere is too dry (H. Mavi and G. Tucker 2004). Air below 0 °C is **air frost**. **Radiation frost** (sometimes called *ground frost*) is a ground-level fog formed by nocturnal radiational cooling of a humid air layer so that its relative humidity approaches 100 %, characterized by a low vertical ceiling of the cold air layer. Gołaszewski (2004) *Acta Agrophys.* **3**, 2 concludes that stratification of cold air temperature on spring nights with radiation frost is much stronger than that on nights with advective-radiation frost. **Hoar frost**, or rime, is a thick coating of white ice crystals on vegetation and other surfaces; Karlsson (2001) *Met. Appls.* **8** finds a correlation between the amount of hoar frost, average wind speed, and difference between dew point and road surface temperature. See also Whiteman et al. (2007) *J. Appl. Met. & Climatol.* **46**, 1.

frost cracking When the frozen ground reaches very low temperatures, it contracts, splitting up to form a pattern of polygonal cracks. Anderson (1998) *Arct. & Alp. Res.* **30**, 4 defines the **frost cracking window**—the optimal thermal range for frost cracking—as between -3° and -10° C. See also Grab, *Geografiska A* **89**, 3. Vegetation and accumulations of organic matter may reduce the likelihood of frost cracking by creating a layer of lower thermal conductivity relative to bare soil, provided the area is not water saturated (Murton and Kolstrup (2003) *PPG* **26**, 2).

frost creep The net downslope displacement that occurs when a soil, during a *freeze–thaw cycle, expands perpendicular to the ground surface and settles in a nearly vertical direction.

((⊕)) SEE WEB LINKS

● The US Permafrost Association glossary.

frost heaving The upward dislocation of soil and rocks by the freezing and expansion of soil water. See Matsuoka et al. (2003) *Geomorph* **52**, 1–2 on frost heaving and sorting.

frost hollow A concentration of cold air in a hollow or valley floor when night-time *terrestrial radiation is greatest on valley slopes. Air over these slopes becomes colder, and hence denser, and flows downslope. Temperatures in frost hollows can be tens of degrees colder than the surroundings. See Gustavsson et al. (1998) *J. Appl. Met.* **37**, 6.

frost pocket *See* FROST HOLLOW.

frost shattering The fracturing of rock by the expansionary pressure associated with the freezing of water in planes of weakness or in pore spaces; 'the major weathering process in cold areas' (Hall et al. (2002) *PPG* **26**, 4). See also Coutard and Francou (1989) *Arct. & Alp. Res.* **21**.

frost thrusting The lateral dislocation of soil and rock by the freezing and expansion of water. See Zhu (1996) *Perig. & Permaf. Procs* **7**, 1.

frost wedging *See* FREEZE–THAW.

Froude number The ratio of the velocity (v) of a river to its *celerity where celerity is the product of the acceleration due to gravity (g) and the mean depth of flow (d). The Froude number (Fe) is calculated from the equation:

$$F_e = \frac{v}{\sqrt{gd}}$$

where F_e is less than 1, deeper flow is tranquil. Where F_e exceeds 1, the flow is turbulent. See Lewin and Brewer (2001) *Geomorph.* **40** on Froude numbers and channel patterns.

fugitive species *See* OPPORTUNIST SPECIES.

fumarole A vent in a volcano through which steam and volcanic gases are emitted. See Hildreth et al. (2003) *GSA Bull.* **115**, 6 and Walker et al. (2006) *GSA Spec. Paper* **412**.

fumigation The trapping of pollutants beneath a stable layer or *inversion, and their transport down towards the ground. See Kim et al. (2005) *J. Atmos. Scis* **62**, 6.

functional linkage The link between industries, including information, components, raw materials, finished goods, and transport links. Functional linkages are at the heart of *agglomeration economies. A firm's linkages are classified by their direction of movement: **backward** or **input linkages** are received by the firm—producer services have large backward linkages (Besser (2003) *Rural Sociol.* **68**, 4). **Forward linkages** are supplied by a firm to another undertaking: 'consumers/firms like to locate close to their suppliers as this decreases their living/production costs' (Ottaviano and Robert-Nicoud (2006) *J. Econ. Geog.* **6**, 2). See Crozet (2004) *J. Econ. Geog.* **4** on the forward linkages that relate labour migrations to the geography of production.

functional region A type of region characterized by its function, such as a city-region or a drainage basin. See Wheeler and Brunn (2002) *PHG* **26**, 4. Coombes et al. in D. Herbert and R. Johnston, eds (1982) define functional regions for the UK Census.

fungi Important in soil science, fungi are a group of simple parasitic plants. Fungi are lacking in chlorophyll and therefore cannot photosynthesize. They attack a wide range of organic residues, such as the woody tissue of plants, and are a major element of the soil-forming processes. Kohler et al. (2006) *Soil Use & Manage.* **22**, 3.

gabbro A coarse-grained, basic, igneous rock, mostly made of calcium-rich plagioclase and pyroxene, formed from the crystallization of basic *magma, usually as an *intrusion.

Gaia hypothesis J. Lovelock (1988) argues that planet Earth— *atmosphere, ecosphere, geosphere, and *hydrosphere—is a single ecosystem/organism, regulating itself by feedback between its *abiotic and *biotic components. Lovelock stresses the overriding importance of Gaia, rather than any individual species. Kleidon (2002) *Climatic Change* **52**, 3 concludes that life on Earth tends to enhance carbon uptake.

Gaia tends to equilibrium, but human agency seems to be overriding its regulatory mechanism. 'We have spread thousands of toxic chemicals worldwide, appropriated 40% of the solar energy available for photosynthesis, converted almost all of the easily arable land, dammed most of the rivers, raised the planet sea level, and now ... are close to running out of fresh water. A collateral effect of all this genetic activity is the continuing extinction of wild ecosystems, along with the species that compose them. This also happened to be the only human impact that is irreversible' (E. O. Wilson 2006).

game theory This concerns making rational decisions under uncertain conditions. In geography, game theory is often used to overcome or outwit the environment. A. Dinar et al. (2008) apply game theory to real-life issues in natural resources and the environment. See also Levinson (2005) *Transp. Res. A* **39** on congestion pricing theory.

gamma index (γ) In a network, a measure of connectivity that reflects the relationship between the number of observed links and the number of possible links. The value of gamma is between 0 and 1 where 1 indicates a completely connected network (extremely unlikely in reality). Where e is edges, and v for vertices (nodes),

$$\gamma = \frac{e}{3(v - 2)}$$

See Xie and Levinson (2007) *Geog. Analysis* **39**, 3.

garden city A planned settlement, as conceived by Ebenezer Howard (1850–1928), with low housing densities, and many parks, open spaces, and allotments; the maximum city size to be about 30 000. The first UK garden city was Letchworth, 1903. Howard's ideas were echoed in the construction of *new towns in the UK. A **garden suburb** is a planned suburban development inspired by Howard's ideas. Garden suburbs were built in the late 19th and early 20th centuries, for example London's Hampstead Garden Suburb, 1907. See P. Hall and C. Ward (1999) on Howard's legacy.

gatekeeper An individual—or possibly a group—able to control access to goods and/or services. This may be a body that vets research (see Castree (2006) *PHG* **30**, 6); an exercise by the US government to reduce illegal immigration between Mexico and the USA (Herbert (2004) *Antipode*, **36**, 1); or men as the gatekeepers of gender equality (Connell (2005) *Signs* **30**).

GATT (General Agreement on Tariffs and Trade) An agreement made in 1974 that was intended to reduce tariff barriers, funded by subscriptions provided by the signatory governments. Since voting power reflected the size of the subscription, and the USA was then by far the biggest economy, it initially dominated GATT (Sarre (2007) *Geog. Compass* **1**, 5). It was replaced in 1995 by the *World Trade Organization.

GaWC A study group and network, with its centre in the Geography Department of Loughborough University, that focuses on the external relations of world cities.

GDP (gross domestic product) The total value of the production of goods and services in a nation over one year. Components for a finished product are not taken into account; only the finished articles (but the definition of 'finished product' varies). GDP is an imperfect measurement of a nation's economy because certain forms of production, especially *subsistence production, are not recorded.

GDP per capita in the most globalized developing countries grew 4.2% annually between 1990 and 2001, 1.9% in the rich countries, and only 0.9% in the least globalized developing countries (D. Dollar and A. Kraay 2001).

G8 A group comprising France, the United States, Britain, Germany, Japan, Italy, Canada, and Russia. The G7/8 Summit deals with macroeconomic management, international trade, and relations with developing countries. Questions of East–West economic relations, energy, and terrorism have also been of recurrent concern. Recently, the summit agenda has broadened considerably to include microeconomic issues such as employment and the information highway, transnational issues such as the environment, crime, and drugs, and a host of political-security issues ranging from human rights through regional security to arms control. A mechanism for an international exchange of views between the major energy consuming nations was agreed at the G8 Gleneagles Summit in July 2005 (Hammond (2006) *Nat. Resources Forum* **30**). 'Britain's first world status is dependent on relationships with the third world underpinning its membership of power groups like the G8' (Noxolo (2006) *Sing. J. Trop. Geog.* **27**, 3). See the University of Toronto's G8 information centre; see also D. Della Porta et al. (2006) on the G8 protests in Genoa, 2001.

gelifluction The downslope flow of soil in association with ground ice, occurring in *periglacial environments, where water cannot percolate through the *permafrost. Gengnian et al. (1995) *Permaf. & Perig. Procs* **6**, 3 find that abundant groundwater, fine grained sediments, and slopes 10°–30° favour gelifluction. Harris et al. (2003) *ESPL* **28**, 12 argue that gelifluction is elasto-plastic in nature.

gelifraction Synonymous with *freeze–thaw.

geliturbation Any frost-based movement of the *regolith, including *frost heaving and *gelifluction.

gemeinschaft society A community on a clear-cut piece of territory, socially homogeneous, and tightly knit. Gemeinschaft is dominated by face-to-face contacts, and said to be a feature of villages and small towns, in contrast to *gesellschaft society. (Both concepts are idealized; in real life there are ingredients of both, in varying proportions, in all societies). 'In taking up Tonnies' classical distinction between Gemeinschaft and Gesellschaft [1887], networking practices in the transient word of projects might be differentiated into network communality, network sociality and network connectivity' (Grabher (2004) *Eur. Urb. & Reg. Studs* **11**, 2).

gendarme On an *arête, an abrupt rock pinnacle which has resisted *frost shattering.

gender While it is generally accepted that sex is biologically determined, societies construct appropriate behaviour for each gender, thus producing local **gender cultures** (Jones (1998) *Geoforum*, **29**, 4). Padmanabhan (2007) *Sing. J. Trop. Geog.* **28**, 1 notes that in rural West Africa the **gendered division of labour** extends to labelling certain crops as 'male' or 'female'; with the introduction of new crop varieties, these gendered plant constructions are renegotiated. Guyat (2005) *NZ Geogr.* **61**, 3, writing on bar staff, sees that when 'the dichotomized gender division [was disturbed], it was extremely contradictory and never straightforward because hegemonic performances of masculinities and femininities are so deeply ingrained'.

  Gendered power relations often stem from poverty and entrenched structural inequality (McEwan (2000) *Pol. Geog.* **19**, 2; see also Budlender (2000) *World Dev.* **28**, 70). Bracken and Mawdsley (2004) *Area* **36**, 3 argue that fieldwork is one of the key sites of gender discrimination for women in physical geography. Bankey (2001) *Gender Place Cult.* **8**, 1 observes that 'the fear of the hysterical woman is also a fear that implies that madness or deviance is rendered visible on and through the body . . . a "logic of visibility" . . . has bifurcated men and women's bodies, spheres and spaces from one another'.

The ordering of space is strongly gendered, and may also reinforce gender stereotypes; when spaces (secluded woodlands, dark alleyways, ill-lit multi-storey car parks) make them feel unsafe, women feel vulnerable, and this will further constrain their movements, so fulfilling the stereotype that women are less adventurous than men. 'While patriarchal structures of inequality often result in the spatial entrapment of women, the spatial boundedness of women's lives can be both enabling and constraining' (J. Lin and C. Mele 2005). Hmm.

Gender may be seen as arising from the norms of (hetero)sex; Robinson (2007) *Gender Place Cult.* **14**, 4 finds that the space of international web-brokered marriages 'is one in which women can be seen as active subjects in a transnational space that allows them to act outside, to certain degrees, of kinship-based power'. See Dua (2007) *Gender Place Cult.* **14**, 4 on the importance of gender in constituting the racialized practices of Canada. See also Heynen (2008) *ACME* **7**, 1.

general circulation of the atmosphere The world-scale system of pressure and winds that transports heat from tropical to polar latitudes, thus maintaining the present patterns of world temperatures. This global circulation, influenced by the *Coriolis force, is driven by intense differences in *insolation between the tropics and the poles. Air moves vertically along the *meridians and horizontally with the wind systems, both at ground level and in the upper atmosphere. See B. Haurwitz (2007). See Yu (2000) *Math. Comput. Simul.* **52**, 5–6 for a general circulation model.

generator cell A localized cyclonic cell, caused by fluctuations in the polar front *jet stream, bringing precipitation. See Trapp et al. (2002) *Monthly Weather Rev.* **129**, 3.

genocide A collective, organized attempt systematically to destroy a politically or ethnically defined group. Ideas of 'difference' and 'other' underpin genocide: 'to truly understand how . . . contemporary genocides and camps remain legitimate within liberal democratic regimes, as well as to understand how sovereign states discriminate between an authentic life and a life lacking political value, we need to theorize biological constructions of difference and their connections to citizenship' (Mitchell (2006) *PHG* **30**, 1). Territoriality is a further component: late 20th-century genocides exhibited a marked spatial pattern, attacks being more frequent in the peripheral areas of the supposed 'ethnically pure' unit: eastern Croatia in the former Yugoslavia, and north-west Rwanda, for example.

Wood (2001) *TIBG* **26**, 1 uses the concepts of Lebensraum, territorial nationalism, forced migration, and ethnic cleansing to explain the production of genocide in Rwanda.

'Ethnic cleansing' (the foul euphemism for genocide) is not achieved solely through mass murder, but through forced expulsions and systematic rape—to 'cleanse' the lineage. 'After a decade of displacement, the legacy of ethnic cleansing endures, forming limits to returns and persistent insecurity for returning communities, thus permanently altering Bosnia's human geography and political future' (Dahlman and Ó Tuathail (2005) *Pol. Geog.* **24**, 5). Typically, the overall ploy moves from segregation to isolation and elimination—which explains the vital role that 'safe areas' can play in countering genocide.

The very term genocide has power; see Campbell (2007) *Pol. Geog.* **26**, 4. Romeo Dallaire, UN commander in Rwanda, 1994, thought exposure to the Western media was worth a battalion on the ground (S. Power 2002). Do see J. Sacco (2003).

gentrification The rebuilding, renewing, and *rehabilitation of depressed areas of the inner city as more affluent families seek to live near to the city centre, trading space and quiet for access to the goods and services of the city centre. (Bourne (2000) *PHG* **24**, 153 does not like the term gentrification; 'it is both pompous and now irrelevant. Even "social upgrading" invokes a sense of class-based superiority.')

S. Zukin (1991) describes gentrification as dominated by signature spaces and lifestyles and new middle-class taste formations that get rid of vernacular traditions. Rofe (2004) *J. Austral. Geogr. Studs* **42**, 2 writes that gentrification provides a cleaner and more positive identity. Walks and Maaranen (2008) *Urb. Geog.* **29**, 4 demonstrate that gentrification is followed by declining, rather than improving, levels of social mix, ethnic diversity, and immigrant concentration within affected neighbourhoods: 'at the same time, gentrification is implicated in the growth of neighborhood income polarization and inequality.' See Latham (2003) *Urban Studies* **40**, 9 on *multiculturalism and gentrification.

Lees (2000) *PHG* **24** and D. Ley (1996) both identify three distinct geographies of gentrification. Lees describes them as international, intra-national, and citywide; while Ley uses international, intra-metropolitan, and urban neighbourhood/intra-urban. Smith (2002) *Antipode* **34**, 3 claims that gentrification, 'which initially emerged as a sporadic, quaint and local anomaly in the housing markets of some command-centre cities, is now thoroughly generalized as an urban strategy that takes over from liberal urban policy'. See Phillips (2004) *PHG* **28**, 1 on rural gentrification.

geochronology The chronology of the Earth generated from geological information; the science of the absolute and relative dating of geological formations and events.

geocomputation 'The art and science of solving complex spatial problems with computers. It is focused on the research, development and application of spatial information technologies to address social, economic and environmental problems at human and geographical scales. Geocomputation represents a conscious attempt to move the research agenda back to geographical analysis and modeling, with or without Geographic Information System (GIS) in tow. Its concern is to enrich geography with a toolbox of methods to model and analyze a range of highly complex, often non-deterministic problems' (Xue et al. (2004) *Future Generation Computer Sys.* **20**, 7).

( ) SEE WEB LINKS

• A survey of geocomputational techniques in spatial data analysis.

geodesy The science of the shape and size of the Earth.

geoecology (landscape ecology) An interdisciplinary science which studies the interactions and interrelations in the environment; a view of geo-ecosystems as dynamic entities constantly responding to changes within themselves, their near-surface environments, and external influences, both geographical and cosmic (R. Huggett 1995). See also Shaw and Oldfield (2007) *AAAG* **97**, 1.

Geographical Analysis Machine (GAM) A *geocomputation technique developed by S. Openshaw (1997) and co-workers. 'While conceptually simple, the GAM represents one of the first attempts to address, in a geographical manner, a clearly geographical problem' (Atkinson (2001) *PPG* **25**, 1).

geographic information systems (GIS) Integrated, spatial, data-handling programmes which will collect, store, and retrieve *spatial data from the real world. They are powerful tools in decision-making, as they can incorporate coordinated data. GIS only contain selected data—the properties which investigators have considered relevant—so that many variables will not be fed into the systems. Miller (2007 *Geog. Compass* **1**, 3) calls for a people-based GIS, built on classic time geography, calculating entities and relationships between transport networks, at high resolution and in virtual space. See C. P. Lo and A. Yeung (2007) on GIS and spatial problem-solving.

geographic scale economies 'Efficiencies that arise within a state as a result of the state's absolute abundance of those endowments that confer comparative advantage in the industry' (Smith (1999) *Rev. Int. Econ.* **7**, 1). These arise as a result of production, are external to firms, and generate own-industry effects that dominate cross-industry effects.

geography 'The science of place and space. Geographers ask where things are located on the surface of the earth, why they are located where they are, how places differ from one another, and how people interact with the environment' (AAAG). 'The subject that bridges the natural and human sciences in understanding societies, places and environments' (Royal Geog. Soc. with IBG). 'A fundamental fascination with, and a crucial method for, understanding the way the world works' (A. Bonnett 2008).

'The nature of geography has meant different things for different people in different places' (P. Hubbard 2002). Thus the Royal Dutch Geographical Society claims that geography 'puts knowing into seeing', while Swedish geography 'manifests itself as a form of scholarly praxis that . . . developed through a process of theoretically informed, hermeneutic reflection concerned with the endogen principles, and history, of geographical writing' (Olweg

(2007) *AAAG* **97**, 4). Shaw and Oldfield (2007) *AAAG* **97**, 1 identify landscape science as a Russian geographical tradition. Bruneau (2005) *Sing. J. Trop. Geog.* **26**, 3 suggests that tropicality has had an epistemologically stronger and more institutionalized relationship with francophone geography than with anglophone geography. Bonnett (2003) *Area* **35**, 1 holds that the spirit and purpose of geography should be 'a militant anti-parochialism and a refusal of ethnocentrism'. Abbott (2006) *Sing. J. Trop. Geog.* **27**, 3 argues that whiteness 'as an analytical framework' challenges the legitimacy of overseas geographical fieldwork. Barnes and Farish (2006) *AAAG* **96**, 4 identify new modes of understanding and representing geography in America that were 'closely wedded to broader geopolitical conditions of war and militarism'. 'The collective work of geographers involves persistently questioning what is left out of the consolidation of any particular geographic account, vision, map, or idea [such as the Iraq War]' (Sparke (2007) *AAAG* **97**, 2). 'War is God's way of teaching Americans geography' (Ambrose Bierce).

Geography is one of the structures of how we understand society and space, practised, *inter alia*, by seeing, dwelling, collecting, travelling, mapping, representing, recording, and narrating; an approach which encourages a focus on complexity, multiplicity, and relational thinking. It would take considerable temerity to find a unifying definition throughout the twists and turns that the discipline has taken. Raper and Livingstone (2001) *TIBG* **26**, 2 believe that 'there is scope for an ecumenical view of geographical thinking, and the constitutive nature of space–time in the production of entities is central to this view'. Couper (2007) *TIBG* **32**, 3, reflecting on the division in geography between physical and human, observes that each discipline has its own system, or language game; parallel ideas may be expressed in different ways in different language games.

geological column Also known as the **stratigraphical column**, this is the separation of geological time into *eras and periods, as below:

ERA	PERIOD	EPOCH	END DATE, MILLION YEARS
Quaternary		Holocene (recent)	
		Pleistocene (glacial)	
Tertiary		Pliocene	2
(Cenozoic)		Miocene	
		Oligocene	
		Eocene	
Secondary	Cretaceous		65
(Mesozoic)	Jurassic		135
	Triassic		190
Primary	Permian		225
(Palaeozoic)	Carboniferous		280
	Devonian		345
	Silurian		395
	Ordovician		440
	Cambrian		500
	Precambrian		570

geomagnetism (terrestrial magnetism) The magnetic field of the earth. The axis runs through the *magnetic poles, whose position varies over time; sometimes the north and south magnetic poles switch places (Maus et al. (2005) *Geophys. J. Int.* **161**, 3). The pattern of the magnetic field at any one time will be preserved in contemporary *extrusions of volcanic rock. The study of past magnetic fields, **palaeomagnetism**, can yield information about the creation of new material at the oceanic ridges, and continental drift; see Chauvin et al. (1991) *J. Geophys. Res.* **96**, B2.

geomorphic indices A tool for analysing the landforms and evaluating the degree of tectonic activity in a given area; see Keller (1986) *Panel on Active Tectonics*, National Academy Press, Washington, DC; see also Guarnieri and Pirrotta (2008) *Geomorph.* **95**, 3–4.

geomorphic systems Open systems where the energy of operation comes from outside, and moves across, the system boundary. Any change in energy or mass causes both the processes and their related landforms to adjust in order to maintain equilibrium in the system. The stability of a geomorphic system will depend on whether the changes are linear, non-linear, or threshold-related, and whether the effects of change are locally damped or propagate through the system (Harvey (2007) *Geomorph.* **84**, 3–4).

geomorphic threshold *See* THRESHOLD.

geomorphological threshold A point at which a landform switches, as in the change from an incision stage to an accumulation stage on the slopes of semiarid areas (Gutiérrez-Elorza and Peña-Monné (1998) *Geomorph.* **23**, 2–4). An **intrinsic threshold** indicates possibly independent changes within the system; see Hugenholtz and Wolfe (2005) *Geomorph.* **70**, 1–2 on the activation of a dunefield, while an **extrinsic threshold** indicates a response to change in an external factor; see Brunsden (2001) *Catena* **42**, 2–4.

geomorphology The science concerned with understanding the form of the Earth's land surface and the processes by which it is shaped, both at the present day as well as in the past (British Soc. Geomorphology). Since many landforms cannot fully be explained by present-day geomorphic processes, geomorphologists also consider the impacts of past events on the present-day landscape; the landscape is a physical system with a history (S. Schumm 2003). B. Rhoads and C. Thorne (1996) show geomorphology as having the potential to change character as it evolves through time. The **process geomorphology** approach focuses upon the dynamic elements of geomorphological, hydrological, geological, and ecological systems; see D. Ritter et al. (2001).

Harrison and Dunham (1998) *TIBG* **23**, 4 stress that landscape change is dominated by uncertainties and probabilities, arguing that an idealist approach to geomorphology (that recognizes the primacy of consciousness) can better appreciate the world's unpredictable and probabilistic nature. Lane (2001) *TIBG* **26** observes geomorphology pays attention to the 'way in which the real becomes contingent to form the actual'; it is place-dependent. Harrison (2001) *TIBG* **26**, 3 argues that process geomorphology is essentially reductionist, but that landforms are not amenable to reductionist explanations.

'With humans becoming one of the Earth's major modifiers, human-altered landscape systems are likely to be a prime focus of future research, tackling environmental issues such as deforestation and pollution. In turn, land-surface processes pose a suite of threats to human resources, through the extreme action of soil erosion, slope instability, and river and coastal flooding, and with humans increasingly impinging on the natural environment, hazards research will be of critical importance in most fields of geomorphology' (Stewart in P. Hancock and B. Skinner, eds 2002).

Rhoads (1999) *AAAG* **89**, 4 discusses the epistemological nature of knowledge in physical geography, the way in which this knowledge is connected with processes of enquiry and reasoning, the metaphysical assumptions and moral valuations underlying enquiry and reasoning, and the important commonalities and differences between human and physical geography, and between geography and other domains of knowledge.

geomorphometry (morphometry, terrain modelling, digital terrain analysis) The quantitative analysis of the morphology of landforms. It uses general geomorphometric measures to capture the basic characteristics of landscapes, like slope, convexity, concavity, aspect, linearity, etc., and **specific geomorphometry**: the use of quantitative measures designed for analysis of specific landscape features. See Pike (2000) *PPG* **24**, 1.

Geomorphometry uses input digital elevation models (DEMs, also known as digital land surface models); see Fisher et al. (2004) *TIBG* **29**, 1. See Pike et al., eds in T. Hengl and H. Reuter, eds (2007) for a 'brief guide'.

geopolitics The view that location and the physical environment are important factors in the global power structure; a discursive practice with the 'spatialization of international politics by core powers and hegemonic states' as its object of study (Ó Tuathail and Agnew in S. Daniels and R. Lee 1996). Larsson (2007) *Pol. Geog.* **26**, 2 broadens the scope of geopolitics to include the 'counter-spatialization' of international political and economic relations by weak states to deny or mitigate claims made by more powerful states on natural resources and populations. Hyndman (2001) *Canad. Geogr./Géogr. canad.* **45** argues that immigration law produces a '**geopolitics of mobility**'.

Critical geopolitics investigates 'the use of geographical reasoning in the service of state power' (Dalby (1996) *Pol. Geog.* **15**), exploring the way the production of geopolitical knowledge about the relationship between states both exercises political power and affirms identity. See J. Sharp (2000) and D. Slater (2004). MacDonald (2007) *PHG* **31**, 5 offers a critique of the application of classical geopolitics to outer space in the form of **astropolitics**.

geopolitik A view of *geopolitics developed in 1920s Germany. Individuals are subordinate to the state which must expand with population growth, claiming more territory—*Lebensraum—to fulfil its destiny. Bassin (1987) *Pol. Geog. Qly* **6** argues that the racial doctrines that defined so much of the Nazi variant of fascism were ultimately incompatible with the environmental determinism of geopolitik.

geosophy The study of geographical knowledge; 'the coalescence during the 1960s of a small number of geographers interested in the social, cultural, and historical foundations of spatial knowledge and behaviour' (Edney (2005) *Cartographica* **40**, 1–2); 'the study of the world as people conceive of and imagine it' (John Kirtland Wright, quoted in Keighren (2005) *J. Hist. Geog.* **31**).

geostatistics Traditional sampling methods are unsuited to *spatially autocorrelated data; in geostatistics, autocorrelations are not a problem and sampling is less restrictive. For a detailed description of most geostatistical methods, see E. Isaaks and R. M. Srivastava (1989). A web search for geostatistics software will yield any number of programmes. Before choosing a method, note that, in a user-friendly paper, Kent et al. (2006) *PPG* **30**, 2 warn that it is vital to specify the scale of interest.

geostrophic wind A theoretical wind, occurring when the *pressure gradient force equals the opposing *Coriolis force (assuming straight/nearly straight isobars; when the isobars are strongly curved, the effect of centrifugal force should be added). The net result is a wind blowing parallel to the isobars. Except in low latitudes, where the Coriolis force is minimal, the actual wind has the same direction as the geostrophic wind. See Danard (1989) *Monthly Weather Rev.* **117**, 6 on calculating the surface geostrophic wind in a local area.
   Supergeostrophic flow describes winds faster than the expected geostrophic wind, and occurs at a jet entry; see Cunningham and Keyser (1999) *Qly. JR Meteorol. Soc.* **125**. **Subgeostrophic flow** describes winds slower than the expected geostrophic wind, occurring when air rotates counterclockwise around a low.

geosyncline A thick, rapidly accumulating sediment body, usually parallel to a plate margin, formed within a subsiding sea basin. See Kashfi (1976) *GSA Bull.* **87**, 10.

geothermal flux (geothermal heat) Heat from the earth's interior generated by early gravitational collapse, and later radioactive decay. **Geothermal energy** is a potential energy source in a volcanically active area; see J. Marti and G. Ernst, eds (2005).

geovisualization An abbreviation of geographic visualization, using cartography, cognitive sciences, computer science, data visualization, *GIS, graphical statistics, human–computer interaction design, and information visualization. Geovisualization differs from *cartography in that it affords direct interaction with the content and presentation of what is seen. See Anselin et al. (2006) *Geog. Analys.* **35**, 1.

Edsall (2003) *Prof. Geogr.* **55**, 2 outlines the advantages of geovisualization systems in exploring complex spatiotemporal information; Kwan (2004) *Geografiska B* **86**, 4 discusses geovisualization in time-geographic research; and Gilbert and Masucci (2006) *Trans. GIS* **10**, 5 find that the geovisualization of information in ways relevant to people's experiences may be more important to them than 'geographic accuracy'.

gerrymandering Redrawing constituency boundaries for political gain. It involves 'careful drawing of constituency boundaries by a party so that either it wins a particular seat or, more generally, it wins more seats than its opponents' (Johnston (2002) *Pol. Geog.* **21**, 1). See Altman (2002) *Pol. Geog.* **21**, 1 on evaluating gerrymandering claims.

gesellschaft society A society characterized by aloof relationships; people merely reside in their neighbourhoods, free from social bonds and ties. *See* GEMEINSCHAFT SOCIETY.

Gestalt theory A theory of perception and social understanding that emphasizes the degree to which individuals understand phenomena as wholes, greater than the sum of their individual parts. 'Gestaltists argue for positive change as a process of mutual and personal discovery' (Hobson (2008) *Geog. Compass* **2**, 1). Gestalt is the perception of a pattern or structure as a whole: Barnes (2001) *PHG* **25**, 4 defines a **Gestalt shift** as a change marked by abrupt disjunctions rather than inevitable progress, thus resembling, or equating to, a paradigm shift. Swyngedouw (in R. Lee and J. Wills, eds. 1997), refers to the **gestalt of scale**—geographical configurations as a set of interacting and nested scales; 'nineteenth-century middleclass women altered the prevailing *Gestalt* of scale by altering the structures and practices of social reproduction and consumption' (Marston (2000) *PHG* **24**, 2).

geyser A jet of hot water and steam, usually as a result of the *geothermal heating of underground water, connected to the surface by a narrow outlet.

ghetto A part of a city, not necessarily a slum area, occupied by a minority group (a **gilded ghetto** is simply a more affluent ghetto). Lifestyles within the ghetto differ distinctly from those of the 'host' population and the prejudices of the host confine the sub-group to particular locations. D. Wilson (2007) describes ghettos in terms of deteriorating levels of poverty, increasing demonization in popular discourse, and their complex relationship with the prison system. 'Any proposition of a particular community in isolation, defined in racial, religious, or regional terms . . . shielded from its powers by fixed boundaries, is destined to end up as a kind of ghetto' (M. Hardt and P. Negri 2000). 'To counter the negative effects of the so-called ghetto life, organizations created alternative spaces that they argued would encourage more appropriate social conduct and would persuade homosexuals to be more supportive of these organizations' but political agenda.' (Nash (2006) *Canad.Geogr./Géogr. canad.* **50**, 1).

giant nuclei *See* NUCLEI.

gilgai Undulations of the soil surface, characteristic of many vertisol soils (Fagan and Nanson (2004) *Geomorph.* **60**, 1–2). Gilgai result from the shrinking and wetting of soils. See Hallsworth in H. C. T. Stace et al., eds (1968).

Gini coefficient In a *Lorenz curve, a measure of the difference between a given distribution of some variable, like population or income, and a perfectly even distribution. More simply, it tells us how evenly the variable is spread; the lower the Gini coefficient, the more evenly spread the variable.

GIS *See* GEOGRAPHIC INFORMATION SYSTEMS.

glacial **1.** Of or relating to a glacier.
   2. An extended length of time during which earth's glaciers expanded widely. The most recent glacial epoch, better known as the Pleistocene glacial epoch, previously the Quaternary, covered at least the last 3 000 000 years. A. Penck and E. Brückner (1909) established the classical sequence of four glacial periods in the Alps: the First Glaciation, at its peak some 550 000 years ago, the Second Glaciation, at its peak 450 000 years ago, the Third Glacial Period (Riss) with a climax 185 000 years ago, and the Fourth Glaciation (Wurm) at its height 72 000 years ago. Ice-sheets grew only during parts of these so-called 'glacials', each of which was followed by an *interglacial period.

glacial abrasion The grinding away of bedrock by fragments of rock incorporated in ice. It can be achieved by bodies of subglacial sediment sliding over bedrock, or by individual clasts within ice (Glasser and Bennett (2004) *PPG* **28**, 3). Abrasion is favoured where effective basal pressures are greater than 1 MPa and where there are low sliding velocities. Landforms of glacial abrasion include striae, grooves, micro-crag and tails, bedrock gouges and cracks. Ice ceases to be an effective agent for abrasion when the weight of the ice is thick enough to bring about plastic flow.
   The Boulton model (1996, *J. Glaciol.* **42**) sees abrasion as controlled by effective normal pressure, and the Hallet model (1979, *J. Glaciol.* **23**) argues that abrasion is highest where basal melting is greatest.

glacial breaching The erosion of a breach between two adjacent valleys when a transfluent glacier flows across the *watershed (divide) between them. See Hall and Jarman in S. Lukas et al. (2004).

glacial deposition The laying down of *sediments which have been removed and transported by a *glacier (see Larson and Piotrowski (2005) *J. Sediment. Res.* **75**). The sediments—known as *till, or drift—are deposited when the ice melts; when *ablation is dominant. Glacial deposition is predominant in marginal areas of present and past ice sheets, such as the Eden

Valley, and much of the English Midlands, but also occurs in uplands, especially
in the form of *moraines. See Escutia et al. in Cooper and Raymond et al. (2007)
Online Procs 10th ISAES on meltwater production and high sediment discharge
in the East Antarctic. *See also* DRUMLINS; ERRATICS; KETTLE HOLES.

glacial erosion Glacially eroded landscapes are moulded by abrasion,
*debris entrainment, the 'conveyor belt' transport of *moraine on top of a
glacier, rock fracturing (Iverson (1991) *J. Glac.* **37**), *plucking/quarrying (Hallet
(1996) *Annal. Glaciol.* **22**), *pressure release (Lewis (1954) *J. Glaciol.* **2**), and
the action of subglacial water pressure water (Iverson (1991) *J. Glaciol.* **37**).
Bougamont and Tulaczyk (2002) *Boreas* **32**, 1 see sub-ice-stream sediment
generation and transport as agents of glacial erosion; Glasser and Bennet
(2004) *PPG* **28**, 1 argue that 'all landforms of glacial erosion provide evidence
for the release of subglacial meltwater'. Glacial erosion can modify the bed of
an ice sheet and alter large-scale ice dynamics and mass balance (Oerlemans
(1984) *Zeitschrift* **20**).

The extent and nature of glacial erosion depend on ice characteristics.
*Glen's Flow Law explains glacier behaviour with respect to strain rates and
shear stress (Tarasov and Peltier (2000) *Annal. Glaciol.* **30**, 1). Alley et al. (2003)
Nature **424** suggest that, if a glacier digs too deep, strong thermodynamic
feedbacks prevent further deepening. For the role of basal velocity in glacial
erosion, see Harbor (1992) *GSA Bull.* **104**; for the thermal regime, see Sugden
(2005) *Geomorph.* **67**, 3–4. Other variables are covered by Brook, Kirkbride,
and Brock (2004) *Geograf. Annal. A* **86**, 3 (rock strength), and Augustinus
(1995) *Geomorph.* **14** (rock stress). Paulsen and Wilson (2004) *Glob. & Planet.
Change* **42** consider the role of geologic structure, and Olymo and Johansson
(2002) *ESPL* **27** provide a useful summary of the roles of rock structure,
lithology, and pre-glacial relief. For the response of glaciated landscapes to
rapid rock uplift, see Brocklehurst (2007) *J. Geophys. Res.* **112**.

In glacially eroded lowlands, relief is confused, weathered rock is stripped
away, and *abrasion and *quarrying of the bedrock are active. Drainage is
rambling, as earlier patterns are disrupted by erosion and the deposition of
*moraines. Lakes and ponds abound, and perched blocks and erratics may be
common. See Le Coeur (1988) *Geografiska A* **70**, 4 and Munro and Shaw (1997)
Geology **25**.

(⊕) SEE WEB LINKS

• Glaciers online's thumbnails of glacially eroded landforms.

glacial meltwater The effectiveness of meltwater as an agent of erosion
depends on: the susceptibility of the bedrock involved (particularly its structural
weaknesses or proclivity to chemical attack); the water velocity and the level of
turbulent flow; and the quantity of sediment in transport (Glasser and Bennett
(2004) *PPG* **28**, 3). A **glacial margin channel** is a stream running between the side
of the glacier and the valley sides. Landforms of glacial meltwater erosion include
both subglacial and ice-marginal meltwater channels.

glacial surge The swift and dramatic movement of a glacier, associated with the growth of ice up-glacier to unstable proportions and with severe *crevassing. Murray and Porter (2001, *Quat. Int.* **86**) link the termination of the surge front on the Svalbard surging glacier Bakaninbreen to the boundary between warm ice and the cold-based snout, and propose that surging takes place by bed deformation and/or sliding in the warm based zone. See Russell et al. (2001) *Glob. & Planet. Change* **28**, 1 on glacier surging as a control on the development of proglacial, fluvial landforms and deposits.

glacial trough A deep linear feature, carved into bedrock by glacial erosion, that is channelled along a trough or valley. Troughs may be cut beneath ice sheets by a combination of glacial *abrasion and *quarrying. The effects of glacial abrasion are more obvious on the smoothed and polished trough walls whilst the effects of glacial quarrying—such as roches moutonnées and rock basins—are most evident on the valley floors. Glasser and Bennett (2004) *PPG.* **28**, 3 see troughs as *palimpsest features*, developed over successive glaciations.

Hebdon et al. (1997) *ESPL* **22** calculate glacial erosion rates as between 1 and 2 mm a^{-1} in Scotland; Harbor et al. (1988) *Nature* **333** cite rates of 10^{-3} ma^{-1} landslide material in a single glacial cycle.

Although the cross-sectional form of a glacial trough is traditionally described as 'U-shaped', their true cross-sectional morphology is better described by empirical power-law functions (Yingkui et al. (2001) *J. Glaciol.* **47**) and by second-order polynomials (James (1996) *ESPL* **21**). Furthermore, this cross-sectional morphology varies with lithology (Harbor (1995) *Geomorph.* **14**), rock mass strength (Brook and Tippett (2002) *Scot. J. Geol.* **38**, 1) and the degree of alluvial modification (James op. cit.).

Modelling studies demonstrate that the cross-sectional morphology of glacial troughs tends towards an equilibrium form (Harbor (1992) *GSA Bull.* **104**). Other characteristics of glacial troughs include classic features of glacial erosion such as hanging valleys and divide elimination (Linton in H. E. Wright and W. H. Osburn, eds 1967), whilst the valley patterns associated with these features have been used to quantify the effects of glacial erosion on landscapes (Haynes (1995) *Boreas* **24**).

glacier A mass of ice which may be moving, or has moved, overland: when enough ice has accumulated, a glacier will start to move forwards. A glacier may be seen to be the result of a balance between *accumulation and *ablation. Glaciers are classified by their location (*cirque glacier, expanded-foot glacier, valley glacier, *niche glacier, *piedmont glacier), by their function (diffluent glacier, *outlet glacier), or by their basal temperature (*cold glacier, warm glacier).

glacier budget The balance in a glacier between the input of snow and *firn, that is, *accumulation, and the loss of ice due to melting, evaporation, *sublimation, and *calving, that is, *ablation. A glacier grows where the budget is positive and retreats when it is negative.

glacier drainage systems In a warm glacier, meltwater will drain into the ice through a network of veins, *moulins, channels, and conduits, which can connect to an *englacial system, discharging at the snout, or to a *subglacial system of channels, linked cavities, subglacial lakes, and water films. These englacial and subglacial systems may become connected during the melt season; in cold glaciers, however, the superglacial and subglacial drainage systems are always separate. Hydrological features related to the evolution of the glacier drainage system include supraglacial routing, moulin types, and chasms (Yde and Knudsen (2005) *Geografiska A* **87**, 3).

glacier hydrology The study of the deposition, distribution, and depletion of snow; the evaluation of hydrologic water balance of basins where appreciable snow occurs; the physics of snow and glacier melt; the storage and transit of liquid water in snow and glaciers, and of possible methods for estimating the rate of stream flow, and the volumes of run-off in basins where snow affects those quantities. See Cooke et al. (2006) *PPG* **30**, 5 on glacier hydrology and supercooling; Hock in U. Huber et al. (2005) on melt modelling; and Hock (2005) *PPG* **29**, 3 for a review of glacier melt.

glacier response time The duration of the morphometric adjustments following change in net mass balance. 'This is a vague concept, because the reaction time may depend on the glacier history in an untransparent way. A distinction should be made between response times for glacier *volume* and glacier *length*. When the specific balance changes, the volume will respond immediately. For glacier length it may take a bit more time, in particular when the changes are more apparent at higher parts. 'Numerical experimentation has indeed shown that response times for volume are typically 30% shorter than response times for length' (J. Oerlemans 2001).

glacier retreat (glacier recession) There is no lack of evidence for glacier retreat: for example, Kulkarni et al. (2007) *Current Science* **92**, 1 report that nearly 500 glaciers in the Chenab, Parabati, and Baspa basins have shrunk more than 20% over a 40-year period; Joughin et al. (2006) *Nature* **432** report on 'catastrophic retreat' in west Greenland; while Owen and Lehmkuhl (2005) *Boreas* **34** comment on Tibet. Barry (2006) *PPG* **30**, 1 provides a useful summary table. See also de Woul et al. (2007) *Geophys. Res Abs.* **9**.

There are reports that some glaciers are extending: Howat et al. (2007) *Climate Dynam.* **28**, 1 report that the Mt Shasta, Calif., glaciers have grown by 30% since 1960.

glacier sliding *See* BASAL SLIPPING.

glaciohydraulic cooling A process that allows water to remain liquid below 0 °C. It is associated with basal water flow through subglacial overdeepenings; as the subglacial melting point changes in response to changing pressure, water may become supercooled, and freeze, but if the adverse bed slope is steep enough, the heat generated will not be enough to match the changed pressure melting point, and it will become supercooled. For more details, see Hooke (1991) *Geol. Soc. Amer. Bull.* **103** and Alley et al. 1998 *J. Glaciol.* **44**). Alley et al. (2003) *Nature* **424** view supercooling as a key link between glacier dynamics, basal ice formation, and subglacial landform evolution. For field evidence of glaciohydraulic cooling, see Hooke and Pohjola (1994) *J. Glaciol.* **40** and Lawson et al. (1998) *J. Glaciol.* **44**.

Glaciohydraulic cooling is associated with subglacial deformation (Echelmeyer and Zhongxiang (1987) *J. Glaciol.* **33**), and ice creep at crystal boundaries (Wolff and Mulvaney (1988) *Ann. Glaciol.* **11**). Cooke et al. (2006) *PPG* **30**, 5 review recent work on glacial supercooling. Knight (2006) *PPG* **30**, 5 suggests that the significance of glaciohydraulic supercooling may have been overestimated.

Glen's Flow Law As temperature increases, the Glen's Flow Law parameter increases; as temperature decreases, towards the top of a glacier, Glen's Flow Law will decrease, and the glacier flow will slow. In using this law to model real ice masses, remember that many simplifications may have been assumed (B. Hubbard and N. Glasser 2005).

gley soils Soils with mottled grey/yellow patches. **Gleying** occurs in waterlogged soils since oxygen supply is limited there. Under these conditions, *anaerobic micro-organisms extract oxygen from chemical compounds, particularly reducing ferric oxide to ferrous oxide. See Zwerman (1973) *Soil Sci. Soc. Am. J.* **37**.

global Smith (2001) *Transnational Urbanism* notes that each member of the global–local binary has been allocated a distinctive set of attributes; global is a synonym for dynamic, thrusting, open, rational cosmopolitan, dominant, masculinized, and economic, while 'local' suggests authentic, closed, static, nostalgic, defensive but defenceless, feminized, and cultural. However, Ley (2004) *TIBG* **29**, 2 argues that 'the global is also the local'.

global circulation *See* GENERAL CIRCULATION OF THE ATMOSPHERE.

global city A central location for the leading industries of globalization; specifically financial and specialist business services, with their own forms of production as 'control centres' of the world economy (Knox in P. Knox and P. Taylor, eds 1995; S. Stassen 2000; see also Cochrane and Passmore (2001) *Area* **33**, 4). Examples include London, New York, Paris, and Tokyo. Scholz (2000) *Zeitschrift* **1** calls these *acting global cities*, which differ from *globalized places* that house only branches/offshoots of the decision-making structures concentrated in the acting global cities (see also Brade and Rudolph (2004) *Area* **36**, 1). Transnational elites have long been associated with the global city (Beaverstock (2002) *Geoforum* **33**, 4). Yeung and Olds's 2001 paper for the IRFD World Forum on Habitat labelled Shanghai an 'emerging global city' as opposed to 'hyper global cities' such as New York and London.

Wei and Leung (2005) *Growth & Change* **36**, 1 argue that the literature on global cities is too dependent upon a perspective based on major Western cities. Global cities are marked by large-scale in-migration: Wong et al. (2005) *J. Housing & Built Env.* **20** highlight the significance of a temporary migrant labour force in the moulding of Shanghai into a global city, and Beaverstock (2002) *Geoforum* **33**, 4 observes that *transnational elites have long been associated with the global city. S. Stassen (1991) has linked global cities with increasing social polarization—a thesis rejected by Nørgaard (2003, *Geografiska B* **85**, 2), who prefers 'inequality' to 'social polarization'.

(((⊕))) SEE WEB LINKS

• Yeung and Olds on the global city.

global commodity chain A network of organizations and production processes resulting in a finished commodity (Ponte (2002) *World Dev.* **30**); a chain of nodes from raw material exploitation, primary processing, through different stages of trade, services, and manufacturing processes to final consumption and waste disposal (Brown et al. (2006) *GaWC Res. Bull.* **236**). Some have argued that the more inclusive language of **value chains** should replace the more specific concept of commodity chain.

Smith et al. (2002) *PHG* **26**, 1 identify limitations to the global commodity chain framework: a limited treatment of both the state and the labour process in governing the organization of productive systems; problems of linearity and dualism; and the problematic conception of the geographies of linkages in commodity chains. Leslie and Reimer (1999) *PHG* **23**, 3 regret the lack of a comprehensive treatment of the spatiality of commodity chains. Coe et al. (2004) *TIBG* **29**, 4 argue that, despite the global focus of the commodity chain approach, it still remains preoccupied with the nation state as geographical scale of analysis, with 'surprisingly little to say about regional and subnational processes'.

(((⊕))) SEE WEB LINKS

• Jennifer Bair on commodity chains and value chains.

global energy balance The difference between the total influx of *solar radiation to the earth's surface and the loss of this energy via *terrestrial radiation, evaporation, and the dissipation of sensible heat into the ground.

globalization The increase in the volume, scale, and velocity of social (and environmental) interactions. Globalization is not new, pre-dating *colonialism. Ash (2004) *TIBG* **29**, 2 describes globalization as a politically driven project, led by the US government, the World Bank, the World Trade Organization, and the G8. He adds that the concept of globalization cannot explain anything, but merely 'redescribes processes then enmeshed in discourses'. Urry (2005) *Theory, Cult. & Soc.* **22**, 5 argues that 'the global order is a complex world, unpredictable and irreversible, disorderly but not anarchic'. Jackson (2004) *TIBG* **29**, 2 prefers to

Global energy balance

Incoming solar radiation 342 Wm⁻²

342

Reflected solar radiation 107 Wm⁻²

107

Reflected by clouds, aerosol and atmosphere 77

77

Absorbed by atmosphere 67

Reflected by surface 30

168 Absorbed by surface

Latent heat 78

24 Thermals

78 Evapotranspiration

Outgoing longwave radiation 235 Wm⁻²

235

Atmospheric window 40

165

30

Emitted by atmosphere

Greenhouse gases

350

40

324 Back radiation

324 Absorbed by surface

390 Surface radiation

think of 'globalizing' rather than 'globalized', suggesting that globalization 'might be better thought of as a site of struggle rather than as a foregone conclusion'. The **strong globalization thesis** stresses the primacy of global economic forces over rational/domestic political ones, emphasizing the decline of the social democratic politics and the limitations of national governments; see U. Beck (2000). Tickell and Peck in J. Peck and H. W. Yeung, eds (2003) argue that globalization is not a monolithic phenomenon, but produces its own geography. Peck and Tickell (2002) *Antipode* **34** differentiate neoliberalism from globalization.

Globalization operates on a number of scales; Swyngedouw in G. Clark et al. (2000) refers to the changes in the relationships between geographical scales, which create 'a new scalar migration, especially of the poor'. Taylor (2007) *GaWC Res. Bull.* **238** claims that, generally, globalization is associated with a rescaling argument in which national domination of social practice is dissipating upwards to the global, and downwards to the local. Ash (2004) *TIBG* **29**, 2 believes that globalization is creating a 'new, topologically and hierarchically structured economic space which is substantially different from the hitherto dominant world system based on territorially organized and state-regulated economies'. But globalization impinges differently on different places, and in locally specific ways (see OECD (2001) *Devolution and Globalization*); in Dicken's view (2004, *TIBG* **29**, 1) bounded political places matter—'global changes are manifested most directly at the local level' (P. Dicken 2003). Taylor in W. Dunaway et al., eds (2003) expects globalization to change 'the existing state-centric view' into a city-centric view of the world. 'The global economy thrives on the specialized differences of countries, regions, and cities. But it does need homogenized standards . . . and . . . it also needs standardized built environments' (Sassen (2008) *Urb. Geog.* **29**, 2).

Martin (2004, *TIBG* **29**, 2) believes that access to key resources and assets are crucial to a nation's (or region's) prosperity under globalizations, while Dicken (op. cit.) underlines the importance of the spatial and territorial range and flexibility of the actors involved.

Globalization has many detractors, who claim that global capital privileges profit over local interests and deplore the 'Westernization' of local cultures and what they see as the negation of local identities and autonomies (Amin and Graham (1997) *TIBG* **22**, 4). 'There is the argument that the global market has made us all the same anyway because there are the same burger bars, coffee shops and fashion outlets in every town centre and in every continent . . . Maybe this immense crumpling of space—the vastly increased connectedness that undoubtedly *has* happened—has given us an *illusion* of knowledge—a presumption that can be both dangerous, I would suggest, and potentially imperial . . . This particular version of the shrinking world is not just a description of the world as it is, however correct or incorrect. It is actually part of a project. It is geographical mythmaking, to convince us of globalisation in its current form' (Massey, OU Radio Lecture, November

2006). Others suggest that globalization is a dialectical process; it invades local contexts of action, but new forms of local existence and expression emerge (Bollywood as well as Hollywood; *see* CREOLIZATION).
Escobar (2001) *Pol. Geog.* **20**, 2 cites *hybridization as proof that culture still 'sits in place'.

In order to rein in excesses of neoliberalism and abuses of state regulation, Falk in J. Dunning, ed. (2003) argues for **globalization with a human face**, based on the consent of citizenry, human rights, participation, and accountability; see also Dicken in J. Peck and H. W. Yeung, eds (2003). 'In the political geographies of globalization, three main themes stand out: dominance, governance, and resistance' (Sparke (2004) *PHG* **28**, 6).

global production network (GPN) The global network of firms, institutions, and other economic agents which shapes and is shaped by the fundamental processes of knowledge and wealth creation, enhancement, and exploitation; corporate, collective, and institutional elements of organizational power; and, spatial and network embeddedness (Henderson et al. (2002) *Rev. Int. Polit. Econ.* **9**, 3). 'As manufactured products are becoming more technologically sophisticated, their production has often involved an increasing number of stages. These production stages can be decoupled and farmed out to a network of geographically dispersed locations to take advantage of lower costs . . . International trade then entails flows of goods belonging to a single industry but at different stages of production. This cross-border multi-staged production process has been facilitated immensely by major improvements in transportation, coordination and communication technologies, and the liberalisation of trade and investment regimes' (G. Nandan (2006) Commonwealth of Australia).

Hesse (2006) *Growth & Change* **37**, 4 relates the 'somehow abstract system dynamics of GPNs' to very particular places, while Coe et al. (2004) *TIBG* **29**, 4 highlight the dynamic 'strategic coupling' of global production networks and regional assets. See also Coe et al. (2004) *TIBG* **29**, 4 on global production systems and intersecting notions of power, value, and embeddedness. The entire issue of (2008) *J. Econ. Geog.* **8**, 3 is dedicated to global production networks. See Rutherford and Holmes (2008) *J. Econ. Geog.* **8**, 4 on the state's role in *actor networks in GPNs.

global space 'Global space . . . is not an isotropic surface of sameness' (Jackson (1996) *TIBG* **21**, 3). Doel and Hubbard in A. Mayr et al. (2002) argue that a world city is a 'bounded place plugged into a global space of flows'. 'Neoliberal policies create a global space in which finance capital can range freely in search of ever-increasing profit' (Peet (2007) *5th Int. Conf. Critical Geog.*). Weyland (in A. Oncu and P. Weyland, eds 1997) finds that 'the global (highly paid) corporate and managerial labour force which sustains the "public" multinational business space epitomizing globalism is itself reproduced by the presence of a female "privatized" global space, often shored up by (unpaid) corporate wives as well as (lowly paid) foreign maids'.

G. Ó Tuathail (1996) notes the early modern horizontal organization of space associated with ideas of state sovereignty and the emerging state system; 'such a "chess-board" vision of global space made it possible to authorise and strategise new campaigns of geopolitical pre-eminence' (Hughes (2007) *Geog. Compass* **1**, 5).

global value chain 'Sets of inter-organizational networks clustered around one commodity or product, linking households, enterprises and states to one another in the world-economy' (Gereffi et al. in G. Gereffi and M. Korzeniewicz, eds 1994). They can be considered as the specific industrial, geographical, social, and institutional environment within which firm-level production networks operate (Palpacuer and Parisotto (2003) *Glob. Networks* **3**, 2), or the entire process of bringing a product from its beginnings to its end use: the extraction of raw materials, design, manufacturing, marketing, distribution, warehousing, delivery, retailing, and customer support. The value chain

	GLOBAL COMMODITY CHAINS (GCCS)	GLOBAL VALUE CHAINS (GVCS)	GLOBAL PRODUCTION NETWORKS (GPNS)
Disciplinary background	Economic Sociology	Development Economics	Relational Economic Geography
Object of enquiry	Inter-firm networks in global industries	Sectoral logics of global industries	Global network configurations and regional development
Orienting concepts	Industry structure Governance Organisational learning Industrial upgrading	Value-added chains Governance models Transaction costs Industrial upgrading and rents	Value creation, enhancement, and capture Corporate, collective, and institutional power Societal, network, and territorial embeddedness
Intellectual influences	MNC literature Comparative development	International business Trade economics	GCC/GVC analyses Actor-network theory Varieties of capitalism

Source: developed from Bair (2005), Table 1.

perspective stresses the importance of activities other than production itself, notably design, logistics, and marketing; see Kaplinsky et al. (2002) *World Dev.* **30**. See also Knorringa and Pegler (2006) *Tijdschrift* **97**, 5 and Sturgeon (2005) *Rev. Int. Polit. Econ.* **12**, 1 on the governance of global value chains.

(⊕) SEE WEB LINKS

• Global value chains.

global warming The increase in global temperatures brought about by the increased emission of greenhouse gases into the atmosphere. 'Many of the indicators of climate change are suggesting that measurable change is broadly apparent globally; with eleven of the last twelve years the hottest on record across the globe, ice sheets and glaciers melting at rates exceeding scientific expectations, and trends in ecosystems resulting from responses to environmental change apparent, such as coral bleaching and the migration of species poleward and to higher altitudes' (Bardsley and Bardsley (2007) *Geog. Res.* **45**, 4.

See Cavan et al. (2001) *Bull. AMS* **82** on western North America, Maheras et al. (2004) *Int. J. Climatol.* **24** and Dünkeloh and Jacobeit (2003) *J. Climatol.* **23** on southern Europe, and Smith et al. (2000) *Int. J. Climatol.* **20** on South Africa, Chile, and south-west Western Australia; see also James J. McCarthy et al., eds (2001).

Nyberg et al. (2007, *Nature* **447**) note that the average frequency of major hurricanes in the North Atlantic Ocean decreased gradually from the 1760s until the early 1990s, but has increased significantly since 1995. This trend has been attributed to anthropogenically induced climate change, and natural variability, but the primary cause remains uncertain. Emanuel (*Program in Atmospheres, Oceans, and Climate*, MIT) agrees that there has been a large upswing in the frequency of Atlantic hurricanes since 1995, but observes that there has been no change in the global, annual frequency of tropical cyclones. However, he adds that there is some evidence that the intensity of hurricanes is increasing.

glocal Describing the seamless integration between the local and global; the comprehensive connectedness produced by travel, business, and communications; willingness and ability to think globally and act locally. 'The concept of glocalization captures the dynamic, contingent, and two-way dialectic between the global and the local' (Swyngedouw (2004) *Camb. Rev. Int. Affairs* **17**). Brown et al. (2007) *GaWC Res. Bull.* **236** show that demand for financial service intermediaries is concentrated in firms operating both at a global and a local scale.

Parnreiter et al. (2005) *Mitteilungen der Österreichischen Geographischen Gesellschaft* **147** attribute the preference for 'glocal' financial institutions to using relatively local banks with sufficient global experience to operate in major financial markets, and to so-called 'soft' factors, such as the initiation of business deals, and an in-depth analysis of the company that sets out to issue bonds or shares. 'Entrepreneurial cities represent key regulatory arenas in which new "glocalized" geographies of national state power are being consolidated . . . the hallmark of glocalizing states is the project of *reconcentrating* the capacities for economic development within strategic subnational sites such as cities, city-regions and industrial districts, which are in turn to be positioned strategically within global and European economic flows' (Brenner in J. Peck and H. W. Yeung, eds 2003). See also Cox (2004) *TIBG* **29**, 2. Bliss (2005) *NZ Geogr.* **61**, 3 argues that responsible **glocal citizenship** makes for a better world.

glurbanization 'Strategies undertaken by cities under conditions of inter-urban competition to secure the most advantageous insertion of a given city into the changing interscalar division of labour in the world economy' (Jessop and Sum (2000) *Urb. Studs* **37**, 12).

gneiss A highly metamorphosed rock of a granular texture, with a banded appearance and possibly metamorphosed from *schists or quartzites. Some may be produced by the interaction of igneous magma with metamorphic rocks.

GNP (gross national product) The *GDP of a nation together with any money earned from investment abroad, less the income earned within the nation by non-nationals. *Compare with* GDP. **GNP per capita** is calculated as GNP/population and is usually expressed in US dollars. It may be used as an indicator of *development.

goal function A function that by current optimization can describe the succession of an ecosystem. 'Ecosystems do not, of course, have goals; but their successional development may be described by a goal function' (Jørgensen and Nielsen (2007) *Energy* **32**, 5). Phillips (2008) *PPG* **32**, 1 reviews theories of evolution and ecosystem development based on goal functions. See F. Müller and M. Leupelt, eds (1998).

Gondwanaland A 'supercontinent': a continuous land surface formed of the now separate units of Africa, Madagascar, Antarctica, Australia, and India. Gondwanaland started in the earliest Paleozoic at the break-up of a Late Proterozoic supercontinent (Veevers (1988) *Geology* **16**, 8), and assembled by 650–570 Ma. See Veevers (2003) *Geology* **31**, 6.

gorge A deep, narrow cleft in an upland area, usually containing a river. Gorges occur in *karst scenery partly as a result of the collapse of caves, or when the incisive power of a river is greater than the processes of valley-wall erosion.

governance The action, fact, or manner of governing (not solely by governments); the various means used to shape society to a desired end. Whitehead (2003) *Area* **35**, 1 sees governance 'as a process whereby formal governing structures are no longer focused primarily on the political realms of public sector government . . . [but] drawn also from the private sector and civil society'. Hooghe and Marks (2003) *Am. Polit. Sci. Rev.* **97** distinguish between governance as the dispersion of authority, and governance as task-specific, intersecting, and flexible.

governmentality A concept of M. Foucault (2003) to encompass the mentalities, rationalities, and techniques used by governments, within a defined territory, actively to create the subjects (the governed), and the social, economic, and political structures, in and through which their policy can best be implemented. In other words, governments try to produce the citizen best suited to fulfil their policies: 'a national form of biopower directed not so much at the management of individual human bodies as at populations and their territories' (M. Hannah 2000). 'Governmentality points at the techniques and rationalities involved in both governing the self and governing others' (Cupers (2008) *Cult. Geogs.* **15**). Governmentality also includes ways in which spaces and places are created, and used, in order to pursue policies; see Airriess (2005) *J. Hist. Geog.* **31**, 4 on politically contested space in Hong Kong's New Territories. **Colonial governmentality** aimed to construct a wage-earning, property-owning, taxpaying, and literate subject (Legg (2006) *Soc. & Cult. Geog.* **7**, 5).

grade In geomorphology, a state of *equilibrium in a system. Thus, the gradient of the long profile of a river declines downstream as a response to changing discharge, bed material size, and sediment transport (Harmar and Clifford (2007) *Geomorph.* **84**, 3–4); see also Lauer and Parker (2008) *Geomorph.* **96**, 1–2. A **graded river** has sufficient *discharge to transport its *load, but not enough energy for erosion; if any of the factors controlling discharge changes, the river will react to re-establish grade; see Eaton and Church (2004) *J. Geophys. Res.* **109**, F03011. Grade over one section of a river course is a **graded reach**. A **graded slope** has an angle at which output, throughput, and input remain in equilibrium.

granite A coarse-grained igneous rock that consists largely of quartz, alkali feldspar, and plagioclase feldspar. Granite is formed by the slow crystallization of deep igneous intrusions but may also be formed by *metasomatism.

granitization *See* METASOMATISM.

granular disintegration (**granular disaggregation**) A form of weathering where the grains of a rock become loosened and fall out, to leave a pitted, uneven surface. André (2002) *Geografiska A* **84**, 3–4 finds that the granular disintegration of crystalline rocks operates at the same rate whatever the environment.

gravel A loose deposit of rounded rock fragments. The lower size limit of gravel is 2 mm but the upper limit varies between scholars. **River gravels** are well rounded (Powell (1998) *PPG* **22**, 1); **solifluction gravels**, produced by *freeze–thaw, are more angular.

gravity anomaly The difference between the actual gravitational force and the calculated force. For example, when a plumb line is set up near a mountain range it will be pulled from the vertical towards, and by the gravitational pull of, the mountains—but by far less than expected from calculations. This is a **negative gravity anomaly**. The gravity anomaly field of continental Ecuador corresponds closely to the physiographic provinces of that country (Feininger and Seguin (1983) *Geology* **11**, 1).

gravity model In human geography, gravity models evaluate, or forecast, the spatial interactions of goods, people, and so on, between origins and destinations. 'The gravity model assumes that the probability of patronizing a facility (facility utility) is proportional to the attractiveness of the facility and inversely proportional to a function of the distance' (Drezner and Drezner (2006) *Geog. Analys.* **38**, 4). These interactions are expressed as:

$$I_{ij} = k \; M_i \; M_j \; Dist_{ij}^{-\beta},$$

where I_{ij} denotes the interaction between two locations i and j; M_i and M_j represent the 'masses' measuring the strength of i and j (usually the population numbers of two settlements); $Dist_{ij}$ stands for the distance between i and j; and k and β are constants (G. Robinson 1998). 'While the gravity model is now accepted by economics as a useful tool (it is said to be all over the place at the International Monetary Fund), it is considered only in an incidental manner in economic analysis that employs it' (Isard (2001) *Int. Reg. Sci. Rev.* **24**, 3). See Rodríguez-Pose and Zademach (2006) *Tijdschrift* **97**, 3 on the gravity model, and mergers and acquisitions.

great circle Any circle on a sphere formed by a plane that passes through the centre of the sphere. The equator and the meridians of longitude are all great circles on the Earth's surface. Any great circle route between two points will represent the shortest line between the two; great circles are therefore relevant to studies of migration and other interactions. Try Zar (1989) *J. Field Ornithol.* **60** for the calculations.

great soil groups The primary classification of global soils into groups; a classification similar to the formations of vegetation. *See* SOIL CLASSIFICATION.

green belt An area of undeveloped land encircling a town, restricting development in the semi-rural areas beyond the built-up zone. The amount of new building is restricted although by no means completely banned. An early green belt was set up in the 1950s around Greater London, in order to limit the spread of suburbs. Other cities have followed this example: in Beijing, all the land in the 'green belt' was forced to keep its status in 2004 without any changes until 2020. 'Under this policy, the trend of central city to form a big pancake would be stopped, while the sub-cities in the region would develop very fast with many small and medium "pancakes" forming outside the 6th Ring Road' (He et al. (2006) *Appl. Geog.* **26**, 3–4). Cullingworth calls green belts 'the first article of the British planning creed', but Nadin argues that they 'have become an outmoded and largely irrelevant mechanism for handling the complexity of future change in the city's countryside' (both in J. Cullingworth and V. Nadin 2002).

greenfield site Areas beyond the city where development can take place unfettered by earlier building and where low-density, high-amenity buildings can be constructed.

greenhouse effect The warming of the atmosphere as some of its gases absorb the heat given out by the earth. Short-wave radiation from the sun warms the earth during daylight hours, but this heat is balanced by outgoing long-wave radiation over the entire 24-hour period. Much of this radiation is absorbed by atmospheric gases, most notably water vapour, carbon dioxide, and *ozone, but also by methane and chloro-fluorocarbons. All of these may be called **greenhouse gases**. Without this absorption, which is also known as *counter-radiation, the temperature of the atmosphere would fall by 30–40 °C.

Through human agency, such as the clearance of rain forest, or the increased rearing of livestock, the concentration of greenhouse gases in the atmosphere is increasing; measurements taken at Mauna Loa, Hawaii, show that the concentration of atmospheric CO_2, for example, increased by 19.4% between 1959 and 2004. It would follow, therefore, that increased concentrations of such greenhouse gases would lead to a rise in global temperatures; see *global warming.

(⊕) SEE WEB LINKS

- The Carbon Dioxide Information Analysis Centre.

green manure A leguminous crop ploughed into the fields after it matures, since it fixes nitrogen from the air and thus improves soil fertility. See Ajayi et al. (2007) *Nat. Resources Forum* **31**, 4.

green revolution The development of high-yielding crop varieties (HYVs) for developing countries began in a concerted fashion in the late 1950s. In the mid-1960s, scientists developed HYVs of rice and wheat that were subsequently released to farmers in Latin America and Asia. The success of these HYVs was characterized as a 'Green Revolution'.

Early rice and wheat HYVs were rapidly adopted in tropical and subtropical regions with good irrigation systems or reliable rainfall. Over the following years, the Green Revolution achieved broader and deeper impacts, extending far beyond the original successes of rice and wheat in Latin America and Asia. 'Productivity growth associated with HYVs had important consequences. Increased food production has contributed to lower food prices globally. Average caloric intake has risen as a result of lower food prices—with corresponding gains in health and life' (Evenson and Gollin (2003) *Science* **300**, 5620). 'In order for an African Green Revolution to happen, it is recommended that the agrarian communities and regions reform their land ownership and land tenure from traditional and/or customary laws, in order to enable the ordinary or poor farmers to access and use the land with ease. The farmers need secure and stable tenancy on the land for them to make the necessary investments in modernization of agriculture. Where there are farmers who are tenants on the land, then there should be reform to make the tenancy long term, so that the farmers have operational legitimacy such as land lease or operational leasehold' (Ngambeki (2003) *SMART/AGRI*).

A. Gupta (1998) describes the mixed reception afforded to the green revolution: 'on the one hand, rural people fused environmental and health concerns with a critique of green revolution technology; on the other hand, they welcomed increased wealth and surpluses flowing from the new technology.' 'The post-World War II green revolution in synthetic chemicals transformed New Zealand agriculture and was interpreted in a progressive way. Today, the same chemicals are often seen in anything but this light. This shift in representation can be seen in the ways we interpret texts and how such re-readings often highlight and contest dominant ideas of the past' (Wildblood-Crawford (2006) *NZ Geogr.* **62**, 1). Wilson (2001) *TIBG* **26** stresses that the need for initiatives and policies that emphasize environmentally friendly agricultural practices has grown as a result of the green revolution. See also G. Robinson (2004).

'Was there a real alternative to the Green Revolution? Did other development programs contribute to the process of producing higher incomes? Certainly, there were many redistributive programs and investments in health and schooling that contributed to welfare... The Green Revolution did raise the productivity of land and water resources. It also raised the productivity of human resources and did lead to higher wages... In short, for the poorest countries, the Green Revolution, late and uneven as it was, was the "only game in town"' (Evenson (2002) *Bureau Econ. Studs Occ. Paper* **19**).

greywacke A sedimentary rock made of angular fragments of quartz, feldspar, and other minerals in a muddy base. Its origin is problematic, since, according to the normal laws of sedimentation, sand and mud should not be laid down together, and some attribute its formation to submarine avalanches, or turbidity currents; see H. G. Reading (1996).

grèzes litées Deposits down a hillslope of *imbricated rock fragments bedded parallel to the slope. It may be that the debris is shattered by freeze–thaw, with larger fragments rolling downwards; wind might be the depositional process (Cailleux (1967) *Biuletyn Peryglacjalny* **17**; Edlund and Woo (1992) *Geol. Surv. Canad. Part B* **92–1B**; Lewkowicz (1998) *Canad. J. Earth Scis* **35**).

gross domestic product *See* GDP.

gross national product *See* GNP.

gross reproduction rate The number of female babies born per thousand women of reproductive age. The **net reproduction rate** also takes into account the number of women who cannot or do not wish to have children; in western Europe, the cohort net reproduction rate is roughly 0.9 children per woman (Goldstein et al. (2003) *Pop. & Dev. Rev.* **29**, 4), and by 2000 the Chinese net reproduction rate had fallen to 0.82 (Young (2005) *Geog. J.* **171**, 1).

gross value added (GVA) The additional value at a particular stage of production, or through image and marketing. It shows the contribution of the factors of production, in increasing the value of a product. GVA measures the contribution to the economy of each individual producer, industry, or sector in a nation. Per capita GVA in the UK for 2003 were:

North East	£10600
North West	£12477
Yorkshire and Humberside	£12459
East Midlands	£12890
West Midlands	£12642
East	£13340
London	£21702
South-East	£15178
South-West	£12873
England	£14620
Wales	£10987
Scotland	£13263
Northern Ireland	£10941

Source: UK Office for National Statistics.

ground failure *See* AVALANCHE; DEBRIS; LANDSLIDE; ROCK FALL.

ground frost *See* FROST.

ground moraine *See* MORAINE.

groundwater All water found under the surface of the ground which is not chemically combined with any minerals present, but not including underground streams. The GroundWaterSoftware.com newsletter is really useful.

growth pole A point of economic growth. Poles are usually urban locations, benefiting from *agglomeration economies, and should spread prosperity from the core to the periphery. Planners have tried to create new growth poles; the best known being in the Mezzogiorno of Italy—which 'signally failed' (G.-J. Hospers 2004). Glasmeier (2002) *PHG* **26**, 2 blames the rampant enthusiasm for growth poles in the USA for rationalizing and supporting concentrated investment in selected places, 'while downplaying the underlying human needs found in adjacent poor places'.

growth triangle A cooperative venture among three or more countries; each has different factors of land, labour, capital, and management, which complement the others, generating mutual advantages in external trade and investment. Growth triangles offer economies of scale, widen potential markets, and may act as buffers against recession; see Landingin and Wadley (2005) *J. Int. Dev.* **17** on growth triangles in Asia.

groyne A breakwater running seawards from the land, constructed to stop the flow of beach material moved by *longshore drift.

Guinea current A warm *ocean current off the coast of West Africa, which transports low-saline, warm waters from the Gulf of Biafra south-west along the north Gabon coast.

gully A water-made cutting, usually steep-sided with a flattened floor. Gullying usually occurs in unconsolidated rock and rarely cuts through bedrock. Gullies usually form quickly as a result of destruction of the plant cover; see Malik (2008) *Geomorph.* **93**, 3–4. It can be stopped by restoring a vegetation cover, by contour ploughing, and by making terraces and small dams across the hillside; see Xiang-zhou et al. (2004) *Env. Sci. & Policy* **7**, 2; Fu et al. (2000) *Catena* **39**, 1.

gust A temporary increase in wind speed, lasting for a few seconds. A typical ratio of gust speed to wind speed in rural areas is 1.6:1, increasing to 2:1 in urban areas due to the effects of high buildings and narrow streets. See Mitsuta and Tsukamoto (1989) *J. Appl. Met.* **28**, 11 on the spatial structure of gusts.

guyot A truncated sea-floor volcano occurring as a flat-topped mountain which does not reach the sea surface. Guyots are thought to be associated with *hot spots.

habitat The area in which an organism can live and which affords it relatively favourable conditions for existence. Keagy et al. (2005) *Restor. Ecol.* **13**, 3 investigate a scenario whereby some higher-quality habitat is destroyed, in order to determine the quantity of low-quality habitat required to maintain the equilibrium population abundance. **Habitat generality** means being able to survive in a multitude of environments. See Brown (2007) *J. Biogeog.* **34**, 12.

habitat rehabilitation The management of a degraded habitat aimed at restoring it to 'an ecologically superior state' (Borg et al. (2007) *Geomorph.* **89**, 1–2). An understanding of the recovery and reassembly of plant communities is imperative for habitat rehabilitation (Thornton (1984) *Ambio* **28**). Borg et al. (op. cit.) argue that rehabilitation projects should provide the appropriate elements for a healthy system; one would then let these elements interact to produce some target state. The problem lies in determining what that target state should be; habitat changes that are beneficial to one species may be detrimental to another; thus, 'in some ways, habitat rehabilitation is based on explicit or implicit species values' (Clough (2003) *Rept. Fish Habitat & Species Recovery Wkshp*).

habitus A socio-cultural milieu in a distinct neighbourhood; the dispositions that shape the actions, bodily movements, tastes, and judgements characteristic of a social class, together with the circumstances that ensure its reproduction. Habitus is improvisatory, operating through human practices rather than through prior conscious thought. See Bridge (2001) *TIBG* **26**, 2 on habitus and *gentrification, and Marshall and Foster (2002) *Canad. Geogr./Géog. canad.* **46**, 1 on habitus and migration.

Hadley cell A simple, vertical, thermally direct, tropical, *atmospheric cell. Air rises at the *inter-tropical convergence zone (ITCZ) and drifts polewards, cooling a little through a net loss of radiation, and thereby transferring heat. As the circumference of the earth decreases with distance from the equator, the moving air has to converge. This combination of cooling and convergence, together with deflection by the *Coriolis force, causes the air to sink around latitude 32°. The air then returns to the equator as surface winds. The basic components of this cell are: weak upper-air easterlies above the ITCZ, subtropical anticyclones at the descending limb, and easterly *trade winds associated with the return of air to the equator, together with *synoptic scale

disturbances, such as easterly waves or *tropical cyclones. The westerly
subtropical *jet is located at the poleward limit of the Hadley cell.

In the tropics, the rotationally confined, thermally direct Hadley circulations
are important for maintaining the observed small meridional temperature
gradients in low latitudes (Held and Hou (1980) *J. Atmos. Sci.* **37**). The
intensification and poleward expansion of the cross-equatorial Hadley cell can
lead to westerly acceleration in the winter subtropics, and enhanced vertical
shear of the zonal wind in the subtropics and mid-latitudes (Hou (1998)
J. Atmos. Sci. **55**, 14).

hail A form of snow made of roughly spherical lumps of ice, 5 mm or more in
diameter, formed when a frozen raindrop is caught in the violent updraughts
found in warm, wet *cumulo-nimbus clouds. As the drops rise, they attract ice,
and as they fall, the outer layer melts, but refreezes when the droplet is again
lifted by updraughts, often causing the hailstones to show an onion-like
pattern of alternating clear ice (glaze) and opaque ice (rime). A hailstone will
descend when its fall speed is enough to overcome the updraughts in the
cloud. Soft hail is white and low-density as it contains air. See Billett et al.
(1997) *Weather & Fcst* **12**, 1 on the prediction of hail size.

halo A ring of light around the sun or, more rarely, the moon, caused by the
refraction of light by ice crystals. A **cloud halo** is a zone of increased humidity
around cumulus clouds; see Lu et al. (2002) *J. Appl. Met.* **41**, 8.

halophyte A plant which can grow in saline conditions: salt marshes,
estuarine environments, and the lower parts of sea cliffs. Halophytes on salt
marshes tend to trap sediment at high tide, thus gradually increasing the
height of the marsh; see Hanslin and Egen (2005) *Seed Sci. Res.* **15**. Controversy
exists over whether halophytes require saline conditions for their existence—
or merely tolerate them (Yeo (1998) *J. Exp. Bot.* **49**; Glenn et al. (1999) *Crit. Rev.
Plant. Sci.* **18**).

ham 1. In Anglo-Saxon place names, a home, as in Birmingham.
2. A water-meadow of rich pasture.

hamada A desert surface of pebbles/gravel/boulders, formed by the washing
or blowing away of the finer material. See Smith (1988) *Geomorph.* **1**, 4.

hamlet A small settlement without services or shops, and usually without a
church.

hanging valley A high-level tributary valley from which the ground falls
sharply to the level of the lower, main valley. The depth of the lower valley may
be attributed to more severe glaciation: 'the height of a hanging valley reflects
the difference in the time-integrated ice discharge in tributary and trunk
valleys and therefore increases as the discharge ratio decreases . . . The rate of
production of hanging valley relief increases strongly as the ratio of the
tributary to trunk ice discharge decreases' (MacGregor et al. (2000) *Geology* **28**,

11). The level of a hanging valley may be used as an indicator of glacier ice thickness (Kuhle (2005) *Island Arc* **14**, 4).

Brocklehurst et al., *Geomorph.* **97**, 1–2, observe that while hanging valleys show glacial modification of a landscape, the overall change in the relief structure of mountain ranges is surprisingly modest.

hardpan A cemented layer in the B *horizon of a soil, formed by the *illuviation and precipitation of material from the A horizon, such as clay (forming a **clay pan**), humus (a **moor pan**), or iron (an **iron pan**). See Lee and Gilkes (2005) *Geoderma* **126**, 1–2.

Harmattan *See* LOCAL WINDS.

Hawaiian eruption A volcanic *fissure eruption where large quantities of basic lava spill out with very little explosive activity. Hawaiian volcanoes evolve through stages that have been delimited by the compositions of their erupted lavas; see Vazquez et al. (2007) *Geology* **35**, 8.

hazard An unexpected threat to humans and/or their property; a hazard includes both the event and its consequences. By this definition, the Indian monsoon is not a hazard, but its failure is. A natural event becomes a hazard through: the social processes which cause some people to be much more at risk from the effects of a hazard; a lack of options to minimize, or escape from, the effects of the hazard; the location of human settlement on potentially dangerous physical sites; the location of human settlement near potentially dangerous economic activity; and activities that aggravate events with natural causes.

'As the range of hazards and vulnerabilities faced by any given community increases, it often becomes possible only to play one kind of risk off against another in search of a "less bad" scenario. Many highly vulnerable communities may deliberately choose to inhabit a hazard prone environment if this reduces other risks, related to income generation for example. Or, should they find themselves in hazard prone zones due to exclusion from formal land markets or for other reasons, they will many times opt to stay in order to maintain those conditions that provide them with the means to reduce daily life risk and vulnerability' (UNDP Expert Group Meeting, Havana, 2002).

Hazard perception is the view which an individual has of a hazard. Robertson (1999) *Geog. Rev.* **89**, 4 compares the reality of the tornado hazard with the portrayal in movies, finding the latter seriously misleading. Walker et al. (2002) *Appl. Geog.* **20**, 2 report on the failure of regulatory practice to distinguish hazards from industrial accidents. *See* COGNITIVE DISSONANCE.

haze A suspension of particles in the air, slightly obscuring visibility. These particles may be naturally occurring—sea salt or desert dust—or man-made, like smoke.

headward erosion The lengthening of a river's course, by erosion back towards its source. A *knick point on a stream profile suggests that the fluvial system has reacted to a lower base level by headward erosion (Schlunegger and Schneider (2005) *Geomorph.* **69**, 1–4). See also Hill et al. (2008) *Geomorph.* **95**, 3–4 on the Grand Canyon, USA.

health, geography of *See* MEDICINE, GEOGRAPHY OF.

heartland A term suggested by H. Mackinder (1904) to indicate the wealthy interior of Eurasia; whoever controls the heartland would be able permanently to dominate the Eurasian landmass, and consequently gain a '*hegemony over the entire world'. Bilgin (2007) *Pol. Geog.* **26**, 7 shows how this concept has been adopted and adapted to help justify foreign and domestic policy-making in Turkey.

heat island In a city, air temperatures are often as much as 3–4 °C higher than over open country. Causes include: the presence of large vertical faces; a reduced sky-view factor; surface materials of high heat capacities and conductivities; large areas of impermeable surfaces; anthropogenic heat fluxes; and atmospheric pollution (Grimmond (2007) *Geog. Rev.* **173**, 1); for more detail, see Yow (2007) *Geog. Compass* **1**, 6.

heavily indebted poor country (HIPC) An extremely poor, *less economically developed country, with huge external debts. 'The HIPC initiative is a comprehensive approach to debt reduction for heavily indebted poor countries pursuing IMF- and World Bank-supported adjustment and reform programs' (IMF). By 2008, debt reduction packages had been approved for 33 countries, 27 of them in Africa, including Uganda, whose economy grew by 6% a year between 1998 and 2004. However, since it started receiving debt relief from the HIPC initiative, debt and debt service indicators have not improved in net present value terms (ADF/BD/WP/2004/26).

((⊕)) SEE WEB LINKS

• The IMF HIPC initiative.

hegemony In terms of power and politics, leadership (especially by one state of a federation); more widely, the way in which one social group will represent its interests as being everyone's interests. This ruling group may keep its grip on society either by **social hegemony**—the use of force to maintain order—or, more commonly, by **cultural hegemony**—producing ways of thinking, especially by subtly eliminating alternative views to reinforce the status quo (Laurie and Bennett (2002) *Antipode* **34**, 1; Hoelscher (2003) *AAAG* **93**, 3). See also S. Zukin (1995) on the cultural hegemony of cities.

  Perreault (2007) *PHG* **31**, 4 sees the core idea of hegemony as 'ideological leadership and popular consent'. Kohl (2006) *Antipode* **38**, 2 discusses neoliberalism as a **hegemonic system**.

Heinrich event During glaciations, at intervals of 8000–10000 years, the release into the ocean of icebergs, with major impacts on oceanic circulation and global climate. See Cacho et al. (2006) *Quat. Sci. Revs* **25**; Mullins et al. (1996) *GSA Spec. Paper* **311**; Kageyama et al. (2005) *Comptes rendus geoscis.* **337**.

helicoidal flow A continuous corkscrew motion of water along a river channel.

henge In the UK, a Neolithic circular earthwork with an encircling bank and inside ditch.

heritage Inherited circumstances or benefits; 'a (re)presentation of the past in cultural forms that can hide some voices and confirm others, and . . . formulate certain national identities through which individuals may come to see themselves in their everyday lives' (Bhatti (2008) *Area* **40**, 1). K. Walsh (1992) argues that 'heritage, in many of its forms, is responsible for the destruction of a sense of *place', but Hubbard and Lilley (2000) *Geog.* **85**, 3 see the **heritage industry** 'as involving conflicts between the different senses of place, with the distinctive character of a town . . . having resulted from different groups seeking to impose their values on the townscape'. Summerby-Murray (2002) *Canad. Geogr./Géog. canad.* **46**, 1 argues that the **heritage discourse** is constructed through the creation of memory and the processes of commodification and consumption. See J. Scarpaci (2005) on heritage tourism in Latin America.

heterogeneous nucleation The freezing of water droplets around a *freezing nucleus. See Khvorostyanov and Curry (2004) *J. Atmos. Scis* **61**, 22.

heteronormativity The societal privileging of heterosexuality (Oswin (2008) *PHG* **32**, 1); the normalization of man/woman as opposites meant to come together within heterosexual relationships that are based on specific class and race-based relations (Browne (2006) *Antipode* **38**, 5). Waitt (2006) *AAAG* **96**, 4 writes on resistance to heteronormativity and Sydney's gay games (Oswin (2008) *PHG* **32**, 1). See the special issue of *Antipode* **34**, 5 (2002).

heterotopia 'The coexistence in an impossible space of a large number of fragmentary possible worlds' (M. Foucault 1967). Cyberspace is a good example. See Soja in S. Watson and K. Gibson, eds. (1995).

heterotrophe An organism which has to acquire its energy by digesting food which has been manufactured by other organisms. Thus all organisms are heterotrophes except the primary *producers which can manufacture their own food usually by photosynthesis.

hide A medieval assessment of land tax liability, see H. C. Darby, 1977.

hierarchical diffusion *See* DIFFUSION.

hierarchy Ordering of phenomena with grades or classes ranked in sequence; *central place theory, for example, posits a hierarchy of settlements from regional capitals to hamlets. In a **nested hierarchy** each level contains/is composed of the level below, with the individuals in that level being smaller-scale sub-systems of the level above. The metasystem (highest level) does not directly respond to the lowest sub-system directly, as the dynamics of the sub-system are on too small a scale to be of any consequence; in other words, information across any three hierarchical levels is *buffered*. This suggests that the sensitivity of a system to change takes a different form, depending on where in the hierarchy the drivers of change originate. Whatever your own focus in geography, you will never find hierarchy expressed more clearly than in Couper (2007) *TIBG* **32**, 3. Smith and Kurtz (2003) *Geogr. Rev.* **93** suggest a hierarchy of coalitions.

high A region of high atmospheric pressure—in Britain, generally over 1000 mb.

high-order goods and services Goods and services with a high *threshold population and a large *range. Examples include furniture, electrical goods, and financial expertise. These goods are usually *shopping goods.

high-technology (high-tech) Any industry concerned with advanced technology, such as biotechnology, computers and microprocessors, fibre optics, aerospace, electronics-telecommunication, and pharmaceuticals (Fischer et al. (2005) *Geog. Analys.* **38**, 3). In theory such industries are *footloose, since the products have a high value to weight ratio. Bernstein and Nadiri (1988) *Am. Econ. Rev.* **78** find that new technological knowledge complements R&D in high-technology industries. Florida (2002) *J. Econ. Geog.* **2** finds a significant, positive correlation between his 'bohemian index' and concentrations of high-tech industry. James (2008) *Geog. Compass* **3** has an interesting paper on the gendered geographies of high-tech economies.

hill farming The extensive farming of an upland area, usually sheep, although some cattle may be kept. British hill farming receives support from the EU, and the dependence of these farms on subsidies has grown (Horton (2005) *Community Dev. J.* **40**, 4).

hill fort A fortified site on a hilltop, usually with a ditch and ramparts. The earliest date from the Iron Age but some British examples were created as late as the Dark Ages. Favoured sites for such structures in southern England seem to be chalk uplands.

hill wave *See* LEE WAVE.

hinge line A line either side of which *isostatic readjustment is uneven; the line separating uplift from subsidence (Sella et al. (2007) *Geophys. Res. Letts.* **34**, L02306). In glaciology, it is the point where a glacier, reaching the coast, starts to float (Debenham (1965) *Geo. J.* **131**, 3). Rignot et al. (1997) *Science* **276**, 5314, refer to the hinge line as the limit of tidal flexing.

hinterland The area serving, and being served by, a settlement. An **information hinterland** is the region for which a city provides the best access to information flows (Wang et al. (2007) *Tijdschrift* **98**, 1). See Baldacchino (2006) *Asia Pac. Viewpt* **47**, 1.

hinterworld The worldwide pattern of connections between a given *world city and other world cities. See Taylor (2001) *Geog.* **86**.

historical geography The study of past human geographies; of past landscapes; 'geographic perspectives on the past' (*Historical Geography* website). 'The study of the organization of space in the . . . context of time' (X. de Planhol 1994). 'The relationship between humans and their environment in particular socio-cultural, spatio-temporal contexts' (U. Fellmeth et al., eds 2007). 'Historical geography may be viewed as being concerned with the historical dimension in geography, and geographical history with the geographical dimension in history' (A. Baker 2003). Gregory and Healey (2007) *PHG* **31**, 5 examine the creation and dissemination of historical GIS databases, and the use of GIS to perform quantitative and qualitative analyses. For 'progress reports' on historical geography, see Naylor (2005) *PHG* **29**, 5; (2006) *PHG* **30**, 6; and (2008) *PHG* **32**, 2.

historical materialism The analysis of history, most closely associated with Marx, which stresses the material basis of society; pointing out that economic systems underlie the development of history and ideas. See D. Harvey (1996).

historiography The study and writing of history. Llewellyn (2003) *Area* **35**, 3 writes on the historiography of architecture, while Maddrell (2008) *TIBG* **33**, 1 comments on the under-representation of women's work in the historiography of geography. See Trimble (2008) *PPG* **32**, 1. Livingstone (2000) *Australian Geog. Studs* **38**, 1 writes that 'the "history of geography" might profitably be reconceptualised as "the historical geography of geography"'.

histosol *See* US SOIL CLASSIFICATION.

Hjulström diagram A diagram showing the relationship in a channel between particle size and the mean fluid velocity required for *entrainment. It shows that an entrained particle can be transported in suspension at a lower velocity than that required to lift the particle initially. When the stream velocity slows to a critical speed, the particle is deposited. Note that higher velocities are needed for the entrainment of clay-sized particles because of the electrostatic forces which bind them together.

Hjulström diagram

holism The view that the whole is more than its parts. In earlier geographies, the region has been seen as having a distinct identity which does not come entirely from its separate parts. See Antrop (1998) *Landsc. & Urb. Plan.* **41**, 3.

hollowing out The reduction of political power; for example, of a member state on joining a supranational body, such as the *European Union; of a local authority by national government; or of a state by global organizations such as the *World Bank or the *International Monetary Fund. The term may also apply to the loss of local government power when national government takes more control (Skelcher (2000) *Pub. Policy & Admin.* **15**, 3), but Radcliffe, Laurie, and Andolina (2002) *Transnat. Communities Prog. Seminar* claim that the state is not hollowed out in this process.

Holocene The most recent geological *epoch, stretching from 12 000 years ago to the present day. See Fisch in H. von Stoch et al., eds (2004).

homelessness, geography of Homelessness is commonly perceived as an urban problem. Lee and Price-Spratlen (2004) *City & Commy* **3**, 1 observe that, in the USA, the homeless, and services for the urban homeless, are mostly located relatively close together, usually in more deprived central city areas, and almost forming a service ghetto. Fagan (2003, *San Francisco Chronicle*) identifies Main Street as the centre of welfare and medical services in San Francisco. For the homeless, such a concentration of services renders services more accessible, but has a number of disadvantages: 'For many, simply getting to the local night shelter is a frightening experience [and this fear has a] gender dimension, with women especially tending to prefer to use services located away from the main "service hub"'. (homeless-research.org.uk). Informal areas—'interstitial niches'—develop in streets, parks, and public car parks, where people find places to set up 'home', sleep, eat, make friends, and look for security, money, and entertainment. These niches are colonized by the homeless at particular times: 'at dusk, the street families lay out boxes, build plastic houses, warm themselves, and cook on small fires. The children rarely have anything. Public space is *their* space only at night' (Morelle and Fournet-Guérin (2006) *Cybergeo, politique, culture, représentations* **342**).

Differentiation is not merely temporal; it is also social. Homeless Research record a division between 'straightheads', 'smackheads', and 'pissheads': 'what for one homeless person might be experienced as a space of care, might for another become a space of fear because of the dominant presence of a particular homeless sub-group . . . thus embodied and performative geographies can become underpinned by specific spatial logic.' See Cloke, Milbourne, and Widdowfield (2003) *Geoforum* **34**, 1 on rural homelessness.

(⊕) SEE WEB LINKS

• Research into homelessness.

homeostasis The maintenance of an ecosystem by self-adjusting mechanisms, under ever-changing conditions (Lau and Lane (2001) *PPG* **25**, 2).

homocline One of a regular series of hills from a large area of rock strata of uniform thickness and dip.

homogeneous nucleation The spontaneous freezing of water droplets at around −40 °C, as clusters of water molecules within a droplet settle, by chance, into lattice formation of ice, making the entire droplet freeze. See Easter and Peters (1994) *J. Appl. Met.* **33**, 7.

homoiotherm A warm-blooded animal.

homonormativity 'A politics that does not contest dominant heteronormative assumptions and institutions, but upholds and sustains them, while promising the possibility of a demobilized gay constituency and a privatized, depoliticized gay culture anchored in domesticity and consumption' (L. Duggan 2003). 'In other words, many gays and lesbians want nothing more than to be considered normal so that they might go on with their day-to-day lives as part of the status quo' (Oswin (2008) *PHG* **32**, 1).

honeycomb weathering The removal of material from sandstone while the cement is preserved, leaving a lacy network; see Rodriguez-Navarro et al. (1999) *GSA Bull.* **111**, 8.

honey-pot A location, such as Shakespeare's birthplace, which is particularly appealing to tourists. Planners in National Parks often develop honey-pot sites with car parks, shops, cafés, picnic sites, and toilets, so that other parts of the Park will remain unspoilt; see Pickering and Buckley (2003) *Mt. Res. & Dev.* **23**, 3 on Mt Kosciuszko, Australia.

'The tourism hotspots Bondi, Byron Bay, Coogee and Manly are being buried under an avalanche of visitors . . . Councils in these honeypot areas are picking up the tab for everything from dumped cars, and closing illegal backpacker hostels, to increased rubbish collection and vandalism. They complain that although tourism may boost the national accounts, it does nothing for the bottom line of council budgets' (Kennedy (2002), cited by Allon (2004) *Space & Cult.* **7**, 1).

horizon (soil horizon) A distinctive layer within a soil which differs chemically or physically from the layers below or above. The **A horizon** or topsoil contains *humus. Often soil minerals are washed downwards from this layer. This material then tends to accumulate in the **B horizon** or subsoil. The **C horizon** is the unconsolidated rock below the soil. These three basic horizons may be further subdivided. Thus, **Ah horizons** are found under uncultivated land, **Ahp horizons** are under cultivated land, and **Apg horizons** are on *gleyed land. The B horizons are also subdivided by means of suffixes: **Bf horizons** have a thin iron pan, **Bg horizons** are gleyed, **Bh horizons** have humic accumulations, **Box horizons** have a residual accumulation of sesquioxides, and **Bs horizons** are areas of sesquioxide accumulation. **Bt horizons** contain clay minerals and **Bw horizons** do not qualify as any of the above. **Bx horizons** or fragipans contain a dense but brittle layer caused by compaction. C horizons are also subdivided: **Cu horizons** show little evidence of gleying, salt accumulation, or fragipan; **Cr horizons** are too dense for root penetration; and **Cg horizons** are gleyed. Additional suffixes may be used. Some soil scientists use the term **D horizon** for the consolidated parent rock.
   In addition to these soil horizons, other layers are distinguished. Thus, the layer of plant material on the soil surface is classified as: the **L horizon** (fresh litter); the **F horizon** (decomposing litter); the **H horizon** (well decomposed litter); and the **O horizon** (peaty). A *leached A horizon is termed an **E horizon** or *eluviated horizon.

horst A block of high ground which stands out because it is flanked by normal faults on each side. See Walsh and Schultz-Ela (2003) *GSA Bull.* **115**, 3. It may be that the block has been elevated or that the land on either side of the horst has sunk.

Hortonian overland flow An overland flow of water occurring more or less simultaneously over a *drainage basin when rainfall intensity exceeds the rate of water infiltration in the soil (Horton (1933) *Trans. Am. Geophys. Union* **14**; Kirkby (1988) *J. Hydrol.* **100**). See also Vigiak et al. (2006) *Geomorph.* **76**, 1–2.

hot desert Located on the west coasts of tropical and subtropical climes, these have average temperatures over 20 °C and annual rainfall under 250 mm. Deserts are too dry for most plant species except for *xerophytes. Desert insects, reptiles, mammals, and birds are all adapted to drought. See M. Evenari et al., eds (1985). See also Nash (2000) *PPG* **24**, 3 on geomorphological processes in arid areas.

Hotelling model A model of the effect of competition on locational decisions, indicating that locational decisions are not made independently but are influenced by the actions of others (Hotelling (1929) *Econ. J.* **39**, 1). Kim (2007) *MPRA Paper* **2742**, U. Munich, develops the Hotelling model.

hot spot An area of localized swelling and cracking of the earth's crust caused by upwelling *magma; deep-seated thermal anomalies fixed with respect to the surrounding mantle (Morgan (1972) *GSA Memoir* **132**). Plumes from the lower mantle rise to the surface, forming volcanoes. As plates move over the Earth's surface, the volcanoes delineate and record the history of plate motion, and the radiometric age of seamounts should show a simple progression away from the present-day hot spot locations. See Wessel and Kroenke (1998) *Earth & Planet Sci. Letts* **158**, 1–2 on the relationship between a hot spot and its seamounts.

hot towers Immensely tall, tropical *cumulo-nimbus, at the rising branch of a *Hadley cell, that siphons sensible and latent heat upwards, to the equatorial *tropopause. See Montgomery et al. (2006) *J. Atmos. Scis* **63**, 1 on vertical hot towers and cyclogenesis.

human capital Talent (Florida (2002) *AAAG* **94**, 2—who finds that the economic geography of talent is strongly associated with high-technology industry location). López-Rodríguez et al. (2007) *Reg. Studs* **41**, 2 observes that, in the European Union, educational attainment levels are higher in those regions with greater market access. 'The presence and concentration of *bohemians in an area creates an environment or milieu that attracts other types of talented or high human capital individuals . . . The geography of bohemia is highly concentrated' (Florida (2002) *J. Econ. Geog.* **2**, 1).

Human Development Index (HDI) A compound indicator of *economic development, established by the United Nations, which attempts to get away from purely monetary measurements by combining per capita GNP with *life expectancy and literacy in a weighted average. 'Human Development is a development paradigm that is about much more than the rise or fall of national incomes. It is about creating an environment in which people can develop their full potential and lead productive, creative lives in accord with their needs and interests. People are the real wealth of nations. Development is thus about expanding the choices people have to lead lives that they value' (UNDP).

(((⊕))) SEE WEB LINKS

• The Human Development Index database.

human ecology An approach whereby the human social system and the ecosystem of the planet are seen as a mesh of reinforcing connections, each influencing and being influenced by the others. See G. G. Marten (2001).

1.

Territory of A   Territory of B

First pattern of market share

2.

Territory of A   Territory of B

Second pattern of market share

3.

Territory of A   Territory of B

Final pattern of market share

Hotelling model. Two vendors, A and B, are positioned in the centre of their own half of the beach. If A moves closer to the centre, he increases his market share, which leads B to also move to the centre, ensuring that market share is equal again.

human geography The study of the patterns and dynamics of human activity on the landscape. Broadly conceived, human geography includes both those areas of geographic study that are more narrowly focused on human activity, and those that are particularly concerned with human-environment dynamics, or the nature–society tradition (W. Moseley et al., 2007). 'That part of the discipline of geography concerned with the spatial differentiation and organisation of human activity and its interrelationships with the physical environment' (R. J. Johnston et al., 2000). 'More than just analyzing and taking note of where people and their possessions are, however, geography studies the interactions and social habits of humans *within* and *across* all spaces. Basically, geographers critically examine how humans organize and identify themselves in space—how we create "places". Thus, the science of human geography does not look at location per se, as much as it does at the mobility of, access to, and barriers against human processes' (Oas (no date) *Implications* **1**, 7). Warf (2004) *Prof. Geogr.* **56**, 1 comments that 'human geography has changed so much and so quickly over the last decade that any prediction of its status ten years hence is undoubtedly doomed to failure'. See M. Phillips, ed. (2005).

human influence index An index of human influence on natural environments: based on population density, land transformation (leading to loss and fragmentation of habitats), human access, and power infrastructure. See Sanderson et al. (2002) *Bioscience* **52**, 10.

humanistic geography Geography centred on human perception, capability, creativity, experience, and values; 'space and time not only have a form, but . . . they also have a meaning' (Ley (2003) *PHG* **27**, 5). It maintains that any investigation will be subjective, reflecting the attitudes and perceptions of the researcher, who may also be an influence on the very field of his study; see Y.-F. Tuan (1974). 'Humanistic geography represents a duality of reason and feeling, science and ethics. As a scientific approach, it is concerned with uncovering the truth regarding people's experiential relationship with place. It does not regard the phenomena under consideration as merely an object of research, rather it bears an ethical message of concern for those objects, be they human beings, nature or place' (Hasson (1984) *Prof. Geogr.* **36**, 1).

Human Suffering Index A compound indicator of distress, compiled by 'adding together ten measures of human welfare related to economics, demography, health, and governance: income, inflation, demand for new jobs, urban population pressures, infant mortality, nutrition, access to clean water, energy use, adult literacy, and personal freedom' (Population Crisis Committee, Washington, DC).

humic acid A complex acid formed when water passes slowly through humus. Humic acid is an example of an organic acid in that it is formed from carbon-based compounds. It is significant in chemical weathering and in the formation of soil. See Yadav and Malenson (2007) *PPG* **31**, 2. *See also* MODER; MOR; MULL.

humidity The amount of water vapour in the atmosphere. It is more exactly defined as the mass of water vapour per unit volume of air, usually expressed in kg m^{-3}. This is **absolute humidity**. **Relative humidity** is the moisture content of air expressed as the percentage of the maximum possible moisture content of that air at the same temperature and pressure. The **humidity mixing ratio** is the ratio of the mass of water vapour in a sample of air to the mass of dry air associated with that water vapour.

humilis *See* CLOUD CLASSIFICATION.

humus A naturally occurring complex combination of substances resultant of the extensive biochemical breakdown of plant and animal residues (Yadav and Malenson (2007) *PPG* **31**, 1). This breakdown is **humification**—the process whereby the simple mineral compounds released by weathering combine with the organic residues to form large, stable organic molecules which bond the soil; see J.-M. Gobat et al. (2004). Humus can form relatively stable complexes and chelates with polyvalent cations, thus increasing assimilation by plants (D. Alexander and R. W. Fairbridge 1999).

hundred An Anglo-Saxon term for part of a *shire or county, perhaps 100 *hides.

hunting and gathering A form of society with no settled agriculture, or domestication of animals, and which has little impact on the environment. The hunting of animals and the collection of edible plants depend on adapting to the environment, rather than changing it. See T. Ingold (2000).

hurricane *See* TROPICAL CYCLONE.

hybrid capitalism A form of capitalism that combines elements of culturally specific imprints—such as employment relations—and globalizing norms—such as international standards of corporate governance (H. W. Yeung (2006) *Econ. Geog.*). The concept of hybrid capitalism offers a way of combining the focus of the global convergence literature on systemic 'drivers' of change with the emphasis on domestic institutional logic found in the 'varieties-of-capitalism' approach (Dyson and Padgett (2005) Anglo-German Foundation). H. W. Yeung (2003) argues that Chinese capitalism will mutate to a form of hybrid capitalism.

hybrid geography Rose (in R. Johnston et al., eds 2000) writes that hybrids 'transgress and displace boundaries between binary divisions and in so doing produce something ontologically new', a notion picked up by Kwan (2004, *AAAG* **94**, 4) who recognizes two major divisions within geography: the partition between physical and human geography—of nature and society; and the separation of spatial-analytical geographies which attempt to create 'a mode of disembodied geographical analysis' from social, cultural, and political geographies. Kwan suggests that geographies based on thematic networks would ameliorate 'the polarizing tendency within the discipline'.

S. Whatmore (2002) agrees, arguing that 'it is both more interesting and more pressing to engage in a politics of hybridity . . . in which the stakes are thoroughly and promiscuously distributed through the messy attachments, skills and intensities of differently embodied lives whose everyday conduct exceeds and perverts the designs of parliament, corporations and labour'.

hybridity The mixture of meanings that emerges when two cultures interact; the new forms that are created when cultures merge. 'Acculturation can be seen as a sort of dialogue between cultures that . . . engenders the creation of new forms of cultural negotiation which themselves materialize as . . . hybridity, creoleness and inbetweeness' (Mitchell (1997) *Antipode* **29**, 2). Identities are not stable, but ambivalent. H. Bhabha (1994) claims that subjects at the margins of cultures, or in a *third space, are best placed to resist a domineering culture/narrative, and to create new hybrid forms.

hydration The incorporation of water by minerals. Hydration is believed to be a major force in weathering coarse-grained igneous rocks; see Hughes (1998) *Geoarchaeol.* **3**, 4.

hydraulic action (hydraulic force) The force of water, including *cavitation and fluvial *plucking.

hydraulic conductivity The ability of a soil or rock to conduct water. The conductivity of dry soil or rock is low, as water must form a film of water around the soil particles before conduction occurs. Tuli and Hopmans (2004) *Eur. J. Soil Sci.* **55**, 1 find that pore geometry and pore-size distribution determine the relations between saturation, hydraulic conductivity, and air conductivity. **Saturated hydraulic conductivity** refers to the maximum rate of water movement in a soil; see Lilly (2000) *Soil Use Manag.* **16**.

hydraulic geometry The study of the interrelationships along the course of a river. *Discharge is linked to the mean channel width, the mean channel depth and slope, the suspended *load, and the mean water velocity, and the wavelength of the meander is related to the radius of curvature; see Singh and Zhang (2007) *Hydraul. Procs* **22**, 2. See also Smith and Pavelsky (2008) *Water Resource Res.* **44**, W03427 on the estimation of hydraulic geometry from space.

hydraulic gradient The rate of change in *hydraulic head with distance. Within a drainage network, topography and valley morphology will accentuate or diffuse the local distribution of flow energy through changes in the hydraulic gradient and constrictions or widening of the river channel (Vandenberghe and Woo (2002) *PPG* **26**, 4).

hydraulic head The pressure exerted by the weight of water above a given point.

hydraulic radius Also known as **hydraulic mean**, this is the ratio of the cross-sectional area of a stream to the length of the *wetted perimeter* (cross-sectional length of a river bed). The hydraulic radius is a measure of the efficiency of the river in conveying water—if the hydraulic radius is large, there is little friction.

hydraulic resistance A measure of the resistance offered by a material, such as vegetation, to a flow of water. The calculation of flow velocities within stream channels and over hillslopes cannot be made without accounting for the hydraulic resistance experienced by flow over different morphologies (Smith et al. (2007) *PPG* **31**, 4). Moore and Burch characterize hydraulic resistance by choosing an appropriate value of the *Manning roughness coefficient, and it can also be characterized using the Darcy–Weisbach friction factor, expressed as a function of the Reynolds number of the flow; see Prosser and Rustomji (2000) *PPG* **24**, 2 for the equations. Fisher et al. (2007) *Glob. Change Biol.* **13**, 11 recommend the introduction of hydraulic resistance to likely flow paths in the management of forests.

hydraulics The study of the mechanical properties of liquids. Smith et al. (2007) *PPG* **31**, 4 summarize the early development of hydraulics. See Boyer et al. (2006) *J. Geophys. Res.* **111**, F04007 on flow turbulence and bed morphology, and Wilcox and Wohl (2007) *Geomorph.* **83**, 3–4 on three-dimensional hydraulics.

hydroecology The study of the influence of dynamic hydrology, *hydraulics, and geomorphology on stream ecosystems, particularly nutrient cycling, primary productivity, and trophic interactions, also known as ecohydrology. See Horn and Richards in P. Wood et al. (2007) on modelling hydraulics in restored floodplains.

hydrograph A graph of *discharge, or of the level of water in a river throughout a period of time. The latter, known as a **stage hydrograph**, can be converted into a discharge hydrograph by the use of a stage-discharge rating curve. Hydrographs can be plotted for hours, days, or even months. A storm hydrograph is plotted after a rainstorm to record the effect on the river of the storm event.

hydrological cycle (water cycle) This is the movement of water and its transformation between the gaseous (vapour), liquid, and solid forms. The major processes are *condensation by which *precipitation is formed, movement and storage of water overland or underground, *evaporation, and the horizontal transport of moisture. The length of time any water stays in the *atmosphere is about eleven days.

Hydrograph

Hydrological cycle

hydrology The study of the Earth's water, particularly of water on and under the ground before it reaches the ocean or before it evaporates into the air. This science has many important applications such as flood control, irrigation, domestic and industrial water supply, and the generation of hydroelectric power.

hydrolysis The chemical reaction of a compound with water. Hydrolysis is an important component of soil formation, and of *chemical weathering—for example, as feldspars in granite decompose to make china clay.

hydromorphic Denoting areas with waterlogged soils. **Gley soils** form in such conditions.

hydrosere A successional sequence of plants originating in water.

hydrosphere All the water on, or close to, the surface of the earth. Some 97% of this water is in the Earth's seas and oceans; of the rest, about 75% is in *ice-caps and sheets, about 25% in surface *drainage and *groundwater, and about 0.03% in the atmosphere.

hygroscopic nuclei *Condensation nuclei which are hygroscopic, i.e. which tend to attract and condense ambient water vapour.

hypabyssal rock An igneous intrusion which has consolidated near the Earth's surface above the base of the crust. Examples include dolerite and quartz porphyry.

hyperconcentrated flow An intermediate condition between debris flow and normal streamflow (Katoka and Natajo (2002) *Sedimentol.* **49**, 2); a liquid–solid two-phase flow, in which the liquid phase is a syrup made of water and fine sediment less than 0.01 mm in diameter, and the solid-phase consists of coarse particles larger than 0.05 mm suspended in the syrup (Chien, ed. (1989), cited by Xu and Yan (2006) *Geomorph.* **73**, 1–2, which see).

hypolimnion The lower layers in a body of water which are marked by low temperatures and insufficient light for photosynthesis. Levels of dissolved oxygen are low.

hyporheic zone A spatially and temporally dynamic mixing zone at the interface between groundwater and surface water. Malcolm et al. (2008) *Geophys. Res. Abstr.* **10**, EGU2008–A–04043 stress the importance of the hyporheic zone to the hydrological and ecological functioning of rivers.

hypsography The study of landforms and processes in physical geography.

hysteresis The impact of its history on a physical system. In many cases hysteresis can be conceptualized as a non-linear response to a causal factor; see Mishra and Singh (2001) *Hydrol. Procs* **15**, 9; Vandenberghe and Woo (2002) *PPG* **26**, 4. Hysteresis occurs where there may be two or more values of a dependent variable associated with a single value of an independent variable (Phillips (2003) *PPG* **27**, 1).

hythergraph A plot of monthly rainfall against monthly temperature over a year. *See* CLIMOGRAPH.

ice When the water content of snow falls below about 8% grain growth occurs by vapour flux of the air in the pores. The resulting growth of ice grains is very slow. In general, growth in wet snow speeds up with increasing water content (D. McClung and P. Schaerer 1993).

ice age A time when there are ice sheets on the continents, within which there may be *interglacial periods of milder climate. Ice ages last for some tens of millions of years, separated by intervals of *c.*150 million years. See Sutcliffe et al. (2000) *Geol.* **28**, 11 on glaciation and Earth's orbit; Easterbrook (1963) *GSA Bull.* **75**, 12 on Pleistocene glacial events and sea-level change; Gore et al. (2001) *Geol.* **29** on East Antarctica (ice free at the last glacial maximum), Bond et al. (1997) *Science* **278** on the Little Ice Age; and Barry (2006) *PPG* **30**, 3 on glacier retreat since then. A. Goudie (2000) is strong on the stratigraphic information from ocean and ice cores, and the causes of climate change.

ice-cap A flattened, dome-shaped mass of ice, similar to an *ice sheet, but under 50000 km^2 in area, such as the Barnes Ice Cap (LeB. Hooke (1973) *J. Glaciol* **12**).

ice contact feature A landform formed in contact with a glacier. For **ice contact terrace**, see *kame terrace. See Hickson (2000) *Geomorph.* **32**, 3–4 on **ice-contact volcanism**.

ice fall An area of *extending flow marked by crevasses, where the gradient of a glacier steepens. The rapid flow at this point causes the ice to thin. See P. Knight (1999).

ice floe A flat section of floating ice which is not attached to ground ice. See Meylan et al. (1996) *J. Geophys. Res.* **101**.

ice flow The major controls on ice flow are temperature and *shear stress. See Joughin et al. (2001) *J. Geophys. Res.* **106**, D24.

ice front The floating ice cliff at the seaward end of an *ice shelf. See Frezzotti and Polizzi (2002) *Annal. Glaciol.* **34**, 1 on ice fronts as indicators of climate change.

ice lens An underground ice segment, often with convex upper and lower surfaces. *See* PALSA.

ice-marginal channel A channel cut by meltwater, running parallel to the ice margin, and found most often in a glacier's ablation area, where rates of surface ablation and meltwater production are highest. Parallel sets of ice-marginal channels may form when the ice margin retreats. See Wadham et al. (2001) *Hydr. Procs* **15**, 12.

ice mound In a *periglacial landscape, a swelling in the ground due to the expansion of an underground *ice lens. *Solifluction displaces the material at the top of the mound, so that when the ice lens melts a depression forms, often surrounded by a rampart. *See* PINGO.

ice polygon (ice wedge polygon) A three- to six-sided polygon of *ice wedges, formed by *ice segregation and the drying and shrinking of sediments, and becoming more regular with age. See Matsuoka (1999) *Polar Geosci.* **12**.

ice segregation The creation of discrete bodies of ground ice in *periglacial conditions. Segregation develops well in silts (which seem to have the optimal pore size), depends on a moderate freezing rate, and often occurs under stones, which have high thermal conductivity. See Hall (2002) *S. Afr. J. Sci.* **98** and Murton et al. (2006) *Science* **314**.

ice sheet A continuous mass of glacier ice with an area over 50 000 km^2. The annual ice-mass loss from Greenland doubled between 1996 and 2005, two-thirds of this from accelerated ice flow (Christoffersen and Hambrey (2006) *Geol. Today* **22**, 3). Alley et al. (2005) *Science* **310**, 5747 report that the Greenland and Antarctic ice sheets, if completely melted, could raise sea level by about 70 m. The resulting freshwater fluxes may affect oceanic circulation.

ice shelf A sheet of ice extending over the sea from a land base, fed by snow falling on it, or from glaciers on the land. Smith et al. (2007) *Antarc. Sci.* **19**, 1 record the constant, but minor, retreat of George VI ice shelf, Antarctica, since the mid-1980s, concluding that while atmospheric warming has been critical in driving this retreat, other mechanisms and feedbacks, such as changes in sea ice extent or oceanography, may be involved.

ice stream Within a glacier or ice sheet, a stream of ice moving faster than, and not always in the same direction as, most of the ice. Ice streams respond to the weight of the ice, and to gravity. They are slowed by side drag from the slower-moving ice that flanks them, and by back pressure from the ice sheet itself (Stokes et al. (2007) *Earth Sci. Revs* **81**, 3–4).

ice wedge In a *periglacial landscape, a near-vertical ice sheet, tens to hundreds of metres long, tapering into the ground to depths of 12 m. Individual wedges usually meet another wedge orthogonally. Matsuoka (1999) *Polar Geosci.* **12** records an expansion of 16 mm across an ice wedge at –17 °C. **Ice wedge casts** are sediment-filled, fossil ice wedges; see Harris et al. (2005) *Geomorph.* **71**, 3–4 for the ice-wedge casting processes.

Ice-wedge networks form as ice fills the tension fractures resulting from cooling. They form in stable, aggrading or sloped surfaces, with winter temperatures from –15 °C to –35 °C. Modelled polygonal networks self-organize through interactions between fractures, stress, and re-fracture (Plug and Werner (2008) *Permaf. & Periglac. Procs.* DOI: 10.1002/).

iconography 'A set of symbols in which people believe, encompassing elements of national feeling, from the state flag to the culture transmitted through the state's schools' (P. Taylor 1993). O'Sullivan (2001, *TIBG* **29**, 3) illustrates the Christian symbolism expressed in late medieval European urban forms. See also Leib et al. (2000) *GeoJournal* **52**.

identity The individuality of a being or entity. National identity may be based on myths; 'like cultural identities, diasporic national identities are actively constructed' (Mavroudi (2007) *Glob. Networks* **7**, 4), but Nairn (2003) *New Left Rev.* **23** observes that 'for growing masses of people ... issues of identity are not metaphorical but [are] treasured, if deplorable, bits of cheap plastic: everyday life and death'. Identity may emerge from the production and reproduction of space; see Saarinen (2004) *Tourist Studs* **4** on tourist destinations and their identity, and try M. Tewdwr-Jones and P. Allmendinger (2006). Dittmer and Larsen (2007) *Soc. & Cult. Geog.* **8**, 5 write that national identity is constructed by the interpellation of subjects from competing discourses that all emphasize different aspects of the lived experience of the inhabitants. The use of the term 'interpellate' is questionable.

In physical geography, identity is established using 'methods of ontology based on meaningful collections of attributes' (Raper and Livingstone (2001) *TIBG* **26**).

identity, geographies of The study of identities and space: 'we are in the midst of a redefinition of space. In the very moment that national and ethnic boundaries are breaking down we encounter paradoxical reinvestments in homeland, territorial integrity, localism, regionalism, and race- and ethnocentrism' (P. Yaeger 1996). Thus, much of the work revolves around the relationship between cosmopolitanism and transnationalism, with an emphasis on migratory movements and the formation of new kinds of transnational social spaces, identities, and relations. 'Consciousness is always in motion and forms of identity such as a subaltern cosmopolitanism reflect particular relationships in particular times and places. Groups and individuals perform their own sense of cosmopolitanism dependent on context—their own particular locations in various axes of power as well as the broader structuring forces of global capitalism and geopolitics' (Cheah in P. Cheah and B. Robbins, eds 1998). For many, forms of citizenship are 'flexible, multicultural and polycultural, and dissenting' (Mitchell (2007) *PHG* **31**, 5).

'Self-perceived liberals often position themselves as fatigued defenders of an identity politics and a form of multiculturalism that has now gone too far. Identity politics is projected (on the left) as either undermining efforts for redistribution or impeding the struggle against a neolib' (Mitchell (2004) *PHG* **28**, 5); see S. Huntington (2004).

idiographic Of, or relating to, the study of anything unique. The idiographic approach has been the cornerstone of *regional geography. In biogeography, **idiographic analyses** are detailed analyses of single species range; see Fattorini (2007) *Divers. & Distrib.* **13**, 6.

igneous rock A rock formed from molten *magma, deep beneath the earth's surface, subsequently coming to the surface as an *extrusion, or remaining below ground as an *intrusion. The plutonic mode of igneous rock formation is characterized by 'intrusive emplacement under compressional tectonic stress regimes. The volcanic mode is dominated by the eruption of large volumes of magma, through fault controlled dyke structures in extensional tectonic stress regimes' (Neuberger and Reuther (1987) *Int. J. Earth. Scis* **76**, 1). The nature of the rock also depends on the rate at which it cooled; intrusions of magma cool slowly, allowing enough time for large crystals to form, while extrusions cool quickly, leaving little time for crystal growth; see R. Le Maitre (1989).

illuviation The downward *translocation from the A *horizon and subsequent precipitation in the B horizon of clay-sized particles in a soil. The **illuvial horizon** is the B horizon, in which there is redeposition or entrapment of matter brought down from above.

image A picture built up by an individual from the social and physical *milieux experienced from birth; see A. Scott's work (2002) on the production of images, symbols, and messages in manufacturing and services. Leung (2006) *Asian Studs Rev.* **30**, 3 shows how images attached to Chinese people in German popular media are exclusionary forces. See A. Buttimer et al., eds. (1999) on texts and images as mediators in the production of spaces.

imagined geographies (imaginative geographies) The ways that other places, people, and landscapes are represented; how these imaginings reflect the preconceptions and desires of their inventors, and the relations of power between subjects and objects thereof (E. Said 1979). Thus, with independence, Singapore's ruling party sought to transform 'a group of people with divergent orientations . . . into a populace of loyal citizens' (L. Kong and B. Yeoh 2003). D. Gregory (2004) argues that, since 9/11, the United States has re-imaged the world into 'us' and 'them', and 'light' against 'darkness': '*orientalism [is] abroad again.' Jeffrey (2007) *Area* **38**, 2 argues that viewing the 1992–5 conflict in Bosnia-Herzegovina as a humanitarian, rather than a political, crisis, justified military and NGO intervention. See Schwartz (1996) *J. Hist. Geog.* **22**, 1 on imaginative geographies of the 19th century.

imagined places 'Imagined places are not fairytale places; they are not just fantasy. In one way or another, imagined places have a connection with a place that exists geographically' (E. van Alphen 2005). P. Shurmer-Smith and K. Hannam (1994) argue that images are not the reality but are our only access to reality. Imagined places are constructed by their representations in art, film, literature, maps, and media; see Gorman-Murray (2007, *MC Magazine* **9**, 3) on Sydney's King Street, imagined as a gay/lesbian place by the gay/lesbian media.

imbricated Of deposits, laid down in overlapping sheets (like roof tiles), as on steep slopes in *periglacial environments (Millar (2006) *Catena* **67**, 1).

IMF *See* INTERNATIONAL MONETARY FUND.

immigration The movement into another country of someone as a permanent resident. Governments may try to restrict immigration by using quotas, discounting family ties, or being reluctant to accept asylum seekers or refugees (Kofman (2002) *Pol. Geog.* **21**, 8). Skenderovic (2007) *Patterns Prej.* **41**, 2 identifies the definition and exclusion of the Other as characteristic of radical-right politics, while Abu-Laban and Garner (2005) *Urb. Aff. Rev.* **40**, 4 compare the Canadian view of immigration (as a national policy) with that of the USA (as individual choice). See Liu (2001) *Soc. & Cult. Geo.* **1**, 2 on space, place, and scale, and the construction of racial-ethnic and immigrant identities. *See* CITY OF REFUGE.

imperialism 'Unequal economic, cultural, and territorial relationships based on domination and subordination' (Jones and Phillips (2005) *AAAG* **95**, 1). Imperialism dominates the external trade of the subordinate nation (E. Arghiri 1972), taking raw materials from the 'colony' and in return selling it finished goods, thus depressing manufacturing industry in the 'colony'; see Zanias (2005) *J. Dev. Econ.* **78**, 1. It can exist without the creation of formal colonies; Domosh (1999) *TIBG* **24**, 4 suggests a multiplicity of imperialisms. The **imperialist gaze** saw subsistence landscapes as empty, despite the presence of indigenous people. Colonized spaces were 'stripped of preceding significations and then re-territorialized according to the convenience of the . . . imperial administration' (D. Harvey 1989).

import penetration The extent to which the country is dependent upon its imports, to reduce which governments may resort to *protectionism (M. Kamiya 2002).

import substitution A strategy that encourages domestic manufacturing in order to reduce imports of manufactures, save foreign exchange, provide jobs, and lessen dependency. Puga and Venables (1999) *Econ. J.* **1109**, 455 find that, while import substitution maintains diversity, trade liberalization generates higher levels of welfare.

inceptisol *See* US SOIL CLASSIFICATION.

incidence matrix An array of a species' incidence in the habitat patches of a system (Maron et al. (2004) *Oikos* **106**, 3). See Patterson and Atmar (2000) *Bonn Zool. Mon.*

incised meander A *meander formed when a *rejuvenated river cuts deeper into the original meander; see O. Catuneanu (2006). An **intrenched meander** is an incised meander with a symmetrical cross-valley profile, an **ingrown meander** one with an asymmetrical profile (Rich (1914) *J. Geol.* **22**).

inclusion The action or state of including or of being included within a group or structure. The principle that binds together the present UK government's (and indeed Europe's) social policy is the reduction in the number of people and places experiencing '*social exclusion' from mainstream societal activities and the promotion of 'social inclusion' into these activities (Social Exclusion Unit 2001), and the Sustainable Development Commission promotes social inclusion and environmental justice (SDC 2002). Cameron (2006) *PHG* **30**, 3 observes that inclusion is mostly defined by *exclusion, and laments the 'general failure to develop a critical understanding of the real and discursive geographies of social inclusion . . . The economic, political and behavioural norms that characterize social inclusion are no longer connected to a particular place or space.' Stewart in P. Askonas and A. Stewart, eds (2000) thinks that an understanding of social inclusion involves 'the recognition and acceptance of the historically situated and therefore contingent character of prevailing institutional realizations of justice and concomitantly privileged conceptions of the good life obtaining at any point in time'. Gray in Askonas and Stewart (op. cit.) argues that the disruption of solidaristic social formations (particularly states) as a consequence of economic globalization renders any meaningful form of social inclusion impossible. I. M. Young (2000) insists that demographic practices depend on inclusion. See also Holt (2008) *PHG* **32**, 2.

index of centrality Many centrality indices are based on shortest paths, measuring, for example, the average distance from other vertices (Brandes and Pich (2006) *Int. J. Bifur. & Chaos* **17**, 7). Simmons and Jones (2003) *Prog. Plan.* **60**, 1 use it as a measure of the external market served by a city, and Channell and Lomolino (2000) *Nature* **403** use it to distinguish between 'central species' and 'peripheral species'. See Beauchamp (1965) *Behav. Sci.* **10** for an 'improved index of centrality'.

index of circulation *See* ROSSBY WAVES.

index of local deprivation This index is calculated for all UK local authorities, although the criteria used vary geographically:

WALES AND ENGLAND (2000) WARD BASED	N. IRELAND (2001) WARD BASED, WITH SOME ENUMERATION DISTRICT AND LOCAL GOVERNMENT DISTRICT INDICATORS
Income deprivation	Income deprivation
Employment deprivation	Employment deprivation
Health deprivation and disability	Health deprivation and disability
Education, skills, and training deprivation	Education, skills, and training deprivation
Housing deprivation	Housing deprivation
Geographical access to services	Geographical access to services
	Social environment

ENGLAND (2004) CENSUS SUPER OUTPUT AREA BASED	SCOTLAND (2003) WARD BASED
Income deprivation	Income deprivation
Employment deprivation	Employment deprivation,
Health deprivation and disability	Health deprivation and disability
Education, skills, and training deprivation	Education, skills, and training deprivation
Barriers to housing and services	Geographical access to services
Living environment	
Crime	

index of segregation A measurement of the degree of residential segregation between two sub-groups inside a larger population. See Johnston et al. (2005) *Urb. Studs* **42**, 7.

index of vitality (I_v) An index to indicate the growth potential of a population:

$$I_v = \frac{(\text{fertility rate} \times \% \text{ aged 20-40})}{(\text{crude death rate} \times \text{old-age index})}$$

indifference curve In economics, a graph of the various levels of *utility achieved at different prices through buying two commodities or services, for example, a mortgage and commuting costs. At a given price, various combinations of the two would yield the same total utility, hence the term 'indifference'. The concept is used in the *Alonso model, and the monocentric city model; see Bertaud and Brückner (2005) *Reg. Sci. & Urb. Econ.* **35**, 2.

indivisibility The minimum scale at which any technology can be used; almost all manufactured capital goods are indivisible. The indivisibility of a tractor makes it uneconomic for many small farmers in the developing world (V. Balasubramanian et al. 1998). It is crucial in finding optimum factory size

Indifference curve

(Stykolt and Eastman (1960) *Econ. J.* **70**, 278), and in economies of scale (Elliethy (2006) *J. Int. Dev.* **4**, 4). Edwards and Starr (1987) *Amer. Econ. Rev.* **77**, 1 hold that **indivisible labour specialization** brings scale economies.

induced demand In *transport geography, the additional traffic resulting from additions to road capacity (Levinson (2008) *J. Econ. Geog.* **8**, 1). Goodwin and Noland (2003) *Appl. Econ.* **35** find that building new roads really does create extra traffic. See also St John and Pouder (2006) *Growth & Change* **37**, 2.

induration The hardening or cementing process that consolidates sediments to form sedimentary rocks. Sedimentary rocks usually become more and more indurated over time.

industrial cluster *See* CLUSTER.

industrial district *See* MARSHALLIAN INDUSTRIAL DISTRICT.

industrial diversification The deployment of investment and labour over a wider range of activities. Wei et al. (2007) *Econ. Geog.* **83**, 4 show that diversification in China is led by local governments, and entrepreneurs embedded in thick local networks.

industrial geography The study of the spatial arrangement of industry; 'the study of production and production systems across all sectors' (Harriman in G. Gaile 2002). 'We are currently in another time of paradigm conflict, with those who focus on the sociology and consumption of production and favor different styles of "theorizing" even while there is a revival of interest in analytical, regional science type research' (Stafford (2003) *Industr. Geogr.*). 'Industrial geography has always been a bellwether sub-discipline reflecting methodological changes within economic geography' (T. Barnes and M. S. Gertler 1999). The 'central question of industrial geography as to why industrial activities grow/decline in particular places' (Watts cited in Stafford, op. cit.). 'Social networks . . . play an important role in determining who can successfully start a new venture in an industry, and where they can start it. Potential entrepreneurs need access to a variety of resources—including knowledge of the key technologies in the industry—to begin operations and to compete successfully; [and] social networks facilitate access to these resources' (Sorenson in D. Fornahl et al. 2003). McCann and Sheppard (2003) *Reg. Studs* **37**, 6–7 argue for the redevelopment of analytical industrial location theory.

industrial inertia The survival of an industry in an area after its original locational factors are gone. It is not unusual in industry, as sunk costs are high, and, over time, enterprises tend to reproduce *external economies (Rota and Vanolo (2007) *EURA Conf. Paper*).

industrialization The development of manufacturing industry from a predominantly agrarian society. Characteristic features of industrialization include the application of scientific methods to solving problems; mechanization and a factory system; the division of labour; the increased geographical/social mobility of the labour force; and capital deepening (Atack et al. (2005) *J. Econ. Hist.* **58**, 3). These are also features of *capitalism, and capitalism is not the same thing as industrialization.

'Industrialization is widely seen as the most important social and economic change of the last 200 years . . . with the power to modify or destroy any pre-existing social arrangement' (A. Janssens 1993). Industrialization encourages urbanization; for example, Chen (2002) *China Econ. Rev.* **13**, 4 finds that those Chinese provinces and municipalities with higher per capita GDP are more urbanized. As industrialization continues, there is a sectoral shift in employment: the sequence is illustrated by the data for Shanghai.

YEAR	1950	1995	1999	2002
Primary industry	3%	4.3%	1.8%	1.6%
Secondary industry	76%	63.8%	48%	47.4%
Tertiary industry	21%	31.9%	50.2%	51%

Source: Zhang (2006) *Urban Affairs Review* **9**, 42.

This change from manufacturing to commerce and services 'to a considerable extent, means [from] more [to] less stable jobs' (Aguilar (1997) *GeoJournal* **43**, 4).

Industrialization is often seen as a solution to poverty; see Fukunishi et al. (2006) *UNIDO*. Yamagata (2006) *Cambodian Econ. Rev.* finds that the Cambodian garment industry offers scope for the poor to increase their wages substantially. But Blumer (1969/70) *Studs Compar. Int. Dev.* **5**, 3 holds that early industrialization brings 'a series of new demands on life' (such as a search for superior economic status) that lead to 'disorganization'.

industrial location policy The neoclassical view of the role of the state in economic activities is 'inherently suspicious of government interference in private sector decision making, and its justification for industrial location policy relies primarily on the promotion of mobility, both with respect to capital and labour. Behaviouralists are more particularly concerned with the effects of industrial incentives on capital mobility, and specifically whether or not industrial location incentives change the location preferences of entrepreneurs in favour of designated regions' (R. Hayter 2004). 'Industrial location policy is ... the result of corporate policies to minimize union strength, to avoid taxes, and to exploit new markets' (T. J. Sugrue 2005).

industrial location theory 'The general critical factors of industrial location are transportation, labour, raw materials, markets, industrial sites, utilities, government attitude, tax structure, climate, and community. In addition, for international location considerations, four general factors are identified: [the] political situation of foreign countries, global competition and survival, [and] government regulations [are critical]' (Badri (2007) *J. Bus. & Public Affairs* **1**, 2—who also reviews industrial location theory). **Industrial organization**—the structure of an industrial unit—is a component of industrial location theory; see McCann and Sheppard (2003) *Reg. Studs* **37**, 6–7.

industrial revolution Although there is some discussion about its timing, the industrial revolution is generally accepted as occurring in Britain in the late 18th and early 19th centuries. K. Pomeranz (2000) explains the industrial revolution though 'Europe's armed exploitation of other resource-rich regions, particularly in the Americas'.

industrial specialization The domination of a limited range of industries in a region. Sölvell et al. (2006) *DRUID Summer Conf.* analyse industrial specialization in terms of absolute number of employees, specialization, and dominance, and identify nineteen 'three-star' regional clusters in the ten new EU states. Svaleryd and Vlachos (2002) *J. Int. Econ.* **57** find that the financial sector has greater impact on industrial specialization among OECD countries than do differences in human and physical capital.

industry A sector of the economy in which firms use similar factor inputs to create a group of related products and/or services. The extraction of natural resources, including agriculture, is **primary industry**; manufacturing is **secondary industry**; **tertiary industry** comprises services, including transportation and communication; **quaternary industry** deals with a range of producer services from banking to retailing to real estate; and **quinary industry** with consumer services such as education, health care, and government information. With economic development, industry moves from primary industry to progressively less material-oriented activities such as textiles and clothing, until, over time, tertiary industries dominate; the sequence is illustrated in the figure.

The relative dominance of the latter three categories over the former two in many industrialized countries has informed discussions of a *post-industrial society, in which Miozzo and Soete (2001) *Tech. Forecast. & Soc.* **67**, 2 identify research and development, design, marketing, distribution, and after-sales maintenance as essential.

Phelps and Ozawa (2003) *PHG* **27**, 5 propose four industrial phases.

	PROTOINDUSTRIAL	INDUSTRIAL	LATE INDUSTRIAL	POST-INDUSTRIAL
Sectorial basis	Agriculture-manufacturing	Manufacturing	Manufacturing	Services-manufacturing
Division of labour	Intra-firm, intra- and inter-sectorial	Inter-firm and inter-sectorial	Intra- and inter-firm, intra-sectorial	Inter- and inter-firm, inter- and intra-sectorial
Source of accumulation	Exports →domestic	Domestic → exports	Domestic → exports	Exports → domestic

Industry. Industrial change in Japan 1990–2002
Source: Ström, 2005, *Soc. Sci. Japan* J **8**, 2

inequality The fact of **global inequality** is not disputed; the world is routinely divided into developed/developing, North/South, more/less/least economically developed, high/medium/low development, and so on. The richest 2% of adults in the world own more than half of global household wealth (J. B. Davies, S. Sandstrom, and A. Shorrocks 2006). There is no consensus about the causes of global inequality, not least because of the difficulties in measuring inequality: if we scale income by some index of health, there is more inequality in the world than if we consider income alone (Deaton (2006) *NBER W. Paper* W12735).

Many writers blame globalization: Ash (2004, *TIBG* **29**, 2) argues that 'TNCs and international banks are deeply embedded in the production of inequality, through their influence over governments, their tax evasion schemes, their poor environmental record, their labour market practices, their locational mobility, their market power, their dumping practices.' Cornia and Court (2001, *Policy Brief 4* World Inst. Dev. Econ. Res.) contend that the liberalization of domestic banking associated with globalization has caused rises in income inequality; Rowson (2000) *healthmatters* **41** agrees.

There is also debate over the impact of *structural adjustment measures on inequality. In 2004, *SAPRIN claimed that structural adjustment polices are the largest single cause of inequality; at the other end of the spectrum, Ferreira (2004) *World Bank Policy Working Paper* **1641** asserts that, under a programme of structural adjustment, the population of Tanzania in poverty fell from 65% in 1983 to 51% in 1991.

Whether global inequality is increasing or decreasing is by no means clear. The UNDP (2005) reports that the ratio of the assets of the world's richest 20% to the poorest 20% increased from 30:1 in 1960 to 78:1 in 1994, yet Firebaugh and Goesling (2004) *Amer. Journal Soc.* **110**, 2 report that global income inequality declined in the last decades of the 20th century.

There are those who hold that there is a relationship between **regional inequality** (as measured by the *Gini coefficient) and level of development (as measured by per capita GDP), but the data in the table do not offer robust support to this view.

REGION	GINI COEFFICIENT FOR PER CAPITA GDP (PPP US$) 2003	GDP PER CAPITA (PPP US$) 2003
Sub-Saharan Africa	72.2	1856
South Asia	33.4	2897
East Asia and Pacific	52.0	5100
Latin America and Caribbean	57.1	7404
High Income OECD	67.0	30181

Source: UNDP (2005) 55 and 222.

National inequality has grown markedly; since 1980, virtually every country in the world has become more unequal (Gilbert (2007) *Geog. Compass* **1**, 3). During the 1980s and 1990s, income inequality rose in the OECD countries, particularly escalated in China, Vietnam, and South Asia, shot up in Latin America, and went ballistic in Russia in the 1990s (2005, *World Bank* **45**). See Alderson and Nielsen (2002) *Amer. Journal Soc.* **107** for an exhaustive study of inequality in sixteen OECD countries.

In Britain, inequality rose in the 1980s and 1990s. Changes since 2000 are less clear, but the proportions of 'exclusive wealthy'—the people with so much wealth that they can exclude themselves from the norms of society—are at their highest since 1990 (Joseph Rowntree Foundation 2007). The rich and poor now live further apart; while in some cities over half of all households are now breadline poor, wealthy households have concentrated in the outskirts and surrounds of major cities, and exclusive wealthy households have increasingly been gathering around London. Since 2000, there seems to have been little progress in reducing geographical polarization.

Does inequality matter? Gilbert (2007) *Geog. Comp.* **1**, 3 argues that inequality creates social ghettos and increased violence, and blocks progress for the poor. 'Inequality is not inherently wrong—as long as three conditions are met: society as a whole is getting richer; there is a safety net for the very poor; everybody, regardless of class, race, creed, or sex, has an opportunity to climb up through the system' (*The Economist*, 15 June 2006). Korpi (2008) *J. Econ. Geog.* **8**, 2 finds a positive link between wage inequality and labour market size, due to increasing upper income levels as labour market size increases.

(((⊕))) SEE WEB LINKS

• The Joseph Rowntree Foundation study of poverty and wealth across Britain.

infant mortality The number of deaths in the first year of life per 1000 children born. Poor sanitation is a major factor: the 'urban penalty'/'urban graveyard effect' (Mercier and Boone (2002) *Hist. Geog.* **28**, 4). Gyimah (2006) *Ethnic. & Health* **11**, 2 finds that mortality is more of a reflection of socio-economic inequalities than ethnic-specific socio-cultural practices, but notes that wider birth spacing and longer breastfeeding also reduce infant mortality. T. Boley (1999) sees infant mortality as an index of social tension and cultural values.

infiltration Water entering soil. Lipiec et al. (2005) *Geophys. Res. Abs.* **7**, 02621 find that tillage increases soil porosity, and thus infiltration. Reduced tillage and stubble retention practices can greatly reduce surface sealing and improve infiltration rates. Soil type is significant—sands take in water faster than loams and clays; see Koelliker and Duell (1990) ASAE Paper 902568 on the effects of burning on infiltration. Infiltration may not occur if the speed of the water is too great.

Infiltration capacity is the maximum rate at which water can infiltrate the soil, and a key determinant of the volume of surface run-off, which occurs when rainfall exceeds a soil's infiltration capacity. **Relative infiltration capacity** is the ratio of mean precipitation intensity to saturated hydraulic conductivity (Berger and Entekhabi (2001) *J. Hydrol.* **247**, 3).

informal sector (informal economy) 'A group of household enterprises or unincorporated enterprises owned by households that includes informal own-account enterprises, which may employ contributing family workers and employees on an occasional basis, and enterprises of informal employers, which employ one or more employees on a continuous basis. The enterprise of informal employers must fulfil one or both of the following criteria: size of unit below a specified level of employment, and non-registration of the enterprise or its employees' (ILO Report of the Fifteenth International Conference of Labour Statisticians, Geneva, 1993).

	% IN NON-AGRICULTURAL INFORMAL EMPLOYMENT	WOMEN AS A % OF NON-AGRICULTURAL INFORMAL EMPLOYMENT
North Africa	48	43
Sub-Saharan Africa	72	84
Latin America	51	58
Asia	65	No data

Source: Chen et al. (2002) International Labour Organization.

'The informal sector plays an important and controversial role. It provides jobs and reduces unemployment and underemployment, but in many cases the jobs are low-paid and the job security is poor. It bolsters entrepreneurial activity, but at the detriment of state regulations compliance, particularly regarding tax and labor regulations. It helps alleviate poverty, but in many cases informal sector jobs are low-paid and the job security is poor' (GDRC). See also B. Guha-Khasnobis et al. (2000).

(((()))) SEE WEB LINKS

• The informal sector.

information city A city whose residents work mostly in education, consulting, banking, services, computer programming, computer manufacturing, and medical equipment. Those in the *creative industries (R. Florida 2002) could be considered as elements of both in some cases. Is Birmingham, UK, an information city? Webster (2007) *City* **5**, 1 thinks not. Is Bangalore? Liang (2005) *Wld Info. City Conf.* says yes, and writes of 'the other information city': 'the proliferation of non-legal media practices ranging from pirated VCD's, DVD, MP3's to grey market mobile phones and pirate book sellers on the streets.'

information society A society built on information storage, retrieval, and transmission, *time–space compression, *post-Fordism, *flexible accumulation, and the advance of finance capital, which is characterized by networking, *globalization, and the flexibility, individuality, and instability of work. 'Information technologies, both as a process and product, contain embedded social knowledge, and thus represent the construction of new social norms and institutions' (Eischen (2001) U. Calif., Santa Cruz, *W. Paper Series* 1009). 'The information society can be usefully characterized as a universe at the intersection of three distinct but interdependent spaces: the geographical space, the social space, and the informational space' (Ekbia and Shurman, *Information Society*, special issue).

infrastructure The framework of communication networks, administration, and power supply; 'the fact that a direct link between investment in transport infrastructure and national or regional economic development has not been shown to exist has not stopped politicians, lobbyists and promoters from asserting that great benefits will accrue from such investments' (Scrafton (2002) *J. Transp. Geog.* **10**, 1). 'In developed countries, where there is already a well-connected transport infrastructure network of a high quality, further investment in that infrastructure will not on its own result in economic growth' (Banister and Berechman (2001) *J. Transp. Geog.* **9**, 3). Nonetheless there is some empirical analysis that supports the idea that infrastructure growing ahead of productive activities would stimulate growth (Reardon et al. (2001) *World Dev.* **29**, 3). Kirkpatrick (2004) *Publish. Res. Quart.* **20**, 3 thinks that African governments' telecommunications monopolies are 'obstructive'.

ingrown meander *See* INCISED MEANDER.

inner city An area at or near the city centre with dilapidated housing, derelict land, and declining industry (also called the *twilight zone*, or *zone of downward transition*). Langois and Kitchen (2001) *Urb. Studs* **38**, 1 use a general deprivation index to locate deprivation in central Montreal (but note that it isn't confined to the inner city). W. J. Wilson (1996) asserts that the decline of US inner cities reflects *deindustrialization, combined with the suburbanization of a small number of middle-class African Americans. 'Following the restructuring process which swept away the traditional manufacturing economy of the inner city twenty-five years ago, new industries are transforming these former postindustrial landscapes. These creative, technology-intensive industries include Internet services, computer graphics and imaging, and video game production, and are integral to the production of the "new inner city" of the 21st century' (T. A. Hutton 2008).

innovation wave The diffusion of an innovation from its point of origin. Beginning at the innovation epicentre, it is adopted by a very few, close to the focus. Through time, the peak of the wave moves away from the source as more distant locations adopt the innovation. Consolidation follows across the whole of the region concerned; finally, as the innovation saturates the region, diffusion slows down until the stage of maximum acceptance.

An innovation wave may occur when the larger cities take up the innovation, which then filters down the urban hierarchy. The largest cities have benefited from many innovation cycles, and, in consequence, have developed a broader diversity of activities, and attained higher levels of social and organizational complexity—thus making them more likely to adopt further innovations (Pumain et al. (2006) *Cybergeo: systèmes, modèl., géostat.* **343**).

input–output analysis A view of the economy which stresses sectorial interdependence—the output of one sector is often the input of another. See R. K. Miller et al. (1989) and Fleischer and Freeman (1997) *Annals. Tour. Res.* **24**, 4 on input–output analyses of tourism.

InSAR (Interferometric Synthetic Aperture Radar) A remote sensing technique which uses radar satellite images. The US Geological Service has an excellent site on InSAR. See Baltzer (2001) *Prog. Phys. Geo.* **25**, 2 for an account of the use of InSAR in forest mapping and monitoring.

inselberg A steep-sided hill composed mainly of hard rock, which rises sharply above a plain in tropical and subtropical areas; see Römer (2007) *Geomorph.* **86**, 3–4. *See also* BORNHARDT.

insolation The *solar radiation hitting the earth's surface. Global variations in insolation are critical in the *general circulation of the atmosphere. 'Climatic changes result from variables in planetary orbits which modulate solar energy emission and change seasonal and latitudinal distribution of heat received by the Earth. Small insolation changes are multiplied by the albedo effect of the winter snow fields of the Northern Hemisphere, by ocean-atmosphere feedbacks, and, probably, by the stratospheric ozone layer' (Kukla (1972) *Boreas* **1**, 1).

instability 1. The condition of a slope just before the initiation of mass movement. See (2008) *Geomorph.* **94**, 3–4, special issue.
2. The condition of a parcel of air which has positive *buoyancy, and thus a tendency to rise through the *atmosphere. It is the temperature of the parcel relative to the *ambient air which is critical, since this affects densities. A parcel remains unstable if it cools more slowly than the surrounding, stationary air.
Absolute instability occurs when an air parcel, displaced vertically, is hastened in the direction of the displacement; such a move will end when the temperature of the parcel is at one with its surroundings. The **potential instability** of a parcel of air can be calculated using a *tephigram. **Conditional instability** occurs when a parcel of air would become unstable if lifted by some other force, such as a mountain, the moistening of lower-level air, or when a lower, stable parcel of air is overlain by an unstable layer. **Convective instability** is a tendency of an air parcel towards instability when it has been lifted bodily until completely saturated, which often occurs via the inevitable stirrings in the atmosphere. The atmosphere is said to be unstable when the fall of temperature with height of the environmental air is more rapid than that experienced by a rising air parcel, so that the parcel continues to be less dense than the ambient air. See E. Fedorovich et al. (2004).

institutional thickness 'The localised conditions that are crucial to the particular articulation of localities and regions within wider scale processes of economic transformation' (Henry (2001) *Env. & Plan. A* **33**). Institutional thickness includes trade associations, voluntary agencies, sectoral coalitions, concrete institutions, and local elites—their effects on local policy, and their consensus institutions: common agreements, shared views and interpretations, and unwritten laws.

E. Sheppard (2002) is sceptical: 'how, exactly, does institutional thickness promote local economic development and prosperity? Further, how do we explain the fact that some very successful regions do not appear to be underpinned by a highly concentrated and co-ordinated matrix of institutions?' Does institutional thickness stimulate growth, or vice-versa? (A. Wood and D. Valler 2004).

integrated assessment (IA) An interdisciplinary process which combines, explains, and communicates knowledge from a range of disciplines in order wholly to weigh up an entire chain of causes and effects. Van Amstel (2005) *Env. Sci.* **2**, 2–3 uses an integrated assessment model to evaluate six methane emission reduction strategies on future air temperature. For a second example, see Mendonça et al. in M. van Asselt et al., eds (2001).

integrated land-use and transportation models Models which attempt to represent human behaviour in a realistic way, measuring and representing the decision-making that leads to the organization of human activities in time and space. These range from daily household decisions— what to do, where and when to do it, and by which mode of transport—to long-term decisions relating to land use and transport networks. See T. Nisbet, H. Orr, and S. Broadmeadow (2004) for using integrated land use planning to resolve problems of sedimentation.

Despite examples such as these, Coucelis (2005, *Env. & Plan. A* **37**, 8) argues that 'despite more than four decades of effort to integrate the two fields, the place of land-use models in planning remains problematic'.

(((()) SEE WEB LINKS)

• The Heartlands Initiative.

integrated water management 'A process that promotes coordinated development and management of water, land, and related resources, in order to not only maximize economic and social welfare, but also ensure equity and sustainability' (Chao (2005) *Env. Informatics Archs* **3**). Radif (1999, *Desalination* **124**) urges that integrated water resources management be based 'on the perception of water as an integral part of the ecosystem, a natural resource and a social and economic good'.

The general objective is the environmentally sound, equitable, and sustainable utilization and development of water resources. This may include supply management, demand management (water conservation, transfer of water to uses with higher economic returns, etc.), water quality management, recycling and reuse of water, economics, conflict resolution, public involvement, public health, environmental and ecological aspects, socio-cultural aspects, water storage (including long-term storage or water banking), conjunctive use of surface water and groundwater, water pollution control, flexibility, regional approaches, weather modification, sustainability, etc. (Bouwer (2000) *Ag. Water Manage.* **45**).

In a study of the *Loess Plateau, China, Chen et al. (2007) *PPG* **31**, 4 list the factors to be considered: 'basic farmland construction, plantation of cash trees, firewood and conservation wood, roads, water banks, agricultural structure modifications, and local people's needs (such as daily fuel, daily drinking water, education, medical care and other factors regarding improving living conditions).' Remedial measures include the use of *check dams; afforestation and the planting of shrubs and grasses on slopes (*see* REHABILITATION ECOLOGY); and land closure, in order for ecosystems to rehabilitate naturally, without human intervention (see Zhang (2005) *Acta U. Agric. Suecia* **126**). 'Natural recovery by land closure is strongly recommended on the steep slopes' (Chen et al., op. cit.).

In natural rehabilitation, the degraded soil may be improved (Jia et al. (2005) *Forest Ecol. & Manage.* **217**), and the biodiversity may be increased steadily over time. However, revegetation usually takes a long time, especially from an extremely degraded state (Bellot et al. (2001) *Land. & Urb. Plan.* **55**). Zhang (op. cit.) estimates that several decades to hundreds of years may be needed in the Loess Plateau.

integration 1. Social integration The process whereby a minority group, particularly an ethnic minority, adapts to the host society and where it is accorded equal rights with the rest of the community. Lemanski (2006) *GeoForum* **37**, 3 reports on social integration in post-apartheid urban South Africa: 'different races are not only living peacefully in shared physical spaces but also actively mixing in social, economic and, to a lesser extent, political and cultural spaces.' But Nagel and Staeheli (2008) *Soc. & Cult. Geog.* **9**, 4 call integration 'a socio-political process by which dominant and subordinate groups negotiate the terms of social membership'. Lemanski notes a social continuum of integration (greeting in the street, visiting homes, inter-marriage) and a continuum of **spaces of integration**, for example physical space (shared neighbourhood), economic space (common employment type), social space (cross-race friendship), political space (common involvement in civic organizations), and cultural space (shared sense of belonging).

2. Economic integration can be the breaking down of trade barriers between nations. 'Economic integration has a direct effect on internationalisation by reducing transaction costs, and information costs are a part of transaction costs. Being in a currency union has been shown . . . to have a significant effect on bilateral trade' (Rose (1999) *CEPR Discussion Paper* **329**, London). The effect of sharing a common currency is significant even when exchange rate volatility is taken into account (Guerin (2006) *World Econ* **29**, 2). **Horizontal integration** occurs across different sectors; **vertical integration** occurs within a hierarchy; see Farrington (2007) *J. Transp. Geog.* **15**, 5.

intellectual property A creative work, or a device or invention that can be patented or copyrighted. Different knowledges have been variously adopted, reworked, and contested at different scales, including attempts to defend and (re)construct more collective notions of knowledge and property (S. Wright, U. Newcastle, UK). 'Endemic healing floras and native intellectual property diffused back and forth between Europe and the colonies, transforming forever the geography of tropical plant pharmacopoeias and folk healing traditions' (Voeks (2004) *AAAG* **99**, 4). Intellectual property rights laws are best understood as particular, culturally defined systems for codifying knowledge, employed to discipline objects, phenomena, and social relations (Parry (2002) *Antipode* **34**, 4).

intensity The size of an earthquake, based on the impact on people, buildings, and the ground surface. Isoseismal lines represent equal earthquake intensity. The Modified Mercalli intensity scale is the world standard; the Medvedev–Sponheuer–Kárnik (MSK) is popular in Europe. Although differing in detail, the MM and MSK scales are much alike. As they are descriptive, and based on recorded damage, **intensity values** can be calculated and isoseismal lines drawn for present day and historic earthquakes (Chester (2001) *PPG* **25**, 3).

intensive agriculture Agriculture with a high level of capital and labour inputs, and high yields. 'It would appear that there is no large complex society without intensified agriculture' (R. C. Hunt, no date). Phillips (1998) *Appl. Geog.* **18**, 3, provides a gloomy list of the effects of agricultural intensification in Italy: declining frequency of crop rotations; increased use of chemical fertilizers and pesticides; more up- and downslope tillage; and enlarged fields. Mechanization has led to a decline in draught cattle numbers, causing loss of agricultural employment, and a fall in soil fertility. See Kasente et al. (2000) *UNRISD Occ. Paper* **12**, on the intensification of non-traditional agricultural exports.

interaction The action between points, each upon the others; 'business concentration clearly facilitates more intensive social interaction, and interaction intensity is argued to be a key element differentiating between types of client-supplier relations' (Bennet et al. (1999) *TIBG* **24**, 4). An **interaction model** describes the reactions of two or more processes or systems as they affect each other; see Eradus et al. (2002) *J. Transp. Geog.* **10**, 2.

interception The retaining of raindrops by plant leaves, stems, and branches. Xiao et al. (2000) *Hydrol. Procs* **14** find that rainfall frequency is more significant than rainfall rate and duration in determining interception. Guevara-Escobar et al. (2007) *J. Hydrol.* **333**, 2–4 evaluate interception around the canopy of a single tree.

interdependence In a heartening review of the term, Smith et al. (2007) *Geog. Compass* **1**, 3 declare that interdependence conveys a sense of relying on and being responsible for others. 'Today we all gather here to declare our interdependence. Today we hold this truth to be self-evident: we are all in this together' (Will Smith, Live8, 2005).

interface The zone of interaction between two systems or processes. Estuaries are interfaces between *fluvial and marine systems (Bate et al. (2002) *Water SA* **28**, 3).

interflow Water that moves downslope through soil pores (as opposed to *throughflow, which is soil water moving downslope along impermeable soil horizons—it's difficult to separate the two processes in the field). Interflow is of major importance for run-off generation and groundwater recharge (Flugel and Smith (1998) *Env. Model. & Sofw.* **14**, 2).

interglacial A long, distinct period of warmer conditions between *glacials when the earth's glaciers have shrunk to a smaller area—the present, Holocene, period is an interglacial. Soon (2007) *Phys. Geog.* **28**, 2 argues that the persistence of insolation forcing at key seasons and geographical locations, taken with closely related thermal, hydrological, and cryospheric changes, suffice to explain transitions in paleoclimates.

intermediate technology Mid-level technology, often the most suitable level for economic development. Chinese aid projects in Africa, for example, consist largely of intermediate technology: Chinese workers are paid at local rates, train their local equivalents, and aim to leave projects in a state where they can be locally managed, operated, and maintained so that even the spare parts could be locally produced (Mawdsley (2007) *Geog. Compass* **1**, 3). The *Intermediate Technology Group* is now called *Practical Action*.

internal deformation A distortion within a material, such as glacier ice: individual ice crystals react to stress by elongating, or by melting and recrystallizing, and ice may also shear along separate shear planes; see Hambrey et al. (2005) *J. Geophys. Res.* **110**, F01006 and Mair et al. (2001) *ESPL.* **23**. See Lewkovicz and Harries (2005) *Geomorph.* **69**, 1–4 on internal deformation in permafrost.

internal migration The temporary or permanent relocation of its population inside a *nation-state. China has the most internal migration; in 2003, the so-called floating population reached 140 million, most of them rural labourers moving from the countryside to cities and coastal areas (Ping and Shaohua (2005) *Chinese Acad. Soc. Scis*). The rapid growth of rural–urban migration in China results from the privatization of agriculture, and the transition to a market economy (Ping and Pieke (2003) *Inst. Sociol., Chinese Acad Soc. Scis*).

international division of labour The allocation of various parts of the production process to different places in the world. 'The world economy is organized through horizontal and vertical linkages of an international division of labour, in which the modes of integration and geographical scopes vary over time' (Appelbaum et al. in G. Gereffi and M. Korzeniewicz, eds 1994). Grinberg (2007) *Glob. Pov. Re. Grp* observes that whilst some countries have tended to concentrate within their boundaries the great bulk of the skilled labour force and therefore of the most complex labour processes (the USA, western Europe, and Japan), others have been mainly transformed into sources of a relatively cheap and disciplined, though less skilful, labour force (mainly East and South-East Asia). A third group of countries have been turned into reservoirs of consolidated surplus population relative to the needs of the accumulation process (for instance, most of Sub-Saharan Africa and parts of Asia, Northern Africa, and Latin America) and, in some cases, have eventually become new sources of cheap and disciplined labour (parts of South Asia, Vietnam, and the Caribbean Basin). See Hardy (2007) *Eur-Asia Studs* **59**, 5 on the new international division of labour and Poland.

International divisions of labour involve transfers of reproductive labour beyond territorial boundaries. Thus, Filipina workers take on the household tasks of middle-class women abroad, while they themselves may hire poorer workers to do their housework in the Philippines (Boyle (2002) *PHG* **26**, 4). *Service offshoring* is an extension of outsourcing as firms apply an international division of labour to service tasks (Bryson (2007) *Geografiska B* **89**). 'To many Marxist scholars, the rise of East Asia as a global factory can often be "read off" as an empirical proof of the working of . . . spatial fixes, particularly in relation to the rise of the new international division of labour' (Cheung (2007) *Econ. Geog.* **83**, 4).

International Fund for Agricultural Development (IFAD) A UN
agency for ending rural poverty in developing countries.

((●)) SEE WEB LINKS

• The IFAD website.

International Monetary Fund 'The work of the IMF is of three main
types. Surveillance involves the monitoring of economic and financial developments, and the provision of policy advice, aimed especially at crisis-prevention. The IMF also lends to countries with balance of payments difficulties, to provide temporary financing and to support policies aimed at correcting the underlying problems; loans to low-income countries are also aimed especially at poverty reduction. Third, the IMF provides countries with technical assistance and training in its areas of expertise. Supporting all three of these activities is IMF work in economic research and statistics. In recent years, as part of its efforts to strengthen the international financial system, and to enhance its effectiveness at preventing and resolving crises, the IMF has applied both its surveillance and technical assistance work to the development of standards and codes of good practice in its areas of responsibility, and to the strengthening of financial sectors. The IMF also plays an important role in the fight against money-laundering and terrorism' (IMF website).

The IMF doesn't get a good press from geographers, largely through its role in promoting *structural adjustment policies: as a result of austerity programmes sponsored by the World Bank and the IMF, 'the 1980s and 1990s were the "lost decades" for development in much of Africa' (Carmody (2008) *Geog. Compass* **2**, 1). Watts et al. (2005) *PHG* **29**, 1 describe the IMF as one of the ungoverned bodies that regulate global capitalism, and D. Harvey (2004) argues that localized financial crises are in some sense intentionally managed by various US-sponsored agencies such as the International Monetary Fund under the guise of neoliberal economics, 'thereby maintaining the United States as the global hegemon'.

SEE WEB LINKS

• The IMF website.

internet A global network of connected computers (Kitchin (1998) *PHG* **22**). Warf (2007) *Sing. J. Trop. Geog.* **28**, 2 identifies costs, wealth, literacy, and telephone penetration rates as determinants of internet uptake. Zook (2002) *Jour. Econ. Geog.* **2**, 2 observes that 'although both the internet and venture capital have been viewed as independent of geography, the development of this industry continues to highlight the continuing relevance of regions and place-based relations'. Thus, Gorman (2002) *Tijdschrift* **93**, 5 finds that website production is dominated by New York and San Francisco.

In a North Carolina study, Palm and Danis (2002) *Tijdschrift* **93**, 5 find that the internet has little impact on housing search patterns. Adams and Ghose (2003) *PHG* **27**, 4 describe the vast, variegated space of international and multicultural communication used by the 'Indian diaspora' and residents of India. See Wharf and Vincent (2007) *Area* **39**, 1 on Arab geographies of the internet, J. Rodgers (2003) on the internet and activism, and Grubesic and O'Kelly (2002) *Prof. Geogr.* **54**, 2 on measuring access to commercial internet sites.

interpluvial A time of increased aridity in deserts; the time between *pluvials.

inverted U Kuznets (1955) *Am. Econ. Rev.* **45** suggests that income equality first increases, and then decreases, with *economic development; List and Gallet (1999) *Rev. Dev. Econ.* **3**, 2 find evidence to support this contention. This concept can be extended to regional inequality between nations; see Yamamoto (2008) *J. Econ. Geog.* **8**, 1 or to industrial concentrations: Grote (2008) *J. Econ. Geog.* **8**, 2 finds an inverted 'U'-shaped concentration of foreign banks in Germany.

intersectionality The interconnections and interdependence of race with other categories (K. Crenshaw et al. 1995). Wills (2008) *Antipode* **40**, 1 believes that geography has real strength as a discipline for the study of intersectionality: 'to understand the ways in which class intersects with other social cleavages, with very different e/affects.' Understanding such intersectionality at the local level can help in mapping larger-scale trends in the geography of class and employment; see A. Ong (2006). G. Valentine (2007) argues that feminist geography needs to use the concept of intersectionality to theorize the complex relationship between and within different social categories: 'the specific debate about intersectionality as a concept has not yet been played out within geography despite its obvious spatial connotations.' See also Valentine (2008) *PHG* **32**, 3.

interstadial A warmer phase within a *glacial which is too short and insufficiently distinct to be classed as an *interglacial. See Phillips et al. (1994) *Geology* **22** on interstadial climatic changes, and Ringberg et al. (2003) *Boreas* **32**, 2 on interstadial events and glacial varves.

inter-tropical convergence zone (ITCZ) That part of the tropics where the opposing north-east and south-east *trade winds converge. It is not a continuous belt, more like a necklace with groups of clouds as the 'beads'; in places there may be two or more 'strings'. The zone is narrower over the oceans, and broader over the continents, where other wind systems may be involved; in West Africa the ITCZ is the convergence of the Guinea *monsoon and the *Harmattan (Miskolczi et al. (1997) *J. Appl. Met.* **36**, 5). See Yancheva et al. (2007) *Nature* **445** on the influence of the ITCZ on the East Asian monsoon.

The ITCZ moves north and south; moving more over land, and arriving in the summer in each hemisphere. Its position is affected by the apparent movement of the overhead sun, the relative strengths of the trade winds, and the changing locations of maximum sea-surface temperatures (R. D. Thompson 1998). This means that the movements of the ITCZ are highly unpredictable. In May 2005, the African portion of the ITCZ was 1° south of the average; a period of drought near the Sahel. In July 2003, it was around 0.9° north of the average; heavy rains fell in Ethiopia, with flooding in Khartoum. The movements of the ITCZ also affect bush/forest fires in South-East Asia (P. Kershaw et al. 2001). Over the oceans, the ITCZ is broad, and often loses its identity.

intervening opportunities theory Stouffer (1940) *Am. Soc. Rev.* **5**, 6 argues that the spatial distribution of competing (alternative) intervening destinations influences decision-makers' choices (Cascetta et al. (2007) *Papers Reg. Sci.* **86**, 4). See Alix et al. (1999) *J. Transp. Geog.* **7**, 3 on Montreal as an intervening transhipment opportunity between Quebec and Europe.

intervention price A guaranteed minimum price, set by a government, for agricultural produce. Should prices fall below this minimum, the government must buy the produce at this price. The intervention price is usually a percentage of the price hoped for in the open market. Try Polet (2007) *GAIN Report* **E47016**. In 2003, the European Union started to lower intervention prices, but raise direct support to farmers.

SEE WEB LINKS

• CAP reform and intervention prices.

intramax methodology The identification of nested functional regions in an information space by aggregating spatial units according to document similarity, using a hierarchical clustering algorithm. Poon et al. (2000) *TIBG* **25**, 4 use the method to analyse the intensity of international trade and foreign direct investment flows; see also Goetgeluk and de Jong (2007) *ENHR Int. Conf. Sustainable Urban Areas*.

intrazonal soil A soil affected more by local factors than by climate, unlike a *zonal soil. For example, waterlogging creates *gley soils, and a limestone bedrock will produce a *rendzina.

intrenched meander *See* INCISED MEANDER.

intrusion A mass of igneous rock which has forced its way, as *magma, through pre-existing rocks, and then solidified below the surface of the ground; hence **intrusive rock**. See Jousset et al. (2003) *J. Volc. & Geoth. Res.* **125**, 1–2.

Igneous intrusions are classified according to their shape and size: *dykes are **discordant**, since they cut across the bedding or fabric of pre-existing rocks, *sills are **concordant**, since they intrude between layers. *Plutons are the largest intrusions, typically spheroidal, and ranging from *stocks to *batholiths.

Intrusion

invasion The establishment of species in ecosystems they are not native to; see, for example, Russell-Smith et al. (2004) *J. Biogeog.* **31**, 8 on the invasion of rain forest into eucalyptus woodland in Australia. Campbell et al. (2002) *J. Biogeog.* **29**, 4 outline a methodology which might be able to be used to predict invasions by alien plant species. 'Humans cause invasions, humans perceive invasions, and humans must decide whether, when, where and how to manage invasions' (Richardson et al. (2008) *PHG* **32**, 2).

invasion and succession **1.** A model of change used in *urban ecology to represent the effects of immigration on the social structure of an urban area. Invasion and succession involve a chain reaction, with each preceding immigrant wave moving outwards and being succeeded by more recent, poorer immigrants (R. E. Park, E. W. Burgess, and D. McKenzie 1925/1974). This model saw immigrant enclaves as transitional stages on the road to eventual acceptance and integration in the larger (American) society (Clark in R. Waldinger and M. Bozorgmehr, eds 1996). Van Kempen(2007) *J. Housing & Built Env.* **22**, 1 argues that divided cities cannot be explained without using concepts like invasion, succession, and filtering.
  2. In plant ecology, the introduction and subsequent spread of a species. Diamond (1974) *Science* **184** provides a dynamic model of successional processes, and R. H. Groves and F. Di Castri (1991) distinguish between invasion and succession.

invasive species A species is invasive if it: is not native to the biogeographical region in question; has adapted to its new location; is becoming more profuse/more widely distributed; is an irritant; and results from anthropogenic activity. Henderson et al. (2006) *PPG* **30**, 1 find that intact natural areas are becoming more invaded and degraded.

inversion The increase of air temperatures with height. (This is the reverse of the more common situation in which air cools with height.) Inversions occur: when strong, nocturnal, *terrestrial radiation cools the earth's surface and therefore chills the air which is in contact with the ground (Lapworth (2003) *Qly J. Royal Met. Soc.* **129**); when cold air flows into valley floors, displacing warmer air (LeMone et al. (20030 *J. Atmos. Scis* **60**); where a stream of warm air crosses the cool air over a cold *ocean current (Vihma et al. (2003) *Boundary-Layer Met.* **107**); where warm air rises over a cold front (Ross and Orlanski (1978) *J. Atmos. Scis* **35**, 3); when air from the upper *troposphere, subsiding in a warm *anticyclone, is compressed and *adiabatically warmed. The boundary between the top of the cold air and the beginning of the inversion is an **inversion lid**. *See also* TRADE WIND INVERSION.

involution **1.** The refolding of two *nappes differing in age so that parts of the younger nappe lie below older rocks.
  2. The convolution of layers of ground under *periglacial conditions. To form thermokarst involutions, ice-rich permafrost must thaw, drainage conditions must be poor, and sediments must vary in texture or composition. In addition, the sediments should be susceptible to fluidization, liquefaction, or hydroplastic deformation (Murton and French (1993) *Perm. & Periglac.* **4**, 3). The repetition of differential heave and soft-loam settlement promotes

decimetre-scale involutions in near-saturated soils subject to deep seasonal frost penetration (Ogino and Matsuoka (2007) *Perm. & Periglac.* **18**, 4).

ionosphere A layer of the *atmosphere containing ions and free electrons. The ionosphere is warmed as it absorbs solar radiation, and it will reflect radio waves. Three bands are recognized: E, F, and F2, at about 110, 160, and 300 km above the earth. The ionosphere is in a state of constant motion, and is affected by tidal forces and by the earth's magnetic field.

irradiance The rate of flow of radiant energy through unit area perpendicular to a solar beam. **Total solar irradiance** is the dominant driver of global climate (Mendoza (2005) *Advances Space Res.* **35**, 5). See also Mishchenko et al. (2007) *Bull. Am. Met. Soc.* **88**, 5.

irrigation The supply of water to the land by means of channels, streams, and sprinklers in order to permit the growth of crops. Without irrigation arable farming is not possible where annual rainfall is 250 mm or less, and irrigation is advisable in areas of up to 500 mm annual rainfall. To some extent, irrigation can free farmers from the vagaries of rainfall and, to that end, may be used in areas of seemingly sufficient rainfall because irrigation can supply the right amount of water at the right time. 'The contrasts in adjustments and attitudes over the use of water for agriculture stem largely from the inconsistent provision of information from the scientific community, and differing ideologies, namely environmental protection versus economic development' (Yang et al. (2006) *Geogr. Res.* **44**, 2). 'Essential complements to water pricing are water distribution rules and technological choices at critical nodes in the delivery system that allow farmers flexibility in conserving water in response to higher prices. Among supporting institutions, water users' associations seem a higher priority than water markets' (Dinar and Mody (2004) *Nat. Resources Forum*, **28**, 1).

Writing on Ecuador, Cremers et al. (2005) *Nat. Resources Forum*, **29**, 1 judge that bottom-up processes of awareness, capacity-building grassroots' claims, collective action, and mobilization are critical in establishing water rights, together with governmental political will and an enabling political environment. In a Malaysian study, Johnson (2000) *Geog. J.* **166**, 3 show that tertiary intervention has increased the capacity of the farmers to unofficially control the distribution and supply of the water resource, resulting in a significant over-supply of water; inefficient water use; and a reduction in yields without a reduction in incomes. Lam (1994), cited by Kurian and Dietz (2004) *Nat. Resources Forum* **28**, 1, finds a negative correlation between land-holding inequality and irrigation management performance. See Reinfelds et al. (2006) *Geogr. Res.* **44**, 4 on sustainable diversion limits in river systems.

isentropic (isentropic surface) In meteorology a surface of constant potential temperature. Isentropic surfaces slope very gently upwards towards the cold air in the presence of a horizontal temperature gradient, and winds flow along these surfaces. See Ambaum (1997) *J. Atmos. Scis* **54**, 4, on the isentropic formation of the tropopause.

island arc An arcuate island chain, mostly of volcanic origin, formed when *oceanic crust plunges into the *mantle where it undergoes *subduction. The *magma thus formed creates a chain of submarine volcanoes—eventual islands; see Jicha et al. (2006) *Geology* **34**, 8 on magma flow in the Aleutian Island arc. See also Takahashi et al. (2007) *Geology* **35**, 3.

island biogeography The relationship between area and species number on islands, as an equilibrium between immigration and extinction. See R. H. MacArthur and E. O. Wilson (1967) and R. Whittaker (1998).

Island biogeography is also relevant to continents, when plant and animal communities are effectively reduced to islands in a sea of cultivation or urbanization; Meadows (2001) *PPG* **25**, 1 writes on *habitat islands*—distinctive communities in a sea of other types of community.

isochrones A line connecting places of an equal journey time to the same location.

isodemographic map A cartogram which represents nations, states, counties, and so on, in proportion to the size of their population. See A. Cliff and P. Haggett (1988).

isogloss A line on a map which represents the geographical boundary of regional linguistic variants. By extension, the term also refers to the dialect features themselves. For example, in the USA north of the Mason–Dixon line, the word greasy is pronounce with an 's'; south of the line, it is sounded with a 'z' (L. Campbell 2004). Isoglosses are usually highly simplified representations and do not depict an abrupt transition. Sometimes, a **bundle of isoglosses** may occur, where a number of isoglosses lie close enough together to indicate a true *dialect boundary; see W. Wolfram and N. Schilling-Estes (1998).

isoline/isopleths Any line on a map joining places where equal values are recorded: an **isohyet** connects points of equal rainfall; an **isophene** connects places with the same timing of similar biological events; an **isotherm** joins places of equal temperature; and an **isotim** is a line drawn about a source of raw materials or a market where transport costs are equal.

isostasy The continental crust of the earth has a visible part above the surface and a lower, invisible one. The balance between these two is isostasy. **Erosional isostasy** is a 'plausible mechanism' (Maddy et al. (2000) *Geomorph.* **33**, 3–4) for differential crustal movement, in which the crust realigns in response to erosional unloading and depositional loading (M. A. Summerfield 1991). The lack of evidence for significant surface denudation may suggest that erosional isostasy may be no more than a positive feedback response to uplift initiated by other influences (Maddy (1997) *J. Quat. Sci.* **12**). **Glacio-isostasy** results from the cessation of pressure from glacier ice on glacier melting, with unloading leading to uplift (Lambeck (1995) *J. Geol. Soc. London* **152**). **Hydro-isostasy** is similar to glacio-isostasy; here, the loading/unloading of the continental shelves due to changes in sea level may have been significant in driving continental uplift.

isosteric Of that part of the atmosphere having uniform density.

isotropic Having the same physical properties in all directions.

jet stream 'A sinuous, relatively narrow ribbon of air that encircles the globe at an altitude of around 11 km, sometimes reaching speeds of 400 kph. They form where there is a substantial variation of air temperature within a relatively short geographical distance' (Meteogroup, 26 Nov. 2006). The westerly jet streams (J) in the northern hemisphere lie where significant breaks occur in the tropopause.

The **polar-front jet stream** is a frontal wind, not necessarily continuous, blowing west to east just below the *tropopause, parallel to the surface fronts and moving with them, and draining the air rising from the fronts. It is strongest at the 200–300 mb level, and swings between latitudes 40 and 60 °N; speed and location vary from day to day with the Rossby waves. This jet marks the polar *front, and has important effects on *convergence and *divergence in the upper air. Divergence is more frequent in the downstream, poleward sector of the jet core, and is associated with extratropical cyclone development. At the 'jet entrance', the pressure gradient steepens, and the wind becomes super-*geostrophic, leading to high-level convergence. The polar-front jets often 'steer' the movement of major low-level air masses.

A strong polar-front jet is associated with rapidly moving *depressions (Galvin (2006) *Weather* **15**, 11), but the jet stream and its accompanying depressions in the surface can be diverted by *anticyclones. This jet splits into the polar front jet and subtropical jet, which develop each year in March through to early Southern Hemispheric spring (Bertler et al. (2006) *Advances Geoscis* **6**, 83386).

The **westerly subtropical jet** is at the poleward limit of the *Hadley cell, around 30° N and S; the northern subtropical jet is strongest at the 200 mb level, and above the Indian subcontinent. This is one of the most powerful wind systems on earth, at times reaching speeds of 135 m s^{-1}, and it follows a more fixed pattern than the polar-front jet. It results from the poleward drift of air in the Hadley circulation and the *conservation of angular momentum. Chen and Yen (1986) *Mon. Wea. Rev.* **114**, 255032570 are among many to link this jet with the South Asian summer monsoon.

The **tropical, easterly jet** develops during the summer months at 15° N, and is strongest at the time of the summer *monsoon. The **stratospheric, subpolar jet stream** blows at a height of 30000 metres, being westerly in winter and easterly in summer.

Jet stream. Identified by J on the chart

joint A crack in a rock without any clear sign of movement either side of the joint.

jökulhlaup 'A "catastrophic" flood, often generated by an ice-dammed lake outburst', for example the volcanic eruption and concomitant catastrophic flooding from Grímsvötn, Iceland, in November 1996 (Tweed and Russell (1999) *PPG* **23**, 1). Alho (2003) *Geograf. Annal. A* **85**, 3–4 illustrates the severe and extensive jökulhlaup-induced process of land degradation.

Jurassic The middle period of *Mesozoic time stretching approximately from 190 to 136 million years BP.

(⊕) SEE WEB LINKS

• An overview of the Jurassic period.

just city Democracy, equity, diversity, growth, and sustainability are indispensable to the creation of just cities (Fainstein (2005) *Urb. Affairs Rev.* **41**): 'I did not attempt a justification for choosing these values but simply assumed agreement on them.' These are problematic: 'illiberal majorities can make democracy indifferent to minority rights; the high cost of achieving equity [can create] resentment . . . diversity can lead to social breakdown; and growth . . . benefits most those who already have the most. Sustainability may diminish growth' (Fainstein (2006) *Lecture on the Just City*). She later writes, 'if the city is composed of diverse communities, conflict arises over whose definition of justice should prevail' ('The Resurgent City'). 'Ethical demands for the cultural recognition of different groups now require that questions of the good life be addressed alongside deontological questions of morality and rights to create meaningful social justice' (Shields (2000) *Distinktion*, **4**).

SEE WEB LINKS

- Susan Fainstein's 'The Resurgent City'.
- MIT's 'Just Jerusalem'.

justice 'Geographers have long been concerned with questions of social justice. In recent years there has been a particular emphasis on the processes through which certain social groups such as women, ethnic minorities, children, lesbians and gay men and disabled people are excluded from, or marginalized within, everyday spaces Research on these topics inevitably has led to questions about how academics can address these issues in ways that are sensitive to, and empower those participating in, such studies and consequently to debates about professional ethics' (Valentine (2003) *PHG* **27**; see also D. Sibley 1995).

Geographical investigations of equity, fairness, and justice have centred on spatial equality, territorial justice, and minimum standards (Hay (1995) *TIBG* **20**, 4): 'what is just at one geographical scale for one set of workers might be unjust for other workers elsewhere.' While aeronautical engineers working in Seattle might feel they are justly treated by Boeing, Mexican workers stitching seat-covers for Boeing aircraft may not enjoy equal pay and conditions. However, 'the logic of capital is... not about... social equity, human happiness, or environmental justice' (N. Castree et al., 2004).

Raustiala (2005) *Fordham Law Review*, rejects the US Federal Court's notion that 'American law is tethered to territory—that simply by moving an individual around in space, the rights that individual enjoys wax and wane... spatial location ought not woodenly foreclose the existence of constitutional rights for noncitizens subject to American power outside the boundaries of the United States'. Osofsky (2008) *Villanova Law School* **53** explores '"wormholes" in the US legal system that transport people... into another timespace in which basic protections are absent'. He is referring to rendition flights.

The Brundtland Report establishes the pursuit of intra- and intergenerational equity as key dimensions of future progress towards sustainability. 'If the ability to live in an environment which meets reasonable standards of quality and tolerable levels of environmental risk is included as a basic need... pursuing environmental justice is... central to the broader framing of sustainable development' (Walker and Bulkeley (2006) *Geoforum* **37**, 5). See A. Dobson, ed. (1999) on *sustainable development and intergenerational justice.

just-in-time system (JIT) A system of production which aims to deliver all the necessary inputs, such as raw materials, components, and labour, just in time for the appropriate stage of production. This is essentially a flexible system of manufacturing, linked with small production runs and flexible methods of production. It is largely credited with being a Japanese innovation. 'Dell [computers] has created a distinctive geography based upon a small number of geographically dispersed, strategically located assembly sites connected into close concentrations of suppliers needed for its just-in-time system . . . "we are not experts in the technology we buy. We are experts in the technology of supply chain integration"' (G. Fields 2004).

Kamchatka current A cold *ocean current off the peninsula of the same name in Siberian Russia. See Cokelet et al. (1996) *J. Phys. Oceanog.* **26**, 7.

kame An isolated hill or mound of stratified sands and gravels which have been deposited by glacial meltwater. Lundqvist (2003) *Geografiska A* **85**, 1 is recommended; see also Bennett et al. (2000) *Quat. Sci. Revs* **19**, 7. **Kame terraces** are flat-topped, steep-sided ridges of similar *fluvio-glacial origin, running along the valley side. They are *ice contact features, formed between the side of a decaying glacier and a valley wall; see Lundqvist (1995) *Geomorph.* **11**, 3 and Bennett et al. (2000) *Quat. Sci. Revs* **19**, 7 on lake-bed kame terraces. Moulin kames form below *moulins (Carlson et al. (2005) *Geomorph.* **67**, 3–4).

karre A collective name (pl. *karren*) for the shallow channels formed by solution on exposed limestone. **Kluftkarren** are enlarged joints; **rillenkarren** are very closely spaced small runnels; small radiating rillenkarren are the overflow of surface solution pans. **Rinnenkarren** are both longer and deeper, as much as 10 m in length and 0.5 m in depth. They may develop as a result of coalescence of small channels. Their walls are sharp, in contrast with the large, rounded hollows known as **rundkarren**. The types of karren can be explained as the results of the solution process under different hydro-dynamical behaviour (Ferrarese et al. (2003) *Geomorph.* **49**, 1–2).

karst (labyrinth karst) The deep canyons of limestone formed by *carbonation. Initially the limestone shows the wide, deep tissues known as bogaz; these widen and deepen into long gorges known as karst streets with other, cross-cutting, lines of erosion. The remnant of this carbonation is **tower karst**. See Zhu and Chen (2006) *Speleogen. & Evolution Karst Aquifers* **4**, 1 on **cone karst**, **tower karst**, and tiankengs (giant dolines), and Harmon et al. (2006) *GSA Spec. Paper* **404** for perspectives on karst geomorphology, hydrology, and geochemistry.

kata- From the Greek *kata-*, down-sinking, as in a *katabatic wind. The term is also used to describe the sinking of air in the warm sector of a *depression at a cold or a warm **kata-front**, bringing about a large-scale inversion of temperature at the fronts, which are fairly inactive. At a kata-warm front, cloud development is limited to cirrus and high stratus, and precipitation is restricted to light rain; at a kata-cold front strato-cumulus is common, and precipitation is similarly moderate.

katabatic wind A gravity-driven atmospheric current, which is forced by cooling air adjacent to a sloped surface (Gallée et al. (1996) *J. Appl. Met.* **35**, 7). Descending, *adiabatically warmed katabatic winds are *föhn winds. Cold katabatic winds result from the slumping down of very cold, and hence dry, air. Coastal Antarctica is dominated by katabatic gales; the gentler katabatic flows of hill slopes produce *frost hollows.

kettle hole Large masses of ice can become incorporated in glacial *till and may be preserved after the glacier has retreated. When one of these bodies of ice finally melts, it leaves a depression in the landscape; a kettle hole. **Kame and kettle topography** is hummocky terrain evolved by melt-out, comprising pitted or kettled outwash (*sandur).

Khamsin *See* LOCAL WINDS.

kinematic wave theory A theory developed by Lighthill and Whitham (1995) *Procs Royal Soc., London, Series A* **229** which has applications in geomorphology. See Howes and Abrahams (2003) *Geomorph.* **53**, 1–2 for a kinematic wave equation for overland flow, and Van de Wal and Oerlemans (1995) *J. Glaciol.* **41** on high-velocity kinematic waves in glaciers, generated by a major accumulation of snow above the *firn line, and causing a *glacial surge when the wave reaches the snout.

kinetic energy The energy of motion; in geomorphology, the energy used by wind, water, waves, and ice. Kinetic energy for channelled flow is defined as:

$$\frac{MV^2}{2}$$

where M is the mass of water, and V is the mean velocity.

knick point (nick point) A point at which there is a sudden break of slope in the *long profile of a river. In areas of uniform geology, the presence of a knick point may be evidence of *rejuvenation; the river is forming a new, lower profile cutting first from the mouth of the river and working upstream as *headward erosion takes place. See Crosby and Whipple (2006) *Geomorph.* **82**, 1–2 and Schlunegger and Schneider (2005) *Geomorph.* **69**, 1–4. A **knickzone** is a steep reach caused by more resistant lithology, by an increase in shear stress downstream of a confluence, or by surface uplift (Bishop et al. (2005) *ESPL* **30**, 6).

knock and lochan topography A glacially scoured, lowland landscape. Low, rounded hills (*knocks*) and roches moutonnées alternate with *striated and eroded hollows, which may be tens of kilometres long, often containing lakes (*lochans*). Drainage is generally chaotic.

knowledge-based organizations Those organizations that successfully create new knowledge as the basis for new products and services, usually associated with large urban agglomerations. The **knowledge-based economy**

(**KBE**) generates information rather than goods and services. M. Castells (1996) sees the production of information and of knowledge-intensive goods and services as characteristics of post-industrial societies; others see it as a product of globalization (see J. Dunning, ed (2004) and Schamp et al. (2004) *Eur. Plan. Studs*, **12**, 50). Larner (2007) *TIBG* **32**, 3, reports on the use of long-standing, informal networks and opportunistic links with expatriate experts in the KBE of New Zealand.

Not everyone sees the rise of the KBE as beneficial. Von Osten in P. Spillmann, ed. (2004) points out that 'controlled access to knowledge goods and information . . . creates new global differences in power, new forms of resistance and subversive practices', and Chi-ang Lin (2006, *Eco. Econ.* **60**, 1) is troubled by the persistent emphasis on knowledge and economic growth at the expense of poverty reduction and environmental conservation. See Hudson (1999) *Eur. Urb. & Reg. Studs* **6**, 1.

knowledge flows. The mobility and migration of the highly skilled in knowledge-based economies is interwoven in complex ways with flows of knowledge. Hughes (2007) *PHG* **31**, 4, summarizes the literature. Park (2003) *Papers Reg. Sci.* **82**, 2 demonstrates that the internet has greatly contributed to the diffusion of codified knowledge and technology to China and the Asia Pacific Rim, but the distribution of knowledge is spatially uneven. Thus, he argues (2004, *PHG* **28**, 3) 'we should not abandon policies to reduce the disparity in the knowledge-based economy'. Egger et al. (2007) *Ann. Reg. Sci.* **41** introduce factor mobility in the **knowledge-capital model** to explain why European nations are less specialized than US regions.

knowledge in geography Harrison (2002) *Geoforum* **33**, 4 considers the way certain presuppositions and idealizations over the nature of understanding and meaning are or have been built into our (social scientific) modes and methods of explanation. For Brown (2004) *TIBG* **29**, 3 the key questions are: what are the motivations behind chosen research topics and questions? What are the reasons for favouring one particular methodology in tackling a given research question?

Lane (2001) *TIBG* **26** stresses the need to achieve the closure required to make things amenable to study; B. L. Rhoads and C. E. Thorn (1966) wonder how place shapes our understanding. Couper (2007) *TIBG* **32**, 3, drawing on Wittgenstein, observes that parallel ideas may be expressed in different ways in disciplines in geography: for example *see* HYSTERESIS; CONTINGENCY. See also Rose (1997) *PHG* **21**, 3 on situating knowledges. Vallance (2007) *Geog. Compass*, **4** detects the emergence of multiple economic geographies of knowledge, the diversity of learning processes in different types of organizational and spatial setting, and the circulation of mobile forms of knowledge across geographical space.

Kondratieff cycle A series of long waves of economic activity, identified by Kondratieff (1935, *Rev. Econ. Stats* **17**). Each cycle lasts 50–60 years and goes through development and boom to recession. The first cycle was based on steam power, the second on railways, the third on electricity and the motor car, and the fourth on electronics and synthetic materials. I think there is now a fifth cycle, based on *knowledge. C. Freeman (1987) suggests that new techno-economic paradigms evolve out of a downswing phase of the previous Kondratieff wave and include various new product and process innovations, which in turn form new industries. This results in major changes in terms of industrial and infrastructure innovations, sources of productivity improvements, and business organization; see Park (2003) *Papers Reg. Sci.* **82**, 2. Alas, Smith (2003) *PHG* **27**, 1 calls the Kondratieff cycle one of the 'big and spectacular accounts of the world', exaggerated in order to simplify, pretending that the partial is the whole.

Kuro Shio A warm *ocean current which, fed by the North Equatorial current, runs from the Philippines to Japan, thence feeding into the North Pacific current.

Kyoto Protocol An international and legally binding agreement to reduce greenhouse gases emissions worldwide, entered into force on 16 February 2005. Kerr (2007) *Area* **39**, 4 examines the efficacy of national climate programmes to meet emissions reduction targets set out in the Kyoto Protocol by considering emissions trends: only four of 21 countries with defined programmes demonstrate improved emissions trends following their inception and in only one is the change statistically significant. 'While there is little doubt that participation in emissions trading will be driven mainly by political (Kyoto targets) and commercial (the cost of domestic actions vis-à-vis trading mechanisms) considerations, European governments that sought to constrain its use in the Kyoto negotiations may prefer to retain allowances for future periods rather than trading, constraining markets, and creating barriers to new market entrants' (Bailey (2007) *AAAG* **97**, 3).

labour The manual or intellectual work which is one of the *factors of production. **Labour intensive companies** seek to find more profitable locations for their activities across the globe. The European Union project (CIT2–CT2004–001695) reports that this delocalization trend has its greatest impact on 'intermediate' countries (such as Greece) which face competition from advanced economies on quality, and from developing countries, on price. S. S. Barney (2001) suggests that China may only be able to maintain its comparative advantages in labour-intensive manufacturing by moving production from eastern coastal regions, where labour costs have risen, to the cheaper western regions.

Sanyal and Menon (2005) *Econ. Dev. & Cult. Change*s **53**, 4 find that **labour disputes** have had detrimental effects on the location of new investment in India. Herod (2002, ESRC Seminar on Geographies of the New Economy) argues that the 'threat of capital flight has allowed corporations to play workers against each other, [undermining their] bargaining power', and Lambert (2005) *J. Indus. Relations* **47**, 3 documents the demise of **organized labour** in Australia.

The **labour market** is the mechanism whereby labour is exchanged for material reward. R. Johnston et al. (2000) argue that all labour markets are territorially constituted, the results of the relationships between home, residential setting and workplace, and between production and reproduction, and that the geographical scales over which they operate have become increasingly intertwined and interactive. R. Martin (with P. Morrison, ed. 2003), argues that 'there is no such thing as the national labour market, instead a mosaic of local and regional markets'; thus, **segmented labour market theory** sees the labour market as composed of self-contained sub-markets. Sunley and Martin (2002, *ESRC*) show that the UK government's schemes for tackling youth unemployment have varied significantly across different parts of the country, depending on local labour markets, and Gordon (1995) *TIBG* **20**, 2 focuses on the significance of labour market segmentation for migration conditions.

Monastinotis (2007) *Area* **39**, 3 presents a set of **labour market flexibility** indicators for the UK. Bande and Karanassou (2006) *ersa06* show that the degree of labour market flexibility differs between high and low unemployment regions.

labour, geographies of The ways in which labour, largely political labour, has influenced the geographical landscape; 'understanding work requires knowledge of the wider economic, political, and cultural context in which employment takes place' (Smith and Butler (2000) *Env. & Plan. A* **32**, 9). Herod (2003) *Int. Labor & Work.-Class Hist.* **64** argues that 'workers make their own geographies, though not under the conditions of their own choosing'.

Major themes are: the role of work and employment in contemporary capitalist society (M. Pacione (2005) and Cumbers et al. (2003) *Env. & Plan. A* **35**, 9); workers' socio-geographical positionality in explaining employment within and beyond the workplace (Smith (2000) *Env. & Plan. A* **32**, 10 on economic practices in eastern Europe, and Herod (2000) *Env. & Plan. A* **32**, 10 on workers, neoliberalism, and globalization); workplace structures and identity (Dowell in T. Barnes et al., eds (2002) and Wills et al., guest editorial (2000) *Env. & Plan. A* **32**, 9).

Gazier (2006) Background Paper, 11th EU–Japan Symposium, discusses the differentiated impact of globalization on work, and Gritsai (2005) *GaWC Res. Bull.* **162** and Winther (2001) *Urban Studs* **38** write on the evolution of a changing spatial division of labour. See Yildirim (1999) *Work, Empl. & Soc.* **13** and Whitelegg (2003) *Antipode* **35**, 2 on spatial changes in employment relations. MacKinnon et al. (2008) *Env. & Plan. A* **40**, 6 detect a major shift in industrial relations 'away from national collective bargaining to a system of localized company bargaining'. See Jones in Valler and Wood, eds (2004) on geographies of local labour regulation, James (2007) *Geog. Compass* **2**, 1 on gendered geographies of labour, Wills (2001) *Antipode* **33**, 3 on geographies of union organization; Wills and Lincoln (1999) *Env. & Plan. A* **31**, 8 on cultures of work; and Jackson and Palmer-Jones (1999) *Dev. & Change* **30**, 3 on the embodiment of employment. See also Lier (2007) *Geog. Compass* **1**, 4 for a review of labour geography.

labyrinth karst Deep canyons of limestone formed by *carbonation. The wide, deep fissures known as bogaz widen and deepen into long gorges (*karst streets*) with other, cross-cutting, lines of erosion. The remnant of this carbonation is tower karst. In the late stages of evolution, labyrinth karst is replaced by limestone towers (Brook and Ford (1978) *Nature* **275**).

laccolith An intrusion of igneous rock which spreads along bedding planes, forcing the overlying strata into a dome. Rocchi et al. (2002) *Geology* **30**, 11 delineate the two-stage growth of laccoliths at Elba Island, Italy.

lacustrine Of lakes, especially in connection with sedimentary deposition. **Lacustrine plains** result from the in-filling of a lake. Soil parent materials are usually fine grained, well sorted, and often *varved. Ground surfaces are level to gently inclined and slightly concave. Landform elements include lakes, playas, some ox-bows, and some lagoons. For **lacustrine terraces**, see Korotkii et al. (2007) *Russ. J. Pacific Geol.* **1**, 4.

lagoon A bay totally or partially enclosed by a spit or reef running across the entrance, known in the Baltic as a *haff*; see Miotk-Szpiganowicz et al. (2007) *2nd MELA Conf. Abstr.*

lag time (lagged time) The interval between an event and the time when its effects are apparent; for example, in the storm hydrograph.

Lag time

lahar A rapidly flowing, high-concentration, poorly sorted sediment-laden mixture of rock debris and water (other than normal streamflow) from a volcano (Lavigne et al. (2007) *Forum Geografi* **21**, 1); 'transient sediment-water flows whose properties are unsteady so that the sediment load fluctuates during the flow' (Lavigne and Thouret (2003) *Geomorph.* **49**, 1). Lahars most commonly occur when a crater lake or an ice-dammed lake suddenly overflows; perhaps because of an eruption, the collapse of a dam, heavy rain, snow melt, or the mixing of a *nuée ardente with lake water. See (28 Sept. 2007) *GeoNet* on the lahar from Mt Ruapehu, New Zealand. Berti and Simoni (2007) *Geomorph.* **90**, 1–2 propose a method for delineating lahar-hazard zones in volcanic valleys.

laissez-faire economics The view that a *market economy will perform most efficiently if it is free from government intervention, and is subject only to market forces. This view takes no account of environmental degradation (see Dean and Lovely (2008) *Economist's View* on China) or social justice (see Antle (1993) *Amer. J. Agric. Econ.* **75**, 3). 'The free-market is not . . . a natural state of affairs which comes about when political interference with market exchange has been removed. In any long and broad historical perspective the free market is a rare, short-lived aberration. Regulated markets are the norm, arising spontaneously in the life of every society. The free market is a construction of state power' (J. Gray, 1998). Gray shows convincingly that laissez-faire capitalism has a very ambivalent relationship to democracy as it demands serious social control (see Wills (2000) *PHG* **24**, 4). In the UK, L. Elliot and D. Atkinson (1998) argue that this control has been achieved partly through the extension of insecurity into the nooks and crannies of everyday life for the majority of the population. Vojnovic (2003) *Geograf. Annal. B* **85**, 1, in a study of Houston, Texas—'the archetype laissez-faire city'—finds that, despite the local laissez-faire rhetoric, government intervention in Houston's growth has been vital. Additionally 'there is a strong contradiction between neoliberal laissez-faire in the East and growing regulative intervention in the West' (M. Redclift and G. Woodgate, eds 1997).

laminar flow A non-*turbulent flow where the movement of each part of the fluid (gaseous, liquid, or plastic) has the same velocity, with no mixing between adjacent 'layers' of the fluid. It may be seen at low velocities in a smooth, straight river channel (see Smith et al. (2007) *PPG* **31**, 4).

land breeze A wind blowing from land to sea (an offshore wind) which develops in coastal districts towards nightfall. Pressure is higher above the land than above the sea, as the land cools more rapidly in the evening, and air therefore moves seawards along the *pressure gradient. See Holmer and Haeger-Eugensson (1999) *Phys. Geog.* **20**, 2.

land capability The potential of land for agriculture and forestry depending on its physical and environmental qualities. The main factor investigated is soil type, but climate, gradient, and aspect are also considered. Present land use is not taken into account.

(((●))) SEE WEB LINKS
• A model of changes in land capability for Tasmania.

land classification Land classification is essential for geographers, planners, and, increasingly, for environmental scientists (Owen et al. (2006) *Landsc. & Urb. Plan.* **78**, 4). There is 'no single correct way to describe reality and solve practical questions' regarding the classification of land cover, and all classifications are subjective. Thus, the quality of the classification depends on the skill of the interpreter (Lofvenhaft et al. (2002) *Landsc. & Urb. Plan.* **58**, 2–4). The Food and Agriculture Organization of the United Nations Land Cover Classification System uses an initial dichotomous phase, where the user derives the main land cover type, and a subsequent modular-hierarchical phase, where a land cover class is defined by determining one classifier at a time.

(((●))) SEE WEB LINKS
• The FAO Land Cover Classification System.

land consolidation A type of *land reform which aims to give each farmer one relatively large plot of land rather than scattered, small parcels of land. Issues in land consolidation include: the consent of the peasants; an assessment of the value of current farmland holdings; consolidation as the promoter of rural development; the assignment of new farmland for each household; the application of modern technologies; the control of corruption; an appeal system; the reimbursement of expenses by local or national government; and, controversially, population control (Zhou (1999) *EUI W. Paper* **99**, 1). Kirit and Nagarajan (2004) *Occ. Paper, Nat. Bank Agric. & Rural Dev.* **31**, record huge benefits from land consolidation in the Indian state of Tamil Nadu.

land cover The observed biophysical cover of the earth's surface. Strictly speaking, it describes vegetation and man-made features, and omits bare rock and water, although, in practice, these elements are often included under this term.

land degradation A noted decrease in, or any adverse influence on, the *carrying capacity of land, revealed by: decreased soil fertility; increased soil erosion; removal of vegetation cover; and general negative changes in land use which compromise the supply and quality of ground and surface water, biodiversity, the carbon-oxygen cycle, and general air quality (anon. (2000) *Mine Policy & Mine Monitoring Unit*). Land degradation is seen as 'an unavoidable by-product of mining'; but see Singh (2004) *EnviroNews* **10**, 1.

In Australia, land clearing and the use of European farming systems have been major causes of severe land degradation, including *salinization and erosion (Pannell and Ewing (2006) *Agric. Water Manage.* **80**, 1–3). Douglas (2006) *Geog. Res.* **44**, 2 links globalization in South-East Asia with land degradation, but finds that local changes may have a greater impact. Davis (2006) *Geog. J.* **172**, 2 finds that land degradation in dryland Morocco is commonly blamed on overgrazing, despite existing evidence that ploughing marginal lands and over-irrigation are the primary drivers: 'all too often the outcomes of neoliberal reforms on the environment . . . include increased pollution of air, earth and water, and land degradation in the form of deforestation, soil exhaustion, salinization and erosion.' Thornes (2007) *Geog. Res.* **45**, 1 emphasizes the need to take forage ecology into account when managing grazing in relation to land degradation.

Mambo and Archer (2007) *Area* **39**, 3 use remote sensing to detect and map susceptibility to land degradation, and Kiunsi and Meadows (2006) *Land Deg. & Dev.* **17** present a land degradation model. See also A. Conacher (2001).

land economics The study of land use and of the factors which influence and shape it. Land values are central to this study as land use and land values are interlinked; see the journal *Land Economics*.

landform segmentation Minár and Evans (2008) *Geomorph.* **95**, 3–4 introduce the concept of elementary forms (segments, units) defined by constant values of fundamental morphometric properties and limited by discontinuities of the properties.

Land Information System A system for the capture, storage, manipulation, analysis, and display of land-use data. See Tabi et al. in A. Bationo et al., eds (2007) on Nigeria, and Karikari et al. (2003) *Prog. Dev. Studs* **3** on Ghana.

land-locked state A nation with no access to the sea. Globally, there are twenty-six land-locked states ranging in size from the tiny Vatican City to Mongolia. Geography has put land-locked countries in a position of disadvantage. To be land-locked is more detrimental for a country's development than at any time in the past; see A. Chowdhury and S. Erdenebileg (2006).

land management The way land is used is driven by the interplay of economic, social, and environmental factors. Land management is about finding the right balance of these, often competing, factors that allows sustainable land use. Whenever land is put to use, it is subject to stresses that can be quite high, as in the case of intensive agriculture. These stresses may not only affect the biodiversity of the land, such as through the loss of hedgerows, but can also have widespread social impacts—for example, by detracting from the land's natural beauty. Research on multifunctional land uses looks at land from a holistic perspective to understand how different patterns of use cause different stresses, what the consequences of these stresses are, and how they can be managed or corrected. For example, Amundson (2001) *Ann. Rev. Earth & Planet. Scis* **29** argues that the store of terrestrial carbon can be increased by proper land management practices. Pressing land management issues make sound ecological information and conservation thinking a major imperative (Foster (2002) *J. Biogeog.* **29**). Bowman (2001) *J. Biogeog.* **28**, 5 argues that ecologists must learn to harness the power of environmental history narratives to bolster land management practices designed to conserve biological heritage. See Eriksen (2007) *Geog. J.* on fire as a savanna land management tool, and Lin and Ho (2005) *AAAG* on the role of the Chinese socialist state in land management.

(((●))) SEE WEB LINKS

• EC Environment Research website.

land reform A sweeping change in land tenure. It usually involves the breaking-up of large estates and the widespread redistribution of the land into smallholdings, but may also be *land consolidation. 'Where issues of justice and equity, [and the] sustainable and productive use of land are addressed in conjunction with one another, land reform can make a major contribution to the future of the Republic of South Africa, to sustainable livelihoods, and to the broader issue of ecologically sustainable development' (Attfield et al. (2004) *Third World Q.* **25**, 2). The redistribution of land is a complex, and slow, business; in 1994 whites owned about 87% of South Africa's farmland. The new government set a target for at least 30% of it to be transferred to blacks by 2014. More than a decade on, only 4% has changed hands; see Oudtshoorn (3 May 2007) *The Economist*.

landscape 1. All the visible features of an area of land, the appearance of an area, or the gathering of objects which produce that appearance: 'the expression of interaction between humans and their environment' (C. Sauer 1925). In a rather grumpy paper, Minca (2007) *PHG* **31**, 2 complains that ideas of the geographical landscape have remained essentially immune to the extraordinary body of critical reflection in the writings of Denis Cosgrove, Stephen Daniels, James Duncan, and other British cultural geographers. 'We may think of individual "landscapes" as being compromised, partial, contested and only provisionally stable as modes of ordering the world and our engagement with it. If so, this suggests that we should not think of individual landscapes as discrete pieces of territory because they are supported by, and help to sustain, the interests of mere sections of any given society. Alternatively, we might think of landscapes as being formed in relation to other landscapes and conceptions of landscape. In that case, perhaps also we should base our analysis in terms of the interconnectedness of landscape, its links with other landscapes, other geographies' (J. Gold and G. Revill 2000). 'Landscape is far from being a neutral concept—[it is] is deeply rooted in a performative and ideological mode of engagement with site' (M. Shanks, site 2006). Revil (2007) *TIBG* **32**, 2 argues for 'landscape as a mode of governance'. Henderson in C. Wilson and P. Groth, eds (2003) argues that landscape in geography can be broken down into landscape as *Landschaft; landscape as social space; the epistemological landscape ('the material revelation of human practice and thought'); and the apocryphal landscape ('landscape as a way of seeing').

    B. Rhoads and C. Thorn (1996) define the geomorphological landscape as 'the totality of surface landforms', referred to by Thwaites and Bell (2004) *SuperSoil*, as the *geoscape*.

    2. An overview of a specific phenomenon, and its structures, which might not be material. Parto (2005, *MERIT-Infonomics Res. Memo* 2005–001) writes that 'the new **landscape of governance** . . . is marked by a departure from territorially defined spaces of the post war nation states . . . governance in this new landscape increasingly has to rely on formal and informal, supra- and sub-national institutions of the nation states if it is to attain a similar degree of legitimacy as governance at the national territorial scale'. See Hashim (OECD 2002) on Malaysia, and Robinson and Shaw (2001) *Reg. Studs* **35**, 5 on north-east England. Parto (op. cit.) contends that 'each subsystem is the product of its "own" institutional landscape'. Make of that what you will.

The term is similarly used in **landscapes of care**; see Moon et al. (2006) *TIBG* **31**, 2; Gleeson and Kearns (2001) *Env. & Plan. D* **19**, 1, and Williams (2002) *Sol. Sci. & Medi.* **55**, 1. See the Health, Place and Society Research Group at Queen Mary College, U. London.

landscape conservation The planning and management of the scenic and wildlife resources in geographical and environmental systems. See Pinto-Correia (2000) *Landsc. & Urb. Plan.* **50**, 1-3 on the possible integration of different interests and policies for nature conservation, and Watson et al. (2005) *J. Biogeog.* **32**, 8 on matrix management in the conservation of fragmented landscapes.

landscape ecology Within this study, a landscape is a 'heterogeneous land area composed of a cluster of interacting ecosystems that is repeated in similar form throughout' (R. T. T. Forman and M. Godron 1986). Landscape ecology specifically addresses the importance of spatial configuration for ecological processes (M. Turner et al. 2001). The central themes are: detecting a pattern, and expressing it in quantitative terms; identifying and describing the agents of pattern formation; understanding the ecological implications of the pattern; spatially and temporally characterizing the changes in pattern and process; and managing landscapes to achieve human objectives. For ten differing definitions of landscape ecology, see the United States Regional Association for Landscape Ecology site.

landscape evaluation An attempt to assess the landscape in objective terms. Sometimes a consensus of views on the landscape is sought so that particular landscapes may be chosen as being outstandingly beautiful. Landscape description studies try to identify important items such as topography or buildings. Some kind of ranking method may be attempted to compare one landscape with another. Using this idea, Leopold attempted to calculate how close a landscape was to being unique. Personal preferences may also be used.

landscape matrix The most extensive and connected landscape type, playing the dominant role in landscape functioning. The characteristics of matrix structure are the density of the patches (porosity), boundary shape, networks, and heterogeneity. The type of landscape matrix surrounding a *patch can mitigate the negative effect of habitat isolation for a given species, according to its degree of matrix habitat use (Barbaro et al. (2007) *J. Biogeog.* **34**, 4). V. Ingegnoli (2002) describes three matrix types: continuous, with a single, dominant element type; discontinuous, with a few co-dominant element types; and web-shaped, with connected corridors of prevailing functions.

landscape morphology In *geomorphology, the form and spatial structure of the *landscape. Landscape morphology has a direct influence on water movements, soil, physical and chemical properties, and on the productivity of the vegetation cover. In *human geography it is the material formation of the landscape, its shaping and reshaping, in which social structures and cultural worlds are enfolded.

landscape pattern The existing pattern in the landscape whose characteristics are filtered by constraints of scale and resolution (in space and time) of the reception system. Technically, this system is defined by physical sensors; the human version is called visual perception (Bartel (2000) *Ecol. Model.* **130**, 1–3). *Landscape ecology seeks to associate ecological processes with landscape pattern (Griffith et al. (2002) *Sing. J. Trop. Geog.* **23**, 1). See Lü et al. (2007) *PPG* **31**, 6 on the interactions between landscape pattern and nutrient transport processes. Misclassification can cause large errors in landscape pattern indices (Foody (2007) *PPG* **31**, 4). See Li and Wu, in J. Wu and R. Hobbs, eds (2007).

landscape preference It is argued that most cultures have a preferred landscape: for example, Eleftheriadis et al. (1990) *Env. Manage.* **14**, 4 find that, out of six European groups, Greeks preferred sea landscapes most, and Italians, Austrians and Yugoslavs (*sic*) least. Wherret (1999) *Landsc. & Urb. Plan.* **45**, 4 uses the internet to detect preferred Scottish landscapes (but doesn't tell us what they are). D. M. Williams (2002) theorizes that landscape preference and sense of place must be understood as existing in juxtaposition to other places and scales. See T. Daniel and R. Booster (1976) on predictive models for estimating landscape preference and scenic beauty. Herzog et al. (2000) *En. & Behavior* **32**, 3 observe that young children display higher landscape preference, since they see landscapes as 'playscapes', while teenagers are more interested in social concerns.

landscape sensitivity The potential for landscape dysfunction as a result of human impacts; the likelihood of change (Miles et al. (2001) *Catena* **42**, 2 landscape health). In reality, landscape sensitivity is difficult to determine, as land uses tend to be confounded with climate type, and assessment of dysfunction is highly related to land use (Hobbs and MacIntyre (2005) *Glob. Ecol. & Biol.* **14**, 1).

Brunsden (2001) *Catena* **42**, 2 states that the landscape sensitivity concept concerns the likelihood that a given change in the controls of a system or the forces applied to the system will produce a sensible, recognizable, and persistent response. Landscape sensitivity (potential instability) may develop as a function of environmental factors acting over a long time period. Inherent landscape sensitivity becomes manifest when change reaches critical thresholds triggered either by extreme climatic events or by cumulative human impact or when both act in tandem (Marker (2003) *Geog. J.* **169**, 1).

landscape studies The examination of the structure and organization of landscapes; of interrelationships between place-making, personal and social memory; of landscape as the location of cultural and ecological patterns,

processes and histories; of the social life of landscapes—their reception by
different societies; of 'how and why places come to be understood and
experienced as thresholds of time-space in different societies, including how
places constitute geographies of belonging through and beyond urban and
national space' (Till (1999) *Ecumene* **6**).

Landscapes may be viewed not simply as 'scenes' into which humans are
inserted, but rather as the products of human activity, shaped through and
shaping cultures (C. Sauer 1925). An important development in the study of
the cultural landscape is the incorporation of social theory: 'the thrust of the
new landscape studies [was] to consider landscapes as part *of a process* of
cultural politics, rather than as the outcome of that process' (B. Bartley et al.
2002). 'Within landscapes are particular sites—monuments or markers—
which facilitate and direct the process of "collective memory" and through
which social groups situate their identities in time and place' (Inwood et al.
(2008) *Soc. & Cult. Geog.* **9**, 4). For the social life of landscapes, see *Site/Lines*,
the journal and online forum of the Foundation for Landscape studies.

SEE WEB LINKS

- *Site/Lines*: the social life of landscapes.

Landschaft A concept of landscape which attempted to classify landscapes,
usually distinguishing between the natural and the cultural landscape. The
term became part of modern geographical currency in Germany towards the
end of the 19th century. 'It did not do so unambiguously, however, and this has
had unfortunate consequences for the translation of the term into landscape in
the English-speaking literature' (A. Baker 2003).

landslide A form of *mass movement where the displaced material retains
its form as it moves. Cruden and Varnes (*USGS Fact Sheet* **2004–3072**) divide
landslides into slides, falls, flows, topples, and lateral spreads. Probably the
world's largest landslide occurred in south-west Iran in 1937, when a segment
of the Kabir Kuh ridge, about 15 km long, 5 km wide, and 300 m thick, slid off
the mountain, with enough momentum to travel 20 km.

Landslides are prompted by an increase in *pore pressure through snow
melt, through precipitation, and through spring action. These all reduce the
friction which binds the mass to the slope. See Hungr. et al. (2008) *Geomorph.*
96, 3–4 on the technique of constructing a regional magnitude-cumulative
frequency curve for landslide prediction.

SEE WEB LINKS

- Worldwide database of empirical rainfall thresholds for the possible occurrence
of landslides.

land surface process models describe the energy, water, carbon, and
nutrient fluxes on a local to regional scale using a set of environmental land
surface parameters and variables. In principle many of these inputs can

be derived through remote sensing (Bach and Mauser (2003) *Trans. Geosci. & Rem. Sens* **41**, 7).

land tenure   The nature of access to land use. Common forms of land tenure are owner-occupied farms, and tenancies which basically involve payment (in the form of labour, cash, or *share-cropping) to the landlord from the tenant. A *plantation is owned by an institution and uses paid labour. Collectives may own land together and work together, sharing any profits. Under traditional communal land tenure systems in Ghana, the direct descendants of a landholder inherit their rights without losing them to a larger group (Asbere (1994) *J. Black Stud.* **24**, 3). Formal land tenure laws and regulations can be conceptualized as 'one of many state simplifications which have the character of maps' (J. Scott 1998). 'In the hands of powerful lawgivers, such legal "maps" not only summarize "reality"; they also have (intended and unintended) transformative effects' (Larsson (2007) *Pol. Geog.* **26**, 7).

land use zoning   The segregation of land use into different areas for each type of use: agricultural, industrial, recreational, and residential. Planners have sought to pursue *sustainable development using more mixed land use zoning in order to reduce the demand for travel, and greater coordination between transport and land use planning, including support for public transport (Haughton and Counsell (2004) *Geog. J.* **170**, 2). See Greed (2006) *Tijdschrift* **97**, 3 on the problems for women in cities divided by traditional land use zoning, and Degg and Chester (2005) *Geog. J.* **171**, 2 on land use zoning in the mitigation of seismic/volcanic hazards.

La Niña   A state of unusually cool sea surface temperatures in the western Pacific, which are associated with lowered precipitation in the southern USA and western South America, but an increase in the frequency of *tropical cyclones in the Atlantic, and heavier rainfall in Indonesia and Australia. La Niña can be seen as the other extreme from *El Niño.

   'There is strong statistical and modelled evidence that persistent, La Niña-like, cooler sea surface temperatures . . . produce multiyear droughts not only in the Great Plains and the Southwest, but also in the Mediterranean region of Europe, the Pampas region of South America, the steppes of Central Asia, and the outback of Western Australia' (Goodrich (2007) *Geog. Compass* **1**).

lapili   *See* PYROCLAST.

lapse rate   In meteorology, the rate at which stationary or moving air changes temperature with a change in height. *See* ADIABAT; ENVIRONMENTAL LAPSE RATE.

latent heat   The quantity of heat absorbed or released when a substance changes its physical state at constant temperature. The release of **latent heat of condensation** in the rising air of a *hurricane is the chief force fuelling that meteorological phenomenon. See Grossman and Rodenhuis (1975) *Monthly Weather Rev.* **103**, 6.

lateral accretion The build-up of sediments, as in a recurved spit (Kumar and Sanders (1974) *Sedimentol.* **21**, 4), or a meander (Brooks (2003) *Geomorph.* **54**, 3–4).

lateral erosion Usually of rivers; erosion of the banks rather than the bed. In a stream or river, it results in undercutting of the banks or terrace formation (Hanson et al. (2006) *Geomorph.* **76**, 1–2).

lateral moraine *See* MORAINE.

laterite Thick, red, and greatly weathered and altered strata of tropical ground. *Horizons are unclear and the nutrient status of the soil is low. See Bourman (2007) *Geog. Res.* **45**, 3. Laterite is soft but gets brick-hard when exposed to the air. **Lateritic soils** (latosols) are tropical/equatorial soils characterized by a deep weathered layer from which silica has been *leached, a lack of *humus, and an accumulation or layer of aluminium and iron sesquioxides. **Laterization** is the formation of lateritic soils, taking place in warm climates where bacterial activity takes place throughout the year. D. Nahon (1991) argues that ferric *duricrusts result from long-term laterization processes and as a consequence are extremely widespread in tropical environments.

latifundium A large farm or an estate (pl. *latifundia*), particularly in Latin America. The estate is farmed with the use of labourers who sometimes lease very small holdings from the landowner. See Simmons et al. (2007) *AAAG* **97**, 3 on the Amazon Land War.

latitude Parallels of latitude are imaginary circles drawn round the Earth parallel to the equator. The parallels are numbered according to the angle formed between a line from the line of latitude to the centre of the Earth and a line from the centre of the Earth to the equator. Those regions lying within the Arctic and Antarctic circles, having values of 66.5° to 90°, are termed **high latitudes**. **Low latitudes** lie between 23.5° north and south of the equator, i.e. within the tropics. **Mid-latitudes**, also known as temperate latitudes, lie between the two.

latosol A major soil type of the humid tropics with a shallow A *horizon but a thick B horizon of clay, sand, and iron and aluminium sesquioxides which, respectively, give it a red or yellow colour. Much of the silica has been leached, and latosols tend to be of low fertility.

Laurasia One of the two original continents which broke from the supercontinent, Pangaea, by *continental drift. As a supercontinent, Gondwanaland lasted much longer than Laurasia, and was hotter (Veevers (1995) *Geology* **23**, 12).

lava *Magma which has flowed over the earth's surface. The *viscosity of lava depends on its silica content, pressure, and temperature. Temperature is the most important factor. **Basic lavas** have a low silica content and flow freely (Arculus (1976) *GSA Bull.* **87**, 4); **acid lavas** are more viscous (Wilson and Grant (2006) *GSA Spec. Paper* **405**). *See* EXTRUSION.

law A theory or hypothesis which has been confirmed by empirical evidence.

law of retail trade gravitation *See* REILLY'S LAW.

law of the sea A framework, agreed to by the majority of maritime nations, for administering the seas. It recognizes seven administrative zones: internal waters, the *territorial sea, the *contiguous zone, the *continental shelf, exclusive fishing zones of up to 200 miles from a nation's coastline, exclusive economic zones of the same extent as the fishing zones, and the high seas. See M. Freeman 1991.

laws of drainage basins *See* DRAINAGE BASIN GEOMETRY.

leaching The movement of water down the *soil profile. This results in the movement of cations, sesquioxides, *clay colloids, and *humus to the lower soil *horizons. Specific types of leaching include: lixiviation (the removal of the soluble salts containing metallic cations); the removal of *chelates; *lessivage; and, in tropical soils, *desilication.

least developed country (LDC) At the time of the 2003 review of the list, the following three criteria were used by the UN: low income, in the light of a three-year average estimate of the gross national income per capita (under $750 for cases of addition to the list, above $900 for cases of graduation); weak human assets, as measured through a composite Human Assets Index; and economic vulnerability, as measured through a composite Economic Vulnerability Index.

(⊕) SEE WEB LINKS

• UNCTAD's Least Developed Countries Report.

Lebensraum Literally, 'living space', the room needed for a nation's expansion. The term was introduced by Friedrich Ratsel (1896) and became a central concept in Nazi propaganda.

lee The side (lee side) sheltered from the wind. Somewhat confusingly, a **lee shore** is the shore towards which the wind is blowing. **Lee depressions** occur when *lee troughs develop into *lows. They are frequent in winter where mountains block low-level air streams. See R. Barry and R. Chorley 1992.

lee trough As a stable air column rises to cross over a ridge, it shrinks in the vertical, therefore *diverging in the horizontal. This gives it a negative relative *vorticity. Accordingly, in the Northern Hemisphere, the flow is deflected anticyclonically: to the right. Since low pressure always lies to the left of an airstream in this hemisphere, the pressure is lower on the lee of the mountain. See Weisman (1990) *Monthly Weather Rev.* **118**, 4.

lee wave A wave motion in a current of air as it descends below an upper layer of stable air, after its forced rise over a mountain barrier, which sets up vertical oscillations (also called a *standing, rotor, hill,* or *mountain wave*). The wavelength is *c.*5–15 km with an amplitude of *c.*500 m. Lee waves are often indicated by higher wind speeds and *lenticular clouds.

lenticular cloud A lens-shaped cloud formed above a peak as air rises over mountain barriers and condensation occurs. *See* LEE WAVE.

lenticularis *See* CLOUD CLASSIFICATION.

leisure, geography of Leisure and consumption have become major forces in contemporary society, with social, economic, and geographical implications (Dodson (2000) *Tijdschrift* **91**, 4). Geographies of leisure study the relationships between leisure and other social practices and behaviours related to human movement; the spaces produced by leisure activities (gardens, heritage sites, parks, theme parks, and so on) and the meanings given to these spaces, together with the spatial patterns of people's behaviour in their free time. See Hubbard et al. (2008) *PHG* **32**, 3 on the urban geography of 'adult entertainment' (yes, really) and Jayne et al. (2008) *PHG* **32**, 2 on geographies of alcohol and drunkenness. D. Crouch (1999) sees leisure, tourism, travel, recreation, time off work, holidaying (however the activity is labelled) as an 'encounter'; between people and places, people and people, people and nature, and people and their past.

less economically developed country (LEDC) A country with low levels of economic development. Indicators include high birth, death, and infant mortality rates (typically over 20; over 30; and over 50 per thousand, respectively); more than 50% of the workforce in agriculture; and with low levels of nutrition, secondary schooling, literacy, electricity consumption per head, and GDP per capita—generally below $US1 000 per capita. The word 'economically' was included in this term in order to redress perceptions of the people of financially poorer nations as uncivilized, irrational, ignorant, and unskilled.

lessivage The translocation of *clay colloids in a soil, with no change in their chemical composition. Characteristic features of lessivage are often observed in the soils without hydrological barriers hampering, or preventing, the vertical migration of soil water and mass transfer processes (Zaidel'man (2007) *Eur. J. Soil Sci.* **40**, 2).

levée A raised bank of alluvium flanking a river. The bank is built up when the river dumps much of its *load during flooding. New Orleans was founded on a natural levée; elevations as high as 19 feet above sea level provided the driest land for settlement. French colonists created their own levées, and ever since permanent settlement began, residents have been expanding the levées. See Turner (2007) *Tech. in Soc.* **29**, 2 on the repetitive disasters resulting from levée failure around, and within, New Orleans.

lexical diffusion In geolinguistics, phonological changes proceeding through the lexicon, word by word. See Britain (1997) *Paper Presented at NWAV-26*.

lichenometry It is possible to estimate the age of previous geomorphological events if the growth rate of lichens is known. Lichen growth rates can also be used to indicate global warming and the status of the environment because lichens are very sensitive to subtle climatic and chemical changes (Brabyn et al. (2005) *NZ Geogr.* **61**). See also Gob et al. (2008) *Geomorph.* **94**, 1–2.

life expectancy The average number of years which an individual can expect to live in a given society, normally derived from a national *life table. Life expectancy is usually given from birth but may apply at any age, and because, in all societies, mortality rates tend to be rather high in the first year of life, life expectancy at birth is usually significantly lower than at 1 year old. Women consistently have a longer life expectancy than men, especially in *more economically developed countries where the risks of childbirth are less than those in *less developed countries.

The lower life expectancies for less economically developed countries generally reflect high *infant mortality rates, but by the age of 70, the years of life remaining to an individual are, globally, very similar. Thus, the strong correlation between *GDP per capita and life expectancy becomes weaker as the age of an individual increases.

life space The limited time and space which an individual has in which to pursue a necessarily limited range of opportunities. Life space is the interaction of the individual with her or his behaviour setting. L. Lofland (1998) notes that the allocation of differing amounts of life space to different realms will vary by the social class of a city's inhabitants.

life table A summary of the likelihood of living from one age to any other. In a life table, a hypothetical *cohort of 100000 births is set up and then the loss by deaths is shown for each year of life. Averages of losses are calculated for a given year, and from this the actual diminution of the cohort is shown.

life world (lifeworld) The taken-for-granted dynamic of everyday experience that largely happens automatically, without conscious attention or deliberate plan (Seamon in T. Mels, ed. 2004).

lift force The upward force produced when fluid rises over a particle. In watercourses, the particle moves up from the bed into the flow when the lift force exceeds the gravitational force provided by the mass of the particle. See Cornelis (2004) *Geomorph.* **59**, 1–4.

light industry The manufacture of relatively small articles (toasters as opposed to girders), using small amounts of raw materials. In consequence, the *material index is low, and such industries are more *footloose than heavy industries.

lightning An emission of electricity from cloud to cloud, cloud to ground, or ground to cloud, accompanied by a flash of light. It results from variations of electrical charge on droplets within clouds and on the earth's surface. This variation may be caused by the break-up of raindrops, the splintering of ice crystals, or differences between splintered ice crystals and soft hail. As a *cumulus cloud develops, the frozen upper layer becomes positively charged, and most of the cloud base negatively charged, with positively charged patches. The negative charges are attracted to the earth, which has a positive charge. When the electrical field strength gets to about 1 MVm^{-1}, the electrical insulation of the air breaks down. The result is a *leading stroke*, or *stepped leader*, from cloud to ground, which creates a conductive path between them. The return stroke, from earth to cloud, follows the same pathway (only millimetres across), with a charge up to 10000 amps. The intense heat of the stroke engenders light, and a violent expansion of the air, making waves, heard as *thunder. Not all the negative charge may be released; there may be several return strokes, each prefaced by a downward dart leader which reactivates the channel. Where the path between ground and cloud is clearly visible, **forked lightning** is seen. The illumination of other clouds by a concealed fork is **sheet lightning**.

De Pablo and Rivas Soriano (2007) *Monthly Weather Rev.* **135**, 7 report on an inverse correlation between lightning flash rates and index values of the *North Atlantic Oscillation. Ball lightning has been described as a sphere of glowing light meandering through the lower air. Little is known about it, but there are a few fascinating 19th-century observations accessible from the American Meteorological Society website.

((⊕)) SEE WEB LINKS

• American Meteorological Society.

limestone A general term for a *sedimentary rock which consists mainly of calcium carbonate. Limestones vary in texture; **oolitic limestone** consists of tiny, rounded grains, **pesolitic** of larger grains, whereas other limestones have a crystalline texture. Limestones also vary in mineral content, and in modes of origin. Limestones are classified according to geological age.

limestone pavement A more or less horizontal, bare limestone surface, cut into by grikes (fissures) running at right angles to each other, leaving clints (upstanding areas of limestone) between them.

limnology The study of all the biological, chemical, meteorological, and physical aspects of freshwater ponds and lakes. See P. O'Sullivan and C.

Reynolds, eds (2004, 2005) and Lau and Lane (2001) *PPG* **25**, 2 on shallow lake ecosystems.

linear city A planned city developed along a single, high-speed line of transport; a city form favoured by Soviet planners in the 1920s and 1930s; see R. A. French (1995). Industry is developed along one side of the link, while shops and offices are located on the other side, with housing beyond them. A **linear village** is an elongated ribbon of settlement usually formed along a routeway such as a road or canal, perhaps reflecting the pattern of *land tenure.

line-haul costs The costs of transporting goods over a route, but not loading or unloading. The selection of any shipment route depends on the total sum of its network access, line haul, interlining, terminal transfer, and network egress costs (Southworth and Peterson (2000) *Transp. Res. C* **8**). Furuichi and Koppelman (1994, *Transp. Res. A* **28**) report that business travellers value time to the airport twice as much as time on the aeroplane (**line-haul time**). The 200-year-long collapse in transport time has now been replaced by a plateau in general **line-haul speeds**, but time/space continues to collapse locally, for example through new fixed links or high speed trains (Knowles (2006) *J. Transp. Geog.* **14**, 6).

line squall A linear, severe storm event, with rapidly rising pressure and wind speed, a sudden temperature fall, low, dark clouds, and often thunder. At certain cold fronts in the central and eastern United States, the advancing wedge of cold air scoops up the warm air before it to form a nearly continuous line of squall-bearing cumulo-nimbus. Bluestone et al. (1987) *Monthly Weather Rev.* **111**, 11 divide line squalls into *broken line*, *back building*, and *broken areal*.

linguistic distance The degree of contrast between two *dialects, measured by the percentage of items differing from a fixed set. See Gooskens (2005) *DiG* **13** on the geographical and demographic determinants of linguistic distance, and Scapoli et al. (2005) *J. Theor. Biol.* **237**, 1 on the link between linguistic similarity and linguistic distance.

link The route or line joining two *nodes.

linkages Flows of inputs and outputs to and from a manufacturing plant in association with other plants. Movements of matter are **material linkages** as opposed to **machinery** and **service linkages** (such as information, advice, and maintenance). Individual plants are also tied together by **forward linkages**—supplying customers, and **backward linkages**—with their suppliers. **Horizontal linkages** occur between plants which are engaged in similar stages of a manufacturing process.

literary geography The combination of the study of literature and a geographical perspective (Ridanpää (2007) *Soc. & Cult. Geog.* **8**, 6). See Sharp (2000) *Area* **32** on fictive geographies, and Kong and Tay (1998) *Area* **30**.

lithification Processes by which sediments are converted into hard rock. These include the expulsion of air, or the suffusion into the rock of *cementing agents in solution, like quartz.

lithology The character of a rock; its composition, structure, texture, and hardness.

lithosphere The earth's *crust, and that upper layer of the *mantle which lies above the *asthenosphere. Defant and Drummond (1993) *Geology* **21**, 6 believe Mount St. Helens to be an example of the partial melting of the subducted lithosphere. The strength of the continental lithosphere is likely to be contained within the seismogenic layer; variations in the thickness of this strong layer determine the heights of the mountain ranges it can support (Maggi et al. (2000) *Geology* **28**, 6). Constantin et al. (1996) *Geology* **24**, 8 find that magmatic intrusion is a fundamental process in the construction of the oceanic lithosphere. See Scoppola (2006) *GSA Bull.* **118**, 1 on the westward rotation of the lithosphere relative to the underlying mantle.

litter The layer of dead and dying vegetation found on the surface of the soil (some soil classifications assign the litter layer to the A *horizon, shown as A00). Cellulose, hemicellulose, and lignin make up 20–30, 30–40, and 15–40% of the total litter mass, respectively; the exact values are governed by plant type and age, onsite conditions, and yearly environmental variations (Yadav and Malenson (2007) *PPG* **31**, 2).

Little Climatic Optimum The time, *c.* AD 750–1200, when Europe and North America were much warmer; see Peterson (1994) *Clim. Change* **26**. See Nunn (2000) *NZ Geogr.* **56**, 1 on the transition between the Little Climatic Optimum and the *Little Ice Age.

Little Ice Age The phase between AD 1550 and 1850 when temperatures were globally lower than they are at present (Thompson et al. (1986) *Science* **234**, 4774). **Little Ice Age glacierization** occurred over about 650 years and can be defined most precisely in the European Alps (*c.* AD 1300–1950) when extended glaciers were larger than before or since. **Little Ice Age climate** is defined as a shorter time interval of about 330 years (*c.* AD 1570–1900) when Northern Hemisphere summer temperatures (land areas north of 20° N) fell significantly below the AD 1961–1990 mean (Matthews et al. (2005) *Geografiska A* **87**, 1).

littoral drift *See* LONGSHORE DRIFT.

lixiviation *See* LEACHING.

load The matter transported by a river or stream. **Solution load** is dissolved in the water. **Suspension load** refers to undissolved particles which are held in the stream. On the river bed, the material of the *bed load jumps by *saltation, or rolls along the bed. The deposits forming a channel bed are known as **bed-material load**. Conventional theory is that meanders develop to an equilibrium form which is related to discharge and sediment load (Hooke (2007) *Geomorph.* **91**, 3–4). See Singh et al. (2007) *Geomorph.* **86**, 1–2 for a three-dimensional model for sediment transportation dynamics.

loam An easily worked, fertile soil, composed of *clay, *silt, and *sand, roughly in a ratio of 20:40:40. A **clay loam** has a clay content of 25–40%, a **silt loam** has more than 70% silt, and a **sand loam** has between 50 and 70% sand. Loams heat up rapidly, drain neither too slowly nor too easily, and are well aerated.

local For a term so frequently employed in geography, 'local' is difficult to define. If we consider 'local' as relating or restricted to a particular area, or one's *neighbourhood, 'particular area' and/or neighbourhood need to be defined, so the problem is compounded. More interesting are the overtones attached to this term. Evans (2004) *Area* **36**, 3 notes 'the suite of values that have become attached to the "local", such as inclusivity, collaborative action, self-determination in terms of policy and action, and responsiveness to real-world needs'. In terms of food production, Goodman (2003) *J. Rural Studs* **19** shows how the 'local' is associated with ecology, differentiation, and quality, with the latter being tied to relational ideas of tradition, trust and 'place' (see also Feagan (2007) *PHG* **31**, 1). Allen's supposed truism needs to be evoked here: 'the local is not everywhere the same' (2003, *J. Rural Studs* **19**).

Most geographers acknowledge a continuum between local and global; thus 'whatever lure the local may hold for landscape geographers it must become one of understanding how particular places, particular landscapes fit into a larger, scalarly complex, mosaic' (Mitchell (2001) *PHG* **25**, 2). And rather than positing the global and the local as a dichotomy, Ç. Keyder and his collaborators (1999) advocate a dialectic in which the global and the local are mutually constitutive.

local climate The climate of a small area such as a moorland or city—a *mesoclimate—falling between a *microclimate and a *macroclimate. At this scale, such variables as local winds, *albedo, relief, slope, and *aspect are of considerable significance. D. Wheeler and J. Mayes (1997) fit local climates into a larger picture for the British Isles. *See* URBAN CLIMATE.

local economic development policy Enterprises promoted by the public sector, or by public–private partnerships, to attract new enterprises and investment. Local economic development is conditional in its geography on land use plans and on the permissions made within the context of those plans (Cox (2004) *TIBG* **29**, 2). Local economic development policy has been identified as amongst the key areas where local authorities can foster more sustainable forms of development (Local Government Management Board 1993; see also Muir et al. (2000) *Area* **32**, 4). Polèse and Shearmur (2006) *Papers Reg. Sci.* **85**, 1 argue against using local economic development strategies as a means of arresting population and employment decline. Nonetheless, diversification into new markets is a key objective of local economic development policy in north-east Scotland (Chapman et al. (2004) *TIBG* **29**, 3). See Bartik (2005) *Growth & Change* **36**, 2 on local economic development incentives.

local food systems (LFS; local food networks) Local food systems include community gardens, community-supported agriculture, farmers' markets, food cooperatives, and community gardens, all of which may be described as direct agricultural markets (Hinrichs (2000) *J. Rural Studs* **16**), or **local food networks** (Jarosz (2000) *Agric. & Hum. Values* **17**). Local food systems will reduce food miles, and should make for fresher food, and support both localized production practices and local crops or livestock.

Hendrickson and Heffernan (2002) *Sociol. Ruralis* **42** describe food in a local system as 'rooted in a space that enables and constrains production and consumption through its own unique characteristics'. Anderson and Cook in J. M. Harris, ed. (2000) argue that local food systems can supplement and complement larger-scale food systems. Bellows and Hamm (2001) *Agric. & Hum. Values*, **18**, 3 concur: 'the realities of a "local food system" necessitate an integration of local and non-local, and conventional and sustainable in local food systems.' Feagan (2007) *PHG* **31**, 1 argues that the local in LFS will have to be contingent on the place—its social, ecological, and political circumstances—going on to warn against xenophobia, 'place purity', and anti-democratic orientations in local food systems.

local governance The institutions around local governments engaged in the design and implementation of economic and social policy: business elites, community leaders, development corporations, training and enterprise councils, and voluntary groups. Local governance, regime theory, and new regionalism literature argue that horizontal networks, consisting of public, private, and non-profit agents, diminish the significance of formal hierarchical political-territorial structures (Kubler and Heinelt in H. Heinelt and D. Kubler, eds 2005). However, Rosen and Razin (2007) *Tijdschrift* **98**, 1 show that they only reshape the intervention of the central state in local development. See Tewdwr-Jones (2003) *Int. Plan. Studs* **8** on place-making as the tool of effective local governance.

locality debate A discussion, underpinned by structuralism, concerning the way general social tendencies are realized locally. Places are understood as local forms of economy constituted in a wider social context; they develop in response to larger-scale processes which bring new elements that may become part of the local layers of development. See Smith (1987) *Antipode* **19**. 'The locality debate taught us *inter alia* that thinking critically about and through scalar categories is essential to the ways that human geographers need to come to terms with the changing world around them, not least because important causal processes seem to operate in a scalar dimension' (Jonas (2006) *TIBG* **31**, 3).

localization economies Advantages arising from the localization together of a number of firms in the same type of industry. Desmet and Fafchamps (2005) *J. Econ. Geog.* **5** distinguish between localization economies—benefits which derive from being located close to other firms in the same industry—and urbanization economies—associated with closeness to overall economic activity. Boschma and Weterings (2005) *J. Econ. Geog.* **5**, 5 list knowledge externalities and a skilled labour market. Viladecans-Marsal (2004) *J. Econ. Geog.* **4** refers to localization economies as associated with a city's specialization.

local winds Local winds blow over a much smaller area than global winds and have a much shorter time span. **Hot winds** originate in vast *anticyclones over hot deserts and include the Santa Ana (California), the Brickfielder (south-east Australia), the Sirocco (Mediterranean), the Haboob (Sudan), the Khamsin (Egypt), and the Harmattan (West Africa).

Cold winds originate over mountains or other snow-covered areas and include the Mistral, funnelled down the Rhone Valley, and the *Bora. Some local winds, such as the Southerly Burster of Australia, are associated with cold fronts. Other local winds include *land breezes and *sea breezes. *See also* MOUNTAIN WIND; FÖHN.

location **Absolute location** is expressed with reference to an arbitrary grid system as it appears on a map. **Relative location** is concerned with a feature as it relates to other features.

Transnational companies can move their production plants, and their call centres, with ease from place to place—from continent to continent indeed—at the slightest whiff of lower costs. 'So, rather than erasing the importance of material location, it could well be argued that current trends in the world economy have in some senses reinforced its significance because as companies can move about more easily they can respond to even the smallest differences between places.' Massey (2006) BBC/OU Open2.net.

locational advantage Although it is true that transport costs have become less significant, and that new computer and telecommunication technologies have reduced the importance of location for production, locational advantage is still important in industrial location; for example, the overwhelming majority of logistics providers possess a geographic speciality (and even the worldwide integrated firms either have branches in, or have acquired firms with, strategic geographic locational advantage). 'This is because transportation is primarily a local industry that requires rich, geographically-specific knowledge, even if the industry deals with international and widely geographically dispersed transactions' (Aoyama et al. (2006) *Prof. Geogr.* **58**, 3). For a second example, see Lee and Rodrigue (2006) *Growth & Change* **37**, 4 on locational advantage and differential growth within the Korean port system.

location-allocation model A mathematical model used to establish optimal locations. The model takes account of the location and demand of the customers, the capacity of the facilities, and operational and transport costs. Rushton (1988 *Econ. Geo.* **64**, 2) provides an easily comprehensible discussion of the use of a location-allocation model to provide 'an explicit framework for diagnosing accessibility problems, measuring the efficiency of location decisions and the current levels of settlement efficiency [and the] generation of viable alternatives for action by decision makers'. A more technical, but still admirably clear, report on the use of location-allocation models in planning health care in Ghana is provided by Møller-Jensen and Kofie (2001) *Geografisk Tidsskrift* **101**. Dejonghe (2004) *Tijdschrift* **95**, 1 uses the model to suggest a restructuring of professional football clubs in Belgium.

location coefficient Also known as the location quotient, this expresses the relationship between an area's share of a particular industry and the national share. Thus, the locational coefficient for a given region equals:

$$\frac{\%\text{employed in a field in a given region}}{\%\text{employed nationally in that field}}$$

See Guimarães et al. (2007) *J. Reg. Sci.* **47**, 4.

location-specific Locational factors bestow *competitive advantages if they offer resources that are not easily transferred to other locations. Such resources are either tangible (like natural resources or labour) or intangible (like expertise, or specialized services).

Location-specific resources influence the locations of transnational corporations (TNCs), especially if they complement the expertise of the TNC; the more easily the TNC can influence the location-specific resources, the more likely it is to invest in the professional development of its human capital. The TNC is then less likely to substitute one location for another one. In a study of the location-specific resources of Vienna, Windsperger (2006 *South East Eur. J. Econ. & Bus.* **20**) identifies geographic distance; transport infrastructure; specific human capital; management know-how; historic ties; a multicultural environment; and a high quality of life, as major factors in location specificity.

location theory A group of theories which seek to explain the siting of economic activities. Various factors which affect location are considered such as localized materials and *amenity, but most weight is placed on transport costs. Witlox (2000) *Tijdschrift* **91**, 2 advocates a relational view on industrial location theory.

Loch Lomond Stadial A stadial characterized by small ice caps and
*cirque glaciers in the Scottish Highlands. See Sissons (1979) *Nature* **280**.

locked zone An area along a *rift where *plates remain attached to each other.
In such a zone, no new crust develops, so that the locked zone stretches, and the
crust thins. Ultimately, the plates will separate. See Wallace et al. (2004)
J. Geophys. Res. **109**, B12 and Hyndman et al. (1995) *J. Geophys. Res.* **110**, B8.

lodgement The release and consolidation of debris from a glacier if the
basal ice reaches its *pressure melting point, as the ice moves. The moving ice
aligns fragments of this debris, known as **lodgement *till**, in the same direction
as the flow of the glacier. Ruszczyńska-Szenajch (2001) *Quat. Sci. Revs* **20**, 4
distinguishes between **hard lodgement till** for deposits released from a glacier
sole mainly due to friction, and **soft lodgement till** for sediments released from
the base of a glacier due to the melting of ice in a more water-saturated
subglacial environment.

loess (löss) Any unconsolidated, non-stratified soil composed primarily of
silt-sized particles. The origin of loess is in dispute. Some writers believe the
deposit to be wind-borne (Mason et al. (1999) *Geomorph.* **28**, 3–4); others note
the occurrence of the soil in *periglacial environments (Pawelec (2006)
Geomorph. **74**, 1–4) or stress the importance of glacial grinding in the
production of silt-sized particles (Sculier et al. (1998) *Geomorph.* **23**, 1).
Sweeney (2005) *Quat. Res.* **63**, 3 points to the frequency of dust storms in
deserts; Smith et al. (1991) *ESPL* **16** stresse the importance of *salt weathering.
The loess sequences of north-central China preserve the longest and most
detailed record of *Quaternary climate change found on land. See Haberlah
(2007) *Area* **39**, 2 for a clear and concise summary on loess. Loess is very fertile
but very difficult to conserve; soil erosion in most of the Loess Plateau is 5000–
10000 Mg/km^2 per year, and 20000 Mg/km^2 per year in some places (Chen
et al. (2007) *PPG* **31**, 4). See also Cheng (2007) *Tillage Res.* **94**, 1.

logit model Model used to represent choice between two mutually
exclusive options; for example, a commuter may decide to drive to work or to
use public transport. The **multinomial logit model** is mathematically simple
(Andrew and Meen (2006) *Papers Reg. Sci.* **85**, 3), and is widely used, but
imposes the restriction that the distribution of the random error terms is
independent and identical over the alternatives, causing the cross-elasticities
between all pairs of alternatives to be identical, and this can produce biased
estimates. The **nested logit model** allows the error terms of pairs or groups of
alternatives to be correlated, but the remaining restrictions on the equality of
cross-elasticities may be unrealistic (Hunt (2000) *Papers Reg. Sci.* **40**, 1). Logit
models which allow different cross-elasticity between pairs of alternatives
include: the **paired combinatorial logit** (Koppelman and Wen (2000) *Transp.
Res.* **B4**); the **cross-nested logit** (Vovsha (1997) *Ann. Transpt. Res. Board,
Washington, D.C.*), and the **product differentiation model** (Filippini (1999)

Int. J. Econ. Bus. **6**, 2). The major weakness of logit models is the implication that the choice between any two groups of alternatives depends solely on the characteristics of the alternatives being compared, and not upon the characteristics of any other alternatives in the choice set.

longitude The position of a point on the globe in terms of its *meridian east or west of the prime meridian, expressed in degrees. These degrees may be subdivided into minutes and seconds, although decimal parts of the degree are increasingly used.

longitudinal data Information on one or more areas over periods of time; for example, on changes in human reproduction (Reher et al. (2007) *Pop. & Dev. Rev.* **33**, 4).

long profile A section of the longitudinal course of a river from head to mouth, showing only vertical changes (Harmar and Clifford (2007) *Geomorph.* **84**, 3–4). See Jain et al. (2006) *Geomorph.* **74**, 1–4 on stream power plots along long profiles. The theoretically smooth curve shown by such a profile may be interrupted by breaks of slope which can result from bands of resistant rock or from *rejuvenation (Larue (2008) *Geomorph.* **93**, 3–4).

longshore drift The movement of sand and shingle along the coast. Waves usually surge onto a beach at an oblique angle and their *swash takes sediment up and along the beach. The *backwash usually drains back down the beach at an angle more nearly perpendicular to the coast, taking sediment with it. Thus there is a zigzag movement of sediment along the coast. **Longshore currents**, initiated by waves, also move beach material along the coast. The term **littoral drift** is synonymous.

long wave *See* EARTHQUAKE.

lopolith A large *intrusion which sags downwards in the centre, forming a saucer-shaped mass. Wilson (1956) *GSA Bull.* **67**, 3 interprets lopoliths as funnel-shaped in cross-section with the most basic rocks at the bottom. See O'Driscoll et al. (2006) *Geology* **34**, 3 on the Great Eucrite intrusion of Ardnamurchan.

Lorenz curve A cumulative frequency curve showing the distribution of a variable such as population against an independent variable such as income or area settled. If the distribution of the dependent variable is equal, the plot will show as a straight, 45° line. Unequal distributions will yield a curve. The gap between this curve and the 45° line is the inequality gap. Such a gap exists everywhere, although the degree of inequality varies. *See* GINI COEFFICIENT.

Lösch model A model of central places developed by A. Lösch (1954) which is less narrow than that of *Christaller, in that it treats the range, threshold, and hexagonal hinterland of each function separately. The resulting pattern of

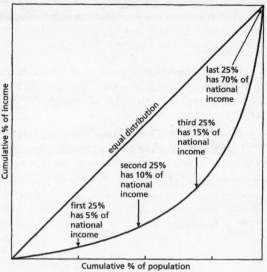

Lorenz curve

central places is much more complex than Christaller's, and yields a continuous, rather than a stepped, distribution of population sizes.

löss *See* LOESS.

low A region of low atmospheric pressure. In Britain, the term low is generally applied to pressures of below 1000 millibars.

low-order goods and services Goods and services with a low *range and a low *threshold population like daily newspapers, and hairdressing. The goods are often convenience goods.

Luminescence Dating (Optically Stimulated Luminescence Dating) (OSL) All sediments include low levels of uranium, thorium, and potassium which produce ionizing radiation, which accumulates over time, and is absorbed and stored within the sediment. If this store is expelled from the sediment, it will generate luminescence, the amount produced being proportional to the accumulated dose. If the annual radiation dose is known, the age of the sediment can be calculated:

$$\text{Age} = \frac{\text{total accumulated radiation dose (Gy)}}{\text{annual radiation dose}}$$

Sediments rich in quartz and feldspar, which cannot be dated by conventional radiocarbon methods, can be absolutely dated ($\pm \sim 10\%$) within a range of

100000 to 200000 years. Aeolian sediments are ideal for luminescence dating, but samples from other sedimentary environments have been successfully dated.

lunette *See* SAND DUNE.

lynchet In earlier, possibly prehistoric times, small belts of uncultivated land were sometimes left between ploughed areas. These now can be seen in the landscape as part of a system of terraces. See Klingelhofer (1991) *Trans. Sm. Philos. Soc.* **81**, 3.

lysimeter A device for the direct estimation of evapotranspiration, composed of a vegetated block of soil 0.5–1 m^3 to which the amount of water added is known, and from which the amount lost as run-off or percolation may be measured. Richards (2004) *PPG* **28**, 1 explains further.

macro- Large-scale. Thus, a **macroclimate** is the general climate of a region extending across several hundred kilometres; and **macrometeorology** is the study of large-scale meteorological phenomena—from *monsoons to the *general circulation of the atmosphere.

macroecology The integration of perspectives over broad spatial and long temporal scales into the study of ecology (Brown and Maurer (1989) *Science,* **243**). Macroecology has moved from the absolute advocacy of small-scale over large-scale studies to an advocacy of some large scales in preference to others; Blackburn and Gaston (2002) *Glob. Change & Biogeog.* **11**, 3 argue that there is no general sense in which one scale of study is better than any other. Although macroecology arose from geographical ecology, it has diverted from a geographical perspective. Ruggiero and Hawkins (2006) *Glob. Change & Biogeog.* **15**, 5 argue that present trends in macroecology adopt a predominantly statistical approach to looking for rules and regularities 'that increasingly ignore the spatial distributional aspects linked to mapping. 'In a vigorous reply, Blackburn and Gaston, 2006 [*J. Animal Ecol.* **75**], stressed that there is much more to macroecology than mapping, and emphasized that 'macroecological patterns will only be elucidated effectively through the formulation of relevant hypotheses and related statistical approaches' (Kent (2007) *PPG* **31**, 6).

macrotidal Having a large tidal range; J. L. Davies (1980) defines *microtidal* as having a range under 2 m, *mesotidal* 2–4 m, and *macrotidal* over 4 m. Areas with a tidal range over 10 m are sometimes distinguished as *hypertidal* (Allen (2000) *Quat. Sci. Revs* **19**). **Macrotidal estuaries** have their dynamics dominated by tidal forcing, as compared to the forcing imposed by the river (Regnier and Steefel (1999) *Geol. & Cosm. Acta* **63**, 9). Reichmüth and Anthony (2007) *Geomorph.* **90**, 3–4 explore the relationship between tidal influence and intertidal bar–trough systems in a macrotidal setting. Macrotidal sandy beaches are thought to represent optimal conditions for coastal dune development (Ruz and Meur-Ferec (2004) *Geomorph.* **60**, 1–2).

magma The molten rock found below the Earth's crust which can give rise to *igneous rocks. Magma occurs in *subduction zones, at continental rift zones, *hot spots and *mid-oceanic ridges, and may rise as **mantle plumes**—see Foulger et al. (2005) *GSA Spec. Vol.* **388**, on plates, plumes, and paradigms. Magma composition is an important control on the geomorphology of lava flows and volcanoes (Carrasco-Núñez (1997) *Geol. Mag.* **134**, 2); see Morgan and Ghen (1993) *Nature* **364** on magma supply. **Magmatic stoping** is the detaching and engulfing of pieces of the country rock, but its role in the location of plutons is disputed: see Žák et al. (2006) *Int J. Earth Sci.* **95** versus Glazner and Bartley (2006) *GSA Bull.* **118**.

magnetic pole reversal The Earth's magnetic field resembles a bar magnet located at the Earth's centre, with its axis emerging at the magnetic poles. The north and south magnetic poles have repeatedly changed places, at irregular intervals, while the axis has stayed in place. When *igneous rocks form, they take up the prevailing pattern of the Earth's magnetic poles. As a result, areas of the ocean bed where *sea-floor spreading has taken place are characterized by **magnetic stripes**—parallel bands of igneous rock with differing magnetic polarity. See Nicolosi et al. (2006) *Geol.* **34** on the Marsili Basin, Italy.

magnitude Of an *earthquake, an expression of the total energy released.

malapportionment The existence of electoral districts with wildly unequal populations (Norman et al. (2007) *Pol. Geog.* **26**, 1). Unintentional malapportionment happens when population growth varies in electoral districts. Redistricting is then needed to even out voter numbers; see Johnson et al. (2006) *Pol. Geog.* **25**, 5.

Malthusianism In 1798, Thomas Robert Malthus (1766–1834) published his *Essay on Population* in which he put forward the theory that the power of a population to increase is greater than that of the Earth to provide food. He asserted that population would grow geometrically, while food supply would grow arithmetically. When population outstrips resources, **Malthusian checks** to population occur: misery, vice, and moral restraint.

Malthus' predictions were not borne out in 18th-century Britain, perhaps because of the increases in food output, and emigration to the colonies; A. Macfarlane (1997) refers to the 'great escape' from the Malthusian trap by large chunks of humanity. P. Wissoker (forthcoming) shows how Malthusian theory is deployed to underpin the theory of European historical superiority by arguing that Europeans, uniquely, have generally (and rationally) avoided the Malthusian disasters of overpopulation while non-Europeans (irrationally) have not done so and therefore not developed as Europe has.

'While old-style Malthusians simply saw an inevitable tendency of human population to outgrow food production, neo-Malthusian scholars offered a more complicated, political argument: overpopulation would cause resource depletion and hunger which in turn would lead to political instability threatening Western interests and world peace' (Flitner and Heins (2002) *Pol. Geog.* **21**, 3). Brander (2007) *Can. J. Econ./Rev. can d'écon.* **40**, 1 sees lower fertility as the central element in achieving sustainable development, 'which is very much a modern translation of what Malthus wrote in 1798'. Furlong et al. (2006) *Int. Interact.* **32**, 1 find that the neo-Malthusian factors are significant, but not dramatic, factors in boundary disputes.

mammatus Breast-shaped lobes of cloud, hanging from the undersurface of a *cumulo-nimbus anvil, signifying negative *buoyancy. **Mammilated** means smooth and rounded.

mandate A statement of empowerment, for example, through an election: 'representing humanity ultimately requires legitimation through some sort of people's mandate' (Dicken (2004) *TIBG* **29**, 1). The term also describes a territory, once part of the German or Ottoman Empires, governed by a member of the League of Nations, 1919–39.

mangrove swamp A number of types of low trees and shrubs, growing on mud flats in tropical coastal areas where the tidal range is slight. Mangroves are significant agents of *progradation along tropical coasts, and are especially well developed in South and East Asia. See Saintilan and Willims (1999) *Glob. Ecol. & Biogeog.* **8**, 2.

Manning's roughness coefficient (Manning's *n*) The resistance of the bed of a channel to the flow of water in it. Representative values of the coefficient are 0.010 for a glassy surface to 0.020 for alluvial channels with large dunes. See Mohamoud (2007) *J. Am. Water Resources Ass.* **43**, 5 for the equations and their application. Smith et al. (2007) *PPG* **31**, 4 note that the application of the Manning equation also requires a measure of resistance. N. Gordon et al. (2004) suggest that n can be thought of as a calibration factor.

Mann's model A model of British urban development which combines the *sector theory with the *concentric zone model. Four basic zones are postulated: middle class, lower middle class, working class, and lower working class, to which are added the *CBD, the transitional zone, a zone of smaller houses, and the outermost zone made up of post-1918 housing.

manor The smallest area of land held in the Middle Ages by a feudal lord, with its own court for minor offences. It usually consisted of a village, the lord's holding (*demesne*), and *open fields farmed on the three-field system. See Z. Razi and R. Smith (1996).

mantle The mantle, with a density of up to 3.3, and a thickness of some 2800 km, lies between the *crust and the Earth's *core. The upper layer, immediately below the *Moho discontinuity, is rigid, forming the lower *lithosphere; the lower layer is the *asthenosphere. Zhou et al. (2002) *Lithos* **62**, 3 see the interaction between continental crust and subcontinental mantle as expressing geodynamic processes deep in the subcontinental lithospheric mantle. See Husson (2006) *Geology* **34**, 9 on upper mantle flow.

manufacturing industry Secondary *industry.

map A cartographic representation of selected spatial information. 'Maps *emerge* through practices and have no secure ontological status . . . maps are of-the-moment, brought into being through embodied, social, and technical

Age of building

■ Transition zone

▨ Pre1918

⊡ Post 1918

Housing

A Upper ⎫
B Lower ⎰ middle class
C Working class
D Poorest houses plus industry

◄— Dormitory towns —►

Mann's model

practices; mapping is a process of constant reterritorialization' (Kitchen and Dodge (2007) *PHG* **31**, 3). We tend to restrict the term to visual maps, but spatial information can be shown on a computer screen, through braille, or a spoken description, and these may also be described as maps. See Pontius et al. (2007) *AAAG* **97**, 4 on the Behavioral Landscape Model and spatial patterns.

Map reading is the interpretation and analysis of map images; an understanding of the physical and psychological processes in map reading helps in map-making. See Lobben (2007) *AAAG* **97**, 1 and Griffin et al. (2006) *AAAG* **96**, 4. *See also* MENTAL MAP.

map generalization Decreasing the detail on a map when reducing its scale.

map projection Both a method of mapping a large area and the result of doing so. The Earth is a sphere; a map is flat, so that it is impossible to produce a map which combines true shape, true bearing, and true distance. 'The usefulness of a particular map projection and the justification behind its transformation from a three-dimensional spherical surface to a two-dimensional plane surface is all too often judged from the esthetic or suspicious eye of a naïve map reader rather than by the science behind its creation' (M. Monmonier 2004). See also Monmonier (2005) *PHG* **29**, 2.

Mercator's projection exaggerates the size of the northern continents; it has been criticized as overemphasizing the importance of Europe and North America, although such was not Mercator's intention. No projection is perfect: for example, Mollweide's and Peters's are **equal area projections** (correct in area), but distort shapes. **Azimuthal projections** show true direction; **gnomic projections** show the shortest straight-line distance between two points; **orthographic projections** convey the effect of a globe. **Interrupted projections** show the earth as a series of segments joined only along the equator. Details of the projection used are given below each map in a good atlas.

Map projection. Mercator's projection is drawn from the shapes and shadows of landmasses that are projected onto a cylinder of paper encircling the earth, when the earth is lit from the centre.

marginal land Land which is difficult to cultivate, and which yields little profit.

margin of cultivation The distance from a market where the revenue from a product exactly equals total cost. If production costs are the same whatever the distance from the market, transport costs determine the margin of cultivation. See J. Plant et al. (2007).

marine erosion Coastal erosion.

maritime Of climates near the sea coast. Such climates have less extremes of temperature, both diurnally and seasonally, than their continental counterparts.

market An arena in which buyers and sellers exchange goods and services, usually for money; it doesn't have to be a physical location. **Market area analysis** is the investigation of how a firm's market area is established. Barker (2000) *Papers Reg. Sci.* **79** presents a shopping model and its application to market area analysis. See also Atsuyuki et al. (2008) *Geogr. Analy.* **38**, 1 and Bruinsma et al. (2002) *Tijdschrift* **91**, 3 on airports.

market cycle A series of periodic—usually one-day—markets such that a trader moves from one location to the next in a weekly cycle.

market economy An economy, characteristic of *capitalism, in which the major parts of production, distribution, and exchange are carried out by private individuals or companies rather than by the government, whose intervention is minimal. Market economies are characterized by specialized production, the freedom to exchange commodities between individuals, and the use of the *market mechanism to determine prices. See Peck (2008) *PHG* **32**, 1 for definitions of the market economy, and social market economy.

marketing principle An arrangement of settlements, termed $k = 3$, such that low-order places are as near as possible to higher-order *central places (W. Christaller 1933, 1966).

market mechanism The interaction of supply, demand, and prices. Page (2005) *TIBG* **30**, 3 compares two approaches to water supply: 'there are those who claim that water is scarce and, like all scarce goods, it is best allocated through a market mechanism. Water must be turned into a commodity to be sold at the full price of its production, including environmental externalities, if it is to be efficiently and equitably distributed and conserved. . . . On the other side are those who argue that water, though scarce, is in its essence not commercial and that such a process of commodification . . . is little more than theft of a common good.'

market potential The intensity of possible contacts with markets. The so-called Harris market potential uses bilateral distances to weight surrounding locations' GDP:

$$MP_{\text{Harris}, i} = \sum_j \frac{GDP_j}{\text{dist}_{ij}}$$

A more sophisticated version of Harris's market potential uses road travel times for lorries as weights, instead of great circle distances:

$$MP_{\text{Traveltime}, i} = \sum_j \frac{GDP_j}{\text{traveltime}_{ij}}$$

See Breinlich (2006) *J. Econ. Geog.* **6**, 5.

marl Clay, usually alluvial, rich in soft calcium carbonate. See Cerdà (1999) *Soil Sci. Soc. Am. J.* **63** on comparative erosion rates in marl, clay, limestone, and sandstone.

Marshallian industrial district A spatial concentration of specialized small and medium-sized firms, supported by a variety of regional institutions for promoting coordination, learning, and innovation (Lee and Saxenian (2008) *J. Econ. Geog.* **8**, 2). Dense interactions among a large number of competing and cooperating firms create an external economy favouring technological innovation and learning (Wang (2007) *J. Dev. Stud* **43**, 6).

Marxist geography A theory and methodology for understanding the political economy of the world—and changing it. It highlights the dialectical processes between the natural environment, spatial relationships, and *social formations* ('layers of relations—economic, social, cultural, political—arranged in space in patterns generally similar, but particularly different, under a given mode of production', Peet (1979) *AAAG* **69**). At all scales, the capitalist mode of production assumes a centre–periphery pattern: 'uneven development is integral to capitalism. Capitalism produces and remakes landscapes and natures of differentiation and is sustained through their continual reworking . . . . Rent, profit (and loss), urbanization (and counterurbanization), devaluation (and speculation), gentrification (and devalorization), imperialism (and revolution), and enclosure (and globalization) are forceful spatial-temporal projects' (Castree and Gregory (2007) *AAAG* **97**, 1).

Perkins (2006) *Antipode* **38**, 1 observes an emphasis within Marxist political thought on the social production of nature, and Castree (1999) *TIBG* **24**, 2 seeks to show how a certain kind of Marxism, and a certain reading of the nature of capitalism and class, can offer the basis on which questions of identity, difference, and otherness can be articulated. While Sage (2006) *Prof. Geogr.* **58**, 1 comments that constant contextualizing through Marxism often seems to limit, rather than widen, the scope of locally derived concepts, Corbridge (2005) *PHG* **29** thinks that the Marxian critique of capitalism has been absorbed and normalized in geography, such that 'we are all political economists now'.

masculinity 'A highly differentiated identity category with its own internal relations of alliance, subordination and dominance' (R. Connell 1995). McDowell (2002) *Env. & Plan. D* **20**, 1 explores the social construction of masculinities among white working-class youth in contemporary Britain; see Malam (2008) *Soc. & Cult. Geog.* **9**, 2 on Thai masculinities.

mass budget The balance between the accumulation and ablation of ice in a glacier. Its components are accumulation, ablation, and iceberg calving. A glacier that is not in balance with the ambient climate loses/gains mass and thereby raises/lowers global sea level (IAMU, 2006–7). The flow of solid ice governs the dynamical response of the glacier to a change in its mass budget. Machguth et al. (2007) *Geophys. Res. Abs.* **9**, 06249 compare the output of three distributed mass balance models of differing complexity.

mass conservation, principle of For a river, sediment continuity. Rivers adjust their width, depth, and slope to keep a balance between the water and sediment supplied from upstream and exported at the outlet; see Dietrich et al. (1999) *Hydrol. Procs* **13**.

mass movement (mass wasting) The movement downslope of rock fragments and soil under the influence of gravity. The material concerned is not incorporated into water or ice, and moves of its own accord, but slides are often triggered by increase in water pressure on rocks and soil. **Granular mass movements** entail transport by large (>0.06 mm) solid grains, mixed with less-dense intergranular liquid or gas, and include rock avalanches, debris flows, pyroclastic flows (Iverson and Vallance (2001) *Geology* **29**, 2). See also Deplazes et al. (2007) *Terra Nova* **19**, 4. A widely used classification of mass movement uses the combination of types of movement (*falls, *topples, *slumps, *slides, and *flows) with the nature of the material (bedrock, *debris, and fine soil); see R. Chorley et al. (1985). Many cases of mass movement include more than one type of movement.

mass strength The strength of a rock's resistance to erosion. Mass strength will vary according to the innate strength of the rock, the jointing/bedding of the rock, and its state of weathering. See Selby (1980) *Geomorph.* **24** on rock mass strength classification.

material index The ratio of the weight of localized materials used in the manufacture of a product to the weight of the finished product. A material index much greater than 1 shows a sizeable weight loss during manufacturing. For many resource-based manufacturing activities where inputs are more expensive to transport than the final product, production will normally be drawn to the source of the material input (Polèse and Shearmur (2004) *Int. Reg. Sci. Rev.* **27**, 431).

mathematical geography The study of the interactions of geography and mathematics, such as the Earth's size and shape, time zones, and motion. See Withers (2006) *PHG* **30**, 6.

matrix *See* LANDSCAPE MATRIX.

maximum sustainable yield The maximum yield of a renewable resource consistent with maintaining its stock. Arlinghaus et al. (2002) *Fish & Fisheries* **3**, 4 review the concept.

meander A winding curve in the course of a river. A *sinuosity of above 1.5 is regarded as distinguishing a meandering channel from a straight one. The dimensions of a meander are related to the square root of water discharge, Q:

$$\lambda = k_1 Q^{0.5};$$
$$A_m = k_2 Q^{0.5};$$
$$w_c = k_3 Q^{0.5};$$

where λ is meander wavelength, A_m is meander amplitude, w_c is the channel width, and k_1, k_2, and k_3 are coefficients whose value varies with location. Bank resistance controls meander wavelength, and is positively correlated with meander wavelength (Abrahams (1985) *ESPL* **10**, 6). Meander bends with a low radius of curvature have deeper pools and more lateral migration; the ensuing bank erosion reduces pool depth via higher sediment inputs. Meander bends with smaller pools have less lateral migration, but use energy vertically, thereby deepening pools and fostering pool-riffle sequences (Hudson (2002) *Phys. Geog.* **23**, 2).

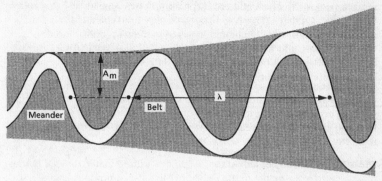

Meander

Eaton et al. (2006) *ESPL* **31** argue that slight initial variations in the shear stress distribution result in local net scour along the banks, increasing local transport capacity. In a positive feedback, this leads to further local net scour, which is mostly vertical until the channel banks begin to fail, at which point bank erosion triggers negative feedback that increases channel sinuosity. Once one initial bend develops, others inevitably follow. See Millar (2005) *Geomorph.* **64** on equations for the meandering-braiding transition. The **meander belt** is the total width across which the river meanders.

mean information field (MIF) In *diffusion, the field in which contacts can occur. It generally takes the form of a square grid of 25 cells, with each cell being assigned a probability of being contacted. The possibility of contact is very high in the central cells from which the diffusion takes place, becoming markedly less so with distance from the centre, that is, there is a *distance decay effect. See Bivand (2008) *J. Reg. Sci.* **48**, 1.

0.0096	0.0140	0.0168	0.0140	0.0096
0.0140	0.0301	0.0547	0.0301	0.0140
0.0168	0.0547	0.4431	0.0547	0.0168
0.0140	0.0301	0.0547	0.0301	0.0140
0.0096	0.0140	0.0168	0.0140	0.0096

Mean information field

means of production The elements needed to produce goods and services: land, labour, and *capital. Any production process depends upon a particular material configuration of the means of production (Hudson (2008) *J. Econ. Geog.* **8**, 3). Marx believed *capitalism to be characterized by a split between the capitalist owners of the means of production, and the proletarians, with only their labour to sell. Akram-Lodhi (2005) *J. Agrar. Change* **5**, 1 notes that the 1980s decollectivization of Vietnamese agriculture has divorced the landless from the means of production.

mechanical erosion Erosion by physical means, with no chemical change. The main agents of physical erosion are wind, water, and ice.

mechanical weathering The splitting up *in situ* of rock, with no chemical change.

medial moraine *See* MORAINE.

medicine, geography of (medical geography) A field of geography that considers the distribution of specific diseases and human characteristics in relation to the geographical and topographical features of regions, countries, or the world. In its initial incarnation, medical geography was an enquiry into the environmental correlations of diseases; see Brown and Moon (2004) *J. Hist. Geog.* **30**. Medical geographers also became interested in the spatial distributions of those medical facilities, from hospitals to clinics to GP surgeries, that society 'invents' to treat diseases (Philo (2004) *NZ Geogr.* **63**, 2). Parr (2002) *PHG* **26**, 2 lists geographical variations in incidences of disease, and the complex intersections between health, illness, place, and space, as key issues, while calling for more attention to the body. Medical geographers map/ represent disease incidence (new cases) or prevalence spatially, to identify social or environmental factors associated with disease and locate areas where risk is particularly high. 'Most medical geographers are well aware of the statistical challenges associated with the analysis and representation of aggregate geographic data—the small numbers problem, the *modifiable areal units problem and the *ecological fallacy are perhaps most often of concern' (Yiannakoulias et al. (2005) *Canad. Geogr./Géogr. canad.* **49**, 3). Philo (op. cit.) argues that 'the human being—living, experiencing, thinking, feeling, often anxious, perhaps suffering, maybe dying—risks obliteration when expressed as "patients", disease entities, figures in graphs, or dots on maps'. See Sui (2007) *Geog. Compass* **1**, 3 on GIS and medical geography, and Smyth (2008) *PHG* **32**, 1 on the geographies of health inequalities.

Mediterranean climate A climate of hot, dry summers and warm, wet winters, characteristic not only of Mediterranean lands, but also in California, central Chile, and the extreme south of Africa. In summer, the climate is dominated by *subtropical anticyclones, and *trade winds prevail. In winter, mid-latitude depressions bring rain. *Local winds are of great significance.

Mediterranean soils, formed in *Mediterranean climates, owe their red colour to the *leaching of clays in the wet winters, which releases iron. Leaching is rare in summer so that a carbonate *horizon often builds up.

megalith Any large stone sited by humans, mostly dating from *c*.3000 to 2000 BC. C. Holtorf (2001) points out that all interpretations of megaliths, whether as tombs or 'mnemonic markers', can support the idea that megaliths were meant for a future time.

megalopolis A many-centred, multi-city, urban area of more than 10 million inhabitants, characterized by complex networks of economic specialization. Vicino et al. (2007) *Int. J. Urb. Reg. Res.* **31**, 2, in a study of the US Bowash Megalopolis, reveal a complex socioeconomic pattern structured by class, education, housing tenure, housing age, and race and ethnicity. They identify distinct clusters: 'affluent places', 'places of poverty', 'Black middle class places', 'immigrant gateway places', and 'middle America places'.

melting pot A concept of a number of ethnic groups, cultures, and religions in a society fusing together to produce new cultural and social forms. Brueckner and Smirnov (2007) *J. Reg. Sci.* **47**, 2 suggest that under mild conditions of interconnectedness, a society proceeds inexorably toward homogenization, with waves of foreign immigration, introducing unconnected groups into the population, providing short-term infusions of heterogeneity. As links form between these new groups and the rest of the population, the homogenization process typically reasserts.

melt-out till Subglacial melt-out *till reflects the amount of sediment in transport at the time of melting; supra-glacial melt-out till is located at the margins of a glacier, and includes lateral and terminal moraines. Larson et al. (2003) *GSA Abstr. with Progs* **35**, 299 suggest that distinctive silt pellets in melt-out tills indicate *glaciohydraulic supercooling.

meltwater Glacial meltwater is produced by the melting of glacier ice at the surface, or by pressure and *geothermal heat at the base, forming *subglacial streams. Some surface meltwater may percolate through the ice to emerge at the base. Both subglacial and proglacial **meltwater erosion** are extremely effective agents of landscape change. Water often flows under pressure within and below decaying ice, pressure that enables meltwater to flow upslope, and carry large quantities of debris, which promotes *abrasion.

   Glacial meltwater landforms form subglacially; at or near the ice margin; and in proglacial settings. **Ice margin meltwater** follows the edge of the glacier when drainage routes have been cut off by ice. **Tunnel valleys** occur below the ice and can cut steep-sided, flat-floored valleys; see Cutler et al. (2002) *Quat. Int.* **90**, 1. **Spillways** are channels cut by streams overflowing from *proglacial lakes; see Thieler et al. (2007) *Pal. Pal. & Pal.* **246**, 1. **Coulees** are canyons which result from the sudden and violent release of water from ice-dammed lakes when the barriers which impound them are breached. See Shaw (2002)

Quat. Int. **90**, 1 on the meltwater hypothesis for subglacial bedforms, and Hannah and Gurnell (2001) *J. Hydrol.* **246**, 1 and Pohl et al. (2005) *Atmos-Ocean* **43**, 3 for models of meltwater action. For meltwater and the reconstruction of former ice sheets, see Clark and Mix (2002) *Quat. Sci. Revs* **21**, 1–3. See also Glasser et al. (2004) *Boreas* **33**, 3 on meltwater processes as evidence for multiple glaciations.

memory We construct meaning about our lives and worlds through the places we create, inhabit, or visit. Till (1999, *Ecumene* **6**) describes places as 'fluid mosaics of memory, metaphor, matter and experience, [which] create and mediate social spaces and *temporalities'. Hoelscher and Alderman (2004, *Soc. & Cult. Geog.* **5**, 3, special issue), observe that 'memory and place conjoin to produce much of the context for modern identities'. Tolia-Kelly (2004, *TIBG* **29**, 3) writes on **re-memory**: 'a conceptualization of encounters with memories, stimulated through scents, sounds and textures in the everyday . . . souvenirs . . . [and] signifiers of "other" narrations of the past, not directly experienced'.

mental map A map of the environment within an individual's mind that reflects the knowledge and prejudices of that individual. A mental map indicates the way an individual acquires, classifies, stores, retrieves, and decodes information about locations. Ojanen and Vuohula (2007) *J. Common Mkt Studs* **45** describe Finland as lacking a prominent presence on the mental map of other EU governments. Walker (2000) *Int. Affairs* **76**, 3 notes that the overall American mental map of Europe is becoming clearer: it includes Turkey, and 'may include Ukraine and even Russia'. Sbragia (2007) *Politics* **27**, 1 describes the EU as a complete puzzle for most Americans. It is not a major actor in international relations, is not represented in the UN, and yet can force Microsoft and other iconic American corporations to change their behaviour if they want access to the European market. 'The American mental map does not have a place for such an entity.'

Mercalli scale A measurement of the intensity of an *earthquake.
Modified Mercalli scale of earthquake intensity:

I. Felt by very few, except under special circumstances.
II. Felt by a few persons at rest, especially on the upper floors of buildings.
III. Felt noticeably indoors, although not always recognized as an earthquake. Vibration like passing lorry.
IV. Felt by many indoors during daytime, but by few outdoors. Some awakened at night. Vibration like lorry striking building.
V. Felt by nearly everyone; many awakened. Some breakages, disturbances of trees, telegraph poles.
VI. Felt by all; many run outside. Some heavy furniture moved.
VII. Everyone runs outside. No damage in well-built buildings; moderate damage in ordinary structures; considerable damage in poorly constructed buildings.

VIII. Considerable damage except in specially constructed buildings. Disturbs people driving cars.
IX. Damage even in specially designed structures. Buildings shifted from foundations; ground cracked; underground pipes broken.
X. Ground badly cracked. Railway lines bent. Landslides considerable.
XI. Few brick-built structures remain standing, if any. Bridges destroyed. Broad fissures in the ground.
XII. Total damage. Waves observed on ground surface. Objects thrown upward into the air.

Adapted from US Geological Survey.

mercantilism The view, current in early modern Europe, that one nation's gain depends on another nation's loss; a trading nation prospers only if it encourages the export, but discourages the import, of manufactures. 'Compared to the US, Japan is thought to actively engage in mercantilism, accumulating substantial foreign reserves through exclusionary trade and industrial practices' (Pantulu and Poon (2003) *J. Econ. Geog.* **3**).

meridian An imaginary circle along the world's surface from geographic pole to geographic pole; a line of longitude. Meridians are described by the angle they form west or east of the **prime meridian**, which has a value of 0° and runs through Greenwich, England.

meridional circulation In meteorology, air flowing longitudinally, across the parallels of latitude. **Meridional flow** occurs in *atmospheric cells, resulting in part from longitudinal temperature variations. See Song et al. (2008) *J. Climate* **21**, 19. The **meridional temperature gradient** is the change in temperature along a *meridian. Von Storch (2008) *J. Climate* **21**, 18 finds that an increase in the meridional gradient of sea surface temperatures strengthens the atmospheric circulation.

mesa A steep-sided plateau or upland, formed by the erosion of nearly horizontal *strata. A mesa resembles a *butte, but is very much wider; see Brookes (2001) *Geomorph.* **39**, 1–4.

mesoclimate *See* LOCAL CLIMATE.

mesometeorology The study of middle-scale meteorological phenomena; between small features, like *cumulus clouds, and large features, like anticyclones. **Meso-scale** describes systems, or patterns of systems between small and *synoptic scale *c.*10–100 km across.

Mesozoic The middle *era of earth's history, from around 225 to 190 million years BP.

(((●))) SEE WEB LINKS
• An overview of the Mesozoic era.

metagovernance A counter-process to *governance whereby economic and political coordination is achieved despite the limitations of governments, states, firms, governance, and clans (Jessop (2004) *Rev. Int. Pol. Econ.* **4**, 3).

metamorphic aureole An area of rock altered in composition, structure, or texture by contact with an igneous *intrusion. See Richards and Collins (2002) *J. Metamorph. Geol.* **20**, 1. **Metamorphic rocks** have been changed from their original form by heat or by pressure beneath Earth's surface; for example, limestone to marble, shale to slate, slate to schist. **Dislocation metamorphism** (a widespread but non-universally recognized concept—see Sobolev et al. (2002) *Bull. Voronzeh State Univ.* 2: *Geology*) occurs through friction along fault planes or *thrust planes. **Regional metamorphism** (also called *dynamic metamorphism*) occurs during an *orogeny; see Patison et al. (2001) *Australian J. Earth Scis* **48**, 2.

metasomatism The change in *country rock brought about by the invasion of fluid; see Storey et al. (2004) *Geology* **32**, 2. Granites are formed at great depths by invading granulizing fluids; see Langer (1966) *Annals NY Acad. Scis* **136**, 1.

meteoric water Groundwater originating from *precipitation.

meteorology The study and science of all aspects of the atmosphere—weather and climate—aiming to understand the physical and chemical nature of the atmosphere, its dynamical behaviour, and its complex interactions with the surface. It includes both short- and long-term weather forecasting, and the determination of past and future climatic change. Meteorology now includes the study of the atmospheres of other planets (S. Dunlop 2001).

metrology The scientific study of measurement, especially the standardization and definition of the units of measurement in science. See Pike (2001) *Prof. Geogr.* **53**, 2.

metropolis A very large urban settlement usually with accompanying suburbs. No precise parameters of size or population density have been established. The structural, functional, and hierarchical evolution of global metropolises is rooted as much in the past as in the present: modern information and communications technology may be more advanced than the 19th-century telegraph, but the processes and outcomes are much the same (Daniels (2002) *PHG* **26**). '[Berlin's] wealth of facilities, as well as their scatter across the metropolis, can be understood only in the light of the city's history and, paradoxically, its troubles' (Grésillon (1999) *Cult. Geog.* **6**).

metropolitan Describing a country or city that is the centre of *colonial or *imperial power.

micelle A microscopic in aggregate of surfactant molecules within a *clay colloid.

microclimates The climates of those parts of the lower *atmosphere directly affected by the features of the earth's surface. The height of this part of the lower atmosphere varies according to the size of the influencing feature, but could be four times the height of the feature (Barry (1970) *TIBG* **49**). Local factors commonly override the influence of larger scale macroclimatic factors (Vitt et al. (1994) *Arct. & Alpine Res.* **26**). See Lindberg (2006) *6th Int. Conf. Urb. Climate* on modelling the urban microclimate.

micro-credit Lending small amounts of money to very poor households at commercial rates, rather than at the 'usurious rates of loan sharks' (S. Buckingham-Hatfield 2000). The classic model is the Grameen Bank, initiated in Bangladesh by Mohammed Yunis. Similar micro-credit schemes have been set up throughout the developing world, and have been enormously effective, not only in alleviating *poverty and improving child nutrition, but also in increasing the voluntary use of contraception. N. Burra et al., eds (2005) assess the impact of micro-credit. Poverty trends would probably not be reversed, 'but booming micro-credit would at least speak for a world where justice would have a greater place' (Santiso (2005) *Int. J. Soc. Sci.* **57**, 185).

(((●))) SEE WEB LINKS

• Grameen Bank website.

mid-latitude depression An area of low atmospheric pressure occurring between 30° and 60°. This low is some 1500–3000 km in diameter and is associated with the removal of air at height and the meeting of cold and warm *air masses in the lower atmosphere. At the fronts between the air masses, a horizontal wave of warm air is enclosed on either side by cold air. The approach of the warm front is indicated by high cirrus cloud. The cloud then thickens and lowers, and rain falls. As the warm sector passes over, skies clear and the temperature rises. The cold front is marked by heavier rain and a fall in temperature. The warm front advances at 20 to 30 miles per hour, whilst the cold front can move forward more quickly at 40 to 50 miles per hour. In consequence, the cold front eventually pinches out the warm air, lifting it bodily from the ground to form an *occlusion. See R. Barry and R. Chorley (2003).

mid-oceanic ridge *See* OCEANIC RIDGE.

migration The movement of people from one place to another. The terms **in-migration** and **out-migration** are used for **internal migration; voluntary migration** refers to unforced movements, **compulsory migration** describes the expulsion of minorities from their country of birth by governments, or by warring factions. In the case of *commuting, migration is a daily act, but, because there is no change of residence, a purist would not call commuting a migration, preferring the term *mobility. In **innovative migration** people move to achieve something new (see Waxman in D. Mittelberg 1999). **Economic migration** is the movement of people from a poor to a richer area; Gerber

(2006) *Int. Mig. Rev.* **40**, 3 argues that market conditions are key determinants of contemporary migration patterns within Russia, for example. Boswell (2003) *Int. Affs* **79**, 3 finds that, in the EU, domestic political-electoral pressures and the institutional context have damaged longer-term strategies of migration management, refugee protection, and relations with third countries. The distinction between economic and betterment migration is unclear. Hoey (2005) *J. Contemp. Ethnog.* **34**, 5 writes on **non-economic migration**. Findlay and Li (1997) *Area* **29**, 1 see the migration act as a manifestation of an individual's identity.

Scott (1998, *IIAS Annual Lecture*) notes that states do not care for 'people who move around'; efforts to prevent people from crossing political borders to seek safety are increasing, giving rise to a new set of safe spaces; see Hyndman (1999) *Prof. Geogr.* **51**, 1.

W. Chan (2001) *Asian Perspec.* **25**, 4 presents an overview of internal migration in China; see also Fan (2003) *Int. J. Urb. & Reg. Res.* **27**, 1 on rural–urban migration and gender division of labour in China. Jones (1998) *Econ. Geog.* **74**, 1, in a study of the impact of migrant remittances on the place of migrant origin in Zacatecas, Mexico, finds that remittances tend first to decrease, and then increase interfamilial income inequalities.

Milankovitch cycles There are three interacting, astronomical cycles in the Earth's orbit around the sun: in the shape of the elliptical orbit (*c.*95 000 years); in the axis of rotation (*c.*42 000 years); and in the date of perihelion (the time of year when the Earth is closest to the sun, about 21 000 years) (M. Milankovitch 1930). There is also a 100 ka cycle; see Berger (1978) *J. Atmos. Scis* **35**.

The medium-scale cycles are poorly known: Campbell et al. (1998) *Geology* **26** recognize up to eight cycles, but cannot explain their origins. Ware (1995) *Fisheries Oceanog.* **4** demonstrates four short-term cycles: 2–3 years, 5–7 years (El Niño–Southern Oscillation), 20–25 years (bidecadal oscillation), and a poorly resolved, very low-frequency oscillation with a 50–75-year frequency. See Harris (2005) *PPG* **29**, 2 for an excellent summary. Imbrie and Imbrie (1980) *Science* **207** show that Milankovitch cycles satisfactorily explain the major fluctuations in ocean temperature.

milieu The sphere in which each individual lives and which they are affected by, including tangible objects and people, social and cultural phenomena, and the *images which influence human behaviour. This is a huge theme in Canadian geography; see *Canad. Geog./Géog. canad.* (2000) **44**, 1; (2002) **46**, 1 and **46**, 4; (2004) **48**, 3; and (2006) **50**, 4. Buciek et al. (2006) *Geografiska B* **88**, 2 consider the ability of the Danish *kulturmiljø* (cultural milieu) to capture and incorporate the 'multiple and often contradictory cultural practices of different groups of actors and, not the least to transgress the often rather static and confined view on local history, which often results from the *heritage perspective'.

In economic geography, Waxell (2005) *Geografiska regionstudie* **62** finds the local milieu to be important for knowledge transfer, and Maskell (2004) *PHG* **28**, 1 questions the view that tacit knowledge transfer is confined to local milieux whereas codified knowledge 'may roam the globe almost frictionlessly'. See also Gareau (2008) *Antipode* **40**, 1 on the *neoliberal milieu.

mimesis 'The belief that we should strive to produce as accurate a reflection of the world as possible' (J. Duncan and D. Ley 1993); 'Less-than-conscious embodied practices; the subconscious inculcation of the specific "rules of the game" of a particular field' (Holt (2008) *PHG* **32**, 2). See Malmberg and Maskell (2002) *Env. & Plan. A* **34** on spatial clusters, spin-offs, and mimesis.

mineralization The breakdown of organic residues in soils; an essential process in humus formation. See Andriesh (2006) *18th World Congr. Soil Sci.*

mineralized zone An enriched zone of mineral deposits around an *igneous intrusion. Rousell and Trevisiol (1988) *Mineralium Deposita* **23**, 2 identify the sequence of events in the mineralized zone of the Wanapitei complex, Ontario.

minimum efficient scale of production (mes. of production) The smallest possible size of a factory compatible with profitable production, below which *economies of scale don't apply. Mes. can be defined in terms of production, or of employment. The turnover of a single product is ultimately bounded by the minimum efficient scale and consumer demand for a specific product (Frenken and Boschma (2007) *J. Econ. Geog.* **7**). A firm may lower its mes. by buying in components, or through technological change (M. Jovanović 2006).

misfit stream A stream that appears to be too small to have eroded the valley in which it flows; a stream whose volume is greatly reduced or whose meanders show a pronounced shrinkage in radius. It is a common result of drainage changes effected by capture, glaciers, or climatic variations (Beaty (1990) *Canad. Geogr./Géog. canad.* **34**, 2).

mist A suspension of water droplets which restricts visibility to between 1 and 2 km.

Mistral *See* LOCAL WINDS.

mixed farming The farming of crops and livestock under unified management. Tiffen (2006) *Expl. Agric.* **42** notes that increasing urbanization in Sub-Saharan Africa (SSA) has a particularly important impact on the livestock element of the mixed farming systems.

mobility A research field that spans 'studies of corporeal movement, transportation and communications infrastructures, capitalist spatial restructuring, migration and immigration, citizenship and transnationalism, and tourism and travel' (Hannam et al. (2006) *Mobilities* **1**, 1). T. Cresswell

(2006) addresses 'how the fact of movement becomes mobility. How, in other words, movement is made meaningful, and how the resulting ideologies of mobility become implicated in the production of mobile practices.' See also Blunt (2007) *PHG* **31**, 5.

The **mobilities paradigm** emphasizes that all places are tied into at least thin networks of connections that stretch beyond each such place and mean that nowhere can be an 'island'. It goes beyond the imagery of 'terrains' as spatially fixed geographical containers for social processes, and calls into question 'scalar logics such as local/global as descriptors of regional extent', and suggests a set of questions, theories, and methodologies 'rather than a totalizing or reductive description of the contemporary world' (Sheller and Urry (2006) *Env. & Plan. A* **38**, 2). Thus, McNeill (2008) *PHG* **32**, 3 considers the hotel as a space 'operating between fixity and flow, locating and refreshing mobile bodies, [and] embedding them in relatively fixed networks within particular cities'.

modal split The varying proportions of different transport modes which may be used at any one time. The choices of modes may be determined by the costs, destinations, capacities, and frequencies of the modes together with the nature of the goods carried and their destinations; Buchanan et al. (2006) *J. Transp. Geog.* **14**, 5 find that, in commuter behaviour, the key variable determining modal split and trip length was the distance the residence was located from the central business district. Titheridge and Hall (2006) *J. Transp. Geog.* **14**, 1 find that location, urban form, levels of car ownership, and the mixture of different social classes in the ward of residence are important determinants of modal split.

model A representation of some phenomenon of the real world made in order to facilitate an understanding of its workings. A model is a simplified and generalized version of real events, from which the incidental detail, or 'noise', has been removed; an 'abstraction of an object, system or process that permits knowledge to be gained about reality by conducting experiments on the model' (Clarke in P. Atkinson et al., eds 2004). Malmberg (1992, *Geografiska B* **74**, 2) defines idealist models as 'designations of theories that lack a specification of how this model is related to processes in time and space'.

In the 1960s, the quantitative revolution in geography saw an emphasis on models: basic mathematical equations and models, such as gravity models, deterministic models, such as von Thünens and Weber's location models, and stochastic models that use probability. The 'cultural' turn—the counter-positivist response from human geography—stressed the weaknesses of models. O'Sullivan (2004) *TIBG* **29**, 3, writes of geocomputational modelling, that 'the compositional model represents a theory about the world, rather than the world itself. The end result is a model . . . whose behaviour may be almost as intractably difficult to account for as the world it represents.' Connecting the model back to the world it represents is difficult for a number of reasons,

principally the *equifinality problem, which makes it impossible to judge the
relative merits of alternative models on purely technical grounds.

model township A planned settlement. A. Mager (1999) describes the
intended model township of Zwelitsha, RSA.

mode of production The way in which society organizes production.
K. Marx and F. Engels (1974) claimed that under the **capitalist mode of
production**, bourgeois democratic society is naturalized, so that the way
people act and feel is universalized as 'human nature', in order to preserve the
status quo. M. Sahlins (1972) identifies a peasant **family mode of production**,
which he describes as pre-capitalist, while F. Ellis (1988) stresses the flexibility
of the family in agricultural and non-agricultural activities. This rather
deterministic view of the critical influence of the mode of production has been
challenged, for example, because of the Eurocentric nature of the argument.
Nonetheless, geographers recognize that each mode of production creates its
own geography: capitalism seems to be inextricably wedded to *uneven
development (see, among many, Walker (1978) *Rev. Radical Pol. Econs* **10**, 3).

moder A humus which is less acid than *mor but more acid than *mull. See
Fons and Klinka (1998) *Geoderma* **86**, 1 on mormoders and leptomoders.

modernity The knowledge, power, and social practices which emerged in
17th-century Europe. Modernity was not associated solely with 'newness', but
also with beliefs in rationality and 'progress', and came to be seen as a
central attribute of Europe, which the rest of the world was expected (or
compelled) to adopt. 'The substitution of new designs for old will be a
progressive move, a new step up the ascending line of human development'
(Bauman in J. Krieger, ed. 2001; read the whole entry). Larsen (2006)
Geografiska B **88**, 3 sees Western modernity as characterized by the binary
oppositions of self–other, nature–culture, and future–past. I. Wallerstein (2004)
argues that modernity is concretely expressed as the modern world system.
Claval (2007) *Tijdschrift* **98**, 2 sees the large city as an expression of 19th-/20th-
century modernity, while Taylor (2007) *TIBG* **32**, 2 sees the subordination of
cities to territorial states as the most important geographical attribute of
modernity. Tomic and Trumper (2006) *Antipode* **38**, 3, writing on Chile, find
that the concept of cleanliness as an expression of the modernity claimed by
neoliberalism for a small part of Chilean society comes 'at a high human cost':
precarious labour under harsh conditions and low wages. Murphy (2006)
AAAG **96**, 1 argues that a core tenet of modernity is an abrupt break with the
past, but D. Harvey (2003) argues against the 'modernity as break' thesis.

modifiable areal units Areas studied in geography, such as states,
counties, census areas, or *enumeration districts, which may be arbitrary units
unconnected with any existing spatial patterns developed—the boundaries of
many African states, for example, cut across cultural and ethnic regions. In
many cases the investigator has to use the units for which data are available

rather than the units which are more suited to the investigation. Guo and Bhat (2004) *Transportation Res. Record* **1898** contend that the fundamental issue is the inconsistency between the analyst's definition of areal units, and the decision-maker's perception of residential neighbourhoods, and propose a way out of the problem.

Moho discontinuity (Mohorovičić discontinuity) The boundary in the earth's interior between the *crust and the upper *mantle, marking a change in rock density from 3.3 (crust) to 4.7 (mantle). Clague and Straley (1977) *Geology* **5**, 3 see the oceanic Moho as the boundary separating partially serpentenized ultramafic rocks, above, and fresh tectonized ultramafic rocks below.

The Moho occurs at about 35 km below the continents (but can be depressed to nearly 50 km depth; see Fuis et al. (2008) *Geology* **36**, 3) and at around 10 km beneath the oceans.

Moh's scale A relative scale of hardness, based on the ability of a harder mineral to scratch a soft one. The ten minerals in the scale are: talc, gypsum, calcite, fluorite, apatite, feldspar quartz, topaz, carborundum, and diamond. Use Moh's scale by testing an unknown mineral against one of these standard minerals. Whichever one scratches the other is harder, and if both scratch each other they are both of the same hardness.

moist coniferous forest Forest found only on the west coast of Canada and the USA where the climate is cool and moist throughout the year. Very large evergreens such as redwoods and Douglas fir grow, and undergrowth is sparse.

moisture index A measure of the water balance of an area in terms of gains from precipitation (P) and losses from *potential evapotranspiration (PE). The moisture index (MI) is calculated thus:

$$MI = \frac{100(P - PE)}{PE}$$

See Smakhtin and Hughes (2007) *Env. Model. & Software* **22**, 6 on drought indices.

mollisols In *US soil classification, soils with a rich humus content, developed under grassland. *Chernozems, *rendzinas, and *chestnut soils fall into this category.

monadnock An isolated peak, the relict of long-term subaerial denudation, now standing above the *peneplain. The term comes from Mount Monadnock, New England; see Fowler-Billings (1949) *GSA Bull.* **60**, 8.

money, geography of The interactions between money, *place, and *space, ranging from the global scale (*uneven development, *capitalism, *circuits of capital), to regional and local, such as variations in access to *financial services. Of considerable interest is the continued predominance, despite *time–space compression, of certain financial centres, such as Tokyo New York, Frankfurt, London, and Amsterdam; see Faulconbridge (2007) *Growth & Change* **38**, 2 on the changing landscape of European financial centres.

'Money is not just an economic entity, a store of value, a means of exchange, or even a "commodity" traded or speculated on for its own sake; it is also a social relation' (Martin (1999) in Stuart Corbridge). 'Value can be thought of as generated through relations and things which, via the material and social practices of the economy (production, exchange and consumption), come to be regarded as socially useful, helpful, uplifting or, more narrowly but generally, as fundamental to everyday life going on "as normal". These flows encompass the exchange of value embodied in products and may involve the exchange of money for work or the capacity to work, which could lead to an augmentation of future production and/or consumption' (Hudson (2008) *J. Econ. Geog.* **8**, 3).

B. Cohen (1998) argues that the globalization of money has undermined the power of sovereign nation-states: 'a system predicated upon the ability to draw a line between the "inside" of the nation state and the "outside" of the international system has been shattered by the growing frequency and volume of monetary flows.' Lee (2002) *PHG* **26**, 3 argues that the decision-makers in specialized investment houses and banks, all too ready to shift hot money around the globe, construct geographies through the intersection of their assessments of internal conditions in emerging markets and their power to engage with them. Lee did not forecast the resultant crash into global depression. Unfortunately.

There is surprisingly little on fraud in the geographical literature. *ID Analytics* examine how US identity fraud rates vary by geography, finding that the highest rates of identity fraud occur in New York state, followed by California, Nevada, Arizona, and Illinois, while the least risky states tend to be in the upper Midwest and upper New England.

monoculture A farming system given over exclusively to a single product. Writing on globalization and agrarian change in the Pacific Islands, Murray (2001) *J. Rural Studs* **17** lists some disadvantages of monoculture: 'increasing pollution, soil degradation and ground water depletion; further concentrating economic power, property ownership and social inequalities; and contributing to urbanization as displaced small growers migrate to towns and cities. Moreover, the Tongan economy has been left more vulnerable to global economic fluctuations.' However, Castree (2003) *PHG* **27**, 3 writes that 'it is assumed to be self-evident that monocultures are undesirable. Yet they are linked to high yields, while their implied opposite—a world of mixed, non-chemical agriculture—might not so readily feed current population numbers.'

monopoly The provision of a good or service by a single supplier, who then has the power to set prices, since there is no competition. See Holmes and

Stevens (2004) *J. Econ. Geog.* **4** on monopoly, geographic concentration, and business size.

monsoon Colloquially, a sudden wet season within the tropics, but, more explicitly, a seasonal shift of air flows, cloud, and precipitation systems. Monsoons have been described in West and East Africa, northern Australia, Chile, Spain, and Texas, but the largest is the south-west monsoon. This Asian monsoon is the atmospheric response, complicated by the presence of water vapour, to the shift of the overhead sun, and therefore zone of maximum heating, from the Tropic of Capricorn in late December to the Tropic of Cancer in late June. Associated with this response are major changes in *jet stream movements and a meridional shift of the rain-bringing inter-tropical convergence zone.

In winter, pressure is high over central Asia; winds blow outwards, and some depressions, guided by upper-air westerlies, move from west to east, bringing rain. During the spring, a *thermal low develops over northern India. Rains, related to a trough in the upper air, enter Burma in April/May, and India in late May/early June. In early summer the direction of the upper air changes, and the tropical easterly jet stream is semi-permanent about 15° N for the rest of the summer. With this change, the monsoon 'bursts', giving heavy rain across the southern half of the subcontinent.

By late June there is a continuous southerly flow of warm, moist air into the monsoon trough lying across northern India. This flow is a continuation of the south-east *trades, altered to south-westerlies by the *Coriolis force as they move north, across the equator, bringing huge quantities of water vapour, and reaching the west coast of India as the south-west monsoon. Here, the rainfall is *orographically enhanced. *Monsoon depressions are formed in association with these air flows, steered by the now easterly jet. Subsiding upper air prevents rainfall in the Thal and Thar deserts of the north-west of the subcontinent. By autumn, the easterly jet stream is replaced by a narrow band of the westerly subtropical jet stream, which follows the southern Himalayas. The south-west monsoon begins to retreat in September. Thereafter, the north-east trade winds dominate, and *hurricanes are common in the Bay of Bengal. See O'Hare (1997) *Geo* **82**, 3 and 4 for an exegesis on the Indian monsoon, and Bhaskara et al. (2001) *Meteor. Applications* **8** on forecasting of the Indian south-west monsoon.

monsoon depression A summer low pressure system affecting South Asia, 1000–2500 km across, with a *cyclone circulation about 8 km. Lasting two to five days, it comes roughly twice monthly, bringing 100–200 mm of rain per day. See Daggupaty and Sikka (1977) *J. Atmos. Scis* **34**, 5.

monsoon forest A tropical forest found in *monsoonal regions. Trees are semi-deciduous or evergreen, in order to withstand drought. Monsoon forest is more open, and has more undergrowth, than *tropical rain forest. See

Russell-Smith and Setterfield (2006) *J. Biogeog.* **33**, 9 on monsoon rain forest seedling dynamics.

Monte Carlo model A simulation model incorporating an entirely random element, and describing a sequence in terms of probability. Change is represented by random sampling from a number of probabilities, and each stage of the model is dependent on the stage before, and subject to chance. See Bivand (2008) *J. Reg. Sci.* **48**, 1.

mor A forest humus characterized by an organic matter on the soil surface in matted horizons, reflecting mycogenous decomposers, and with an abrupt boundary between the organic horizon and the underlying mineral soil. See Olsson (2005) *Glob. Change* **11**, 10.

moraine The rocks, boulders, and debris that are carried and deposited by a glacier or ice sheet. Moraine is often partly stratified, since some may have been formed under water. **Ablation moraine** is common on retreating glaciers, and coarse, because meltwater has washed out finer particles. As the glacier shrinks, lateral moraines are deposited at the side of the glacier. Where two lateral moraines combine, a central, **medial moraine** may be formed. **Ground moraine**, also called a *till sheet, is a blanket covering the ground. Other moraines have been moulded by ice parallel to the direction of ice movement. These include **fluted moraines** which are long ridges, possibly formed in the shelter of an obstruction; see Benn (1994) *Sedimentol.* **41**. *Drumlins are streamlined moraines. Boone and Eyles (2001) *Geomorph.* **38**, 1 suggest that the growth and decay of ponds in repeated cycles of subglacial till failure and landform development can produce hummocks, but Lukas (2006) *PPG* **30**, 6 notes that the term **hummocky moraine** can be problematic.

End moraines/terminal moraines are till ridges, usually less than 60 m high, marking the end of a glacier. In plan, they are crescentic, corresponding with the lobes of the glacier; a well-developed end moraine shows that the ice front was there for some time. Krzyszkowski (2002) *Sed. Geol.* **149**, 3–4 argues that end moraines often constitute alluvial fans, formed at the ice margin by the redeposition of supraglacial material. See Dyke and Lavelle (2000) *Canadian J. Earth Scis* **37** on major end moraines in the Canada.

Recessional moraines mark stages of stillstand during the retreat of the ice; a transverse moraine is a recessional if the up-glacier surface shows streamlining. **Rogen moraines** are fields of transverse moraines, 10–30 m in height, up to 1 km long, 100–300 m apart, and often linked by cross-ribs. Their origin is uncertain. **De Geer moraines** form where a glacier meets its *proglacial lake, and consist of till, layered sand, and lake deposits (Evans et al. (2002) *J. Quat. Sci.* **17**, 3). **Push moraines** occur when a glacier is retreating in the melt period but advancing in the cold season; see Krüger et al. (2002) *Norwegian J. Geog.* **56**. Glaciohydraulic supercooling may aid the development of moraines created by melt-out from basal ice (Lawson et al. (1998) *J. Glaciol.* **44**).

moral geography Studies of the interactions and variations within morality, place, and space, such as the analysis of global variations in moral beliefs and practices; 'perspectives on human geography with a normative emphasis' (D. Smith 2000). Smith stresses the need for geographers to engage with moral issues, asking: does distance diminish responsibility? Should we interfere with the lives of those we do not know? Is there a distinction between private and public space? Which values and morals, if any, are absolute, and which cultural, communal, or personal? And are universal rights consistent with respect for difference?

Castree (2007) *Int. J. Reg. Res.* **31**, 4 argues that all workers, knowingly or not, operate with moral geographies: 'moral geographies matter because they are the ethical basis for all worker solidarity and division, at whatever geographical scale happens to interest us.' Whitehead (2003) *Ethics, Place & Env.* **6**, 3 argues that, in addition to geographical space, geographical scale also plays a crucial role in the construction and maintenance of moral frameworks. See also R. Lee and D. Smith, eds (2004).

morbidity Ill health. **Morbidity rates** are of two types: the *prevalence rate* (the numbers suffering from a specific condition at any one time), and the *incidence rate* (the numbers suffering from a particular condition within a given time period, usually a year).

more economically developed country (MEDC) A country with: low birth, death, and infant mortality rates (characteristically around 10; around 10; and under 12 per thousand, respectively); less than 10% of the workforce in agriculture; high levels of nutrition, secondary schooling, literacy, electricity consumption per head; and per capita GDP generally above $US20. 'Economically' was included in this term some 30 years ago, to redress the perceptions of financially richer nations as more civilized, and more rational.

morphodynamics The study of the evolution of landscapes and seascapes in response to the erosion and deposition of sediment (Parker (2006) *U. Illinois*); 'the interaction between the form and the processes' (I. Livingstone and A. Warren 1996). See Stephenson (2006) *PPG* **30**, 1 for a review of coastal morphodynamics, and H. Viles (2001). For morphodynamic modelling, see Abad and García (2004) *Wld Water & Env. Resources Congr.* and Federici and Seminara (2003) *J. Fluid Mechs* **487**.

morphogenetic region 'A region resulting from processes that strongly reflect regional climate, geology, vegetation, soil, and hydrology' (Vale (2003) *Phys. Geog.* **24**, 3).

morphological mapping Mapping landform segments and facets using breaks of slope and inflections (Savigear (1965) *AAAG* **55**). 'Landforms are discrete units that can readily be defined and verified at different scales by proven techniques, are therefore acceptable integrated classifiers of the landscape, and can be used to divide it into discrete segments' (Bocco et al. (2001) *Geomorph.* **39**, 3–4).

Morphological mapping

morphology 1. The study of form; the term is now used as a synonym for the form itself, as in the morphology of the landscape or of the city.

2. In geolinguistics, the phonological shapes of words that adapt to special grammatical functions. See U. Ammon (2005).

morphometry The process of measuring the external shape and dimensions of landforms (for example, Federici and Spagnolo (2004) *Geografiska A* **86**, 3 on cirques), or living organisms (for example, Wilfert et al. (2006) *J. Biogeog.* **33**, 11 on termites).

morphotectonics 'Morphotectonics is a rather poorly defined term, but most geomorphologists would probably agree that it concerns causal linkages between topography and underlying tectonic or structural features within the crust' (French (2006) *PPG* **30**). Morphotectonic features may reveal recent seismic activity (Caputo et al. (2006) *Terra Nova* **18**, 3); can provide an insight into past sediment flux to depositional basins (Jones et al. (2004) *Terra Nova* **16**, 3); or illuminate drainage basin evolution (Ghassemi (2005) *Basin Res.* **17**, 3). A. Scheidegger (2004) claims that morphotectonics is fundamental to the understanding of landscape evolutions.

mother cloud A cloud from which another cloud develops; for example strato-cumulus cumulo-genitus (from cumulus). See Curic et al. (2002) *EGS XXVII Abstract* #3084.

moulin (glacier mill) A rounded, often vertical hole within stagnating glacier ice. Meltwater, heavily charged with debris, swirls into the hole. 'Moulins are ephemeral features in such an environment' (James (2003) *Geomorph.* **55**, 1–4). Some of this debris settles out at the base of the moulin, forming a **moulin kame** after the ice retreats.

mountain building The creation of uplands by movements of the Earth's crust. This is often, but not necessarily, associated with an *orogeny. Possible causes of the uplift of mountains are: a decrease of the density of the crust causing it to rise, possibly above a *hot spot; thickening of the crust at *collision margins; subduction below continents, causing them to rise; overthrusting of sedimentary rocks during collisions; and compression at *converging or at *conservative plate margins.

mountain ecology A change in altitude up a mountain equals a latitudinal shift away from the equator, and vegetation types change accordingly. The sequence in the San Francisco mountains is hot desert cactus at 200 m; oak scrub at 800 m; coniferous forest at 2000 m; and alpine tundra at 4000 m (Woodward (1993) *Ecol. Applications* **3**).

mountain meteorology Upland areas are cooler than lowlands of the same latitude, since temperatures fall by 1 °C/150 m. The heating of a valley floor may cause *anabatic winds, and colder and heavier air at the peaks may spill down as a *katabatic wind. When warm, moist air rises over mountain barriers, *orographic rain falls; when this air has passed over the barrier, it descends and is adiabatically warmed, which may bring *föhn winds. The warmed air can now hold the remaining moisture, so a 'rain shadow' of drier air develops in the lee of mountains. Lee depressions form downdraught of mountains, and major barriers such as the Himalayas affect the upper atmosphere. See C. D. Whiteman (2000).

mountain wind A wind formed when dense, cold air flows downslope from mountain peaks, or from a glacier. *See* KATABATIC WIND; BORA.

mud flat An accumulation of mud in very sheltered waters. Mud is carried into *estuaries and sheltered bays and settles at low water. See Proffitt et al. (2005) *J. Ecol.* **93**, 2.

mud flow The *flow of liquefied clay. Debris of all sizes, including large blocks of rock, may be transported by mud flows. In a study of northern Italy, Luino (2005) *Geomorph.* **66**, 1–4 finds that mud flows in small drainage basins are usually triggered when the rainfall has reached a value equal to 10–20% of the local mean annual rainfall (MAR). Mud flows in larger basins are triggered when rainfall is equivalent to 15–30% of the local MAR.

mudslide A mass movement of *debris lubricated by melting ice and snow, or by heavy rain. Chandler and Brusden (1995) *ESPL* **20**, 3 illustrate *dynamic equilibrium in a study of the Black Ven mudslide complex, Dorset, UK. *See also* JÖKULHLAUP; LAHAR.

mull In soil science, a mild humus produced by the decomposition of grass or deciduous-forest litter. Earthworms and other soil fauna mix this humus thoroughly with the mineral content of the soil. See Ampe and Langohr (2003) *Catena* **54**, 3.

multicell storm A severe local storm comprising a succession of convective cells. The first cell is formed by a powerful updraught of moist air. As the ensuing downdraught stifles this updraught, a new updraught develops to the right of the first. This sequence continues until the surface air cools. See Fovell and Tan (2000) *QJ Royal Met. Soc.* **126**, 562.

multiculturalism 'The political accommodation by the state and/or a dominant group of all minority cultures, defined first and foremost by reference to race or ethnicity; and more controversially, by reference to nationality, aboriginality, or religion' (Modood in I. McLean and A. McMillan, eds 2003). Nash (2005) *Antipode* **37**, 2 describes multiculturalism as 'the diverse policies and ways of thinking about societies characterised by cultural plurality'.

Liberal multiculturalism focuses on cultural diversity, celebrating ethnic variety, and teaching tolerance. Howard-Hassman (1999) *Canad. Pub. Policy/ Analyse de pol.* **523** observes that the public policy of liberal multiculturalism encourages private, individual choices of identity; 'paradoxically, this liberal policy also encourages identification with Canada and Canadian citizenship'. Nash (2003) *PHG* **27**, 5 contends that neoliberal multiculturalism 'delegitimates the politics of class and redistributive justice and provides opportunities for more overtly reactionary versions of nationhood to emerge'.

Critical multiculturalism is concerned with 'majorities' as much as 'minorities', and considers the institutions and practices forming the whole society. It sees inequalities of power, and racism, as central, and emphasizes recognition and rights. 'The continuous invocation of "oppression" in critical multiculturalism, and related calls for "counter-hegemonic" action, look incongruous: oppression becomes a thoroughly discursive, contestable and pluralized modality, and "hegemony" itself is an impossible attempt to constitute imaginary unity out of palpable diversity' (McClennan (2001) *Ethnicities* **1**, 3).

multinational corporation *See* TRANSNATIONAL CORPORATION.

multinational state A state which contains one or more ethnic groups, as identified by religion, language, or colour. Bell (1999) *Cult. Geogs* **6**, 183 encounters iconic references to the pre-Revolutionary cultural traditions of Muslim central Asia in Tashkent, with the aesthetic of a modernized, multinational (yet Russianized) society, arguing that political elites have a difficult task in rallying citizens around symbols of national autonomy, while not undermining their own legitimacy, which still rests in large part on Soviet-era institutions.

multiple nuclei model A representation of urban structure based on the idea that the functional areas of cities develop around various points rather than just one in the central business district. Some of these nuclei are pre-existing settlements, others arise from urbanization and *external economies. Distinctive land use zones develop because some activities repel each other; not all land users can afford the high costs of the most desirable locations, and some require easy access.

Outlying business district

CBD

Low class residential

Medium class residential

High class residential

Light industry

Heavy industry

Industrial suburb

Residential suburb

Multiple nuclei

multiplier The economic consequence of an action, intended or otherwise. Once a location becomes established as a production centre, economies of agglomeration (which increase with increasing regional size) tend to give it permanent cost advantages over other locations. These scale and cost advantages are reinforced by the relatively higher wages paid to workers in the scale-intensive industries, which are thought to act, by means of a Keynesian income multiplier, as a stimulus to local markets, resulting in additional scale economies, which, in turn, lead to further growth of regional exports (Leichenko (2000) *Econ. Geog.* **76**, 4).

Holloway and Kneafsey (2000) *Sociologia Ruralis* **40** argue that farmers' markets provide a local multiplier effect. In services, Glückler (2005) *Env. & Plan. A* **37**, 10 observes that continuous collaboration with a client may produce multiplier effects through client referrals and provide cheap access to new potential clients and future business. See Peck (2002) *J. Econ. Geog.* **2** on the multiplier effects of major factory closures: reduced demand, spending power, savings, and investment.

multi-scale landscape morphometry Digital images of landscape features may be captured in a sequence of successively coarsened resolutions. Hierarchy theory (see O'Neill et al. in M. G. Turner and R. H. Gardner 1991) suggests that patterns observed at any given resolution will, to some extent, weaken the patterns observed at finer resolutions and will themselves be constrained by coarser resolution patterns. Establishing the range of scales over which landscapes show hierarchically nested spatial patterns makes it possible to identify the scales at which fine-scale processes affect global scale patterns and vice versa. The selection of appropriate scales of measurement is fundamental to this process, and is generally guided by principles of hierarchy theory, since hierarchy incorporates relational links between nested levels. Bailey (1987) *Landscape and Urban Planning* **14** suggests a hierarchy of criteria for multi-scale ecosystem mapping and Tay et al. (2005) *Geosci. & Rem. Sens. Letts, IEEE* **2**, 4 provide a simple framework to generate multi-scale digital elevation models and extract topologically significant multi-scale geophysical networks.

mythical geography 'Symbolic places that celebrate history and invest locations with mythical meaning provide a sense of identity in place and time; they fuse history and geography in terms of myth and memory' (Azaryahu and Kellerman (1999) *TIBG* **24**, 1). 'The Greeks used the mythical geography of their land to re-create a physical view of the past that their poets, priests and politicians used as a paradigm for contemporary behaviour, and they drew upon the world around them, not just to illustrate that past, but also in many ways to create it' (J. Boardman 2002). 'Mythical geography is inherently holographic, in the sense that its interpretation is not tied to the identification of specific physical locations; each human being has the universe within himself and myths are meant to be discovered and understood by every one as if they had been created for him alone' (C. Carpentier de Gourdon 2002).

naive geography The incorporation of the knowledge that people have about their surroundings into *GIS and cartography, in order to make spatial information and decision support more accessible (here, naive means instinctive, not stupid, geography). Its creed is: 'topology matters, metric refines' (Egenhofer and Mark in M. Rumor et al., eds 1996). For a comprehensive review of naive geography, see Egenhofer et al. in A. Frank and W. Kuhn, eds (1995).

nappe Part of a broken recumbent fold which has been moved forward over the rock formations beneath and in front of it, finally covering them; see van Hinsbergen et al. (2005) *Geology* **33**, 4 on nappe stacking in the Aegean region.

nation 'Governable space' (Watts (2004) *TIBG* **29**). The ethnic or cultural nation has ethnicity as the basis of membership, with culture as a key cohesive element. The cultural interpretation regards 'nation' as something organic, evolving on the basis of an aesthetic, primordial community, whereby ethnic group solidarity is based on ties passed down from the past, and hence the modern nation may be related to historical ethnicity. Culture and language are crucial cohesive elements of this community. In contrast, the territorial or civic nation has territory as the basis of membership and citizenship as the cohesive force (Robinson et al. (2001) *Pol. Geog.* **20**, 8). Mavroudi (2007) *Glob. Netwks* **7**, 4 shows that the nation is not a 'given' entity; see Uimonen (2003) *Glob. Netwks* **3**, 3. Although many *multinational states exist, E. Gellner (1983) argues that a mismatch between national and political units engenders nationalist movements; see Sekulic (1997) *Nats & National.* **3**, 2. L. Kong and B. Yeoh (2003) hold that, in post-independence Singapore, the creation of a fixed national identity has denied the population's innate hybridity; see also Chua in J. S. Kahn, ed. (1998).

nationalism The feeling of belonging to a group linked by common descent, language, and history, and with a corresponding ideology which values the *nation-state above everything else, expecting supreme loyalty from its citizens. 'Nationalism understood as a modern concept, is a project of political and social engineering, which works through the invention of history and the reproduction of geography, space and architecture. The "geography of nationalism" is the spatial expression of strategies of exclusion, displacement and dispossession of the externalized "other", as well as of strategies of re-construction and re-production for the sovereign and hegemonic "self" of the nation' (ÖKTEM (2003) Socrates Kokkalis Graduate Workshop Paper). Writing on the break-up of Yugoslavia, Robinson et al. (2001) *Pol. Geog.* **20**, 8 observe that in recent times nationalist sentiments have been inextricably linked to demands for statehood (especially by the Serbs, Croats, and Slovenes). 'Thus nationalism in this context has exhibited political, ideological and cultural dimensions, but strongly linked to the ethnic characteristics embodied within the various nationalist enterprises.' Jones and Desforges (2003) *Polit. Geog.* **22**, 3 illustrate the local, as well as the national and global, processes creating nationalist sentiment; 'the national scale is itself *socially constructed*'.

National Park An area less exploited by humans, containing sites of particular scenic or scientific interest, and protected by a national authority. 'Geography, and our identification with it, give us a sense of place. Geography also affects our national identity, and for many National Parks is the fundamental reason for their establishment as parks' (United States National Parks Authority).

There may be conflicts between the conservation of the natural environment and public access; 'where this happens, priority must be given to the conservation of natural beauty' (Lord Sandford 1974). Imrie and Edwards (2007) *Geog. Compass* **2** record protests by locals over the Maya Biosphere Reserve, Guatemala. See also Milian (2007) *Cybergeo* on the politics of 'sites naturels' in the Pyrenees.

Attempting to resolve post-colonial tensions, the management of many national parks in the developing world is shared between indigenous communities and the non-indigenous; see Waitt et al. (2007) *TIBG* **32**, 2, on the Uluru National Park. For more on management, see T. Prato (2005) and for the economic benefits of a national park, see S. L. Patiño's study of the Madidi National Park, Bolivia.

nation-state A *nation whose boundaries coincide with the boundaries of the state which governs the territory of that nation; a condition which rarely occurs in practice. Indeed, Walby (2003) *Sociol.* **37**, 3 argues that the nation-state is more mythical than real: there are more nations than states; several key examples of presumed nation-states are actually empires; there are diverse and significant polities in addition to states, including the EU and some organized religions; and polities overlap, and rarely politically saturate the territory where they are located. Nonetheless, D. Slater (2004) argues that as the nation-state is an important crossroads between the global and the local, it is the critical site for geopolitical study. Focusing on 'the hybrid entity of states linked by hyphens to nations', M. Sparke (2005) suggests that the hyphens are deserving of much more attention than they usually get. Mahtani and Mountz (2002) *Res. on Immigration in the Metropolis W. Paper* 02-15 note the conflicting roles played by the nation-state as facilitator and enforcer; see

Mountz (2003) *Antipode* **35**, 3 on a poststructuralist approach to geographies of the nation-state.

natural area 'A geographical area characterized both by a physical individuality and by the cultural characteristics of the people who live in it.' (Zorbaugh in J. Lin and C. Mele 2005). Most cities contain numbers of natural areas which are delimited by informal boundaries, such as canals, parks, railways, rivers, and main roads. 'Culture is the agent, the natural area the medium, the cultural landscape the result' (C. Sauer 1925).

natural increase The surplus or deficit (negative natural increase) of births over deaths.

natural region A region unified by its physical attributes, especially its latitude, relief and structure, and location. Lane et al. (2004) *Australian Geog. Studs* **42**, 3 argue that the 'natural' region is a highly contested concept. Even so, in Canada, federal and provincial park and protected area system plans adopt natural region representation approaches (Lemieux and Scott (2005) *Canad. Geog./Géog. canad.* **49**, 4).

natural resource Any property of the physical environment, such as minerals, or natural vegetation, which is exploited by humans. Natural resources may be classified as *renewable and *non-renewable. The journal *Natural Resources Forum* covers technical issues of extraction, development, and efficient exploitation of resources, and international, multidisciplinary issues related to sustainable development and management of natural resources, particularly in developing countries and countries in transition.

natural resource evaluation Assessing the economic value of the natural environment. Key considerations include: an assessment of the importance of each resource to human life in a particular context, how far one resource can be replaced by another, whether the resource is a stock or a flow, and whether it is exhaustible or renewable. For the analytic tools used in natural resource evaluation, see S. Schueller et al. (2006).

natural resource management (NRM) The management of natural resources to promote development that is economically viable, socially beneficial, and ecologically *sustainable; S. Moore and S. Rockloff (2006) analyse the Australian government's attempts to effect nationwide changes in the management of agricultural areas and rangelands; see also Robins and Dovers (2007) *Geogr. Res.* **45**, 3 on NRM in Australia, and Perreault (2006) *Antipode* **38**, 1 on the exclusion of Bolivia's poor resulting from the restructuring of water and gas resources. For the methodologies used in natural resource management, see Viergever et al. (2006) *Geomatics World* **4** and G. Herath and T. Prato (2006).

natural vegetation In theory, the grouping of plants which has developed in an area without human interference; since most landscapes have been changed by humans, it is argued that there are relatively few areas of truly natural vegetation left; many biogeographers use, instead, the concept of potential natural vegetation—the vegetation that would exist today if humans were removed from the scene and if the plant succession after their removal were telescoped into a single moment. The time compression eliminates the effects of future climatic fluctuations, while the effects of humans' earlier activities are permitted to stand. The potential natural vegetation reveals the biological potential of all sites (US Env. Protect. Agency).

nature 'Nature, in the common sense, refers to essences unchanged by man; space, the air, the river, the leaf' (R. W. Emerson 1836); 'a leisure-time concept . . . something to be consumed' (D. Harvey 1985). Bakker and Bridge (2006) *PHG* **30**, 1 observe that resource and environmental geography have conceptualized nature in predominantly physical terms: as an assemblage of things independent of (although related to) society, whose properties (responsiveness to technology, geographical location) and social utility are revealed by science. 'In a world with multiple realities of "nature" that are constructed by diverse groups with differing motivations and access to power, decision-making in environmental management can become a contest over whose knowledge is "right"' (Pedynowski (2003) *PHG* **27**, 6).

'Contemporary privatization is remaking nature–society relations as property. Privatization innovates and proliferates new forms of property such as patents for genetic information, markets for water, and tradable credits for polluting. In so doing, privatization transforms people's relationships to themselves, each other, and the natural world' (Mansfield (2007) Lecture, Ohio State University). See Castree (2003) *PHG* **27**, 3; see Mrozowski (1999) *Int. J. Hist. Archaeol* **3**, 3 on the commodification of nature.

nearest neighbour analysis The study of settlements in order to discern any regularity in spacing by comparing the actual pattern of settlement with a theoretical random pattern. For the methodology, see Mak (2003) *Geog.* **516**, 4.

nebkha *See* SAND DUNE.

nehrung A synonym for *barrier beach.

neighbourhood A district forming a community within a town or city, where inhabitants recognize each other by sight. See C. Moughtin (2005). Neighbourhoods evolve their own distinctive characteristics, or subculture; see Shaw et al. (2004) *Urb. Studs* **41**, 10. Schnur (2005) *Tijdschrift* **96**, 5 shows that local social capital can make the crucial difference in neighbourhood development.

neighbourhood effect The power of the environment to structure social contacts, plus the empirical fact that contact across class boundaries makes a consensual impact on partisan choice (Miller (1978) *Brit. J. Pol. Sci.* **8**). McAllister et al. (2003) *Brit. J. Pol. Sci.* **31** provide 'very strong' evidence that members of each social class in the UK were much more likely to vote Labour than Conservative in the low-status than in the high-status areas. 'This is entirely consistent with the concept of the neighbourhood effect, but alternative explanations are feasible.' Johnston et al. (2004) *Polit. Geog.* **23**, 4 find that neighbourhood effects are among the strongest influences on voting patterns.

neighbourhood unit An areal unit of between 5000 and 10000 people, each unit having its own low-order centre supplying convenience goods, medical facilities, and primary education, all within walking distance. Through traffic is discouraged. See Schulyer (2006) *J. Reg. Sci.* **46**, 5 and C. Moughtin (2005).

neoclassical economics Neoclassical economics focuses on the behaviour of individual consumers and firms. Consumers and firms interact in markets, which set the prices of goods and services. Taking into account their limited incomes and the prices set in the market, consumers buy the goods and services that maximize their well-being. Taking into account the prices set in the market, firms make decisions, including location decisions, that maximize their profits. Firms compete with one another, and this competition ensures that they will seek out and take advantage of profitable opportunities (Cortright (2006) Brookings Inst. Met. Policy Prog.).

Neoclassical economics suggests that surplus labour and *capital will move to areas where labour and capital are in short supply, so that uneven development will be eradicated. (Taylor (2007) *Globalizations* **4**, 4). Other important ideas include *cluster development and trickle-down economics— but see Basu and Mallick (2007) *Camb. J. Econ.* **32**, 3.

neo-colonialism The control of the economic and political systems of one state by a more powerful state, usually the control of a developing country by a developed one. It is marked by the export of *capital from the least economically developed countries (LEDCs) on the periphery to the controlling more economically developed countries (MEDCs) at the core, a reliance by the LEDCs on imported manufactures from the MEDCs, and adverse *terms of trade for the periphery or satellite nations (Blattman et al. (2004) *Harvard Inst. Econ. Res. Disc. Paper* 2040). The means of control are usually economic, including trade agreements, investment, and the operations of *transnational corporations, which are often seen as the primary instruments of neo-colonialism; but see Law (2000) *east asia cult. crit.* **8**, 1. Forstorp (2008) *Studs Philos. & Educ.* **27** argues that 'globalization' and 'knowledge society' are expressions of neo-colonialism.

neo-conservatism A particular version of conservatism that holds that security is best attained by using state power to spread freedom and democracy, if necessary by force and without international cooperation; a view strengthened in the USA after 9/11. Elden (2007), *Int. Politics* **44**, 1 writes that the war on Iraq violated territorial sovereignty, even though neo-conservatives see territorial sovereignty as an absolute. As a result, in the USA, funding programmes addressing 'natural' hazards have been cut to fund counter-terrorist strategies (Graham (2006) *Int. J. Urb. & Reg. Research* **30**, 2); government spending on public services has decreased (McGary (2004) *Philosophy & Geog.* **7**, 1); and public–private sector partnerships replace traditionally public-sector activities (O'Neill et al. (2002) *Health Prom. Int.* **15**, 2). See also Yamazaki (2002) *Geopolitics* **7**, 1 on neo-conservatism in Japan; Peunova (no date) *Grad. Stud. Rev.* **2**, 2 on a Russian retro-empire; and Mahs (2001) *Urb. Geog.* **22** on neo-conservatism and urban homelessness.

Neogene The current geological period of the *Cenozoic era, consisting of the Miocene, Pliocene, *Pleistocene, and *Holocene epochs. The Neogene began 23 million years ago (E. Martin and R. Hyde 2008).

neoglacial A time of increased glacial activity during the *Holocene. Lamoureux and Cockburn (2005) *Holocene* **15** suggest that increased moisture, perhaps driven by shifts in oceanographic conditions, were important in North American neoglacials.

neoliberalism It seems generally to be agreed that this term refers to the liberalizing of global markets associated with the reduction of state power: state interventions in the economy are minimized; privatization, finance, and market processes are emphasized; capital controls and trade restrictions are eased; free markets, free trade, and free enterprise are the buzzwords. Beyond that, definitions become more partial; D. Harvey (2005), for example, speaks of 'the doctrine that market exchange is an ethic in itself, capable of acting as a guide for all human action . . . [the reduction of] the obligations of the state to provide for the welfare of its citizens . . . unfettered individual rights'. A. Tickell and J. Peck (2006) argue that neoliberalization refers to 'the process of political-economic change'. Neoliberalism, they contend, 'is not a monolithic phenomenon, but produces its own geography'. 'Neoliberalism, in its changing forms, is playing a part in the reconstruction of extralocal relations, pressures, and disciplines' (Peck and Tickell (2002) *Antipode* **34**, 3). Peck (2002) *J. Econ. Geog.* **2**, 2 argues that 'both the institutional durability and the political tenacity of neoliberalism may have been underestimated'. D. Harvey (2005) argues that privatization writ large, financialization (*sic*), the skilful manipulation of crises practically and discursively to spread 'market rule', and state redistributions to the private sector and the rich have been neoliberalism's modus operandi. Do see Roberts et al. (2003) *Antipode* **35**, 5.

Case studies of the impacts of neoliberalism abound; see, for example: Budds (2004) *Sing. J. Trop. Geog.* **25**, on Chile; McCarthy (2006) *AAAG* **96**,

1 on British Columbia; Ryan and Herod (2006) *Antipode* **38**, 3 on Australia and Aotearoa; Gökarkisel and Mitchell (2005) *Global Networks* **5**, 2 on Turkey; Fisher (2006) *Soc. Justice* **33** on race, neoliberalism, and 'welfare reform' in Britain; and Emery (2006) *Soc. Justice* **33**, 3 on contesting neoliberalism in South Africa. See Liverman and Vilas (2006) *Ann. Rev. Env. & Resources* **31** on neoliberalism and the environment.

Attention has also been paid to **governance/governmentalities of neoliberalism**: see Larner and Butler (2005) *Studs Polit. Econ.* **75** on New Zealand; McCarthy (2004) *Geoforum* **35**, 3. **Spaces of neoliberalism** are addressed by N. Brenner and N. Theodore, eds (2003), INITIAL Basu (2004), W. Larner and W. Walters (2004); and **scales of neoliberalism** by Kohl and Warner *Int. J. Urb. & Reg. Research* **28**, 4.

neotectonics The study of the causes and effects of the movement of the Earth's crust in the neogene, i.e. the late *Cenozoic era. See Székely et al. (2002) EGU *Stephan Mueller Spec. Pub. Sers* **3** on neotectonic movements and their geomorphic response.

nesting The way in which one network fits into a larger one. For example, Inkpen et al. (2007) *Area* **39**, 4, see scale as a **nested hierarchy**. Species communities often exhibit **nested assemblages**, the species found in species-poor sites representing subsets of richer ones (Schouten et al. (2007) *J. Biogeog.* **34**, 11).

net primary productivity In ecology, the amount of energy which primary producers can pass on to the second *trophic level; see, for example, McNaughton et al. (1989) *Nature* **341**, 6238.

net radiation The balance of incoming solar radiation and outgoing terrestrial radiation (also called *net radiative balance*).

(((●))) SEE WEB LINKS

● Michael Pidwirny on net radiation and the planetary energy balance.

network A system of interconnecting routes which allows movement from one centre to the others. Most networks are made up of nodes (*vertices*), which are the junctions and terminals, and links (*edges*) which are the routes or services which connect them. 'Networks constitute the new social morphology of our societies, and the diffusion of networking logic substantially modifies the operation and outcomes in the processes of production, experience, power and culture' (M. Castells 1996). Dicken (2004) *TIBG* **29**, 1 describes **situated networks** as a generic form of social organization. Brown et al. (2008) *GaWC Res. Bul.* **236** see the world economy as based upon economic nodes connected as chains/networks. See Glückler (2007) *J. Econ. Geog.* **7**, 5 on economic geography and the evolution of networks.

Network connectivity is the extent to which movement is possible between points on a network—cities' housing firms with high service values will record large measures of network connectivity. Taylor et al. (2002) *esrc* attempt to measure the global **network connectivity of cities.** See also Taylor and Catalano (2001) *GaWC Research Paper* **61** for global, and banking, network connectivity.

network society The interaction between the revolution in information technology, the process of globalization, and the emergence of networking as the predominant social form of organization (Castells (2002) *Tijdschrift* **93**, 50); a society whose social structure is made of networks powered by microelectronics-based information and communication technologies (M. Castells 2004). D. Barney (2004) doubts whether a new network politics has emerged from the new network technology and economy, disputing that the terrain of politics and the operation of power has shifted wholesale into a politics of information management; Fuchs (2007) *21st Cent. Soc.* **2**, 1 argues that this term obscures 'the continuity of domination and capitalism'.

névé An alternative term for *firn.

new economy 'The economic frenzy of the 1990s . . . [typified by] new styles of production, novel forms of consumption and organization in everyday life, new horizons of planning and logistics of mobility, [and] new forms of materiality and sensibilities' (Löfgren (2003) *Glob. Netwks* **3**, 3). It is being built around smaller enterprises distributed in a much more dispersed geography, relies on the old economy for physical capital and a portion of demand, and the location of business activity in the new economy is far less tied to the cost-based logic that has shaped and reshaped the distribution of industry in the old economy. It relies to a growing extent on telephone-based networks for production and delivery, but also has leading-edge layers that require face-to-face contact (Beyers (2002) *J. Econ. Geog.* **2**, 1). For J. Bryson et al. (2004) the key players in the new economy are service activities, although Chiang (2008) *Geog. Compass* **2**, 1 focuses on 'the critical role of manufacturing', which spurs innovation spillover and generates multiplier effects. However, for Wood (2002) *J. Econ. Geog.* **2**, 2 the term signifies that things are not what they used to be 'even if we are not certain what they are'. Ettlinger (2007) *PHG* **32**, 1 argues that 'the popular characterization of the new economy in terms of "time-space compression" requires revision to consider the complex content of *what* is being compressed over time and across space'.

new industrial spaces Agglomerated production systems that coordinate interfirm transactions and the dynamics of entrepreneurial activity (Moulaert and Sekia (2003) *Reg. Studs.* **37**, 3). See Hansen and Winther (2007) *Danish J. Geog.* **107**, 2 on Copenhagen.

new international division of labour (NIDL) A global division of labour associated with the growth of *transnational corporations and the *deindustrialization of the advanced economies. The most common pattern is for research and development in *more economically developed countries, and cheap, less skilled labour in *less economically developed countries. See

Marin (2006) *J. Eur. Econ. Ass.* **4**, 2–3 on outsourcing. The impacts of the NIDL have been uneven: between nations, where some benefit more than others, and within nations, in locally specific ways; Martin (2004) *TIBG* **29**, 2 refers to the latter as a 'new sub-national development paradigm'. See OECD (2001) *evolution and globalization*.

newly industrializing country (NIC) A country which has seen rapid industrialization since the 1970s. The economic success of newly industrializing countries has been attributed to shared values and high levels of mutual trust within cultural groups (Batley in V. Desai and R. Potter 2002). See Sum in H. Beynon et al., eds (2001) on the 'embedded exportism' of East Asian newly industrializing countries. See also N. Forbes and D. Wield (2002).

new political economy The methodology of new political economy rejects the split between agency and structure characteristic of classical political economy. Instead, it builds on structuration theory, institutional economics, and *public choice theory, to develop an integrated analysis which combines theories which analyse agency in terms of instrumental rationality with contextual theories which analyse structures institutionally and historically.

new regionalism A territorially fluid, process-based, diverse, collaborative, and open mode of governance: see Smith (2007) *Reg. Studs Ass.* Lisbon; Wallis (2002) *Munimet*; I. Sagan and H. Halkier, eds (2005); and Rossi (2004) *Int. J. Urb. Reg. Res.* **28**, 2).

E. LeSage and L. Stefanick (2004) provide a useful summary table:

NEW REGIONALISM	OLD REGIONALISM
*Governance	Government
Process	Structure
Open	Closed
Collaboration	Coordination
Trust	Accountability
Empowerment	Power

Lovering (1999) *Int. J. Urb. & Reg. Res.* **23**, 3 differentiates between the 'new regionalism in thought'—'that sub-set of ideas in policy related economic geography which converge on the claim that "the region" is displacing the nation-state as the "crucible" of economic development'—and the 'new regionalism in practice'—an economic analysis 'overwhelmingly dominated by a "productivist" and micro-economic focus. That is, it treats regional economic activity as if the driving influences on the regional economy *as a whole* can be understood by looking at only one type of regional economic actor.' For other criticisms of new regionalism, see Bull (1999) *3rd World Qly.* **20**; Rossi (2004) *New Int. J. Urb. & Reg. Res.* **28**, 2; and Sandstrom (2001) *J. of Urban Affairs* **23**, 5 (2001).

new town A newly created town, planned to relieve overcrowding and congestion in a major conurbation. In the UK, housing was arranged in *neighbourhood units of around 5000 people with their own facilities such as shops, schools, and medical centres, and housing densities were low—about five houses per hectare. All British new towns were planned to give a balance of social groups. See S. Ward (2002) on planning the 20th-century city; D. Frantz and C. Collins (1999) on Celebration, a master-planned city near Disney World; and Popke (2000) *Growth & Change* **31**, 2 on a new town in South Africa.

niche A set of ecological conditions which provides a species with the energy and habitat which enable it to reproduce and colonize. A niche is usually identified by the needs of the organism. See Chave et al. (2002) *Am. Nat.* **159** on niche differentiation and the maintenance of species diversity. Etienne and Olff (2004) *Am. Nat.* **163** develop a simple model uniting a neutral community model with niche-based theory.

niche glacier A small patch of glacier ice on an upland slope; see Allen (1998) *Geomorph.* **21**, 3–4. Unlike *cirque glaciers, niche glaciers have little effect upon topography.

nimbo-, nimbus Referring to clouds bringing rain, as in nimbo-stratus clouds.

nitrogen cycle The cycling of nitrogen and its compounds through the *ecosystem. Most plants obtain the nitrogen they need as inorganic nitrate from the soil solution. Animals receive the required nitrogen they need for metabolism, growth, and reproduction by the consumption of living or dead organic matter containing molecules.

(⊕) SEE WEB LINKS

• Michael Pidwirny on the nitrogen cycle.

nitrogen fixation The alteration of atmospheric, molecular nitrogen to nitrogen compounds. The fixation mechanisms are: biological micro-organisms (such as those in the root nodules of leguminous plants), *lightning and other natural ionizing processes, and industrial processes. See Vitousek et al. (2002) *Biogeochem.* **57–8** on modelling nitrogen fixation.

nivation The effects of snow on a landscape. The main processes are backwall failure, sliding and flow, niveo-aeolian sediment transport, supra- and ennival-sediment flows, niveo-fluvial erosion, development of pronival stone pavements, accumulation of alluvial fans and basins, and pronival solifluction (Christiansen (1998) *ESPL* **23**, 8). For the factors influencing the effects of nivation, see Raczkowska (1995) *Geograf. Annal. A* **77**, 4. Nivation may produce the shallow pits known as **nivation hollows**. In time, these hollows may trap more snow and may deepen further with more nivation so that *cirques or *thermocirques are formed; see Kariya (2005) *Catena* **62**, 1.

node Accessibility of a node is given by the travel cost between a node and another node, or a weighted average of travel cost to a number of nodes in the network (Holl (2007) *J. Transp. Geog.* **15**, 4). Transport nodes show a distinct hierarchy, and the importance of a transport node appears to correlate strongly with its effects on the surroundings (van den Berg and Pol (1998) *Env. & Plan. C* **16**, 4). Factors that determine the importance of a transport node are: the situation of the transport node with respect to major metropolitan regions; the number and type of the converging transport modes; the number of destinations that can be reached from the transport node, and the frequency of the transport moves. The more important a transport node, the greater the potential regional economic development (van den Berg and Pol (1997) *Eur. Reg. Sci. Assoc.* 37th Eur. Congr.). See also de Langen and Visser (2005) *J. Transp. Geog.* **13**, 2. See Devlin et al. (2008) *J. Transp. Geog.* **16**, 1, on **arc-node topology**—the topological data structure ArcInfo uses to represent connectivity between arcs and nodes.

nomadism A form of social organization where people move from place to place. Nomadism incorporates the advantage of mobility; traditionally nomadic groups were able to exploit natural resources at dispersed locations. 'In the discourse of modernization and social change nomadism's place is usurped by agriculture' (Kreutzman (2003) *Geog. J.* **169**, 3). As international boundaries are increasingly well defined, with border guards, nomadism declines: N. M. Shahrani (1979) uses the term **closed frontier nomadism**, with sedentarization and confined migration cycles.

Governments try to immobilize nomads for the purposes of taxation as well as to improve their health and literacy; the mobility of capital and labour in the age of empire produced a spatial order where nomadism had to be seen as a cultural problem (Noyes (2000) *Cult. Geogs* 7). Conversely, Drakakis-Smith (2007) *Mobilities* **2**, 3 shows how local authorities keep some families mobile and excluded. Mouffe (quoted in Pugh (2007) *Area* **39**, 1) writes: 'against those who . . . advocate "nomadism" I am convinced that radical politics cannot avoid "territorialization" and that all forms of territorialization should not be perceived as machines of capture.' B. Jordan and F. Düvell (2003) argue that **global economic nomadism** requires a redefinition of citizenship beyond national borders which involves shared duties for those who have access, and rights for those who remain on the outside.

non-basic Concerned with serving the city alone; not bringing wealth into the city.

nonconformity A series of sedimentary strata overlying an igneous or metamorphic rock.

non-ecumene The uninhabited or very sparsely populated regions of the world. It is not easy to draw boundaries between the *ecumene and the non-ecumene as regions of dense occupation merge into sparsely populated regions. If there is a boundary, it is not static.

non-linear Lacking proportionality between two related variables, such as input and output; describing the unpredictability of those systems which do obey 'laws' but are hypersensitive to variations in the initial conditions; they may exhibit *chaotic behaviour. See Phillips (2003) *PPG* **27**, 1 on non-linearity in geomorphic systems.

non-renewable resource A finite mass of material which cannot be restored after use, such as natural gas. (S. Barr and P. Chaplin (2004) argue that cultural heritage is a non-renewable resource.) Each non-renewable resource has a theoretical depletion time—a ratio of its stock and rate of use. Non-renewable resources may be sustained by *recycling, which Jenkins (2000) *Econ. Ecol.* **34**, 3 claims tends to be built into traditional cultures.

non-representational theory Non-representational theorists consider how researchers might 'represent' what they encounter in their fieldwork, since, they argue, 'representational' theory generates an unwavering, deadened picture of the world. They emphasize knowing through connection and participation; the spotlight is on the process, rather than the outcome—'it ain't what you get but the way that you got it'. This means that academics have to move beyond mere representation. M. Doel (1999) provides a helpful parallel. The task of a painter, he says, is not to paint an object, or even to represent it—but to be that object's effects. This does not mean that descriptions are no longer admissible. 'Close descriptions can still be offered of particular encounters, attending in the process to the situated, embodied sense-making work being (unavoidably) undertaken by the peoples involved, [the work] that makes those encounters what they are' (Laurier and Philo (2006) *Area* **38**, 4).

That this theory has been taken up by human geographers owes a very great deal to the work of Nigel Thrift, whose major work on the topic was published in 2007. This 'folded mix of the witnessed and witnessing world' (Dewsbury (2003) *Env. & Plan. A* **35**, 11) chimes with the concept of *hybrid geographies. Lorimer (2008) *PHG* **32** waxes lyrical about non-representational theory, but geographies based on this concept are painfully few. Thrift's use of the term 'non-representational theory' has been seen as problematic, and Lorimer (2005) *PHG* **29**, 1 offers the term 'more-than-representational'.

North Following the terminology of the Brandt report (1979) a portmanteau term used to describe the advanced economies/more economically developed countries.

North American Free Trade Agreement (NAFTA) A *free trade area, established between Canada and the USA in 1989, Mexico joining in 1993. NAFTA aims to remove import tariffs and *quotas on most raw materials and manufactured goods, and restrictions on trans-border financial services. The agreement also has proposals on health, pollution, and the use of renewable resources. The North American Free Trade Agreement is the most comprehensive free trade pact (short of a common market) ever negotiated between regional trading partners, and the first reciprocal free trade pact between a developing country and industrial countries (G. C. Hufbauer and J. Schott 1993).

((⊕) SEE WEB LINKS)

• The text of NAFTA.

North Atlantic Drift   The Gulf Stream is a warm surface ocean current originating in the Gulf of Mexico, which widens to several hundred kilometres, and slows to less than 2 km/h, and then divides into several sub-currents, one of which is the North Atlantic Drift (Hayward and Williams (2005) *Geol. Today* **21**, 4). The intensified westerlies in the northern North Atlantic and strengthened northerlies along the western part of the Nordic Seas force the North Atlantic Drift to follow a more easterly path, giving an asymmetric sea surface temperature response in the northern North Atlantic, and thereby maintaining the properties of the Atlantic Water entering the Nordic Seas (Otterå et al. (2004) *Tellus A* **56**, 4). See Rasmussen and Thomsen (2004) *Pal. Pal. & Pal.* **210**, 1 on the role of the North Atlantic Drift in millennial glacial climate fluctuations. Bryden et al. (2005) *Nature* **438** calculate that the North Atlantic Drift has slowed by 30% over the past 50 years.

North Atlantic Oscillation (NAO)   A large-scale alternation of atmospheric mass between the Icelandic low- and the Azores high-pressure area. When the pressure difference is large, with a deep Icelandic low and a strong Azores high, the NAO is said to be high or positive, and said to be negative when the pressure difference is less (or occasionally negative) as a result of persistent blocking in the Iceland–Scandinavia area. Strong positive phases of the NAO tend to be associated with above-normal temperatures in the eastern USA and across northern Europe, and below-normal temperatures in Greenland and across south-eastern Europe and the Middle East. Opposite patterns are observed during negative phases (Perry (2000) *PPG* **24**, 2—an excellent source). Kingston et al. (2006) *PPG* **30**, 2 find generally positive NAO/ Arctic Oscillation streamflow relationships in north-west Europe and north-east USA, and positive and negative streamflow relationships in parts of eastern Canada. Kuszmina et al. (2005) *Geophys. Res. Letts.* **32**, L04703–L04703 suggest that the NAO may intensify with further increases in greenhouse gas concentrations.

North Pacific Current   An eastward current, originating east of the Emperor Seamounts (170° W) and forming the northern part of the North Pacific subtropical gyre. It maintains the Arctic Polar Front with the Pacific Subarctic current. See Dinniman and Rienecker (1999) *J. Phys. Oceanog.* **29**, 4.

nuclear family The small family unit of parents and children. 'The institutionalization of the nuclear family household ... emerges in interaction with the processes of capitalism, as well as in relation to existing (patriarchal) processes' (Lier (2007) *Geog. Compass* 1, 4). D. Lal (1998) claims that 'capitalism is not inevitably connected with the nuclear family or even individualism, although its genesis depended on both'.

nuclear power A form of energy which uses nuclear reactions to produce steam to turn generators. Naturally occurring uranium is concentrated, enriched, and converted to uranium dioxide—the fuel used in the reactor. This undergoes nuclear fission, which produces large amounts of heat. Some of the highly radioactive spent fuel may be reprocessed while the bulk must be disposed of. Both are costly and hazardous undertakings. The main advantage of nuclear power is the low CO_2 emissions: 'nuclear power is a low-carbon form of electricity generation that can make a significant contribution to tackling climate change. Ruling out nuclear as a low carbon energy option would significantly increase the risk that the UK would fail to meet its CO_2 reduction targets' (Meeting the Energy Challenge: A White Paper on Nuclear Power URN:08/525).

The major disadvantages are very high construction and decommissioning costs; – 'it would be irresponsible for us to benefit from nuclear power and leave it to later generations to deal with the waste' (N. Lenssen, worldwatch institute)—and the major problems which may arise with any accident, such as the Chernobyl disaster of April 1986. Furthermore, nuclear power stations have a short lifespan.

nucleated settlement A settlement clustered around a central point, such as a village green or church. 'The problem is to determine where dispersed settlement finishes and nucleated settlement begins' (B. Roberts 1996). Nucleation is fostered by defence considerations, localized water supply, the incidence of flooding, or rich soils so that farmers can easily get to their smaller, productive fields while continuing to live in the village. 'Nucleated settlement does not in itself necessarily induce a sense of community ... nucleated settlement in itself is only a part of the process of constructing and maintaining a sense of common interest' (C. Dalglish 2003).

nuclei Minuscule solid particles (sing. nucleus) suspended in the *atmosphere. Three types of atmospheric nuclei are distinguished: **Aitken nuclei**, of radii less than 0.1 μm (Heggs et al. (1991) *J. Geophys. Res.* **96**, 18), **large nuclei**, radii 0.2–1 μm, and **giant nuclei**, with radii greater than 1 μm. See Clarke (1992) *J. Atmos. Chem.* **14**.

These nuclei may be scraps of dust, from volcanic eruptions or dust storms, or salt crystals, or given off when bubbles burst at the surface of the sea. Atmospheric nuclei can scatter sunlight enough to lower temperatures, if enough are present, and play an important role in *cloud formation; see Kulmala et al. (2001) *Tellus B* **53**, 4.

nuée ardente 'A glowing cloud of volcanic ash, *pumice, and larger *pyroclasts which moves rapidly downslope, typically starting as large blocks, and then fragmenting to a mixture of dust through to house-sized particles' (Mursik et al. (2005) *Rep. Prog. Phys.* **68**). The material from a nuée ardente consolidates to form ignimbrite (*welded tuff*).

numerical modelling 'Numerical models can be useful for explaining poorly understood phenomena or for reliable quantitative predictions. When modelling a multi-scale system, a 'top-down' approach—basing models on emergent variables and interactions, rather than explicitly on the much faster and smaller-scale processes that give rise to them—facilitates both goals. Parameterizations representing emergent interactions range from highly simplified and abstracted to more quantitatively accurate. Empirically based large-scale parameterizations lead more reliably to accurate large-scale behaviour than do parameterizations of much smaller-scale processes. Conversely, purposely simplified representations of model interactions can enhance a model's utility for explanation, clarifying the key feedbacks leading to enigmatic behaviour. For such potential insights to be relevant, the interactions in the model need to correspond to those in the 'real' system in some straightforward way. Such a correspondence usually holds for models constructed for predictive purposes, although this is not a requirement' (Murray (2007) *Geomorph.* **90**, 3–4).

nunatak A mountain peak which projects above an ice sheet, generally angular and jagged due to *freeze–thaw, contrasting with the rounded contours of the glaciated landscape below. Confirming a former nunatak is generally difficult (Goodfellow (2007) *Earth Sci. Revs* **80**, 1–2). See Kleman and Stroeven (1997) *Geomorph.* **19**, 1–2.

nuptiality The frequency of marriage within a population, usually expressed as a marriage rate. R. Woods (2002) suggests that the role of nuptiality in influencing fertility during the late 19th century was geographically variable and could not be attributed to some overarching explanatory factor. Aria (2001) *Int. Mig. Rev.* **35**, 2 considers *assimilation as a cause of falling nuptiality in Cuban Americans.

NUSAP A diagnostic and analytic tool of uncertainty in science and policy. Brown (2004) *TIBG* **29**, 3 sees a need for geographers to collaborate in developing uncertainty methodologies or 'strategies for openness'.

nutrient cycle The uptake, use, release, and storage of nutrients by plants and their environments; the transformation of inorganic compounds into organic forms due to biological uptake, and then back to an inorganic state (D. Allen and M. Castillo 2009). The nutrient cycle is the main functional process maintaining the stability and production of an ecosystem, and land cover is the source and sink of the material and energy supporting the biosphere.

Schematic nutrient cycle

Nutrient cycling processes

A. Sources (inputs)
    4. Atmospheric inputs (eg. dust, aerosol, biological fixation, precipitation)
    5. Weathering from parent material (primary minerals from rocks)
    6. Deposition of minerals through the movement of animals (eg. movement of ungulates in Serengeti), soil (down slope), and seepage (drainage waters)

B. Storage capacities (pools)
    1. Living biomass
    2. Detritus (eg. litterfall, dead/decaying roots, fallen logs and branches)
    3. Soil

C. Internal ecosystem cycling
    10. Litter inputs (above- and below-ground)
    11. Decomposition and mobilization processes
    12. Uptake, utilization, and redistribution processes by higher plants (eg. trees primary sink for nutrients – ⅓ of the total annual requirement of nutrients acquired from internal redistribution. Excluding 70% of N removed from the leaves before litterfall)

D   Losses (outputs)
    7. Volatilization to the atmosphere
    8. Harvest transport, and erosion
    9. Leaching below the rooting zone

Nutrient cycle

oasis A watered spot in an arid area, occurring when the *water-table reaches the surface. See Woo and Guan (2006) *Permaf. & Perigl. Procs* **17** on a **polar oasis** environment.

object-field A model that attempts to integrate the field- and object-view in a single, combined, integrated data model; see Raper and Livingstone (2001) *TIBG* **26**, 2 for a brief explanation and useful references.

occlusion A stage in a *mid-latitude depression where the cold front to the rear catches up with the leading warm front, lifts the wedge of warm air off the ground, and meets the cold air ahead of the warm front. If this overtaking air is colder than the cold air ahead of it, it will undercut it, forming a **cold occlusion**. If it is warmer than the cold air which is ahead of it, it will ride over it, in a **warm occlusion**. See B. Haurwitz (2007); see also Martin (1998) *Month. Weather Revs* **126**, 2 and (1999) **127**, 10; and Levizzani et al. (1989) *Il nuovo cimento C* **12**, 2 (online), on a three-dimensional single radar analysis of an occluded front.

Power and Sidaway (2005) *Soc. & Cult. Geog.* **6**, 6 use the term **occluded geographies** in describing Lisbon's Expo '98 to signify a reworking/ re-presentation.

occupancy rate The number of people dwelling in a house per habitable room (except for kitchens and bathrooms). A rate of 1 per room is taken as acceptable; more than that represents overcrowding. The 2001 UK Census finds that 41.2% of black African, and 40.7% of Bangladeshi, households were overcrowded, but only 6.3% of white British.

occupational mobility The ability of the individual to change jobs after the acquisition of a new skill. Using 1970 US census data, Lin and Christiadi (2006) *Papers Reg. Sci.* **85**, 4 develop a set of loglinear models that synthesize gravity models of interregional mobility and loglinear models of occupational mobility, finding that occupational mobility was greater for males than for females. Monso (2006) *INSEE Premières* **1112**, finds that upward occupational mobility in France greatest in the 30–34 age group, and Power and Rosenberg (1995) *Fem. Econ* **1**, 3 spring no surprises in noting that white women in the USA are more occupationally mobile than are black.

ocean–atmosphere interaction (ocean–atmosphere coupling)
Oceans and the atmosphere constantly interact with each other; surface winds

Occlusion

drive the *ocean currents, moving warm water polewards and cold water equatorwards, and evaporation from warm oceans removes latent heat from the atmosphere. This latent heat is released when the vapour condenses with height. See Smith et al. (2006) *J. Climate* **19**, 18 for a low-resolution coupled ocean–atmosphere general circulation model, and Ciasto and Thompson (2008) *J. Climate* **21**, 6 on the Southern Hemisphere. **Ocean–atmosphere oscillations** are interactions which switch suddenly from one phase to another; the best known being the El Niño–Southern Oscillation. See McCabe et al. (2004) *Proc. Nat. Acad. Sci.* **101** on ocean influences on drought.

ocean current A permanent or semi-permanent horizontal movement of unusually cold or warm surface water of the oceans, to a depth of about 100 m. Ocean currents are generated by the thrust of the wind, the spin of the Earth, and the Moon's gravity, and are also affected by variations in water temperature and density, and by the *Coriolis force. These currents are an important factor in the redistribution of heat between the tropics and the polar regions; *see* THERMOHALINE CIRCULATION. The Southern Ocean, for example, is 'a crucial cog in the global heat engine' (anon. (2002) *Australian Arct. Mag.*). Saetra et al. (2008) *Tellus A* **60**, 1 believe that an increase in the sea-surface temperatures of 1–2 °C in the North Atlantic current may intensify cyclones. See also Ridgway and Dunn (2007) *Geophys. Res. Letts* **34**, 13.

Cold currents can modify temperatures up to 100 km inland. As tropical air streams move over these currents, advection *fog forms over the sea, stripping the air streams of most of their moisture; onshore winds are therefore dry; see T. T. Warner (2004). Conversely, **warm currents** originate in tropical waters, bringing unusually warm conditions to coastal areas in higher latitudes. The *North Atlantic Drift, for example, creates a difference of around 4 °C between the winters of coastal Canada and coastal western Europe of the same latitude (Bryden in G. Siedler and J. Church, eds 1999).

oceanic core complexes (OCC) Massifs in which lower-crustal and upper-mantle rocks are exposed at the sea floor, formed at mid-ocean ridges through slip on detachment faults rooted below the spreading axis. The formation of these large, shallow sea-floor features appears to be an episodic outcome of plate rifting and accretion at slow spreading ridges. An outstanding example of an oceanic core complex has been mapped at the eastern intersection of the Mid-Atlantic Ridge and the Atlantis transform fault; see Smith et al. (2006) *Nature* **442**. See also Ohara et al. (2007) *Island Arc* **16**, 13.

oceanic crust That portion of the outer, rigid part of the earth which underlies the oceans. It is made mostly of basalt layer, rich in iron and magnesium, average density around 2.9, and 5–6 km thick, but see Rodger et al. (2006) *Geology* **34**, 2.

oceanic ridge An underwater mountain range developed where *magma rises up through a cracking and widening ridge in the oceanic crust. Some magma cools below the crust, some forces into fractures, and much flows out to form new crust, which is then pushed away from the ridge. McCarthy (2004) *J. Geophys. Res.* **112**, B03410 explains that the disparity in spreading rates between the Northern and Southern Hemispheres is correlated with the rate of latitudinal circumferential extension.

As new crust is created each side of the ridge, it takes up the prevailing *magnetic polarity of the earth, which reverses from time to time. As a result, symmetrical bands of crust, with alternating polarity, develop on either side of the ridge. These magnetic patterns are used to calculate the rate of the *sea-floor spreading; see Michaud et al. (2006) *Geology* **34**, 1. The term **mid-oceanic ridge** properly refers to the ridge at the centre of the Atlantic Ocean, which comes to the surface at points such as Tristan da Cunha and Ascension Island. Other ridges, such as the Pacific–Antarctic ridge, are not truly at the centre of the ocean.

oceanography The study of the oceans. This covers the shape, depth, and distribution of oceans, their composition, life forms, ecology, and water currents, and their legal status. A. P. Trujillo and H. V. Thurman (2007) provide a good general introduction.

ocean trench A long, narrow, but very deep depression in the ocean floor where two adjacent plates converge, and one descends beneath the other. For this reason ocean trenches are also known as convergent plate boundaries. 'The arc like structure of many ocean trenches can be qualitatively understood by the ping-pong ball analogy. If a ping-pong ball is indented, the indented portion will have the same curvature as the original ball, that is, it will lie on the surface of an imaginary sphere with the same radius as the ball. The lithosphere as it bends downward might also be expected to behave as a flexible but inextensible thin spherical shell' (D. L. Turcotte and G. Schubert 2002). An alpinotype orogeny is also termed an **ocean trench orogeny**.

offshore 1. In meteorology, moving seawards from the land, as in an offshore wind.

  2. In geomorphology, the zone to the seaward side of the breakers.

  3. Outside the regulatory controls of nation-states; see Donaghy and Clarke (2003) *Competition & Change* **7**, 1. *See also* OFFSHORE FINANCIAL CENTRE.

offshore bar *See* BAR.

offshore financial centre (OF) Better known as 'tax havens', offshore financial centres—such as the Cayman Islands or Panama—are 'jurisdictions that oversee a disproportionate level of financial activity by non-residents. Financial activity in them is usually dominated by the provision of intermediation services for larger neighbouring countries . . . By providing competition for domestic banks, offshore banks also (inadvertently) lower the costs for anyone using the domestic financial system. For example, simply by being close to Andorra and Monaco, France has a more competitive banking system, which provides more credit at lower interest rate spreads. Such indirect competitive benefits of offshore financial centres more than offset their costs' (Rose and Spiegel (2007) *Econ. J.* **117**, 523, which see).

ogives Bands of dark and light ice that stretch across glaciers, in the shape of a pointed arch. Multiple shear zones are formed, through which basal ice is uplifted to the glacier surface to produce the dark, foliated ogive bands (Goodsell et al. (2002) *J. Glaciol.* **48**, 161).

oil crisis (oil price shock) In 1973, the *OPEC countries, headed by Libya, quadrupled the price of oil, and many member states nationalized the oilfields in their own countries. Oil companies began to develop previously uneconomic alternative resources; attention shifted to alternative sources of energy, energy conservation, and energy-efficient technology; and countries lacking alternative sources of supply, or the technology to develop alternatives, found the cost of their oil imports rising painfully.

okta A measure of cloud cover, from 1 okta (scant) to 8 oktas (complete). Oktas are shown on *synoptic charts by a circle which is progressively shaded in as cloud cover increases.

oligotrophic Poor in nutrients, usually of lakes, soils, and peat bogs.

onion weathering *See* EXFOLIATION.

ontology The study of being. Each discipline has its own ontology: 'economists tend to think about labour markets in terms of wages and human capital: workers locate where they receive the highest returns for their skills . . . in contrast, geographers typically see labour location as a result of worker attachments to complex, fixed social institutions and networks' (Ashby and Monk (2007) *Oxonomics* **2**, 1). See Boschma and Martin (2007) *J. Econ. Geog.* **7**, 5, on an ontology for evolutionary economic geography. A debate on the ontologies of scale occupies much of the (2007) *TIBG* **32**, *passim*. See also Deng (2007) *PPG* **31**, 4 on the ontological conceptualization and methodological treatment of a terrain surface.

((()) SEE WEB LINKS

• UK Ordnance Survey ontologies.

ooze A deposit either of algae, or fine inorganic sediments, developed on the sea floor, or at the bottom of lakes (Dasey et al. (2005) *Marine & Freshw. Biol.* **56**). See H. Armstrong and M. Brasier (2005).

OPEC The Organization of Petroleum Exporting Countries was established in September 1960 by Venezuela, Saudi Arabia, Iran, Iraq, and Kuwait. Other countries which qualified for membership and joined later were Qatar (1961), Indonesia and Libya (1962), Abu Dhabi (1967, later transferred to the United Arab Emirates, 1974), Algeria (1969), Nigeria (1971), Ecuador (1973), and Gabon (1975). Ecuador and Gabon withdrew in 1992. The San Francisco Bay Area-based collective, Retort, argues that since the 1960s relatively high oil prices have benefited both the oil majors and military contractors and manufacturers, construction companies, engineering and industrial design firms, and the banking and finance sectors as OPEC states recycled and spent their abundant petro-dollars. Watts (2005) *PHG* **29**, 360 writes that the removal of Saddam Hussein 'obliterated at a stroke one of the most assertive voices for an aggressive OPEC price policy, and set the stage for clipping the wings of another; Hugo Chavez'. See Bridge (2008) *J. Econ. Geog.* **8**, 3.

opencast mining A system of mining which does not use shafts or tunnels and is, hence, cheaper. The layer above the mineral seam is removed and the exposed deposit is extracted using earth-moving machinery. The overburden may then be replaced. See Filcheva et al. (2000) *Ecol. Engineer.* **15**, 1–2 on afforestation on surface coal-mine spoils.

open field A distribution of farmland associated with *feudalism in Europe. Each manor had two or three large open fields and each farmer was awarded a number of strips within each field, so that no one had all the good or all the poor land. 'From late in the seventeenth century both the open fields and land held in common . . . began to be enclosed and to be allocated in compact parcels to each of those who had possessed grazing or other rights in it' (Pounds (2005) *J. Hist. Geog.* **31**, 3).

open system A system that is not separated from its environment, but exchanges material or energy with it. Trudgill and Briggs (1977) *PPG* **1** see soil as an open system; see Nordstrom (1994) *PPG* **18** on an open system model of beach change, and Douglas (1981) *PPG* **5** on the city as 'an integrated open system of living things interacting with their physical environment'.

opportunist species Species which survive by the rapid colonization or recolonization of a *habitat: for example, the roach in Ireland (Caffrey and McLoone, *Environ 2003*). Such species have very good powers of dispersal in order to seek new habitats. Life is short, fecundity and reproduction are high, and these species are able to withstand difficult conditions. The term *fugitive species* is also used. Holling (in C. Folke et al. 1995) sees the proliferation of pioneer and opportunist species as the first stage of the adaptive process.

opportunity cost The second choice that you give up when choosing the best; for example, in choosing a better-paid job further away, commuting time is the opportunity cost. Carlino (1985; cited in Desmet and Fafchamps (2005) *J. Econ. Geog.* **5**, 3) argues that, though the cost of moving goods has gone down through improved technology, the cost of moving people—as measured by the opportunity cost of time—has gone up. Baldwin and Okubo (2006) *J. Econ. Geog.* **6**, 3 show that regional policies tend to attract the least productive firms since these have the lowest opportunity cost of leaving an agglomerated region; 'the result is a "sorting effect"'. See Willis et al. (1998) *Transp. Res. D* **3**, 3 on land prices and **social opportunity cost**.

optimal city size There are benefits as a city grows—for example, cheaper transport and the more economical provision of services—up to, and until, the optimum size is reached. Zheng (2007) *Urb. Studs* **44**, 5–6 shows that the optimal city size of Japanese metropolitan areas in 2000 was about 18 million, 'which means that the Tokyo metropolitan area, with nearly 32 million inhabitants, is obviously too large'.

optimization model A model used to find the best possible choice out of a set of alternatives. A simple example might be finding the most efficient transport pattern to carry commodities from the point of supply to the markets, given the volumes of production and demand, and transport costs. See Ward et al. (2003) *Annal. Reg. Sci.* **37** on evaluating urban change. Kuby et al. (2005) *Adv. Water Resources* **28**, 8 introduce a new class of combinatorial-optimization models to appraise dam removal.

optimizer A decision-maker who seeks to maximize profits; in choosing an industrial location, the profit maximizer is assumed to know all the relevant factors, including the cost of assembling materials and distributing products, the price of labour, and *agglomeration and *external economies. Among many who find optimizer a static and simplistic concept are Feola and Binder, *Tropentag 2007. Compare with* SATISFICING.

optimum location The best location for a firm in order to maximize profits; 'in many circumstances, the optimum location of the firm and the optimum production relationships can be shown to be inter-dependent' (P. McCann 2002). See also Bradshaw (2001) *Australian Geogr.* **32**, 2.

optimum population A theoretically perfect situation where the population of an area can develop its resources to the greatest extent, and achieve maximum output while enjoying the highest possible standards of living. The concept of optimum population is grounded in the concept of absolute limits (Burke (1996) *Pop. & Env.* **17**, 6).

ordnance datum (OD) In the UK, mean sea level at Newlyn, Cornwall, from which all other spot heights on Ordnance Survey maps—and hence all *contours—are established.

Ordovician A period of *Palaeozoic time stretching approximately from 500 to 430 million years BP.

(⊕) SEE WEB LINKS

• An overview of the Ordovician period

organic acid Acid compounds of carbon, such as acetic acid, which are produced when plant or animal tissues decompose. Bormann et al. (1998) *Biogeochem.* **43**, 2 show that organic weathering plays a significant role in the release of nutrients to plants. See also Jones et al. (2003) *Plant & Soil* **248**, 1–2.

organic weathering The breakdown of rocks by plant or animal action or by chemicals formed from plants and animals, such as *humic acids.

Organization for Economic Cooperation and Development (OECD) Constituted in 1961, this is a group of nations, comprising most of western Europe, together with Australia, Canada, Japan, New Zealand, and the USA, and formed to develop strategies which will boost the economic and social welfare of the member states. See OECD (2005) *Regions at a Glance* (ISBN Number 9789264018631). An OECD report of October 2007 (ISBN 9789264037809) concludes that 'region-specific advantages—embedded in specialized firms, skilled labour, and innovation capacity—will maintain the competitiveness of OECD regions, thus seeming to contradict the hypothesis that globalization reduces the importance of geographical proximity in business'.

Organization of African Unity (OAU) An association of independent
African states constituted in 1963, and designed to encourage African unity, to
discourage *neo-colonialism, and to promote development.

(⊕) SEE WEB LINKS

• The African Union website.

Organization of American States (OAS) An association of Latin
American states with the USA. Constituted in 1948, it is designed to encourage
American solidarity, to aid cooperation between members, to maintain
present boundaries, and to arbitrate in disputes between members.

(⊕) SEE WEB LINKS

• The OAS website.

orientalism A description of the West's depiction of the cultures of the
eastern hemisphere, orientalism became a political term through the work of
Edward Said (1978), for whom orientalism means European academic and
popular discourse about the Orient, expressed through colonial bureaucracies
and styles, doctrines, imagery, scholarship, and vocabulary. (In this, he was
heavily influenced by Michel Foucault, and his concept of *imagined
geographies.) Said argued that 'the Orient has helped to define Europe (or the
West) as its contrasting image, idea, personality, experience'. In other words,
to Western eyes, orientalism is what the West is not; orientalism is *other. By
representing the East as an exotic, bizarre, and inferior attachment, the
Orientalists made colonial conquest a natural and logical extension of the rise
of the West. Samuel Huntington (1993) *Foreign Affairs* **72**, 3 for example,
divides huge sections of the earth into 'civilization groups', with Western
civilizations at the top of the hierarchy, and D. Gregory (2004) shows how the
Western powers have reproduced and extended the orientalist scripting of the
Middle East in current-day Afghanistan, Palestine, and Iraq.

R. Guha and G. Spivak (1988) criticize Said's work for ignoring voices of
resistance within the Middle East, while H. Bhabha (1995) dismisses a clear
separation between colonizers and colonized, recognizing 'the ambivalence,
*hybridity, and mimicry found in colonial representations of Orientalised
places and, at the same time, to the same phenomena in the self-
representations of colonized peoples'.

orogeny A movement of the earth which involve the folding of sediments,
faulting, and metamorphism. A **cordilleran orogeny** begins with
sedimentation at a passive continental margin, ranging from coarse sand and
silt near the shore, to limestone reefs in tropical seas. Fine-grained clastic
sediments accumulate on the deeper continental slopes, forming shales and
greywackes. The deposition of these *geosynclinal deposits is followed by
*subduction and compression. Folds, thrust faults, and a volcanic arc form.
Lateral growth continues through igneous activity, and metamorphism, uplift,

and deformation result from continuing plate convergence. See Şengör (1999) *GSA Spec. Paper* **338**.

Initially, a **continental collision-type orogeny** is similar to the cordilleran type. However, when two continental plates collide, both are too thick and too buoyant for subduction. Consequently, the plates are welded together to produce a large mountain chain. See the excellent Andersen et al. (2002) *Abstr. & Procs Norw. Geol. Soc.* **22**, 12–14 (available online, 2008).

orographic precipitation (orographic rainfall) Also known as *relief rainfall*, this forms when moisture-laden air masses are forced to rise over high ground. The air is cooled, the water vapour condenses, and precipitation occurs. See Gray and Seed (2000) *Meteorol. Applics* **7** on the characterization of orographic rainfall. The step-duration **orographic intensification coefficient method** may be used to estimate the effects of topography on storm rainfall; see Lin (1989) *Atmospheric Deposition, IAHS Publ.* **179**.

orthogonal At right angles. Orthogonals plotted through the crests of waves in plan illustrate the process of *wave refraction.

osmosis The passage of a weaker solution to a stronger solution through a semi-permeable membrane. In soils, the more dilute soil moisture passes by osmotic pressure into plant roots. *Salinization causes **reverse osmosis** in soil, where higher concentrations of salt in the soil water draw water from the plant roots, causing crops to wither, and perhaps die. See Kotzer (2005) *Desalinisation* **185**, 1–3.

otherness Barnett (2005) *PHG* **29**, 1 finds two quite distinct understandings of otherness. One is based upon the empirical analysis of the ways in which identities are constructed through exclusion or denigration. In the second, the movement between identity and difference is understood to be ontologically constitutive of any and all subjectivity (P. Osborne 2000). 'By running these two understandings of otherness together, the uncovering of exclusion becomes the taken-for-granted manoeuvre of critical analysis for poststructuralized geography, according to which the cultural other is necessarily posited as the unstable foundation of hegemonic identities.' Kuus (2004) *PHG* **28** finds 'a broadly orientalist discourse that assumes essential difference between Europe and Eastern Europe and frames difference from Western Europe as a distance from and a lack of Europeanness'. J. Wedel (2001) observes that the assumption of otherness is not only inscribed on east-central Europe by the West but is also appropriated by east-central Europeans themselves.

outer space The physical universe beyond the earth's atmosphere. MacDonald (2007) *PHG* **31**, 5 aims to establish outer space as a mainstream concern of critical geography. 'More than half a century after humans first cast their instruments into orbit, contemporary human geography has been slow to explore the myriad connections that tie social life on Earth to the celestial realm. My starting point is a return to an early-modern geographical

imagination that acknowledges the reciprocity between heaven and earth.'
Warf (2007) *Tijdschrift* **98**, 3 reminds us that, although satellites circulate in
outer space, their origins and impacts are very much on the ground. See
T. Sandler (1980) on the exploitation of outer space.

outlet glacier A glacier streaming from the edge of a body of ice located on
a plateau. For an example, see Glasser and Hambrey (2002) *Sedimentol.* **49**, 1.

output geographies Geographies based on areal units designed by the
researcher, rather than by a *census body. See Martin (1998) *Int. J. GIS* **12**, 7.

outwash (outwash sands) and gravels Sorted deposits which have been
dropped by *meltwater streams issuing from an ice front. The material
deposited is coarse near the ice front, becoming progressively finer with
distance from it. Outwash fabric tends to be well *bedded, and *current
bedding is common; *see* SANDUR. The finest elements of the outwash may be
deposited in *proglacial lakes to form *varves. A series of streams may produce
several alluvial fans which coalesce to form an **outwash plain** (valley train).

overburden The layer of rock and soil overlying a particular rock *stratum.

overdeepening A phenomenon found in *cirques and in the steps of
*glacial troughs, where the middle section(s) of the feature are lower than the
mouth. One suggestion is that the overdeepened section is the zone of
maximum ice thickness, another that overdeepening coincides with less
resistant rocks. Evans (1999) *Annals Glaciol.* **28** ascribes overdeepening to
rotational flow, while Hooke (1991) *GSA Bull.* **103**, 8 describes a positive
feedback process, as overdeepening causes crevassing at the surface, resulting
in erosional forces that accentuate that overdeepening.

overflow channel A steep-sided and relatively narrow channel cut by
meltwater from a *proglacial lake as the level of the lake rises and spills over
the relief barrier which contained the water. See Reheis et al. (2002) *GSA Bull.*
114, 8.

overland flow The flow of water over a hillslope surface. It may be laminar,
turbulent, and transitional or consist of patches of any of these flow states
(A. J. Parsons and A. D. Abrahams 1992). Horton (1945) *GSA Bull.* **56** suggests
that overland flow is likely to be laminar near watersheds, but with distance
downslope overland flow becomes more turbulent (Smith and Bretherton
(1972) *Water Resources Res.* **8**). Horton (op. cit.) explains that these changes
result from the microtopography of the soil surface: run-off gathers into
depressions; which increase in size downslope. As the depressions deepen
they capture more flow through the cross-grading of the hillslope surface
(Dunne et al. in J. Costa and P. J. Fleisher, eds 1995). Surface run-off deepens
as a result of this 'micro piracy'; see Croke et al. (2005) *Geomorph.* **68**, 3–4.

The Darcy-Weisbach, Chézy, and Manning equations are used to predict
overland flow velocity (Nunnally (1985) *Env. Manage.* **9**, 5). Smith et al. (2007)

PPG **31**, 4 review the problems of measuring overland flow, and evaluate some solutions.

overpopulation Too great a population for a given area to support. 'What might be termed overpopulation in one context might not in another because of differences in the standard of living aspired to by different populations . . . Overpopulation is often examined in relation to the term "demographic pressure". This is a difficult concept to define since pressure is perceived in various ways, depending on whether countries are developed or less developed. In developed countries, which have reached a high standard of living, demographic pressure . . . is perceived as something that degrades the environment . . . Demographic growth, overpopulation and depopulation cannot be studied in the same way everywhere, because there are countless demographic systems, each of which is nested as a subsystem interlinked with other territorial, economic, cultural and social systems' (Faus-Pujol and Higueras-Arnal (2000) *Applied Geog.* **20**, 3).

Perhaps the foremost diagnoses of overpopulation come from P. Ehrlich (with A. Ehrlich 1990). Neo-Malthusians support the argument; see Lempert (1987) *E. Afr. Econ. Rev.* **3**, 1. Marxists, however, view overpopulation as the result of the maldistribution and underdevelopment of resources (Marx, *Capital*, vol. iii; see also Y. S. Brenner 1969). In the developed world, some would suggest that pollution and the desecration of the countryside are indicators of overpopulation. B. Lomborg (2001) provides a basic overview of the topic.

overspill 1. The population which is dispersed from large cities to relieve congestion and overcrowding, and, possibly, unemployment. It occurs with redevelopment in the city where new building is at much lower densities so that some people—the overspill—cannot be housed in the city.

2. The condition of a fluvio-glacial lake when water levels rise above the barrier impounding it; see Kaiser et al. (2007) *Earth Surf. Procs Ldfms* **32**, 10.

3. The dispersal—whether spontaneous or planned—of population in response to congestion, overcrowding, and high property prices. Overspill in Canada reflects the ongoing search for cheaper accommodation, by those who have been hardest hit by the crisis in affordable housing that has characterized Canada's largest and fastest-growing cities since the 1980s (Mitchell et al. (2004) *Canad. Geogr./Géogr. canad.* **48**, 2).

overthrust A nearly horizontal fold subjected to such stress that the strata override underlying rocks; see Price (1998) *GSA Bull.* **100**, 12 and Hubbert and Rubey (1978) *GSA Bull.* **70**.

over-urbanization Population growth in an urban area which outstrips its job market and the capacity of its infrastructure; 'urbanization without industrialization' (M. Davis 2006). In 1988, 100000 migrants from mainland China flooded into Hainan. The provincial government was swamped by more than 180000 applications for only 30000 jobs, with the result that those unable to find work turned to street hawking. Meanwhile, in-migrating peasant workers congregated in sprawling urban 'village' settlements that lacked basic amenities. The infrastructure of urban centres in Hainan sagged under these immense pressures (Gu and Wall (2007) *Sing. J. Trop. Geog.* **28**, 2).

ox-bow lake A horseshoe-shaped lake once part of, and now lying alongside, a meandering river with a narrow 'neck' between meander loops. When the river breaks through this narrow stretch of land, the old meander becomes a temporary lake; ox-bow lakes quickly fill up and become hollows in the landscape. See Gautier et al. (online 2006, *ESPL* **32**, 2).

oxidation The absorption by a mineral of one or more oxygen ions. Oxidation is a major type of chemical weathering, particularly in rocks containing iron; see Price and Vebel (2003) *Chem. Geol.* **202**, 3. Oxidation occurs in soils when minerals take up some of the oxygen dissolved in the soil moisture; see Nayak et al. (2007) *J. Env. Qual.* **36**.

oxisol A soil of the *US soil classification. *See* FERRALLITIZATION.

ozone A form of oxygen, and an atmospheric trace gas, made by natural photochemical reactions associated with solar ultraviolet radiation. Ozone has three atoms of oxygen combined in one molecule, rather than two atoms, as in free oxygen. The proportion of ozone in the atmosphere is very small, but it is of vital importance in absorbing solar ultraviolet radiation. The **ozone layer** (*ozonosphere*), is an ozone-rich band of the atmosphere, at 10–20 km above the earth, but is at its most concentrated between 20 and 25 km. See Schiermeier (2007) *Nature* **449**. When the ozone layer thins (the 'hole' over Antarctica is an example) increased solar ultraviolet radiation reaches the surface of the Earth, with consequent damage to human health. See European Space Agency 2004, 5 October.

o

pacific coast A coastline where the trend of ridges and valleys is parallel to the coast. If the coastal lowlands are inundated by the sea, a coast of interconnected straits parallel to the shore may result.

pack ice Large blocks of ice on the surface of the ocean, formed when an ice field is broken up by strong waves. Since 2002, the East Greenland pack ice is much less, or even non-existent (Brooks (2005) *Geol. Today* **21**, 6).

pahoehoe A type of lava flow which spreads in sheets, associated with highly fluid, basic lava, such as that ejected from *Hawaiian volcanoes. The surface is a glassy layer which has been dragged into ropy folds by the movement of the hot lava below it. Whether the lava is aa or pahoehoe depends mainly on viscosity and strain rate, which in turn depend on crystallinity, dissolved gas content, temperature, bubble content, slope, eruption rate, and lava composition. Self (1998) *Annual Rev. Earth & Planetary Scis* **26** writes on the importance of pahoehoe in emplacing large lavas.

palaeoclimate The climate of a particular period in the geological past, before historical records or instrumental observations. Palaeoclimatology uses evidence from glaciers and ice sheets, geology, sediments, and tree rings; even the micromorphology of leaves (Haworth and McElwain (2008) *Pal., Pal. & Pal.* **262**, 1–2) to determine Earth's past climates.

palaeoecology The reconstruction of past environments from fossils evidence. See Seppä and Bennett (2003) *PPG* **27**, 4 on progress in palaeoecology, and Meadows (1999) *PPG* **23**, 2.

palaeoenvironmentology The study of prehistoric and historic environments. Traditionally, the most widely applied technique in Quaternary studies has been that of *palynology. 'There appears to be no limit to the potential diversity of proxy types of evidence [for palaeoenvironmentology], and Quaternary palaeoecologists are . . . prepared to investigate virtually any organism . . . accumulating in sediments as to its palaeoenvironmental value' (Meadows (1999) *PPG* **23**, 2).

palaeohydrology The study of the structure and dynamics of the water cycle in the geological past. See Schettler et al. (2006) *Boreas* **35**, 4 and A. Issar (2003), both on the *Holocene. The extraction of hydrological information from geomorphological evidence in drainage basins is not straightforward; erosion and sedimentation complicate the alluvial record.

palaeo-ice streams See Stokes and Clark (2001) *Quat. Sci. Rev.* **20** and Golledge and Stoker (2006) *Boreas* **35**, 13 on ice streaming in the British Ice Sheet.

palaeomagnetism *See* GEOMAGNETISM.

palaeosol A soil exhibiting features which reflect past conditions and processes. The presence of a palaeosol may have **palaeoclimatic significance**; see Chlachula et al. (2004) *Boreas* **33**, 2, for example, on using loess-palaeosols in southern Siberia.

Palaeozoic The era stretching approximately from 570 to 225 million years BP.

• An overview of the Palaeozoic era.

Palmer Drought Severity Index (PDSI) A drought index that uses temperature and precipitation values from the climate division dataset of the National Climatic Data Center along with other components of the water balance equation to measure the departure of soil moisture supply from normal. The output of the PDSI consists of positive (wet) or negative (dry) values centred on 0 (normal) with values above +4.0 or below –4.0 generally considered extreme (Goodrich (2007) *Geog. Compass* **1**, 4). See Hessel and Graumlich (2002) *J. Biogeog.* **29**, 7 on tree-ring reconstructions of this index.

palsa 'There is no general agreement about the meaning of the word "palsa". Usage and recent suggested definitions indicate that the word is chiefly used for cryogenic mounds covered by peat that were formed by an accumulation of segregation ice in the discontinuous permafrost zone' (Pissart (2002) *PPG* **26**, 4).

palynology The study of pollen grains as an aid to the reconstruction of past plant environments. Kiage and Liu (2006) *PPG* **30**, 5 combine evidence from palynology, lake sediments, and associated records, to provide a more accurate and complete assessment of the paleoenvironmental changes. 'One drawback of palynological data is that lakes accumulate pollen from a large geographic area making it difficult to obtain information on a specific region from lake pollen' (anon. (1997) *LandOwner Resource Center*). The other weakness is that most of the pollens found come from wind-pollinated species, and animal-dispersed pollen is under-represented.

pan A large, shallow, flat-floored depression found in arid and semi-arid regions. Pans may be flooded seasonally or permanently. See Viles and Goudie (2007) *Geomorph.* **85**, 1–2.

panarchy The structure in which natural, human, and human-natural, systems are interlinked in continual adaptive cycles of growth, accumulation, restructuring, and renewal (L. Gunderson and C. Holling 2001). The initial steps in the process involve a dominant species accumulating biomass, capital, and wealth. However, if conditions change then there is creative destruction and reorganization in the system backloop. The result is that biomass, capital, and wealth is released to create irreversible and often unpredictable change to the system (Holling (2001) *Ecosystems* **4**). Panarchical relations suggest that both top-down and bottom-up interactions occur (Walker et al. (2006) *Ecol. & Soc.* **11**, 1).

Pangaea A supercontinent comprising all of Earth's landmass, before it was split by *continental drift. Crowley et al. (1989) *Geology* **17**, 5 find that very high summer temperatures and a large annual range are major features of supercontinents; a finding with implications for Permian extinctions.

parabolic dune *See* SAND DUNE.

paradigm The prevailing pattern of thought in a discipline or part of a discipline; perhaps the most powerful Western paradigm has been the 'scientific method'. T. Kuhn (1972) argued that the evolution of a new paradigm marks a new stage in thinking. This paradigm persists until too many anomalies occur, when it is replaced by a new paradigm that is able to explain the anomalies. In geography, for example, the regional geography paradigm was superseded by the *quantitative revolution, and then the *cultural turn.

Lomolino (2000) *Glob. Ecol. & Biogeog.* **9**, 1 offers a new paradigm for island biogeography; Essletzbichler and Rigby (2007) *J. Econ. Geog.* **7** find that economists are still far from formulating a coherent research paradigm; Chang (2007, *IGU Comm. Gender & Geog.*) looks for 'a new paradigm within which various aspects of transmigration can fit together'; Pompili (2006, *ERSA Conference Papers*; ersa06923) sees inter- and intra-city networks as 'paradigms for urban development'. See also Poloz (2007) *Export Dev. Canada*.

paraglacial processes The non-glacial Earth-surface processes, sediment accumulations, landforms, landsystems, and landscapes that are directly conditioned by glaciation and deglaciation (Ballantyne (2002) *Quat. Sci. Revs* **21**). This distinguishes it from the term 'periglacial' which is defined as 'cold, non-glacial' and is applied to environments in which frost-related processes and/or permafrost are either dominant or characteristic. See Iturrizaga (2008) *Geomorph.* **95**, 1–2 on paraglacial debris landscapes.

parallel drainage *See* DRAINAGE PATTERNS.

parallel slope retreat The evolution of a hillslope when the slope angle remains constant. Ellis et al. (1999) *Basin Res.* **11**, 1 argue that parallel slope retreat requires some mechanism to remove weathered material from the foot of the slope in order to maintain the existing topography. See Leeder et al. (2002) *Basin Res.* **14**, 1 on parallel slope retreat by earthquake-induced collapse failure. Jackson et al. (2005) *Sedimentol.* **52**, 5 find parallel slope

371 **particulates**

retreat on beaches, when breaking waves on the upper foreshore are lower; enhanced when waves approach at large angles to the beach.

parent material The rock or deposit from, and on, which a soil has been formed. The nature of the parent rock will largely determine the nature of the *regolith, and hence the soil texture—basalt tends to produce clay soils; granite sandy soils. S. Smith (1996) *PPG* **20**, 4 links parent material with bimodality in fluvial beds. See also Mahaney (1993) *PPG* **19**, 3.

Pareto optimality A situation where it is impossible to improve the economic lot of some without making others worse off. Sen (1970) *J. Polit. Econ.* **78** shows that, within a system of social choice, it is impossible to have both a commitment to minimal liberty and Pareto optimality.

park Originally an enclosed area used for hunting, in the 18th century the term applied to the grounds of a country house. It now refers to open land used for recreation in a town or city. Within the UK, parks were a feature of 19th-century *urbanization, in order to relieve the 'artificiality' of urban living. However, Brainard et al. (2006, *CSERGE*, ECM-2006-05) argue that parks have usually been sited with little regard for the geography of where different social groups live: within Birmingham, UK, 'even within the most deprived areas, whites have better access to park areas than non-whites'. Conlon (2004) *Sexualities* **7**, 462 describes Christopher Park, Greenwich Village, New York, as a specific space that 'highlights the relationships amongst queer theory, identity and material public spaces'.

participatory GIS 'A more socially aware type of *GIS which gives greater privilege and legitimacy to local or indigenous spatial knowledge . . . context- and issue-driven rather than technology-led and [seeking] to emphasize community involvement in the production and/or use of geographical information. A Participatory GIS celebrates the multiplicity of geographical realities rather than the disembodied, objective and technical "solutions" which have tended to characterize many conventional GIS applications' (Dunn (2007) *PHG* **31**, 5).

participatory research A collaborative and non-hierarchical approach to research. 'The keystone of PR is that it involves those conventionally "researched" in some or all stages of research, from problem definition through to dissemination and action. Ownership of the research is shared with participants, who negotiate processes with the academic researcher. Education and knowledge building are also often viewed as important outcomes' (Pain (2004) *PHG* **28**, 3).

particulates (particulate matter) Atmospheric particles—dust, pollen, soot, salt, sulphur, etc. See Zhang et al. (1994) *J. Atmos. Scis* **33**, 7 on the impact of dust on particulate nitrate and ozone formation.

passive margin In plate tectonics, a margin between *continental and *oceanic crust contained entirely within a single *plate. Tectonic activity here is minimal. See Leroy et al. (2007) *Geophys. J. Int.* on the uplift and evolution of passive margins.

patch An area of vegetation that has a uniform structure and composition, and differs from the surrounding vegetation. Watson (2002) *J. Biogeog.* **29**, 5–6 distinguishes between fragments—remnants of a previously widespread habitat—and islands, which have always been restricted and isolated in their spatial extent.

path In *behavioural urban geography, a channel people move along, within a city.

path-dependent 'Path dependence exists when the outcome of a process depends on its past history, on a sequence of decisions made by agents and resulting outcomes, and not only on contemporary conditions' (Baláž and Williams (2007) *Int. Mig.* **45**, 2). 'The future development of an economic system is affected by the path it has traced out in the past' (G. Hodgson 1996). This term seems, to some extent, to resemble *hysteresis, or *contingencies. See also Martin and Sunley (2006) *J. Econ. Geog.* **6**. 'The strength of the path dependence model lies in the insistence of its practitioners on the importance of the specific sequence of micro-level historical events' (Ruttan (1997) *Econ. J.* **107**, 444). Greener (2002) *Manage. Decisions* **40**, 6 explains path dependence as a **lock-in** (the entry of a system into a trap from which it can't escape without the involvement of an outside force/shock) by historical events. See also Peck (2007) *PHG* **31**, 6.

patriarchy A gendered system of social control that pervaded all aspects of human existence, including politics, industry, the military, education, philosophy, art, literature, and civilization itself (K. Millett 1970). This includes the 'private' spheres of love, sexuality, marriage, and children—the family is politicized as the foundation of patriarchal power; see Bennett (2006) *Gender, Place & Cult.* **13**, 2. 'Feminist writings extended the original commonsense meaning of [this] term to apply to a situation of male dominance in a society. This immediately raises the question of whether the term is purely descriptive, or whether it refers to an explanatory theory. It is important to appreciate this difference as many sophisticated feminist arguments use the term in its descriptive sense' (Blackburn et al. (2002) *Brit. J. Sociol.* **53**, 4).

'Patriarchy and masculinity are not eternal but (like everything else) constructed products, and open to change' (Massey (2001) *TIBG* **26**, 2). Erman (2001) *Int. J. Urb. & Reg. Res.* **25**, 1 shows how an urban community reproduces traditional patriarchal authority over in-migrants through the social construction of female labour within the framework of familialism, and the housewife ideology. Adams and Ghose (2003) *PHG* **27**, 4 observe that Bollywood idols reflect not only patriarchy but subethnic and colonial power relations. 'It is a sad fact that one of the few profoundly non-racial institutions in South Africa is patriarchy' (A. Sachs, no date). Hern (2006) *Pol. Geog.* **28**, 6 defines **transpatriarchies** as forms of **patriarchy** extending across and between national societies. 'Remasculinization of public rhetoric and privileging economic and military concerns act to defend patriarchy diffusely (Connell (2005) *Signs* **30**, pp. 1816–17), and have sharp impacts on diverse women's livelihoods and political spaces' (Radcliffe (2006) *PHG* **30**, 4). See Waitt (2006) *AAAG* **96**, 4 on white gay patriarchy.

patterned ground Most commonly in *periglacial areas, the arrangement of stones into polygons, isolated circles, and concentrations of circles known as nets, steps, and stripes. Polygons and circles are more common on level surfaces, stripes generally form on slopes; the patterns are made of coarser stones, separated by much smaller stones (*fines*).

George et al. (1989) *Geophys. & Astrophys. Fluid Dynam.* **46**, 3 argue that convection currents within the *active layer produce patterned ground; Grab (1997) *ESPL* **22** cites desiccation and the formation of needle ice as major causal processes. Matsuoka (2003) *Geomorph.* **52**, 1–2 attributes patterned ground to the upward injection of slow-freezing, waterlogged silts at particular points. Kessler and Werner (2003) *Science* **299**, 5605 note *self-organization in patterned ground.

peak land-value intersection (PLVI) The point in a *CBD, often, but not always, at a road intersection, where land values are at a maximum.

peasant A farmer whose activities are dominated by the family group. The family provides all the labour and the produce is for the family as a whole; occasional surpluses are sold in the open market. Landholdings are small, sometimes owned by the family, but often leased; see Kull (2006) *AAAG* **96**, 3 on the perceived rights of peasants versus state control. 'Peasants are not a declining class but rather are undergoing a variety of transformations. In some settings they will indeed disappear, in others experience stresses that lead to their semi-proletarianization and, in yet others, peasants were, and still are, being created by capitalism' (Johnson et al. (2005) *Antipode* **37**, 5). Although peasants have been characterized as backward and resistant to change, peasant strategies can be highly rational in a society where there is little margin for error; see Faminow et al. (1999) *World Animal Rev.* **93**. See Murray (2006) *J. Peasant Studs* **33**, 4 on neo-feudalism in Chile.

peat A mass of dark brown or black, partly decomposed plant material, formed in waterlogged ground, where temperatures are low enough to slow down the decomposition of plant residues. It is characteristically found in cold climates, and as a relict feature in temperate zones. See Turetsky et al. (2007) *Glob. Change Biol.* **13** on climate change and peatland. Peat may be used as fuel, and is widely used by gardeners; so much so that peat bogs are now under threat. See M. Evans and J. Warburton (2007).

ped In a soil, an aggregate of silt, sand, and clay of characteristic shape, resulting from the formation of hydrogen and ionic bonds between soil particles. Peds may be further developed by plant roots, by *polysaccharide gums secreted by soil fauna (Piccolo and Mbagwu (1999) *Soil Sci. Soc. Am. J.* **63**), and by alternate freezing and thawing, or wetting and drying (Hart et al. (1985) *Eur. J. Soil Sci.* **36**, 3). Ped properties can be reflected in the structure of pore space (Pachepsky and Rawls (2003) *Eur. J. Soil Sci.* **54**, 3). See SOIL STRUCTURE.

pedalfer Any soil high in aluminium (Al) and iron (Fe), and from which the bases such as calcium and magnesium carbonates have been *leached. Pedalfers generally occur in regions with an annual rainfall of more than 600 mm. H. Jenny (1992) notes that C. Marbut's differentiation of soils into pedocals and pedalfers (1935) applies only to mature soils.

pediment A low, *concave element at the foot of a hillslope (also called a *concave*, or *waning*, slope. See Strudley and Murray (2007) *Geomorph.* **88**, 3–4 on the numerical simulation of pediment development.

pedocal A soil high in calcium carbonate and magnesium carbonate..

pedogenesis The formation of soils. **Pedogenic processes** are soil-forming processes. See Foss (2006) *Soil Sci.* **171**, 6, Suppl. 1 on milestones in pedogenesis. The chief pedogenic factors are time, *hydrology, *parent rock, climate, fauna and flora, and terrain, but terrain variables may be ineffective predictors of soil characteristics where a dense canopy hides a non-uniform erosional environment (Williamson et al. (2006) *Soil Sci. Soc. Am. J.* **7**). **Pedology** is the science of soils: their characteristics, development, and distribution.

pedon A small sample of a soil large enough to show all the characteristics of all its horizons.

pelagic Of marine life, belonging to the upper layers of the sea.

peneplain An erosional plain, near to base level. 'Despite more than a century of effort, no convincing example of a contemporary peneplain has been identified' (Phillips (2002) *Geomorph.* **45**, 3–4).

perception The manner by which we make sense of the world; a process shaped by a cultural script, or filter, which moulds our take on life. One such filter is the available technology; where once disasters were interpreted as 'Acts of God', today they are perceived as the result of human irresponsibility or malevolence; see Furedi (2007) *Area* **39**, 4. Perception also changes with age: Tapsell et al. (2001) *Area* **33**, 2 come to the unsurprising conclusion that children's perceptions differ from those of adults.

Of particular interest is **hazard**, or **risk, perception**; 'what science deems to be an acceptable level of risk may not match the social perception of acceptability. When the differences between social and scientific notions of risk become acute, then the outcome is a *social amplification of risk* by the public' (Herrick (2005) *Area* **37**, 3). However, Crozier et al. (2006) *Area* **38**, 2 find that people in high-risk earthquake zones tend to be fatalist, or at least resigned to the consequences. Perception also shapes economic decision-making; see Farley (2007) *AAAG* **97**, 4 on the perception of 'good' forests, and Warren (2007) *PHG* **31**, 4 on the clash between perception and ecology; '[the] popular perception of "cultural belonging" conflicts with scientific judgements that such species do not belong ecologically'. *See* ENVIRONMENTAL PERCEPTION.

perched water-table A partly saturated, isolated, confined *aquifer underlain by an impermeable rock, with the main *water-table below the two. See Bryan and Rockwell (1998) *Geomorph.* **23**, 1–2 on perched water tables, rill incision, and sediment transport.

percolation The filtering of water downwards through the *bedding planes, *joints, and pores of a permeable rock.

percolines An underground network of water seepage zones: old root channels, soil cracks, and animal burrows enlarged by *interflow. **Percoline flow** is 'laterally concentrated diffuse interflow perhaps including pipeflow' (A. Lerman and M. Meybeck 1988).

perennial stream A watercourse which flows throughout the year.

perfect competition. Under conditions of perfect competition, there are many suppliers; a perfectly elastic supply of the *factors of production; no collusion between suppliers; and buyers and sellers are fully aware of the prices being charged throughout the market. This is an unlikely state of affairs; imperfect competition is much more common. Try Mossy (2003) *Chaos, Solutions & Fractals* **18**, 3: technical, but interesting.

perforation kame *See* KAME.

performance In the social sciences, the modes of constructing 'the now' by non-textual means; Thrift in I. Cook et al., eds (2000) writes of the possibilities of drawing on 'street theatre, community theatre, [and] legislative theatre'. Latham (2003) *Env. & Plan. A* **35**, 11 argues that social action can be viewed as performance, and that reframing the research process as a performance is productive.

performativity The citational practices by which discourse produces the effects that it names, and which at the same time enable and discipline subjects and their performances. (Gregson and Rose (2000) *Env. & Plan. D* **18**, 4); the 'reiterative power of discourse to produce the phenomena that it regulates and constrains' (J. Butler 1993). In other words, what we say and do about something shapes what it is. A good example is given by Waitt (2008)

Soc. & Cult. Geog. **9**, 1, who observes the perfomative production of gender as exhibited by Australian surfers. See also Powell and Cook (2006) *Int. J. Sociol. Social Policy* **26**, 7–8 on performativity, patriarchy, and the elderly in China.

In a performative perspective, all subjects are subjects-in-process. To study a subject is to 'analyse the practices by which it is produced as a self-constituting subject' (Kuus (2007) *TIBG* **32**, 1).

peridotite A coarse-grained igneous rock, made mostly of olivine and pyroxene, and thought to be the main constituent of the *mantle.

periglacial Referring to the processes and landforms of any area with a *tundra climate, or where frost processes are active and *permafrost occurs in some form. **Periglacial climates** are arid, with temperatures below 0 °C for at least six months, and summers warm enough to allow surface melting to a depth of around 1 m. See Vandenberghe (2006) *Geomorph.* **54** on the transition between temperate to periglacial climates, and vice versa.

Periglacial processes include *abrasion, *freeze–thaw, *nivation, and *solifluction, and are responsible for the formation of new deposits, the alteration of existing unconsolidated deposits, and the modification of existing landforms by *mass movement; see Millar (2005) *Geomorph.* **72**, 1–4 and Braun (1989) *Geomorph.* **2**, 1–3. Hall et al. (2002) *PPG* **26**, 4 identify snow fall as a crucial factor in periglacial weathering. See Vandenberghe and Thorn (2002) *PPG* **26**, 4 for progress in periglacial research.

periodic market A market place where buyers and sellers converge at regular intervals (Ghosh (1981) *Prof. Geogr.* **33**, 4).

periphery The edge, or margin. Murray and Challies (2006) *Asia Pacif. Viewpt* **47**, 3 describe New Zealand as a **resource periphery** and Chile as **semi-peripheral**. 'Urban peripheries organize what remains of the center' (Dear (2005) *Urb. Geog.* **26**, 248).

permafrost Areas of rock and soil where temperatures have been below freezing point for at least two years. Permafrost need contain no ice; a sub-zero temperature is the sole qualification. **Continuous permafrost** is present in all *periglacial areas apart from small, localized thawed zones, while **discontinuous permafrost** exists as small, scattered areas of permanently frozen ground. A **frost table** marks the upper limit of permafrost, which is overlain by the *active layer. Ice develops mainly in upper levels in **epigenetic permafrost**; in **syngenic permafrost** it is regularly distributed throughout the whole thickness of the permafrost; go to the US Army Permafrost tunnel for both types.

Permafrost aggradation (growth) decreases the thickness of the active layer, is responsible for the formation of *pingos, and may be caused by the freezing of *taliks. **Permafrost degradation** (decline), which plays a key role in the development of *thermokarst; see Jorgensen et al. (2001) *Clim. Change* **48**, 4. Permafrost is a very sensitive system; small mistakes in constructing buildings in this environment can have catastrophic effects; see Couture et al. (2000) *Geol. Soc. Canada*, 2000–B2.

Permian The latest period of *Palaeozoic time, stretching approximately from 290 to 248 million years BP.

((⊕)) SEE WEB LINKS
• An overview of the Permian period.

persistence (geomorphological persistence) The persistence of a landform varies with the magnitude and recurrence interval of the event which brought it about. Calver and Anderson (2004, *TIBG* **29**, 1) suggest the terms **static persistence** where 'spatial positioning is constant, [and there is] virtually no change in the feature over time', and **dynamic persistence** where 'the type of feature remains in evidence . . . but individual representations are formed and/or degrade'. See also Wilcock and Iverson, eds (2003) *Geophys. Monog. Ser.* **135**.

personal construct theory This suggests that humans are continually constructing and testing their own, individual images of reality, and that investigations should be based on the personal constructs of the people involved, not on those of the researcher. See Pomery and Green in P. Dearden and B. Sadler (1989); see also Hudson (1980) *PPG* **4**.

personal space The zone around individuals—usually that reserved for a normal conversational voice and for friendly interaction—which they reserve for themselves. The extent of a personal space around an individual is reckoned to be 1–1.5 m for an Anglo-Saxon (Beaulieu (2004) *J. Applied Social Psych.* **34**, 4). Empowerment strategies provide a conduit for women to claim personal space (George (2007) *Gender, Space & Cult.* **14**, 6). Barker (2000) *Area* **32**, 4 only hints at the 'globalization of personal space', as does McCormack (2003) *TIBG* **28** at the 'personal space of emotion'.

p-forms and micro-channel networks Small-scale landforms of glacial erosion—usually less than 1 m in width and depth—on glaciated bedrock surfaces. Sichelwannen are sickle-shaped bedrock depressions, with an open end pointing in the direction of ice flow, and are the most common (Glasser and Bennett (2004) *PPG* **28**, 1). See Shaw (1994) *Sed. Geol.* **91**. The term 's-form' applies if a glacial meltwater origin can be established.

phacolith An elongated dome of intrusive igneous rock usually located beneath the crest of an anticline or the trough of a syncline.

phase space **1.** Very simply, a continuum representing two variables: A and B. The continuum ranges from 100% A, 0% B, to 0% A, 100% B. See, for example, Hooke (2003) *TIBG* **28**, 2 on the phase space of meander movement and bend curvature. Melton (1958) *J. Geol.* **66** describes a mature drainage basin phase space with four variables.

2. Very much less simply, an abstract concept that captures all the possible spaces in which a spatio-temporal system might exist *in theoretical terms*; not just what happens but what might happen; 'the space of the possible' (J. Cohen and I. Stewart 1994). 'The geography of phase space is flexible, but not totally arbitrary: the main possibilities are "already there", constrained by contextual realities' (I. Stewart and J. Coen 1997). This use of the concept is not totally distinct from 1; there are just many more factors involved. Martin (in Harrrison et al. (2006) *Area* **38**, 4) draws on Richards in J. Matthews and D. Herbert (2004) to see phase space as 'a middle road between relational space and more fixed notions of spatiality'.

Hooke (2007) *Geomorph.* **91**, 3–4 uses **phase space plots** to help uncover emergent behaviour in meandering rivers.

phenomenon instance In *geomorphology, an entity whose form and behaviour varies over time and space. Thus, for example, a coastal spit may only 'act as such' at certain stages of the tide cycle, at other times more resembling an embryonic *sand dune (Raper and Livingstone (2001) *TIBG* **26**, 2). *See the* EMERGENCE.

phenotype The outward appearance, physical attributes, or behaviour of an organism, which develop through the interaction of nature and nurture. For **phenotypic variation**, see Robertson and Robertson (2008) *J. Biogeog.* **35**, 5.

Phillips curve A negative exponential curve demonstrating the relationship between the percentage change in wages and the level of unemployment. The theory is that high wages cause high inflation, and the lower the rate of unemployment, the higher the rate of inflation. Fujiki and Wall (*St. Louis Fed. W. Papers* 2006-57) adjust the Phillips curve for Japan, by taking account of the geographical dispersion of labour-market slack.

photic zone Those upper levels of a water body that are penetrated by light. See Håkanson (2006) *Glob. Ecol. Biogeo.* **15** on suspended particulate matter and the photic zone.

photochemical smog Nitrogen dioxide (NO_2) is emitted from petrol engines. Ultraviolet light splits this into nitric oxide (NO) and monatomic oxygen (O). The hydrocarbons emitted from the burning of *fossil fuels react with some of the monatomic oxygen to form photochemical smog. Photochemical smog is most common where the sunshine is strong and long-lasting and where car use is high; see Zhang and Oanh (2002) *Atmos. Env.* **36**, 26 on photochemical smog in Bangkok. Photochemical smog damages plants (Arbaugh et al. (2003) *Env. Int.* **29**, 2–6), and irritates eyes and lungs (Fukuoka (1997) *Int. J. Biometeorol.* **40**, 1).

photogrammetry The use of aerial photographs and images from *remote sensing to measure terrestrial features. See *J. Photogrammetry & Remote Sensing*.

phreatic Describing *groundwater below the *water-table. The **phreatic zone** is permanently saturated; see Bekele et al. (2006) *Aus. J. Earth Sci.* **53**, 6 on a phreatic aquifer, and Lajura et al. (2006) *Env. Forensics* **7**, 4 on a coastal phreatic terrain. A volcanic **phreatic eruption** has *meteoric water mixed with the lava, and may create steam, or a *geyser (Fontaine et al. (2003) *Earth & Planet. Sci. Letters* **210**, 1–2). A **phreatophyte** is any desert plant with roots down to the phreatic zone; see Ridolfi et al. (2007) *J. Theoret. Biol.* **248**, 2.

phylogenetic Referring to evolutionary development. The **phylogenetic diversity** (PD) of a species is a measure of its taxonomic distinctness and can be estimated by looking at the phylogenetic relationships among taxa. Species-specific metrics on PD can then be used to determine conservation priorities at various biogeographical scales (McGoogan et al. (2007) *J. Biogeog.* **34**, 11).

phylum The second highest category (of seven) in the scientific system of classification for organisms (pl. *phyla*), consisting of one or several similar or closely related classes. Hence, **phylogeography**—the geography of phyla; see Richards et al. (2007) *J. Biogeog.* **34**, 11.

physical geography 'Physical geography focuses upon the character of, and processes shaping, the land-surface of the Earth and its envelope, emphasizes the spatial variations that occur and the temporal changes necessary to understand the contemporary environments of the Earth. Its purpose is to understand how the Earth's physical environment is the basis for, and is affected by, human activity. Physical geography was conventionally subdivided into *geomorphology*, *climatology*, *hydrology* and *biogeography*, but is now more holistic in systems analysis of recent environmental and Quaternary change. It uses expertise in mathematical and statistical modelling and in remote sensing, develops research to inform environmental management and environmental design, and benefits from collaborative links with many other disciplines such as biology (especially ecology), geology and engineering' (K. Gregory 2002).

Between 1850 and 1950, the main ideas that had a strong influence on the discipline were uniformitarianism, evolution, exploration and survey, and conservation (G. P. Marsh 1864). In the 1960s, 'a new type of physical geography began to emerge that accentuated a concern with dynamic processes of earth systems. This new approach, which has evolved to the present, is founded on basic physical, chemical, and biological principles and employs statistical and mathematical analysis. It has become known as the "process approach" to physical geography ... Over the past fifteen years, physical geographers, who have always acknowledged that the systems they study are complex, have turned to emerging ideas in the natural sciences about nonlinear dynamical systems and complexity to explore the relevance of these ideas for understanding physical-geographic phenomena' (Rhoads (2004) *AAAG* **94**, 4). 'Advances in remote sensing, geographical information systems and information technology have enabled a more global approach; a second new development has been the advent of a more culturally-based approach throughout many branches of physical geography. By 2000 a series of issues can be identified including the increasingly holistic trend, greater awareness of a global approach and of environmental change problems, and of the timely opportunities which can arise from closer links with human geography and with other disciplines' (Gregory (2001) *Fennia* **179**, 1).

physical quality of life index (PQLI) An attempt to measure the quality of life or well-being of a country. The numerical value is derived from the adult literacy rate, infant mortality rate, and life expectancy at age 1, all equally weighted. It is one of a number of measures created because of dissatisfaction with the use of GNP as an indicator of development.

physiological drought When soil water, although present, is unavailable, as when it is frozen, or when *evapotranspiration exceeds plants' uptake of water. See Susiloto and Bellinger (2007) *Silva Fennica* **41**, 2 on morphological and physiological drought.

phyto- Of a plant; relating to plants; hence **phytogeography**. See DeForest Safford (2007) *J. Biogeog.* **34**, 10 and Kuentz et al. (2007) *J. Biogeog.* **34**, 10.

piedmont glacier A glacier formed from the merger of several *alpine glaciers as they emerge from the mountains; see Hall and Denton (2002) *Holocene* **12**.

pingo **1. open system pingo, hydrostatic pingo** A large *ice mound formed under *periglacial conditions, so called because it is formed from an unfrozen pocket confined by approaching *permafrost. Mackay (1979)

1. Open system pingo (East Greenland type)

a. Permafrost advances

Ice
'Free' water
Talik
Permafrost
Saturated ground

Pingo

b. 'Free' water freezes and expands, lake floor mounds up, lake water drains away

2. Closed system pingo (Mackenzie type)

3. Collapsed pingo (open or closed system)

Pingo (continued)

Geographie physique et quaternaire **33** argues that open system pingoes develop on former lakes.

2. closed system pingo An ice mound, as above, but formed as water, under *artesian pressure within or below permafrost, which causes it to buckle upwards. MacKay (op. cit.) argues that not all such pingoes form in a truly closed system. Gurney (1998) *PPG* **22**, 3 proposes a third category: polygenetic (or 'mixed') pingoes.

pioneer community The earliest *sere developing on a raw site. A **pioneer species** is a plant species dominating a community in the first stage of succession.

pipe **1.** A volcanic channel or conduit filled with solidified *magma.
Sometimes the hard pipe rock is exposed after erosion. See Barnett and Lorig
(2007) *J. Volcanol. Geotherm. Res.* **159**, 1–3. For **pipe eruption**, *see* CENTRAL
VENT ERUPTION.

2. In *hydrology, a subsurface channel, often near horizontal, through which
water passes. Pipes can transfer water underground as a rapid route for
subsurface storm flow. See Carey and Woo (2000) *J. Hydrol.* **233**, 1–3.

pipkrakes In *periglacial environments, needle-like ice crystals which form
when frosty nights alternate with morning thaws over several days of this cycle.
They prise off the material above them, and aid in downhill creep. See
Rosenfeld (1999) *Geomorph.* **31**, 1.

place Staeheli in J. Agnew et al., eds (2003) distinguishes five
conceptualizations of place: place as physical location or site, place as a
cultural and/or social location, place as context, place as constructed over
time, and place as a social process. 'If we make space through interaction at all
levels . . . then those spatial identities such as places, regions, nations, and the
local and the global must be forged in this way, too, as internally complex,
essentially unboundable in any absolute sense, and, inevitably, historically
changing' (Massey (2004) *Geografiska B* **86**, 1). H. Raffles (2002) understands
places as 'spatial moments [that] come into being and continue being made at
the meeting points of history, representation, and material practice'.

'A national ideology about women's place . . . is filtered through
geographical differentiation at the local level' (McDowell (2003) *TIBG* **28**, 1).
The world's cultural geography can be viewed as a 'mosaic' (Crang et al. (2003)
PHG **27**). See also Nijman (2007) *Tijdschrift* **9**, 2. Waterman (1998) *TIBG* **23**
comments on the relationships between the art produced at a specific place
and the place itself: 'place becomes a metaphor for social trends.' Castree
(2004) *Pol. Geog.* **23**, 2 has a different take: 'it is useful to reflect on what is lost
when we are encouraged to see place-bound movements as tendentially
"regressive"; trans-local organising as largely "progressive" (not to mention
pragmatically necessary); and an openness to other peoples and places as the
mark of a proper (sic) place politics.' A **relational reading of place** works with
the ontology of flow, connectivity, and multiple geographical expression, to
imagine the geographies of cities and regions through their plural spatial
connections (Amin (2004) *Geografiska B* **84**, 1). Martin (2003) *AAAG* **93**, 3
argues that **place frames**—the physical conditions of the neighbourhood and
the daily life experiences of its residents—are used to unite residents for
collective action.

Within physical geography, there remain many possibilities that are
determined by local, specific, factors linked to particular places and times—
that is, by *contingencies (Phillips (2007) *Geomorph.* **84**, 3–4). Contingencies
are not necessarily complications that can eventually be described by general
laws. They may be irreducibly place dependent, and as important as, or even

more important than, general laws in determining how the world works (Phillips (2004) *Prof. Geogr.* **56**, 1).

place dependent Temporally and spatially bound, by, for example, fixed *capital, institutions, labour skills, and shared conventions. 'Products and commodities take on the qualities of the places from which they come. In their tendency to persist even in the context of geographically homogenizing forces, place differences permeate the artifacts whose creation they stimulate' (Molotch (2002) *Int. J. Urb. Reg. Res.* **26**, 4.

place identity 'Place identity is subjected to an imposed, created and manufactured place construction, while sense of place remains personal, found and grounded in the lived experiences of many residents' (Carter et al. (2007) *Soc. & Cult. Geog.* **8**, 5). See Hall (1997) *Ecumene* **4** on repositioning place identity; see McGuirk and Rowe (2001) *Australian Geog. Studs* **39**, 1 and Schwartz (2002) *Pol. Geog.* **25**, 1.

placelessness E. Relph (1976) claimed that, with mass communication, and increasingly ubiquitous high technology, places become more and more similar, so that locations lose a distinctive 'sense of place'. 'Placelessness dehumanizes the world and because dehumanized places have less or no human attachments, the people in these placeless places become even more vulnerable to more dehumanization' (I. Vogeler 1996).

Dicken in J. Peck and H. Yeung, eds (2003) argues that 'geography and place are profoundly important to the ways firms are produced and behave. Firms are embedded in specific places, both in the place(s) in which they operate, and more so in their place of origin.' Dicken (2004) *TIBG* **29**, 1 thinks the claim that TNCs are becoming 'placeless' is 'a nonsense'. And Gertler in Peck and Yeung (op. cit) demonstrates the limited ability of head offices to impose uniformity across their operations overseas.

(⊕) SEE WEB LINKS

• Ingolf Vogeler's lecture notes about placelessness.

place names The study of the early forms of present place names may indicate the culture which gave the name together with the characteristics of the site. In the UK, for example, **ey** meaning a dry point and **ley** meaning a forest, wood, glade, or clearing appear in many place names such as Chelsea and Henley-in-Arden.

Place names label, define, and represent places and people; 'a place name sometimes fills up its territory with sense of place and homogenizes it. Fixing a boundary on a map reinforces a real differentiation and might segregate the dwellers' (Okamoto (2000) *PHG* **24**, 328). In the Hawaiian Islands, place names change 'from being reflections of Hawaiian geographic discourse, to being encoded within Western approaches to knowledge, commodification of the environment, and control of territory' (Hermann (1999) *AAAG* **89**, 1). Place names mark the spatiality of power relationships (Myers (1998)

Tijdschrift **87**, 3); see also Usher (2003) *Canad. Geogr./Géog. canad.* **47**, 4 on aboriginal land claims in Canada. Kearns and Berg (2002) *Soc. & Cult. Geog.* **3**, 3 consider pronunciation to be an important element of the cultural politics of place naming within post-colonial societies.

It is common for revolutionary regimes to create new state symbols to 'remove evidence of the deposed regime and to establish an identity for the usurper'. During the Soviet period a special framework of laws and instructions was created to regulate a renaming policy. With the fall of the Soviet Union, renaming began in earnest; in Armenia, for example. By 1988 some 598 out of 980 place names in the Armenian SSR had been renamed.

(⊕) SEE WEB LINKS

• Arseny Saparov on the alteration of place names in Soviet Armenia.

places of memory Sites where the symbolic imaginings of the past interweave with the materialities of the present; places of memory are both constructed and contested (Rose-Redwood (2008) *Soc. & Cult. Geog.* **9**, 4). K. Till in J. Agnew et al., eds (2003) explores places of memory, topographies and sites of social memory, and conflicts over national places of memory. See Forest and Johnson (2002) *AAAG* **92**, 3 on Soviet memorials, and Alderman (2003) *Area* **35**, 2 on street names.

place utility The significance a place has to a particular function, or people. Damm and Rosholm (2005) *CReAM Discuss. Paper Series* **0605** argue that 'instead of mandatory dispersal policies of refugees, voluntary spatial dispersal will increase place utility'. When people evaluate places, they don't necessarily look for the one with the highest overall utility (Lin-Yuan and Kosinski (1994) *Int. Migr.* **32**, 1).

planation surface *See* EROSION SURFACE.

planetary boundary layer The lowest 500 m of the troposphere; the layer most strongly influenced by the land, or sea, beneath it. See Amarydis et al. (2007) *Atmos. Chem. Phys. Discuss* **7**, 13537–660.

planetary winds The major winds of the earth such as the westerlies, *trades, etc., as opposed to local winds. See Civiet et al. (2005) *ADS* on measuring planetary winds.

planeze One of a series of triangular facets facing outward from a conical volcanic peak. The planezes are separated by radiating streams which run down the flanks of the cone. Try Coltorti et al. (1999) *Geol. Soc. Special Publication* **162**.

plankton Minute organisms which drift with the currents in seas and lakes. Plankton *algae, various animal larvae, and some worms; the animals are **zooplankton** and the plants are **phytoplankton**. See W. Lampert and U. Sommer (1997) for a discussion of the 'top-down' versus 'bottom-up' plankton controversy.

planning As practised by local or national government, the direction of development. See S. Ward (2004) for a well-presented and easily read history of planning in the United Kingdom. S. Campbell and S. Fainstein, eds (2003) provide a valuable stimulant for constructive debate and reflection.

Planning blight is the adverse effect of a proposed development, such as a motorway, which could cause a drop in house prices. If the landowner cannot dispose of the property, or cannot make as much use of it as was previously possible, he or she may serve a purchasing notice on the planning department of the local authority. See R. Eddington (2006).

plant **1.** In a system, the buildings, machinery, and land that inputs enter, and output exits.

2. In industrial geography, an individual factory producing power or manufactured goods.

plant community An assembly of different species of plants growing together in a particular habitat; the floral component of an *ecosystem. The concept of community can be applied to a range of scales from a small pond to the Amazon rain forest. Kent et al. (1997) *PPG* **21**, 3 summarize developments in the conceptualization of the plant community. Burns (2007) *J. Biogeog.* **34**, in a study of the assembly of an island plant community off the west coast of Vancouver Island, BC, finds that his results provide evidence for both the deterministic and individualistic viewpoints on community structure, and suggest that both chance and determinism contribute to the assembly of this island plant community. Duckworth et al. (2000) *PPG* **24**, 4 argue that plant functional types can aid in the understanding of the assembly and stability of plant communities.

plant succession The gradual evolution of a series of plants within a given area. This occurs in a roughly predictable order, while the habitat changes. **Primary succession** is 'ecosystem development on barren surfaces where severe disturbances have removed most biological activity' (L. Walker and R. de Moral 2003). **Secondary succession** is the replacement of a community after a disturbance; see Haplern et al. (1997) *Ecology* **70** for a summary of the variables influencing secondary succession. In **autogenic succession** the plants themselves are the genesis of change (Ellis and Coppins (2006) *J. Biogeog.* **33**, 9); in **allogenic succession** the changes are driven by forces outside the ecosystem. The two are not mutually exclusive; see Francis (2006) *Area* **38**, 4.

plantation A large agricultural system, generally a monoculture, employing labour on a large scale to produce tropical and subtropical crops. Early stage processing often takes place on site. Plantations may be seen as a spatial expression of *imperialism, and *capitalism (B. Warren 1980). See Caney and Voelks (2003) *PHG* **27**, 2 on landscape legacies of plantations in Brazil, and P. Curtin (1990) on the rise and fall of the plantation.

plastic flow Movement of material, especially rocks and ice, under intense pressure, when it flows like a very *viscous substance and does not revert to its original shape when pressure is removed. As it moves, it *shears. In ice, plastic flow is due to pressure at depth; a thickness of at least 22 m is needed for plastic flow *temperate glaciers. Plastically flowing ice will flow around and over an obstacle, which may cause deposition in the lee of the obstruction: this is **plastic moulding**; see R. LeB. Hooke (2005). **Plastic deformation** is an irreversible change in the shape of material, without fracturing, resulting from compression or expansion. See Moeyersons et al. (2006) *Geomorph.* **76**, 3–4 on plastic deformation in a *vertisol.

plate A rigid segment of the earth's crust which can 'float' across the heavier, semi-molten rock below. The plates making up the continents—**continental plates**—are less dense but, at up to 35 km deep, are thicker than those making up the oceans—the **oceanic plates**—which are up to 5 km deep. A plate is a part of the *lithosphere which moves over the plastic *asthenosphere. The boundary of a plate may be a *constructive, *destructive, *conservative, or, more rarely, a *collision margin. The theory of **plate tectonics** submits that the earth's crust is made up of six large plates: the African, American, Antarctic, Eurasian, Indian, and Pacific plates, and a number of small plates, the chief of which are the Arabian, Caribbean, Cocos, Nasca, Philippine, and Scotia plates. Hansen (2007) *Geol.* **35**, 12 postulates that the ability of a terrestrial planet to evolve plate tectonics is a function of planet size.

The movement of plates causes global changes, such as *continental drift and a remodelling of ocean basins and the creation of major landforms: *oceanic ridges, *fold mountains, *island arcs, and *rift valleys, together with *earthquakes and *volcanoes, which occur at a destructive plate boundary where one plate plunges below another. See Cloos (1993) *GSA Bull.* **105**, 6 on lithospheric buoyancy and subduction of oceanic plateaus, continental margins, island arcs, spreading ridges, and seamounts. Plates are driven, at least partly, by forces from the convecting mantle. Oceanic driving forces are exerted when a lithospheric plate is being carried by a faster-moving asthenosphere. *Ridge-push* is the gravitational sliding of a plate off a mid-ocean ridge, in response to the *ridge-push force*, caused by the horizontal spreading of the near-surface asthenosphere at constructive margins; see Wilson (1993) *J. Geol. Soc.* **150**. Anderson (2001) *Science* **293** holds that only upper mantle convection may be driven by plate tectonics; Bokelman (2002) *Geology* **30**, 11 finds that, in the case of the North American plate, the mantle plays an important role in driving the plates. The sinking of a cold, dense slab of oceanic plate provides a *slab-pull force* (see Schellart (2004) *Geophys. Res. Letters* **31**, L07611). Conrad and Lithgow-Bertelloni (2002) *Science*, **298** estimate the relative importance of 'pull' versus 'suction', finding that the pull from lower mantle slabs is not coupled to the surface plates. Scoppola et al. (2006) *GSA Bull.* **118**, 1 observe that slab pull is 'notoriously considered one order of magnitude larger than ridge push'.

Plates

1 Juan de Fuca   5 Turkish
2 Cocos   6 Arabian
3 Adriatic   7 Iranian
4 Aegean

--- uncertain plate boundary
ΤΤΤΤΤΤΤ destructive margin
—— constructive margin
▬ ▬ ▬ collision zone

→ movement of plates
② rate of movement in cm/year

PACIFIC PLATE

ANTARCTIC PLATE

EURASIAN PLATE

INDO-AUSTRALIAN PLATE

AFRICAN PLATE

NORTH AMERICAN PLATE

SOUTH AMERICAN PLATE

NAZCA PLATE

Collision zone

p

Eagles et al. (2009) *Tectonophysics* offer an animation of the Southern convergent margins, and M. Summerfield, ed. (2000) has a collection of papers on global tectonics and macroscale landform development. N. Oreskes, ed. (2001) brings together the reminiscences the major players in development of plate tectonics theory, and C. Ollier (2000) analyses current problems related to mountain tectonics.

plateau An extensive, relatively flat upland. Some plateaus are formed structurally, from resistant, horizontal rocks; some are *erosion surfaces; some made of **plateau lavas**; see Bondre et al. (2004) *Procs Ind. Acad. Scis, Earth Planet. Sci.* **113** on the Deccan Plateau.

playa A flat plain in an arid area found at the centre of an inland drainage basin, such as the Salinas Grandes, Jujuy Province, Argentina. Within such an area, lakes frequently form. Evaporation from the playa is high and alluvial flats of saline mud form. The term is also used to describe a lake within such a basin. The deepening of the basins by deflation is favoured by salt weathering processes. See Goudie and Wells (1995) *Earth-Sci. Revs* **38**; and Gutiérrez-Elorza et al. (2005) *Geomorph.* **72**, 1–4.

Pleistocene An *epoch of the Quaternary period, also known as the Ice Age, from 1.8 million to ~10000 years ago.

(⊕) SEE WEB LINKS
• An overview of the Pleistocene epoch.

plinian eruption A highly explosive volcanic eruption with dense clouds of gas and *tephra propelled upwards for many kilometres. *Fissure eruptions are common, as are the destruction of the crater walls, and the fragmentation of solidified lavas. 'The magma mass discharge rate governs the eruption magnitude of plinian eruptions, and depends strongly on conduit geometry' (Varekamp (1993) *J. Volc. & Geothermal Res.* **54**, 3–4).

plinthite A hard capping or crust at the surface of an unconsolidated soil. See D. Steila and T. Pound (1998) on the stages of plinthite development.

plucking See QUARRYING.

plug flow See BLOCKSCHOLLEN FLOW.

plume An upwelling of molten rock through the *asthenosphere to the lower *lithosphere. The *hot spot thus formed shows up in volcanic activity at the surface. As continents move over the hot spot, there forms a chain of volcanoes at the surface, since the plume is stationary with respect to the *mantle. See Constantini et al. (2007) *J. Volc.*

plunge pool A pool at the base of a waterfall, formed by *eddying, *hydraulic action, *cavitation, and *pothole erosion. Nott and Price (1999) *Earth & Planet. Letters* **171**, 2 use plunge-pool palaeoflood records to predict the frequency and magnitude of extreme floods.

pluralism **1.** Any situation in which no particular political, cultural, ethnic, or ideological group is dominant (R. Dahl 1961). When individuals affected by a given issue in the same way band together, the group empowers the individual by aggregating resources. 'Each group competes, negotiates, and compromises with other groups for influence, resources, such as money, or public support' (Eisenberg (1996) *Soc. Sci. Info.* **35**, 2).

2. The cultural diversity of a plural society. Clifford (2006) *Area* **38**, 4 speaks of 'pluralism, which has been much talked about but rarely practised', and Smith (2005) *Antipode* **37**, 5 has a merry tilt at pluralism. Ethington and Meeker in M. Dear and S. Flusty (2002) suggest that the urban is itself plural, such that 'any attempt to view the city other than through a tolerance of multiple views cannot suffice' (Moore et al. (2003) *Area* **35**, 2). R. Eversole and J. Martin (2005) observe the underlying assumption in Australia that 'a recognition of pluralism may be able to overcome potentially conflicting political interests in the regions, or actually sustain some communities that may be becoming economically and environmentally non-viable or socially dysfunctional'. See Nash (2005) *Antipode* **37**, 2 on pluralism in Northern Ireland. Mouffe uses **agonistic pluralism**—in which *the political* refers to the antagonism that is constitutive of society and *politics* refers to practices, discourses, and institutions that order and organize society but that are always open to the possibility of antagonisms because they are embedded in the political (Mouffe (2000) IHS). M. Kenny (2004) discusses the **new pluralism**, finding that the preservation of difference tends to become a condition of any group's inclusion in the political community in a way that 'dethrones the principles of democratic citizenship'.

pluralist geography An 'academic debate that allows distinct approaches to engage with one another without presuming either that one must eventually "win" by dominating the other, or that anything goes' (Plummer and Sheppard (2006) *J. Econ. Geog.* **6**). A. Salanti and E. Screpanti, eds (1997) call for methodological pluralism in geographical economics, and Stokke (no date) *Tidsskrift for geografene i Bergen* depicts a pluralism of 'neo-perspectives' in development geography. Within radical geography, Amin and Thrift (2005) *Antipode* **37**, 2 'map a pluralist and forward-looking position whose guiding principle is the promotion of emergence through the process of disagreement'. Smith (2005) *Antipode* **37**, 5 sees pluralism as 'the core politics of neo-critical geography'.

pluton A solidified, underground, igneous mass, varying in size from *batholiths to *sills and *dykes. See McNulty et al. (2000) *GSA Bull.* **112**, 1 on the Mount Givens pluton, Calif., and Stevenson et al. (2007) *J. Geol. Soc.* **164**, 1 on the Eastern Mourne pluton, Ulster. A **plutonic rock** is an igneous rock which has cooled and crystallized slowly, at great depth.

pluvial A time of heavier precipitation than normal. See Pinson (2008) *Geoarchaeology* **23**, 1 on evidence of more than one pluvial during the last million years. Pluvials have created current desert landforms, including wadi accretion terraces (A. Alsharhan 1998).

pneumatic action Of, or acting by means of, wind or trapped air, for example, of the air compressed between a sea wave and a cliff face. *See* CAVITATION.

podzol (podsol) A soil characteristic of the coniferous forests of Russia and Canada. **Podzolization** occurs when severe *leaching leaves the upper *horizon virtually depleted of all soil constituents except quartz grains. Clay minerals in the A horizon decompose by reaction with *humic acids and form soluble salts. The leached material from the A horizon is deposited in the B horizon as a humus-rich horizon band or as a hard layer of sesquioxides. See Sauer et al. (2007) *J. Plant Nutrit. & Soil Sci.* **170**, 5 for a review of the present knowledge on the development, functions, and threats of podzols in 2007.

poetics The linguistic, literary, and theoretical resources used in *representation; hence **geographical poetics**—the varieties of language in which geography has been, continues to be, and ought to be, written. See Kong (2001) *PHG* **25**, 2 for an example.

poikilotherm An organism which has its body heat regulated by the temperature of its surroundings. See Lindsey (1966) *Evolution* **20** on latitudinal variations in the body sizes of poikilotherm vertebrates.

point-bar deposit The accumulation of fluvial sediment at the *slip-off slope on the inside of a *meander.

point pattern analysis For an examination of the method, potential, and limitations of two classes of point pattern analysis, see Bishop (2007) *Area* **39**, 2.

polar Applying to those parts of the earth close to the poles. The **polar area**, in particular the Arctic, is considered to be the forerunner of climate change (Schiermeier (2006) *Nature* **441**, 146. **Polar air masses (P)** originate in the mid-latitudes (40°–60°) and are characteristically cold.

polar front The discontinuous, variable *front which forms over the North Atlantic and North Pacific, where polar maritime air meets tropical maritime air. The formation of *mid-latitude depressions at the polar front is connected with the development of troughs in the **polar front jet stream**, a band of high-velocity winds in the wider *Rossby waves. The polar front *jet stream, moves southwards in winter and northwards in summer. During an *El Niño event, it shifts south, bringing cooler air, which might suppress tornado activity in the US Midwest. Mid-latitude aridity can result from the effect of mountains on the polar jet stream; see W. G. Whitford (2002). See Dupont et al. (2005) *Geol.* **33**, 6 on the sensitivity of the Namibian upwelling to latitudinal shifts in the Southern Ocean polar front zone; Ingvaldsen (2005)

Geophys. Res. Letter **32**, LI6603 on the link between the polar front and the North Cape Current; and Hayes and Zenk (1977) *NOAA Technical Report ERL. PML* **28** on the Antarctic Polar Front.

Polar easterlies blow east–west, from *polar highs to sub-polar lows. A **polar high** is a mass of cold, heavy air, centred at the poles, and produced by downward, vertical air currents from the *polar vortex, bringing high pressure at high latitudes. For **polar low**, s*ee* COLD LOW.

polar vortex A rotating polar high pressure system, formed as a distinct column of cold air in the middle to lower stratosphere, developed during the long polar night; wind speeds may reach 100 metres s^{-1}. Upper *troposphere air is sucked into the vortex, then sinks to cause *polar highs. See Chshyolkova et al. (2007) *Angeo* **25** on the Arctic Polar vortex.

polder Land reclaimed from the sea, lakes, or river deltas; the land is bounded by a dyke, drained, and is maintained by pumping. **Calamity polders** are overflow areas, to be flooded during extreme discharges; see Roth and Warner (2007) *Tijdschrift* **98**, 4.

political ecology The study of the relationships between nature and society: how political, economic, and social factors affect, and are affected by, environmental issues. These issues may address: 'mutually causal' interactions between society and land-based resources, for example that poverty is brought about by poor management, which leads to environmental degradation, which, in its turn, deepens poverty. However 'it cannot be assumed that the poor have an intrinsic propensity to degrade environmental resources' (*International Fund for Agricultural Development); the influence of political, cultural, or economic systems; for example, the cultural *eutrophication of Lake Erie (D. Kaufman and C. Franz 1993); the role of external structures; for example, the deepening poverty of Masai herders when blocked from traditional grazing (P. Blaikie 1985); the reinforcement and misrepresentation of existing economic, gender, and racial inequalities: see Moore (2000, *Jour. Political Ecol.* **7**) for an analysis of 'law's patriarchy' confronting global inequalities; see Bryant (1991, *Glob. Ecol. & Biogeog. Letts* **1**, 6) on the role of social structures: in Somaliland, for example, customary mechanisms of conflict resolution have helped rebuild local society.

Walker (2005) *PHG* **29**, 73 notes that 'while political ecology has thrived, its coherence as a field of study and its central intellectual contributions remain the subject of sometimes contentious debate'. One of the recurrent, and unresolved, questions has been 'Where is the ecology in political ecology?' Vayda and Walters (1999) *Hum. Ecol.* **27** argue that 'some political ecologists do not even deal with literally the influence of politics in effecting environmental change but rather deal only with politics'. See also P. Blaikie et al. (2004).

(⊕) SEE WEB LINKS

• An overview of the potential violent conflict in pastoral communities.

political economy An approach to economics based on social institutions and their implications, including insights from geography, history, politics, psychology, and sociology. See the review essay 'From international political economy to global political economy' (1997) *Polit. Geog.* **16**, 1. Political economists observe markets in the context of the class structures and power relationships they are embedded in, and while orthodox analyses are based on the concept of 'economic man', who makes rational choices, political economists argue that human behaviour cannot be quantified; Castree (1999) *TIBG* **24**, 2 proposes theoretical foundations for a renewed Marxian political economy in geography.

political geography The study of the geographical and spatial aspects of politics; the spatial expression of political ideals and the consequences of decision-making by a political entity. Of particular interest are the geographical factors which influence voting patterns; for example, Morrill et al. (2007) *Pol. Geog.* **26**, 5 on the red (conservative/republican) and blue (liberal/democrat) divide in the United States, or Buhaug (2006) *Pol. Geog.* **25**, 3 on local determinants of African civil wars. M. Castells (2000) concludes that power 'is no longer concentrated in institutions (the state), organizations (capitalist firms), or symbolic controllers (corporate media, churches). . . . Power, as such, is the fulcrum of identity.' Agnew (2005) *AAAG* **95**, 2 agrees: 'we cannot meaningfully apply the orthodox conception of sovereignty to the conditional exercise of relative, limited, and partial powers that local, regional, national, international, and non-territorial communities and actors now exert.' However, Houston (2005) *Pol. Geog.* **24**, 1, writing on Turkey, finds that the production and signification of space by the nation-state is not cancelled out by the global logic of capital. For clear reviews, see Sparke: (2004) *PHG* **28**, 6; (2006) *PHG* **30**, 3; (2008) *PHG* **32**, 3.

Studies on quantitative methodologies and spatial analyses based on GIS include Buhaug and Lujala (2005) *Pol. Geog.* **24**, 4, who find that the relation of a conflict affects its duration. For critical, feminist, and popular geopolitics, see Hyndman (2004) *Pol. Geog.* **23**, 3; for peace and conflict studies, see Sene (2004) *African Gepopol.* 15–16, on the federating role of soccer, and territoriality.

politics of place An emphasis on the importance of place and locality as arenas through which the forces of capitalist forces and the state structures that support it are challenged; 'far from being bounded, homogenous, and stable entities, localities are actively produced through a politics of place that may simultaneously involve political economic processes (capital, and localized modes of political regulation), and cultural processes (contested meanings and senses of place' (Elmhirst (2001) *Sing. J. Trop. Geog.* **22**, 3; a clear and helpful paper); see also D. Massey (2005) and (2004) *Geografiska B* **86**). Amin (2004) *Geografiska B* **86**, 1 (unclear and verbose), detects a new politics of place, characterized by demand for localized decision-making; 'a restricted democracy'. See also Cumbers et al. (2008) *PHG* **32**, 2.

polje In *karst terminology, a flat-floored, steep-sided, enclosed basin up to 65 km long and 10 km wide. Streams can be ephemeral or permanent; usually the water drains into *streamsinks. Most poljes are aligned with underlying structures such as folds, faults, and troughs. See Nicod (2003) *Acta Carsalogica* **32**, 2 and Miller (2006) *GSA Sp. Paper* **406**.

pollen analysis. *See* PALYNOLOGY.

pollution 'As far as the geographical distribution of pollution is concerned: pollution is concentrated in agglomerations, as in [major] Third World cities . . . environmentally harmful activities are located in peripheral regions, as is the case with nuclear power stations in industrialised countries . . . [and] in some cases . . . pollution is rather uniformly distributed over the country' (Rauscher and Bouman siti.feem.it/worldcongress/abs2/rausch2.html). See Lange and Quaas (2007) *BEJ Econ. Analys. & Policy* **7**, 1 on the effect of environmental pollution on agglomerations.

Biodegradable pollutants, like sewage, cause no permanent damage if they are adequately dispersed; see Zagorc-Koncan and Somen (1999) *Water Sci. & Tech.* **39**, 10 on biodegradable industrial pollution. Land cover change and the expansion of modern agricultural practices has significantly increased leaching of chemicals to the surface waters, leading to the degradation of aquatic ecosystems worldwide (Donner (2003) *Glob. Ecol. & Biol.* **12**, 4).

Non-biodegradable pollutants, such as lead, may be concentrated as they move up the food chain; this is *biological magnification*—see J. Nathanson (2002). Within western Europe, air pollution, associated with basic industries such as oil refining, chemicals, and iron and steel, as well as with motor transport, is probably the principal offender, followed by water and land pollution. Other forms include noise, and thermal pollution.

Present-day problems of pollution include *acid rain and the burning of *fossil fuels to produce excessive carbon dioxide. The **pollution haven** (or ecological dumping) hypothesis states that if free trade occurs between countries with different environmental standards, countries with lower standards will develop a comparative advantage in relatively 'dirty' industries (Rauscher (1994) *Oxf. Econ. Papers* **46**).

pollution dome A mass of polluted air in and above a city or industrial complex which is prevented from rising by the presence of an *inversion above it. See Victor et al. (2004) *Annal. NY Acad. Scis* **1023**. Insolation responds to pollution domes—Tomalty and Brazel (2000) *Procs Urb. Envs*, Conf. AMS, Davis, record inner-city values as 15% lower than outside values. Winds may elongate the dome into a pollution plume; see Nemecek et al. (1995) *Int. J. Rock Mech. &c.* **32**, 7 on pollution plume containment.

The term has been extended to describe **light pollution domes.**

polygon A **sand wedge polygon** is a micro-relief landform, 7–14 m in diameter, 'devoid of vegetation and . . . outlined by textural changes in the soil' (Péwé (1959) *Am. J. Sci.* **257**). **Ice-wedge polygons** are formed after a temperature drop so extreme that 'the ground contracts and cracks, forming a narrow void. During the winter, snow blowing across the tundra surface fills in the crack and freezes to form a thin ice vein. The following spring the ice vein thaws, but it refreezes the following winter and the ground preferentially cracks again along the same ice vein. New snow blows into the crack and freezes, gradually "wedging" the ground open' (Morgan (1972) *Canad. J. Earth Scis* **9**, 6).

polysaccharide gum The sticky by-product of the decomposition of roots by micro-organisms, which can bind soil minerals into aggregates, such as *peds.

pool and riffle The alternating sequence of deep pools and shallow riffles along the relatively straight course of a river. The distance between the pools is 5–7 times the channel width. 'If alternate bars are present, the pools deform into meander bends and the riffles become points of inflection . . . however, in this paper it is suggested that pools correspond to the points of inflection in a meander system' (Leopold in R. D. Hey et al. 1998). See also Wohl and Merrit (2008) *Geomorph.* **93**, 3–4.

population A group of individuals of the same species within a *community. Life tables provide a picture of survival and mortality in populations. Cohort life tables—whereby individuals are tracked from birth—are appropriate for plants and sessile organisms. Static life tables—which record age at death within a certain period of time—are appropriate for mobile and long-lived organisms. For the laws of population ecology, see Turchin (2001) *Oikos* **94**, 1.

population density The ratio of a population to a given unit of area. Lagerlöf and Basher (2006) *MPRA Paper* **369** make the bold claim that population density correlates with per capita income in Canada and the United States. On firmer ground is Sato (2007) *J. Urban Econ.* **61**, 2, who argues that migration from areas of lower, to higher, population density in Japan maintains regional variations in fertility. **Crude population density** is simply the number of people living per unit area and is not necessarily a development indicator.

population dynamics The study of the numbers of populations and the variations of these numbers in time and space. The main determinants are birth and death rates and migration; which, in turn, reflect levels of development, and socio-economic and cultural factors. See Millington et al. (1999) *App. Geog.* **19**, 4 on population dynamics and socio-economic change; Bhandari and Grant (2007) *J. Rural & Comm. Dev.* **2** on population dynamics and land use; Ahn in H.-R. Kim and B. Song (2007) on population dynamics and urbanization in Korea. See also H. Macbeth and P. Collinson, eds (2002); try L. M. Hinter (2000) on the environmental implications of population dynamics. For geolinguists, see N. Komarova and M. Nowak (2002) on the population dynamics of grammar acquisition.

population equation The future size of a population depends on a range of functions. Thus:

$$P_{t+1} = P_t + (B - D) + (I - E)$$

where Pt and P_{t+1} are the sizes of population in an area at two different points in time, t and $t + 1$ are those points, B is the birth rate, D is the death rate, I is the immigration, and E is the emigration. For an example of its use, see Roos (2005) *J. Econ. Geog.* **5**, 616.

population geography The study of human populations; their composition, growth, distribution, and migratory movements with an emphasis on the last two. It is concerned with the study of demographic processes which affect the environment, but differs from demography in that it is concerned with the spatial expression of such processes. *Population, Space and Place* is the journal of the UK Population Geography Research Group.

population growth The rate of increase of the world's population has become ever faster:

YEAR	POPULATION IN BILLIONS
1850	1.2
1900	1.6
1950	2.55
2000	6.1
2050 (estimated)	9.2

'Future population growth is highly dependent on the path that future fertility takes. In the medium variant, fertility is projected to decline from 2.6 children per woman today to slightly over 2 children per woman in 2050. If fertility were to remain about half a child above the levels projected in the medium variant, world population would reach 10.6 billion by 2050. A fertility path half a child below the medium would lead to a population of 7.6 billion by mid-century. That is, at the world level, continued population growth until 2050 is inevitable even if the decline of fertility accelerates (Population Division, Dept. Econ. & Soc. Affairs, UN Secretariat, *2004 Revision*). *See* POPULATION PROJECTION.

Wolfgang Lutz et al. (2004) discuss expert argument-based probabilistic forecasting of population; Bongaarts and Bulatao (1999) *Pop. & Devt. Rev.* **25** quantify the relative contributions of fertility, mortality, migration, and age structure in projections for all countries of the world.

population potential The accessibility of people from a given point.

$$\text{population potential, } V_1 = \sum_{j=1}^{n} \frac{P_1}{d_{ij}} = \frac{P_1}{d_{i1}} + \frac{P_2}{d_{i2}} + \frac{P_3}{d_{i3}} \ldots + \frac{P_n}{d_{in}}$$

where the population potential (V_1) at point i is the summation (Σ) of n

populations (*j*) accessible to the point *i* divided by their distance (d_{ij}) to that point. Transport costs may be used instead of distance.

population projection The prediction of future populations based on the present age–sex structure, and with the present rates of fertility, mortality, and migration. Wilson and Rees (2005) *Pop. Space Place* **11** review recent developments in population projection methodology.

Population projection. Population of the world 1950–2050 by projection variants

population pyramid *See* AGE–SEX PYRAMID.

pore In geomorphology, a minute opening in a rock or soil, through which fluids may pass. Porous rocks allow water to pass through or be stored within them. **Pore water pressure** is the pressure applied by water in the pores to soil and rock particles. When the rock or soil is saturated, pore water pressure can be so great that slope failure results; see Simon et al. (2000) *Geomorph.* **35**, 3–4, and Darby et al. (2007) *J. Geophys. Res.* **112**, F03022.

pork-barrel effect Government expenditure aimed at gaining votes. Government contracts can be of vital importance to employment within a region and may increase the popularity of an administration, as elected members of a government divert spending to the areas they represent. Investigations of the geography of federal investment, for example in the allocation of clothing contracts to the US Army in the immediate post-war period, have revealed the significance of the pork-barrel effect for certain local economies. See Hagen (2007) *Southeastern Geogr.* **47**, 2.

porosity The ratio of the volume of pores to the volume of matter within a rock or soil, expressed as a percentage. Iverson et al. (2000)

Science **20** conclude that relatively small variations in porosity can influence landslide behaviour profoundly. Koulouri and Giourga (2007) *Catena* **69**, 3 use soil porosity as one variable in determining soil erodibility.

Chemical weathering increases rock porosity under virtually all circumstances (Gordon (2005) *Physical Geogr.*). Damody et al. (2008) *Geomorph.* **94**, 1–2, use the porosity of granite to date tors. Porosity is one of the best indicators of soil structural quality.

positive discrimination A policy designed to favour some deprived region or minority and to redress, at least in part, *uneven development. Policies of positive discrimination have been criticized for treating the effects of inequality rather than tackling its causes. Other criticisms made are that not all of the minority or region needs help, and that many deprived people are outside the catchment area. Nonetheless, the EU, for example, still uses schemes of positive discrimination, through the European Regional Development Fund; see Roberts (2003) *Prog. Planning* **59**, 1. Bosher et al. (2007) *Dev. & Change* **38**, 4 find examples of positive discrimination in Andra Pradesh, India.

post-colonialism A sub-discipline of geography which awaits a consensus on its definition; some definitions and observations follow: 'The geographically dispersed contestation of colonial power and knowledge' (A. Blunt and C. McEwan, eds 2000). 'No coherent project of postcolonial development geography has emerged' (Radcliffe (2005) *PHG* **29**, 3). 'Much of what passes for postcolonial theory in British geography reinforces new forms of colonial epistemologies and colonial hierarchies, while destabilizing their older forms' (Gilmartin and Berg (2007) *Area* **39**, 1). 'Post-colonialism studies might best be regarded now as a term for a body of diverse and often contesting formulations of the cultural production of colonized people rather than as a discipline or methodology *per se*' (B. Ashcroft et al. 2002). See R. Phillips (2006). **Green post-colonialism** insists on 'the factoring of cultural difference into . . . ecological and bioethical debates' (D. Curtin 1999).

postdevelopment A political acknowledgement of the injustices and unequal power relationships of the practice of development, and a view that development cannot offer sufficient tools for transformation. Instead, it suggests resistance to *globalization provided by grassroots organizations and non-governmental organizations, alternative forms of livelihood, and local forms of knowledge and practice. See A. Ziai, ed. (2007). See also Sidaway (2007) *PHG* **31**, 3.

post-Fordism A system of production and *accumulation characterized by flexibility of both labour and machinery, the vertical break-up of large corporations, small-batch production, better use of links between firms so that subcontracting is increasingly used, *just-in-time production, niche consumption, commodified culture, and the popularity of monetarist ideologies among governments. Post-Fordism is associated with *agglomeration, which will simplify interaction between linked forms of economic activity. See A. Amin (1994). Phelps (2002) *Antipode* **34**, 2 considers the lineage of post-Fordist work practices. *See* FLEXIBLE ACCUMULATION; FLEXIBLE SPECIALIZATION.

post-industrial city A city exhibiting the characteristics of a *post-industrial society. Service industries dominate with a strongly developed *quaternary sector and *footloose industries abound, often on pleasant open space at the edge of the city. Post-industrial cities are also characterized by large areas of office blocks and buildings for local government administration. These cities often exhibit marked inequality of income distribution because of the contrasts between those who are appropriately skilled—professionals, managers, administrators, and those in high-technology service industries— and the poorly paid service workers who look after their needs, together with the unemployed. The former can afford high house prices, and, in fact, contribute to them; the latter cannot. Phelps and Ozawa (2003) *PHG* **27**, 5 present a taxonomy of forms of agglomeration in proto-industrial, industrial, and post-industrial urban contexts, commenting that contemporary, post-industrial agglomerations are larger than preceding forms, with an increasingly complex pattern of specialization within and between urban areas. *See also* ENTREPRENEUR; INFORMATION CITY.

post-industrial society A post-industrial society has five primary characteristics: the domination of service, rather than manufacturing, industry, the pre-eminence of the professional and technical classes, the central place of theoretical knowledge as a source of innovations, the dominating influence of technology, and levels of urbanization higher than anywhere else in the world. Mahon (2006) *TIBG* **31**, 4 argues that childcare arrangements are central to post-industrial societies.

postmodernism A philosophical stance which claims that it is impossible to make grand statements about the structures of society or about historic causation, because knowledge is partial and situated, and no one interpretation is superior to another. Postmodernism in geography is expressed in the recognition that space, place, and scale are social constructs, not external givens. Of particular interest is the way that *time and space have been 'compressed' by modern transport systems, especially by jumbo jets; hence **postmodernity**: the ever swifter movements of capital and information in postmodern society. This development has reduced the *friction of distance for most *advanced economies. D. Harvey (1989) sees *time–space compression as a fundamental social feature of postmodernism. Some geographers claim that postmodernism challenges the dominance of time and history in social theories; instead stressing the significance of geography and spatiality (see E. Soja 1989). The postmodern tradition also stresses, and indeed champions, difference.

Ley (2003) *PHG* **27**, 5 notes that interest in postmodernism is waning, postmodernism as a form of local knowledge, characterized as knowledge inevitably framed through the 'here' of our collective presence and the 'now'. He then extends his argument to modernism and postmodernism in the built environment.

post-productivist transition A transformation in rural economies since the mid-1980s characterized by the production of food within an increasingly competitive international market; the progressive withdrawal of state subsidies for agriculture; reductions in food output; growing concerns over food quality; the growing environmental regulation of agriculture; and the creation of a more sustainable agricultural system. Wilson and Rigg (2003) *PHG* **27**, 6 ask whether notions of post-productivism can be applied to the developing world, concluding that the concept needs to be adapted and developed to address specific conditions in the rural South, whilst Holmes (2002) *TIBG* **27** judges that Australia's rangelands are moving from a post-productivist agricultural occupance to a multifunctional rural occupance.

In the UK the *Countryside Stewardship Scheme attempts to bring about a transformation from a *positivist to a more post-positivist approach to agriculture. *See* THIRD FOOD REGIME.

post-socialist city Discussions of the post-socialist city can founder on the definition of post-socialism (see Andrusz (2001) *Anthropolis*). Lin (2004) *Eurasian Geog. & Econ.* **45**, 1 argues that the interaction of the socialist legacy of industrialization 'with the new forces of marketization and globalization has given rise to a peculiar pattern of simultaneous industrialization and tertiarization, differing from the Western norm of linear progression'. See also Hirt (2007) *Urb. Geog.* **28**, 8; try K. Stanilov (2007).

poststructuralism An extension and review of structuralism which emphasizes plurality and deferral of meaning, and rejects such binary oppositions as *representation/'reality', discursive/material. For geographers, it has prompted a re-examination of *cartography, and confirmed the *spatiality of social life and the social construction of *space. See J. Munroe (2006) on the amalgam of socio-biological-physical processes that generate the 'geographies' that are commonly discussed as 'geography'.

potential evapotranspiration (PET) The maximum continual loss of water by evaporation and transpiration, at a given temperature, given a sufficient supply of water. PET often outstrips actual evapotranspiration. Trudgill and Briggs (1981) *PPG* **5**, consider the significance of potential evapotranspiration in soil and land potential.

pothole Loosely, a vertical cave system; more precisely, a more or less circular hole in the bedrock of a river. The hole enlarges because pebbles inside it collide with the bedrock as the water swirls. Potholes carved into streambeds can be important components of channel incision. Springer et al. (2005) *Geomorph.* **82**, 1–2 demonstrate that the pothole dimensions of radius and depth are strongly correlated.

poverty 'A pronounced deprivation in well being.' A. Sen (1985, 1993)
argues that what is important to well-being is the 'capability to function in
society'. These capabilities include access to food, shelter, clothing, education,
security, and good health: capabilities which the poor lack. The poor also lack a
voice in society, are vulnerable to risk, and are powerless.

'Over the past 30 years, the world has undoubtedly registered impressive
gains in the average life expectancy of its population—from 60 years at birth, to
67 years. Infant mortality has decreased substantially—from 96 to 56 per 1,000
live births' (Bidwai (2006) *Development Dialogue* **47**, 31). The 2005 UN Human
Development Report explains that these changes are partly a product of falling
child death rates, improvements in access to water and sanitation, and global
immunization. There is also evidence that literacy levels have risen.

There have been setbacks: eighteen countries registered lower scores on the
HDI in 2003 than in 1990. Twelve of these are in sub-Saharan Africa and the
former Soviet Union accounts for the other six (ibid.) Chuhan (*Glob.
Monitoring Re. 2006*) adds that 'over three-quarters of a billion people, many of
them children are malnourished. Each year nearly 11 million children die from
malnutrition or disease before reaching their fifth birthday . . . and over half a
million women die every year in childbirth from lack of appropriate health
care, malnourishment, or disease. More than 100 million children of primary
school age, the majority of them girls, do not attend school.'

Percentage of people living on less than $1 and $2 per day

	\$1.08 per day at 1993 PPP							
	1981	1984	1987	1990	1993	1996	1999	2001
East Asia	57.7	38.9	28.0	29.6	24.9	16.6	15.7	14.9
of which China	63.8	41.0	28.5	33.0	28.4	17.4	17.8	16.6
Eastern Europe and Central Asia	0.7	0.5	0.4	0.5	3.7	4.2	6.3	3.7
Latin America and Caribbean	9.7	11.8	10.9	11.3	11.3	10.7	10.5	9.5
Middle East and North Africa	5.1	3.8	3.2	2.3	1.6	2.0	2.6	2.4
South Asia	51.5	46.8	45.0	41.3	40.1	36.6	32.3	31.3
of which India	54.4	49.8	46.3	42.1	42.3	42.2	35.3	34.7
Sub-Saharan Africa	41.6	46.3	46.8	44.6	44.0	45.6	45.7	46.9
Total	**40.4**	**32.8**	**28.4**	**27.9**	**26.3**	**22.8**	**21.8**	**21.1**

	\$2.15 per day at 1993 PPP							
	1981	1984	1987	1990	1993	1996	1999	2001
East Asia	84.8	76.6	67.7	69.9	64.8	53.3	50.3	47.4
of which China	88.1	78.5	67.4	72.6	68.1	53.4	50.1	46.7
Eastern Europe and Central Asia	4.7	4.1	3.2	4.9	17.2	20.6	23.7	19.7
Latin America and Caribbean	26.9	30.4	27.8	28.4	29.5	24.1	25.1	24.5
Middle East and North Africa	28.9	25.2	24.2	21.4	20.2	22.3	24.3	23.2
South Asia	89.1	87.2	86.7	85.5	84.5	81.7	78.1	77.2
of which India	89.6	88.2	87.3	86.1	85.7	85.2	80.6	79.9
Sub-Saharan Africa	73.3	76.1	76.1	75.0	74.6	75.1	76.0	76.6
Total	**66.7**	**63.7**	**60.1**	**60.8**	**60.1**	**55.5**	**54.4**	**52.9**

Source: Chen & Ravaillon, (2004)

Poverty is not restricted to the developing world: in inner London, 51% of
children are currently living in poverty (2006) and US Census Bureau records
show that the proportion of Americans living in poverty rose between 1974 and

2004, from 11.2% to 12.7%. For the period 2000–2005 the Cuban infant mortality rate was 8/000; the same as for the USA (*World Mortality Report*, 2005). 'The only positive interpretation that can be given to this picture of poverty and inequality in the developed world is that many of the poor are recent arrivals from even poorer places in Africa, Asia and Latin American countries' (Gilbert (2007) *Geog. Compass* **1**, 3).

(⊕) SEE WEB LINKS
- The Human Development Report.

power The power structure of a society is reflected in its social organization and in its economy: *capitalism is controlled by a minority who dominate the *factors of production. In a *centrally planned economy it is the state that dominates, and these have their own spatial expression; see Boudreau (2007) *Env. & Plan. A* **39**, 11 on making new political spaces.

   Geographies of power study the spatial distribution of power across a multiplicity of geographical scales, from the body to the globe. The mix of scales changes dynamically, as territories are continually subjected to resistance, contestation, renegotiation, *deterritorialization, and reterritorialization; see, for example, Mawdsley (2007) *Geog. Compass* **1**, 3 on China and Africa as emerging challenges to the geographies of power.

   Themes include state formation, where power is transmitted through social channels to territorial ones; see, for example, Jessop (2007) *Pol. Geog.* **26**, 1 and the relationship of minority cultures and empowerment, as in Mavroudi (2008) *Pol. Geog.* **27**, 1., on Palestine. See Kobrin (2001) in A. Rugman and T. Brewer (2003).

prairie A large area, found outside the tropics, with grassland and occasional trees as natural vegetation, as in the prairies of North America, the South American pampas, the Russian *steppes, and the South African veld. A **prairie soil** is a soil of the wetter prairies, resembling *chernozem in its high humus content and its development under grassland. Prairie soils are *leached of calcium, and are slightly acid. See Barrett and Randall (1998) *Soil Sci.* **163**, 6.

Pre-Boreal A warmer division of the *Flandrian. For the rapid climate change at the Younger Dryas-Pre-Boreal transition, see Starkel (2003) *Glob. & Planet. Change* **35**, 1.

Precambrian The oldest *era in earth's history dating from about 4600 million years BP. 'To speak of "the Precambrian" as a single unified time period is misleading, for it makes up roughly seven-eighths of the Earth's history' (UCMP).

(⊕) SEE WEB LINKS
- The divisions of Precambrian time.

precipitation In *meteorology, the deposition of moisture from the atmosphere onto the Earth's surface in the form of *rain, *hail, *frost, *fog, sleet, or *snow. Initially, cloud droplets grow around nuclei through condensation and diffusion. In warmer clouds the larger droplets then grow by collision and coalescence with the smaller ones. In colder clouds the *Bergeron–Findeisen mechanism is thought to operate, probably in conjunction with the growth of ice crystals: through *accretion, as supercooled water droplets freeze on impact with the ice; and aggregation, as smaller ice crystals stick to larger ones. Much precipitation begins in the form of ice crystals, but melts as it falls, to become rain. See Zhang et al. (2007) *Nature* **448** on *anthropogenic precipitation; Holmer (2007) *Geograf. Annal.* **89**, 4 on rainfall change; Halfon and Kutiel (2007) *Eur. Geosci. Union* **9** on precipitation mapping; and Malby et al. (2007) *Hydrol. Scis J.* **52**, 2 on long-term variations in *orographic rainfall.

predator–prey relationships In theory, there should be an equilibrium between predators and prey. Thus, when predators are scarce, the numbers of prey should rise. Predators would respond by reproducing more and, possibly, by changing their hunting habits. As the population of predators rises, more prey is killed and their numbers fall. Many of the predators then die; thus numbers of predators and prey oscillate between two extremes. See Gilpin and Rosenzweig (1972) *Science* **177**, 1358 and Rosenzweig and MacArthur (1963) *Am Nat.* **97** for the classic model.

The **keystone predator effect** predicts that prey that are strong resource competitors are expected to dominate at low levels of potential productivity; augmenting prey resources permits less efficient competitors but more predator-resistant prey to invade and dominate. The underlying trade-off among prey can be termed the 'keystone trade-off' (Leibold (1996) *Am. Nat.* **147**). See Steiner (2003) *Oikos* **101**, 3.

pre-industrial city A model of the pre-capitalist city, proposed by G. Sjöberg (1960). The city centre is occupied by small elite. The lower classes occupy the concentric zone surrounding the centre, and the outcasts are consigned to the outer edges of the city. Abbott (1974) *Amer. Soc. Rev.* **39**, 4 finds that the distribution pattern of socio-economic status in pre-revolutionary Moscow were consistent with Sjöberg's model.

preservation The protection of human features in the landscape, as opposed to *conservation which is concerned with the protection of the natural landscape. This distinction is not always made; see, for example, Shafer (2004) *Lands. & Urb. Plan.* **66**, 3 on the voluntary preservation of America's natural heritage.

pressure gradient The rate of change in *atmospheric pressure between two areas, providing a force which moves air from *high to *low in an effort to even up the unequal mass distribution of the air. On a global scale, the most powerful pressure gradients are in a *meridional direction, caused by meridional disparities in *insolation. See Bacon and Carter (1993) *Int.*

J. Climat. **13** on the connection between mean wave height and atmospheric pressure gradient in the North Atlantic, and McPherson (2007) *PPG* **31**, 3. The **pressure gradient wind** is the movement of air in response to pressure differences, blowing from high to low. It is modified, however, by the action of the *Coriolis force. The term is also used in fluvial geomorphology; see Powell (1998) *PPG* **22**, 1.

pressure melting point In glaciology, the temperature, often well below 0 °C, at which ice under pressure will melt. In basal ice, flow at the pressure melting point is partly from basal sliding and partly from shear deformation (Zwally et al. (2002) *Science* **297**). Warm glaciers have bases at or above the pressure melting point of their ice.

pressure release The expansion of a rock formed under pressure when that pressure is released (also known as *dilatation*). A glacier may remove the *overburden, and the revealed rock 'bursts' open; see the classic Lewis (1954) *J. Glaciol.* **2**. Twidale (1973) *Rock Mech. & Rock Eng.* **5**, 3 attributes sheet jointing to pressure release.

price mechanism The regulation of economic activity by supply. Hayek (1945) *Amer. Econ Rev.* **35** described the price mechanism as 'that division of labour on which our civilization is based'. The spatial impossibility theorem tells us that price mechanism does not work in a homogeneous spatial economy (Ottaviano and Thiesses (2005) *Env. & Plan. A* **37**, 10. **Pricing policies** are the arrangements whereby prices of commodities to the consumer are determined; see, for example, Arentze and Timmermans (2007) *J. Transp. Geog.* **15**, 1.

primary industry (primary activity) Economic activity, such as fishing, forestry, and mining and quarrying, concerned with the extraction of natural resources.

primate city The largest city within a nation which dominates the country not solely in size—being more than twice as large as the second city, as in London and Birmingham, UK—but also in terms of influence. Tammarau (2000) *Tijdschrift* **91**, 1 explores this concept in centrally planned, developed, and developing economies.

process domain Spatially identifiable areas characterized by distinct suites of geomorphic processes. This may be on the basis of observed precedence, or involve superposition of information layers describing the distribution of causal factors (Preston in A. Lang et al., eds 2003). See J. Phillips and W. Renwick, eds (1992) and Montgomery (1999) *J. Amer. Water Resources Ass.* **35**, 2.

proclimax Vegetation which appears instead of the climatic climax (F. Clements 1934), because of repeated disturbances, for example, from fire, or overgrazing.

producer An organism which can fix energy from the sun and transform it by photosynthesis into food. See Olff et al. (2000) *Glob. Ecol. & Biogeog.* **9**, 4 on producers and consumers.

production In ecology, the increase of body mass as food is converted into new living material.

productivist Describing an agricultural system which is *intensive, expansionist, and based on the expansion of world trade in food, ever-increasing farm sizes, and the use of technology to increase output; factory farming and *agribusinesses are both examples. Walford (2003) *J. Rural Studs* **19**, 4 reports that commercial farmers continue to intensify and specialize; in Brittany, for example, the removal of hedges has decreased *infiltration and *groundwater storage (Douguet and Schembri (2000) CRITINC *W. Paper* **13**). Annually, some 30000 tonnes of nitrates are leached into British rivers (Harvey (1998) *TIBG* **4**), and slurry from livestock has added to pollution; see Willis (2001) *Asia-Pac. Viewpt* **42**, 1. See also Jay (2007) *Food Policy* **32**, 2.

productivity The output of an economic activity, in terms of the economic inputs. Fischer et al. (2007) *SSRN*, illustrate the 'central link between productivity and knowledge capital', and Broersmar and van Dijk (2007) *J. Econ. Geog.* **8** find that investing in highly dense regions increases congestion, thereby reducing productivity. Some argue that social and environmental 'costs' must also be considered; see Compassion in World Farming Trust (1999) on factory farming, for example, and the cost of raw materials is also the depletion of finite resources—see Ehui et al. (1990) *J. Env. Econ. & Manage.* **18**, 2. A **productivity rating** is an estimate of an area's ability to support plant growth; see Yang et al. (2005) *Rev. Agric. Econ.* **27**, 1 on conservation reserve enhancement.

profit surface Variations in profit shown as a three-dimensional surface, derived from subtracting the relevant *cost surface from the corresponding revenue surface, a calculation of the utmost difficulty. See N. Smith (1984).

proglacial Situated in front of a glacier. A **proglacial lake** is formed between the terminus of the ice and the higher ground which is often in the form of a terminal *moraine. Glacial Lake, Wisconsin, was a large proglacial lake. Glacial recession caused the ice dam to fail catastrophically and the lake drained rapidly (Clayton and Knox (2008) *Geomorph.* **93**, 3).

progradation The accumulation of beach material when there is an excess in the supply of sediment; see Goy et al. (2003) *Geomorph.* **50**, 1–3. Progradation is a feature of, for example, *delta and *mangrove coasts; Hori et al., *Geomorph.* **41**, 2 find that the average progradation rate of the Yangtze delta rose from 50 to 80 km/year about 2000 years BP.

programming region A region which is designed to serve a particular purpose, such as the Italy/Slovenia **Programming Region**—see EUROREG, FP6–506019.

propulsive industry A fast-growing high-tech industry with expert management, and an extensive market, as in semiconductors (Meardon (2001) *Amer. J. Econ. & Soc.* **60**, 1).

project In economic geography, 'a temporary system with institutionalised termination' (Lundin and Söderholm (1995) *Scand. J. Manage.* **11**); a 'one-off' activity, such as shipbuilding, film-making. See Grabner (2002) *Env. & Plan. A* **34**, 11 and Kloosterman (2008) *J. Econ. Geog.* **8**, 4 on **project-based organizations**.

protalus rampart A thin ridge, about a metre high and many metres long, composed of rock fragments, running along the foot of a mountain face and thought to be produced from rock fragments sliding over a snow surface (Curry et al. (2001) *Procs Geols Ass.* **112**, 4).

protection The imposition of *quotas or *tariffs on foreign imports by a government: to become self-sufficient, to protect new industries, or as a bargaining tool. See Bai et al. (2003) *W. Davidson W. Paper* **565**, and Young (2000) *Quart. J. Econ.* **115**, 4 on China.

proximity Nearness in space, time, or relationship. Oerlemans et al. (2001) *Tijdschrift* **92**, 1 explain the **proximity effect** as benefiting from localized ties. Boschma (2005) *Reg. Studs* **39** claims that geographical proximity per se is neither a necessary nor a sufficient condition for learning to take place. Nevertheless, it facilitates interactive learning, most likely by strengthening the other dimensions of proximity. However, proximity may also have negative impacts on innovation due to the problem of lock-in. Accordingly, not only too little, but also too much proximity may be detrimental to interactive learning and innovation. Ponds et al. (2007) *Papers Reg. Sci.* **86**, 3: geographical proximity plays 'a significant, yet minor role for collaboration between academic organisations within the Netherlands'—but a quick study of the scholars writing on geographical proximity shows a preponderance of Dutch/Flemish surnames!

psychic income The enjoyment, which cannot be measured in financial terms, that people derive from location or particular place (G. Becker 1964). See Reardon and McCorkle (2002) *Int. J. Retail & Distribution Manage.* **30**, 4 on psychic income and shopping. Garnham (1996) *Festival Manage. & Event Tourism* **4** argues that major events improve self-image and raise morale, but Olds (1998) *Festival Manage. & Event Tourism* **11** suggests that the psychic income of a hosting community is not always positive.

public administration, geographies of The study of the roles of space and place in contemporary public administration; the implications for public policy-making of the reconfigurations of spaces and places; and the use of information and communications technology to redesign organizations (for example, the e-Europe Action Plan). See Baltodano (1997) *J. Pub. Admin. Res.* **7**, 6. Historically, the organization of government activity has been based on co-location of actor (public administrator), act (administration), and geographical setting (administrative unit); see Painchaud (1977) *Publius* **7**.

Globalization brings about new spatial forms of organization and processes in a 'virtual world', and these new spatial forms of organization and processes represent new 'public spaces'; see Toonen (1998) *Public Admin.* **76**, 2. The key criteria in determining the appropriate sizes of administrative units may well be a sense of geographical identity for the 'citizens', and equity in resource distribution and access to services. See S. Crate (2003) and Ahas et al. (2005) *Futures* **37**, 6.

public choice theory The behaviour of public-sector bureaucrats is at the heart of public choice theory. Hospers (2005) *Intereconomics* **40**, 6 discusses two types of policy-maker: offensive and future oriented, or traditional and preserving, and plumps for the latter.

public goods Goods freely available, either naturally, like air, or from the state, like education in most developed countries. See R. Baldwin (2005).

publics A scarcely bearable term for small groups of people who follow one or more particular issue very closely. Different kinds of publics occupy different kinds of spaces: 'for example, a public space that would foster a liberal-economistic version of publicity might be one that facilitates state functioning and makes it easier for the state to safeguard and regulate individual rights and to promote economic development and growth' (Staeheli and Mitchell (2007) *PHG* **31**, 6). 'Publics are deeply embedded in social and machinic complexes involving the mobilities of people, objects, and information' (Sheller (2004) *Env. & Plan. D* **22**, 1). For Fuller and Askins (2007) *Antipode* **39**, 4 publics, while multiple and in flux, are also created.

public sector That part of a national economy owned and controlled by the state, including nationalized industries, national and local government services, and public corporations. Mrinska (2007) *Northern Economic Agenda*, paper 2, concludes that public intervention is necessary to revive the economy of economically depressed regions, resulting in increased employment and spending, which should stimulate growth in the private sector.

public space Public spaces, including parks, squares, and markets, are 'co-produced', and the key principles for their development include leaving room for self-organization, diversifying activities to encourage diverse people to participate, and maintaining access and availability (M. Mean and C. Tims 2005). K. Worpole and K. Knox (2007) conclude that strategies intended to 'design out crime', such as cutting down bushes, installing vandal-proof street furniture, and closing public toilets, affected their attractiveness and damaged the usefulness of public spaces; see also Klauser (2007) *Eur. Urb. & Reg. Studs*, on using CCTV to regenerate public space. Veronis (2006) *Env. & Plan. A* **38**, 10 finds that the use of public space is central to the cultural politics of migrants in Toronto; Ortiz et al. (2004) *GeoJournal* **61**, 3, report on women's use of public space in Barcelona; and Mackintosh (2007) *J. Hist. Sociol.* **20**, 1–2 writes on the engagingly titled bourgeois geography of bicycling.

pull factor A positive factor exerted by a locality that people move to. These have included: new land grants for farmers (the Great Plains; Gutmann (2000) *Climatic Change* **44**, 3), assisted passages and other government inducements (for Australia, see P. Burroughs and A. J. Stockwell 1998), freedom of speech or religion (see N. Philbrick 2006), and material inducements (Hong Kong; see Lin (2002) *Asia Pac. Viewpt.* **43**, 1).

pumice A very light, fine-grained, and cellular rock produced when the froth on the surface of lava solidifies. See Calcaterra et al. (2007) *Geomorph.* **87**, 3. A **pumice raft** comprises pumice fragments floating on the sea; one was spotted in 2006 (*Geol. News*, Nov. 2006).

pumped storage scheme Electricity cannot be stored, so when demand is low, at night, some can be used to pump water from a lower to a higher reservoir. At peak demand, the water is allowed to fall back to the lower level, passing through turbines which turn generators. See Bruch (2007) *UNEP Dams and Development*, **92–3** on the Palmiet Pumped Storage station, South Africa.

push factor In *migration, any adverse factor which causes out-migration. Examples include: famine, changes in land tenure (the Highland Clearances, 1790–1850, A. MacKenzie 1999), political persecution (Tamil separatists, Sri Lanka, Stokke (2000) *Growth & Change* **31**, 29), and mechanization which made agricultural workers redundant and which made factory products cheaper than those of cottage industry (see Pisani and Yaskowitz 2002, *Soc. Sci. Quarterly* **83**, 2 on rural depopulation in Portugal). Relatively few migrations are spurred by push factors alone. See Barnett and Adger (2007) *J. Polit. Geog.* **15**, 4, on climate-change induced migration.

push moraine *See* MORAINE.

puy The French term for a volcanic neck, revealed by differential erosion. The type location is the Puy de Dôme. The Hohentweil, Hegau, Germany, is a further example.

pyramidal peak Synonym for horn or aiguille.

pyramid of numbers (ecological pyramid) A diagram of a *food chain which shows each *trophic level as a horizontal bar, drawn in proportion to the *biomass. There is a large fall from *producers and primary *consumers;

thereafter the decreases are smaller. The animals on the higher levels are generally larger and rarer than animals lower down the pyramid. See Kent (2000) *PPG* **24**.

pyroclast A fragment of solidified lava, ejected during explosive volcanic eruptions. Classification of pyroclasts is by size. Fragments less than 4 mm across are ash; compacted ash is *tuff*. Material between 4 and 32 mm is *lapili* and fragments larger than 32 mm are *blocks*. Collectively, these fragments are *tephra. Pyroclasts formed from lava produce volcanic bombs and volcanic breccia. **Pyroclastic flows** are also known as *nuées ardentes. They result from the bursting of gas bubbles within the magma, which fragments the lava. Eventually, a dense cloud of fragments is thrown out to form a mixture of hot gases, volcanic fragments, crystals, ash, pumice, and shards of glass. See Adas et al. (2006) *GSA Bull.* **118**, 5.

qualitative Concerned with meaning, rather than with measurement. The emphasis is on subjective understanding, communication, and empathy, rather than on prediction and control, and it is a tenet that there is no separate, unique, 'real' world. All qualitative researchers are positioned subjects; as such, the rigour of their research depends not only on the suitability of the methodology, the use of multiple methods, and the inclusion of verbatim quotations, but also on its credibility and transferability. Thomas (2007) *Prof. Geogr.* **59**, 4 suggests that researchers must avoid psychoanalysing research subjects: 'this limits the ways in which scholars can "read" personal narratives for unconscious processes.' Bradshaw (2001) *Area* **33**, 2 outlines 'some choices which might help address possible difficulties in qualitatively researching the powerful'. For methods in qualitative research, see Davies and Dwyer (2007) *PHG* **31**, 2; 2008, *PHG* **32**, 3; and for a checklist for evaluating qualitative research, see Baxter and Eyles (1997) *TIBG* **22**, 4. See Suchan and Brewer (2000) *Prof. Geogr.* **52**, 1 on qualitative methods in cartography.

quantification The numerical measurement and analysis of processes and features.

quantitative revolution In geography, the intellectual movement beginning in the 1950s that explicitly introduced to the discipline scientific forms of theorizing and techniques of empirical verification (Barnes (2004) *PHG* **28**, 5), transforming geography into an analysis-oriented scientific discipline. The quantitative revolution is generally considered to have emerged from a general dissatisfaction with regional geographic study, and a consequent shift in focus towards more systematic and specialized approaches (Keylock (2003) *TIBG* **28**, 2). Barnes (1998) *Env. & Plan. A* **30** argues that the success of this new geography was due to the manner in which it was able to collapse an initially disparate set of concerns into a single framework (the regression or multiple regression equation), to which Bracken and Wainwright (2006) *TIBG* **31**, 2 would add the use of concepts 'the Western modernist psyche finds easy to accept'. See Johnston and Sidaway (2004) *PHG* **28**. Quantification began, flourished, and was, in turn, criticized: 'the dominant paradigm at the time was that the western experience of urban-industrial development held the key to progress everywhere, and the challenge was to overcome any obstacles to the diffusion of ideas and innovations that would hinder this transformation ... much that characterised the quantitative revolution was about theorising and employing statistical procedures in the office to generalise about change everywhere' (Bedford (2005) *Asia Pac. Viewpt* **46**, 2). Quantification was attacked for being unrealistic and bloodless, turning humans into automata, for being too deterministic, and for ignoring the importance of subjective experience. T. J. Barnes (1998) attributes the reduction in emphasis upon quantification in more recent human geography to the growing awareness that techniques built upon an assumption of statistical independence are not strictly applicable to situations where spatial interdependence is apparent. Was the quantitative revolution a conspiracy? No, says Johnston (2007) *TIBG* **32**, 3.

quarrying 'The removal of pre-loosened bed material and/or material resulting from bed failure' (Rea and Walley (2006) *Earth Surf. Procs & Landfms* **19**, 7). Since the tensile strength of ice is low, *glacial quarrying is not possible unless the rock is shattered.

Glasser and Bennet (2004) *PPG* **28**, 1 observe that there is now general agreement that quarrying is favoured beneath thin, fast-flowing ice, and Drake and Shreve (1973) *Procs Royal Soc. London, Ser A* **332** propose a heat pump effect: water that is melted in high-pressure areas flows away and does not refreeze at the immediately adjacent low-pressure area, and this leaves cold patches that advect downglacier. Fluctuations in basal water pressure also play an important role in the formation of glacially quarried landforms; see Glasser and Bennet (2004) *PPG* **28**, 1. Landforms of glacial quarrying include roches moutonnées, rock basins, and zones of *areal scouring.

Quaternary (Pleistogene) The most recent period of geological time, covering the last 2 million years, and including the *Pleistocene and the *Holocene.

quaternary industry (quaternary activity, quaternary sector) Economic activity concerned with information: its acquisition, manipulation, and transmission. Into this category fall law, finance, education, research, and the media. See D. Ley (1996) on the middle classes and the expansion of quaternary industry.

queer geographies A major concern of queer geographies has been the critical role of place and space in the production of sexual identities, practices, communities, subjectivities, and embodiments, and the scholarship of queer space includes its location, nature, and definition, its history, memories, events, and subcultures, and the relationship between space and social justice. This is examined by looking at spaces such as closets, communities, and cruising grounds; 'finding or creating spaces in which to know ourselves and become known to others' (Sember (2003) *Space and Culture* **6**, 3). 'Through repeated acts of "coming out", gay men and lesbians map where and how to be 'out' and when and where not to be. These environments are largely concealed and require careful reading' (Space, Sexualities and Queer Working Group of the RGS). Correspondingly, there is a focus on the centrality of sexualities in constituting social space. See Gorman-Murray (2007) *Soc. & Cult. Geog.* **8**, 1 on rethinking queer migration through the body.

Podmore, in a study of Montreal lesbian territory (2006, *Soc. & Cult. Geog.* **7**, 4), asserts that, although gay men have often produced clearly visible territorial enclaves in inner-city areas (as in Manchester or San Francisco), lesbian spaces have been comparatively invisible 'since their communities are constituted through social networks rather than commercial sites'. Further, Nast (2003, *Antipode* **34**) argues that 'gay white patriarchies' exist, which depend 'structurally and implicitly upon white supremacy and heteropatriarchy'. Oswin (2004) *Acme* **3**, 2, also, warns against an empiricist, cognitive reading of the affluent, white, gay male. Binnie and Valentine (1999) *PHG* **23**, 2 believe that queer geographies need to move away from simple mapping of lesbian and gay spaces towards a more critical treatment of the differences between sexual dissidents. Knopp (2007) *Prof. Geogr.* **59**, 1 focuses on the potential of feminist-inspired and allied queer geographies to rethink a variety of spatial, and other, ontologies. See also Brown and Knopp in Anderson et al., eds (2003, *Antipode* **34**). Hubbard (2000) *PHG* **24**, 2 writes that 'everyday space' is experienced as aggressively heterosexual by lesbians and gay males; heterosexuality has served to create and justify other forms of oppression and confinement in Western cities.

Queer theory challenges the idea of the preconstituted sexual subject and understands power as productive rather than simply oppressive: 'yet critical geographers generally depict queer spaces as spaces of gays and lesbians or queers existing in opposition to and as transgressions of heterosexual space' (Oswin (2008) *PHG* **32**).

quickclay Clay whose structure collapses completely on remoulding, and whose shear strength is thereby reduced to almost zero (Eilertsen et al. (2008) *Geomorph.* **93**, 3–44).

quickflow That part of a storm rainfall which moves quickly to a stream channel. Amit et al. (2002) *Ground Water* **40**, 5, compare quickflow—probably through cracks—with baseflow—the slow flow through the porous medium. 'Quickflow and slowflow refer to relative time constants for characteristic components of observed basin-scale response' (Schwartz (2007) *J. Am. Water Resources Ass.* **43**, 6).

quota A limit on the import or export of a particular product imposed by a government. Import quotas may be imposed as protectionism; export quotas may be imposed in countries that depend on the export of a particular raw material, as a means of stabilizing prices. Quotas are usually controlled by the issue of licences. Tokatli (2007) *J. Econ. Geog.* **7**, 1 explains that *transnationals can relocate production from one country whose annual quota is exhausted to another whose quota has not been filled or that is not bound by quotas, in the process playing off different producers and governments against each other.

The Multi-Fibre Arrangement, which has governed the international trade in textiles since 1974 through sets of trade restrictions and quotas, terminated in 2005. In consequence, Hale and Burns in A. Hale and J. Wills, eds (2005) expect a large-scale shift in manufacturing to the lowest-cost areas of production, such as China and India.

q

radial drainage *See* DRAINAGE PATTERNS.

radiation Energy travelling in the form of electromagnetic waves: X-rays, ultraviolet, visible, infra-red, microwaves, or radio waves. In urban areas, a reduction in the *sky view factor decreases the loss of terrestrial radiation thereby contributing to the creation of *urban heat islands (Grimmond (2007) *Geog.* **30**, 3).

radiation fog *See* FOG.

radiative forcing The increase in the trapping of outgoing *terrestrial radiation by greenhouse gases. In a study of the capacity of agriculture to reduce the emission of greenhouse gases, Robertson et al. (2002) *Science* **289** find that only soil carbon accumulation in non-tillage systems came close to mitigating radiative forcing.

radical geography An analysis of the processes by which inequalities—in race, class, gender, or age—are produced and maintained. Radical geographers hold that studying the visible geography of spatial relationships is not enough; the power relationships, specifically the political and economic structures, are fundamental. For example, any study of industrial location would be worthless, unless the operations of *transnational corporations, and tariff and trade agreements, are taken into account (Byrne (1995) *Capital & Class* **56**).

  Radical geography includes attempts to change the situation on the ground for everyday people in everyday places. See D. Fuller and R. Kitchin (2004) for an exploration of the role of the radical academic in society, and Lawson (2007) *AAAG* **97**, 1 on geographies of care and responsibility.

radioglaciology The use of ice-penetrating radar to investigate the thickness, roughness, motion, and debris, of a body of ice, and to measure and to detect crevasses and sub-ice lakes. See Plewes and Hubbard (2001 *PPG* **25**, 2) for a detailed account of the investigative processes.

radiometer A passive remote sensor, sensitive to *terrestrial radiation of one or more wavelengths of the visible and infra red. Advanced Very High Resolution Radiometer currently flies aboard many US and European satellites, and its data may be analysed using the AVHRR Hydrological Analysis system, together with ILWIS.

(⊕) SEE WEB LINKS

- AVHRR Hydrological Analysis system.

radiosonde A free-flying balloon carrying meteorological instruments. The balloon climbs to a height of 20–30 km above mean sea level, sending information from these sensors to ground stations, whereupon it bursts, returning the equipment to the ground.

rain A form of *precipitation consisting of water droplets ranging from 1 to 5 mm in diameter. The type of rain produced reflects the circumstances in which it formed. A mass of warm air rising at a warm front will develop layered clouds and produce steady rain. Air forced to rise quickly at cold fronts will bring heavier rain. These are both examples of frontal rain. *Convection rain occurs when warm, unstable air rises rapidly. Air forced to rise over mountains may form *orographic (relief) rain. *See also* BERGERON–FINDEISEN THEORY; COALESCENCE THEORY.

raindrop erosion The dislodging of soil particles by large drops of rain. See Wainwright et al. (1995) *ESPL* **20**, 277–91, on raindrop erosion and desert pavements. See also Patrick (2002) *PPG* **26**.

rainfall intensity The rate at which rain falls, usually measured in millimetres per hour. Intense rainfall is associated with **convectional rain**, notably in thunderstorms and tropical regions, where intensity may be over 100 mm per hour. (British rainfall intensity is normally of the order of 2 mm per hour.) The intensity of rainfall is normally inversely proportional to its duration. Denault et al. (2006) *J. Am. Water Resources Ass.* **42**, 3 examine the potential impacts of increases in short duration rainfall intensity on urban infrastructure.

rainfall run-off The overland and downslope flow of rainwater into channelled flow when the rock or soil is saturated. In **rainfall run-off modelling**, hydrological models are categorized as *lumped*—which treat the catchment area as a single unit—or *distributed*—which represent the catchment as a system of interrelated subsystems, both horizontally and vertically. For **deterministic rainfall models**, see Ramos et al. (1995) *Water Resources Res.* **31**, 6. For **fuzzy conception rainfall models**, see Ozelkan and Duckstein (2001) *J. Hydrology* **253**, 1. For a very clear guide through a **parametric rainfall-run-off** simulation model see Dawdy, Lichty, and Bergmann (1972) *US Geol. Surv., Prof. Paper* **506**-B. For the use of *artificial neural networks to relate meteorological variables to run-off, see Antar et al. (2006) *Hydrol. Procs* **20**, 5.

rain forest An area of luxuriant forest, found where rainfall exceeds 1000 mm yr^{-1}. Although rain forest does develop in temperate latitudes most of the world's rain forest is tropical. Puyravaud et al. (2003) *J. Biogeog.* **30**, 7 applaud the use of vegetation thickets as restoration tools in rain forests.

rain shadow An area of relatively low rainfall to the lee of uplands. The incoming air, forced to rise over the high land, causes precipitation on the windward side. The descending air is subject to *adiabatic warming, which increases its capacity to hold much of the remaining water vapour, further reducing rain on the lee side. See Ralph et al. (2002) *Conf. Mount. Meteo.* on the impact of a prominent rain shadow in California.

rainsplash The impact of raindrops on the soil may break down soil peds, loosen soil particles, and cause *turbulence in the *sheet wash of water flowing downslope. 'Rainsplash . . . is not a single process, but a combination of several discrete but interacting soil particle detaching and [soil particle] transport mechanisms' (Terry (1998) *Austr. J. Soil Res.* **36**, 3). As overland flow becomes deeper, the effect of raindrop impact on the soil decays significantly (Mosley (1973) *Zeitschrift* **18**). Furbish et al. (2007) *U. Vanderbilt* argue that soil grain transport by rainsplash is largely an *advection-dispersion process.

raised beach A former *beach, recognizable by beach deposits and marine shells, often accompanied by rock platforms and dead *cliffs, which now stands above sea level some metres inland. At sites of *isostatic rebound, several raised beaches may be seen at different levels; the raised beaches of Scotland range from 6 to 14 m OD. Raised beaches result from tectonic uplift; see van Vliet-Lanoe et al. (2000) *J. Geodynam.* **29**, 1.

ramparts As *ice wedges grow, they push up the soil at the edges into ramparts; see M. Shahgedanova (2002). **Protalus ramparts** are coarse, angular, non-stratified thin ridges of rock fragments, about 1 m high, lying along the base of a mountain face. They are associated with persistent snow banks (Sass (2006) *Geomorph.* **80**, 1–2). Shakesby et al. (1999) *Geografiska A* **81**, 1 show that they can be formed by snow-push.

range of a good or service The maximum distance/travel time an individual will travel to obtain a given good or service; rare speciality shops have a larger range than shops selling *convenience goods. D. Harvey (2001) noting that heavy, low-value items are now traded over such long distances, argues, controversially, that this concept now makes little sense.

ranker An *intrazonal soil, not yet fully developed. This soil is shallow, with a fibrous O *horizon, and an A horizon directly on top of loose, non-calcareous rock.

ranking of towns Various methods have been used to devise a *hierarchy of towns, including comparing the number and type of functions, using multivariate analysis, measuring the interaction between a settlement and its field, ranking settlements by the size of their *spheres of influence; and using graph theory. Recent rankings have been based on quality of life (Rogerson in F. T. Seik et al., eds (2000); Berger et al. (1987) *Soc. Sci. Quart.* **68**) or on urban competitiveness (Rodinelli and Vastag (1997) *Econ. Dev. Q'ly* **11**). The 2007 China Urban Competitiveness Study (Hong Kong) found, unsurprisingly, that Hong Kong was at the top of the Chinese hierarchy.

rank-size rule This 'rule' predicts that, if the settlements in a country are ranked by population size, the population of a settlement ranked *n* will be 1/*n*th of the size of the largest settlement. When settlement size is plotted against rank, on normal graph paper, a concave curve results; plotted on logarithmic scales, a straight line emerges—this is the *rank-size pattern*.

Many developing countries show *primacy*: a sharp fall between the largest, *primate city and the other cities. The *binary pattern*, found mostly in federal countries such as Australia, shows a concave curve. The *stepped order* pattern shows a number of settlements at each level, with each place resembling others in size and function. Fonseca (1989, Institute of Mathematical Geography) explains why the rank-size rule 'works'.

Taylor (2007) *GaWC Res. Bull.* **238** notes that national urban hierarchies assume that the national urban system is a closed one; relations between cities in different countries are factored out. 'However, vibrant dynamic cities are always cosmopolitan, [and] to treat New York as US only is to severely under-estimate its economic significance . . . the national bounding of cities was [always] a critical weakness . . . it is even more nonsensical today.'

Raoult's law This states that the presence of a solute will lower *saturation vapour pressure. *See* CONDENSATION NUCLEI.

rapids Areas of greatly disturbed water across a river, rapids have a continuous and relatively gentle slope, rather than a sudden vertical drop.

rates A form of local taxation in the UK superseded by the community charge or poll tax, which was, in turn, replaced by the council tax.

rational choice theory A theory arguing that individual self-interest is the fundamental human motive, and that individual actors pursue their goals efficiently (see Clarke in M. Tonry and N. Morris, eds 1983). Strauss (2008) *J. Econ. Geog.* **8**, 2 notes that economic rationality finds expression in rational choice theory. Cornish in R. Clarke and M. Felson, eds (1993) argues that criminals make travel decisions based on rational choice theory: everything else being equal, criminals choose closer targets rather than those that are further away; see also Herbert and Brown (2006) *Antipode* **38**, 4.

rationality The process of using reason or logic to solve a problem. 'The insights of behavioural economists have allowed social scientists to acknowledge certain unique traits specific to humans as a species as rationality without succumbing to an ontological position that posits *utility maximisation as an *a priori* foundation of human behaviour . . . In the UK, for example, strong assumptions are being made about the economic rationality and decision-making competence of individuals in areas of government policy such as health, pensions and education' (Strauss (2008) *J. Econ. Geog.* **8**, 2).

Ravenstein's 'laws' of migration E. Ravenstein's laws of migration, 1889 (1995, *J. Royal Stat. Soc.* **48**) introduced the notion that people move in order to better themselves economically. In this view, migration is considered as the individual's response to regional differentials in economic development. Ravenstein's fifth 'law' of migration suggests that women are more mobile than men, at least across short distances—a contention supported by Boyle and Halfacree (1995) *Reg. Studs* **29**, 1 and Faggian et al. (2007) *J. Reg. Sci.* **47**, 3.

reaction time The time between any kind of change and the response it elicits in a system. In geomorphology it is the interval between a disturbance or change in controls and observable morphological change. In hydrology, the bigger a catchment, the longer its reaction time. Harrison (2003) *Am. Geophys. Union* **C22A-01** extends this term to glaciers, but Oerlemans (2007) *J. Glaciol.* **53**, 182 thinks that reaction time is not a well-defined concept, 'because the reaction time may depend on the glacier/climate history in a non-transparent way'.

recession The decline in river flow after a storm event has passed. On a *hydrograph, the **recession limb** records the fall in *discharge after the river has reached peak flow due to a storm event. The slope of the recession limb reflects the amount of water stored in the basin and the way it is held in the *catchment area. See Stravs et al. (2008) *Geophys. Res. Abstr.* **10**, EGU2008–A-02978 on recession limbs and flow forecasts.

recessional moraine *See* MORAINE.

reclamation The process of creating usable land from waste, flooded, or derelict land.

recovery rate The time taken for a *diffusion wave to re-form having been blocked by a diffusion barrier. The recovery rate of a wave front is directly related to both the type and size of the barrier it encounters. See Cliff and Hagget (2006) *J. Geogr. Syst.* **8**, 3.

recreation *See* LEISURE, GEOGRAPHY OF. **Recreation carrying capacity** is the amount of recreation which a site can take without any deterioration of its qualities. See, for example, W. Theobold (2005).

rectangular drainage *See* DRAINAGE PATTERNS.

recurrence interval The length of time between events of a given magnitude. Anderson and Calver (2004) *TIBG* **29**, 1 note that 'landscape form depends on the overall effect of past frequency distributions of sizes of geomorphic events, and on the weather conditions' return'. In hydrology, the return period has an inverse relationship with the probability that the event will be exceeded in any one year. See Keylock (2005) *Adv. Water Resources* **28**, 8 on flood recurrence intervals. See also Srikanthan and McMahon (1985) *Hydrol. Scis* **30**, 2 on calculating the recurrence interval of drought events through the stochastic analysis of rainfall and streamflow data.

recycling The reuse of *renewable resources in an effort to maximize their value, reduce waste, and reduce environmental disturbance. A *cycle economy*, based on the economical and responsible treatment of limited resources (Seliger et al. (2001) *CIRP Anns.—Manuf. Ind.* **51**, 1) is a key concept of today's Japanese environmental policy (Kenichi (2002) *Mining & Materials Processing Inst. Japan* **118**, 9). Williams and Kelly (2003) *Resources, Conserv & Recycl.* **38**, 2 evaluate public perception of local authority recycling, and Asomi-Boateng (2007) *J. Plan. Ed. & Res.* **27**, 2 reports that waste-based urban cultivation in Ghana could minimize waste collection and disposal costs.

redevelopment The demolition of old buildings and the creation of new buildings on the same site. Redevelopment can solve existing problems of congestion and poor design but, for residential areas in particular, it is seen to be wasteful of resources, destroying communities. Wu and He (2005) *Tijdschrift* **96**, 1 find that the root cause of such problems was the way local officials and local developers 'utilised their privileges inherited from a planned economy to benefit from the emerging real estate markets'. Fang and Zhang (2003) *Asia Pac. Viewpt* **44**, 2 also detect 'the hidden agenda of local elites' during urban redevelopment.

redlining The identification by a mortgage lender of a specific neighbourhood for which people cannot get a mortgage; 'a form of place-based social exclusion that particularly disadvantages low-income neighborhoods and ethnic-minority neighborhoods' (Aalbers (2008) *AMIDSt UvA*). However, Harris (2003) *Canad. Geogr./Géogr. canad.* **47**, 3 argues that the suburban origin of redlining in Canada 'had little to do with the distribution of ethnic minorities. As long as it lasted, roughly until the late 1950s, it perpetuated social class diversity in Canadian suburbs.' Eisenhauer (2001) *GeoJournal* **53**, 2 describes the location of out-of-town superstores as 'supermarket redlining'.

red rain The washing out of fine dust particles over mid-latitudes. For Europe, the major sources of dust are the Sahara and the fringes of the Sahel. This dust is picked up by air streams which rise into the upper *troposphere, and carried north and west until it is washed out by precipitation. McCafferty (2008) *Int. J. Astrobiol.* **7** examines historical and mythical accounts of red rain.

reduced-complexity modelling/approaches Physical description matched to computational parsimony (Brasington and Richards (2007) *Geomorph.* **90**, 3–4, special issue).

reduction The loss of oxygen from a compound. For example, the sesquioxide ferric oxide can be reduced to the monoxide ferrous oxide by bacteria. *See* GLEY SOILS.

reflective beach A beach which reflects a major part of the incoming wave. A reflective beach has a convex beach profile, without nearshore bars, is steeper than the wave it reflects, and is typically caused by *accretion. Reflection may generate *edge waves on beaches. See Miles and Russell (2004) *Cont. Shelf Res.* **24**, 11.

reforestation Replanting a previously wooded area that has been felled; the implicit assumption is that the restoration will succeed reforestation. See Vallauri et al. (2002) *Restoration Ecol.* **10**, 1.

refugee 'A person who, owing to a well-founded fear of persecution for reasons of race, religion, nationality, membership of a particular social group or political opinion, is outside her or his country of nationality and who is unable or unwilling to return' (UN Protocol 1976). Australia's refugee programme seeks to provide a humanitarian response and protection to individual refugees; participate responsibly in the international community; honour its Convention obligations; further the interests of the people of Australia; meet high standards of administration; and acknowledge as much as possible changes in refugee populations (King (2001) *Int. Mig.* **39**, 1). 'Though attempts to de-naturalize the relationship between people and places have been important for how the refugee experience is conceptualized, there has been too much focus on imagination accompanied by a neglect of the local perspective of migrants and displaced people' (Brun (2001) *Geografiska B* **83**, 1). *See also* ASYLUM MIGRATION.

reg A North African term for *desert pavement.

regime theory A set of empirical equations relating channel shape to discharge, sediment load, and bank resistance. The theory proposes that dominant channel characteristics remain stable for a period of years and that any change in the hydrologic or sediment regime leads to a quantifiable channel response (such as erosion or deposition). Valentine and Haidera (2005) *Water Manage.* **158**, 2 propose a modification to the Wallingford rational regime theory. See also Eaton and Millar (2004) *Geomorph.* **66**, 1–2. Eaton and Church (2007) *J. Geophys. Res.* **112**, F03025 find that classical hydraulic geometry represents only an approximation of the variation in channel forms.

region Any tract of the Earth's surface with either natural or man-made characteristics which mark it off as being different from the areas around it. 'The formation of any given regional map is reflective—and indeed constitutive—of an unevenly developing, often overlapping and superimposing mosaic' (Martin and MacLeod (2004) *TIBG* **29**, 4).

Paasi (2001, *Eur. Urb. & Reg. Studs*) writes of regions as 'collective institutional structures . . . regions are not independent actors; they exist and "become" in social practice and discourse'.

Many geographers have attempted to distinguish regional boundaries; for example, Slaymaker (2007) *Sing. J. Trop. Geog.* **28**, 1 discusses whether or not South-East Asia is a legitimate physical geographical region. However, as

Shields (2007, Putting Region in its Place Conf., U. Alberta) points out, 'topography is only one component of geographical identities . . . a place might be said to realize or to embody a regional character, but these statements are always made with a view to contrasting some quality of the region with the qualities of other regions.' Jones (2006) *AAAG* **96**, 2 finds international region building 'a messy, problematic, and highly contested activity for parceling, regulating, and representing geopolitical space'.

regional development **1.** The provision of aid and other assistance to regions which are less economically developed.

2. The differential in economic outcomes. Ezcurra and Rapún (2006) *Eur. Urb. & Reg. Studs* **13**, 4 explore the relationship between regional *inequality and economic development level in fourteen western European countries for the period 1980–2002, finding that, beyond a given level of per capita GDP, regional inequality can be seen to decrease. McCoe et al. (2004) *TIBG* **29**, 4, in a study of BMW in Germany and Thailand, argue that the 'strategic coupling' of the global production networks of firms and regional economies ultimately drives regional development. Liard-Muriente (2007) *Area* **39**, 2 highlights the potential opportunities and pitfalls when designing incentive strategies.

Regional Development Agency (RDA) Eight regional development agencies were established in England in 1999. These are:

- the North West RDA—North-West
- OneNorthEast—North-East
- Yorkshire Forward—Yorkshire and Humberside
- East Midlands Development Agency—East Midlands
- Advantage West Midlands—West Midlands
- East of England Development Agency—East of England
- South West of England Development Agency—South-West
- South East England Development Agency (SEEDA)—South-East

The London Development Agency was established in 2000. Their remit is to promote sustainable economic development. Each has stakeholders from the local public, private, and voluntary sector community, and each RDA produces a 'regional economic strategy' which outlines the planned economic development and regeneration of the respective region.

(⊕) SEE WEB LINKS

- Regional Development Agencies online.

regional geography The study in geography of *regions and of their distinctive qualities. A precondition of this study is the recognition of a region, its naming, and the delimitation of its boundaries. One approach has been to identify 'natural' regions while another was to establish economic regions based on agriculture and/or industry. Often there was an intimation of a link between the two types of region. Once the keystone of geography, the status of regional geography has been in decline: 'regional geography survives—if not in most of our major universities, then at least in the public eye. Believe it or not, the informed public still thinks that geographers have particular and unusual perspectives and insights on foreign areas as well as domestic issues, and that geographers can explain "the way the world works" because they have a comprehensive view of it' (Murphy et al. (2005) *PHG* **29**, 2). 'The great number of feasible ways of dividing space into regions and time into periods opens the door for alternative narratives, including those which challenge conventional Eurocentric interpretations' (Wishart (2004) *PHG* **28**, 3).

regional inequality A disparity between the standards of living applying within a nation. It is difficult to quantify the prosperity or poverty of a region, but there are two basic indicators: unemployment (which has been used in Britain as a symptom since the 1920s), and per capita income, which in England generally falls to the north and west: the North–South divide was 'produced essentially by the . . . metropolitan South as a way of differentiating itself from northern "others"' (A. Baker and M. Billinge 2004). Some would assert that economic development brings about regional inequality, but Fan and Sun (2008) *Euras. Geog. & Econ.* **49**, 1 reveal that both interregional and intraregional inequalities in China have declined since 2004. *See also* UNEVEN DEVELOPMENT.

regionalism A move to foster or protect an indigenous culture in a particular region. ECLAC (1994) defines open regionalism as 'a process of growing economic interdependence at the regional level, fuelled both by preferential integration agreements and other liberalization and deregulation policies, in order to increase the competitiveness of the region's countries and to consolidate the foundations for a more open and transparent international economy'. Poon (2001) *Area* **33**, 3 sees the major goal of open regionalism as overcoming the social and psychological barriers that stop businesses networking. Jones and MacLeod (2004) *TIBG* **29**, 4 write on **spaces of regionalism**, which (re) assert 'national and regional claims to citizenship, insurgent forms of political mobilization and cultural expression and the formation of new contours of territorial government'; see J. Agnew (2002), for example.

regionalization The demarcation of regions. E. Ginzburg (1977) suggests that regionalization is 'a slippery concept that varies from writer to writer', but that, regardless of definition, the intention of regionalization programmes is to improve access, quality, cost, and equity. Parr (2002) *Int. Reg. Sci. Rev.* **25** sees the advantages of economic activity being organized around a well-defined localization, urban concentration, or activity complex (or some combination of these) realized with a regional spatial structure which is more dispersed. Wishart (2004) *PHG* **28**, 3 thinks regionalization should be seen as part and parcel of the art of *representation.

regional multiplier In the framework of a simple regional multiplier model, the operation of a new enterprise creates additional regional income due to its regional exporting activity, which may be considered as the first round of impact. In a second round, the additional expenditures of the firm in the local economy will create more income (Skuras et al. (2005) *Eur. Urb. & Reg. Studs* **12**, 4). See Keet in J. Adésínà et al. (2006) and A. Hirsch (2005), both on African topics, and Bathelt (2007) *Geog. Compass* **1**, 6 on a knowledge-based multiplier model of *clusters.

regional policy A policy, adopted by government, aimed at redressing *uneven development within a country. Pritchard (2005) *Geog. Res.* **43**, 1 argues that the neoliberal approach to regional policy 'advances a relatively simplified conception of regional economic and social life that pays little regard to their richness, complexity and grounded realities'. Gleeson (2003) *Austral. Geog. Studs* **41**, 3 critically analyses the European Union's regional policy framework, particularly the advocacy of 'economic normalization' as a leading aim. 'While regional policy traditionally draws upon regional assets to produce competitiveness, the value of this becomes questionable in the era of globalization' (Bathelt (2006) *PHG* **30**, 2).

regional science An interdisciplinary study which concentrates on the integrated analysis of economic and social phenomena in a regional setting. It seeks to understand regional change, to anticipate change, and to plan future regional development, and draws heavily on mathematical models. See Bailly and Gibson, and Boyce, both (2004) *Papers Reg. Sci.* **83**, 1.

regolith A general term for the unconsolidated, *weathered, broken rock debris, mineral grains, and superficial deposits which overlie the unaltered bedrock. The depth of the regolith varies with the intensity and duration of the weathering process; within the tropics it may be hundreds of metres deep. Soil is simply regolith with added organic material.

rehabilitation ecology The establishment of a community which is similar to the original when it is impossible to restore a site to its original condition. Soil and water loss may be controlled effectively by planting trees, shrubs, and grasses on the slopes (Xu et al. (2004) *Env. Sci. & Policy* **7**). See Christian and Wilson (1999) *Ecology* **80**.

Reilly's law The principle that the flow of trade to one of two neighbouring cities is in direct proportion to their populations and in inverse proportion to the square of the distances to those cities (W. Reilly 1929). See Brodsky (2003) *Prof. Geogr.* **55**, 2.

rejuvenation The renewed vigour of a once active process. The term is generally applied to streams and rivers which regain energy due to the uplift of land through *isostasy or by a fall in the base level (Schlunegger and Schneider

(2005) *Geomorph.* **69**, 1–4). Rejuvenation may also apply to tectonic movements (Valdiya (1993) *Current Sci.* **64**).

relationality Relational means indicating relation(s) or relationship; concerning the way in which two or more people or things are connected. It would follow, therefore, that relationality means 'being related to'; 'in a relationship'; 'connectedness'; but try Yeung (2005) *TIBG* **30**, 1, who describes relationality as an 'essential quality embedded in an iterative process of drawing interconnections between two or more discrete categories and phenomena that may not necessarily be binaries'.

Smith et al. (Open University) shed some light: 'one of the ways that geographers explore relationality is not simply by making assertions or pronouncements, but by actually tracking the practices through which geographically extensive linkages are established and sustained. Following flows through specific networks and nodes, researchers "flesh out" the causal chains that implicate distant lives.' Hopkins and Pain (2007) *Area* **39**, 3 also help: 'relationality does not just pose interesting questions about age, but marks a fundamental change in the way we approach and think about it, focusing more widely on families, generations and interactions'; and Sparke (2008) *PHG* **32**, 3 writes that local regions and places should be studied 'with attention to the ways in which they are interarticulated with global space-spanning ties and traces'.

(((⬡))) SEE WEB LINKS

• Smith et al. on interdependence.

relational space A perception of space as unbounded and formed through non-territorial flows and networks. Ash (2004) *Geografiska B* **86**, 1 sees regions as 'places of overlapping—but not necessarily connected—relational networks'. Hudson, too (2005, *PHG* **29**, 5), sees regions as 'constituted from spatialized relations stretched out over space and materialized in various forms, and representational narratives'. Jones (2005, *Reg. Studs Ass. Conf.*) points out that regions and networks should not be seen as non-spatial and without 'geographical anchors', nor should territories and scales be viewed as closed and static. There have been criticisms. Yeung (2006) *W. Paper Series* 2006-15 stresses that regions 'are not some kind of autonomous actors capable of effecting spatial change', and Tomany (2007) *Env. & Plan. D* **25**, 2 challenges those advocates of the relational region 'who see places as simply the local articulation of global flows and who present any concern with local culture and identity as atavistic and archaic'.

relative humidity (U) The ratio of the actual vapour density (which indicates the amount of water vapour present in the air) to the theoretical maximum (saturation) vapour density at the same temperature, expressed as a percentage. This may be expressed as:

$$U = 100 \; e/e'w$$

where *e* is actual vapour pressure and *e′w* saturation vapour pressure with respect to water at the same temperature.

Saturated air has a relative humidity of 100%. Air with a relative humidity in excess of 100% is said to be **supersaturated**. Relative humidity varies both diurnally, with a dawn maximum and an afternoon minimum (Priante-Filho et al. (2004) *Glob. Change Biol.* **10**, 5), and, less conspicuously, annually, both variations being in opposition to the pattern of temperatures.

relaxation time The time taken by a geomorphological system to adjust to a sustained change in the nature/intensity of an external process. For the Yellow River, Wu et al. (2008) *Geomorph.* **100**, 3–4 find a mean relaxation time of about five to six years for channel adjustment.

relict landform A geomorphological feature which was formed under past processes and climatic regimes but still exists as an anomaly in the changed, present-day conditions. Only a few very young landforms are the result of currently operating geomorphic processes (Bloom (2002) *Geomorph.* **47**, 2–4). See Stroeven et al. (2002) *Geomorph.* **44**, 1–2 on tors, boulder fields, and weathering mantles in north-eastern Sweden, and Gutiérrez-Elorza et al. (2005) *Geomorph.* **72**, 1–4 on blow-outs and playas in the Duero Depression, Spain. **Relict surfaces** contain information on past surface processes and long-term landscape evolution (Goodfellow et al. (2008) *Geomorph.* **93**, 3–4).

relief rain *See* OROGRAPHIC PRECIPITATION.

relocalizaton The preservation of local values and services; an alternative to *globalization. See C. Hines (2000). DuPuis and Goodman (2005) *J. Rural Studs* **21**, 8 hold that 'resistance to the agro-food distanciation' is in some key ways at the core of relocalization appeals within local food systems; see also Feagan (2007) *PHG* **31**, 1. See Hendrickson and Heffernan (2002) *Sociol. Ruralis* **42** on opening spaces through relocalization. Renting et al. (2003, *Env. & Plan. A* **35**) argue that advocacy for shortened food chains is directly tied to respatialization and localization.

rematerialization The grounding of geographical analysis in the concrete world of actual physical (Jackson (2004) *TIBG* **29**, 2); 'the way that the "material" and the "social" intertwine and interact material in all manner of promiscuous combinations' (Bakker and Bridge (2006) *PHG* **30**, 1). Kearnes (2003) *Soc. Cult. Geog.* **4**, 2 takes issue with the rematerialists, particularly with Jackson's assumption that matter is a universally undifferentiated conditionality. He argues, instead, that social and cultural geography must account for the 'wayward expressiveness of matter'.

re-memory *See* MEMORY.

remote sensing The scanning of the earth by satellite or high-flying aircraft in order to obtain information about it. Wooster (2007) *PPG* **31**, 1 provides an updated situation report with regard to some of the more recent sensors and systems, and a brief tour of their current applications.

rendzina An *azonal soil rich in humus and calcium carbonate, developed on limestone. The A *horizon is dark and calcareous, but usually thin; the B horizon absent, and the C horizon is chalk or limestone.

renewable resource A recurrent *resource which is not diminished when used but which will be restored, such as *wind energy. Renewable resources may be consumed without endangering future consumption as long as use does not outstrip production of new resources, as in fishing. In principle, wood is a renewable resource, but in the absence of well-planned management, short-term exploitation can induce environmental impacts or conversion to other uses yielding results better likened to mining than sustainable use (Berlik et al. (2002) *J. Biogeog.* **29**). The EU proposes a mandatory target of 20% for renewable energy's share of energy consumption by 2020 and a mandatory minimum target of 10% for biofuels. It also proposes creating a new legislative framework to enhance the promotion and use of renewable energy.

((())) SEE WEB LINKS
• The Renewable Energy Roadmap.

rent gap The gap between the actual rent paid for a piece of land and the rent that could be collected if the land had a 'higher' use. The idea is central to, but does not entirely explain, *gentrification; see N. Smith (1996).

rent gradient The decline in rents with distance from the city centre, reflecting the cost of transport from the outlying districts to the centre. See Wu in L. Ma and F. Wu (2005).

replacement rate The fertility rate required to hold the population constant in the absence of immigration or changes in longevity (Greenspan (2004) *Pop. & Dev. Rev.* **30**, 4).

representation The ways in which meanings are formed, conveyed, and shared among members of social groups. These representations can be defined as *culture, and cultural forms, notably language, and to some extent shape the reality they represent. Given that it is a working, creative, language-driven process, representation cannot be neutral or without power relations. Del Casino and Hanna (2000) *PHG* **24**, 1 argue that tourism maps and other representations play an important role in the production of tourism spaces. Bhattacharyya in M. Balshaw and L. Kennedy, eds (2000) considers the many difficult stories of Birmingham and the city's desire to reshape its image. *Landscapes may also be representations, from which people choose, rearrange, and specify their meanings; 'representation frequently draws on cultural imaginings of places and a past that may never have existed . . . such representations are responsive to the geographical and historical specificities of the context within which they are created and consumed' (Markwick (2001) *Geog.* **86**, 1).

research and development (R and D) Research on new discoveries for industrial processes. Rutherford (2004) *PHG* **28**, 4 observes that, while R and D *transnational corporations remain home-nation oriented, and in many cases locally embedded, there are forces eroding this orientation. 'TNCs may innovate domestically, but the need for maximizing returns on investment means that R+D will spill over internationally. Home base has become insufficient to generate technological advantage.' In 2004, 1.9% of GDP was spent on research and development in the EU-25 (1.95% in the EU-15), compared to 2.6% of GDP (NFS data) for the USA (Crescenzi et al. (2007) *J. Econ. Geog.* **7**, 6).

reserves The proportion of resources, notably mineral resources, which can be extracted using the prevailing technology.

residential differentiation (residential segregation) The evolution of distinct neighbourhoods, recognizable by their characteristic socio-economic and/or ethnic identity. *See also* SEGREGATION; ETHNIC SEGREGATION.

residential mobility Newbold and DeLuca (2007) *Urb. Geog.* **28**, 7 suggest that whereas mobility is high, new arrivals primarily remain in their initial destination with little difference in the overall distribution. Residential moves are associated with various individual and household characteristics, along with neighbourhood effects and the type of housing initially occupied.

resistance In hydrology, the friction force counterbalancing the downslope gravitational force (L. B. Leopold et al. 1964). Resistance is a function of the surface area in contact with the flow, and so flow becomes more efficient when the ratio of the flow cross-sectional area A to the wetted perimeter P is high (defining the *hydraulic radius R). See Smith et al. (2007) *PPG* **31**, 4 on applying flow resistance equations to overland flows.

resource An available supply of something that is valued because it can be used for a particular purpose, usually to satisfy particular human wants or desires (C. Park 2007). Giordano (2003) *AAAG* **93**, 2 develops a typology classifying common resources into one of three categories—open access, fugitive, and migratory—based on spatial relationships between resources and resource users. *See also* NATURAL RESOURCE.

resource allocation The assessment of the value of a resource or of the effects of exploiting a resource. Spatial decision support systems (SDSS) are designed to make complex resource allocation problems more transparent and to support the design and evaluation of allocation plans (Jeroen et al. (2003) *Trans. GIS* **7**, 2). K. Kyem et al. (2001) *Trans. GIS* **5**, 2 describe a decision support algorithm for managing resource allocation problems involving competing interests.

resource-frontier region A newly colonized region at the periphery of a country which is brought into production for the first time. See Gwebo (2006) *Tijdschrift* **97**, 4 on Botswana; *see also* CORE–PERIPHERY MODEL.

resource management The allocation and conservation of *natural resources. The main emphases are on: an understanding of the processes involved in the exploitation of resources; the analysis of the allocation of resources; and the development and evaluation of management strategies in resource allocation. Sustainable development and environmental protection are major goals; see Barton (2006) *Asia Pac. Viewpt* **47**, 3 on fishery resource management in Chile and New Zealand. See Johnson et al. (2004) *Nat. Res. Forum*, **28**, 3 on gender analysis in resource management *See also* NATURAL RESOURCE MANAGEMENT.

response time The speed of a reaction to a change in environment. See Dyke and Peltier (2000) *Geomorph.* **32**, 1–4 on response times and sea-level changes. For land ice, the dynamic response time is proportional to the size of the ice mass (W. Paterson 1994). J. Bamber and A. Parsons (2004) have a table of estimated response times for various components of the climatic system. *See also* REACTION TIME; RELAXATION TIME.

restructuring A change in the economic make-up of a country. It may involve: reordering production to achieve *economies of scale; a switch of investment from one sector to another (*see* DEINDUSTRIALIZATION); a change in the spatial distribution of industry; or a change in the economic system; see D. Harvey (2005) on the restructuring of capitalism. 'The tendential "ecological dominance" of neoliberal restructuring strategies . . . has profoundly disrupted inherited scalar hierarchies, while setting in train new rounds of creative institutional destruction across multiple scales' (Peck and Theodore (2007) *PHG* **31**, 6; see also N. Brenner et al. (2005) and Jessop (2000) *Int. J. Urb. & Reg. Res.* **24**). 'Firms emphasizing the social at the expense of the material may require dramatic restructuring during times of crisis when debt, for example, becomes unwieldy' (Ettlinger (2008) *PHG* **32**, 1). See Clarke et al. (2006) *Env. & Plan. A* **38** on retail restructuring and consumer choice. Restructuring may be imposed by an authority, like the World Bank, to improve the ailing economy of a nation which is in debt to that authority. *See also* STRUCTURAL ADJUSTMENT.

retailing, geography of The study of spatial patterns of retail and consumer behaviour. This includes the analysis of retailing within a city; hierarchies of retail centres based on *central place theory, and the relationship between out of town malls and city shopping centres. Models are used to forecast retailing and consumer decisions, mostly at the intra-urban scale. These methodologies treat space as neutral, or independent.

In contrast, the '**new economic geographies of retailing**' (N. Wrigley and M. Lowe eds, 1996) view the relations between space and retail activity as mutually constitutive. In this way, retail capital can structure spaces, defining our urban spaces with shopping streets, markets, and malls, but is also configured by socio-spatial processes. Key themes include: the reorganization of working corporate structures, retailer–supplier interfaces and labour; the social relations of production; the organization and technology of retail distribution; and the workings of retail capital. 'The potential of retail geography is that categories such as "economy" or "culture" are constantly being shattered. The two seem mutually implicated' (Nick Blomley, Simon Fraser University).

In the developed world, it has becoming increasingly difficult to distinguish between consumption and entertainment: 'not only is shopping melting into everything, but everything is melting into shopping' (S. T. Leong 2001). Retailing is now part of the expression, construction, and contestation of identity; consumption may define who we are. At the same time, much of current thinking is that consumption is intrinsically evil, and morally corrupting.

Perhaps the supreme incarnation of shopping as leisure is the out-of-town shopping centre/shopping mall. Malls were built on a large scale and pioneered by Victor Gruen (1973), who aimed to 'restore the lost sense of commitment and belonging . . . [to] counteract the phenomenon of alienation, isolation and loneliness and achieve a sense of identity'. Shopping malls offer consumers a safe and climate-controlled alternative to the perceived dangers and unpredictability of city centre shopping, transforming our public spaces into sanitized, privatized, serially reproduced zones. G. Ritzer (2004) argues that 'these malls, corridors, consumers and shops could be almost anywhere—Los Angeles, Singapore, Moscow, Rio or Johannesburg', and M. Sorkin (1992) claims that 'the globalisation of retailing is creating the ageographical city, a city without place attached to it'. 'The mall is essentially a pseudoplace which works through spatial strategies of dissemblance and duplicity' (J. Goss (1991) *AAAG* **81**, 1). See Birkin, Clarke, and Clarke (2002).

retail TNC 'The conceptualization of the retail *TNC draws attention to the necessarily high levels of territorial *embeddedness which retail TNCs must invest in, and to the fact that the network relationships of the retail TNCs are simultaneously shaped by the institutional/regulatory/cultural contexts of the home economies/societies from which they have emerged, and the host economies/societies that they enter in the emerging markets' (Wrigley and Lowe (2007) *J. Econ. Geog.* **7**, 4). Currah and Wrigley (2004) *Glob. Netwks* **4**, 1 argue that the competitiveness of the retail TNC 'increasingly rests upon its ability to adapt the portfolio of retail formats to different and rapidly changing business environments by mobilizing and blending knowledge from multiple locations'.

re-territorialization The restructuring of a territory/place or that has experienced *deterritorialization. See Bunnell et al. (2006) *Int. J. Urb. Reg. Res.* **30**, 1 on the complex re-territorializations on the Indonesian island of Bintan.

return flow Water which has seeped through the soil as *interflow but which backs up the hillslope when it has reached a saturated layer. See Guebert and Gardner (2001) *Geomorph.* **39**, 3–4.

return period The length of time between events of a given magnitude.

reurbanization The development of new homes, businesses, and community facilities within existing urban areas. Ogden and Hall (2001) *TIBG* **29**, 4 provide strong evidence for re-urbanization in ten French cities, 1990–1999. Pietry (2000) *Cybergeo: écon. et statis.* **336** attributes some of this to the preference of French young people for city-centre living.

revegetation Re-establishing and developing plant cover, either by artificial or natural means; see, for example, Li et al. (2007) *Geomorph.* **88**, 3–4. Owen et al. (2001) *J. Biogeog.* **28**, 5 report that most revegetation following anthropogenic and/or environmental interference is through vegetative reproduction. One of the key components of any restoration programme is to adequately and efficiently identify potential sites for revegetation. As with any restoration activity, economic cost plays a major role (Lawson et al. (2007) *Geogr. Res.* **42**, 4).

Reynolds number (R$_e$) Four factors combine to determine whether the flow of water within a channel is *turbulent or *laminar: the density, velocity, and viscosity of the water, and the hydraulic radius of the channel. Since the density of water is 1, the Reynolds number expresses this combination as:

$$R_e = \frac{VR}{\mu}$$

where V = velocity of the liquid, R = hydraulic radius, and μ = viscosity of the liquid. The Reynolds number is a dimensionless quality; in streams, the maximum number for laminar flow is between 500 and 600, depending on temperature, and at high Reynolds numbers, above 2000 to 25000, flow is turbulent. Smith et al. (2007) *PPG* **31**, 4 note that the flow is laminar for Reynolds numbers below 2300 (in circular pipes), and above 4000 in turbulent. 'An ill-defined region between these two limits is known as the transitional zone. The critical Reynolds number transition to turbulent flow depends on the exact flow configuration and should be determined experimentally.'

ria The seaward end of a river valley which has been flooded as a result of a rise in sea level. The name is from the type location in Galicia, Spain.

ribbon development A built-up area along a main road running outwards from the city; a location that combines cheaper land with high accessibility, and the chance of attracting trade from passing traffic. See Milward (2003) *Canad. Geogr./Géog. canad.* **46**, 1 and Murdoch and Lowe

(2003) *TIBG* **28**, 3 on the campaign by the Council for the Preservation of Rural England (CPRE) against ribbon development.

Richter scale A scale of the *magnitude of earthquakes, ranging from 0 to (in theory) 10. Richter (1935) *Bull. Seism. Soc. Am.* **25** equated the magnitude of an earthquake to the base 10 logarithm of the ground motion in millimetres, measured on a certain type of seismograph, plus a correction factor related to the distance of the earthquake. The distance is calculated from the difference in arrival time for different types of waves that travel at different speeds. So, for a constant distance between an earthquake's hypocentre and the seismograph, the ground motion has to increase by a factor of 10 to cause an increase of 1 on the Richter scale. Professor Richter was trying to characterize the energy of a seismic event, not damage. Newer measurements, referred to as the 'seismic moment' and 'moment magnitude', have been developed to address some of the Richter scale's shortcomings (Johnson (2004) US Dept Energy). *See also* MERCALLI SCALE, for measurements of earthquake *intensity.

ridge and furrow A set of parallel ridges and depressions formed during the period of strip cultivation in the Middle Ages. As the land was ploughed always to the same pattern, the plough threw up earth to make ridges which often survive in the present landscape. See Whittington and Brett (1979) *J. Hist. Geog.* **5**, 1. Ridge and furrow landscapes are still clearly visible in the landscapes of midland England.

rift valley A long strip of country let down between normal faults, or between a parallel series of step faults (also called a *graben*). There is a maximum rift valley relief, controlled by the competition between isostasy and lithospheric geometry (Quin and Buck (2005) *J. Geophys. Res.* **110**, B03404). The biggest terrestrial rift valley system is the East African system, at 3000 km long. *Plate tectonic theory suggests that rift valleys are the result of large-scale doming above a mantle *plume, followed by fracturing along the crest of the dome as plates diverge (Moucha et al. (2008) *Geology* **36**, 6). Many rifts, like the Rhine rift valley, have the Y-shaped pattern characteristic of a triple junction, indicating that they arise from plate separation; see Horner-Johnson et al. (2003) *Eos* (*Trans. Am. Geophys. Union*) **84**, 46.

right to the city Writings on 'the right to the city' stress the intrinsic worth of cities to their inhabitants, not merely their instrumental or economic value. 'The right to the city is the right to make full use of the city and to live a richly urban life' (Painter (2005) *ICRRDS, U. Durham*). Varsanyi (2008) *Urb. Geog.* **29**, 1 draws heavily upon the legal geographical literature on 'the right to the city' to explore the way in which cities are deploying local public space and land use ordinances to constrain the behaviour of undocumented day labourers seeking work and living within their jurisdictions.

rills Small *channels, between 5 and 2000 mm in width, and very closely
spaced. They develop well in areas with heavy rainfall, especially upon weaker
rocks, such as volcanic ash. Sediment yields increase with slope gradients and
flow rates (the former having more effect than the latter) and greater flow rates
cause more rill erosion and soil loss under the same slope gradient (Lei et al.
(2006) *Eur. J. Soil Sci.* **57**, 3). Torri et al. (2006) *Geomorph.* **76**, 3–4 find power
relationship exists between width and total flow discharge in rills. Rills may
widen and deepen to form *gullies.

rime *See* FROST.

ring dyke *See* DYKE.

riparian Relating to a river bank. Riparian vegetation dynamics are driven by
allogenic hydrogeomorphological factors, with autogenic (plant-induced)
influences affecting both plant dynamics and the river environment from the
earliest stages of plant establishment, and becoming more important as
landform stability is achieved (Francis (2006) *Area* **38**, 4).

rip current A strong current moving seawards in the near-shore zone. Munk
(1949) *Ann. NY Acad. Sci.* **51**, 3 suggested that rip currents were formed by
onshore wave mass transport and that water piled up against the beach, which
provided a hydraulic head for a rip current (an offshore return flow). Brander
and Short (2001) *J. Coast. Res.* **17**, 2, explain velocities in a rip channel by the
continuity equation, in which the discharge out of the rip channel is equal to
incoming wave-induced transport over the transverse bar. See also MacMahan
et al. (2005) *Marine Geol.* **218**, 1–4.

rising limb The section of a *hydrograph from the start of increased
discharge to the maximum flow.

risk The likelihood of possible outcomes as a result of a particular action or
reaction. Technically speaking, the likely outcomes of risks can be assessed as
a series of different odds, while there is no calculation of probabilities in
*uncertainty. 'Social scientists have long argued that risk is socially
constructed . . . the interpretation of physical threats is not just a subjective
process engaged in by individuals but is also strongly affected by mores,
norms, values, institutions, and other influences on choice that are held in
common by members of social groups. By these means the almost infinite
number of physical risks that inhabit our world is prioritized to facilitate
collective action' (Mitchell (2007) *AAAG* **97**, 2). See J. Kasperson and
R. Kasperson (2005). Herrick (2005) *Area* **37**, 3, in a study of risk perception
and GM food, writes that 'what science deems to be an acceptable level of risk
may not match the social perception of acceptability. When the differences
between social and scientific notions of risk become acute, then the outcome is
a "social amplification of risk" by the public.' Wakefield and Elliott (2003)
Prof. Geogr. **55**, 2 suggest that risk messages are chosen and shaped by

journalists on the basis of the pressures they themselves are under, and that, while newspapers were a major source of risk information, their impact was mitigated by readers' distrust, and access to their own personal information networks. Watson and Stratford (2008) *Soc. & Cult. Geog.* **9**, 4 recognize three socio-spatial orderings of risk (displacement, replacement, and reorientation). **Risk analysis** identifies the level of hazard in an area, and estimates the probability of occurrence of future risk (Chung and Fabbri (2008) *Geomorph.* **94**, 3–4). See Hungr et al. (2008) *Geomorph.* **96**, 3–4 on **quantitative risk analysis**. *See also* UNCERTAINTY.

river 'Rivers are truly perceived not as "things in space" but as "processes through time"' (Goodwin (1999) *Stream Notes*).

river basin management 'The process of coordinating conservation, management and development of water, land and related resources across sectors within a given river basin, in order to maximise the economic and social benefits derived from water resources in an equitable manner while preserving and, where necessary, restoring freshwater ecosystems' (Global Water Partnership (2000) *Backgr. Paper* **4**). Restoration, rehabilitation, river basin management, and their derivatives are practised in at least 21 different countries in response to the exploitation and subsequent deterioration of the riverine environment (Wheaton (2006) *Area* **38**, 2). 'Interventions on hydro/ecological systems by different categories of stakeholders characterized by different political, decision-making, and discursive power, and varied access to resources, tend to generate costs, benefits, and risks that are distributed unevenly across spatial and temporal scales and across social groups. This is due to the interconnectedness of users through the hydrologic cycle entailed by their dependence upon the same resource' (Molle (2007) *Geog. J.* **173**, 4). See Hirsch and Wyatt (2004) *Asia Pac. Viewpt* **45**, 1 on the Se San River

river capture *See* CAPTURE.

riverine Relating to, similar to, or formed by, a river. Ward et al. *Freshwater Biol.* **52**, 1 argue that hydrological connectivity (the exchange of matter, energy, and biota via water) plays a major role in sustaining **riverine landscape diversity**.

river restoration Assisting the recovery of ecological integrity in a degraded watershed system by re-establishing hydrologic, geomorphic, and ecological processes, and replacing lost, damaged, or compromised biological elements (Wohl et al. (2002) *Landsc. Ecol.* **17**). Restoration activities include: cutting a new, meandering course, creating stepped banks to provide diverse ecological niches, removing dams to restore fish movement, reconnecting flow by cutting linking channels, setting back or breaching levees, and dredging. McDonald et al. (2004) *TIBG* **29**, 3 argue that the purpose of restoration has shifted from simple utilitarian needs for flood and erosion/sedimentation control towards the incorporation of ecological concerns.

See Millington and Sear (2007, *Earth Surf. Procs Landforms* **32**, 8) on the geomorphic and ecological effects of reintroducing woods in river restoration schemes.

((⊕)) SEE WEB LINKS

- The UK River Restoration Centre has case studies.
- The European Centre for River Restoration has a useful bibliography.

river terrace A bench-like feature running along a valley side, roughly parallel with the valley walls. Bridgland and Westaway (2007) *Geomorph.* argue that uplift is essential for the formation of river terraces; 'their cyclic formation, however, almost invariably seems to have been a response to climatic fluctuation'. Bridgland (2002) *Quat. Sci. Revs* **19**, 13 suggests a six-stage model:

1. the incision phase in which terrace generation occurs, occurring at the transition to interglacial conditions (discharge is high as a result of melting permafrost);
2. an aggradation phase, again at the glacial–interglacial transition, seen mainly in the lower reaches of valleys;
3. the interglacial phase, in which fine-grained sedimentation (rarely preserved) is predominant;
4. a further phase of incision at the interglacial–glacial transition;
5. the main aggradational phase, at the interglacial–glacial transition, as a result of considerable sediment being liberated by the decline of vegetation;
6. a phase of glacial climate during which there is relatively little activity, much of the potential discharge being locked up in permafrost.

river training The attempt to 'flood-proof' a river/channel. This may include the straightening, narrowing, and shortening of a river course, removing river gravels, installing new flood banks and river-bank walling, and improving floodplain drainage (McDonald et al. (2004) *TIBG* **29**, 3). See Korpak (2007) *Geomorph.* **92**, 3–4 on the geomorphological impacts of river training in mountainous channels.

road pricing A strategy to reduce *urban *congestion, first used in Singapore in 1978. An electronic road pricing (ERP) system can electronically monitor, and track, vehicles entering a restricted zone to control the flow of inbound traffic. The system is capable of automatically imposing a demand-sensitive congestion toll on every vehicle without requiring them to slow down or stop, when the congestion level in the restricted zone exceeds a preferred threshold level.

The ERP system was first tried out in Hong Kong from July 1983 to March 1985, but not implemented due to public rejection, arising from concerns over the privacy of movement. In Singapore, on the first day of its implementation (1 April 1998), the usual morning rush hour traffic along one of the heavily congested highways decreased by 17% (Goh (2002) *J. Transp. Geog.* **10**, 1). Road pricing varies in coerciveness depending on households' economic resources, and the public perceive road pricing as unfair, given that wealthy households are not forced to adapt (Jones in J. Schade and B. Schlag, eds 2003). Jakobsson et al. (2000) *Transp. Policy* **7**, 2 find that lower-income groups find road pricing less acceptable than high-income groups because it affects their perceived freedom and justice negatively.

rock avalanche The mass movement of a minimum volume of 500 000 m³ of rock (Scheidegger (1973) *Rock Mechanics* **5**) derived from the collapse of competent masses of coarse-grained, hard bedrock. Hewitt (2006) *PPG* **30**, 3 examines rockslide-rock avalanches in mountain landscapes, and the landforms associated with them, backed up with useful pictures and really clear explanations.

rock fall *See* FALL.

rock flour Silt- and clay-sized particles of debris formed from grinding due to *abrasion within and at the base of a glacier, and carried by *meltwater streams.

rock glacier A very slowly moving river of angular rock debris, characteristic of high mountain permafrost and made up of debris and ice that creeps on hillslopes at a typical rate of a few decimetres per year (Serrano et al. (2006) *Geomorph.* **74**, 1–4). Ikeda and Matsuoika (2006) *Geomorph.* **73**, 3–4 distinguish between **boulder rock glaciers**, with an active layer of matrix-free boulders derived from crystalline rocks, and massive limestone; and **pebbly rock glaciers**, made of matrix-supported debris derived from less resistant shales, and platy limestone.

Rossby waves Long ridges and troughs in the westerly movements of the upper air, with a wavelength of around 2 000 km, discovered by C. J. Rossby in 1939. Four to six waves girdle the Northern Hemisphere at any one time. Some are a response to relief barriers, like those east of the Rockies, and east of the Himalayas (O'Kane and Frederiksen (2005) *ANZIAM J.* **46** (E), C704–C718).

Rossby waves are thought to be a reaction to the unequal heating of the earth's surface, and are intimately connected with the formation of *cyclones (Li et al. (2006) *J. Atmos. Scis* **63**, 5), *anticyclones (Chen and Newman (1998) *J. Climate* **11**, 10), and *mid-latitude depressions. As air in the middle latitudes of the Northern Hemisphere travels west–east into a trough, it slows down, and piles up, causing *convergence just ahead of the ridge which follows. Convergence in the upper air causes a downflow to the ground, creating high pressure systems at ground level, below and just ahead of troughs in the Rossby waves. As air leaves the trough, and its passage straightens out, air speeds pick up, and the air moves very fast as it swings round the outer arc of the ridge. Air then diverges just ahead of the next trough. *Divergence in the upper air causes low pressure systems at ground level below and just ahead of ridges.

At times, the waves are few, and shallow; a pattern known as a high zonal index. In other cases, the flow becomes markedly *meridional; a pattern known as a low *zonal index. Upon occasion, the waves break down into a series of cells (**Rossby wave-breaking**; *see* BLOCKING). Woollings and Hoskins (2008) *J. Atmos. Scis* **65**, 2 suggest that the low-frequency variability of the *North Atlantic Oscillation results from variations in upper-level Rossby wave-breaking events.

Rossby wave

rotational slip The semicircular motion of a mass of rock and/or soil as it moves downslope along a concave face. Evidence for rotational slip in *cirques comes from the dirt bands observed in cirque ice, which become progressively steeper from the back wall, but then flatten towards the cirque mouth.

r-selection, k-selection Two major strategies may be adopted for the survival of a species. r-selected plants (where r stands for maximum increase) respond swiftly to favourable conditions with most of their energies devoted to rapid maturity and reproduction. *See* OPPORTUNIST SPECIES. k-selected plants survive by putting their energies into persistence. *See* EQUILIBRIUM SPECIES.

rubification The change of soil colour to yellow or red. In warm climates, intense weathering liberates iron, which, combined with some *lessivage, rubifies the soil.

rudaceous Coarse-grained sedimentary rock, either consolidated as in *conglomerate or unconsolidated as in *till.

rugosity Roughness, wrinkliness; see Donovan (2002) *Geol. Today* **18**, 4 on rugosity in karst landscapes, and Knudby et al. (2007) *PPG* **31**, 4 on coral reefs. In ecology, more rugosity will provide more available habitat for colonization, foraging, and shelter.

(((●))) SEE WEB LINKS

• Monitoring of coral reef ecosystems.

run-off The movement over ground of rain water.

rural In, of, or suggesting, the country. Bell (2007) *J. Rural Studs* **23**, 4 distinguishes between 'first rural' (basically, low population density), and 'second rural'—'a rural of associations. 'It calls upon the connections we have long made between rural life and food, cultivation, community, nature, wild freedom, and masculine patriarchal power, and the many contradictions we have also so long associated with the rural, such as desolation, isolation, dirt and disease, wild danger, and the straw-hatted rube.' Halfacree (1995) *J. Rural Studs.* **11**, 1 makes the case for defining 'the rural' as 'an abstract social representation; a set of rules and resources existing out of space and time which are drawn upon in both discursive and non-discursive actions. The precise form that this representation takes in these actions is highly contextualised and depends upon its precise usage.' **Rurality**, through its strong cultural entanglement with 'nature', is seen as a repository of more embedded and authentic values (see R. Mabey 2005). McCarthy (2006) *PHG* **29**, 6 argues that 'rurality' 'remains at least as much a product of divisions of labor within the academy and social contexts as a category defined by particular sectoral mixes, land uses, densities, or other empirical descriptors'. M. Woods (2005) concludes that 'it seems that the real power in the British countryside is the very idea of rurality itself'.

Halfacree in P. Cloke et al., eds (2006) notes that *productivist agriculture was to be the essence of 'modern' rurality. Tonts and Grieve (2002) *Australian Geog. Studs* **40** examine the destructive tendencies associated with the **commodification of rurality** in some of Australia's more scenic and accessible rural areas. See Bell (2000) *Rural Sociol.* **65**, 4 and Little (2002) *PHG* **26**, 5 on rurality and sexuality.

rural depopulation The decrease in population of rural areas, by migration, or through falling birth rates as young people move away. In 2004, Ioffe et al. (*AAAG* **94**, 4) reported that, as a result of farmland abandonment, no fewer than 20 million hectares of arable land were already deserted in European Russia. However, Stockdale (2002) *Int. J. Pop. Geog.* **8** asserts that the historical trend of rural depopulation is being replaced by a repopulation of the countryside.

rural geography The study of the rural landscapes of the developed world. It includes the origin, development, and distribution of rural settlement, *rural depopulation, the causes and consequences of agricultural change, patterns of recreational use of the countryside, tourism, planning, and the growing influence on rural areas of urban dwellers. For a review of the sub-discipline, see Roche (2005) *PHG* **29**, 3. McCarthy (2006) *PHG* **29**, 6 argues that recent developments 'call into question the geographic categories and concentrations of rural geography'.

rural–urban continuum The belief that between the truly rural and the truly urban are many 'shades of grey'; if we actually look along a scale from the single isolated farm all the way to the *megalopolis, we do not find any clear boundaries between hamlets, villages, towns, and cities. Sheppard and Nagar (2004) *Antipode* **36**, 4 state that 'it is important to reconceptualize the urban as rural–urban continuum in attending to processes of domination, co-optation and resistance, especially in Asia and Africa where populations are predominantly rural and historical patterns of colonial and postcolonial capitalist development and labor migration connect the lives in the urban, peri-urban, and rural areas in intricate and inseparable ways'.

rural–urban fringe The transition zone between the city and its suburbs, and the countryside. Certain types of land use are characteristic of this zone: garden centres, country parks, riding stables, golf courses, sewage works, and airports are common, and these are neither truly urban nor truly rural uses. They do, however, give an urban air to the countryside, which can be cited as an argument for further development—since the zone is not really 'countryside', it need not be preserved. Thus, C. Bryant et al. (1982) define the **inner urban fringe** as 'land in the advanced stages of transition from rural to urban uses—land under construction, land for which subdivision plans have been approved—in short, land where there is little doubt over much of its area about its urban oriented function and ultimate conversion to urban uses'.

Irwin and Bockstaehl (2002) *J. Econ. Geog.* **2**, 1 think that fragmented patterns of development in rural–urban fringe areas could be due to negative externalities that create a 'repelling' effect among residential land parcels. Lagarias (2007, *Cybergeo, systèmes, modél., géostat.*, **391**) presents a fractal analysis of urbanization at the periphery of Thessaloniki, Greece. Bunker and Houston (2003) *Australian Geog. Studs* **41**, 3 report that the fringe is becoming increasingly complex due to multifaceted demographic change, a broadening economic base, and demands for better environmental management, 'all within the context of an evolving understanding of sustainability'. Qviström (2007) *Geografiska B* **89**, 3 objects to the characterization of the rural–urban fringe as a *process* rather than a *place*; questions representations informed by a dichotomous way of thinking; and hopes to 'bring forward more nuanced accounts of fringe landscapes'.

rustbelt *See* SNOWBELT.

Sahel The **Sahelian zone** borders the southern Sahara. Vegetation—scattered grasses, shrubs, and trees—increases in density towards the southern margin. Annual rainfall is 200–400 mm, but unreliable. Severe droughts bring losses of livestock, crop failures, and famine. See *Glob. Env. Change* (2001) **11**, 1, special Sahel issue.

saline Salty. A **saline soil** is a non-*sodic soil containing enough salt to adversely affect crop growth. Saline soils get their salts from natural, or anthropogenic *salinization. **Salinity** is the degree to which water contains dissolved salts, usually expressed in parts per thousand.

salinization The accumulation of unusually high concentrations of dissolved sulphates and chlorides of sodium and calcium, often as a result of large-scale irrigation in semi-arid areas; as irrigation water is evaporated or transpired, the salt ions are largely left behind to accumulate in the soil (Saysel and Barlas (2001) *Ecol. Model.* **139**, 2–3). Salinization can reduce plant growth rate and yield, and, in severe cases, cause total crop failure. In certain circumstances, excessive salinity in soils cannot be reduced over time by routine irrigation and crop management practices (Qadir et al. (2000) *Land Degrad. & Dev.* **11**).

In coastal areas, salinization can be associated with the over-extraction of groundwater, which can lower the water table and lead to the invasion of sea water (Cardonna et al. (2004) *Env. Geol.* **45**, 3). Tsunamis can cause severe salinization problems.

((())) SEE WEB LINKS

• European Land Management and Natural Hazards Unit on soil salinization.

saltation The bouncing of material from and along a river bed or a land surface. The impact of a falling sand grain may splash other grains upwards so that a chain of saltating particles may be set up. Saltation upwards is the result of *lift forces (see J. S. Bridge 2003); the downward movement occurs when lift is no longer effective and the particle is subject to drag and gravity. See Stout (2004) *ESPL* **29**, 10 on the relationship between saltation activity, and relative wind strength; see also Bauer et al. (2004) *Geomorph.* **59**, 1–4 on shear velocity. The **saltation–abrasion model** represents river incision by saltation; see Sklar and Dietrich (2006) *Geomorph.* **82**, 1–2.

salt dome A large mass of evaporate minerals that has pierced, and risen through, denser, overlying *sedimentary rock, forming a dome-shaped arch with salt at its core. See Autin (2002) *Geomorph.* **47**, 2 on salt domes in Louisiana.

salt marsh Tidal salt marshes are found along the low-energy coastlines of mid- to high-latitude regions of the major continents (except Antarctica), and lie at the *ecotone between terrestrial, freshwater, and marine system (Greenberg et al. (2006) *BioSci.* **56**, 8). As the marsh develops, *halophytes (such as marsh samphire and sea aster in Britain) pave the way for less hardy species, and the marsh becomes part of the coastland; see van Wijnen et al. (1997) *J. Coastal. Cons.* **3**, 1.

salt water weathering (salt weathering) A form of weathering, especially important in hot deserts, which combines crystal growth with *hydration. See Smith and Warke in D. Thomas et al., eds (1997) on the importance of the nature of salt and moisture delivery to weathering in arid areas. Salt weathering is closely linked to groundwater controls in enclosed arid depressions (Aref et al. (2002) *Geomorph.* **45**, 3–4). See Viles and Goudie (2007) *Geomorph.* **85**, 1–2 on salt weathering in coastal salt pans. Smith et al. (2005) *Geomorph.* **67**, 1–2 evaluate recent simulations of salt weathering. Landforms associated with arid salt weathering include columnar weathering networks (Goudie et al. (2002) *Zeitschrift Geomorph.* **46**) and *tafoni.

sand Particles of rock with diameters ranging from 0.06 mm to 2.00 mm. Most sands are formed of the mineral quartz. Sandy soils are loose, non-plastic, and permeable, and have little capacity to hold water.

sand dune A hill or ridge of sand accumulated and sorted by wind action. Sand will settle on a dune because the friction of the sandy surface is enough to slow the wind, which then sheds some of its *load. Dunes formed in the *lee of an obstacle are **topographic dunes**: *lunettes* form in the lee of a *deflation hollow, *nebkhas* in the lee of bushes, and **wind shadow dunes** in the lee of hills and plateaux. **Crescentic dunes**—sand mounds, barchans, barchanoids, and transverse ridges—are restricted to areas with unidirectional high winds and minimal vegetation (Bishop (2001) *ESPL* **26**).

Transverse dunes form when sand supply is abundant, and wind direction constant (Walker and Nickling (2002) *PPG* **26**, 1). As the sand supply is reduced, the dunes transform into **parabolic dunes**—hairpin-shaped with the bend pointing downwind (Mitasova et al. (2005) *Geomorph.* **72**, 1–4). Where the direction is very changeable, **star dunes** form. The linear **seif dunes** form when two prevailing winds alternate, either daily or seasonally; the leeward slope is a zone of deposition, as well as a zone of erosion (Tsoar et al. (2004) *Geomorph.* **57**, 3–4). When sand is limited, **barchans** form with the horns pointing downwind. The height of the slipface of a barchan is proportional to the width of the horns (Wang et al. (2007) *Geomorph.* **89**, 3–2). Barchans may

Sand dunes. Arrows indicate wind direction

be changed into dome dunes, and vice versa; completely gradational forms exist between the two (Fryberger et al. (1984) *Sedimentol.* **31**). **Barchanoid** dunes are undulating, continuous cross-wind dunes which may grade into long transverse dunes; see R. Cooke et al. (1993). Dune fields have recently come to be recognized as self-organizing systems that can be seen progressing from states of disorganization or randomness to uniformity; see Wilkins and Ford (2007) *Geomorph.* **83**, 1–2.

sand dune dynamics The processes of dune movement; see Wiggs (2001) *PPG* **25**, 1.

sand dune stabilization Techniques designed to prevent the erosion and deposition of sand include: the establishment of shelterbelts; mass tree revegetation using a synthetic water absorbing polymer; developing live hedges; and mechanical fencing. Shelterbelts were the most effective; mechanical fencing the least (Raji et al. (2004) *Env. Model. & Assess.* **99**, 1–3). Schwendiman (1997) *Int. J. Biomet.* **21** describes the three-stage approach in temperate climates, and Gadgil and Ede (1999) *Land Deg. & Dev.* **9**, 2 outline sand dune stabilization methods in New Zealand.

sand flux The flowing of sand. Dong et al. (2004) *Geomorph.* **57**, 1–2 find that sand flux over a sandy surface increases with height in the very near surface layer, but then decays exponentially; see also Wang et al. (2008) *Geomorph.* **96**, 1–2.

sand ramp A sloping accretion of *aeolian or *fluvial material, sometimes including *palaeosols, up- or downwind of an obstacle. Sand ramps can provide evidence of *palaeoclimates; see Thomas (1997) *Quat. Res.* **48**, 2.

sandstone A sedimentary rock made of compacted and cemented quartz and feldspar.

sandur A sheet, or gently sloping fan, of *outwash sands and gravels (pl. *sandar*); see Marren (2004) *Sed. Geol.* **164**, 3–4, and Magilligan (2002) *Geomorph.* **44**, 1. Sandar *fans are commonly found in glacial troughs (Thomas and Chiverrell (2006) *Quat. Sci. Revs* **25**, 21–2).

sand-wedge polygons *See* FROST CRACKING.

Santa Ana *See* LOCAL WINDS.

sapping The breaking down and undermining of part of a hillslope, triggering small slips. Sapping is significant in valley formation (Nash (1996) *ESPL* **21**, 9). See Twidale (2007) *Revista C. & G.* **21**, 1–2 on sapping at a river bluff. For **sapping erosion**, see Fernández et al. (2008) *Geomorph.* **95**, 3–4.

SAPRIN The Structural Adjustment Participatory Review International Network is a global network established to strengthen the organized challenge to structural adjustment programmes.

((⊕)) SEE WEB LINKS
• The SAPRIN website.

saprolite Weathered rock *in situ* formed by deep weathering to depths of tens or hundreds of metres (Ollier et al. (2007) *Geomorph.* **87**, 3).

saprophyte A largely surface-dwelling, primarily free-living, organism, using enzymes to break down dead organic matter. Most are fungi or bacteria—fungi play vital roles in the nutrient cycles forest ecosystems, acting as mutualists, parasites, and saprophytes of virtually all plants (Czederpiltz et al. (1999) *McIlvainea* **14**).

sapropel An unconsolidated mud in estuaries, lagoons, lakes, and shallow seas, its formation depending on heavier precipitation (Rohling and Hilger (1991) *Geol. Mijnbouw* **70**, and to variations in the Earth's orbit—and are thus useful in reconstructing palaeoclimates (Larrasoaña et al. (2006) *Phys. Earth & Planet. Interiors* **156**, 3–4).

sastrugi Ridges of ice particles lying across an ice sheet, formed by strong winds, and oriented at right angles to them. The sastrugi alter the surface roughness, and this alters the resistance to wind. Consequently, more drift snow builds up; this self-enhancing process thus forms larger sastrugi (Herzfeld et al. (2003) *Hydrol. Procs* **17**, 3).

satellite town Satellite towns are established around big cities to attract population surplus and control population growth. See Ziari (2006) *Cities* **23**, 6 on satellite towns in Iran.

satisficing The setting of sub-optimal targets to which people aspire. Satisficing has been widely observed among residential developers: 'for developers, the most important rule is to choose locations that meet a minimum profit criterion—a rule that results in satisficing . . . The main consequence of satisficing is that developers prefer projects on greenfield sites that take a shorter time to build and sell' (Mohamed (2006) *J. Planning Ed. & Res.* **26**).

saturated adiabatic lapse rate (SALR) *See* ADIABAT.

saturated mixing ratio lines Lines of constant humidity mixing ratio plotted against height and pressure on a *tephigram.

saturated zone overland flow (saturation overland flow) *See* OVERLAND FLOW.

saturation (saturated air) There is a constant exchange of water molecules between liquid water, or ice, and the air, as evaporation and *condensation take place. **Saturation vapour pressure** (*es*) depicts a balance in the air between condensation and evaporation. *es* is greater in water droplets than in sheet water, and lower in impure water than in pure. Where *es* in an air parcel is greater than *e* (the *ambient *vapour pressure) there will be net evaporation; where *e* is greater than *es*, there will be net condensation.

savanna Broad belts of tropical grassland flanking each side of the equatorial forest of Africa and South America. Furley (2006) *PPG* **30**, 1 reviews the determinants of savanna: climate and water, fire, soil–vegetation relationships, grazing. See also Boone et al. (2002) *Afr. J. Ecol.* **40**. The savanna belts are associated with the sinking of high-level equatorial air on its return to the *inter-tropical convergence zone. Rainfall is therefore slight, and trees are modified to minimize water loss, with small leaves, and often thorny (Chidumayo (2001) *J. Veg. Sci.* **12**).

The boundaries of the savanna are far from clear; there is a gradual change from tall grasses, 1–3 m high with scattered trees, to grassy woodland, and finally to the rain forest. However, where savanna vegetation has been repeatedly burnt (Balfour (2002) *Afr. J. Range & Forest Sci.* **19**, 1) there can be a sharp division between this and the equatorial forest, which is less easily fired.

scale A level of representation; in cartography, the ratio between map distance and distance on the ground. Geographical difference is expressed at all scales, from the inter-personal to the institutional, and from the national to the international. However, plotting geographical information on too small a scale can conceal information which only becomes apparent at smaller scales and higher resolutions. The map of US persistent poverty purports to show poverty at county level; and shows no poverty in Oregon. The map of Oregon, beneath it, also at county level, tells a different story.

Conversely an examination of geographical information at too large a scale may conceal the bigger picture. Cash and Moser (2000) *Glob. Env. Change* **10**, 2 advise on matching the scale of the assessment with the scale of management, thus avoiding **scale discordance**. Vincent (2007) *Glob. Env. Change* **17**, 1 describes the development of two empirical adaptive capacity indices for use at different scales of analysis.

'Scale is both a methodological issue inherent to observation—its epistemological moment—and an objective characteristic of complex interactions within and among social and natural processes—its ontological moment' (Sayre (2005) *PHG* **29**, 3). 'Every scale brings a new set of problems that coincide with a new nature/society interface' (Dolidon (2007) *Cybergeo: Env. Nat. Paysage* **363**). Sassen (2000) *Pub. Cult.* **12**, 1 observes that the national and the global are not mutually exclusive scales but 'overlap and interact in ways that distinguish our contemporary moment', and Biggs et al. (2007) *Ecol. & Soc.* **12**, 1 produce a dialogue on multi-scale scenarios.

The **politics of scale** describe the ways in which scale choices are constrained overtly by politics, and more subtly by choices of technologies, institutional designs, and measurements; see Lebel et al. (2005) *Ecol. & Soc.* **10**, 2, and Silvey (2004) *PHG* **28**, 4. Haarstad and Fløysand (2007) *Pol. Geog.* **26**, 3 criticize the debate on politics of scale 'for leaving several central questions relatively unexplored'.

For modelling scale in GIS, see E. Wentz (2003) and Florinsky and Kuryakova (2000) *Int. J. GIS* **12**.

scarp retreat A synonym for *parallel retreat.

U.S.: persistent poverty

Scale. Oregon: persistent poverty

schist A metamorphic rock, characteristically with broad, wavy bands, which are not *bedding planes, but sorted zones of minerals such as mica. Schists form from the metamorphosis of slates and shales, and are foliated.

science park An area of industrial development set up in collaboration with, and in close proximity to, a centre of higher education, in order for industry to capitalize on academic research, for jobs to be created locally, and for the educational institute to generate income. Knowledge from foreign companies

is also important; see Lefner et al. (2006) *Env. & Plan. A* **38**, 1 on the
Zhongguancun Science Park, Beijing.

Scirocco *See* LOCAL WINDS.

scoria A volcanic rock made of sharp rock fragments and full of air pockets
once occupied by gases. **Scoria cones** are constructed by the accumulation of
ballistically ejected clasts from discrete, relatively coarse-grained Strombolian
bursts, and subsequent avalanching (Valentine et al. (2005) *Geology* **33**, 8).

scour Erosion; see, for example, Saynor et al. (2004) *Geografiska A* **86**, 2 on
scour holes and gullying. The depth of a scour hole in a river bed is strongly
correlated with *discharge, with scouring on the rising limb of the
hydrograph, and filling as the rate of transport through the pool declines (Borg
et al. (2007) *Geomorph.* **89**, 1–2). See also Leopold in L. Leopold et al.
(1964). **Scour and fill** is the erosion and subsequent filling of a channel; see
Xu (2004) *Geomorph.* **57**, 3–4, and Fuller and Hutchinson (2007) *NZ Geogr.*
63, 3 on measuring scour and fill. For **river bed scours**, see Eilertsen and
Hansen (2008) *Geomorph.* **93**, 3–4.

scree Shattered rock fragments which accumulate below and from free rock
faces and summits. Scree is formed by *freeze–thaw; see Statham (1973) *TIBG*
59. Its formation requires jointed rocks that allow water to penetrate.

sea breeze When coastal land is warmed by the morning sun, the air above
it rapidly becomes warmer. This rising air is replaced by cool, moist air, drawn
in from above the sea. See J. E. Simpson (2007). The **sea breeze front**—the
boundary between the sea breeze flow and its opposite flow—can mark sharp
contrasts in temperature and humidity, and may move tens of kilometres
inland (Prado (1997) *Boundary-Layer Met.* **82**, 1).

sea-floor spreading The creation of new crust as *magma rises up at a
*constructive *plate margin. Sea-floor spreading can be detected through the
presence of magnetic anomalies; see Roest and Srivastava (1989) *Geology* **17**.

sea level Modern analyses of sea-level changes due to glacial *isostatic
adjustment are based on the classic sea-level equation by Farrell and Clark
(1976) *Geophys. JR Astr. Soc.* **46**; Mitrovica and Milne (2003) *Geophys. J. Int.*
154, 2 present a further equation. See Kendal et al. (2005) *Geophys. J. Int.* **161**, 3
on theoretical approaches to computing gravitationally self-consistent sea-
level changes due to ice growth and *ablation.

For something like 227000 of the last 250000 years, sea levels were around
10 m below their current stand (Voris (2000) *J. Biogeog.* **27**). Nunn (2000) *NZ
Geogr.* **56**, 1 demonstrates the 'possible and/or likely' effects of a fall in sea
level around AD 1220–1510 on Pacific islands: falling water-tables; the
emergence of reef surfaces and the consequent reduction of nearshore water
circulation; the emergence of reef islets; and the conversion of tidal inlets to
brackish lakes. Heany et al. (2005) *J. Biogeog.* **32**, 2 argue that late Pleistocene

low sea-level shorelines, rather than current shorelines, define patterns of genetic variation among mammals on oceanic Philippine islands.

N. Bindoff et al. (2007) conclude that sea levels rose at an average rate of 1.7 ± 0.5 mm yr^{-1} during the 20th century. They estimate future sea-level rise of between 0.18 and 0.59 m for the 21st century. 'These simple figures conceal substantial spatial and temporal variability' (Church et al. (2006) *Glob. & Planet. Change* 53); see Edwards (2007) *PPG* 31, 6. For many atoll island states, sea-level rise is more than just a threat to their tourism; it also determines their survival (Wong (2003) *Sing. J. Trop. Geog.* 24, 1). These countries also have low adaptive capacity to sea-level rise, heavily dependent as they are on fragile environmental goods and services, such as coral reefs and fisheries (Nurse and Moore (2007) *Nat. Resources Forum* 31, virtual issue). Sea-level rise in Tuvalu is likely to contribute to saltwater intrusion, long-term loss of land, and damage to ecosystems, agriculture, and livelihoods (J. McCarthy et al., eds 2001). See Dawson in C. Gopalakrishnan and N. Okada, eds (2008) on sea-level rise and the Thames estuary.

seamount A relatively small-scale (up to 250 km across) rise, sub-circular in plan view. The date a seamount is formed, and the tectonic setting they erupt into (on-ridge or off-ridge), reflect the way the lithosphere interacts with the fluid mantle beneath. For the radiometric dating of seamounts, see Hillier (2007) *Geophys. J. Int.* **168**. McClain (2007) *J. Biogeog.* **34**, 11 challenges the view that seamounts are oases/biodiversity hot spots.

search behaviour The way an individual, or entity, selects a location. The choice depends on the information available, and on the decision-maker's ability to evaluate the information. Try McColl-Kennedy and Fetter (2001) *J. Services Marketing* **15**, 2.

seasonal forecasting A forecast of weather conditions for a period of three to six months, based on, in particular, sea surface temperatures. Murphy et al. (2001) *Nat. Hazards* **23**, 2–3 find prospects for the early warning of climatic hazards in the tropics and subtropics 'particularly promising'.

seawater incursion (**seawater intrusion**) In coastal aquifers, lighter fresh water lies above a wedge of more dense salt water, and is separated from it by a mobile transition zone. If aquifers are pumped at rates exceeding their natural capacity to transmit water, seawater is drawn into the system to maintain the regional groundwater balance. See Qahman and Larabi (2006) *Hydrogeol. J.* **14**, 5 on Gaza.

secession The transfer of part of the territory and population of one state to another, whether pre-existing, or newly created. 'The last act of the European geopolitical drama has yielded the greatest "advantage" for the former minorities in Central and Eastern Europe, but, as calculations show, the nation-state model became a less effective tool to solve ethnic problems, since each new secession engenders the appearance of new minorities' (Kotossov and Treivish (1998) *Polit. Geog.* **17**, 5). Writing on Biafra's abortive secession in 1967, Diamond (2007) *Dialec. Anthrop.* **31**, 1–3 observes that 'Nigeria is not a nation but a political idea imposed by force of foreign arms. The variety of peoples within its borders will continue to seek accommodation.' See Le Breton and Weber (2001) IMF WP/01/176, on 'how to prevent secession'.

secondary air mass An *air mass which has been modified by the passage of time or by its movement to a different area. A *k* air mass is colder than the surface it moves over, and is usually *unstable; a *w* air mass is warmer, and usually stable.

secondary industry (secondary sector) The creation of finished products from raw materials; that is, manufacturing. See Ito (2002) *EPTD Disc. Paper* **91** on the significance of secondary industry in interregional imbalances in China.

second home A property occasionally used by a household whose normal residence is elsewhere. Second homes are usually found in rural areas; the 'compensation hypothesis' describes the association between urban density and the consumption of second homes (Módenes and López-Colás (2007) *Tijdschrift* **98**, 3). Second homes can bring increased custom to rural areas, but may drive up house prices beyond the pockets of local residents. Stressing that he is referring to Sweden only, Marjavaara (2007) *Tourism Geogs.* **9**, 3 observes that second homes become a convenient scapegoat for rural decline: second home owners are often targeted as holders of alien values; locals are seen as fragile victims.

sector principle The use of meridians to demarcate boundaries or claims on land and in the oceans (Forbes and Armstrong (1995) *IBRU Boundary & Security Bull.*). This is the principle on which claims to territory in the Arctic and Antarctic are made; see J. O'Brien (2001).

sector theory The view that housing areas in a city develop in sectors along the lines of communication, from the *CBD outwards. High-quality areas locate along roads; industrial sectors develop along canals and railways, away from high-quality housing; working-class housing locates near industry. 'Basic urban pattern such as concentric zone, sectors and multi-nuclei . . . can be interpreted as "ideal" conditions or basic structure at local scale, and the real pattern could be viewed as a composition of multiple structures in the urbanization process' (Xu et al. (2007) *Landsc. Ecol.* **22**, 6).

Sector theory

sedentary Fixed, not moving, as in **sedentary agriculture** where the farmer and the fields are permanently settled.

sediment 1. Material which has separated and settled out from the medium—wind, water, or ice—which originally carried it. For fluvial sediments the ability of a river to carry sediment depends on particle size as well as the river discharge. *See also* LOAD.

2. A **sediment budget** is the tally of inputs and outputs for a specific *open system, over a given time period; the classic paper is Trimble (1983) *Am. J. Sci.* **283**. A **balanced sediment budget** is an equation of *mass conservation in which the sediment fluxes related to the sources, sinks, and storages are balanced. The concept is important in environmental management: see Nordstrom et al. (2002) *Geomorph.* **47**, 2–4.

sedimentary rock A rock composed of sediments, usually with a layered appearance (*bedding plane). Sedimentary rocks may be subdivided by size into: *argillaceous, *arenaceous, and *rudaceous. **Sedimentation** is the deposition of sediment from a state of suspension in water or air, also called *siltation*.

sediment cell A length of coastline, and its associated nearshore area, within which the movement of sand and shingle is largely self-contained. The littoral sediment cell concept 'has become synonymous with pro-active coastal management practices in England and Wales' (Cooper and Pontee (2006) *Ocean & Coastal Manage.* **49**, 7–8).

Sediment yield is the total mass of sediment that reaches the exit of a drainage basin. Tamene et al. (2006) *Geomorph.* **76**, 1–2 find that pronounced terrain steepness, easily detachable slope material, poor surface cover, and gullies promote high sediment yields; de Vente and Poesen (2005) *Earth Sci. Revs* **71**, 1–2 propose a general relation between basin area, active erosion processes, sediment sinks, and total sediment yield; and Verstraeten et al. (2003) *Geomorph* **50**, 4 develop an index model to explain variability in area-specific sediment. **Sedimentation** is a term often used to describe the blocking of an aquatic system by the deposition of sediment; see Bennett et al. (2005) *Water Resources Res.* **41**, W01005, and Wang et al., *Water Resources Res.* **41**, W09417. See Flemming (1965) *J. Sed. Res.* **35**, 2 on **sedimentary particles**, their transport history, and their depositional environments.

segmented labour market A *dual labour market may be composed of skilled and unskilled segments. 'Welfare, tax and labour-market institutions continue to facilitate and maintain a segmented labour market with a wide range of employment systems and practices, making it virtually impossible for the growing proportion of the labour force in low-wage, service-dominated jobs to achieve independent living on an individual basis' (McDowell (2004) *PHG* **28**, 2). See also Hedberg, SULCIS *W. Paper* 2008:1.

segregation The separation of a large population into subgroups—based on grounds of income, race, religion, or language—located in distinct residential areas. Blumen and Zamir (2001) *Env. & Plan. A* **33** find spatial segregation to be intrinsic to the capitalism-patriarchy, while Ellis et al. (2004) *AAAG* **94**, 3 write on prejudice and racial segregation. Schnell and Yoav (2001) *AAAG* **91**, 4 suggest a socio-spatial isolation index of segregation. Goldhaber (2007) *Urb. Geog.* **28**, 6 presents an empirical model for testing ethnic segregation in the context of spatial perception. *See also* INDEX OF SEGREGATION.

seiche A stationary or *standing wave in an enclosed body of water, resulting in major shifts in wind magnitude or direction (Luettich et al. (2002) *C. Shelf Res.* **22**, 11–13).

seif *See* SAND DUNE.

seismic Of an *earthquake. The **seismic focus** is the point of origin of the earthquake within the crust; see Deji (1999) *Eng. Geol.* **51**, 3 on seismic foci around the Three Gorges Project. The velocity of **seismic waves** varies with the density of the material through which they travel. This property is used in **seismic tomography**—the investigation of the interior of the Earth; see W. K. Lee (2003).

selective logging The felling, at intervals, of the mature trees in a forest of mixed age. Selective logging (combined with large tree retention) is recommended for minimizing ecological impacts; see Rüger et al. (2007) *Forest Ecol. & Manage.* **252**, 1–3.

selective migration This may be spontaneous, as when a particular age-sex group constitutes most of the migrants; 'age-selective migration tends to dichotomize areas between younger and older, with higher and with lower natural increase' (Morrill (1995) *Annal. Reg. Sci.* **29**, 1). 'Age-selective migration is systematically draining the entry labour pool in rural America' (W. Kandell and D. L. Brown 2006). Boyle (2004) *PPG* **28**, 6 argues that selective migration 'will alter our interpretation of health and mortality inequalities'.

A leading cause of selective migration is the increasing skill-focus of immigration policy in a number of industrialized countries—'a trend that is likely to intensify as rich countries age, and competitive pressures build in knowledge-intensive sectors' (Kapur and McHale (2005) Center for Global Development). Vigdor (2002) *Rev. Econ. & Sta.* **84**, 4 considers selective migration and economic outcome.

self-organization In geomorphology, the patterns that arise autogenically; independently of any external forcings (Chin and Phillips (2007) *Geomorph.* **83**, 3–4). Examples include the formation of evenly spaced, nearly uniform ripples on sand dunes or stream beds, or of *patterned ground in periglacial landscapes. 'Nonlinear, dissipative interactions among the small- and fast-scale constituents of a system give rise to order at larger spatial and longer temporal scales' (Phillips (2003) *PPG* **27**, 1). 'Those not convinced by arguments of self-organisation and inherent behaviour as an explanation in itself commonly ask the mechanism of the changes. Even for those who support the theory, questions still arise of how the self-organisation takes place and how the changes are communicated from one part of the system to another such that order emerges' (Hooke (2007) *Geomorph.* **91**, 3–4).

Self-organized criticality describes a system in dynamic equilibrium near a threshold condition The concept arises from Bak et al. (1987) *Phys. Rev. Letts* **59**, 4, who develop a simple *cellular automaton model in which sand is added, grain by grain, to a surface to form a pile. When local slopes become too steep a collapse occurs, moving sediment to neighbouring cells, which too can collapse if the adjusted slopes are too steep. See Coulthard and Van De Wiel (2007) *Geomorph.* **91**, 3–4.

semi-globalization A term used by Ghemawat (2003, *J. Int. Business Studs*, **34**, 2) to describe incomplete cross-border integration, and the location-specificity of certain enterprises. 'Semi-globalization covers the range of situations in which neither the barriers nor the links among markets in different countries can be neglected.' Ghemawat argues that semi-globalization is the underlying structural condition most conducive to thinking carefully about competition across multiple locations, because complete market insulation is a less challenging condition.

semiotics The ways in which signs and meanings are created, decoded, and transformed; 'cultural landscapes . . . are a collection of symbols to be deciphered—an exercise in semiotics' (Knudsen et al. (2007) *Tourism Geog.* **9**, 3).

sense of place Either the intrinsic character of a place, or the meaning people give to it, but, more often, a mixture of both. 'Sense of place . . . is an umbrella concept that includes all the other concepts—attachment to place, national identity, and regional awareness . . . The messages transmitted are not neutral, but rather they reflect the subjective senses of the beholder or the perceptions of "society"' (Shamai (1991) *GeoForum* **22**, 3). Robinson (2006) *UNESCO Univ. & Heritage, 10th Int. Seminar*, argues that it is 'the intangible characteristics of places which are central to our recognition of the more perceptually and socio-culturally founded notion of sense of place'. Planners consciously create or preserve memorable and singular structures to make a space distinctively different—such as the UK Angel of the North.

sensible temperature Not the temperature recorded on a thermometer, but the temperature as felt by the individual.

serac An ice pinnacle, or series of pinnacles on the surface of a glacier resulting from tensional failure in the more rigid upper crust when the glacier: moves over a slope; spreads out over a plain; or passes round a bend in its valley. See Bishop et al. (1998) *Geomorph.* **21**, 3-4 on spectral variability in serac fields.

sere One of a sequence of developmental stages in plant succession. A primary stage is a **prisere**; a **hydrosere** develops in water; a **psammosere** on a sand dune; and a **xerosere** in an arid location—see Dimeyeva (2007) *Basic & Appl. Dryland Res.* **1**. If a plant succession is interrupted, the resulting secondary stage is a **sub-sere**.

service industry Those economic activities (also called *tertiary industry) concerned with the distribution and consumption of goods and services. Proximity to clients is the major locational factor. Other factors promoting the *clustering of service firms and workplaces are: agglomeration economies, supply of infrastructure, supply of labour, and new firm formations.

Some services—particularly less specialized household services—are fairly evenly located in relation to settlement patterns. Those less evenly located include specialized household services, central government functions, and professional or business services, which concentrate in major city regions, or within regional clusters. With information and communication technology, service activities can decentralize from urban cores to the suburbs, and to more peripheral sites, both nationally and internationally (Hermelin and Rusten (2007) *Geografiska B* **89**, 1).

set-aside grant Within the EU, a gift of money from the European Parliament to persuade a farmer to take land out of agricultural production, in order to cut down on the creation of surpluses; 'in the arable sector, set-aside is interpreted as delivering *extensification through an overall lowering of output from individual farms' (Evans et al. (2002) *PHG* **26**, 4). Robinson and Lind (1999) *Tijdschrift* **90**, 3 find that limited conservation gains have been realized through EU set-aside. In 2007, European Union agriculture ministers suspended set-aside payments, recognizing that current EU farm policy is helping to drive up cereal prices.

settlement Any form of human dwelling—from a single house to the largest city.

settlement hierarchy A division of settlements into ranks, usually by population size. Milbourne (2007) *J. Rur. Studs* **23**, 3 notes that, when rural poor and homeless groups relocate to places with better job, housing, and welfare prospects, they tend to move short distances up the rural settlement hierarchy, rather than to the more distant cities.

settlement pattern The spatial distribution of settlements. Some settlement patterns may be seen as a reflection of cultural traditions: a dispersed settlement pattern in Norway has acquired a symbolic meaning (Cruickshank (2006) *Norsk Geograf. Tidssk.* **60**, 3). *Nearest neighbour analysis may be used to test for regularities in settlement patterns.

sexuality, geographies of The spatialities of sexualities, intersecting discussions of sexualities with issues such as development, race, gender, and other forms of social difference (K. Browne et al., eds 2007). For example, C. Hemmings (2002) examines bisexual spaces as places that are defined by both geographical boundaries and cultural significance, showing how and why safe places have developed for gay, lesbian, and bisexual communities. In 1999, Binnie and Valentine (*PHG* **23**, 2) argued for a move away from a simple mapping of lesbian and gay spaces towards a more critical treatment of the differences between sexual dissidents. Consult Hubbard et al. (2008) *PHG* **32**, 3 and Oswin (2008) *PHG* **32**, 1 to see if their call has been answered, while Del Casino (2007) *Health & Place* **13**, 4, discusses geographies of sexual health. *See also* QUEER GEOGRAPHIES.

shadow price The shadow price of any good is the increase (or decrease) in social welfare brought about by providing one more (or one less) unit of that good. Thus, the shadow price of carbon is defined as the social cost of emitting a marginal tonne of carbon (or the social benefit of abating a tonne). See Dietz (2007) DEFRA. This concept is used in *cost–benefit analysis to value intangible items like *amenity.

shale A fine-grained *sedimentary rock formed when layers of clay are compressed by the weight of overlying rocks. Shales are easily split along *bedding planes.

shanty town *See* SQUATTING.

share-cropping A farming type where the tenant pays his rent in produce rather than in cash. In the USA, lack of capital left the cropper dependent upon others for subsistence until market time. Storekeepers furnished the cropper with food and clothing during the year, the accumulated debt to be repaid, with interest, at harvest time. The store owner was often also the buyer of the crops (M. P. Conzen 1994). 'Share-tenant relations are . . . compromises between risk and incentive . . . Given a choice, better-off workers would seek tenant relations, while poorer workers or recent arrivals would have to enter sharecropping arrangements' (Brannstrom (2001) *Polit. Geog.* **20**, 7).

shear (shearing) The deformation of a material so that its layers move laterally over each other. Shearing bends, twists, and draws out rocks along a fault or *thrust plane. A **shear plane** is the face along which shearing occurs. **Shear strength** is the ability of a rock or soil to withstand shearing; soil shear resistance is proportional to the tangent of the friction angle and therefore soils with larger friction angles exhibit a higher shear strength. Slope failures are the consequence of loading conditions that exceed soil shear strength (Simoni et al. (2008) *Hydrol. Procs* **22**, 4). **Shear stress** is the load referred to above; see Namikas (2000) *J. Sed. Res.* **72**, 2 on measuring bed shear stress during aeolian saltation; Sime et al. (2007) *Water Resources Res.* **43**, W03418 on estimating **river** bed shear. The **basal shear stress** of ice is related to the basal shear strength (Tulaczyk et al. (2000) *J. Geophys. Res.* **105**, B1).

sheet erosion A very slow-acting form of erosion whereby a thin film of water—**sheet wash**—transports soil particles by rolling them along the ground. See Chaplot and Bisonnais (1998) *Soil Till. Res.* **46** and A. Goudie (2004).

sheeting The separation and peeling away of the outer layers of a rock mass (Wakasa et al. (2006) *Hydrol. Procs* **31**, 10). *See also* EXFOLIATION.

shield The very old, rigid core of relatively stable rocks within a continent. See Stein (2003) *Precamb. Res.* **123**, 1–2 on the formation of the Arabian-Nubian Shield, and Lidmar-Bergström (1995) *Geomorph.* **12**, 1 on the Baltic Shield.

shield volcano A volcano formed of successive eruptions of free-flowing *lava which create a gently sloping, broad dome tens of kilometres across and more than one kilometre high. Mount Mazama, Oregon, is composed of several overlapping composite and shield volcanoes (Bacon and Lanphere (2006) *GSA Bull.* **118**, 11).

shifting cultivation A farming system based on the clearance of land, cultivation until soil fertility declines, followed by fallowing for up to twelve years, in order to restore the vegetation to its initial condition. Vasconcelos et al. (2002) *Appl. Geog.* **22**, 2 argue that shifting cultivation is ecologically sustainable for the subsistence of small groups. 'Shifting cultivation is a truly *sustainable system, perhaps the only one for humans. It can maintain agricultural production indefinitely with little reliance on external inputs when population levels are low—roughly no more than one person per 2 hectares of agricultural land' (C. A. S. Hall 2000).

shift-share analysis A method of estimating the relative importance of different elements in any growth or decline of regional industrial employment. See Fotopoulos (2007) *Growth & Change* **38**, 1 for the methodology and for the application of shift-share analysis to demography. See also Hanham and Banasick (2000) *Growth & Change* **31**, 1.

shingle Pebbles on a beach, rounded by *abrasion, and reduced in size by *attrition. The abrasion rate of shingle is linked to, in rank order: mean wave height, pebble type, and pebble weight (Dornbusch et al. (2002) *J. Coastal Res.* **SI36**).

shopping, geography of *See* RETAILING, GEOGRAPHY OF.

shopping centre M. Ibrahim and P. McGoldrick (2003) examine the relations between the use of shopping centres and the transport modes used or available, together with policies that attempt to reduce the negative impacts of decentralizing retail facilities. The hypermarket is 'dedicated to creating a space of control that has taken its departure from the classical regime of surveillance' (D. B. Clarke 2003).

shopping goods In the process of selecting and buying things, those goods which the customer compares on such bases as suitability, quality, price, and style—such as furniture and clothing (AMA Report Definition Committee (1948) *J. Marketing*, **13**, 2). Makers of shopping goods can limit the number of wholesalers and retailers in a geographic area in order to maintain channel control and enhance the image or exclusivity of their products (Gable et al. (1995) *J. Retail. & Consum. Services* **2**, 4).

shore The land adjoining a large body of water/the sea. The **backshore** is normally above the high water mark, but still influenced by the sea; the **foreshore** is the area between high and low tide marks and is exposed at low tide. The **nearshore** is seaward of the foreshore, ending at the breaking point of the waves; the **offshore** is the zone seaward of the breakers, but in which material is moved by the waves.

shore platform A very gently sloping platform extending seaward from the base of a cliff, and subject to *salt weathering, alternate wetting and drying, *water-level weathering, and *quarrying, *hydraulic action, *pneumatic action, and *abrasion. Trenhaile (2005) *ESPL* **30** finds that platform gradient increases with tidal range; with slow downwearing, platform width increases with tidal range (but decreases with fast downwearing); and platform gradient decreases and width increases with wave energy, decreasing rock resistance, and platform roughness. However, Stephenson and Kirk (2000) *Geomorph.* **32** argue that waves on platforms are not capable of causing erosion because of their high loss of energy in these shallower waters.

sial The material of the continental crust, mostly of silica- and aluminium-rich minerals.

Sichelwannen Glacially eroded, crescent-shaped depressions, up to 10 m long and about 5 m wide; the horns facing down-glacier.

silcrete A *duricrust cemented with silica.

sill An *intrusion of igneous rock which spreads along bedding planes in a nearly horizontal sheet. Sills are mostly concave, and can be fed from dykes, or the steep climbing portions of deeper sills. Both sills and dykes can provide magma to overlying volcanic fissures, and sills can feed shallow laccoliths (Thomson (2006) *Bull. Volc.* **66**, 4).

silt Fine-grained minerals, 0.002–0.06 mm in diameter, ranging in size between *clay and *sand in size. Silts are often deposited by rivers when the flood water is quiet.

Silurian A period of *Palaeozoic time stretching approximately from 430 to 395 million years BP.

(((●))) SEE WEB LINKS
• An overview of the Silurian period.

sima The material of the lower part of the continental crust and the oceanic crust, dominated by silica and magnesium.

sink hole *See* DOLINE.

sinter Mostly silica and sulphate minerals, precipitated in layers from volcanic activity. Sinters provide evidence of past environments (Lynne et al. (2006) *Geology* **34**, 9). **Snow sintering** is the bonding of snow particles, either by the diffusion of water molecules, or by pressure-induced creep; see D. McClung and P. Schaerer (1993).

sinuosity The extent to which a river bends. The **sinuosity ratio** is the distance between two points on the stream measured along the channel, divided by the straight line distance between those points; it is used to determine whether a channel is straight or meandering (Aswathy et al. (2008) *Env. Monit. Assess.* **138**, 1–3).

site The position of a structure or object in physical, local terms, such as a river terrace.

Site of Special Scientific Interest (SSSI) A site in the UK which is of particular importance because of its geology, topography, or ecology. SSSIs are graded in terms of importance from 1 to 4.

(⊕) SEE WEB LINKS

• The Board of Natural England's SSSI website.

situation The location of a phenomenon, such as a town, in relation to other phenomena, such as other towns. *Compare with* SITE.

skid row An inner-city district characterized by 'a stigmatised resident population that is predominantly poor, street-entrenched, addicted, alcoholic and/or mentally ill, the presence of a local drug market and a concentration of bars, low-rent hotel rooms and social service facilities' (Huey and Kemple (2007) *Urb. Studs* **44**, 12).

sky view factor A measure commonly used by urban climatologists to quantify the openness of a site within an urban setting. It has major implications for incoming and outgoing solar and terrestrial radiation (Grimmond (2007) *Geog.* **30**, 3).

slaking In geomorphology, the disintegration of fine-grained rocks. Slaking is one of the main rock-weathering mechanisms (Stefano and Ferro (2002) *Biosys. Engineer.* **81**). **Slake durability** is the rate and amount of strength reduction in a rock after soaking (Morgenstern and Eigenbrod (1974) *J. Geotech. Eng. Div., ASCE* **101**), and is an important parameter in slope stability (Yilmaz and Kalacan (2005) *Env. Geol.* **47**, 7).

slash and burn An agricultural system whereby the trees are felled, the land cleared of most of the trunks, and the remaining vegetation fired. See Tschakert et al. (2007) *Ecol. Econ.* **60** on carbon offsets in small-scale slash-and-burn agriculture.

slate A weak sedimentary rock, easily split along thin bedding layers, and formed by the compression of shales by the overlying rocks.

slickenside A set of linear marks on a fault or bedding plane caused by the frictional movement of one rock body against another. See Ree et al. (2003) *Island Arc* **12**, 1.

slide A form of mass movement in which material slides in a relatively straight plane, generally breaking into many blocks as it moves. Slides usually have a length far greater than the depth of the moving material, and are triggered by high water pressure (Dai et al. (1999) *Eng. Geol.* **53**, 3–4), but the competence of the rock is also significant; see Warburton et al. (2003) *ESPL* **28**, 5.

slip-off slope The relatively gentle slope at the inner edge of a meander.

slope elements The differing parts of a slope. These may be convex, straight, or concave; the shape of a slope is an expression of the predominant processes acting on it.

Slope elements

Convex slope elements are usually gentle, generally at the top of a slope, and formed by soil *creep and *rainsplash. Downslope from the convex segment, there may be a **straight slope element** of bare rock. This is the *free face (fall face). Other straight slopes develop where mass movement is the dominant geomorphological process. Slopes with shallow debris and steep slopes undercut by rivers or waves both tend to have straight segments; *talus slopes are often straight. **Concave slope elements** are either due to increased water erosion downslope or to larger pieces of debris rolling further than the rest. **Compound slope profiles** may exhibit some or all of these segments and may show a repetition of certain segments.

slope wash The downslope transport of sediment by an almost continuous film of water. It can be seen as three erosional processes: sheet flow, rill erosion, and gullying.

slope winds *Anabatic and *katabatic winds.

slum An area of poor housing, characterized by multi-occupance and overcrowding; see McFarlane (2008) *Env. & Plan. A* **40**, 1 on sanitation in Mumbai's informal settlements. Rao (2006) *Int. J. Urb. & Reg. Res.* **30**, 1 rails against the use of 'slum' as shorthand for 'the distortion of urban substance into a dysfunctional stage for violence, conflict, and the iniquitous distribution of resources'. Durch et al. (1998) *Cities* **15**, 4 review housing policies implemented in Mumbai; from slum clearance and the construction of high-rise apartment blocks to a range of self-help strategies and privatized market-led schemes. **Slum clearance** is the demolition of substandard housing, usually accompanied by *rehabilitation and *redevelopment. At the time of writing, history's most ambitious slum-clearance project is nearing completion, with the elimination of Dharavi, a Mumbai slum that houses between 600 000 and a million people.

slump A form of *mass movement (also called a single rotational slide) where rock and soil rotate backwards with movement down a concave face: 'a more or less rotational movement, about an axis that is parallel to the slope contours, involving shear displacement along a concavely up-ward facing failure surface' (D. Varnes 1978). Slumps are most common in thick *regoliths (Nwadije (1998) *Nat. Hazards* **1**, 2) and large mudstone rock units (Plint (1986) *Sedimentol.* **33**, 3).

smog A combination of smoke and fog. The fog occurs naturally; the 'smoke' is introduced into the atmosphere by human agency. See R. Kosobud (2006) on the cost-effective control of urban smog. *See also* PHOTOCHEMICAL SMOG.

(⊕) SEE WEB LINKS

• Details of the air pollution in Mexico City from the International Development Research Centre.

SMSA A Standard Metropolitan Statistical Area in an urban area of the USA. This can be a town of 50 000 people, or two towns, each with more than 15 000 people, and together totalling more than 50 000; or a county with more than 75% of its population working in industry. Areas which seem, by employment, commuting, or population density, to be urban rather than rural are also included.

snow Frozen atmospheric water. A snow crystal is ice, up to 5 mm across, and variously shaped as a prism, plate, star, or needle. Snow crystals fall from stratiform clouds when the low-level air is degrees below freezing point, and the air above it is colder (Mason (1998) *QJ Royal Met. Soc.* **124**, 545). When the low-level air is near 0 °C, snow crystals aggregate to form snowflakes (Fujiyoshi and Wakahama (1985) *J. Atmos. Scis* **42**, 15).

snowbelt The states of the north-east USA, such as Michigan, Illinois, Indiana, Ohio, New Jersey, New York, and Pennsylvania. Discussion of a population shift from the snowbelt to the *sunbelt was frequent in the 1970s, but Essletzbichler (2004) *AAAG* **94**, 3 maintains that the snowbelt/sunbelt dichotomy collapsed in the 1990s.

social area analysis The analysis of a city to define **social areas** (urban areas which contain people of similar living standards, ethnic background, and lifestyle). 'Social area analysis tends to view the city as constituted of a mosaic of socially homogeneous containers' (Schnell and Benjamini (2005) *Urb. Studs* **42**, 13).

social capital The relations that exist between individuals within both families and communities, together with the educational, cultural, and social resources and expertise of a society (J. Coleman and F. Thomas 1987). 'At the macro level, formal institutions, such as state constitutions and legal norms, make up the social capital of a society . . . at the micro level, social capital can be conceptualized as personal trust which exists in a particular community or social network' (Batheld and Glückner (2005) *Env. & Plan. A* **37**). However, Clarke (2008) *Pol. Geog.* **27**, 1 claims that 'many scholars would agree that an emphasis on community as the site for generating social capital and citizenship may be misleading', and Antoci et al. (2007) *J. Socio-Econ.* **38**, 1 emphasize the negative welfare consequences of social capital. For an overview of geographical accounts of social capital see Holt (2008) *PHG* **32**, 2.

social cleavage The spatial division of society into distinct groups; as manifestations of divisions within society, social cleavages are held to be linked to geographic patterns of electoral politics (Secor (2001) *Pol. Geog.* **20**, 5; see also Perepechko et al. (2007) *Pol. Geog.* **26**, 2). However, Blondel (*U. Siena & Eur. U. Inst, Occ. Paper* 16/2005) argues that 'the time has come to consider whether it is realistic to continue to regard as correct the view that social cleavages are the paramount . . . factor accounting for the relationship between "modern" parties and their supporters'.

social cohesion The interdependence between the members of a society, shared loyalties, and solidaity (Durkheim, quoted by Jensen (1998) *Canad. Policy Res. Netwk Stud.* **F03**). 'A cohesive community is a community that has naturally many cross-links, where people from different race, age, background, feel free and happy to mix together in housing, in education, [and] in leisure facilities' (UK Housing, Planning, Local Government and the Regions Sixth Report). Nagel and Staeheli (2008) *Soc. & Cult. Geog.* **9**, 4 see the social cohesion agenda as 'marked by the assertion that the host society should define the terms of integration, and that the primary responsibility for integration lies with immigrants and minorities'. Yuval-Davis et al. (2005) *Ethnic & Racial Studs* **28** see the concern with social cohesion as an attempt to limit the exclusion of groups.

(((⊕))) SEE WEB LINKS

• UK Housing, Planning, Local Government and the Regions Sixth Report.

social costs In economics, the total costs of any action. **Private social costs** are met by the individuals concerned; Williams and Balaz (2002) *J. Ethn. & Mig. Studs* **28**, 4 note that Slovaks working in Austria reduce their social costs through commuting rather than migrating. Indirect costs, such as the displacement of local communities because of large infrastructure projects, are borne by third parties.

social Darwinism The application of the concept of evolution to the development of human societies over time, emphasizing the struggle for existence of each society, and the survival of the fittest of them. Such ideas have been used to justify power politics, imperialism, and war; see Weikart (2003) *German Studs Rev.* **26**, 2.

social distance The perceived distance between social strata (different socio-economic, racial, or ethnic groups), usually measured by the amount of social contact between groups; see White et al. (2005) *Sociol. Methods & Res.* **34**, 2. Agrawal et al. (2006) *J. Econ. Geog.* **6**, 5, analysing knowledge flows, find that geographic proximity works to overcome social distance. Autant-Bernard et al. (2007) *Papers Reg. Sci.* **83**, 3 define this term as the number of links between any two firms: 'firms with a higher number of partners and a smaller social distance are more likely to be engaged together in a research project . . . in this context, social distance matters more than geographical distance.'

social exclusion The exclusion of part of society from 'any of the social, economic, political and cultural systems which determine the social integration of a person in society' (A. Walker and C. Walker 1997). Kenyon et al. (2003) *J. Transp. Geog.* **10**, 3 see social exclusion as 'the denial of access, to an individual or group, to the opportunity to participate in the social and political life of the community, resulting not only in diminished material and non-material quality of life, but also in tempered life chances, choices and reduced citizenship'.

Social exclusion is a more complex concept than poverty: the Transport Research Group notes that 'it is possible to be excluded without experiencing poverty; and it is possible to experience poverty, yet feel included'. Exclusion may come about through a combination of problems such as income poverty or unemployment; geographical isolation or reduced accessibility to social networks, facilities, goods, and services unemployment; poor housing; high crime environments; family breakdown; or the denial of citizenship rights and freedoms. See the DFID Policy Paper *Reducing Poverty by Tackling Social Exclusion*.

((⊕)) SEE WEB LINKS

• Transport Research Group summary table on social exclusion.

social formation The prevailing pattern of class structure which goes hand in hand with a particular *mode of production; try Bruneau (1981) *Antipode* **13**, 3. J. Osterhammel and N. Petersson (2005) list qualifying criteria: the social interaction between over two people; the longevity of these interactions; and their reinforcement by institutions.

social geography Originally this was defined as the study of the spatial patterns of social, as distinct from political and economic, factors; see, for example, Meijering et al. (2007) *Tijdschrift* **98**, 1. Social geography may now be subdivided into three categories. The first is the spatial expression of capitalism: 'private property . . . is a spatial expression of power and surveillance that we have obediently internalized to make the actual exercise of power unnecessary' (Gruenewald (2003) *Am. Educ. Res. J.* **40**, 3). The second stresses the 'alternative' view of human geography; for example, studying the economically disadvantaged rather than the successful: D. Darton et al. (2003) point out that the distribution of disadvantage in the UK has a strongly geographical dimension. A third category emphasizes *welfare geography.

socialism A social system based on equality and social justice, once linked with common ownership of the *means of production and distribution. Hensman (2001) *Antipode* **33**, 3 argues that the opposition to globalization is 'national socialism, socialism or state socialism [but] still a form of capitalism'. 'In the mid-1970s, Zygmunt Bauman, 1976, described socialism as intrinsically utopian. Contemporary critical geographers, however, are better at mapping current dystopias than imagining utopic alternatives' (Blomley (2007) *PHG* **31**, 1). 'There is a moral responsibility to take ownership of what we propose, implicitly or explicitly, when we counter capitalism in the name of socialism or (today) postdevelopmentalism' (Corbridge (2005) *PHG* **29**, 5).

social justice 'Putting it simply, geographers interested in justice focus on who gets what, who misses out and where all this occurs' (Bauder (2006) *PHG* **30**). These are questions that can be asked on global, national, regional, and local scales. 'Social justice requires the removal of the obstacles that inhibit individuals and groups from developing their potential . . . contrary to the orthodox view, increasing social equality improves economic efficiency' (Kitson et al. (2000) *Camb. J. Econ.* **24**, 4). Also of interest to geographers is the way in which moral systems vary spatially; 'humans seek not only to improve technical means but also to use these means to reflect upon, regulate, and place value on their own conduct' (Collier and Ong (2003) *Current Anthrop.* **44**, 3). *See also* JUST CITY.

social network The relatives, family, and neighbours an individual, or family, is connected to. See Boyd and Ellison (2007) *J. Computer-Mediated Comm.* **13**, 1 on MySpace, Facebook, Cyworld, and Bebo.

social physics The use of concepts from physics to illuminate aggregate human behaviour; for example, the *gravity model. In general, 'new social physics falls well short of sociological adequacy' (Crossley (2005) *Soc. Rev.* **53**, 2).

social polarization The results of *segregation within a society; 'a widening gap between individuals, households or groups of people in terms of their economic and social circumstances and opportunities' (Dorling and

Woodward (1996) *Prog. in Plan.* **45**, 2). 'Differentiation as such is not necessarily a problem, but polarization between the have and the have-nots is. It becomes even more of a problem if polarization increases' (Brattbakk et al. (2000) *Paper, 2nd Int. Conf. Critical Geog.*).

social reproduction The processes through which the fundamental structures and relations of society continually recreate the existing mode of production—a mode in which the structures of dominance of one group over another are maintained (Althusser (1971) cited in Mitchell et al. (2003) *Antipode* **35**, 3). Katz (2001, *Antipode* **33**, 4) writes that 'social reproduction is the fleshy, messy, and indeterminate stuff of everyday life . . . secured through a shifting constellation of sources encompassed within the broad categories of the state, the household, capital, and civil society . . . It is also a set of structured practices that unfold in dialectical relation with production, with which it is mutually constitutive and in tension . . . the production and reproduction of the labour force calls forth a range of cultural forms and practices that are also geographically and historically specific, including those associated with knowledge and learning, social justice and its apparatus, and the media.'

'Distinctive national, regional and local patterns in the ways in which men and women divide paid labour and caring work result in an uneven geography of the total work of social reproduction' (McDowell (2003) *TIBG* **28**, 1). Even childcare arrangements are integral to social reproduction in post-industrial economies (Mahon (2006) *TIBG* **31**, 4), while in developing countries 'traditional means of social reproduction, including children's participation from an early age in family businesses and agricultural endeavours, is giving way to oversight by federal and state government agencies' (Jennings et al. (2006) *Area* **38**, 3).

social space The combined use and perception of space by distinct social groups, as opposed to *personal space. 'Social space is produced by societies according to the spatial practices that exist within the society. The produced space is a set of relations between objects within the space' (Carter (2004) *Am. Ass. Geogrs, Mid. Atlantic Div. Annual Meeting*). Social space provides an environmental framework for the behaviour of the group; it is flexible/networked (Peck and Tickell (2002) *Antipode* **34**, 3), and multi-layered—'vulnerability is a multi-dimensional and multi-layered social space defined by political, social and institutional capacities' (Watts and Bohle (1993) *PHG* **17**).

social well-being A situation where income levels are high enough to cover basic needs, where there is no poverty, and where there is easy access to social, medical, and educational services. To achieve this in developing countries, El-Ghannau and Ragab (2002) *Perspec. Glob. Dev.* **1**, 1 conclude that population growth must slow down.

sodic soils Soils that are traditionally defined as having an exchangeable sodium percentage greater than 6%—'but there are many soils with lower sodium levels which exhibit similar behaviour' (S. Raine and R. Loch 1995).

soil The naturally occurring, unconsolidated, upper layer of the ground, made of humus and weathered rock. Major factors affecting soil formation are: climate, relief, parent material, vegetation, and time; see Svoray and Shoshany (2004) *Rem. Sens. Env.* **92**, 2.

soil association A cluster of soils dominated by a single soil taxon (US Dep. Agric. (1993) *Agric. Handbook* **436**).

soil classification A naming system to describe and identify soils.

(((●))) SEE WEB LINKS

• National soil classifications.

soil creep *See* CREEP.

soil erosion The removal of the soil by wind and water and by the *mass movement of soil downslope. The wind erosion is by *deflation; water erosion takes place in gullies, *rills, or by *sheet wash; downslope mass movement ranges from soil *creep to *landslides.

Accelerated soil erosion is erosion increased by human activity. Causes include: wind, tillage (*NSR Report for DEFRA*); and continuous cropping (Jaiyeoba (2003) *Soil & Till. Res.* **70**, 1). However, Ives (1991) *New Scientist* **1764** disputes a link between deforestation and soil erosion, and Rowntree et al. (2004) *Land Degrad. & Dev.* **15**, 3 doubt the claim that overgrazing leads to soil erosion. War (as a result of agricultural intensification and the use of marginal land, or of 'scorched earth policies'), fire, urbanization, and mining also accelerate soil erosion; see Robichaud (2005) *Int. J. Wildland Fire* **14** and Limartin in M. Haigh, ed. (2000) on mining. Loughran and Balog (2008) *Geogr. Res.* **46** record a drop in soil erosion through changing from intensive tillage to none at all. However 'there is still some way to go before the cause and effect links are sufficiently understood to form a basis of tools for the management of grazing in a way that would minimise erosion and maximise the benefits' (Thornes (2008) *Geogr. Res.* **46**).

soil horizon *See* HORIZON.

soil moisture Moisture is held in the *capillary soil pores. Soil moisture controls many important processes in seedling emergence, evapotranspiration, mineralization of the soil organic fraction, surface run-off, leaching, and crop yield; the extent of these processes cannot be quantified without knowing the moisture status of the root zone (Akinrem and McGinn (1996) *Can. J. Soil Sci.* **76**).

The **soil moisture budget** is the balance of water in the soil, resulting from precipitation and *potential evapotranspiration in combination. Mintz and

Serafini (1992) *Clim. Dynam.* **8** give a simple water budget model. *See also* MOISTURE INDEX.

soil pore Any open space within the soil. **Soil porosity** is the percentage of pore space. Water will not drain freely through the fine **capillary pores** (average diameter < 0.03 mm), which retain water through surface tension, but drains freely through the larger non-capillary pores. With 2–5% soil porosity, most water flows directly to the subsurface.

soil profile A vertical series of soil *horizons from the ground surface to the parent rock. A soil is classified according to the arrangement of its horizons. The Soil Profile Analytical Database of Europe of Measured Profiles was created to provide a common structure for storing harmonized information on typical soil profile properties of European soils (Hiederer et al. (2006) *Geografisk Tidsskrift* **106**, 1).

soil science 'Historically, soil science has followed a circuitous path in its evolution from a discipline with foundational roots in geology, to an applied agricultural and environmental discipline, and now to a bio- and geo-science through the Earth's Critical Zone investigations. This closes the loop or spiral, but along the way, soil science has become more comprehensive, extensive, integrative, analytical, and quantitative . . . now is a golden era for soil science to integrate its expertise more closely with other bio- and geo-sciences. This will significantly enhance the opportunity to obtain extramural funding and public support, as well as the advancement of soil science. As such, soil science needs to vigorously become more interactive and extend its role beyond traditional agriculture. The knowledge of spatial soil diversity and landscape dynamics is a fundamental underpinning critical to the success of this venture. Pedology, as a unique sub-discipline of soil science, contributes inordinately to earth science, including, for example, elucidation of field variability, surficial weathering processes, earth system dynamics, and vadose zone flow and transport. With the blooming of hydrogeosciences, hydropedology is a timely addition in this era of interdisciplinary, multidisciplinary, and systems approaches for developing comprehensive prioritization of science and applications in earth science. Hydropedology has a niche in this march with other bio- and geo-sciences in addressing global earth science priorities. Soil scientists support changing paradigms and favour closer linkages with the bio- and geo-sciences community. In this regard, hydropedology has a unique role to play'. Wilding and Lin (2006) *Geoderma* **131**, 3–4.

soil series A group of soils formed from the same parent rock and having similar *horizons and *soil profiles, but with varying characteristics according to their location.

(((⊕))) SEE WEB LINKS

- Soil Survey Staff, Natural Resources Conservation Service, United States Department of Agriculture: Official Soil Series Descriptions.

soil structure The arrangement of soil particles into larger particles, or clumps. Platey structures are formed of thin, horizontal layers; prism-like structures are called columnar where the tops are rounded, and prismatic where the tops are level; blocky peds are bounded by flat or slightly rounded surfaces; and spheroidal structures are called crumbs if highly water absorbent, and granular if only moderately so. See Levine et al. (1996) *Ecol. Model.* **92**, 1.

soil texture The proportions of sand, silt, and clay in any soil. Twelve different textural classes are recognized, and the structure of the soil can be determined when the percentage of these three soil constituents is plotted on a ternary diagram. TAL is a group of programs to determine soil texture classes based on major classification schemes.

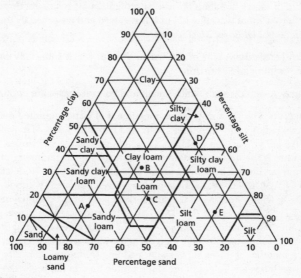

Soil texture

(⊕) SEE WEB LINKS

• TAL software.

solar constant The rate per unit area at which *solar radiation reaches the outer margin of the earth's atmosphere. This fixes the energy supply for the *atmospheric heat engine. Gribben (1991) *New Scientist* **132**, 1796 discusses variations in the solar constant.

solar energy 1. Of all the sunlight that passes through the atmosphere annually, only 51% is available at the Earth's surface to do work. This energy is used to heat the Earth's surface and lower atmosphere, melt and evaporate water, and run photosynthesis in plants. Of the other 49%, 4% is reflected back to space by the Earth's surface, 26% is scattered or reflected to space by clouds and atmospheric particles, and 19% is absorbed by atmospheric gases and clouds. The geography of a place determines how much solar energy is absorbed by the earth, how much is stored and how readily it is released to the atmosphere; see R. Thomas and R. Huggett (1980) and Martin et al. (1998) *Geophys. Res. Lett.* **25**, 23 on the distribution of solar energy at the Earth's surface.

2. Any energy source based directly on the sun's radiation. See Zweibel et al. (2008) *Scientific American* for a wide-ranging review.

solar radiation The electromagnetic waves emitted by the sun, varying in wavelength from long-wave radio waves, through infrared waves and visible light, to ultraviolet waves, X-rays, and gamma radiation. Earth gets only 0.0005% of the sun's radiation Most solar radiation passes straight through the *atmosphere without warming it, but it is received and absorbed by the Earth. See Allen (2006) *QJ Royal Met. Soc.* **84**, 362.

sol brun lessivée A type of *brown earth from which some clay has been *leached, forming an E *horizon.

solfatara A vent through which steam and volcanic gases are emitted.

solifluction The slow, downslope movement of water-saturated debris in *periglacial regions and other areas with cold climates. Rates of movement may be 0.9 cm yr^{-1} on gentle slopes, 12–25 cm yr^{-1} on steeper slopes. Temperature is the most important factor controlling morphometry of solifluction landforms, explaining about 22% of the variability (Ridefelt (no date) ISSN 1650–6553, 70). Solifluction may develop in areas without *permafrost; winter freezing of the sub-surface layer may be sufficient (Matsuoka (2001) *Earth Sci. Revs* **55**, 1–2). Solifluction terraces are step-like, linear, patterned-ground formations, made of fine-grained material (Walsh et al. (2003) *Geomorph.* **55**, 381). Heights and widths are about 1 m, and, with very fluid material, terraces show rough bedding (*tumultuous bedding*). **Solifluction lobes** are bulges in the slope profile that do not break the surface (Matsumoto and Ishikawa (2002) *Geografiska A* **84**, 3–4). The major environmental controls on solifluction lobes are gradient and soil moisture (Ridefelt and Boelhouwers (2006) *Perm. & Periglac. Procs* **17**).

solonchak An *intrazonal saline soil found in hot, arid climates. Evaporation of soil moisture brings saline groundwater to the surface where it, too, evaporates. The sodium and calcium chlorides and sulphates which have been translocated remain as a grey surface crust.

(⊕) SEE WEB LINKS

• A solonchak succession.

solonetz An *intrazonal, formerly saline soil. Periodic rainfall has leached the salts from the surface layer and these accumulate in the B *horizon. Deep ploughing destroys the solonetz horizon (Bogomolova (2008) *Geophys. Res. Abstr.* **10**, EGU2008–A-10603).

solum In a soil, the layer above the parent material. The solum is strongly influenced by vegetation (Schaetzl (1991) *Eur. J. Earth Sci.* **42**, 3) and climate. Alvarez and Lavado (1998) *Geoderma* **83**, 1–2 find precipitation as a key influence in Pampa and Chaco soils.

solution In geomorphology, the picking up and dissolving of particles by a fluid, usually water or carbonic acid. *See also* CARBONATION; AGGRESSIVITY.

sorting In geomorphology, the deposition of sediments in size order. 'Grain diameter indicates the strength of the transporting environment' (Potocki and Angiel (2004) *Miscellanea Geographica* **11** on *aeolian sorting). River beds become finer grained downstream through a combination of grain abrasion, size-selective transport, and deposition (Ferguson (2003) *Geology* **31**). *See also* VARVE.

Southerly Burster *See* LOCAL WINDS.

sovereignty 'Exclusive political authority exercised by a state over a given territory' (Agnew (2005) *AAAG* **95**, 2, who proposes **sovereignty regimes**). See D'Arcus (2003) *Pol. Geog.* **22**, 4 on sovereignty, power, and authority in the Wounded Knee conflict, and Kuus (2002) *Pol. Geog.* **21**, 3 on the Estonian sovereignty discourse. 'Absolute sovereignty ought to be trumped by an "ethics of care"' (N. Crawford 2002). 'Sovereignty in the modern era of multi-tiered governance is often murky and shifting' (Rodden (2004) *PSOnline*, on **fiscal sovereignty**). 'There is an urgent need for the global community to develop a fresh approach towards the concept of sovereignty, and norms and rules of conduct in the international system that will hopefully replace the current system which is ad hoc and open to abuse by hegemonic power(s)' (Tekin (2005) *Futures* **3**). Ong (2000) *Theory, Cult. & Soc.* **17**, 4 defines **graduated sovereignty** as the state's differential treatment of the population based on ethno-racial differences and the dictates of developmental programmes; see also Park (2005) *Pol. Geog.* **24**, 7.

space The extent of an area, usually expressed in terms of the Earth's surface. The term **spatial** derives from this meaning; and spatial relationships are at the heart of geography. '"Space" refers to the operational dynamics of a network: across space, over time, at a variety of scales, from the body to the neighborhood, to the region, and across nations' (Ettlinger and Bosco (2004) *Antipode* **36**, 2). D. Massey (2005) sees space as lived experience: 'one important aspect of space is that it is the dimension of things (and people) existing at the same moment. If *time* is the dimension of change then *space* is the dimension of simultaneity. Space is the dimension of all these things happening at once. Space is the dimension, then, that presents us with the existence of others.'

Y.-F. Tuan (1997) contends that a space requires a movement from a place to another place. Similarly, a place requires a space to be a place. 'Hence, the two notions are co-dependent.' Thrift in S. Holloway et al. (2003) writes that 'place refers to the process whereby spaces are ordered in ways that open up affective and other embodied potentials'. See also Sanders (2008) *PHG* **32**, 2.

'**Absolute space** is fixed and we record or plan events within its frame . . . the **relational view of space** holds there is no such thing as space or time outside of the processes that define them . . . . Processes do not occur *in* space but define their own spatial frame' (D. Harvey 2006). Raper and Livingstone (1995) *Int. J. GIS* **9** distinguish between absolute space, which is geometrically indexed, and **relative space**, which is object oriented; ' "space" refers to the operational dynamics of a network: across space, over time, at a variety of scales, from the body to the neighborhood, to the region, and across nations' (Ettlinger and Bosco (2004) *Antipode* **36**, 2).

Mark and Freundschuh (1997) *Trans. GIS* **2**, 4 identify 'at least five types of spaces . . . because for most if not all human cultures, there are different kinds of spaces, applied to different situations or phenomena, that are conceptualized in different ways'. Every society uses its space differently, both artistically and technologically: 'Europeans have a notion of time and space that is generally assumed by them to be universal. This gratuitous assumption is naive, arrogant, and wrong' (Hopgood (1993) *AGOCG Tech. Report* **24**). 'Space is used as a site and mechanism for gendered relationships of power' (Corroto (2001) *J. Arch. Educ.* **55**, 2).

space–time 'Space and time are treated as sticky concepts that are difficult to separate from each other' (Dodgshon (2008) *Geografiska B* **20**, 1). We need to be able to understand *time* to tell the story of how an individual place developed, and to understand *space* to understand the complexity of the way different places develop. Galton (2004) *Spat. Cogn. & Comput.* **4**, 1 explains that in the real world there is rarely a clean separation between (spatial) objects and (temporal) events. 'For example, a meteorologist tracking a tropical cyclone will naturally view it as an object which comes into existence at a certain place and time and moves along a well-defined path, eventually fading away and ceasing to exist. From the point of view of an inhabitant of a town in its path, the cyclone is much more like an event: the sudden onset of strong winds and rain, bringing destruction to the town and then dying away leaving the inhabitants to pick up the pieces.'

Massey (1999) *TIBG* **24**, 3 argues that space 'is not static, not a cross-section through time; it is disrupted, active and generative. It is not a closed system; it is constantly, as space-time, being made.' Lane (2001) *TIBG* **26**, 2, following Massey's argument (op. cit.) that the sort of space–time model that we adopt needs to be informed by the entities that we study, concludes that some of the space–time models that Massey critiques 'may still be fundamental to what we do, and in no sense necessarily ahistorical'.

For *geomorphologists, Raper and Livingstone (1995) *Int. J. GIS* **9** propose a **relational space–time,** since the identity of a geomorphological phenomenon is generated by the spatio-temporal relations between causal processes and the local nature of the environment. 'The relational notion of space-time implies internal relations; external influences get internalized in specific processes or things through time' (D. Harvey 2005). Neutens et al. (2007) *Int. J. GIS* **21**, 10 visualize space and time in three dimensions using a hybrid (CAD/GIS) system.

spatial To do with geographic space; with distribution or location across a landscape or surface. See B. Warf and S. Arias (2008) on the **spatial turn.**

spatial analysis (spatial data analysis) A geographical analysis which seeks to explain patterns of human behaviour and their spatial expression in terms of mathematics and geometry; see L. Sanders (2008). *See also* GEOCOMPUTATION.

spatial autocorrelation A clustering pattern in the spatial distribution of some variable which seems to be due to the very fact that the occurrences are physically close together, that is, that they are in geographical proximity. Spatial autocorrelation is widespread: rich people move to areas where other rich people live; people only go to parties because other people go, and so on. In any exercise using sampling, the problem with spatial autocorrelation is that if observed values are spatially clustered, samples will not be independent; see T. Bailey and T. Gatrell (1995) and Ehlschlaeger (2000) *Computers Env. & Urb. Sys.* **24**, 5 on measuring spatial autocorrelation.

spatial covariation The study of two or more geographic distributions which vary over the same area, such as unemployment and crime, or vegetation cover and gradient. A close fit shows that the phenomena are associated by area.

spatial differentiation A. Knowles (2002) is probably not alone in defining geography as 'the study of spatial differentiation'. Within geography, spatial differentiation is seen as:

- a process (see Krevs (2004) *Dela* **21** and Li and Wu (2006) *Housing Studs* **21**, 5),
- a causal mechanism (see Walker (1978) *Rev. Radical Polit. Econs* **10**, 3),
- a response (see Spark and Burt (2002) *Soil Sci. Soc. Amer. J.* **66**).

The identification and measuring/mapping of spatial differentiation is fraught with difficulties. The European Union has proposed seven criteria for the differentiation of the European territory (Study Programme on European Spatial Planning—Final Report, 2001). There you will find the list of methods, together with the problems.

spatial diffusion A phenomenon spreading through geographic space, such as a language. Silvestre and Campoy (2002) *Studia Anglica Posnaniensia*

38 conclude that models may give adequate statistical explanations for the volume, distance, and direction of geographical diffusion flows, but they do not reveal causal factors.

spatial-interaction theory The view that the movement of persons between places can be expressed in terms of the attributes, such as population or employment rates, of each place. See Greenwood and Hunt (2003) *Int. Reg. Sci. Rev.* **26**, 1 for an excellent review.

spatiality The effect of space on actions, interactions, entities, and theories. 'Spatiality is a social construct, not an exogenously given, absolute coordinate system . . . but a product of the political economic system' (Sheppard (2004) *Antipode*, **36**, 3). 'Spatiality is constitutive of the particular ways in which the different modalities of power take effect. The two major dimensions of spatiality are reach and intensity' (Clegg (2006) *Area* **38**, 1). 'From the perspective of spatiality, space and society do not gaze at each other but rather are mutually embedded' (Ettlinger and Bosco (2004) *Antipode* **36**, 2). The term is also used as a synonym for distribution, or spatial expression.

spatial margin The points at a distance from a factory where costs are equal to revenue and no profit is made; 'when firms choose, intentionally or by accident, a location that falls within the so-called spatial margin of profitability, they have a better chance to survive and prosper' (Boschma and Frenkma (2006) *J. Econ. Geog.* **6**, 3). See also Boschma and Lambooy (1999) *J. Evol. Econ.* **9**.

spatial monopoly The monopoly of a good enjoyed by a supplier over a marked area where no competitor exists. 'In the literature on spatial monopoly, it is common to assume that the firm produced (or sold) only one good. This assumption, however, departs from the real world' (Peng (2004) *Southern Econ. J.* 1 Jan.). 'Spatial monopoly leads to the spatial impossibility theorem: no competitive equilibrium involving trade across locations exists in a homogeneous space' (Stafford (2003) *AAAG* **93**). Spatial monopolies happen when firms agree to split the market spatially, or through the privatization of public utilities, as in the case of the British Water Boards.

spatial ontology The many ways people make geography. Hayes and Laforte (IHMC, U. W. Florida) identify some particularly 'geographical' ontological concepts: continuant—physical entity with space-like part; occurrent—physical entity with time-like parts; location—a piece of physical space; terrain—a piece of geographical space (consisting of locations suitably related to each other); and history—the spatio-temporal region. Knopp (2007) *Prof. Geogr.* **59**, 1 focuses on the potential of feminist-inspired and allied queer geographies to rethink a variety of spatial ontologies.

spatial preference The choice of one spatial alternative, such as a housing area, or a shopping centre, rather than another. Meester and Pellenbarg (2006)

a. costs vary spatially; revenue constant

b. revenue varies spatially; costs constant

c. costs and revenue vary spatially

Spatial margin

Tijdschrift **97**, 4, in a study of Dutch entrepreneurs, find that their criteria are potency, activity, and evaluation, and that spatial preference does not change much over time.

spatial randomness A distribution pattern of some variable where the observed spatial pattern is no more, nor less, likely than any other; 'complete spatial randomness is a neutral condition often thought to reflect the dominance of external, stochastic processes, for example disturbance in ecosystem development' (Cutler et al. (2008) *J. Ecol.* **36**, 1).

spatio-temporal fix Capitalism is an imperfect system, which creates problems that need fixing. For example, over-accumulation can result in surpluses, *dumping, idle capacity, lack of investment opportunities, and unemployment. The capitalist system fixes these problems by developing new (*temporal*) sites (*spatial*) that can absorb surplus labour and/or capital. MacDonald (2007) *PHG* **31**, 5 thinks that so much money and time is spent on outer space that 'capitalism will never reach wealth saturation'. MacDonald also has a fascinating account of GPS and sat-nav.

special economic zone (SEZ) An area within the territory of a *state where normal regulatory practices, such as foreign exchange and remittance restrictions, and export levels, do not apply. By 2002, there were more than 5000 SEZs in China (Chen (2008) *Urb. Studs* **45**, 2). Li et al. (2005) *Pacif. Econ. Rev.* **10**, 4 find that foreign investment has played an important role in Chinese SEZs, but that the future of the Chinese open door policy will depend on improving the Chinese political system. *See also* FREE TRADE ZONE.

speciation The concept of ecological speciation invokes Darwinian natural selection as a means of generating biodiversity and, specifically, reproductive isolation (Rundle and Nosil (2005) *Ecol. Letts* **8**). Patten (2008) *J. Biogeog.* **35**, 1 argues that speciation and specialization, 'long studied as distinct disciplines, one evolutionary, the other ecological', could well be 'one and the same thing'.

species–area relationship 'Along a gradient of ecosystems of increasing size, the numbers of species inhabiting those ecosystems increases; rapidly at first, but then more slowly for the larger ecosystems' (Lomolino (2001) *J. Biogeog.* **27**, 1). Losos and Schluter (2000) *Nature* **408** explain the relationship as the outcome of the effect of area on immigration and extinction rates. Báldi (2008) *J. Biogeog.* **35**, 4 argues that habitat heterogeneity overrides the species–area relationship.

specific humidity The actual mass of water vapour present in a kilogram of moist air; in general terms, the mixing ratio. See Chapman and Thornes (2003) *PPG* **27**, 3 on the use of GIS to compare specific humidity with wind flow.

speleothem A collective noun for depositional features such as *stalactites and *stalagmites. Most are made of calcareous rock. See Mattey et al. (2008)

Geophys. Res. Abstr. **10**, EGU2008–A–00461 on seasonal controls on speleothem growth.

sphere of influence The field/hinterland, or area of control; see Bilgim (2007) *Pol. Geog.* **26**, 7 on Turkey's 'natural sphere of influence'. The term is also used for the *urban field of a city; see Ma (2005) *Pol. Geog.* **24**, 4.

spheroidal weathering The weathering of jointed rocks by water along the joints, such that shells of decayed rock surround isolated, unaltered corestones. Sarracino and Prasad (1989) *Geojournal* **19**, 1 see the concentric patterns formed by spheroidal weathering as evidence of *self-organization.

spillover In economic geography, an overflow, or spreading, of information from elsewhere, whereby 'the mysteries of the trade become no mysteries, but are as it were in the air' (A. Marshall 1920), such as spillovers as that between university *research and development, and firms (although Boders (2004) *J. Econ. Geog.* **4** reports that only regions with low R&D density benefit from interregional spillovers). **Knowledge spillovers** mean that firms can acquire knowledge and information created by others through expenditure on industrial R&D without paying for it. Movements of labour between firms, informal contacts, business meetings, close customer–producer relations, and face-to-face contacts are all mechanisms fostering spillovers; see Engelkstoft et al. (2006) *Papers Reg. Sci.* **85**, 1. Adams (2002) *J. Econ. Geog.* **2** finds that academic spillovers are more localized than industrial spillovers. *See* PROXIMITY.

spit A ridge of sand running away from the coast, usually with a curved seaward end. Breaking waves produce a turbulent flow of water that dislodges sediment from the beachface, and create the *longshore currents that transport the sand along the beachface. 'Without breaking waves, the littoral current and the drift of sand ceases' (Oertel and Overman (2004) *Geomorph.* **58**, 1–4). Since the water beyond the end of the spit is usually deep, waves do not break, but are refracted and diffracted around the spit, curving it. Clemmensen et al. (2001) *Holocene* **11**, 3 find that spit growth is closely linked to sea-level variation.

   Ciavola (1997) *Catena* **30** concludes that 'the strategy of coastal management that seems more appropriate to the spit-tidal flat sedimentary system is a "do-nothing option". The sea has to be left free to overwash the spit, supplying sand to the tidal flat. The spit will start again to migrate uniformly westwards, even if this means that for some time it will be transformed into an island.'

spodosol A soil of the *US soil classification. *See also* PODZOL.

spontaneous settlement An unsatisfactory term for a *squatter settlement, since it implies that squatter settlements are unplanned. They often are planned, although rarely legally.

spread (lateral spread) The *slumping of clayey sediment at the edge of *ice sheets.

spread effect The filtering of wealth from central, prosperous areas, to *peripheral, needier areas. The spread effect is the spatial equivalent of trickle-down economics.

spring tide *See* TIDE.

squall A storm characterized by sudden and violent gusts of wind. A **squall line** is a cluster of storm-bearing *convection cells along a non-frontal line, or belt; see Wang and Carey (2005) *Monthly Weather Rev.* **133**, 6. Squall lines form along an *occlusion, along an *inversion created by a flow of maritime air over a cool surface, or below an upper-air trough (López (2007) *Atmos. Res.* **83**, 2–1). Mohr et al. (2003) *J. Hydromet.* **4**, 1 define squall line as a convective line with a trailing stratiform cloud anvil.

squatting The unlawful requisition of housing or land. **Squatter settlements** are illegally occupied, which necessarily involves insecurity of tenure, irrespective of the quality of the buildings and the prevailing physical conditions. Squatter settlements are not necessarily *slums because not all houses in a squatter settlement have to be substandard for it to be a classical squatter settlement (Matovu (2000) Municipal Dev. Prog., Eastern and Southern Africa). *Compare with* SLUM.

Squatter settlements grow because demand for cheap housing outstrips supply. Houses are often flimsy, and sanitation is grossly inadequate (in Indian urban areas, scarcely 20% of the population has access to water/flush toilets; see Raghunaathan (2006) *Hafta Magazine*), power may be unavailable, roads are not metalled, and education and medical facilities are severely limited. Brazil's squatter settlements are characterized by a lack of access to credit and funding for urban infrastructure, and alarming levels of violence, particularly in poor areas; see the movie *City of God*.

It's heartening that Klaufus (2000) *J. Housing & Built Env.* **15**, 4 focuses on the inhabitants of squatter settlements as individuals; similarly, Datta (2007) *Gender, Place & N. Cult.* **14**, 2 describes the way different women in a Delhi squatter settlement experience different forms of spatial control, using multiple modes of resistance to different spatial controls. Patel and Arputham (2007) *Env. & Urbs* **19**, 2, warn of 'the disruptions that the slum dwellers will bring if they are not involved in the planning and implementation of such redevelopments'.

stability The state of a parcel of air which, if displaced vertically, will return to its original position. Thus, if a parcel of air cools more on rising than the air which surrounds it, it becomes denser than its surroundings and therefore sinks. The atmosphere is **absolutely stable** when the environmental *lapse rate is less than both the dry and saturated *adiabatic lapse rates. Atmospheric stability is reinforced by *inversions. See Peppier (1988) *SWS Miscellaneous Publication* **104**.

stable ecosystem One which will maintain, or return to, its original condition after any disturbance. 'The stable ecosystem in equilibrium is a myth' (Bowker in A. Ong and S. Collier, eds 2005).

stable population A population where fertility and mortality are constant. A stable population will show an unvarying age distribution and will grow at a constant rate. Stable population theory is based upon measuring the reported age–sex distribution against that of an appropriately chosen stable population, in order to validate census information (*UN Publication* ST/ESA/STAT/SER. M/67/Rev. 2).

stack An isolated islet or pillar of rock standing up from the sea bed, close to the shore. Initially, caves develop; some ultimately meet to form an *arch. With the collapse of the roof of the arch, a stack is left.

stadial A time when glaciers advanced and periglacial conditions extended, but not as significantly as in a *glacial. Renssen et al. (2005) *Vrije Universiteit Amsterdam*, maintain that stadial–interstadial transitions in the North Atlantic were most probably caused by changes in the thermohaline circulation in the North Atlantic Ocean.

stage In geology, a stratigraphic division of rocks formed at the same time, usually with the same fossil assemblage.

stalactite A limestone column hanging from a cave roof. It grows as an underground stream deposits its dissolved load of calcium carbonate, and may extend far enough to meet a *stalagmite, thus forming a continuous column. Stalactites retain a record of environmental conditions; see Mayer et al. (2003) *Geophys. Res. Abstr.* **5**, 12952.

stalagmite A limestone column, formed on a cave floor when underground water evaporates, depositing the calcium carbonate dissolved within it.

standard industrial classification A grouping of industries classified by a government.

(((●))) SEE WEB LINKS

• British government standard industrial classification.

standard of living, index of An assessment of living standards using indicators such as access to health care, standard of education, house ownership, car ownership, take-home pay, and employment rates. As the standard-of-living index is a constructed measure, it does not have an absolute interpretation, so it's better to use it hierarchically (Subramanian et al. (2006) *Am. J. Public Health* **96**).

standardized mortality ratio The ratio of observed to expected deaths. Julious et al. (2001) *J. Public Health Med.* **23**, 1 criticize this statistic.

standing crop The *biomass present at a given time in a given area.

standing wave A wave where the surface oscillates vertically each side of a fixed, motionless node, with alternating up and down motion each side of it. The wavelength of a standing wave is the length of the water body it develops in (R. Davis and D. FitzGerald 2004). The most common standing wave is the *seiche.

star dune *See* SAND DUNE.

state A territorial unit with clearly defined and internationally accepted boundaries, an independent existence, and responsibility for its own legal system. 'Most social science has taken state territoriality for granted, equating it with homogeneous bounded space neatly dividing the domestic from the foreign' (N. Brenner et al., eds 2003). 'The state thrives through simplification, making social relations legible so they can be controlled' (Taylor (2007) *TIBG* **32**, 2). B. Jessop (2002) argues for a new state form: 'Schumpeterian competition replaces Keynesianism, workfare substitutes for welfare, postnational multiple scales are imposed upon the national mosaic, and a governance regime blurs the private–public division that defines the state.' Wacquant (2001) *Eur. J. Criminal Policy* **9**, 4 argues that neoliberal ideology has brought about the erasure of the economic state, the dismantling of the social state, and the strengthening of the penal state: 'an integral part in the consolidation of the penal state in France has been the state's practices that produce spaces.' 'Processes of globalization increasingly breach state borders and have led to questioning of the degree to which "fixed" territorial structures of the modern state are adapted to organizing, describing and analysing the contemporary world.... The spatial foundations of the state are also undermined by formal international integration processes, most notably those of the supranational European Union' (Wise (2006) *Area* **38**, 2).

State theorists analyse the state's role in regulating capitalism; fearlessly, Purcell and Nevins (2005) *Pol. Geog.* **24**, 2 hold that the current analysis of the state in critical state theory 'is not in a position to think comprehensively about the imperatives that affect the state and shape state restructuring'.

steady state A system where input is balanced by output; see Willet and Brandon (2002) *Geology* **30**, 2 on steady states in mountain belts. *See also* DYNAMIC EQUILIBRIUM.

steam fog A shallow, wispy, smoke-like fog, formed when cold air passes over warmer water, and is rapidly heated. Convection currents carry moisture upwards, which quickly recondenses to form fog. *See also* ARCTIC SEA SMOKE; FOG.

stem flow *See* INTERCEPTION.

step An abrupt, but short, break of slope, especially in a *glacial trough. Steps tend to be smoothed and striated upstream, as the ice grinds the bedrock, and craggy downstream, because of *quarrying; see R. Huggett (2003).

steppe The wild grasslands of central Europe and Asia. The natural vegetation has by now been removed or much altered by cultivation and grazing.

stepped leader *See* LIGHTNING.

stepwise migration A type of migration which occurs in a series of movements, for example, moving to a town larger than the home town, but not directly to one of the city regions (Nilsson (2003) *Urb. Studs* **40**, 7).

sticky Referring to those cities and regions which can 'anchor' income-generating activities. For the conditions under which some places manage to remain 'sticky', see Markusen (1996) *Econ. Geog.* **72**, 3.

sticky spot An area of high drag at the glacier bed; see Mair et al. (2001) *J. Glaciol.* **47**, 9–20, and Kavanaugh and Clarke (2001) *J. Glaciol.* **47**, 158 on basal motion precipitated by the gradual failure of a sticky spot.

stillstand A time of tectonic inactivity: see Everest and Kubik (2006) *J. Quat. Sci.* **21**.

stochastic Governed by the laws of probability. Frenken and Boschma (2007) *J. Econ. Geog.* **7** reformulate Simon's model of stochastic growth.

stock An irregular igneous *intrusion, cutting across the strata of the *country rock. A stock is less than 10360 hectares in area (Buddington (1969) *GSA Bull.* **70**).

stocking rate The number of livestock per unit area, a rate used to measure the ecological pressure of livestock (K. Zimmerer and T. Bassett 2003).

stone pavement In a *periglacial landscape, pavements of large boulders on saturated land in valley floors. Wells et al. (1995) *Geology* **23** argue that stone pavements remain at the surface because of vertical inflation caused by windblown dust.

stone stream A linear arrangement of rocks, seemingly developed through *freeze–thaw; see Hall (2002) *S. African J. Sci.* **98**, 78.

stone stripes *See* PATTERNED GROUND.

stoping The assimilation at depth of *country rock by an igneous *intrusion. The heat of the intrusion melts the country rock which then mingles with the *magma. Žák et al. (2006) *Int. J. Earth Sci.* **95**, 5 hold that roof pendants, stoped blocks, and discordant intrusive contacts are evidence of magmatic stoping.

storm beach A beach ridge situated well above the normal limit of high tides, formed as unusually large waves fling ashore shingle, cobbles, and boulders.

storm surge As a *depression passes over the sea, the water is subject to lower *atmospheric pressure than its surroundings, causing the water level to rise and the surrounding water to sink; a fall in pressure of 1 mb will produce an increase in height of almost 1 cm. Such rises in the sea surface can be compounded by high winds.

Storm surges associated with *hurricanes are common in the relatively confined locations of the Gulf of Mexico and the Bay of Bengal. Davis and Sheng (2003) *Int. J. Numer. Meth. Fluids* **42** propose, and illustrate, a parallel storm surge model.

stoss and lee Glacial landforms with a pronounced asymmetric profile—an abraded slope on the up-ice side (**stoss**), and a steeper rougher quarried slope on the down-ice side (**lee**) that has both abraded and quarried surfaces. Individual examples are commonly known as roches moutonnées (Glasser and Bennet (2004) *PHG* **28**, 1).

strategic environmental assessment *See* ENVIRONMENTAL IMPACT.

stratified Showing distinct layers. Glacial till is often stratified through sorting and redeposition by *meltwater. *See also* SORTING.

strato-, stratus Layered cloud. *See also* CLOUD.

stratosphere A layer of the Earth's atmosphere, above the *troposphere, 50 km thick. See Cohen et al. (2007) *J. Climato.* **20**, 21 on stratosphere-troposphere interaction events.

stratovolcano A cone-shaped volcano with a layered internal structure; see Hildreth et al. (2003) *GSA Bull.* **115**, 6.

stratum In geology, a layer of distinctive deposits with surfaces roughly parallel to those above and below (pl. *strata*).

streamer Aeolian sand transport occurs primarily in the form of streamers—'wriggles' of sand—that are driven by distinct eddies travelling down toward the surface through the boundary layer and scraping across the bed while exciting saltation along their trail (Baas (2007) *Geomorph.* **91**, 3–4).

streamflow Discharge that occurs in a natural stream channel. Streamflow velocity varies with distance from bed; with distance from banks, downstream, and over time. Channel geometry affects turbulence, and thus velocity, and hydraulic jumps and drops affect velocity and depth. Variations in streamflow affect the amount and type of geomorphic work a river can do. The driving force is gravity, which varies with channel gradient; the resisting forces are the viscosity of the water; the *Reynolds number; *Manning's roughness coefficient; and the *Froude number.

See Rantz et al. (1982) *US Geol. Surv. Water-Supply Paper* **2175**, 631 on the measurement and computation of streamflow.

stream order The numbering of streams in a network. Ng (2006) *Geomorph* **76**, 1–2 finds a negative relationship between stream order and landslide density. *See also* BIFURCATION RATIO.

streamsink *See* DOLINE.

strength In geomorphology, the resistance of a rock mass to rupture under stress. Strength varies with, in order of importance: the spacing of joints, the cohesion and *frictional force of the rock, the *dip of any fissures, the state of weathering of the rock, the width of fissures, the movement of water in or out of the rock mass, the continuity of the fissures, and the amount of infilling of soil within the fissures. See Jackson (2002) *GSA Today* **12**, 9 on the 'jelly sandwich' view of the continental lithosphere; see also Townend and Zoback (2000) *Geology* **28**, 5 on the way faulting keeps the crust strong.

stress The force applied to a unit area of a substance measured in newtons per square metre. **Compressive stress** crushes the rock; see Bunds (2001) *GSA Bull.* **113**, 7. **Tensile stress** is a force which tends to pull a rock or soil apart and which may cause fractures and pores to open; see Wiprut and Zoback (2000) *Geology* **28**, 7.

stress–response model A model based on environmental stress–response, including stress-dependency, frequency–recovery relationships, environmental heterogeneity, spatial hierarchies and linkages, and temporal change (Gabriel and Kreutzwiser (2000) *Env. Manage.* **25**, 1).

striations Long scratches on a rock, a few millimetres across, formed by glacial abrasion, usually in clusters, often well developed on the stoss side of roches moutonées. Striations can show the direction of glacier movement (Fjellanger (2006) *Geomorph* **82**, 3–4).

strike The direction of a sloping stratum, perpendicular to the *dip.

strike-slip fault A horizontal displacement fault; surfaces on opposite sides of the fault plane have moved horizontally relatively to each other, and parallel to the strike of the fault (also called transcurrent fault).

strip mining The removal of the overburden to extract minerals. See E. Reece (2006) on strip mining and the devastation of Appalachia.

Strombolian eruption Volcanic activity, relatively frequent and mild, in which gases escape at intervals, producing small explosions.

structural adjustment Changes to a nation's economy usually imposed upon debtor nations by their institutional creditors; for example the

*International Monetary Fund. Changes include: slimming down of government employment, cuts in taxation to stimulate investment, a reduction in public services, privatization of the economy, the promotion of exports; liberalization, and a reduction in government subsidies in order to bring domestic prices more in line with world prices (Clarke and Howard (2006) *Geogr. J.* **172**, 2) and combat inflation (with great success in Bolivia, for example: Kohl (2006) *Antipode* **38**, 2). However, in countries that have undergone structural adjustment programmes, the majority of citizens have seen their standard of living fall (M. Kurtz 2004). 'It is clear that structural adjustment policies have negative human rights consequences for loan recipients, [but] these bad outcomes probably have been unintended' (Abouharb and Cingarelli (2006) *Int. Studs Qly* **50**, 2). Clarke and Howard (loc. cit.) note mixed effects in Jamaica: while formal employment has fallen in Kingston, the investment in education and housing undertaken by the middle and upper classes has helped them to sustain the dislocation introduced by structural adjustment.

Structural adjustment policies may be adopted unilaterally; see Wong (2007) *China & World Econ.* **15**, 2.

subaerial Occurring on land, at the Earth's surface.

subduction The transformation into *magma of a denser *plate as it dives under another, less dense, plate. A **subduction zone** occurs where rocks of an oceanic plate are forced to plunge below much thicker continental crust. As the plate descends it melts and is released into the *magma below the earth's crust. Tectonic erosion of the overriding plate by the downgoing slab is believed to occur at half the Earth's subduction zones; see Vannucchi et al. (2008) *Nature* **451** who provide excellent diagrams. The subduction of features such as ridges and seamounts increases basal erosion and the subsidence of the accretionary wedge (Morgan (2008) *Geophys. Res. Abstr.* **10**, EGU2008-A-00560). See also Chiarabba et al. (2008) *Earth and Plan. Sci. Letts* **268**, 3–4.

subglacial water The principal water inputs to a glacier are the bergschrunds in cirques and crevasse fields lower on the glacier. Subglacial water flows are maintained by high water pressure, rather than by melting of the subglacial channel walls. See Siegert (2005) *Ann. Rev. Earth & Plan. Sci.* **33** and Boulton et al. (2007) *Qly. Sci. Rev.* **26**, 7–8.

subgraph In *network analysis, the graphs, or networks, forming an unconnected part of a whole graph or network.

sublimation Sublimation is *not* the direct conversion of solid ice to vapour—as the liquid on the surface of the solid lump has a high partial pressure it rapidly evaporates, so that it appears that the solid phase is changing directly into the gas phase (Holland (2004) *Weather* **59**, 5). **Latent heat of sublimation** is released in this way.

sublittoral zone The zone from the lowest mark of ordinary tides to the end of the *continental shelf.

submarine slide A form of underwater *mass movement. 'Earthquakes are among the most frequently suggested triggering mechanisms for submarine slides . . . sliding in various forms is a major agent in the evolution of passive continental margin morphology' (Bugge et al. (1987) *Geo-Marine Letts* 7). See Yamamoto et al. (2007) *Island Arc* **16**, 4; and Dawson and Stewart (2007) *PPG* **31**, 6 on tsunamis generated by submarine slides.

sub-optimal location A satisfactory, but not *optimal, location.

subsequent streams Rivers running down the *strike of usually weak *strata, or along the line of a fault, at right angles to *consequent streams.

subsistence farming A farming system where almost all of the output feeds and supports the household. See Pandi et al. (2007) *Ag. Sys.* **92**, 1–3 for a detailed, practical analysis.

subtropical anticyclones Areas of high pressure formed when air, which has risen in the tropics, subsides in subtropical areas. The air is warmed adiabatically as it descends; therefore rainfall is unlikely. The subtropical anticyclone over the North Atlantic, often called the 'Azores' or the 'Bermuda high', influences weather and climate over the eastern United States, western Europe, and north-western Africa (Davis et al. (1997) *J. Climate* **10**, 4). Miyasaka and Nakamura (2005) *J. Climate* **18**, 23 suggest a local land–sea–atmosphere feedback loop associated with a subtropical high.

suburb A one-class community, located at the edge of the city, with low rates of housing per hectare. 'Suburbs were originally characterised by housing types not entirely different from those found closer to the urban core, but the "garden suburb" tradition came to dominate Western cities during the twentieth century . . . Individual suburbs became more regulated, more "planned" and more middle class. The widespread adoption of the garden suburb model transformed streetscape into landscape' (Sharpe (2005) *Canad. Geogr./Géog. canad.* **49**, 4). Suburb-making in the USA is 'the most geographically extensive version of suburbanization' (Martin (2007) *Area* **39**, 1). 'The current structure of financing of communities and municipalities is mainly responsible for growing settlements . . . a community can increase its revenue by providing plots of land for interested families and businesses' (Jetzkowitz et al. (2007) *Sociol. Ruralis* **47**, 2). Peach (2000) *PHG* **24**, 4 reports on immigrants in 'leafy suburbs': 'instead of the traditional central concentration of the first generation, the new pattern achieves instant suburbanization but remains fairly segregated.'

succession (plant succession) A series of complexes of plant life at a particular site. Plant succession is viewed as the development of plant life on originally bare earth, in a definite sequence; see Leal and Lorscheitter (2007) *Acta Bot. Bras.* **21**, 1.

sunbelt In the USA, the southern and western states, such as New Mexico, Arizona, and Florida. Guhathakurta and Stimson (2007) *Int. Plan. Studs* **12**, 2 report on sunbelt coalitions that form to save the environment, 'supported by the same businesses that had earlier supported new development at the expense of existing neighbourhoods'. *See* SNOWBELT.

sunspot A dark patch on the surface of the sun, usually occurring in clusters, and lasting about two weeks. Sunspot activity fluctuates in an eleven-year cycle. It has been suggested that the sun is 1% cooler when it has no spots, and that this variation in solar radiation affects the climates of the earth; see Schrier and Vesteeger (2001) *Geophys. Res. Letts* **28**, 5.

superadiabatic lapse rate A *lapse rate over the 9.8 °C/1000 m dry *adiabatic lapse rate, rare in the free *atmosphere, but common just above land surfaces emitting strong terrestrial *radiation. When the environmental lapse rate is superadiabatic, the density of the atmospheric parcel relative to its surroundings results in a buoyancy force in the same direction as original vertical movement (Slonaker (1996) *Weather & Forecast.* **11**).

superimposed drainage If a drainage pattern develops on a non-resistant mass covering a buried, resistant bedrock, and if base level falls enough, the drainage will cut into the bedrock in the established pattern (Douglass and Schmeeckle (2007) *Geomorph.* **84**, 1–2).

Super Output Areas (SOAs) A new geographic hierarchy designed to improve the reporting of small area statistics in England and Wales. The layers of SOA are:

Lower layer	Minimum population 1000; mean 1500. Built from groups of OAs (typically 4 to 6) and constrained by the boundaries of the Standard Table (ST) wards used for 2001 Census outputs.
Middle layer	Minimum population 5000; mean 7200. Built from groups of lower layer SOAs and constrained by the 2003 local authority boundaries used for 2001 Census outputs.
Upper layer	To be determined; minimum size *c*.25000.

(⊕) SEE WEB LINKS
• Names and codes for Super Output Area geography.

supraglacial On the surface of a glacier. See Benn et al. (2000) *Int. Ass. Hydrol. Scis* **264** on the early stages of **supraglacial lake** growth.

surface boundary layer *See* BOUNDARY LAYER.

surface process model *See* LAND SURFACE PROCESS MODELS.

superterrane *See* TERRANE.

surge phenomenon *See* EDDY.

surging glacier A faster-flowing glacier. Normally, glaciers flow at 3–300 $m^{-1}yr^{-1}$, whereas a surging glacier may reach 4–12 $km^{-1}yr^{-1}$. Some glaciers always flow at surging speed; others surge periodically, suddenly accelerating to 10–100 times their normal velocity. Surging glaciers are closely crevassed, with compressional flow at the leading edge and *extending flow at the rear. The net result is a deepening of the long profile of the glacier. Kamb et al. (1985) *Science* **227** and Kamb (1987) *J. Geophys. Res.* **92** suggest that surging is initiated by a switch from a tunnel system to a distributed system that facilitated sliding. The surge will continue as long as high water pressure sustains rapid sliding.

susceptibility The degree to which an environment, landform, or organism is sensitive to a particular factor. A landslide zonation map divides the land surface into zones of varying landscape susceptibility, based on the estimated significance of causal factors in inducing instability. Such maps identify and delineate potentially unstable areas, so that suitable mitigation strategies can be used, and help planners choose favourable locations for siting development schemes (Tangestani (2004) *Australian J. Earth Scis* **51**, 3). See Sandercock et al. (2007) *PPG* **31**, 2 on a **vegetation-susceptibility index**, defining the force required to bend vegetation to 45° from the vertical; see Märker et al. (2008) *Geomorph.* **93**, 1–2 on **land degradation susceptibility**.

suspension The state in which small, insoluble particles are evenly distributed within a fluid. Particles are carried upwards when *turbulence outstrips the force of gravity.

sustainable development Article 9 of the UN Declaration of Human Rights (1948–66) states that all people 'should promote sustainable development all over the world to assure dignity, freedom, security and justice for all people'. Sustainable development 'meets the needs of the present without compromising the ability of future generations to meet their own needs' (1987 UN Convention on Environment and Development). 'Sustainable development requires a balance between economic prosperity, social equity and the environment to ensure quality of life now without damaging the planet for the future' (UK DEFRA). For the economic geographer, development is sustainable 'if it does not decrease the capacity to provide non-declining per capita utility for infinity' (Neumayer (2003) *Env. & Res. Econ.* **15**).

Advocates of '**strong sustainability**' argue that natural capital—the range of functions the natural environment provides for humans and for itself—should be afforded special protection, while those who espouse '**weak sustainable development**' hold that natural capital can be substituted by other forms of capital, especially produced capital. 'The weak interpretations are bound to an anthropocentric world-view, and have typically phrased the essence of sustainable development in terms of an environmental question, rather than also a developmental one ... The strong interpretations refuse to treat humans and nature as separate from each other and wish to maintain life opportunities by reconciling sound human development and environmental integrity ... while the weak interpretations tend to have faith in authoritative and coercive structures, such as market forces, the strong interpretations hold that more participatory, transparent and democratic processes are required' (Kallio et al. (2007) *Sust. Dev.* **15**).

*Agenda 21 concludes that 'indicators of sustainable development need to be developed to provide solid bases for decision-making at all levels and to contribute to a self-regulating sustainability of integrated environment and development systems'.

'A **sustainable community** is one in which improvement in the quality of human life is achieved in harmony with improving and maintaining the health of ecological systems; and where a healthy economy's industrial base supports the quality of both human and ecological systems (Indigo Development). A **sustainable drainage system** aims to mimic as closely as possible the natural drainage of a site to minimize the impact of urban development on the flooding and pollution of waterways (Jones and MacDonald (2007) *Geoforum* **38**, 3).

((⊕)) SEE WEB LINKS

- UK government sustainable indicators.
- Indigo Development.

sustained-yield resource A view of resource extraction as a mutually advantageous long-term partnership with nature (N. Langston 1995). Forestry may be managed as a slow-growing but renewable resource of fuel and timber and a sustained-yield recreational resource. McCarthy (2006) *AAAG* **96**, 1 chronicles the failures of forestry sustained-yield policies to achieve their stated goals. The implications of sustainable yield resource input depend on the relative importance of capital and resource inputs in production (M. Common and S. Stagl 2005).

swallow hole *See* DOLINE.

swash The water moving up a beach from a breaking wave. The **swash zone** involves a moving land–water boundary travelling across the beach; at least some part of the beachface is usually unsaturated. Water depths in the swash zone are very shallow, and sediment is often transported as complicated, single-phase, granular-fluid flows (Horn (2002) *Geomorph.* **48**, 1–2).

swidden cultivation *Shifting cultivation.

symbiosis An association of two participants whereby both partners benefit, as when flowering plants rely on insects for pollination and the insects feed on their nectar. **Industrial symbiosis** describes organic relationships between dissimilar industries; see Lyons (2005) *Local Env.* **10**, 1, on the use of waste products from one industry as input to another. See Wu and Wall (2005) *Int. J. Tourism Res.* **7** on economic and environmental symbiosis in tourism.

symbolic capital 'The various forms of distinction and prestige acquired through cultural recognition . . . the form that the various species of capital assume when they are perceived and recognized as legitimate' (P. Bourdieu 2005). Forset and Johnson (2002) *AAAG* **92**, 3 discuss the role of symbolic capital in Moscow in the transformation of national identity. See Matos-Wasem (2005) *J. Alpine Res.* **93**, 1 on 'good' air—a slightly diagonal usage of the term symbolic capital.

symbolic places Places invested with meaning: 'symbolic place is the keeper of national consciousness and promoter of communality. The national consciousness and self-esteem are fed by tradition and ethnic symbols' (G. Taylor and S. Spencer 2004). Missingham (2002) *Urb. Studs* **39**, 9 describes Government House, Thailand, in a time of popular protest by the poor: 'the streets outside Government House are both particular and symbolic. The symbolic and metaphorical meanings of mobilising on the streets and pavements served to highlight the poor's exclusion from the places of authority and power within the walls of Government House.'

sympatric Living in the same region; a term used in ecology to specify separate species whose territories overlap. **Sympatric speciation** is the emergence of two species from a single, mixed-parent population, without the aid of geography. Jiggins (2006) *Current Biol.* **16**, 9 finds evidence of sympatric speciation in isolated island habitats.

synchronic analysis The study of the internal linkages of a system at a given time, as in cross-sections in *historical geography (Conedera et al. (2007) *J. Hist. Geog.* **33**, 4).

syncline and **synclinorium** *See* FOLD.

synekism The causal power of urban agglomeration (E. Soja 2000). See Blake (2002) *J. Soc. Anthrop.* **2**, 2 for Soja's explanations.

synoptic Overall, or relating to a large-scale or broad view. Synoptic meteorology is primarily concerned with large-scale weather systems, such as extratropical cyclones, but not on a global scale. See C. N. Hewitt and A. Jackson (2003).

system Any set of interrelated parts. Systems can be classified as open, closed, or isolated. **Open systems**—such as the ocean—allow energy and mass to pass across the system boundary. 'Capitalism emerges as an inherently open system which, through abstract and concrete labour, is constantly infused by its putative "exteriors"; differences of nationality, gender, sexuality, geographical location and so on are constantly gathered together in the

domain of concrete labour and . . . forcibly articulated into a global system'
(N. Castree 1999).

A **closed system** allows energy but not mass across its system boundary. The
Earth system *as a whole* is a closed system (the boundary of the Earth system is
the outer edge of the atmosphere). The Japanese *keiretsu* model represents a
relatively closed system of vertically networked firms; see Gross et al. (2005)
J. Japanese Int. Economies **19**.

In a **cascading system**, a series of small sub-systems are linked from one
system to another; 'terrestrial hydrology involves a dynamic, nonlinearly
lagged and cascading system consisting of complex feedforwards and
feedbacks, with any perturbations to inputs or stores interacting' (Schultze
(1997) *PHG* **21**).

systematic geography The study of a particular element in geography,
such as agriculture or settlement, seeking to understand the processes which
influence it and the spatial patterns which it causes.

tafone A hollow on a sheer face, produced by localized *weathering, mainly through *granular disintegration (pl. *tafoni*). Canopy-shaped, overhanging cave roofs are typical. Norwick and Dexter (2002) *ESPL* **27** find that the rate of tafoni enlargement decreases over tens of thousands of years. Hejl (2005) *Geomorph.* **64**, 1–2, describes a prehistoric wall painting which displays stages of tafoni development.

taiga The predominantly coniferous forest, south of the *tundra in northern continents, also known as boreal forest. See Eugster et al. (2000) *Glob. Change Biol.* **6**, 1 and Kuris and Ruskule (2006) *Baltic Env. Forum* on management of boreal forest habitats

talik Within a *permafrost zone, the layer of unfrozen ground that lies between the permafrost and the seasonally thawed *active layer. Talik most often occurs below rivers and lakes, or where groundwater upwelling occurs along fractures (Delisle (1998) *J. Quat. Scis* **13**). See Grosse et al. (2007) *Geomorph.* **86**, 1–2. For the role of talik in the formation of pingoes, see Gurney (1998) *PPG* **22**, 3.

talus A *scree slope formed of *frost shattered rock *debris which has fallen from the peaks above. A talus slope is usually straight, and at an angle of 34–35°. Slope angle in a talus cone reflects the balance between the supply of sediment from upslope (usually by physical weathering of bedrock) and the processes moving the sediment downslope while spreading it out. Lithology is also an important control. Grain flows are frequent occurrences; see Croasta et al. (2007) *Geophys. Res. Abstr.* **9**, 07610 on grain size, and Pech and Kotarba (2002) *Studia Geomorph. Carpatho-Balcanica* **36**, on talus slopes in the high Tatra. Obanawa and Matsukura (2006) *Computers & Geosci.* **32**, 9 propose a mathematical model for the topographic change of a cross-section of a talus landform. See van Steijn et al. (2002) *PPG* **26**, 4 on **cliff-talus systems.**

The slow, downslope movement of talus is **talus creep**, initiated either by the shock of new fragments falling on the scree or by the movement of individual particles resulting from heating and cooling.

tariff A list of duties or customs to be paid on imports. Tariffs may be imposed to increase the cost of imported goods in relation to domestic production, thereby reducing the volume of imports and keeping the *balance of payments in credit, or to protect domestic industry from foreign competition. **Preferential**

tariffs reduce import duties on products of a certain type or origin. The impact of tariffs may be less than expected: Behrens (2006) *Papers Reg. Sci.* **85**, 3 shows that decreasing transport costs are more likely to trigger agglomeration than decreasing tariffs, and Medvedev (2006) *World Bank Policy Research Working Paper* **4038** finds that preferential trade agreements do not much increase trade flows. In a highly technical paper, Mai et al. (2006) *CIRJE-F-***435** show that, when the transport cost is sufficiently small, tariff competition coupled with firm migration leads to a core–periphery economy.

taxocene *See* TROPHIC LEVEL.

taxon *See* TROPHIC LEVEL.

Taylorism A system of production characterized by the division of factory work into the smallest and simplest jobs, while closely coordinating the sequence of tasks in order to achieve maximum efficiency, as, for example, on a production line. Skilled managers and technicians oversee semi-skilled or unskilled workers who are engaged in simple, repetitive chores. Ellem (2005) *Asia Pacif. J. Hum. Resources* **43**, 2 observes that 'most people would say that the biggest change in work towards the end of the last century was the abandonment of Taylorism'. However, Lorenz and Valeyre (2005) *J. Indus. Relations* **47** observe that traditional Taylorist forms of work organization are holding their own in certain nations and sectors of the EU.

tear fault A *fault characterized by lateral movement, transverse to the *strike of the rocks. See Huang et al. (2004) *Marine Geophys. Res.* **25**, 1–2.

technological revolution The development in the last 25 years in information and telecommunications technology. It has made possible the worldwide integration of economic processes while allowing organizational structures to respond rapidly to changes in markets and technology. Information technology has pushed forward globalization and its overall effects (anon. (2006) *Pop. & Dev. Rev.* **32**, 4). Unilateral advantages do not last long; backwardness can sometimes be advantageous if it creates opportunities for technological 'leapfrogging', as 'late-comer' institutions and capital are not locked into older technologies, enabling quicker uptake and diffusion of new ones (Carmody (2008) *Geog. Compass* **2**, 1). S. Amin (2003) even refers to the 'obsolescence of capitalism' due to the technological revolution, making labour redundant. See Harris in S. Danzigler and R. Haveman (2001) on anti-poverty policy after the technological revolution.

technopole A global, rather than local, *growth pole. Technopoles include three major types: industrial complex of high-tech firms, emerging spontaneously, and linking research and development with manufacturing. See Wang et al. (1998) *Asia Pac. Viewpt* **39**, 3 on the development of technopoles in China, and Fontan et al. (2004) *Canadian J. Urb. Res.* **13**, 2 on the Angus Technopole in Montreal: 'a new type of community initiative which combines endogenous and community development with business oriented projects.'

tectonic Of, or concerned with, the processes acting to shape the Earth's crust.

teleconnections Causal links between patterns of weather in two locations, or between two atmospheric occurrences, which are very far apart, such as the *El Niño–Southern Oscillation, where unusually high sea surface temperatures off the west coast of South America can be connected with droughts in Indonesia. Teleconnections are usually expressed statistically by the amount of shared variance between the two locations or systems. See Simpson and Erkal (2004) *Papers Applied Geog. Confers* **27** on using teleconnections to predict temperatures.

teleology The study of purposes, goals, or ends; the theory that acts, objects, states of affairs, and so on can only be justified in terms of the ends—'rightness' is not intrinsic in an action or process, but is dependent on the consequences of the action or process. A. Memmis (1957) describes the teleology of colonial power and its resistance. H. Obenzinger (1999) claims that the field of 'Holy Land' studies has been controlled by colonialist teleology in much American and Israeli scholarship.

temperate Describing those locations and climatic types falling between subtropical and subarctic.

temperate glacier A glacier with basal temperatures of around 0 °C. Warmth is provided by friction with the bedrock, by shearing within the ice mass, and from *geothermal heat. Glacier flow is rapid because warm glaciers are lubricated at their base, and velocities of 20–30 m per day have been recorded. Warm glaciers are the most effective agents of *glacial erosion. See D. Evans (2003) on active temperate glacier margins, and Lingle and Fatland (2003) *Annal. Glaciol.* **36**, 1 on temperate glacier surges.

temporality In contrast to the measurable and calculated notion of time/chronology, temporality is concerned with the way in which a sequence of events, a kind of history, is physically experienced by those who live through them or experience them. Thus the passing of time is treated not as a neutral dimension but rather as being constituted by social practices (T. Darvill 2002). Academic geographers do not express—or do not want to express—this concept very clearly: 'time and space do not exist singly, but only as a hybrid process term' (J. May and N. Thrift 2001); 'the predominant poststructuralist understanding of temporality is in terms of a series of successive moments of pure contingency, tied together by nothing other than the force of an imposed convention or act of vitalistic will' (Barnett (2004) *Pol. Geog.* **23**, 5); 'to take on board the coevality of space is to refuse that flipping of the imaginative eye from modernist singular temporality to postmodern instantaneity' (Massey in O. Eliasson 2003), and Brown (2007) *Canad. Geogr./Géogr. canad.* **51**, 1 writes of 'the implicit linear temporality of the two-dimensional planes that define the Western agricultural landscape'.

tephigram An *aerological diagram, of two axes aligned at 45° to each other. The nearly horizontal lines show *atmospheric pressure, together with the heights at which they are found; the lines at 45° to the isobars show temperatures, running from bottom left to top right. Superimposed upon these two are three sets of guidelines: those running from bottom right to top left show the dry *adiabatic lapse rate (DALR); those running from bottom left to top right, at about 60° to the base, show constant humidity (mixing ratio lines); and convex, curved lines indicate the saturated adiabatic lapse rate (SALR). With the help of these lines, the behaviour of a rising parcel of air may be predicted. First, the readings from a *radiosonde are plotted to show the environmental lapse rate. Next, from the same source, dew point temperatures are plotted against height. From the temperature and height of the air parcel under investigation, a plot is made of its temperature fall at the DALR, parallel to the DALR guidelines. Similarly, the dew point change of the air parcel with height is plotted, parallel to the mixing ratio lines. Where these two plots intersect, the parcel will begin to cool at the SALR, and a plot is made of the parcel parallel to the SALR guidelines. At the point where this plot intersects the environmental lapse rate line (above), the height of the top of any cloud formed by the parcel may be established.

tephra A deposit made of fragments of rock—from so-called 'bombs' to dust and *ash—shattered by an explosive volcanic eruption. The coarser, heavier particles fall out close to the volcano vent, while the finer dust may be carried hundreds of kilometres. See Hall and Mauquoy (2005) *Holocene* **15** on tephra dating, and Xia et al. (2007) *Env. Geol.* **51**, 8 on correlating tephra layers.

terminal moraine *See* MORAINE.

terms of trade The ratio of the index of export prices to the index of import prices. If export prices rise at a faster rate than those of imports, then there has been an improvement in the terms of trade.

'Since the early 1980s, all developing countries taken together have been experiencing a downward trend in their net barter terms of trade . . . From a development perspective; the use of the additional income resulting from terms-of-trade changes is of crucial importance. For example, if terms-of-trade gains resulting from higher export prices accrue in the form of higher company profits, and if these are reinvested, the medium-term impact on growth will be much greater than in a situation where the gains accrue to the government through transfers from State-owned enterprises, which are used to service the public debt, or in a situation where they accrue to workers in the form of higher wages that are spent for consumption. Similarly, a deterioration in the terms of trade resulting from higher import prices or lower export prices can lead, *inter alia*, to either a reduction of investment, an increase in government

Tephigram

indebtedness or higher unemployment and wage compression if it is not counterbalanced by productivity and export volume growth' (UNCTAD Trade and Development Report 2005).

Chung (2007) *Int. Econ. J.* **21**, 1 finds that labour-augmenting technological progress turns the terms of trade against the growing country while capital-augmenting technological progress shifts them in favour of the growing country.

terrace *See* RIVER TERRACE.

terracette A small terrace, about 50 cm across and closely spaced with other terracettes. They rise above each other in steps of less than a metre. Bielecki and Muelle (2002) *Geomorph.* **42**, 1 propose soil creep as the most likely cause of terracettes.

terracing The construction of a series of horizontal levels built on a hillside, especially in upland or piedmont zones, in order to retain water and reduce soil erosion (C. Barrow (1999), Chen et al. (2007) *PPG* **31**, 4); terracing technology is intimately related to irrigation and the availability of water. Ye and Yang (2002, *J. Chongqing U., Nat. Sci. Ed.* **25**, 6, in Chinese, with English abstract) suggest that slopes of less than 8° are more suitable than steep slopes for changing into terraces. The terrace is a highly variable and only sometimes distinctive cultural artefact (Field (1966) *DTC* AD0679468).

terrain analysis (digital terrain analysis) The use of *remote sensing satellite data for mapping various aspects of terrain, such as land cover, land use, and soils. Software may then be utilized to derive terrain parameters, such as aspect, catchment area, and wetness index, which are then used to describe the morphology of the landscape and the influence of topography on environmental processes. See J. Wilson and J. Gallant (2000).

terrain attribute A quality in the landscape. **Primary terrain attributes** can be computed directly from topographic data and **secondary terrain attributes** combine primary attributes (see I. D. Moore et al. 1991). Schmidt and Dikau (in R. Dikau and H. Sourer, eds 1999) argue that this primary–secondary distinction is insufficient in representing subtle, but fundamental, differences among topographic characteristics. The function of topography is usually scale and landscape dependent (see W. H. Zhang (2000) and Florinsky and Kuryakova (2000) *Int. J. Geog. Info. Sci.* **12**).

terrains of resistance The material and/or symbolic ground upon which action by social groups takes place. 'A terrain of resistance is simultaneously both metaphoric and literal' (Routledge (1996) *Pol. Geog.* **15**, 6/7). See also Routledge (2000) *GeoJournal* **52**, 1. R. Neumann (1998), for example, shows how Arusha National Park, Tanzania, 'embodies all the political-ecological dilemmas facing protected areas in Africa'.

terrain surface The landscape, which may be simulated in a terrain surface display system. Smith and Mark (2003) *Env. & Plan. B* **30** identify six factors as topographic existences: elevation, terrain surface shape, topographic position, topographic context, spatial scale, and landform.

terrane In geology, a fragment or block of crust with an individual geological history that differs from the surrounding areas, and is usually bounded by faults. **Accreted terranes** are those that become attached to a continent by tectonic processes. If a relatively small crustal fragment is accreted to a larger continent, it is an **exotic terrane** (originating elsewhere). **Superterranes** are composite terranes, composed of individual terranes and other assemblages, which share a distinctive tectonic history; see Mezger et al. (2001) *J. Metam. Geol.* **19**, 2.

SEE WEB LINKS

• Cordilleran accretionary terranes.

terra rossa A red intrazonal soil developed in Mediterranean regions on weathered limestone.

terrestrial magnetism *See* GEOMAGNETISM.

terrestrial radiation The heat radiated from the earth. Short-wave *solar radiation reaching the earth does not heat the atmosphere it passes through, but does heat the earth's surface. In turn, and particularly on clear nights, much of this heat is radiated out from the earth. The earth also absorbs terrestrial radiation reflected from the overlying opaque atmospheric layer. It is by long-wave terrestrial radiation that the *atmosphere is heated. Almost one-third of the solar radiation intercepted by Earth is radiated back into space. In urban areas, a reduction in the *sky view factor decreases the loss of terrestrial radiation, thereby contributing to the creation of *urban heat islands (Grimmond (2007) *Geography* **30**, 3).

terrier A written survey of an estate, recording the value and extent of the land. The earliest English example dates from AD 900, but they were extensively produced from the 16th century onwards. See Fletcher (1998) *TIBG* **23**, 2.

territoriality The need by an individual or group to establish and hold an area of land is an urge, fuelled by aggression, to define a territory for mating and food supply. In human beings, territoriality is an organization of space in order to make sense of it. The individual needs security and identity, and this is shown most clearly in relation to the home, and the community requires a suburb or small town with which to associate. The importance of territory extends to larger units; the reorganization of the counties of Britain always causes distress, and the term is most often associated with nation-states which have the formal power to demarcate and control their borders. Hensel in J. Vasquez, ed. (2000) notes that, between 1816 and 1992, over half of all disputes began between neighbours, over a quarter involved territorial issues, and both contiguity and territory were involved in over half of all full-scale wars. See D. Storey (2001) for theories of territoriality and the geographical outcomes of territorial control, Johnston (2001) *Polit. Geog.* **20**, 6 on territoriality in political geography, and Oosten (2003) *Acta Politica* **38**, 2 on territoriality and globalization. Heeg and Ossenbrügge (2002) *Geopolitics* **7**, 3 provide a concise overview of the history of the EU territoriality; see also K. Cox (2002). See Johnson (2001) *Pol. Geog.* **20**, 6 on theories of territoriality.

Agnew (2005) *AAAG* **9** defines **political territoriality** as the administration of infrastructural power, and identifies four 'sovereignty regimes' that result from distinctive combinations of central state authority and degree of political territoriality. Lewis et al. (2002) *PHG* **26**, 4 describe a **territorial point of entry**, which 'emphasizes local social institutions, local government, the territoriality of resources, the articulation of local and global knowledge, and the spatial strategy of the enterprise'.

territorial justice The application of ideas of *social justice to an area of territory; that is, the identification by a government of areas of need, followed by a deliberate policy of redressing an imbalance. Ideas of social justice vary according to the *mode of production and the prevailing ideology, and so territorial justice varies globally. See Morgan (2006) *Publius* **36** on territorial justice and the North–South divide.

territorial seas (territorial waters) The coastal waters together with the sea bed beneath them and the air space above them, over which a state claims *sovereignty. Most countries claim twelve nautical miles. In 1983, the Law of the Sea Convention proposed a 200-nautical-mile exclusive economic zone, but most nations are less than 400 nautical miles apart, so a median line is drawn between the baselines of the states concerned.

territory **1.** In ecology, the living space of an animal which it will defend from the forays of other territorial animals. Animals need space in which to reproduce and their territory can be some or all of the following: a source of food, a source of mates, and a breeding area. Pakeman (2001) *J. Biogeog.* **28**, 6 investigates possible links between simple migration rates and herbivore territory size.

   2. A socially constructed division of space. A 'national territory'—the area of land seen to be inhabited by a nation—is based on claims to a particular space, usually backed up by references to history (actual or invented) and central to the nation's being. 'Territory comprises ancestors, knowledge, the use of plants, their evolution, the perception of the cosmos, customs and community and living history' (Ceceña (2004) *Antipode* **36**, 3). Jones (2008) *PHG* **32**, 1 argues that 'it is not territory itself that produces spatial differences between "non-economic" influences but the social practices enacted across and through them. Places are social constructions that are themselves potentially, infinitely "translocal".'

   Nonetheless, control over territory is a key political motivating force, and competing territory claims are a major source of conflict. In a paper that well rewards careful study, Alatout (2006) *Pol. Geog.* **25**, 6 shows that different perceptions of territory between Palestinians and Israelis lead to forms of resistance that are focused on property rights and questions of sovereignty in Palestine and on concerns over quality of life of the population in Israel. M. Toft (2003) proposes a theory of ethnic violence, civil wars, terrorism, and many interstate wars. See Elden in M. Margaroni and E. Yiannopoulou, eds (2006)

terroir 'An area or terrain, usually rather small, whose soil and microclimate impart distinctive qualities to food products' (Barham (2003) *J. Rural Studs* **19**), and which is tied in with the marketing and cultural branding of food through its association with place. Ilbery and Kneafsey (1998) *Eur. Urb. Reg. Studs* **5** find that 'terroir' represents a more embedded set of relations between producers and consumers. Gade (2004) *AAAG* **94** uses the French term *patrimonialisation* to describe 'this mesh of authenticity, heritage and food as manifested in regional cuisine, [and] the protection of rural landscapes'.

tertiary The first period of the *Cenozoic era, from about 65 to 1.8 million years ago, subdivided into the Palaeocene, Eocene, Oligocene, Miocene, and Pliocene *epochs.

tertiary industry Economic activity concerned with the sale and use of economic goods and services; service industry. Economically advanced countries have a high proportion of employment in services: USA 75.3%; Britain 72.6%, France 72.4%, Germany 69.2%, Japan 68.1%. See Daniels, O'Connor, and Hutton (1991) *Growth and Change* **22**, 4 on the planning response to service-sector growth in urban areas.

China's employment statistics indicate that industrialization in developing countries may also be accompanied by a percentage rise in services.

	1978	1989	1997	2002
% employed in primary industry	70.5	60.1	49.9	50.1
% employed in secondary industry	17.3	21.6	23.7	21.4
% employed in tertiary industry	12.3	18.3	26.4	28.6

Source: National Bureau of Statistics of China, *China Statistics Yearbook* (2003).

The corresponding decline of primary and secondary industries in advanced economies has ushered in the *post-industrial era.

Tethys An ocean which developed during Palaeozoic and Mesozoic times, running from the coast of southern Spain to South-East Asia. Great thicknesses of sediment were formed since the sea floor kept subsiding at the same rate as deposits were laid down. These sediments were compacted, subjected to *vulcanism, and then uplifted and deformed by the earth movements; see Pivnik and Wells (1996) *GSA Bull.* **108**, 10 on the transition from Tethys to the Himalayas, and M. C. Şengör (1989) on the formation of the Dinaric Alps.

text A group of practices for signalling meaning(s). This commonly means written texts, but has recently included economic, political, and social institutions, paintings, landscapes, and maps. Anthropologists view culture as a text. These texts have been exposed to the *semiotic analyses associated with structuralism, and the *deconstruction and *discourse analyses of *poststructuralism. The main foci of texts in geography have been the interpretation of maps as cultural texts (see Kokkonen (1998) *Cartographia* **35**, 3–4 on cartography in the Baltic Sea) and the nature and usefulness of texts as a metaphor: '[a] metaphorical understanding of geography is popular because it suggests an alternative to both relativism and to naturalist conceptions of mimetic truth. These alternatives often involve rather different understandings of geography' (Demeritt and Dyer (2002) *Area* **34**, 3).

thalweg The line of the fastest flow along the course of a river. This usually crosses and recrosses the stream channel. Channel morphology can undergo rapid transitions along the thalweg, depending on changes in topography, bedrock structure, water, and sediment supplies; see Ramonell et al. (2002) *Zeitschrift* NF **192**. For techniques to quantify thalweg profiles, see Bartley and Rutherfurd in F. Dyer et al. (2002).

thermal erosion *See* THERMO-EROSION.

thermal expansion (insolation weathering) The rupturing of rocks and minerals mainly as the result of large daily temperature changes. The exterior of the rock expands more than the interior. Whether thermal expansion is effective in an environment with no water is open to question; see Koch and Siegesmund (2002) *Spec. Publ.-Geol. Soc. Lond.* **205**. See also Smith et al. (2005) *Geomorph.* **67**, 1–2 and Wigley and Raper (1987) *Nature* **357** on the thermal expansion of sea water associated with global warming.

thermal low An intense, low-pressure system caused by local heating of the Earth's surface, and leading to the rising of air by convection. Heavy rainfall will result if the air rises and cools enough for condensation to occur. The Iberian thermal low is thought to be the most prominent example of its type; see Portela and Castro (2006) *Qt. J. Royal Met. Soc.* **122**, 529.

thermal pollution The contamination of cold water by warm water. Sources of heat include water used for cooling in electricity stations, the *urban heat island, and the construction of reservoirs. That warmer temperatures lower dissolved oxygen in the water, increase respiration rates of organisms, and increase fish and wildlife susceptibility to disease, parasites, and toxic chemicals is generally agreed; see B. Phillips, ed. (2001) on thermal pollution in the Murray-Darling Basin Waterways.

Interestingly, Beser's 2007 paper (*Digital Repository U. Maryland*), in a study of the effects of the thermal effluent in Chesapeake Bay, finds that thermal effluent does 'not cause diversity differences between submersed aquatic macrophyte communities in different areas of the thermal regime'.

thermal stress The strains set up in a result of temperature changes. Hall (1999) *Geomorph.* **31**, 1–4 outlines the importance of thermal stress in cold regions; see also Hall and Andre (2001) *Geomorph.* **41**, 1 on weathering by thermal stress in Antarctica. The term is also used to describe the impact of temperature changes on humans: Auliciems and Kalma (1979) *J. Applied Met.* **18**, 5 propose a climatic classification of human thermal stress in Australia, and Yan (1997) *Sing. J. Trop. Geog.* **18**, 2 analyses climatic thermal stress on the people of Hong Kong.

thermal wind Not a real wind, but an expression of *wind shear for a given layer of *atmosphere; the *vector expressing the difference between the *geostrophic winds at the bottom and top of the layer. It is proportional to the thickness of the layer, and is directed along the isotherms, with cold air to the left in the Northern Hemisphere, and to the right in the Southern. However, the term is used to denote a wind developing as follows: the *pressure gradients which produce surface winds may be due to the presence of cold and warm air masses. The fall in pressure with height is rapid in cold air, and much less rapid in warm air. Thus, at height, air pressure in the cold air will be less than that in the warm air. This creates a high-level pressure gradient and, therefore, a wind, often described as the 'thermal wind'. The strength of this wind is a function of its height and the temperature difference between air masses; the greater the difference, the stronger the wind.

Since there is a marked *meridional temperature gradient in the *troposphere, influenced at height by a powerful westerly factor, thermal winds are very strong at the point where the temperature gradient is greatest; at the polar *front. The result is the polar front jet. The force of a thermal wind may be strengthened by any pressure gradient at ground level. See Chu-shih (1963) *Acta Met.* **33** on thermal wind in a *baroclinic atmosphere, and Roth et al. (1999) *Boundary-Layer Met.* **92**, 2.

thermocirque A large hollow on a hillside formed from the coalescence of *nivation hollows. See Kizyakov (2005) *Atlas Conferences Inc, Document #* capx-22., for the formation and retreat of thermocirques in the Arctic.

thermo-erosion The combined thermal and mechanical activity of running water in *periglacial conditions. See Etzelmüller (2000) *Zeitschrift Geomorph.* **4**, on thermo-erosion in proglacial areas.

thermohaline circulation Also known as the **global thermohaline conveyor belt**, the Atlantic thermohaline circulation begins when the Gulf Stream (and its extension, the *North Atlantic Drift) bring warm, salty water to the north-east Atlantic, warming western Europe. The water cools, mixes with cold water coming from the Arctic Ocean, and becomes so dense that it sinks, both to the south and east of Greenland. This current is part of a larger system, connecting the Atlantic with the Indian and Pacific Oceans, and the Southern Ocean. Further sinking of dense water occurs near to Antarctica. Water from the two main sinking regions spreads out in the subsurface ocean affecting almost all the world's oceans at depths from 1000 m and below. The cold, dense water gradually warms and returns to the surface, throughout the world's oceans. These surface and subsurface currents, the sinking regions, and the return of water to the surface form a closed loop.

Model simulations (for example, Srokozs (2003) *Royal Society*) suggest that a sudden shutdown of the thermohaline circulation will cause global climate changes. For an assessment of the risk of a collapse of the Atlantic thermohaline circulation, see Yin Yohe et al. (2005) *Change* **54**).

Thermohaline circulation

thermokarst A landscape characterized by irregular hummocky topography and interconnected linear ridges, caused by the irregular heaving and melting of ground ice under periglacial conditions (Costard and Baker (2001) *Geomorph.* **37**, 3-4). The exact form of the depressions depends on the original distribution of ice segregations, the subsurface movement of water during warmer periods, and the presence or absence of water in the hollows, since water-filled hollows tend to perpetuate themselves. (The term 'karst' is used to indicate the numerous features formed by subsidence, and does not imply the presence or development of a limestone landscape.) See Kokelj and Lantz (2007) *Annual Conf. Geol. Ass. Canada*, on thermokarst slumps and global warming, and Seppälä (1997) *Geomorph.* **20**, 3-4 on piping in thermokarst.

thermosphere That part of the atmosphere, starting at about 85 km above the earth (the top of the mesosphere), extending to the uttermost fringe of the atmosphere. Here, temperatures increase with height. See C. Lagos (1967) on the dynamics of the thermosphere.

Thiessen polygon A subdivision of a drainage basin, containing a rain gauge. Polygons are constructed by first siting the rain gauges, plotting them on a base map, and connecting the sites by straight lines. The lines are bisected with perpendiculars, which meet to form the polygons. The areas of the polygons are calculated and expressed as fractions of the total area. Each fraction is multiplied by the precipitation recorded by its rain gauge. The sum of these calculations represents total precipitation over the catchment area. For an account, see Yoo et al. (2007) *J. Hydrol.* **335**, 3–4.

Thiessen polygon

thin region As compared to a region with *institutional thickness, a less favoured region. A region may be 'thin' if it is located far from relevant knowledge organizations; for example, the Finnish sub-region of Seinäjoki (Kosonen, *ERSA Conference Papers* No. 3, 112). Functionally, it may have small industrial milieux, a lack of relevant organizations, or its firms may lack collective learning; see Kira and Frieling (2007) *J. of Wkplace Learn.* **19**, 5. This functional thinness can be a consequence of a region's decision-making powers, financial resources, or policy orientation (Tödtling and Kaufmann (1999) *Eur. Plan. Studs.* **7**).

third food regime A regime in which 'strategies for profit capture are built around expediting internationally co-ordinated flows of production, commodities and capital' (Pritchard (1998) *Econ. Geog.* **74**, 1). The third food regime is highlighted by the increasing globalization of production and consumption markets reflected, in part, by corporations which are 'geared to the global or regional markets that are replacing national markets' (McMichael (1992) *Sociolog. Perspectives* **35**, 2), and characterized by flexible production and adaptation to niche markets, aided by research and development. Other researchers see the third food regime as centred around the expansion of the fresh fruit and vegetable complex (LeHeron and Roche (1995) *Area* **27**, 1). The development of organic food production (and its creation of social and economic space) is an expression of the third food regime. See Mckenna et al.

(2001) *Geoforum* **32**, 2. The third food regime is also termed the *post-productivist transition.

third space theory E. Soja (1996) proposes a different way of thinking about space and spatiality. First and second spaces are two different, and possibly conflicting, spatial groupings where people interact physically and socially: such as home (everyday knowledge) and school (academic knowledge). Third spaces are the in-between, or hybrid, spaces, where the first and second spaces work together to generate a new third space. 'Soja is anxious to avoid the common dualities of the social and the individual, culture/nature, production/reproduction, the real versus the imagined, (which pervade geographical analysis, arguing "there is always another way"' Wharf (2006) *PHG* **30**, 819). Linehan and Gruffudd (2004) *TIBG* **29**, 1 apply the term to the tented camps set up in inter-war Wales for retraining miners ('isolated from their wives and families, and separated from the rural society around them'); and, in a cheeky usage, Starbucks propose their premises as 'third space experience: the place between work and home'. See Moje et al. (2004) *Reading Res. Qt.* **39**, 1 on third space in an educational context.

third world Originally, a synonym for those nations that aligned themselves with neither the West nor with the Soviet bloc during the Cold War. Today, the term is used to denote nations with the smallest UN Human Development Indices (HDI). There is no objective definition of Third World or 'Third World country', and these countries are also referred to as 'the South', developing countries, and least developed countries. H. Arendt (1970) argues that 'the Third World is not a reality but an ideology'. D. H. Aldcroft (2007) refers to 'Europe's third world' when considering the puzzling persistence of economic backwardness outside the north-western European core; but see Sidaway (2007) *PHG* **31**, 3, on the decline of third worldism.

(((⊕))) SEE WEB LINKS

● The Third World Network.

thread A river channel. *Anastomosing (braided) river channels are multi-threaded. Richardson and Thorne (2001) *Geomorph.* **38**, 3–4 argue that the division of the water in a single-channel river into multiple velocity threads is a prerequisite for development of *braiding.

threshold In a system, the 'tipping point' which brings about lasting, non-reversible changes, after which the system must develop a new equilibrium condition, adjusted to the characteristics of the altered controlling factors. If, thereafter, the system reverts to the original, balanced condition, it has not crossed a threshold. Schumm (1979) *TIBG* **4**, 4 notes that the definition has been broadened to include abrupt landform change as a result of progressive change of external controls. Ritter et al. (1999) *Geomorph.* **29**, 3 argue that 'if thresholds are accepted as boundaries between equilibrium and disequilibrium, they are required to be time-dependent phenomena. Thus, a steady-time cut and fill of a stream channel during the passage of a flood event is a different type of threshold crossing than the sustained, graded-time entrenchment of a channel due to distinct climate change.'

Most work on geomorphic thresholds has been on fluvial processes (see, for example, Ritter et al. (1999) *Geomorph.* **29**, 3 and Tucker and Stokes (2004) *Amer. Geophys. Union*, Abstr H53E-) and mass movement (see Caine (1980) *Geograf. Annal. A* **62**, 1–2 and Friend (2000) *Prof. Geogr.* **52**, 2). 'If the threshold conditions can be recognized, not only will different explanations for some landforms emerge, also the ability to identify incipiently unstable landforms and to predict their change will be of value to land managers and engineers' (Schumm, op. cit.).

threshold population The minimum population needed to justify the provision of a certain good or service. This may be expressed crudely, in population numbers, but purchasing power may be a better yardstick for commercial goods or services. Note that threshold populations may vary regionally, and certainly vary nationally; the threshold population for a baker's shop is far lower in France than in the UK, for example. The actual evaluation of the threshold population for most goods and services is difficult. However, Coon and Leistritz (2002) *North Dakota State U., Agribus. & Appl. Econ. Misc. Rept.* **191** have, with many caveats, attempted to do so.

throughfall *See* INTERCEPTION.

throughflow The movement diagonally downslope of water through the soil, as opposed to the vertical movement known as *percolation. It may follow natural *percolines in the soil. Throughflow is a major factor in the hydrology of a drainage basin where the rocks underlying the soil are impermeable. See Kirkby and Chorley (1967) *Bull. Int. Assoc. Sci. Hydrol.* **12**. See also Borga et al. (2002) *Hydrol. Procs* **16**.

throw Of a *fault, the vertical displacement of strata along a fault line.

thrust A movement causing the formation of a reverse *fault of a very low angle. The **thrust plane** is the low-angle fault face over which movement occurs. Rubin et al. (2001) *Eos* **82**, 47 write on Taiwanese **thrust faults**. A **thrust belt** is a geological formation caused by compressional tectonics, which ultimately results in the formation of large mountain ranges. The Patagonian fold and thrust belt is an example (Suarez et al. (2000) *Geol. Mag.* **137**, 4). Kwon and Mitra (2004) *Geology* **32**, 7 present a model of a fold-thrust belt wedge. Thrust belts present enormous potential for tapping into oil and gas deposits; see Johnson et al. (2007) *The Leading Edge* **26**, 2.

thufur A low hummock which forms part of a polygonal pattern in *periglacial, or cool areas. They are related to a dynamic snow cover in areas of open birch stands or tundra, and their development is polygenetic, with alternating frost disturbance and phases of humification or podzolization

(Vliet-Lanoe (2002) *Holocene* **12**, 2). Several hypotheses have been proposed for the genesis of earth hummocks, including the 'cryoexpulsion' of clasts, hydrostatic and cryostatic pressure, cellular circulation, and differential frost heave; see Grab (2005) *PPG* **29**, 2.

thunder When a stroke of *lightning passes through the atmosphere, the air becomes intensely hot, perhaps to 30 000 °C. The violent expansion thus caused makes a shock-wave heard as thunder.

thunderstorm A storm including strokes of *lightning, which cause the thunder, and draw off electrons earthward as part of the atmospheric electrical cycle.

tidal energy Energy based on the motions of the tide. See the UK government *World Offshore Renew. Energy Rept. 2004–2008*.

tidal prism The volume of water exchanged between a lagoon or estuary and the open sea in the course of a complete tidal cycle. McBride et al. (2007) *Geomorph.* **88**, 3–4 argue that tidal prism and the quantity of littoral sediment transport are the most important factors controlling inlet stability. The tidal prism is also significant in the size and configuration of ebb-tidal deltas (Fitzgerald et al. (2002) *Geomorph.* **48**, 1–3).

tidal wave *See* TSUNAMI.

tide Tides result from the gravitational pull of the moon and sun on the Earth (which also affects land masses, but the reaction of the water is greater and more apparent). The greatest effect of this pull is on each side of the Earth as it faces the moon. The moon 'pulls out' two bulges of water, which are fixed—the Earth moves through them, causing high water twice daily. The sun also attracts water. When the effects of both sun and moon coincide, twice monthly in the second and fourth quarters of the moon, high **spring tides occur**. When the sun and moon seem to be at right angles to each other from the Earth, the forces of moon and sun are opposed to each other, and lower, **neap tides** result.

till The unsorted sediment deposited directly below a glacier, which exhibits a wide range of particle sizes, from fine clay to rock fragments and boulders. The rate of **till flow** beneath a glacier is thought to depend on basal shear stress and on the effective pressure N, defined as the difference between overburden pressure and the interstitial pore water pressure within the sediments. See Evans et al. (2006) *Earth Sci. Revs* **78**, 1–2.

time-space compression The impact of new systems of transport and communications as experienced by the individual, and especially the emphasis given by modernity to 'the shock of the new', immediacy, and simultaneity. The major periods of time-space compression are from about 1850 to 1914, and from the late 20th century onwards, when there occurred 'a radical restructuring in the nature and experience of both time and space . . . with a dissolution or collapse of traditional spatial co-ordinates' (J. May and N. Thrift 2001). Janelle (2003) UCSB colloquium, provides the following data for Los Angeles to Santa Barbara:

- 500 minutes apart in 1901
- 100 minutes apart in 2001.

time–space convergence Places are separated by absolute distance and by time. With improvements in communication systems and methods of transport, this time-distance diminishes. In essence, time–space convergence means that the *friction of distance—a concept fundamental to conventional *central place theory, *diffusion theory, and *location theory—is lessening. In 1992, R. O'Brien predicted 'the end of geography'. He was wrong. Simson (2005) *J. Transp. Geog.* **26**, 2 writes that 'speed permeates the history of transport and modernity, but it does so in multiple ways. The time–space compression narrative is not adequate to encompass this history.' Furthermore 'there are different kinds of technologies, different levels of technological change, and even the same technology can be used differently by different people in different places' (N. Coe et al. 2007).

time-space geography Time–space geography provides a method of mapping spatial movements through time. Hägerstrand (in T. Carlstein 1978) expressed the time and space needed for events, as a web model, based on four propositions:

- space and time are scarce resources which individuals draw on to achieve their aims;
- achieving an aim is subject to *capability constraints, *coupling constraints, and *authority constraints;
- that these constraints interact to demarcate probability boundaries;
- that choices are made within these boundaries.

Ahmed and Miller (2007, *J. Transp. Geog.* **15**) acknowledge Hägerstrand's conceptual framework, but argue that 'rigorous, analytical definitions of basic time geography entities and relationships do not exist'. Accordingly, they propose 'analytical formulations for basic time geography entities and relations, specifically, the space–time path, prism, composite path-prisms, stations, bundling, and intersections'. Raubal et al. (2004) *Geografiska. B* **86** also propose a spatio-temporal theory of location-based services, which they claim to be closer to the individual user and more plausible with respect to her or his daily life. See also Miller (2005) *Geog. Analysis* **37**, 1. A **time–space prism** is a representation of the constraints limiting the time within which the individual can act.

TNC *See* TRANSNATIONAL CORPORATION.

tolerance The ability of an organism to survive environmental conditions. The prefixes *eury-* and *steno-* refer to wide and narrow ranges of tolerance respectively. An organism can be widely tolerant of one factor, such as temperature, but narrowly tolerant of another, such as salinity. The **climatic tolerance thesis** is that boundaries to the distribution of life forms and species often coincide with isometric lines of climatological variables (Grace (1987) *New Phytologist* **106**, 1); see Ruggiero (2001) *J. Biogeog.* **28**, 10.

tombolo A spit, resulting from *longshore drift, which joins an offshore island to the mainland. See S. Codrington (2005).

topographic(al) map A map which indicates, to scale, the natural features of the Earth's surface, as well as human features, with features at the correct relationship to each other.

topological map A map designed to show only a selected feature, such as the stations on the London Underground. Locations are shown as dots, with straight lines connecting them. Distance, scale, and relative orientation are not important. Try Remolina and Kuipers (2004) *Artific. Intell.* **152** on a general theory of topological maps.

topology 'The mathematics of connectivity or adjacency of points or lines that determines spatial relationships in a GIS. The topological data structure determines exactly how and where points and lines connect on a map by means of nodes (**topological junctions**). The order of connectivity defines the shape of an arc or polygon. The computer stores this information in various tables of the database structure. By storing information in a logical and ordered relationship, missing information, e.g., a line segment of a polygon, is readily apparent. A GIS manipulates, analyzes, and uses topological data in determining data relationships' (Foote and Huebner (1996) Dept. Geog. U. Texas at Austin, which see).

topophilia The feeling of affection which individuals have for particular places; 'the term **topophilia** couples sentiment with place' (Y.-F. Tuan 1974); an 'affective bond between people and place' (Y.-F. Tuan 1990). Environmental designers have long exploited the basic ideas of topophilia to create presumably attractive surroundings that restore mental health based on the use of materials, sensory stimuli, and arrangements that remind people of the place and environmental settings that are comforting and/or associated with healing potential; see A. Carlson (2000). Try also Ogunseitan (2005) *U. Cal. Postprints* **623**. 'Topophilia can be employed to map national territories in a more nuanced manner that reflects nations' own understandings of their territories' (White (2007) Fac. Env. Earth, & Resources, U. Manitoba).

topple The forward rotation, out of the slope, of a mass of soil or rock about a point or axis below the centre of gravity of the displaced mass (Cruden (2003) *Geomorph.* **55**, 1–4).

topsoil The cultivated soil; the surface soil as opposed to the subsoil. Xiao et al. (2006) *Int. J. Rem. Sensing* **27**, 12 propose a new index for detecting topsoil

grain size composition, and Katra et al. (2007) *Geomorph.* **86**, 3–4 outline
dynamics of topsoil moisture. **Topsoil erosion** occurs when the topsoil layer is
blown or washed away; it takes approximately 500 years for one inch of topsoil
to be deposited, but there are 25 billion tons of topsoil lost each year. See
Lovett et al. in E. Buhmann et al., eds (2006) on modelling and visualizing soil
erosion, and Evans (2006) *CABReviews* **1**, 30 for a review of sustainable
practices to limit soil erosion.

tor An upstanding mass of rocks or boulders which rises above the gentler
slopes which surround it. Tors were probably formed as a result of deep
weathering and stripping in preglacial times, with additional sediment
removal by mass movement under periglacial conditions in interglacial and
interstadial times (Stroeven et al. (2002) *Geomorph.* **44**, 1–2; see also
Hätterstrand and Stroeven in the same issue). See Knight (1987) *U. Aberdeen,
Dept. Geog., discuss. paper* **9**, on structural controls on tor location.

tornado A destructive, rotating storm under a funnel-shaped cloud which
advances over the land along a narrow path. The tangential speed of the
whirling air may exceed 100 m s^{-1}, the core may perhaps be 200 m across, the
duration of the storm about 20 minutes. Within a tornado, the central pressure
is around 100 mb below that of the exterior; this may cause buildings to
explode. This type of storm is very common in 'tornado alley', extending from
northern Texas through Oklahoma, Kansas, and Missouri, with as many as 300
tornadoes a year. The exact mechanism of its formation is not fully
understood, but tornadoes are associated with intense local heating coupled
with the meeting of warm, moist air from the Gulf of Mexico and cold air from
the *basin and range area of the western United States. See H. B. Bluestein
(2006).

(((∰))) SEE WEB LINKS

• The enhanced Fujita scale, which uses 28 tornado damage indicators.

total fertility rate *See* FERTILITY.

tourism Making a holiday involving an overnight stay away from the normal
place of residence. Tourism has grown to become the world's second largest
industry, directly accounting for 3.8% of global GDP (World Travel and
Tourism Council, 2005). The mushrooming of **international tourism** may be
explained by high levels of disposable income and longer holidays in *more
economically developed countries; the development of package holidays,
which reduce risk; cheap, mass air transport; and place *myths, which
persuade the tourist that the local culture they see represents the 'real thing'.
Of critical importance has been the internationalization of finance: credit and
debit cards, travellers' cheques, and hotel vouchers. **Creative tourism** means
allowing visitors to participate in events specific to the city or area visited; see
G. Richards and J. Wilson (2007).

Kweka (2004, *Conf. Growth, Poverty & Human Dev. Africa*) finds that tourism expansion has a substantial impact on the Tanzanian economy, while Mitchell and Ashley (2006, *ODI Briefing Paper*), point out that the most important pro-poor impact of tourism is local employment: 'tourism employment relatively favours the marginalized sector as a higher proportion of women workers relative to other similar industries are employed.' However, the tourism industry generally exhibits a high degree of leakage—whereby part of the foreign exchange earnings generated by tourism is either retained by tourist-generating countries or repatriated to them in the form of profits, income and royalty remittances, repayment of foreign loans, and imports of equipment, materials, capital, and consumer goods (Diaz (2001) *Symposium on Tourism Services Paper*).

Dixon et al. (2001) *World Bank, Env. Dept Paper* **80** list some of the environmental impacts of tourism: the generation of substantial amounts of solid and liquid waste, high demand on energy and freshwater resources, and the damage to environmentally sensitive areas. Peeters et al. (2007) *J. Transp. Geog.* **15**, 2 find that climate change generates more than half of the externalities of tourist transport. Lee et al. (2006) *Annals Tourist Res.* **33**, 2 study environmentally friendly practices in the Vietnamese hotel businesses; Singh (2007) *Annals Tourist Res.* **34**, 4 describes environmental management strategies for damage control in India's mountainous areas; and Holden (2005) *Annals Tourist Res.* **32**, 3 considers the environmental ethics of tourism. Socio-cultural impacts include increased crime rate, higher levels of prostitution, and sex tourism; see Shah and Gupta (2000) Overseas Development Institute, UK. Shepherd (2002) *Tourist Studs* **2**, 2 reflects on commodification, culture, and tourism, and Clarke (2004) *TIBG* **23**, 4 thinks tourism makes the world and its people 'modern'. For part 1 of a review of geographies of tourism, see Gibson (2008) *PHG* **32**, 3.

tower karst Limestone towers, 30–200 m high, with nearly vertical walls and gently domed or serrated summits. The towers stand above large, flat *flood plains and swamps and show undercutting from rivers and swamps. Tower karsts are thought to represent the last remnant of a limestone outcrop. See Zhu and Chen (2006) *Cave & Karst Sci.* **32**, 2–3 on the tower karst (*fenglin*) in southern China.

traction load *See* BED LOAD.

trade The movements of goods from producers to consumers. The classic explanation for trade is expressed in terms of *comparative advantage. Greenaway and Torstensson (1997) *TU. Inst. Econ. Res. W. Paper* **144** support comparative advantage models, observing that an abundant human capital endowment, as well as a large domestic market, increases the quality of OECD countries' manufacturing exports, with human capital increasingly important over time. Forslid and Wooton (2003) *Rev. Int. Econ.* **11** find that lowering trade costs leads initially to increased concentration and then to dispersion of production. When a pattern of comparative advantage exists, integration may lead to international specialization of production. 'This may be good news for peripheral countries, which may be able to retain industry despite the attraction of the core.'

Baier and Bergstrand (2005, *Atlanta W. Paper* 2005–3) answer the question 'do free trade agreements increase members' international trade?' with a *yes!* Bernard et al. (2007) *CEP Discuss. Paper* **795** urge that the number of products firms trade as well as the number of countries with which they trade are central to understanding the role of distance in dampening aggregate trade flows. See Hummels and Hillberry (2005) *NBER W. Paper* **11339** on the impact of distance on trade.

trade bloc A large area of *free trade, formed through tariff, tax, or trade agreements, such as the EU. Buckley et al. (2007) *World Econ.* **30**, 2, examine the link between globalization and the growth of trade blocs, and Glebe (2006) *Technische U. München, Env. Econ. & Ag. Policy Gp, Discuss. Paper* 022006, analyses the way the enlargement of a trade bloc will affect national welfare. *See also* BLOC.

trade wind inversion A temperature *inversion, found in the tropics at heights of between 3000 and 2000 m, caused by descending air of the *Hadley circulation. This inversion acts as a ceiling for pollution; *see* PHOTOCHEMICAL SMOG.

trade winds The tropical *easterlies, blowing towards the equator from the *subtropical anticyclones at a fairly constant speed. They are at their strongest around the equatorial flank of these *highs. Trades are not true *zonal winds; their zonal character is weaker in summer because of the south-west *monsoon. They are most regular around 15°, and are associated with fine weather resulting from the *anticyclonic subsidence of the Hadley cell. Over the equator and the western oceans, **trade wind weather** is rainier; see Snodgrass et al. (2006) *12th Conf. Cloud Phys.* Liu (2002) *Science Daily* suggests that shifts in Pacific winds may support *El Niño formation.

traffic The movement of people and vehicles along a route. **Traffic capacity** is the maximum number of vehicles which can pass over a route in a given time, while **traffic density** is the existing number of vehicles. See Barthelemy et al. (2007) *Advances Complex Sys.* **10**, 1 on the role of geography and traffic in complex networks.

traffic segregation The subdivision of towns and cities into certain units where road traffic is restricted and pedestrians predominate. Beckmann (2001) *Env. & Plan. D* **19** comments that 'the automobile turns into a structural prerequisite for the organization of everyday life, while at the same time the variety of forms of everyday action becomes the structural prerequisite for the expansion of the automobile.' See Bickerstaff and Walker (2005) *Urb. Studs* **42**, 5 on local transport planning.

tragedy of the commons G. Hardin (1968) described an increase in the use of common land by a number of graziers, with each grazier continually adding to his stock of animals for as long as the marginal return from each animal is positive, even though the average return for each animal is falling, and even though the quality of the grazing deteriorates. Hardin used this metaphor to describe any situation where the interests of the individual do not coincide with the interests of the community, and where no organization has the power to regulate individual behaviour. Giordano (2003) *AAAG* **93**, 2 attempts to apply scale and space to an understanding of the tragedy of the commons.

transferability The capacity of a good to be transported. Transferability is largely determined by transport costs: as economic distance increases, transferability decreases. Since *economic distance and *intervening opportunities vary, transferability may change over time. See Chang (2004) *J. Geographical Sys.* **6**, 1.

transfer costs Total transport costs involved in moving a cargo including extra costs such as tariffs and insurance. Transfer costs are highest for people, because of the very steep cost of insurance. If there is just one market and just one material source, transfer costs can be minimized by locating the processing unit at one of those two points and not at any intermediate point; see McCann (2005) *J. Econ. Geog.* **5**, 3.

transform fault *Faults which are parallel to the arc of sea-floor spreading. See Greg et al. (2006) *Geology* **34**, 4.

transfrontier zone A space which straddles a land or sea frontier— whether political, environmental, or socio-economic. See Bunnell et al. (2006) *Int. J. Urb. Reg. Res.* **30**, 1 on transfrontier zones and resistances in Bintan, Indonesia, and Duffy (2006) *Polit. Geog.* **25**, 1, on the politics of **transfrontier conservation** areas.

transnational capitalist class (TCC) 'That segment of the world bourgeoisie that represents transnational capital; the owners of the leading worldwide means of production as embodied in the transnational corporations' (W. Robinson 2003); an '"ungrounded and deterritorialized" class' (Ley (2004) *TIBG* **29**, 2).

Carroll and Meindert-Fennema (2002, *Int. Sociol.* **17**) claim that 'transnational corporate interlocking is less about intercorporate control than it is about the construction of an international business community', and Robinson and Harris (2000, *Sci. & Soc.* **64**, 1) argue that 'the TCC became politicized from the 1970s into the 1990s and has pursued a class project of capitalist globalization institutionalized in an emergent trans-national state apparatus and in a "Third Way" political program'. 'Transnational elites have long been associated with the global city: embedded at work, disembedded at home' (Beaverstock and Bostock (2000) *GaWC Res. Bull.* **27**).

transnational corporation (TNC) A firm which owns or controls production facilities in more than one country, through direct foreign investment. Transnationals are made possible by improved international communications which provide rapid containerized trans-shipment and foreign travel; easy communication of information; and international mobility of capital. TNCs have the potential to 'take advantage of geographical differences in the availability and cost of resources, and in state policies, and to switch and re-switch operations between locations' (Dicken (2004) *TIBG* **29**, 1), and when one market is saturated, the transnational can rapidly develop others, since foreign investment cuts transport costs, and makes possible a rapid response to local markets. TNCs are probably the major force affecting worldwide shifts in economic activity, since the largest have a turnover greater than the *GNP of many *less economically developed nations.

Theoretical approaches to TNCs include Pitelis (2001) *Glob. Business & Econ. Rev.* **3**, 1, on the resource-based theory of the TNC, and Wrigley et al. (2005) *PHG* **29**, 4 on the retail TNC. Hudson (in J. Peck and H. Yeung, eds 2003) argues that 'rather than seeing a dichotomy between "global outposts" and "embedded branch plants", we need to conceptualise these as opposite ends of a continuum'; A. Isaksen (1996) agrees: 'TNCs may act as interfaces between global production networks and regional economies.'

Discussions of the advantages and disadvantages of TNCs abound. UNCTAD's *World Investment* states that, since host developing countries tend to use more labour-intensive industries and technologies than do developed countries, the location of a TNC in the former is likely to have greater employment-generating potential than in the latter. Ivarsson (2002, *J. Econ. Geog.* **2**, 2) finds that foreign-located affiliates of TNCs generate technological competencies, 'both internally as well as through organized cooperation with external business partners in the host country'.

However, the ability of transnational corporations to move capital across frontiers, and to manipulate the transfer prices at which their component firms exchange goods and services, has been held against them. Transnationals have also been criticized for undermining national cultures through intensive use of advertising to substitute their synthetic and standardized goods for natural and distinctively local alternatives; the powdered milk controversy comes to mind. Smith (1998, *Eur. Urb. Reg. Studs* **5**) in a study of east and central European economies argues that reliance upon inward investment may not be the 'godsend' that many suggest it will be, and Herrmann (1995, *Hum. Ecol.* **23**, 2) examines the 'culpability' of TNCs for human environmental crises.

Some claim that TNCs undermine states, or, indeed, the state system, but Dicken (2004, *TIBG* **29**, 1) argues that 'states still have power over what they permit'.

transnationality There seems to be no general agreement on the definition of transnationality; its aspects include international cooperation on a level other than exclusively between nation states; cooperation not only with neighbouring countries, but rather with regions that suffer from similar difficulties and problems; and relating to a transnational region, such as the North Sea Region (see the Northern Periphery Programme).

For an individual, Mau et al. (2008) *Glob. Netwks* 8, 1 use a **transnationality index** composed of the number of private transnational relations, the number of times an individual had been abroad in the twelve months before the time of the interview, and the total time the individual had lived abroad for a longer period. For a company, the UNCTAD transnationality index (an indicator of the share of a company's activities taking place in a country other than its own) is the unweighted average of the following ratios: foreign assets/total assets, foreign sales/total sales, and foreign employment/total employment. In 2003, the most transnationalized economy was Hong Kong China, followed by Ireland.

SEE WEB LINKS
• The Northern Periphery Programme.

transnational relations theory A field of study focusing on cross-boundary political space, arguing that states are not the only significant actors. This theory emphasizes the direct linkages across international boundaries, where at least one actor is non-state; for example, a non-governmental organization or transnational corporation. **Transnationalism** describes activities that cross state boundaries, such as the flows of ideas, information, money and credit, and people. Radcliffe, Laurie, and Andolina (2002 *Transnat. Communities Prog. Seminar*) argue that 'the crossing of scales (body, local, national, regional, international) is constitutive of transnationalism. In [the] political transnationalism of Andean indigenous groups, there are multiple scales of interaction, with no fixed location within which actors or practices are found. In other words, transnationalism is as much about discontinuous space as *relational space.' N. Foner (2005) argues that transnationalism did not just appear with recent formulations of late capital or globalization, but rather has existed as a constitutive, dynamic part of migrant experiences; see also Ashutosh (2008) *Urb. Geog.* **29**, 3. For P. Jackson et al. (2004) transnationalism 'encompasses all of those engaged in transnational cultures, whether as producers or consumers. It includes not just the material geographies of labour migration or the trading in transnational goods and services but also the symbolic and imaginary geographies through which we attempt to make sense of our increasingly transnational world.' See also Blunt (2007) *PHG* **31**, 5 and Ley (2004) *TIBG* **29**, 2.

transport costs Krugman (1991) *J. Pol. Econ.* **99**, 3 suggests that so long as transport costs are low, firms can benefit from concentrating in a single location and delivering products to farmers in another location. Alonso-Villar (2005) *J. Econ. Geog.* **5**, 5 finds that the effects of cost reductions in transporting final goods are different from those in intermediate goods. For **feeder costs** (the costs of transferring freight from large ships/vehicles/aircraft to smaller ones) and

handling costs (the costs of physically shifting the freight), and marginal cost/revenue analysis, see Konings (2007) *J. Transp. Geog.* **15**, 6. Baier and Bergstrand (2001) *J. Transp. Geog.* **53**, 1 find that, for the growth in world trade, 1992–2000, trade liberalization was three times more important than transport costs.

Knowles (2006) *J. Transp. Geog.* **14**, 6 adds the concept of **cost/space convergence** to *time–space convergence: 'time/space and cost/space have both collapsed differentially and especially in the last 50 years. Small but incremental changes can have significant consequences for time/cost–space convergence or divergence whilst response lags and inefficient decision making retard change. Collapses in time/space and cost/space are due to innovations in air, rail, road and sea transport technology, communications and handling technology. The differential spatial effects reflect a more unequal world and are largely due to unequal investment in modal capacity, routes and terminals at local, national and international scales. This tends to enhance the importance of the largest demand centres in developed countries and of nodal centrality and intermediacy.' Knowles summarizes the result as a 'shrunken, misshapen world'. Combes and Lafourcard (2005) *J. Econ. Geog.* **5**, 3 show that transport technology and market structure are responsible for most of the transport cost decrease. 'Infrastructure improvements only condition the spatial distribution of the gains.' See Link (2005) *J. Transp. Geog.* **13**, 1 on the **social costs of transport**: pollution, noise, and accidents among them.

transport geography A branch of human geography concerned with: transport policy practice and analysis, especially the impacts of deregulation, privatization, and subsidy control; infrastructure impact on trip-making, the spatial economy, and regional development; technological innovation in transport and telecommunications and global and regional economic integration; the growing mobility gap between rich and poor and differential accessibility to jobs and services; transport, environment, and energy; travel, recreation, and tourism; and spatial and behavioural aspects of modelling transport demand. See Woudsma and Andrey (2004) in the special issue on transport geography, *Canad. Geogr./Géog. canad.* **48**, 4. See also Keeling (2007) *PPG* **31**, 2 and (2008) *PPG* **32**, 2. Try B. Hoyle and R. Knowles, eds (1998); no longer new, but very well reviewed.

travel-to-work area (TTWA) In theory, a self-contained labour market area is one in which all commuting occurs within the boundary of the area. In practice, there are entirely separate labour market areas. For those involved in labour market analysis and planning, it is useful to be able to define zones in which the bulk of the resident population also work, so, by applying a multi-stage allocation process, the UK Office for National Statistics has defined 'Travel to Work Areas' as approximations of self-contained labour markets. The basic criteria used for defining a TTWA are that, of the resident economically active population,

- at least 75% actually work in the area,

• of everyone working in the area, at least 75% actually live in the area.

The definitive minimum working population in a TTWA is 3500, but many are much larger—indeed, the whole of London and surrounding area forms one TTWA.

For the methodology, see appendix 1, Office for National Statistics 1991 Travel-to-Work Areas, and for an update, see M. Coombes, S. Raybould, and C. Wymer (2005).

(⊕) SEE WEB LINKS
• Office for National Statistics 1991-based Travel to Work Areas.
• National Statistics, Travel to Work Areas.

travertine A pale, dense, banded limestone derived from the evaporation of hot springs. The term is sometimes used for *stalactites and *stalagmites.

trellised drainage *See* DRAINAGE PATTERNS.

triangulation In the social sciences, a form of cross-checking; the equivalent of 'validity and reliability' in scientific paradigm research (that is, using multiple people's perspectives, or research gathering methods, to check that your interpretation is robust and not wholly a subjective construct) (Ernest (2000) *U. Exeter*). Forms include data, investigator, method, and theoretical triangulation; see K. Hoggart et al. (2002) on researching human geography, and Farmer et al. (2006) *Qual. Health Res.* **16** on developing and implementing a triangulation protocol. Hobson (2006) *Area* **38**, 3 uses the term for interdisciplinary work: 'triangulation from the human sciences literature with human geographers' arguments could strengthen both sub-disciplines' ability to have the viable policy inputs that both have missed out on in recent times, as it may, in the eyes of policymakers at least, add legitimacy through inter-disciplinarity.'

Triassic The oldest period of *Mesozoic time stretching approximately from 225 to 190 million years BP.

(⊕) SEE WEB LINKS
• An overview of the Triassic period.

trigger In geomorphology, an event which puts an end to the balance between driving and resisting forces and thus sets the mass moving (T. Erismann and G. Abele 2001). P. Fookes et al. (2005) consider the main predisposing and triggering factors of geomorphological surface processes, including climate and weathering, sedimentology, tectonics, and stratigraphy. Jone and Omoto (2000) *Sedimentol.* **47**, 6 outline the criteria that can be used to identify a **seismic triggering** agent. See also Moro et al. (2007) *Geomorph.* **89**, 3–4. For an example of atmospheric triggering, see Miniscloux et al. (2001) *J. Appl. Climat.* **40**, 11. *See also* THRESHOLD.

trope A figurative or metaphorical turn of phrase, used in social constructions of gender, race, imperialism, and so on.

trophic level An individual layer on the *pyramid of numbers which represents types of organisms living at parallel levels on *food chains. All herbivores live at one level, all primary carnivores on the next level, all secondary carnivores on the next level, and so on. The animals on each level are remarkably distinct in size from those on other levels; there is a clear jump in size between an insect and a bird, for example.

The trophic biology of a **taxon** (species/family/class) fundamentally constrains its ability to convert productivity into individuals. Predator taxa require at least an order of magnitude more *net primary productivity (NPP) to support an individual than the trophic level upon which it feeds; the more trophic links between NPP and a consumer **taxocene** (taxonomically related set of species within a community), the more energy is necessary to support viable populations of its individuals. If taxocene abundance is energy limited, consumer taxocenes that occupy broad NPP gradients should show a predominance of lower trophic levels at low NPP. Higher trophic levels should accrue as NPP increases, and taxocenes comprised largely of predators should be under-represented at lower NPP (Kaspari (2001) *Glob. Ecol. & Biogeog.* **10**, 3).

Trophic level

tropical cyclone Also known as a hurricane, or tropical storm, this is a disturbance about 650 km across, spinning about a central area of very low pressure, with winds over 140 km per hour. The violent winds are accompanied by towering clouds, some 4000 m high, and by torrential rain; 150 mm (6 inches) frequently fall within the space of a few hours. There is, as yet, no complete understanding of how these storms develop; they can begin when air spreads out at high level above a newly formed disturbance at low levels. The upper-level outflow acts rather like a suction pump, drawing away the rising air at height and causing low-level air to be pulled in. The winds spiral into the centre because they are affected by the earth's rotation. The intense energy of these storms comes from the warmth of the tropical seas over which they develop. Thus, an extensive ocean area with surface temperatures of over 27 °C is necessary for hurricane formation.

The source regions must be far enough away from the equator—5° at least—for the *Coriolis force to have an effect. The removal of air at height may be along the eastern limb of an upper air trough. Moisture-laden air

spirals into the centre and rises, condensing to form a ring-like tower of
*cumulo-nimbus clouds. With this condensation, *latent heat is released
which causes the air to rise further and faster. The condensation also causes
torrential rain. In the upper *troposphere water droplets freeze and form
*cirrus clouds, which are thrown outwards by the spin of the storm.

At ground level, the temperature at the centre, or eye, of the storm is only
slightly warmer than that at the margins, but, at heights of around 5000 m, the
centre can be 18 °C warmer than the margins. This warm core maintains the
low pressure which drags in the winds. *See also* HURRICANE.

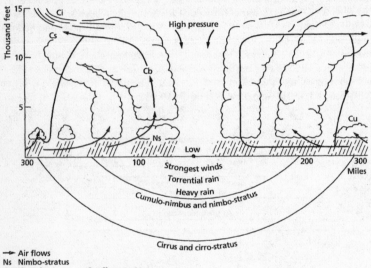

→ Air flows
Ns Nimbo-stratus
Cu Cumulus   Cs Cirro-stratus
Ci Cirrus   Cb Cumulo-nimbus

The structure of a tropical hurricane

tropical dry forest A tropical forest type that goes through an annual dry
season. Drier soils sometimes produce pockets of tropical dry forest
within a tropical rain forest. R. Robichaux and D. Yetman, eds (2002) provide
the ur-text; see also Miles et al. (2006) *J. Biogeog.* **33**, 3 for an overview
of the conservation of tropical dry forests.

tropical forest Forested areas which often extend beyond the tropics, and
consist of *tropical rain forest and *mangrove forest.

SEE WEB LINKS
• The European Tropical Forest Network.

tropical montane forest Forest found in mountainous regions of the
tropics, where rainfall can be heavy and persistent. The trees are typically short

and crooked, and sites for mosses, climbing ferns, lichens, and *epiphytes. See Bruijnzeel and Veneklaas (1998) *Ecology* **79**, 1 on climatic conditions and tropical montane forest productivity. Martin et al. (2007) *J. Biogeog.* **34**, 10 examine relationships between climate-disturbance gradients, vegetation patterns, and *ecotones on subtropical mountains.

tropical rain forest Tropical forest of trees characterized by buttress roots, long, straight, lower trunks, and leathery leaves (see I. Turner 2001). The vegetation shows distinct layering: the canopy, or upper layer, at around 30 m; the intermediate layer at 20–25 m; and the lower layer at around 10–15 m. Undergrowth is poorly developed but *epiphytes and lianas are common. Davidar et al. (2005) *J. Biogeog.* **32**, 3 indicate that seasonality, and not annual rainfall, is the variable that drives tropical rain forest tree diversity. The range of plant and animal species is immense. Hill and Hill (2001) *PPG* **25**, 3 evaluate the following theories explaining the species richness of tropical rain forests: cumulative evolution, productivity, genetic drift and isolation pleistocene refugia, intermediate disturbance, *niche packing, *habitat heterogeneity, seedling and foliar herbivore, recruitment limitation, and temporal resource partitioning.

The felling of the rain forest causes soil erosion and the destruction of potentially useful species (Voeks in D. S. Edwards et al. 1996), and the reduction of oxygen from photosynthesis, while smoke from burning logs increases the quantity of *aerosols and carbon dioxide in the atmosphere (Stein (1982) *GeoJ.* **6**, 2). Clark et al. (2003) *PNAS* **100**, 10 report strong reductions in tree growth and large inferred tropical releases of CO_2 to the atmosphere during the record-hot 1997–8 *El Niño; 'these and other recent findings are consistent with decreased net primary production in tropical forests in the warmer years of the last two decade.' Lettau et al. (1979) *Monthly Weather Rev.* **107**, 3 find that modest deforestation in the Amazon basin will bring slightly warmer, moister weather with drier soils, but that more radical deforestation will result in a drier, cooler climate. See E. Guhardja, ed. (2000) on human impacts on tropical rain forest in Kalimantan.

tropical storm An intense low-pressure cyclonic system that does not contain *fronts between warm and cold air masses, developing when the upper 50 m of tropical ocean warms to over 26.5 °C in regions over 500 km from the equator. See DiMego and Boshart (1982) *Monthly Weather Rev.* **110**, 5 on the transformation of a tropical storm into a *tropical cyclone.

tropics The *Tropic of Cancer* lies approximately along latitude 23° 30′ N. Around 21–2 June, the sun's rays are perpendicular to the ground along this line and the sun exerts its maximum strength in the Northern Hemisphere. Conversely, the sun is overhead at the approximate latitude of 23° 30′ S, the *Tropic of Capricorn*, on 22–3 December when the sun's heat is at its maximum in the Southern Hemisphere. Between these two lines of latitude lie the tropics.

In America this zone is known as the *neotropics* and in Africa and South-East Asia as the *palaeotropics*.

tropopause The upper limit of the *troposphere, above which very few clouds, except for the nacreous and noctilucent, form.

troposphere The lower layer of the *atmosphere, extending to 16 km above ground level at the equator, 11 km at 50 °N and S, and 9 km at the poles. Most clouds and *precipitation, and, indeed, weather events, occur within this layer. Increasingly, it is understood that air movements in the upper troposphere greatly influence weather systems in the lower troposphere. *See also* JET STREAM; ROSSBY WAVES.

Within the troposphere, temperatures of rising air fall, at varying *lapse rates, because the air expands and, therefore, cools. Sinking parcels of air experience a corresponding heating. Most of the water vapour in the troposphere is concentrated in the lower, warmer zone; there is little where the temperature falls below about –40 °C because moist air, rising through *convection or *turbulence, condenses out as ice crystals form.

truncated spur A steep bluff on the side of a *glacial trough, protruding between tributary, possibly *hanging, valleys. This landform is the result of mainly vertical *glacial erosion.

tsunami A huge sea wave, or series of waves. Most are formed from earthquakes of 5.5 or more on the *Richter scale. Other causes include the eruption of submarine volcanoes, very large landslides off coastal cliffs, or the calving of very large icebergs from glaciers in fiords. Fault-generated tsunamis may spread thousands of kilometres across ocean basins with velocities up to 800 km h^{-1}. In mid-ocean the height of a tsunami may be no more than 30–50 cm, but this can increase up to 500 m when the tsunami hits the shore. The most active source region of tsunamis between 1900 and 1983 was along the Japan–Taiwan island arc, where over a quarter of all tsunamis were generated.

The earthquake which caused the tsunami on 26 December 2004 had a magnitude of 9.0, with an epicentre under the Indian Ocean, near the west coast of Sumatra. This earthquake was caused when a portion of the Indian tectonic plate slid under the Burma plate. The US Geological Survey estimates that the ensuing rupture was more than 1000 km long, displacing the sea-floor above by perhaps 10 m horizontally. The great volume of the ocean displaced along the line of the rupture created the tsunami, which hit the coastline of eleven Indian Ocean countries, with wave heights up to 15 m. Estimates of the death toll range between 188000 and 320000.

The largest recorded earthquake in the world occurred in May 1960, with a magnitude of 9.5. The damage caused by the resulting tsunami caused:

- approximately 1655 deaths and $550 million damage in southern Chile;
- 61 deaths, $75 million damage in Hawaii;
- 138 deaths and $50 million damage in Japan;

- 32 dead and missing in the Philippines;
- $500 000 damage to the west coast of the United States.

These statistics reflect the population densities in the affected zones, the economic value of the properties affected, and the importance of warning systems and evacuation planning (Keys et al. (2004) *Antipode* **38**, 2).

tsunami deposits Many former tsunamis have left sedimentary fingerprints; see Dawson and Stewart (2007) *PPG* **31**, 6. See also Scheffers and Kelletat (2003) *Earth-Sci. Rev.* **63**, 1–2 and Scheffers and Scheffers (2007) *Earth Sci. & Planet. Letts* **259**.

tufa A deposit of calcium carbonate found in deserts along a line of once-active springs. The presence of tufa in areas which are now arid points to a *pluvial.

tundra The barren plains of northern Canada, Alaska, and Eurasia. Temperatures and rainfall are low, so tundra vegetation is restricted by the intense winter cold, insufficient summer heat, and waterlogged soil, and characteristically comprises herbaceous perennials with scattered trees, mosses, lichens, and sedges. Arctic tundra ecosystems are warmth-limited and can respond to temperature dynamics on a relatively short time scale. They therefore are highly sensitive to both environmental fluctuation and directional climate change, and play a significant role in biospheric feedbacks to global climate (Gensuo et al. (2006) *Glob. Change Biol.* **12**, 1).

tundra soil A dark soil with a thick, peat layer of poorly decomposed vegetation, which is usually underlain by a frozen layer of soil. Translocation is limited and there is, therefore, little development of *horizons. Tundra soils range from *brown earths in the more humid areas to polar desert soils in arid areas. Climate change may cause tundra soil to dry—accelerating oxygen uptake and hence heterotrophic respiration, which would result in increased CO_2 and methane emissions (Grant et al. (2003) *Glob. Change Biol.* **9**, 1).

tunnel valley A large, elongate, *overdeepened depression cut into bedrock or glacigenic sediment, up to 4 km wide, over 100 m deep, and 30–100 km long. In long profile they contain enclosed hollows or isolated and elongated basins, sometimes displaying convex-up or undulating longitudinal channel profiles and anabranching channel networks (Beaney (2002) *Qt. Int.* **90**). Ó'Cofaigh (1996, *PPG* **20**) attributes their formation to the creep of deformable subglacial sediment into a subglacial conduit, followed by the removal of this material by meltwater flow.

turbosphere That part of the *troposphere where the mixing of the atmosphere is achieved more by *convection than by *diffusion. Its upper limit is the **turbopause**, an ill-defined layer about 100 km above the Earth's surface.

turbulence (turbulent flow) A gustiness in the three-dimensional flow of a fluid, irregular in both space and time, and characterized by local, short-lived rotation currents known as vortices. Turbulence is hierarchical; large *eddies produce smaller ones, and so on, down a series of smaller and smaller scales. In meteorology, it develops because of disturbances in air flows, the most important of which is *wind shear; air parcels caught in a wind shear tend to roll over and over. The *Reynolds number for turbulent flow is 2000 to 2500. In geomorphology, turbulent flow is classified according to the *Froude number of a stream.

turbulent boundary layer *See* BOUNDARY LAYER.

ubac That side of a valley which receives less *insolation; the shaded side, termed **back ridge** in Jamaica (Beckford and Barker (2007) *Geog. J.* **173**, 2).

ultisol A soil of the *US soil classification. *See also* FERRUGINOUS.

umland *Sphere of influence, *catchment area, tributary area, or urban field.

uncertainty A term which has subtle differences in meaning according to its academic context; F. Knight (1921) distinguishes between *risk* (randomness with knowable probabilities) and *uncertainty* (randomness with unknowable probabilities). The measurement of uncertainty is imperative for the production of knowledge; as Goodchild in S. Shekhar and H. Xiong (2008) notes: 'all geographic information leaves some degree of uncertainty about conditions in the real world.' The two broad questions of uncertainty in geography are the estimation of experimental error, and the question of 'how we come to know'.

The first theme is less problematic; D. W. Hubbard (2007) provides useful guidelines for assessing the value of information, and Schroeder (2007) *Geogr. Analysis* **39**, 3, for example, outlines a model used to assess uncertainty in analyses of US census tracts, using aggregated data from incompatible zonal systems.

The second theme relates to the production and status of 'knowledge'. 'In assessing uncertainty, there is a need to question both "what is known" in geography, where knowledge and reflections about the "status" of knowledge differ between individuals and groups of scientists, and "how we came to know", where knowledge and uncertainty are closely related to research tradition and practice' (Brown (2004) *TIBG* **29**, 3). Reed and Peters (2004, *ACME* **31**, 1) draw on contemporary systems ecology to design research practices which 'embrace the uncertainty and partiality of knowledge creation'—one of their wise practices is 'preparing for surprise'. See also Ekinsmyth et al. (1995, *Area* **27**, 4) on the uncertainty of *economic geography, G. Foody and P. Atkinson's e-book on *GIS and uncertainty, Murphy et al. (2006) *Area* **38**, 1 on uncertainty and climate change, and Hunsaker et al. eds (2001) on spatial uncertainty in *ecology. *See also* NUSAP.

underdevelopment The original meaning of the term indicated that existing resources had not been exploited. The word is now close in meaning to 'poverty', although some oil-rich underdeveloped countries have high incomes which are enjoyed by the few. Indicators of underdevelopment include: high birth rates, high infant mortality, undernourishment, a large agricultural and small industrial sector, low per capita *GDP, high levels of illiteracy, and low *life expectancy.

Mawdesly and Rigg (2003) *Prog. Dev. Studs* **3**, 4 identify, between 1978 and 2001, a gradual weakening in the view that development is a global responsibility, with increasing emphasis placed on the choices (and shortcomings) of national governments; 'the international donor community has conveniently emphasised national governance as the principal . . . explanation for underdevelopment, rather than factors such as unfair trade and agricultural dumping of subsidised products (global governance), in which they are implicated' (Carmody (2008) *Geog. Compass* **2**, 1). Power and Sidaway (2004) *AAAG* **94**, 3 provide a review of explanations for underdevelopment; easy to read with useful references and footnotes. Desmet and Ortán (2007) *Scand. J. Econ.* **109**, 1 describe what they call **rational underdevelopment**: 'the advanced region gives transfers to protect itself against low-wage competition, thus preventing the backward region from taking off, and ensuring its dominant position; the backward region accepts these transfers—and rationally chooses to remain backward— because even without transfers it is not sure to benefit from the new technology.'

underfit stream (misfit stream) A stream which flows with narrower meander belts and shorter meander wavelengths than are appropriate to the valley. The valley was initially formed by a river of much greater *discharge, so that an underfit stream is evidence of climate change (Davis et al. (2006) *Geomorph.* **75**, 3–4).

underpopulation A situation where there are too few people to develop fully the economic potential of an area or nation; a larger population could be supported on the same resource base (see M. Schouten (1998) on Minahasa). Christophers (2007) *Geografiska B* **89**, 1 finds a tendency for underpopulation being espoused, or more often predicted, mostly by individuals with a political, ideological, or economic agenda: 'it can never be "proven"; but that does not diminish its potential power.' Underpopulation is a largely hypothetical concept; 'the concept of underpopulation ordinarily relates numbers of people not to physical resources but to economic systems. In short, the argument goes, as fertility falls and populations age, the primary "inputs" to national economic systems (the generation of wealth and of taxation receipts by those of working age) decline, making it increasingly difficult to fund the public expenditure "outputs" demanded of those systems, *particularly* those that inflate in line with an ageing population' (Christophers (2007) *Geografiska B* **89**, 1).

unemployment Being involuntarily out of work. **Structural unemployment** occurs when the labour market no longer requires a particular skill; in the case of **fractional unemployment**, jobs are available but not taken up because of immobility or the lack of information. Suedekum (2005) *Papers Reg. Sci.* **84**, 2 shows that a core–periphery in real wages is associated with and magnified by regional unemployment disparities. For Italy, M. F. Cracolici et al. (2007) explain spatial variations in unemployment differentials in terms of spatial equilibrium and disequilibrium factors, and a significant degree of spatial dependence among labour markets at the provincial level in Italy. Additionally, provinces marked by high, or low, unemployment tend to be spatially clustered. Writing on Spain, López-Bazo et al. (2002) *Papers Reg. Sci.* **81**, 3 also find evidence of clustering. *Core–periphery? Hämäläinen (2002) *Papers Reg. Sci.* **81**, 4 suggests that regional measures to reduce unemployment may cut out-migration, but only during an era of low unemployment. Christopoulos (2004) *Papers Reg. Sci.* **83**, 3 confirms Okun's law: that unemployment falls with increases in the ratio of real output (GDP) growth to *potential* real output growth. The **unemployment rate** is the number unemployed as a percentage of the total population of working age.

uneven development On a multitude of scales, the condition of an economy which has not benefited equally from development either in a spatial sense and/or within classes in society. Henderson et al. (2000) *World Bank Working Paper* WPS 2456: 'The most striking fact about the economic geography of the world is the uneven spatial distribution of economic activity, including the coexistence of economic development and underdevelopment. High-income regions are almost entirely concentrated in a few temperate zones, half of the world's GDP is produced by 15 percent of the world's population, and 54 percent of the world's GDP is produced by countries occupying just 10 percent of the world's land area. The poorest half of the world's population produces only 14 percent of the world's GDP, and 17 of the poorest 20 nations are in tropical Africa.' 'The world . . . *cannot* become smaller while such grotesque inequalities divide us' (D. Massey, Open2Net).

Global uneven development may be seen to be a result of capitalism, based as it is on competition and *accumulation, but inequality is not unique to capitalism. For a national case study, see Hsu and Cheng (2002) *Reg. Studs* **36**, 8 on uneven development in post-war Taiwan. Gallup et al. (1998) *NBER Working Paper* **6849** find that location and climate have large effects on income levels and income growth, through their effects on transport costs, disease burdens, and agricultural productivity, among other channels, and geography seems to be a factor in the choice of economic policy itself. 'Many geographical regions that are not conducive to modern economic growth have a combination of dense population and rapid population increase. This is especially true of populations that are located far from the coast, and thus that face large transport costs for international trade, as well as populations in tropical regions of high disease burden. Furthermore, much of the population

Uneven Development. World per capita GDP

GDP per capita

- Not available
- Less than $1,000
- $1,000 to $5,000
- $5,000 to $10,000
- $10,000 to $20,000
- More than $20,000

Source: World Bank

Uneven Development in China

increase in the next thirty years is likely to take place in these geographically disadvantaged regions' (Gallup et al. (1999) *Harvard U. CID W. Papers*, **1**).

Business development is still very uneven within the UK, because of local transaction costs, variations in *institutional thickness, *agglomeration, *external and urbanization economies, and the continued importance of face-to-face contacts. Transport costs, despite dissenting views, are still important (Bennett, Graham, and Bratton (1999) *TIBG* **24**, 4). Peripheral location appears to affect economic development strongly in China. Wei has shown that the per capita gross value of industrial output of the coastal provinces, such as Guangdong, was double that of provinces deep inland, such as Xizang (2006, *J. Planning Lit.* **20**). J. Y. Lin (2005, East Asian Bureau of Economic Research *Development Economics Working Papers* **656**) proposes that flawed development strategy is responsible for the increasing disparities in economic development among Chinese provinces.

uniformitarianism The view that the interpretations of earth history can be based on the present-day evidence of natural processes. From this comes the maxim 'the present is the key to the past'. See Ager (1989) *J. Geol. Soc.* **146**, 4.

unloading The removal, by erosion, of rock or, by ablation, of ice. See Bonow et al. (2006) *Geomorph.* **80**, 3–4; and Cossart et al. (2008) *Geomorph.* **95**, 1–2 on paraglacial stress release in the immediate aftermath of glacial unloading.

Upton Warren An *interstadial, between 43 000 and 40 000 years BP. Faunal and floral evidence suggests that the landscape was devoid of trees. *See also* PLEISTOCENE.

upwelling The rise of sea water from depths to the surface, bringing nutrients for plankton. Many of the world's best fishing grounds are located at such points. See Ramp and Bahr (2008) *J. Phys. Oceanog.* **38**, 1 on upwelling-favourable winds. *See also* THERMOHALINE CIRCULATION.

urban In medieval Europe, for example, there was a clear-cut boundary between urban and rural, usually in the form of a wall. Subsequently, towns have progressively overflowed their municipal boundaries. In the 1960s, UNO introduced the concept of the **urban agglomeration** (UK—urban area; France—*unité*, Germany—*Verdichtungsraum*) in order to harmonize the statistical definitions of urban settlements from one country to the next. The extensive suburban sprawling based on mass car ownership led to the concepts of *daily urban systems, or *Standard Metropolitan Statistical Areas (*SMSAs). Later, the French statistical system defined the *aire urbaine*: a contiguous urban core and its commuter belt. About 360 *aires urbaines* were defined (Le Jeannic (1996) Cybergeo *écon. & stat.* **294/295**).

urban assets The urban talent in the knowledge economy, believed to 'energize a new capitalist epoch as well as urban economic renewal' (Amin and Thrift (2007) *PHG* **31**, 2).

urban climate Urban centres are subject to climatic conditions that represent a significant modification of the pre-urban climatic state, arising as a result of the modification of radiation, energy, and momentum exchanges resulting from the built form of the city, together with the emission of heat, moisture, and pollutants from human activities. 'For the increasingly urbanized population of the world these effects are already of similar to, or of greater magnitude than, the climatic changes predicted to occur at larger scales as a result of the enhanced *greenhouse effect' (McKenry (2002) *PPG* **37**, 4).

Increasingly, the focus of urban climate research is on understanding the fundamental processes that generate urban climates, not just the resultant effects. 'While the stated rationale for much urban climate research is human health and well-being or energy and water consumption, urban climatologists often note the lack of communication of new knowledge and its implications to end-users, such as planners, architects and engineers' (Souch and Grimmond (2006) *PPG* **30**, 2). See Assimakopoulos et al. (2003) *Atmos. Env.* **37**, 29 on pollution in 'street canyons'. Mills (2006) *Theoretical and Applied Climatology* **84**, 69–76 provides a useful summary of ways to modify urban climates, including outdoor landscaping, street orientation, zoning, and transport policy. *See* URBAN HEAT ISLAND.

urban containment Limiting *urban sprawl. New Zealand's Regional Growth Forum has developed a strategy for regional urban containment, matched by urban intensification (Memn et al. (2007) *NZ Geogr.* **63**, 1). Urban containment has been promoted by US state governments since the 1980s; Wassmer (2006) *J. Reg. Sci.* **46**, 1 finds that only statewide growth management programmes, with the stipulation that local plans must coincide with a state plan or with geographically contiguous local plans, have been effective at reducing the square mile size of US urban areas.

urban design A. Cuthbert (2006) sees urban design as primarily concerned with social interaction and communication in the public realm, using architectural elements and ambient space as its basic vocabulary. Yow (2007) *Geog. Compass* **1**, 6 believes that smart urban design makes it possible to create sustainable cities. See Charney (2007) *Area* **39**, 2 on very tall buildings in London.

urban development corporation (UDC) A government-sponsored enterprise with the task of facilitating urban regeneration in specific areas. J. Susser and I. Schneider (2003) believe that urban development corporations can be seen to offer 'the promise of reconstruction and the spectre of further destruction at the same time'. Dodman (2008) *Geog. J.* **174**, 1 examines the history, role, and function of urban development corporations in Jamaica, Antigua and Barbuda, and Trinidad and Tobago: 'urban development corporations have been able to achieve large-scale transformations of the urban infrastructure and to improve many aspects of the urban environment. However, an over-reliance on *governance structures of this type creates a distinct risk of limiting participation and marginalising large numbers of poor urban residents. Whilst representing a transformation in the role of governments in the management of urban areas, the transfer of some aspects of urban planning and development to these organisations necessarily results in reduced direct accountability to urban residents.'

urban diseconomies The financial and social burdens arising from an urban location. These include constricted sites, high land prices/local taxation/

commuting costs, traffic congestion, and pollution. Firms concentrate geographically in order to exploit benefits of *agglomeration which are traded off against urban diseconomies in determining city sizes (Black and Henderson (2003) *J. Econ. Geog.* **3**, 4).

urban ecology The application of the principles of *ecology to a study of urban environments. *See also* HUMAN ECOLOGY. Urban ecologists look at individual areas of the city in the context of the whole city, and focus on the way a population organizes itself, and the way it adapts to change. Just as ecologists study the way in which an ecosystem seeks to re-establish equilibrium after a sudden alteration, so urban ecologists assume that people will try to re-establish equilibrium after sudden change. Urban ecology has been criticized by many; among them E. Fowler (1996), but Longley (2005) *PHG* **29**, 1 detects some rehabilitation in urban ecology and social area analysis.

urban geography The spatial study of the city, including urban policy; race, poverty, and ethnicity in the city; international differences in urban form and function; historical preservation; the urban housing market; and provision of services and urban economic activity. Pacione (2003) *Urb. Geog.* **24**, 4 writes that 'it is now difficult, if not impossible, to place analytical boundaries around the city. Even those . . . who maintain the value of a distinctly urban focus for their studies must integrate their inquiries into broader social processes . . . The strengths of urban geography lie in its unique spatial perspective and its integrative approach to urban analysis.'

R. Johnston (2000) observes that many of the concerns formerly encapsulated within urban geography are now studied under different banners, but Lees (2002) *PHG* **26**, 1 argues that urban geography has taken on board the interpretative turn, engaging with language, discourse, culture, the immaterial, and so on. Batty (2000) *Env. & Plan. B* **27**, 4 calls for a **new urban geography** of the third dimension: 'the time is nigh for urban geography to reassert itself based on new data sets, on data which are temporally as well as spatially focused, amidst a growing concern to embed such data and associated theory into urban design.' Possibly in contradistinction, Wolch (2007) *AAAG* **97**, 2 argues for 'an urban geography that is not solely a human urban geography but an urban biogeography, an urban geomorphology, an urban hydrology, and an urban GIScience'.

urban governance The exercise of political, economic, and administrative authority in the management of a city. Proudfoot and McCann (2008) *Urb. Geog.* **29**, 4 discuss how street-level bureaucrats negotiate the constraints and pressures inherent to their practice while also exercising a degree of discretion, arguing that these micro-level concerns are important to understanding how cities are produced, but must also be linked with analyses of wider processes that shape contemporary urban development.

urban heat island (UHI) The temperature differentials between urban and rural areas, occurring on the ground, at the surface, and at various heights in the air. The UHI may actually be described in multiple ways with various methodological approaches to investigate each type, and is divided into: the **subsurface heat island** (skin surface temperatures) (Allen et al. (2003) *Geothermics* **32** state that shallow aquifers existing below many northern European cities have groundwater temperatures 3–4 °C higher than in surrounding rural areas) and urban heat islands in the air. Each city's unique land cover patterns will influence the spatial character of its UHI (Yow (2007) *Geog. Compass* **1**, 6). Spatial and temporal patterns of CO_2 concentrations are useful in understanding urban climate; see Wentz et al. (2002) *AAAG* **92**, 1. See Atkinson (2003) *Boundary-Layer Met.* **109** on the numerical modelling of urban heat-island intensity.

urban hydrology The study of the consistent transformation of hydrological regimes that results from urban development. These changes have been attributed to the construction of impervious surfaces and the compaction of pervious surfaces, which promote rapid surface run-off from rainfall, and the efficient drainage of surface run-off through storm sewers to the river network. The percentage urban land cover, percentage impervious cover, or percentage connected impervious cover within a catchment provide simple but very effective predictors of changes in streamflow characteristics resulting from urban development (A. O. Akan and R. Houghtalen 2003). See Gurnell et al. (2007) *Geog. Compass* **1**, 5 for an excellent review of the topic.

urbanism Wirth (1938, *Am. J. Soc.*) used this term to discuss a way of life associated with urban dwelling (suggesting that urban dwellers follow a distinctly different way of life from rural dwellers). Thus, Kern (2007) *Urb. Geog.* **28**, 7 defines **neoliberal urbanism** as the production of self-governing, consumption-oriented, autonomous urban citizens. Wilson (2008) *Urb. Geog.* **29**, 3 finds that a key underpinning of neoliberal urbanism is 'the widespread acceptance of distinctive, simple sensibilities about people, places, and processes that underpin extensions of neoliberalism into urban redevelopment'. These include the beliefs that hierarchical power arrangements in cities and beyond are natural; that power arrangements reflect differences in people's levels of aspirations, abilities, and desires to determine their own destinies; and that human nature is prone to conflict spurred by the 'naturalness' of racial, ethnic, and class differences. This is an illuminating paper. See Ward (2003) *Area* **35**, 2 on **entrepreneurial urbanism**.

However, it is clear from the literature that urbanism can mean *urbanization in North America: E. Talen (2005) identifies four traditions in the United States: incrementalism, the urban plan-making culture, planned communities, and regionalism (the last two outside the existing infrastructure of the city). Talen suggests that new urbanists, 'unlike practitioners of past

planning movements', are attempting to decipher and implement the
'successes' of these four.

urbanization The increase in the proportion of the population residing in
towns, brought about by migration of rural populations into towns and cities,
and/or the higher urban levels of natural increase resulting from the greater
proportion of people of childbearing age in cities. R. Bilsborrow (1997)
discusses the problems in the definition of urbanization, and the difficulties of
making international comparisons. Start with Solecki and Leichenko (2006)
Environment **48**, who give an excellent overview, while also discussing
urbanization and sustainability. See also T. Hall (2006).

Urbanization indicates a change of employment structure from agriculture
and *cottage industries to mass production and *service industries. This backs
up the view that urbanization results from, rather than causes, social change.
This is most notable in the development of *capitalism and its attendant
*industrialization (see Fields (1999) *Berkeley Plan. J.* **13**; try E. Soja 1989).
Others argue that urbanization is the inevitable result of economic growth,
with the rise of specialized craftsmen, merchants, and administrators
(Fafchamps and Shilpi (2001) *U. Oxford Econ. Series W. Papers* **139**). A further
view stresses the importance of *agglomeration economies; cities offer
markets, labour, and capital with a well-developed infrastructure, all of which
increase their *comparative advantage. J. Jacobs (1969) argues that
urbanization economies have been the primary driving force for the
geohistorical development of human societies over the past 12 000 years.

N. Brenner and R. Keil, eds (2006) state that globalizing cities in the
'south', as with those in the 'north', although being distinctly different
have a number of common traits, in particular: a colonial legacy; imposed
neoliberalism and an associated socio-spatial polarization often driven by
international organizations such as the World Bank and United Nations,
and mass urban migration, creating extreme pressure on infrastructure
and service provision.

T. Champion and G. Hugo (2003) argue that there is no longer any
difference between the urban and the rural; instead, there are new forms
of urbanization. Gandy (2005) *Int. J. Urb. & Reg. Res.* **29**, 1 argues that an
emphasis on **cyborg urbanization** (a 'dialectically conceived version of
urban metabolism relating technical developments to a broader political
and cultural terrain') 'extends our analysis of flows, structures and
relations beyond so-called "global cities" to a diversity of ordinary or
neglected urban spaces'.

urbanization curve A model of the progress of urbanization, based on
empirical evidence from Europe. In a traditional society, urbanization is below
20%, and the rate of urbanization is slow, so the curve starts gently. With
industrialization, and a rise in the importance of manufacturing and services, the
pace of urbanization quickens, but the curve slackens after about 75%. While

most developed countries have reached this third stage, the countries of the developing world are still on a rising curve of urbanization, often with a steeper gradient than is characteristic of advanced economies.

Some geographers believe that *counter-urbanization constitutes a fourth stage, when urban populations percentages fall; others that counter-urbanization is temporary.

Urbanization curve

urbanization economies Advantages gained from an urban location (see J. Jacobs 1996). These include proximity to a market, labour supply, good communications, and financial and commercial services such as auditing, stockbroking, advertising, investment, industrial cleaning, and maintenance. Desmet and Fafchamps (2005) *J. Econ. Geog.* **5**, 3 distinguish between *localization economies*, which derive from being located close to other firms in the same industry, and urbanization economies, associated with closeness to overall economic activity. Larger cities have a greater *comparative advantage than small, and the relationship between city size and urbanization economies is non-linear. Demand-side explanations for this argue that the bigger the city, the more likely a match between specific demands and suppliers (an explanation with echoes of *threshold population), while supply-side explanations cite the provision of traded, and non-traded, inputs which increase innovation and variety, and lower costs.

urban morphology The form, function, and layout of the city. Geographic attributes—including shape, density, and pattern of land use categories—all reflect the outcome of the *urbanization process (Longley and Mesev (2002) *Papers Reg. Sci.* **81**, 1). C. Boone and A. Modarres (2006) provide an account and explanation of urban morphology through time, noting that it was within the constraints of the city-state that urban morphology had its genesis, setting up the foundation for socially defined classes and the related hierarchy.

urban planning An attempt to manage the city, often in order to avoid, or alleviate, common urban problems such as inner-city decay, overcrowding, traffic and other forms of congestion; 'the management of spatial interdependence' (Castells in Dunford and Pickvance (2007) *PHG* **31**, 4).

McCann (2001) *Prof. Geogr.* **53**, 2 argues that, in order to understand the role of visioning in contemporary urban politics and in policy-making outcomes, 'we must recognize the sociospatial context in which it is deployed'. Soja, quoted in Warf (2006) *PHG* **30** saw urban planning as the attempt by the state to contain the contradictions of *capitalism.

urban political ecology If one component of *political ecology is the study of political struggles for control over natural resources, urban political ecology is easy to define, and Swyngedouw and Heynen are its begetters: 'the political programme of urban political ecology is to enhance the democratic content of socio-environmental construction' (Swyngedouw and Heynen (2003) *Antipode* **35**, 5). N. Heynan (2006) calls for 'more equitable distribution of social power and a more inclusive mode of the production of nature'; see Myers (2008) *Urb. Geog.* **29**, 3.

urban regeneration (urban renewal, urban revival) The attempt to reinvigorate a run-down urban area, through strategies including the redevelopment of *brownfield sites, rehabilitation of the existing building stock, and enhancement of public spaces and infrastructure (see Temelová (2007) *Geografiska B* **89**, 2). In the *slums of the developed world, substandard housing is broken up by urban regeneration projects, and low-income households, 'unable to gain access to the shrinking stock of public rental housing find themselves increasingly restricted to pockets of low quality housing in the private rental sector' (Randolph and Holloway (2005) *Urb. Policy & Res.* **23**). See Forster (2006) *Geog. Res.* **44**, 2.

Hall (2007) *Geog. Compass* **1**, 6 notes that public art 'has become wedded' to urban regeneration in developed world cities, and Bell (2007) *PHG* **31**, 1 finds that hospitality has become central to regeneration 'as cities increasingly rebrand themselves as pleasure'. Podagrosi and Vojnovic (2008) *Urb. Geog.* **29**, 4 report that, in Houston's urban revival, physical upgrading has been accompanied by the displacement of the community's traditional population and the destruction of an historic neighbourhood.

urban social geography The study of the social patterns and processes of cities; the social interpretation of the spatial structure and functioning of the urban environments. See P. Knox and S. Pinch (2000).

urban sprawl The extension of the city into the countryside, particularly associated with improvements in mass transport. Torrens (2006) *AAAG* **96**, 2 argues that population growth is one of the most important engines of urban sprawl, and Irwin and Bockstaehl (2002) *J. Econ. Geog.* **2** think that sprawl may be caused by negative *externalities between developments. The implication is that the undeveloped land that is adjacent to sprawl is less valuable in residential use and less likely, *ceteris paribus*, to be developed residentially in the future. See J. Wolch et al. (2004) on urban sprawl in California, and Glaster and Cutsinger (2007) *Urb. Geog.* **28**, 6 on sprawl and racial segregation.

urban system Any network of towns and cities, and their hinterlands, which can be seen as a system, since it depends on the movements of labour, goods and services, ideas, and *capital through the network. Freestone et al. (2003) *Tijdschrift* **94**, 2 illustrate the way Australian urban systems reflect broader economic, demographic, and social trends. Abdel-Rahman (2002) links *J. Reg. Sci.* **42**, the structure of an urban system with income disparities: economies that are dominated by a large metropolis, with a number of low-technology cities, will have large income disparities.

urban theory Theorizations of *urbanization which stress the logic of capital in *uneven urban development, appraise the tensions between modernity and postmodernity as conceptual frameworks for understanding 'the urban', explore the *globalization and the concept of world cities, and analyse the complexities of multiple and hybrid urban identities. P. Hubbard (2006) concludes that urban theory can only progress 'by taking the distinctive and excessive materiality of cities more seriously'. A. Amin and N. Thrift (2002) stress instead the 'significant banality' of everyday life. Try Beauregard (2003) *Antipode* **35**, 5, although he is somewhat diffuse.

urban village A residential area of the city containing a cluster of individuals of similar culture and/or interests. Ghazali (2003) *Area* **35**, 2 notes that many rural villages in Malaysia have been absorbed by urban expansion and become urban villages.

U-shaped valley Most U-shaped valleys—valleys with a parabolic cross-section—are *glacial troughs. Brook et al. (2008) *Geomorph.* **97**, 1–2 find that valley cross-profiles become U-shaped with *c.*400 000–600 000 years of glacial occupancy. When no longer supported by ice, the steep sides of U-shaped valleys are prone to rock fall (Blair (1994) *Mountain Rse. & Dev.* **14**). Valleys with this form are also encountered in non-glaciated chalk topography.

US soil classification The US Department of Agriculture recognized ten major soil groups in the Seventh Approximation. Alfisols are relatively young and acid soils with a clay B horizon. Aridisols are semi-desert and *desert soils. Entisols are immature, mainly *azonal soils. Histosols are primarily organic in content, developing in marshes or peat bogs. Insectisols are young soils with weakly developed horizons. Mollisols are characteristic of grassland, high in bases, and with a thick, organically rich A horizon. Oxisols occur in tropical and subtropical areas. They are well weathered and often have a layer of *plinthite near the surface. Spodosols have been *podzolized. Ultisols develop where summers are wet and winters are dry. They are quite deeply weathered and are often reddish-yellow in colour. Vertisols are clay soils characterized by deep, wide cracks in the dry season.

utility The satisfaction given to an individual by the goods and services used.

vadose Describing the zone immediately below the ground surface and above the *water-table in which the water content varies greatly in amount and position.

SEE WEB LINKS

• The *Vadose Zone Journal*, online

valley train *See* SANDUR.

valley wind An anabatic air flow generated as a valley floor is heated by the sun, and the warm air moves upslope. Valley winds are at their strongest in valleys of a north–south orientation. See Weigel et al. (2007) *Boundary-Layer Met.* **123**, 1.

value The importance, worth, or usefulness of something; its material or monetary worth. 'The crucial distinction between poor and rich countries is in the relative value of the commodities produced in each area' (Gereffi et al. in G. Gereffi and M. Korzeniewicz, eds 1994). Gidwani and Chari (2004) *Env. & Plan. D* **22**, 4 describe the **governmentalization of value**: the proliferation of tactics (programmes, laws, codes, standards, and institutionalized knowledges) that 'interpellate subjects of value who are in subjection to the Darwinian logic of value, yet are animated by its almost spiritual promise of self-betterment' ('interpellate' means 'formally question'). Smith et al. (2002) *PHG* **26**, 1 argue that a focus on *value chains and networks* is more useful than a focus on *commodities* ('which implicitly embody that value') *per se*, in understanding territorial economies in increasingly integrated macro-regions.

vapour pressure In meteorology, the pressure exerted by the water vapour in the air. This is a partial pressure since pressure is also exerted by the air itself. Moisture content is commonly measured by *relative humidity. The absolute value of the water content is not usually the most important issue but its value compared with saturation vapour pressure; thus *relative humidity is more relevant than specific humidity.

varve A pair of coarse and fine deposits which reflect seasonal deposition in *proglacial lakes. In summer, *meltwater streams deposit silt- to sand-sized coarse sediments; in winter the lake surface is frozen, the water is calm, and the fine deposits settle out in a thinner layer than the summer sediments.

A **varve couplet** represents the total *fluvio-glacial deposition on a lake floor for one year. Analysis of a series of varves may help in the reconstruction of climatic changes during glaciations; see Snowball et al. (2002) *Holocene* **12**, 1.

veering Of winds in the Northern Hemisphere, changing direction in a clockwise motion, e.g. from westerly to northerly. The *Coriolis effect causes freely moving objects to veer toward the right in the Northern Hemisphere and to the left in the Southern Hemisphere.

vegetation index The most common way to measure Earth's vegetation from space is by using a vegetation index that describes the relative 'greenness' (a comparison of amounts of visible and near-infrared sunlight that are absorbed and reflected by the plants) of Earth's vegetation on a scale of minus one to plus one. In physical terms, it's a way to describe the net result of canopy coverage, leaf area, and canopy architecture in a vegetated area: the number and size of leaves, and how they are arranged horizontally and vertically. Low values on the index mean less vegetation (*Earth Observatory*, 2003). See Liang et al. (2007) *PPG* **3**, 5.

velocity reversal hypothesis An explanation of the occurrence of coarse sediments in riffles, and fine sediments in pools by variations in *discharge. At high discharge, high velocities would move coarse sediment rapidly through the pools, while relatively low velocities in riffles would cause its deposition there. The fine particles are winnowed from riffles and deposited in the pools. Booker et al. (2001) *ESPL* **26** think that velocity reversal is an inadequate explanation of pool-riffle sequences.

vent In geomorphology, an opening in the *crust through which volcanic material flows. **Vent conglomerate** consists of rounded blocks of juvenile material, derived from the collapse of the cone walls during a volcanic eruption, filling a volcanic vent. See Oppenheimer (2003) *PPG* **27**, 2.

venture capital Private equity (finance) offered to outside investors in new/ growing businesses. Venture capital investments are usually high risk, but offer the possibility of above-average returns. Mason and Harrison (2002) *TIBG* **27**, 4 find that, in the UK, venture capital investments were concentrated in Greater London and the South-East, in the 1980s, but that this regional concentration has been greatly reduced since the 1980s. Martin et al. (2002) *J. Econ. Geog.* **2**, 2 find that, although it has recently experienced a marked increase in activity, the venture capital industry is much less developed in Europe than it is in the USA. Zook (2002) *J. Econ. Geog.* **2**, 2 thinks a major reconfiguration is under way and the financial collapses of 2008 seem to be proving him right.

vertex (pl. *vertices*) In *network analysis, this is the place joined by two or more *links (also called *node*).

vertisols Soils of the *US soil classification found in regions of high temperature where bacteria destroy organic residues; hence the humus content is low. Alternate wet and dry seasons lead to the alternate swelling and shrinking of these soils. By these processes, horizons become mixed or inverted.

viscosity The resistance to flow exhibited by a material; the parameter describing the flow characteristics of a fluid. Easy flows result from low viscosity.

vital rates Rates of those components, such as birth, marriage, fertility, and death, which indicate the nature and possible changes in a population.

volcanic ash Finely pulverized fragments of rock and lava which have been thrown out during a volcanic eruption. Volcanic ash forms a thick coating on the landscape adjacent to a volcano, and can travel great distances. See Ort et al. (2008) *GSA Bull.* **120**, 3. Atwater et al. (2004) *Holocene* **14**, 4 use volcanic ash as a dating method. A **volcanic bomb** is a block of lava ejected into the air from a volcano. As it is thrown out, it cools and spins; see Francis (1973) *GSA Bull.* **84**, 8, on cannonball bombs.

volcanism (vulcanicity, vulcanism) All the processes associated with the transfer of magma and volatiles from the interior of the Earth to its surface. *Magma beneath the crust is under very great pressure. Deep in the crust, faults and joints develop downward, reach the magma, and allow it to rise up and intrude the crust. The magma then rises in conduits, depressurizes, forms bubbles, and can give rise to explosive volcanism; see Gaonac'h et al. (2003) *Geophys. Res. Letters* **30**, 11, but also see Gonnermann and Manga (2003) *Nature* **426**. **Intrusive volcanicity** describes magma injected into the crust which fails to reach the surface, and cools slowly underground forming *batholiths, *laccoliths, *dykes, *pipes, and *sills. **Extrusive volcanicity** describes magma which forces its way to the Earth's surface and erupts as *lava, from volcanoes or fissures forming cones or plateaux.

Current volcanicity is confined to regions of the Earth where lithospheric *plates converge, diverge, or pass over possible mantle *hot spots. In a *subduction zone, water is trapped in the descending plate. As the plate is heated, water is released, rising into the wedge of mantle above the plume, causing it to melt. This process forms chains of volcanoes; the cordillera of North America is an area of volcanism related to the subducting eastern margin of the Pacific Ocean. Where one oceanic plate is subducted beneath another, a chain of volcanic island forms; Japan is an example (Bunbury in Hancock and Skinner, eds 2000). When a mantle plume rises through the lithosphere, and eventually erupts at the surface, it is often called *hot-spot volcanism.* About 10% arise through hot-spot activity. Mass movements on volcanic islands may generate large magnitude tsunamis; see Whelan and Kelletat (2003) *PPG* **27**, 2. Hillier (2007) *Geophys. J. Int.* **168**, in a fairly technical paper, notes that *seamounts constitute some of the most direct evidence about intra-plate volcanism, reflecting the ways that the properties of the lithosphere interact with magma generation in the fluid mantle beneath.

In recent years the focus of volcanology has shifted from the physical aspects of why, when, and how volcanoes erupt to incorporate studies on the short- and long-term effects volcanic eruptions have on society and the atmospheric, acquatic and terrestrial environments; see J. Marti and G. Ernst, eds (2005).

volcano An opening in the crust out of which *magma, ash, and gases erupt. The shape of the volcano depends very much on the type of lava. *Cone volcanoes are associated with thick lava and much ash. *Shield volcanoes are formed when less thick lava wells up and spreads over a large area, creating a wide, gently sloping landform. Most volcanoes are located at *destructive or *constructive plate margins.

von Thünen models Johann von Thünen (1783–1850) had two basic models: the first postulates that the intensity of production of a particular crop declines with distance from the market; the second is concerned with land use patterns as they relate to transport costs. See Barnes (2000) *PHG* **27**, 1 for a summary.

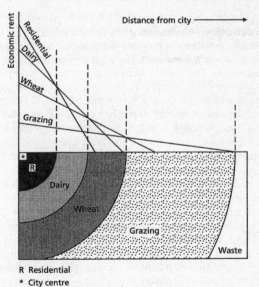

von Thünen model

vortex (pl. *vortices*) A whirling mass of fluid or air. See Walker and Nickling (2002) *PPG* **26**, 1 on helical vortices in dunes; Wiggs (2001) *PPG* on the roll-vortex theory of dune formation; Haines (1982) *J. Applied Met.* **21** on horizontal roll vortices and crown fires; and Davies and Thorne (2005) *J. Geophys. Res.*

110, C05017 on waves in the vortex ripple regime. Huang et al. (2000) *Atmos. Env.* **34** simulate a stable canyon vortex, and MacDonald (2000) *Boundary-Layer Met.* **97** has a numerical model to describe the spatially averaged wind speed profile within the urban canopy layer.

Power (2003) *PPG* **27** explains that the strong **polar vortex** in the Antarctic atmosphere isolates the polar air mass for much of the year. Over the Arctic region the polar vortex is weaker—resulting, in part, from the uneven distribution of land and water. Perlwitz and Graf (1995, *J. Climate* **8**) show that the *North Atlantic Oscillation tends to be positive during winters with an anomalously strong **stratospheric polar vortex**.

vorticity In meteorology, a measure of the local spin of a part of the *atmosphere. The local spin of the atmosphere relative to the Earth has opposite signs in *cyclones and *anticyclones; conventionally the cyclonic direction is taken as positive. A major principle governing the vorticity change of flowing air is the *conservation of angular momentum: as air spreads out horizontally, the rate of spin falls; as it contracts horizontally, its rate of spin rises. Bornstein et al. (1996, *J. App. Met.* **35**) and Schayes et al. (1996, *J. App. Met.* **35**) describe and evaluate a general-purpose vorticity-mode mesoscale model.

vulcanian eruption Also known as a *Vesuvian eruption*, this is marked by periodic lulls during which gas pressure builds up behind the lavas that clog the vent. This blockage is removed by an explosion which throws off masses of *pyroclasts.

wadi In a hot desert, a steep-sided, flat-floored valley, very occasionally occupied by an intermittent stream. Wadis were probably cut during the rainier *pluvials; see Morin et al. (2007) *Adv. Geosci.* **12** on wadis. See also F. El-Baz, ed. (2003).

Walker circulation 'One of the world's most prominent and important atmospheric systems, closely linked to the mean-state of the equatorial Pacific Ocean. It extends across the entire tropical Pacific Ocean, encompassing: the trade winds blowing from east to west; air forced to rise over the western Pacific, south-east Asia and northern Australia through enhanced convection; winds blowing counter to the trades aloft, and air descending over the eastern Pacific Ocean' (Power and Smith (2007) *Geophys. Res. Abstr.* **34**, L18702). Changes in the Walker Circulation are linked to the *El Niño–Southern Oscillation; see Power et al. (1999) *Climate Dyn.* **15**.

wandering river A fluvial channel with alternate stable single-channel reaches, and unstable multi-channel sedimentation zones. This term has also been applied to channels which are midway between meandering and braiding. Wang et al. (2007) *Geomorph.* **91**, 1–2 describe changes from braided to wandering in the Weihe River, China.

waning slope *See* PEDIMENT.

war on terrorism (war on terror) An operation initiated by the United States government under George W. Bush, using legal, military, personal, and political actions to limit the spread of terrorism after 9/11. Graham (2006, *Int. J. Urb. & Reg. Res.* **30**, 2) describes the reworking of US cities in order to construct them as 'homeland' spaces, re-engineered in the interests of 'national security', while Dalby (2007, *Geopolitics* **12**, 4) outlines the 'dichotomous mapping of the world into civilized core and dangerous periphery'. Thobani (2007) *Feminist Theory* **8**, 2 examines the practices that constitute whiteness, triggered by the war on terror, and Macrae and Harmer (2003) *Overseas Dev. Inst.* provide a clear review of the issues.

wash board moraine Synonymous with De Geer *moraine.

water balance The balance at any location between the input (*precipitation, P), and the outputs: *evapotranspiration (E) and *run-off (R):

$$P = E + R$$

If water balance is computed for a number of years, *groundwater storage (S) is held to be constant, but for a single year, groundwater fluctuations are taken into account:

$$P = E + R \pm S$$

Storage and run-off tend to be higher in winter, and evapotranspiration is higher in summer. See Arnell (1999) *J. Hydrol.* **217** on a simple water balance.

waterfall A site on the *long profile of a river where water falls vertically. Waterfalls may be found at a band of more resistant rock (Alexandrowicz (1994) *Geomorph.* **9**, 2), at a *knick point (Crosby and Whipple (2006) *Geomorph.* **82**, 1-2), or where deposition has occurred. Hyakawa and Matsukora (2003) *ESPL* **28**, 6 note that the principal factor in determining the rate of waterfall recession is the ratio of the erosive force of stream to the bedrock resistance, and derive an empirical equation connecting the force/resistance ratios and the rates of waterfall recession.

water framework directive *See* EUROPEAN WATER FRAMEWORK.

water-level weathering The development and enlargement of tidal pools by *weathering, and by the action of rock-grinding animals. Water-level weathering is seasonal, with higher rates in the summer months. See Inkpen (2007) *Area* **39**, 1.

water quality This can be registered as a measure of the ecological health of the water, determined by the level of indicator species in relation to a predetermined critical threshold, derived from toxological data; the biochemical oxygen demand can also be used. Land cover change and the expansion of modern agricultural practices has significantly increased leaching of chemicals to the surface waters, leading to the degradation of aquatic ecosystems worldwide (Donner (2003) *Glob. Ecol. & Biol.* **12**, 4). See Beasley and Kneale (2002) *PPG* **26**, 2 on pollution from urban surface run-off.

watershed In the UK, the boundary between two river systems. Horton (1945) *GSA Bull.* **56**, 3 uses the term *watershed line*, and refers to drainage basins as separate entities, but watershed is now commonly used in North America to mean drainage basin; 'the Mississippi River has the largest of all North American watersheds' (Galler and Anderson (2008) *GSA Bull.* **120**, 3).

water-table The level below which the ground is saturated; the upper surface of the *groundwater. See Quade et al. (1995) *GSA Bull.* **107**, 2 and Romanelli et al. in Massimo Greco et al., eds (2004).

wave A ridge of water between two depressions. As waves approach a shore, they curl into an arc and break. The energy of surface waves is responsible for the *erosion of the coast. Short and Brander (1999) *J. Coastal Res.* **15**, 3 suggest that distinct morphodynamic scaling exists between various wave energy environments.

Waves are the moving force in *longshore drift; Aagaard et al. (2004) *Geografisk Tidsskrift* **104**, 1 show that longshore transport rates relate to changes in the wind climate, and to human intervention. Aagaard et al. (2002, *J. Coast. Res.* **18**) describe a model of sediment transport on a barred beach. See Watt et al. (2005) *Procs Coast. Dynam.* on surface sediment distribution patterns in response to wave conditions. The height of a wave is generally proportional to the square of wind velocity.

wave-cut platform *See* SHORE PLATFORM.

wave energy Energy generated by the force of ocean waves. The University of Hull has a group researching the flow dynamics of wave and tidal power devices, and analysing marine energy resources; see Hardisty (2006) *J. Mar. Sci. and Env.* **C4**.

wave refraction The change in the approach angle of a wave as it moves towards the shore. As water becomes shallow, waves slow down. This change in speed causes the *orthogonals of a wave to 'bend' so that the line of the wave mirrors the submarine contours. Refraction causes waves to converge on

Wave refraction

headlands and diverge in bays. This means that the energy of the waves is concentrated on the headlands rather than on the beaches. See Bristow and Pucillo (2006) *Sedimentol.* **53**, 4 and Gamito and Musgrave (2002) *Computers & Graphics.* **26** on calculating shallow-water wave refraction.

waxing slope The *convex element at the foot of a hillslope.

weathering The breakdown, but not the removal, of rocks. **Chemical weathering** involves chemical change, and includes *carbonation, *hydration, *hydrolysis, *oxidation, and some *organic weathering. See A. Turkington et al. (2005); see also the special issue of *Geomorphology* **67**, 1–2. **Mechanical weathering** is the physical disintegration of the rock, as in *pressure release, crystal growth, *salt weathering, *thermal expansion, and some organic weathering such as *chelation, and bacterial reduction as in *gley soils. See Trenhaile in D. E. Smith and A. Dawson, eds (1993) on freeze–thaw and shorelines. Hall et al. (2002) *PPG* **26**, 4 find that in summer, and often in winter, rock temperatures support mechanical *and* chemical weathering, if water is present. Thorn (2004) *Polar Geog.* **28**, 1 summarizes periglacial weathering research. **Biological weathering** is the disintegration of rocks through the chemical and/or physical agency of an organism. Etienne (2002) *Geomorph.* **47**, 1 considers biogeochemical processes in the morphogenic system of periglacial environments.

Weathering rates vary with the physical structure of the rock (particularly on crystalline rocks—see Borrelli et al. (2006) *Geomorph.* **87**, 3); bond strength (Velbel (1999) *Am. J. Sci.* **299**), and chemical composition (Yokoyama and Matsukura (2006) *Geology* **34**).

The **weathering front** is the zone of contact of the *regolith with the underlying rock. In humid environments, increasing soil thickness increases weathering rates, primarily due to moisture storage and biological effects as soil develops. Eventually the rock weathering rate slows down with increasing distance from surface moisture (Gracheva et al. (2001) *Quat. Int.* **78**). **Weathering pits** are depressions on a flat surface, usually on very soluble rocks, and varying in shape and ranging in size from a few centimetres to several metres in width (Domínguez-Villar (2006) *Geomorph.* **76**, 1–2). **Honeycomb weathering** is a grouping of many, closely spaced pits (Pye and Mottershead (1995) *Int. J. Rock Mechan. Mining Scis & Geomech.* **33**, 8). *Tafoni are weathering pits which are cut into near vertical rock faces. **Weathering rind** is the chemically altered 'skin' of rock around an unaltered core (Etienne (2002) *Geomorph.* **47**, 1).

Weber's theory of industrial location A model proposed by Alfred Weber (1909, trans. 1929). Barnes (2003) *PHG* **27**, 1 reviews a number of locational models.

welding The bonding of grains; see, for example, Anthony et al. (2006) *Geomorph.* **81**, 3–4 on dune accretion.

welfare While welfare might be equated with well-being, within human geography it refers mostly to factors within the control of societies: environmental quality, security, and access to commodities and services. It therefore incorporates income, standard of living, housing, employment, and access to educational, health, and social services; see Sen's **welfare index** (1976) *Rev. Econ. Studs* **43**. Ezcurra et al. (2006) *Tijdschrift* **98**, 4 find that, within the EU, regions with high welfare tend to have advanced services, high productivity, and good human capital. Low-welfare regions tend to higher unemployment, and a larger agricultural sector.

DeVerteuil (2007) *AAAG* **97**, 1 thinks that 'the current decline of the welfare state within a context of *neoliberalism in most developed nations is a key factor in exacerbating the situation of those worst-off, thereby widening the socioeconomic gaps even further'.

Welfare capitalism is the provision of welfare by corporations aiming to foster loyalty among the workforce. S. Jacoby (1998) provides a comprehensive history of 20th-century welfare capitalism.

Welfare geography studies *inequalities in *social well-being and *social justice, looking at the areal differentiation and spatial organization of human activity from the point of view of the welfare of the people involved. It focuses on those factors which affect the quality of human life: crime/lack of crime, poverty/wealth, housing/ homelessness, and the provision/lack of educational, health, leisure, and social services. The first line of enquiry examines inequality in the distribution of welfare indicators, which provides a base from which to evaluate the impact of past or proposed changes; see Falah (1999) *Urb. Geog.* **20**, 5 on Israel's Arab citizens. From this, the second aim is an attempt to explain these inequalities. Current explanations link inequality with *uneven development as an inevitable consequence of *capitalism; or with an uneven spatial pattern of service provision; see DeVerteuil (2001) *PHG* **24**, 1 and I. M. Young (1990). The third aim is the formulation of measures to bring about a fairer distribution of resources and opportunities. Haylet (2001) *TIBG* **26**, 1 argues for a welfare geography attuned to 'the languages and practices through which dominant systems of social and economic distribution are constituted'.

westerlies (westerly winds) Winds blowing from the west, most often in mid-*latitudes.

wetland Any land which is intermittently or periodically waterlogged, including *salt marshes, tidal *estuaries, marshes, and bogs. Wetlands are rapidly disappearing habitats; see M. Williams (1993) and Liu et al. (2004) *Ambio* **33**, 6 on the loss of wetland and *biodiversity.

In the United States a **constructed wetland** is intentionally created from non-wetland sites for wastewater or storm water treatment; the Constructed Wetland Association (CWA) website has a reading list. The concept of a **designer wetland** emphasizes the life history strategy of species as the major

Welfare variations in the E. U.

Legend:
First quartile (lowest)
Second quartile
Third quartile
Fourth quartile (highest)

factor in restoring a wetland type, with no fixed endpoint; see M. Reuss (1998). For **wetland restoration**, see Tooth and McCarthy (2007) *PPG* **31**, 1 and Holden et al. (2004) *PPG* **28**, 1; see also Van Lonkhuyzen et al. (2004) *Env. Management* **33**, 3. A **wetland mitigation bank** site is a property purchased and developed by a public agency or utility to earn credits to compensate for adverse impacts to wetlands due to development activities of other agencies, utilities or, in specific instances, private-sector developers; see Gilman (1997) *Ocean & Coastal Manage.* **34**, 2.

((()) SEE WEB LINKS

• The Constructed Wetland Association reading list.

wetness index (*wi*) Also called the *compound topographic index*, the wetness index describes the effect of topography (slope and aspect) on the location and size of areas of water accumulation in soils. In its simplest form, the wetness index is calculated as: $wi = \ln (\alpha/\tan \beta)$, where α is upslope contributing area and β the local surface gradient. See Sulebak et al. (2000) *Geografiska A* **82**, 1.

wetted perimeter In a cross-section of a river channel, the line of contact between water and bed. *See also* HYDRAULIC RADIUS.

whaleback A streamlined rock knob with symmetrical longitudinal profiles caused by abrasion of both stoss and lee sides. Small whalebacks can form under only a few hundred metres of ice; larger ones under deep ice streams. In theory, whalebacks tend to form in areas of deep ice with rapid flow velocities (James (2003) *Geomorph.* **55**, 1–4). See also Glasser and Harrison (2005) *Geograf. Annal. A* **87**, 3.

wilderness An area which has generally been affected more by natural forces than by human agency; a region little affected by people. Some 77000 km^2 of the USA have been designated as **wilderness areas**, under the Wilderness Act of 1964. These are, ideally, areas which have never been subject to human manipulation of the ecology, whether deliberate or unconscious, and which are set aside as nature reserves to which human access is very severely restricted. Roads, motor vehicles, aircraft (except in an emergency), and any economic use are all forbidden.

The idea of wilderness is socially constructed: Baker (2002) *Space & Cult.* **5** argues that 'the spatial production of wilderness is enmeshed in late capitalist economics and politics, whereas its consumption is structured to reproduce class distinction'. Hintz (2007) *Eth. Place. Env.* **10**, 2, writes that 'wilderness, we are told, can no longer be seen as a scenic playground for weary humans—it is, rather, an ecological necessity for the conservation of biodiversity'.

Some constructions of wilderness omit the works of people. P. Gobster and B. Hull, eds (2000) write of 're-wilding' the landscape, an ideal shared by Noss (2000) USDA, *Rocky Mountain Research Station Proceedings* **15**, 1: 'the goal is not just to save existing wilderness, but to 're-wild' much of what has been lost.' This view has led to the expulsion of indigenous people from designated wilderness areas (W. Cronon 1995); Warren (2007) *PHG* **31**, 4 observes that 'native species and wilderness areas are accorded high value precisely because of their naturalness, and "natural" typically carries the force of a moral imperative: this is the way things *ought* to be. But the belief that nature, to be natural, must be human-free, is rooted in a binary separation of nature and culture which is now widely regarded as false.' Whatmore (1998) *TIBG* **23**, 4 observes that animals have not been helped by the term 'wild': 'Even as they are caught up in the assemblage of global regulatory networks designed to "protect" them, they find themselves objectified again in the urgent business of "wildlife management".'

Wilson cycle A description of the aggregation and break-up of supercontinents. Initially, a *hot spot rises up under a *craton, heating it, causing it to swell upward, stretch, and thin, not only splitting a continent in two, but also creating a new divergent plate boundary. A new ocean basin is generated between the two new continents, and as this widens (possibly by thousands of miles), wedges of divergent continental margins' sediments accumulate on both new continental edges. All of the above is the opening phase of the Wilson cycle. The closing phase of this cycle begins when a new *subduction zone forms at the margin of one of the continents.

Once this zone is active, the ocean basin begins to be subducted, causing the two continents to draw nearer. Magma is generated deep in the subduction zone, rising to the surface to form a cordillera of volcanoes, accompanied by *metamorphism, folding, and faulting. When the two continents finally collide, the closing phase of the Wilson cycle is technically over. Because the subduction zone acts as a ramp, the continent with the subduction zone slides up over the edge of the other one. After this collision, the cordillera will be eroded to a *peneplain. Most of the upper continent will be eroded away, and the lower continent will eventually return to the Earth's surface. See Russo and Silver (1996) *Geology* **24** for further explanation on cordillera formation and the Wilson cycle.

((⊕)) SEE WEB LINKS

• Information about the Wilson cycle.

wind energy Power generated by harnessing the wind, including electricity generation. Drawbacks include the inconstant nature of the wind, construction difficulties, and finding a suitable site; see Wolsink (2006) *TIBG* **31**, 1. 'Installed wind energy capacity has been increasing rapidly worldwide over the last 5 years and now exceeds 70 GW. Wind energy contributes up to 18.5% of electricity consumption in Denmark and about 3% in Europe. Wind energy is economically competitive at good wind sites, can be rapidly deployed compared to conventional sources, and contributes to the goal of meeting increasing electricity demand while reducing emissions of greenhouse gases' (Barthelmie (2007) *Geog. Compass* **2**). Globally, wind power had the fastest rate of growth of any source for electricity (Heiman in S. Majumdar et al., eds 2000). See also Heiman and Solomon (2004) *AAAG* **94**, 1.

wind shear The local variation of the wind vector, or any of its components, in a given direction. A change in wind speed and/or direction with height is the **vertical shear**, given by the *thermal wind equation, if the wind is *geostrophic. See Shepherd and Knutson, (2007) *Geo. Compass* **1**, 1, and Vecchi and Soden (2007) *Geophys. Res. Letts* **34**, L08702 on vertical wind shear and hurricanes.

Windermere interstadial An interstadial of about 13000 to 11000 years BP, including the *Bølling, Older *Dryas, and *Allerød chronozones of Scandinavia. See Holloway et al. (2002) *Scot. J. Geol.* **38**, 4.

window In *meteorology, the ability of radiation of wavelength around 10 μm to escape absorption by the earth's *atmosphere. This wavelength nearly coincides with Earth's peak radiation, and allows some of the outflow of terrestrial radiation to be lost to space, thus, in part, upholding the thermal equilibrium of the atmosphere, which is also achieved by *convection. See, for example, Gube et al. (1996) *J. Climatol.* **3**, 4.

Wirthian theory of urbanism L. Wirth (1938) suggested that urbanization has psychological and social consequences, expressed, at its extreme, in loneliness, mental illness, deviant behaviour in city dwellers, and *anomie.

work, geographies of The geography of work, whether at global, national, regional, or local scale, varies over time and space with the nature of the economy, the level of technology, and spatial variations in *class structure. 'To stake an agenda for "geographies of work" rather than for "labor geographies" means, in the most general sense, mounting a critique of the putative unitariness of value on three fronts: its basis—ostensibly, the production process; the homogeneity of its extraction—ostensibly, the wage/ labor relation; and its frequently posited autonomy from use-values— ostensibly, because the qualities of things, including labor, are effaced when values truck as commodities or exchange-values' (Gidwani and Chari (2004) *Env. & Plan. D* **22**, 4). 'We are encouraged to extend geographies of work beyond the workplace, to contest definitions of work, to repeatedly and insistently connect work to lives beyond work and to see workers as parents, partners, consumers, activists and much more besides' (Stenning (2008) *Antipode* **40**, 1). See James (2008) *Geog. Compass* **2**, 1 on gendered geographies of labour, and Jennings et al. (2006) *Area* **38**, 3 on children's geographies of work. See Ward et al. (2007) *Geoforum* **38** on living and working in working-class communities; see also A. Herod (2001).

World Bank The International Fund for Reconstruction and Development, popularly known as the World Bank, and now generally concerned with aid to the developing world. The Bank acquires its funds partly through government subscription, but mainly through borrowing, and voting power is proportional to capital subscription, so the bank is effectively controlled by the rich countries. It has made loans of over $US 200 billion. Try Bunnell (2007) *Pol. Geog.* **26**, 5 on the International Monetary Fund and World Bank in Singapore, Schec (in D. Moore, ed. 2007), on the World Bank and gender justice, Coleman's (2002, *Pol. Geo.* **21**, 4) charge that, with reference to the debt crisis, and structural adjustment, the Bank 'struggles to explain contemporary developments in the geography of money', and Shriar (2005) *J. Lat. American Geog.* **42** on the impact on Jamaica of structural adjustment, trade liberalization, and economic globalization.

SEE WEB LINKS

• The *World Bank Economic Review* and the *PovertyNet Newsletters*.

world city A city that has a substantial, straightforward effect on world affairs: economically, socially, and politically. Key descriptors include: centrality—a concentration of international activities; externality—a relatively high global role, compared to its national role; and connectivity—as measured by the numbers and intensity of transactions. 'World cities are not merely outcomes of a global economic machine; rather, they are key structures of the world economy itself' (Taylor et al. (2002) *Urb. Studs* **39**, 13; see also P. Taylor 2003). Gugler (2003) *Int. J. Urb. & Reg. Res.* **27**, 3 notes that world cities in the developing world are more subject to the economic, political, and cultural impact of powerful foreign actors: governments, corporations, international organizations, universities, the media. The resources they can draw on are much more limited, even as they have to make heavier investments in infrastructure to meet the standards required by foreign investors.

Beaverstock et al. (1999) *Cities* **16**, 6 propose a hierarchy of world cities: 'alpha world cities', such as London, New York, and Frankfurt; 'beta world cities', such as Zurich, São Paulo, and Mexico City; and 'gamma world cities' such as Buenos Aires, Montreal, and Santiago. Ewers (2007, *Geog. J.* **68**, 2–3) also examines measurement issues and the methodological problems in creating indices of world cities. See Brown et al. (2007) *GaWC Research Bulletin* **236** on *world city networks, and Ewers (loc. cit.) on the role played by world cities in global migrations. Smith (2003, *PHG* **27**, 1) holds that further progress in the study of world cities and their networks can be made through actor-network theory and *non-representational theory. See Taylor (2007) *GaWC Res. Bull.* **238** for an admirably clear summary of GaWC's work on world cities.

(((())) SEE WEB LINKS

- GaWC Inventory of World Cities.

world city network An interlocking network with cities at a nodal level, the network at inter-nodal level, and business service firms as a sub-nodal level. These last 'interlock' cities through their myriad networks of offices. In this new city system, it is the functions of a city that largely determine its role and importance nationally, regionally, and globally (Taylor et al. (2002) *Urb. Studs* **39**, 13). See Beaverstock et al. (2000) *Applied Geog.* **20** on the conceptualization and measurement of the network connectivity of cities.

world-systems theory A materialist description of global social structures, developed by I. Wallerstein (1974, 1983). The *core* consists of the most economically developed countries, the *periphery* the less economically developed countries, and the nations of the *semi-periphery,* such as the newly industrialized countries, have aspects of both core and periphery—actively part of the capitalist world system but with limited authority therein. These three economic classes are continually changing, and nations move up and down in the system. See A. Ash (2003) and Sheppard (2005) *Antipode* **37**, 5.

This theory has been criticized, in turn, for representing the global economic system as a predominantly capitalist phenomenon, for neglecting the historical and cultural differences of developing countries, and for overlooking the core–periphery inequalities existing within nations. 'World systems approaches . . . tend to theorize hegemony chiefly on a global scale' (Sparke (2004) *PHG* **28**, 779).

World Trade Organization (WTO) An international organization established in 1995 to replace *GATT. Its stated aims are: expanding free-trade concessions equally to all members; establishing freer global trade with fewer barriers; making trade more predictable through established rules; and making trade more competitive by removing subsidies (see Mullins (2004) *J. Econ. Geog.* **4** on the WTO's dismantling of the protocols that gave Caribbean, African, and Pacific ex-colonies preferential access to EU markets). Since 2005, WTO members have been unable to place quotas on imports of textiles and garments from other members (Thomsen (2007) *J. Econ. Geog.* **7**).

Potter and Tilzey (2005) *PHG* **29**, 5 argue that 'the entrenched position of neoliberal ideas and framings within the WTO means that the ideology of the free market is coming to define the terms of international trade'. Wade (2003) *Rev. Int. Polit. Econ.* **10**, 4 claims that the 'development space' and 'self-determination space' for diversification and upgrading policies in developing countries is being shrunk behind the rhetorical commitment to universitization and privatization. Woods (2007) *PHG* **31**, 4 argues that conclaves of the WTO have created new sites of political authority for the global countryside. Uitermark (2002) *PHG* **26**, 6 contends that global institutions, such as the WTO, have facilitated competition between different nations such that the nation-state has been 'localized'. As against these tirades, Zhu and Sarkis (2006) *J. Cleaner Prod.* **14**, 5 confirm that *globalization and China's entry into the WTO have helped promote green supply chain management practices.

xeroll A soil of the *US soil classification. *See also* CHESTNUT SOIL.

xerophyte A plant which is able to grow in very arid conditions because it has adapted to restrict any water loss. Such adaptations include dense hairs or waxy leaves and shedding leaves at the start of the arid season. **Succulent xerophytes** incorporate water into their structure.

xerosere A plant succession developed under dry conditions such as bare rock, or sand.

yardang In a *desert landscape, a long ridge which has been isolated by the removal of rocks on either side. Yardangs can be 100 m or more in height and can stretch for many kilometres.

yazoo A tributary stream which does not join the main stream directly but runs parallel to it for some distance, usually because it cannot breach the *levées which flank the main stream.

Younger Dryas A phase of reduced temperatures and glacial advance in north-west Europe between 11 000 and 10 000 years BP. There is evidence that other regions, such as the Mediterranean, Russia, and North America, also underwent climatic deterioration. During this phase, these non-glaciated areas experienced periglacial conditions. This climatic cooling has been attributed to increased *precipitation, and therefore cloudiness, resulting from changes in the locations of the polar and oceanic Atlantic fronts.

young fold mountain *See* FOLD MOUNTAIN.

Zelinsky *See* MOBILITY.

zero population growth The ending of population growth when birth and death rates are equal. This would require an average number of 2.3 children per family. In 1993, 58 members of the World's Scientific Academies stated that 'there is no doubt that the threat to the [global] ecosystem is linked to population size and resource use... Family planning could bring more benefits to more people at less cost than any other single technology... Success in dealing with global social, economic and environmental problems cannot be achieved without a stable world population... We must achieve zero population growth within the lifetime of our children' (1993, Royal Society, London). This is a contentious subject.

zero-sum game A *formal game whereby, on choosing a particular strategy, one competitor's gain is his opponent's loss. Liard-Muriente (2007) *Area* **39**, 2 writes that 'regional development policies could be described as a zero-sum game, with local job reshuffling as the outcome. After all, if one area accomplishes growth, it will be at the expense of another area. Thus, overall welfare is unchanged.' Polèse and Shearmur (2006) *Papers Reg. Sci.* **85**, 1 agree: 'stable national populations means a zero-sum demographic game for interregional systems... Assuming (as most migration models do), that population movements largely follow the spatial distribution of economic opportunities, the key factor becomes the spatial distribution of employment. In such an environment, if shifts in employment continue, on balance, to favour the centre over the periphery, the latter must decline.'

zeuge (pl. *zeugen*), An upstanding rock in a *desert landscape capped with a harder *stratum and undercut by wind at the base, indicative of *differential erosion.

zonal Referring to phenomena occurring in bands roughly parallel with lines of *latitude.

zonal index Meridional displacements of the zonally averaged zonal jet (Feldstein and Lee (1998) *J. Atmos. Scis.* **55**, 19). Robinson (2005) *J. Atmos. Scis* **57**, 3 proposes a *baroclinic mechanism for the positive eddy feedback on the zonal index. *See also* ROSSBY WAVES.

zonal model *See* CONCENTRIC ZONE MODEL.

zonal soil A soil where differences in local rock formation and *lithology are largely masked by the overriding effects of climate. The major zonal soils are *tundra soils, *podzols, Mediterranean soils, *chernozems, *chestnut soils, and *ferallitic soils.

zonal winds Winds, such as the *trade winds or the *westerlies, which are associated with particular *latitudinal zones.

zone of assimilation The area which increasingly develops the functions of the *CBD; the CBD of the future, characterized by whole scale redevelopment of shops, offices, and hotels. The **zone of discard** is that area that was once a part of the CBD but is now in decline and characterized by low-status shops and warehouses, and vacant property.

zone of overlap **1.** An area served by more than one urban centre, i.e. within two or more different urban fields.
    2. A transition zone between two floral/faunal communities. See, for example, Wolf and Flamenco-S (2003) *J. Biogeog.* **30**, 11.

z

Oxford Paperback Reference

A Dictionary of Psychology
Andrew M. Colman

Over 10,500 authoritative entries make up the most wide-ranging
dictionary of psychology available.

'impressive ... certainly to be recommended'
Times Higher Educational Supplement

'Comprehensive, sound, readable, and up-to-date, this is probably the
best single-volume dictionary of its kind.'
Library Journal

A Dictionary of Economics
John Black

Fully up-to-date and jargon-free coverage of economics. Over 2,500
terms on all aspects of economic theory and practice.

A Dictionary of Law

An ideal source of legal terminology for systems based on English law.
Over 4,000 clear and concise entries.

'The entries are clearly drafted and succinctly written ... Precision for the
professional is combined with a layman's enlightenment.'
Times Literary Supplement

Oxford Paperback Reference

A Dictionary of Chemistry

Over 4,200 entries covering all aspects of chemistry, including physical chemistry and biochemistry.

'It should be in every classroom and library ... the reader is drawn inevitably from one entry to the next merely to satisfy curiosity.'

School Science Review

A Dictionary of Physics

Ranging from crystal defects to the solar system, 3,500 clear and concise entries cover all commonly encountered terms and concepts of physics.

A Dictionary of Biology

The perfect guide for those studying biology – with over 4,700 entries on key terms from biology, biochemistry, medicine, and palaeontology.

'lives up to its expectations; the entries are concise, but explanatory'

Biologist

'ideally suited to students of biology, at either secondary or university level, or as a general reference source for anyone with an interest in the life sciences'

Journal of Anatomy

More Social Science titles from OUP

The Globalization of World Politics
John Baylis and Steve Smith

The essential introduction for all students of international relations.

'The best introduction to the subject by far. A classic of its kind.'
Dr David Baker, University of Warwick

Macroeconomics
A European Text
Michael Burda and Charles Wyplosz

'Burda and Wyplosz's best-selling text stands out for the breadth of its coverage, the clarity of its exposition, and the topicality of its examples. Students seeking a comprehensive guide to modern macroeconomics need look no further.'
Charles Bean, Chief Economist, Bank of England

Economics
Richard Lipsey and Alec Chrystal

The classic introduction to economics, revised every few years to include the latest topical issues and examples.

VISIT THE COMPANION WEB SITES FOR THESE CLASSIC TEXTBOOKS AT:

www.oup.com/uk/booksites

CONTENTS

For my mother and late father, Mary and Jim Riddell

ABOUT THE AUTHOR

David qualified as an EFL teacher in 1984 and became an approved trainer in 1987. At the time of writing, he both teaches students and trains new teachers in London. Since 1994, he has also been a Senior Assessor . In one capacity or another his work has taken him to ten countries. As much as he enjoys teaching, he still hopes to win the lottery and retire overnight.

Acknowledgements

I would like to thank:

Thomas Grossmann, for contributing so much, and for being a trainee who brought a smile to everybody's face day after day.

Bertrand Enjalbert for his great support and friendship.

Zsigmond Sipos, for his friendship and support, and for being the ideal student!

Guillaume Tellier, for his encouragement and support throughout.

Jeff Mohamed and John Shepheard, for many years of personal and professional inspiration. And to Jeff and his wife Deanne for being the perfect employers in the perfect city of San Francisco!

Sue Hart, my Editor at Hodder, for her consistent support and encouragement, and for giving me the opportunity to write this book.

INTRODUCTION

Despite the title of this book, I do not think it is possible for someone to 'teach' himself or herself how to teach, without actually doing it, just as you cannot learn from a book how to use a computer if you do not have a computer to use. However, I do know from my experience as a teacher trainer that there are EFL/ESL teachers who have no qualifications or training. This is not good for the profession, and it is certainly not good for the students. But it is a fact. I also know that even those people who **do** undertake a training course need as much practical help and support as they can get, not only during the course, but also – and maybe particularly – after the course, as they begin their first teaching assignment. So, you cannot teach yourself to be a teacher, but you can learn and improve your skills as you go along.

This book, therefore, is intended for these three types of reader:

■ teachers teaching without having had any training
■ teachers who have recently trained and who lack experience
■ trainee teachers doing a training course

By 'teachers' I refer to those in any country, teaching to monolingual and multilingual classes, to students whose first language is not English, and to students who are classified as *adult learners* (generally defined as 16+). Since the book has such a broad focus it inevitably deals with the general, rather than the specific, teaching context. This is no bad thing. The book's purpose is to provide new teachers with the basic teaching skills, background knowledge and awareness that will subsequently enable them to develop and fine tune what they do in the classroom.

I have seen situations where too much is expected of new/ training teachers. Many of my ex-trainees have complained at the lack of support and help from some first-time employers who sometimes seem to expect students to know everything from having done a four-week training course, or similar. This is unrealistic. Equally, tutors on training courses should not forget that most of their trainees have never been in front of a class before in their lives and that they need support and encouragement as well as expert training.

Fortunately, most employers and tutors do provide such support. However, they are not always available when needed by the new or training teacher. I hope this book will fill that space.

This book should not be seen as the be-all-and-end-all of EFL/ESL teaching. To many experienced teachers reading the book (a book not intended for such readership!) much of what I have written will be seen as prescriptive, just as the typical training course and most course books are prescriptive. But I will always argue, from long experience, that a level of prescriptiveness is needed (and welcomed by trainees) at the beginning of their teaching career. I make no apologies for being prescriptive in this book – once teachers have more experience and confidence, they can develop their awareness and methodology. The models of teaching that I outline here can be applied successfully in most contexts, giving new teachers a much-needed degree of security. Please note, when referring to teachers in general, I have used 'she'/'her' throughout.

This book is not intended for experienced teachers or for those teaching young learners. It is not intended to apply in all parts to all teaching contexts. It certainly does not cover all aspects of teaching. Instead, it aims to raise awareness of the role of teachers and students teaching and learning English, and to give practical ideas for teaching English as a foreign or second language to adults.

This is effectively, a teaching 'handbook' which can be referred to at short notice for suggestions, examples and reassurance. It is just the beginning of the long process of becoming an effective and successful teacher. It has been written to provide help and encouragement. I hope it does.

Good luck, and enjoy the ride! It won't be dull.

David Riddell
davi77@hotmail.com

1 | BEING A STUDENT

My own memories of my schooldays and the teachers that taught me are very vivid. As a learner of French I experienced two very different styles of teaching – one where I was taught primarily in English, and one where I was taught mostly in French. I found the second much more intimidating, but also much more effective. The first was safe, comparatively easy, but did little to make me a competent user of the language. Just as it was no good for someone to *show* me how to change a spark plug (I had to *do* it to learn), so it was no good for me to learn a new language without getting more opportunity to use it. That is one reason why our students in EFL are mostly taught entirely in English, regardless of their level. Allow a couple of students to start using their own language and you will quickly hear loud complaints from the other students – they know that this is not helping their learning.

Teachers have differing views on the use of L1 (the student's mother tongue) in the classroom, and it depends to an extent on whether you are teaching a monolingual or a multilingual class. In a multilingual class, lessons are taught almost entirely in English, partly for the reason already given, and partly because this 'total immersion' approach encourages real communication as opposed to rule learning and the completion of exercises. Of course, there may be times when the occasional quick translation may be helpful, but if either the teacher or other students start using non-English to any extent beyond this, the students are very likely to become unhappy.

In a monolingual class it is inevitable that the students will speak their own language more, and perhaps the teacher will too (if she knows it!). But if the teacher starts teaching in that language rather than English, the *authenticity* of the target language will be lost.

All teachers, especially of 'General English', know that different students prefer to learn in different ways. Let's be honest here – as much as teachers believe in the effectiveness of the 'communicative approach', there are still many students who prefer to listen to explanations of grammar, learn rules, and do practice exercises. It is *safer* – just as when my French teacher taught in English. On the other hand, there are those students who only want to 'talk', being more concerned with fluency than with accuracy. There are students who want learning to be 'fun', and those who perceive 'fun' as 'not learning'.

It is obvious, then, that 'being a student' is a complicated existence. We, as teachers, have to appreciate this and be as aware as we can of our students' different attitudes to learning.

 Task

Here are some questions I used in a survey done with some of my students during January 2000. Before I report the findings, answer the questions yourself based on your own experience of new-language learning.

1 What makes a good student?
2 What helps you when you are learning a new language?
3 How important is learning grammar when you are learning a language?
4 Should learning be communicative, or should you learn from explanations, rules, and practice exercises?
5 How important is it for a student to have some responsibility for their own learning and not to rely exclusively on the teacher?
6 Is it a good thing to be in a class with students from your own country (of course, in non-English-speaking classes students will all be from the same country, so this question does have rather limited value!)?
7 Is it a good thing for the teacher to use your language?
STOP here and consider your answers before reading on!

The students I asked were from eleven different countries – Japan, Mexico, Brazil, Russia, Turkey, Italy, Israel, France, Korea, Algeria, and Spain. They were intermediate and upper intermediate level students studying in London with ages ranging from 18 – 45. Most had experience of learning another language besides English, and all had experienced more than one learning style.

I also put the same questions to Thomas Grossmann, a Swiss German teacher who passed his Proficiency exam in December 1998. Six months later, he was a qualified EFL teacher and was teaching shortly after that. This remarkably quick transformation made Thomas an ideal person to ask about both learning and teaching.

But first the student reactions:

1 What makes a good student?
"respect your teacher"
"know that you're not the only student in the class"
"you must have a desire to study"
"listen"
"respect other students"
"be interested"
"socialise with other students and teachers"
"speak in class"
"pay attention"
"do your homework"
"spend time outside class using English"
"go over what you have learned in class"
"know why you are learning English"
"respect the culture of the country"
"be attentive, disciplined, active in class"

'Respect' is a word that came up several times, and some cultures are especially respectful of the role of the teacher. Others have had strained relationships with teachers from their own schooldays and are perhaps still a little suspicious of teachers in general. Most, however, have opinions about what the role of the teacher should be, something we will look at in the next unit.

Having a 'desire to study' – an 'interest' in what is being learned – is crucial, in my view. The inquisitive student, the one who wants to know 'why?' and 'how?' and who asks lots of questions, may be the student that new and learning teachers dread the most, but in the long run these are the students who thrive and make visible progress and who keep the teacher challenged and rewarded.

'Pay attention' and 'listen' are very important points, too. 'Listen' should be followed by 'respond' because to listen is not enough. Think of any conversation you have had with a person who, you know full well, is not really listening to what you are saying. It is frustrating. Their thoughts are elsewhere and you do not have their attention. Students whose minds wander, or who are content only to listen (back to that 'safe' option!) suggest that their desire to study, to learn, is limited. This was certainly the view of the students I spoke to, some of whom were themselves 'guilty' of this at times!

'Go over what you've learned', 'do your homework', 'use English outside class' are points which cannot be emphasised too much. I should add that all my students stressed the ideal of learning English in an English-speaking country, if possible. This obviously makes it much more possible to use English outside school, as well as making the learning process more authentic. But, of course, this is not always possible.

However, it is important for students not to perceive their lessons as the be-all-and-end-all of their learning. Yes, they should do their homework. Yes, they should revise what they have studied. But most of all they should take whatever opportunity they can to *use* their English – speaking, listening, reading, writing. As teachers we should not only encourage them to do so, but also tell them *how* to do it (e.g. suggesting English newspapers, magazines, books, etc. to read). The student who limits his 'learning' to inside the classroom will make very little real progress.

2 What helps you when you are learning a language?
   "be in the country of the language"
   "don't be shy"
   "watch TV"
   "a balance and variety"

"use only English when possible outside class"
"have a part-time job if possible"

Most of these have already been mentioned. The point made about the part-time job only applies to work in an English-speaking country, but this situation certainly plunges the student into using English authentically.

Watching TV (equally, watching films, listening to the radio, reading newspapers, etc) underlines what has already been said about using English outside class.

3 How important is learning grammar when you are learning a language?

"grammar, grammar, grammar is very boring"
"quite important"
"very important"
"gives meaning to conversation"
"not too much"
"it is important because if you learn grammar you can speak properly" (whatever 'properly' means!)

Predictably, different views were expressed, reflecting, in part, students' own language-learning experiences from earlier education. The very word 'grammar' has negative connotations for most people, and can be a real turn-off in the classroom, much as the word 'homework' can. But it is a word with an undeserved reputation! Ask most people what they understand by 'grammar' and they will usually come up with "rules", "exercises", "copying from the board", "boring", "not useful" etc. Say to a class "Now we're going to do grammar" and you won't be greeted with a sea of smiling faces – though some students still view the learning of rules as 'real learning'. We will look at this issue in Unit 5.

4 Should learning be communicative, or should you learn from listening to explanations and making notes and doing practice exercises?

"beginners need more practice exercises, after that learning should be communicative" "it should be more communicative because if French schools did that we would speak better English"

"communicative, otherwise it is boring"

This 'communicative' approach is partly the result of using only English in the classroom, but more than this, it is giving students the opportunity to use the language they are learning in authentic contexts. We will look at this much more later, but it is the overwhelming view of students I have taught that this communicative approach is more successful, more *real* than the approach of rule learning.

5 How important is it for a student to have some responsibility for their own learning and not to rely on the teacher alone?
"very important"

"teacher helps to show you the right way"

"trains you for 'real life'"

"it is important to have some responsibility for your own learning"

"a good student has some responsibility outside the class, but they have to be guided"

The student who comes to class but never uses his English outside it (more likely when they are learning in their own country), or who never studies outside it, will make very little genuine progress. The classroom is not "real life" (as mentioned above). It is *artificial*. Everything is controlled by the teacher and how she has planned the lesson.

We have to encourage our students to read English, speak English, hear English, write English. We also have to encourage learner independence *within* the classroom. Unit 5 will explore these issues.

6 Is it a good thing to be in a class with students from another country?
(this is mostly relevant to classes in English-speaking countries)
"more interesting"

"different cultures"

"yes, it makes you use English"

"it is good to open your mind, and we have to speak English"

7 Is it a good thing for your teacher to use your language some of the time when teaching?
"teacher must first try and explain in English"
"never, never"
"only with low-level students"
"it is not necessary"

These comments support the views expressed at the start of this unit and show how aware most students are of the need to immerse themselves in English in the classroom.

Comments from Thomas Grossmann, from his perspective of a student who became a teacher, are also very relevant. His views on what helped him gain the Cambridge Proficiency of English in a relatively short time cover many of the points already mentioned, as well as one or two others:

"a positive attitude towards the language, the country where it is spoken, and its people"
"absorbing as much language as possible (especially vocabulary, phrasal verbs and idioms) by watching TV, talking with natives, listening to audiobooks"
"reading books, magazines, newspapers"
"discussions in class"
"being able to figure out a grammar rule for oneself" (described by Thomas as the 'aha' effect!)
"not being afraid of making mistakes"
"a good reference book"
"trying to get the 'feel' and 'rhythm' of the language"
"copying the pronunciation from natives and slightly exaggerating it"
"the iron will to learn the language (motivation)"
"keeping penfriends"

Thomas also says: "I think an advantage I had was that I learned English in England where I could use it every day. I imagine that if I had learned it in Switzerland it would have been much harder and less close to my heart. My views on effective learning since becoming a teacher haven't altered much. Every person has his or her individual way and pace of learning and, whatever that is, it should be encouraged. Another point which I think is very important (especially if one doesn't learn English in a country where it is spoken), is to use as much *real* language as possible."

Review

- Our students usually have different favoured learning styles, influenced by their previous learning experiences.
- Most prefer, or accept the advantages of, a mostly communicative approach to their learning.
- Students have to accept some responsibility for their own learning and not rely only on the teacher.
- Learning has to include authentic practice, and has to continue outside the classroom.
- Students need motivation and commitment.
- Students prefer – and expect – English to be the only language spoken in the lessons, except with very low-level classes.

Everything which has been mentioned in this unit inevitably affects the role of the teacher and how she plans her teaching, and this is what the next unit will look at.

2 BEING A TEACHER

Most people remember at least some of their teachers from school-days, whether positively or negatively. Teachers make an impression. And students have expectations. If being a student is a 'complicated existence', then being a teacher is trying to meet the many varying expectations that students have, and that the teacher herself has, of the teacher's role. Perhaps we should say 'roles' because 'teaching' is only part of what the teacher has to do inside – and outside – the classroom.

 Task

1 Think back to your own schooldays and the teachers you had. Which ones did you consider to be 'good' and which were 'bad'? Why?

2 If you studied a language at school, how did that teacher approach his lessons? What was effective and not effective and why? How did you learn? Was anything lacking in your lessons?

3 On the basis of your previous answers, what do you think constitutes an effective and popular teacher of English to speakers of other languages? Make a list of the most important qualities.

STOP – do this task before reading on

Many people have experienced the 'listen and take notes' approach to language learning, followed by doing practice exercises from a book, and with a strong emphasis on remembering what has been learned, as opposed to *using* the language. It must be stated that some students do prefer this approach (which is not necessarily to say that it is an effective approach). This method implies a very limited responsibility for the teacher whereas, in reality, the effective and popular teacher has a multitude of responsibilities.

The survey I did with students highlighted not only this multitude of responsibilities, but also the expectations that students have of their teacher. Students, of any subject, are a critical and discerning audience, which can make a teacher's job hard. However, if teachers can respond to their students' personalities, attitudes, needs and backgrounds, they will become better teachers.

What makes a good teacher?

I asked the same students featured in Unit 1 "What makes a good teacher?" Of all the questions I asked, it was this one that produced the most opinions. In no particular order, these are some of the qualities mentioned in their replies:

"they must be passionate about teaching, and patient, adapting to the level of the students"
"a good teacher must be strict, friendly, interesting"
"should be passionate and intelligent"
"should teach grammar clearly"
"should not rely on the course book only"
"must know students' individual needs"
"help students make progress"
"give interesting lessons"
"be kind"
"not show a distance"
"be funny"
"have an original teaching style"
"have energy"
"have lots of patience"
"adapt the course book"
"treat students as individuals"
"be stimulating"
"know the students' names"
"you must know your students"
"must care"
"it is not enough to know your subject"

'Knowing your subject is not enough'

Just from these comments it is obvious that a teacher has to be more than a 'teacher'. That last comment – "it is not enough to know your subject" – underlines the point that it is not enough for a teacher to go into a classroom and effectively 'give a lecture' while students make notes. Of course, the teacher has to know her subject – this is essential – but it is not enough.

For new or training teachers, however, 'knowing your subject' feels more than enough! What that strikes terror into new teachers' hearts is the 'G-word' – grammar. The most common cry I hear is "I don't know grammar – I never did it at school." Or, if grammar was learned at school, it was in the form of so-called rules of language, and terminology that has all been forgotten. Well, 'grammar' is, or should be, much much more than this. We will see in a later unit what is involved in learning 'grammar', and how it can be 'taught'. However, that is the 'teaching' not the 'knowing'. Yes, as teachers we *do* have to know how the present perfect simple is formed and how it is used, we need to know what a phrasal verb is and the different kinds that exist, we need to know why we can say "if I *left* now" rather than "if I *will* leave now", and we need to know why *may* has different meanings in the sentences "You may leave now if you want" and "Luc's late, he may have missed the bus". The 'mistake' new or training teachers sometimes make is to think that they have to know everything straight away. Well, relax. First, no teacher knows everything about the language they are teaching and never will.

Second, teachers starting out are students themselves, students of the language they are teaching. You learn as you go along and you build up your knowledge. This is how it has to be. So don't read a grammar book from cover to cover. Instead, concentrate on understanding the language that you are going to be focusing on next in the classroom. Little by little your confidence will grow.

The teacher also has to know *how* to teach, she has to have an awareness of effective and successful methodology. The comments about the teacher not relying on the course book, giving interesting lessons, having an original teaching style, and being stimulating, all confirm that students want – expect – their teacher to teach in a much more challenging and interesting way than simply setting them work from the course book. The main part of this book deals with this very area – the methodology.

'Give interesting lessons'

Students get bored easily! They expect the teacher to plan her lessons so that learning becomes interesting, challenging and varied. They expect the teacher to teach in a way that is appropriate to the level being taught, and to the individuals in the class. They expect the teacher to adapt and supplement the course book being used and definitely not to rely on it as the sole resource. In short, they expect the teacher to know *how* to teach effectively, and for the teacher to "be stimulating".

We saw in Unit 1 that students generally respond better, and benefit most from a communicative approach to learning, and it is this approach that we will focus on here. Very frequently, people who have undertaken training courses in teaching English in the communicative way have told me they wish they had learned a language in the same way. I certainly wish I had.

'Know your students'

But still that is not enough. A teacher may know her subject, she may know how to teach it, but she has to know the people that she is teaching – and I don't just mean 'know their names', though that is important, too! So, find out about the students – the people – that you are teaching: what they like and are interested in; and what they don't like and are not interested in. Find out about their previous learning experiences. Find out what their expectations are of the course you are going to be teaching. Find out what their aims are. Ask them about themselves: their country, their culture, their interests. Speak to them outside class. Speak to them inside class! *Listen* to them. Show an interest in what they are saying. Show an interest in them as human beings. Be approachable. Help and support them. Be friendly. *Respond* to their needs. Be tolerant. Be patient.

In my opinion, no teacher is complete if they do not go beyond 'teaching the lesson'. The personal qualities required by the teacher described here can be summed up in the phrase 'show an interest'. If you do not want to show an interest in students as human beings, you will only have limited success as a teacher.

'Be passionate and patient'

Thomas Grossmann, remember, can look at the roles of the teacher from both perspectives. When he gave me his views on being an English teacher he had only recently started his first job, and only a short time before that had passed his Cambridge Proficiency of English exam. He trained in England and went on to teach in his home country, Switzerland. This is what he says:

"I think to be a successful teacher you have to like the job. One should be able to motivate the students and make lessons fun. A dash of humour would be good, too. The ability to explain something complicated in a clear way is important. Having patience, creativity, and good communication skills are an advantage, too.

I started teaching and it's great fun. I was really nervous before the first lesson but everything went smoothly. The worst thing that has happened to me as a teacher was that I wanted to teach them too much – I had three fifty-minute lessons and prepared a lesson plan for each lesson. But after two lessons I was still doing material which I had prepared for the first and in the end I hurried them just to get through everything. Looking back, I think that was a mistake."

You have to "like the job". Yes, you do. Of course, at times all teachers want to scream and do *anything* but go and teach their 25th hour of the week. But there are few jobs, if any, which are always positive. With this job, you meet people from all over the world, you have the scope for creativity (mentioned by Thomas), and can have a very good time. As one of my students said, a teacher needs to have energy – you have to always appear enthusiastic about what you are doing, even at 9 am on Monday. If the teacher appears uninterested, the students will become that way themselves.

But, as Thomas says, you *do* need enormous patience at times. And when, inevitably, the time comes when you do want to scream, just think back to how difficult you found it to learn a new language, or to remember it the next day. In other words, keep thinking of how the student is feeling – maybe tired, maybe frustrated, maybe stressed – and if you then show impatience, how will that affect them thereafter?

Thomas also mentions humour. This doesn't mean the teacher has to be an entertainer, but that they should inject some fun, some light relief from time to time. Most humour arises naturally from the different situations that occur with the various students in your class. The sound of students laughing is (usually!) a very welcome sound, so encourage it when appropriate.

'Help students make progress'

Teachers must know how to plan their lessons appropriately (see Units 17–20), and how to assess objectively the strengths and weaknesses of their own teaching in general, and of specific lessons. This is all part of professional development (Unit 25).

If there is not this critical self-evaluation, there will be no development or progress. New or training teachers have a tendency to dismiss most of their lessons as 'disasters', but this usually revolves around something terribly minor like "I forgot to give the instructions before handing out the papers". As we shall see later, one of the first questions the teacher should ask herself is "Did I achieve my aims?". The better the planning, the more likely the lesson is to be effective.

Review

1 Teachers have many different roles inside and outside the classroom.
2 They need to know their subject.
3 They need to use appropriate teaching methodology.
4 They need to find out about their students; know them; respond to them.
5 They need to demonstrate various personal qualities.
6 They need to show interest in their work and their students, and they must have limitless energy!
7 They need to plan lessons according to the needs of their class.
8 They need to know how to properly evaluate the success of a lesson, and evaluate their own strengths and weaknesses.

As this whole book is about 'being a teacher', this chapter has been no more than an introduction. One point not mentioned in this unit – but another vital aspect of the teacher's role – is classroom manner and management. This is what we will focus on in the next unit.

3 | CLASSROOM MANAGEMENT AND MANNER

We saw in the previous unit that a teacher has many roles, including managing her classroom and students. We shall look at how she does this, and at the manner she adopts when teaching. Effective classroom management is an essential ingredient in successful teaching. For the purposes of this unit we will focus on the following:

Teacher talk   the grading of the teacher's language and the amount spoken, instruction giving, directions

Gestures   how gestures can help to clarify what a teacher wants her students to do

Interaction and management   factors to consider, depending on who is interacting with whom in the classroom and in the teacher's management of the class

Manner   the importance of an appropriate teaching manner

Teacher talk

The teacher has to use a level of language which will be understood by the students in the class. This level will be influenced, of course, by the level of the students. This does not mean speaking at an unnaturally slow pace, or raising your voice – do not patronise your students. If a student has a poor level of English this does not mean they lack intelligence, so never give that impression to them. Instead, *simplify* the language being used. For example, rather than saying "work in pairs", to a very low-level class, say "work in twos" (accompanied by a gesture pairing students off). There is no point in saying "work in pairs" when they might not know what 'pairs' means.

Rather than just saying "underline" (e.g. "underline the correct answer"), *demonstrate* visually or by a gesture what you mean.

The same applies to the teacher's language in general conversation. Rather than saying "What did you get up to last weekend?", say "What did you do…?". And rather than saying "Do you want me to go through your job application with you?", say "Do you want me to look at…?" So, *simplify* your language.

When teaching, this *grading* of language needs to be accompanied by an appropriate level of *teacher talk time* (TTT). This does not mean 'try not to talk' (I have seen new teachers struggling desperately to avoid speaking at all having been told to keep their TTT low!). It means 'don't use too much *unnecessary* language, because the more you say, the more likely you are to be misunderstood. There are times when the teacher has to do a lot of talking. There are times when lessons need to be teacher centred. But what we should be aiming for is *economy* of language coupled with appropriate grading. It's a little like the advice for packing a suitcase. First pack it, then reduce the contents by half to what you *really* need. As a teacher, it is easy to start saying far more than you have to, and to run the risk of confusing your students. This is especially the case, the lower the level of the class. Clearly, a class of advanced students will understand most of what you say, and you can be less careful with your language. But if you are teaching an elementary class, for example, you need to think about everything you say.

 Task 1

Look at the following examples of teacher talk. They are a combination of instructions, directions, and general chat (either as part of the lesson or before or after it) given to an intermediate level class. Try to simplify the language, and reduce the TTT to an appropriate amount. In some cases you might also want to question how the teacher's choice of words may impact on the students' motivation level. When you're done, check the key on page 25.

  1 "In a moment I'm going to ask you to do this exercise in which you have to answer ten questions. You can do it by yourself if you prefer, or work with your partner. When you finish I'll put

you in larger groups for you to check your answers."

2 "I want you to do a mingling activity now, which I'm sure you're going to hate!"

3 "Now, erm, what you need to do, er, is to, erm, get into pairs ... no, erm, make that threes ... wait a minute, there's thirteen of you, better make it threes and one four. Javier, do you want to work in a '3' or a '4'?"

4 "That's a jolly nice scarf. It must have cost you a bomb."

5 "I'm handing out a piece of paper…"

6 "This is going to be boring but it has to be done."

7 "This is going to be easy..."

8 "Did you enjoy that?"

9 "I want you to go through Exercise 1 and work out the answers. You must also justify your answers."

10 "Would you like to stand up, please?"

✳ *Tips for successful instruction-giving*

1 Plan your instructions – even script them at first if you feel it will help you.

2 Having planned them, review them and edit as necessary.

3 Only say what you have to.

4 Use simplified language, but do not over-simplify for the level.

5 Don't patronise (e.g. talking unnaturally slowly or loudly).

6 If you are getting blank faces it probably means they are not understanding. So adjust your language as you speak, if necessary.

7 When giving instructions, make sure the students are looking at you and listening to you.

8 Don't give out papers before your instructions unless you have to – you will lose the students' attention.

9 Stand front and centre. Keep still.

10 Project your voice *slightly* so your instructions stand out.

11 Maintain eye contact.

12 Break up the instructions: "Do questions 1–5. Say if the sentences are true or false. Give reasons for your answers. Work together. You have five minutes. Go."

13 Only give instructions for what they have to do at that time, not later.

14 Give an example of what they have to do / do Question 1 with them.

15 Give a rough idea of how long they have.

16 Make it clear if you want them to work individually, in pairs, or in groups.

17 Don't ask lower-level students "So, what do you have to do?" – they haven't got the language for this, and confusion will result. Once your instructions are over, tell them to begin, then check everyone is doing the activity correctly and deal with individual problems.

18 If one person says "I don't understand", help that person individually, don't hold up the rest of the class, and the lesson.

19 Don't *invite* the students to do something; politely *tell* them. So, don't say "Would you like to stand up?", say 'Stand up, please".

20 Students like – and expect – the teacher to be polite but direct in her language. They need to be told clearly what they have to do and how you want them to do it. Don't be afraid to do this! But don't tell the students they are going to find the activity boring, or difficult, or interesting, or easy!

Gestures

Everyone uses their hands and face to express or clarify meaning. Teachers do it more than most because they quickly realise how much meaning their body language can add to the words they are using. At the beginning, you may have to remind yourself to do a 'writing' gesture with your hand when you say "I want you to write…" to a low-level class; but after a while you will use gestures all the time quite automatically. You can work out for yourself what appropriate gestures you can use to clarify meaning – but make sure your choice is clear, appropriate, and consistent. You will find that your students' understanding will be greater as a result. You will also find yourself gesticulating more than you need to when you are outside the teaching context!

Task 2

Consider what gestures you could use to indicate the following. There are no 'right' answers; the aim of this task is to get you thinking what gestures *you* might use in the classroom.

1 "Work in pairs / threes."
2 "Get in a circle / straight line."
3 "Stand up / sit down."
4 "Listen / Read / Write / Speak."
5 "Tick the right answer."
6 "Think about..."
7 "Look at me."
8 What gesture could you use to indicate the student is wrong in his answer? Only partly right?
9 "Past tense" / "Now" / "In the future".
10 What type of gesture could you use to clarify the meaning of word/sentence stress?

✳ *Tips*

1 Be clear/unambiguous.
2 Be consistent.
3 Don't over-exaggerate to the extent that you appear patronising.
4 Don't use inappropriate gestures (e.g. pointing is considered rude by most people).

Interaction

There is a huge variety of different teaching contexts worldwide. Whether you are teaching a class of six sitting together in a semi-circle ('horseshoe') formation in a multilingual class, or a monolingual class of fifty students sitting behind rows of desks, you must consider how best to manage your classroom and the people in it. Because there is a huge range of possibilities, there are very few 'rights and wrongs', but lots of questions for the teacher to consider.

Here is a selection of questions to think about. The answers sometimes depend upon the situation! With some questions, you may have no choice (e.g. seating arrangements). With other questions, it is possible to give general answers.

- Where do you want the students to sit for a particular activity?
- Do you want them to work individually? In pairs? In groups?
- If you need them to work in pairs, what will you do if there is an odd number of students?
- If you are teaching a multilingual class, do you want to separate students who speak the same language?
- Where are you going to place yourself? What does it depend on?
- Will you always stand?
- How can you try and be sure all the students can hear you?
- Where will you stand when you use the board?
- How does your distance and position affect your 'relationship' with the class?
- How important is eye contact and why?
- How important is it to use your students' names?
- What will you be doing when the students are working individually/together?
- Should every activity last until all students have finished it?
- How can you attract the class's attention at the end of an activity?
- How can you politely stop two students from talking when they should be listening to you?

What should you check in your classroom before the lesson begins?

✳ *Tips*

1   Plan in advance what type of interaction you want for each stage of the lesson (teacher to class / pairs / groups etc).
2   If you need even numbers (A working with B) and you get an odd number of students, two students can work together as 'one' (A, A and B). Don't participate yourself because then you will not be able to monitor other students.

3 In a multilingual class try, if possible, to have students from different nationalities working together to encourage only English in the lesson. In any case, vary who works with whom from time to time.

4 If you are doing 'teacher to class' work, stand where you can be seen by everyone (usually front and centre). If students are working together, go around and monitor their work, helping individuals as necessary and generally just 'keeping an eye' on how everyone is doing. If someone finishes an activity quickly, give them something else to do. You should aim to keep everyone busy all of the time, otherwise pace drops and students get bored (especially at higher levels).

5 Sometimes it might seem more appropriate to sit. If you have a smallish class and are just 'chatting', for example. This would not be appropriate, though, if you have a class where students are forced to sit in rows.

6 When using the board, try not to block students' view. Stand to one side. If you are still blocking someone's view, move out of their way as often as you can so they can see what you are writing.

7 Writing on the board is not easy … it needs practice. Make sure your board looks neat, well organised and clear. Students copy what you write so it needs to be a proper 'record' for them to copy and refer to in the future.

8 Maintain eye contact with your class. This will keep you aware of how students are reacting to each part of your lesson (for example, you can usually tell if someone is not understanding what you are saying just from their facial expression). This will help you to remain in control of the lesson. Also, as in any conversation, it is polite to look at the person you are talking to!

9 Learn your students' names as quickly as you can and use them! Again, this is just polite. But it is very difficult to attract a student's attention in class if you do not know their name. "Excuse me, you…" (or equivalent) is simply not acceptable. Students will expect you to know who they are and show an interest in them. For the first two or three lessons it is perfectly understandable if you cannot remember all the names, but after that it gets embarrassing. Do a 'name-game' activity on the first day, perhaps. If the class is all new, they too will want to learn names.

10 If students are working together or individually on an activity for which you have given them "ten minutes", this does not mean *literally* ten minutes, it is just a rough idea of how long they have. Maybe they only need five minutes, or maybe you realize they are having unexpected difficulty and need a few extra minutes. So be flexible. Do not feel that you have to wait for every single person to finish what they are doing before you stop the activity – very rarely does everyone have time to finish. Otherwise, you are running the risk of bored students sitting twiddling their thumbs, waiting for the lesson to continue. Do not be afraid, then, to say "Okay, stop there, don't worry if you haven't finished, we'll go through the answers together."

11 Devise a way, which works for you, to attract the attention of the class when you want them to stop working together and listen to you again. Projecting your voice (not shouting!) may be enough; tapping the board with a pen works for some; perhaps raising your hand. The bigger the class, the more important this will be.

12 If two students are chatting to each other when they should be listening (to you or another student), stop them, otherwise it becomes very distracting. If you are teaching adults, treat them as adults – so just a gentle and polite request for them to listen.

13 Before the lesson begins, check you have everything you need – board pens/chalk, books/photocopies, tapes (cued up already). Check you have any technology that is required – cassette player, overhead projector etc. – and that you know how to use it. Do not wait for the lesson to begin to check all this since if you hit a problem you will appear unprofessional.

Manner

In my opinion, however good you might be at teaching, however well you might know your subject, however popular you might be with you colleagues, if you do not treat your students with personal respect then you are not a good teacher. I have seen and heard teachers shouting at adult students before (for being late to class, for example). I have seen teachers tell students to go away and not disturb them during their tea break. I have seen teachers treating students with remarkable rudeness. And I always shudder when I witness this (only very occasionally, thank goodness). I would rather employ an 'average teacher' who is friendly and kind to students, than a highly skilled teacher who shouts at students or has no patience with them. Yes, it can be a very frustrating job. Yes, some students will want to make you scream at times. Yes, you will teach students with very different cultures from your own. But if you have limited patience or tolerance levels, then this isn't the job for you. If you are not friendly and patient with your students then they will not respect you, whatever the standard of your teaching.

So, be friendly. Show an interest in the students as people. Be patient. Be tolerant. Be understanding. Have a joke with them. Never shout at an adult student. Talk to your students outside class. If you can actually like them, all the better. Think of how *you* want to be treated by a teacher if you were learning a language in a foreign country, and of how you expect others to behave towards you in daily life.

Key to Task 1

1 *"Answer the ten questions. Work together."*
Avoid unnecessary talk ("in a moment"), unnecessary instructions ("when you've finished") which can be saved for later, complex words ("avoid").

2 *"Stand up and walk around..."* (accompanied by a gesture).
'Mingling' is teacher language, avoid it. If you say "you're going to hate it" then they probably will!

3 *"Work in twos (plus gesture)".*
These were clearly unplanned instructions from a teacher who hadn't thought about how the activity was going to work!

4 This is the teacher chatting outside class. Poor grading ("jolly"), inappropriate and overly-colloquial language ("cost you a bomb") even for a high-level class.

5 Don't give a running commentary! You don't need to tell them that you are handing out paper. Also, never give instructions

while doing something else at the same time.

6 Again, don't be negative when setting up an activity!

7 If you tell them it's going to be easy and it isn't, how will they feel? If it is easy they may wonder why you gave it to them in the first place.

8 And if they say "No"?

9 Avoid phrasal verbs ("go through"). Say "do". And "justify" is too difficult. Say "give reasons".

10 *"Stand up, please."*
   If you ask them if they want to stand up and they say "No", then what?

Review

- Teachers need to adust their language to suit the level of the class they are teaching.
- This is done not by talking loudly or unnaturally slowly, but by simplifying the language they use.
- With lower levels especially, teachers should restrict the amount of language they use.
- Meaning can be clarified by the use of gestures and other body language.
- Instructions should be clear, relevant and precise – planned in advance if necessary.
- Class management is an important part of the teacher's responsibility and it needs to thought about before and during the lesson.
- Teachers should adopt a supportive, friendly and encouraging manner.

4 | TEACHING GRAMMAR VIA A SITUATIONAL PRESENTATION

What is grammar?

> *'An abstract system of rules whereby a person's mastery of their native language can be explained'*
> (adapted dictionary definition)

Different dictionaries define 'grammar' in different ways, but they all effectively talk about the 'rules of the language'. I don't think it really matters what definition we give the word, anyway, as we all have an idea in our minds of what the word means to us. However, do not assume, as many do, that 'grammar' equals only 'tenses' or verb forms (e.g. present perfect, past continuous/progressive). Grammar is language and how we use it; tenses are just a fraction of our language.

Personally, I don't like the word 'grammar' as it has too many negative connotations. For many it implies the rather dry learning of rules and then controlled practice of these rules with little or no communicative practice. Grammar doesn't have to be like this, and it shouldn't be. Despite my own dislike of the word, I will continue to refer to 'grammar' lessons, or components of lessons, because it is not a word English teachers can avoid!

What is much more important than a definition of the word 'grammar' is knowing what students need to know when they are learning language. We also need to know different ways to 'teach' grammar. And there's another awkward word: 'teach'. 'Teach', like 'grammar', has connotations. It implies a lot of teacher talk and explanations, initial student ignorance of what is being covered, a beginning and an end (they didn't 'know' it before, they did after), and it implies a completeness (they now know everything there is to know about this). You don't hear a teacher going into class saying "I'm going to teach them relative clauses today," but teachers do know what their aims are in any lesson. I prefer 'clarification' of language to 'teaching' of language, but like 'grammar', 'teach' is a useful, all-embracing word that has to be used, and will continue to be used here. But do not think that language is just a series of unrelated items that can be dealt with one by one!

There are other words and terms that are, let us say, controversial in the profession, but at this stage of your career it is more important to concentrate on how to 'teach' than to be concerned with terminology.

What do students need to know?

They need to understand **meaning** within the given context.

They need to know the natural **pronunciation** of the language.

They need to know how to **form** the language (how it is constructed).

This is often referred to as **MPF**. As well as knowing the MPF, students need to use the language – **practice.**

This is a simplistic answer to the question, but for new and training teachers it is a very helpful one, especially when it comes to assessing if the lesson you have just taught was successful in terms of your language aims. We will look at each of the MPF areas in the remaining part of this and later units.

Situational presentations

'Presentation' is another controversial word! For many it implies, as does 'teach', that language can be broken up into convenient chunks which can be taught in isolation and built upon brick by brick until students 'know English'. Of course, it does not work this way. Experienced teachers know how best to develop a student's awareness of the language without falling into the "Today we're going to do conditional sentences" approach. But until experience has been gained, it is very difficult to avoid dealing with language in chunks. Too much should not be expected of a trainee teacher, but it often is! Certainly, no person can become a teacher in four weeks – but they can develop their teaching skills and awareness sufficiently to be able to go in and deliver a reasonably successful lesson. That is where the real training begins.

So I make no apologies for talking about 'situational presentations', or 'a lesson on the 2nd conditional', or whatever. Rightly or wrongly this is what trainees are exposed to on training courses, and this is how the majority of course books deal with teaching and language. Once you gain real experience, you will be able to re-evaluate your approach.

Language needs **context** because context provides meaning. One way to illustrate meaning is via a situation. This can be done in a number of ways, including pictures. This situational presentation tends to be quite controlled by the teacher, at least initially, but with the students involved at all times. Let's take the word *should* – a modal auxiliary verb which, like other modals, can have different meanings depending on the context. If a student asks "What does *should* mean?" you cannot usefully answer them. In isolation it is meaningless. If we put the word into a sentence:

You should wear a suit, it is better because the word now fits in grammatically to a sentence – the **form** is illustrated. But meaning is still very limited. We don't know who is speaking, who they are speaking to, and why they are speaking. In short, there is no situation, thus no context, thus no meaning. One way to provide context whilst maintaining student involvement is as follows. This is for an elementary/pre-intermediate class.

Stage 1

Spend a few minutes on a class discussion about unemployment in the students' countries, how people go about finding work, the stages of applying for a job – leading to tips for a successful interview. The teacher asks questions and maximizes student talking time.

Stage 2

The teacher says she has a friend, Jason who is going for an interview tomorrow for a job at the bank. She elicits from the students how Jason should appear when he goes for the interview (appropriate clothes, shaved, hair combed etc). She then shows a picture of how Jason typically appears and asks if this would be appropriate for his interview. She highlights his dirty shoes, uncombed hair, unshaven face, and casual clothes.

Stage 3

The teacher says she saw Jason earlier. She emphasizes that Jason is her friend and she really wants him to get this job but is not confident because the interviewer is likely to expect interviewees to be dressed in a certain way. So, pointing to Jason's casual clothes the teacher asks if the students think it would be a good idea for Jason to go to the interview dressed in this way. Having established that it wouldn't, and still pointing to Jason's clothes, the teacher says "So what did I say to Jason when I saw him?" If possible, the teacher could also have a picture of someone wearing a suit to make the contrast with Jason's clothes and to make it clearer to the students what she is thinking. At this point, the teacher is testing to see if any students are familiar with the language being presented, wondering if anyone will reply "*You should wear a suit*". This is the teacher's model/marker sentence containing the language being targeted. Maybe someone will know it, or may say the sentence with mistakes. Or maybe no-one has any idea. This is not important, you are just seeing if anyone knows it. If no-one does, the teacher says the sentence to the class, making sure everyone is listening!

At this stage of the lesson the main focus has been on **meaning**. The model sentence *You should wear a suit* now has a context, and the 'function' of 'giving friendly advice'. The function is basically the reason for speaking – other examples being 'inviting', 'apologizing', 'refusing', 'expressing surprise' etc. What the teacher has therefore done is to *illustrate* meaning by means of a pictorial situation and introduced a *model sentence* containing the *target language* (subject + *should* + base verb – this is the **form**). The function of giving friendly advice has been established via the situation. Depending upon the precise level of the class and what you have taught them before, you may need to have begun this lesson by checking they understood essential vocabulary (*suit, polish, shave, comb, suggest*) so as not to have the lesson 'stall' at a critical point. See Unit 7 for ways of presenting vocabulary.

So far in this lesson the only aim that has been achieved is that meaning has been illustrated. However, it is essential that the teacher checks that the students really have understood the meaning. The fact that one student, maybe, has given you your model sentence does not mean that everyone has understood its meaning in the given context. Before you go any further in this unit, therefore, you should read Unit 8, then return to complete this unit.

Read Unit 8 now!

Now you have read about how to check students' understanding of new language, the **meaning** of **MPF** has been covered for this example lesson. As for the **pronunciation**, you should later read

Unit 9, but for now it is enough to say that the students must hear you say the model sentence several times naturally and then be given the chance to say it themselves. This is not as straightforward as it may sound, as we shall see later. But if we are teaching language for active use then the students need to know how to say it!

The **form** of the language has already been implicitly established via the giving and repetition of the model sentence. But this needs to be checked, just as the meaning needed to be checked. In the 'Jason' lesson it is easy to elicit further examples of *should*: *You should polish your shoes*, *You should comb your hair*, *You should shave*. These sentences can then be put onto the board (keeping the students participating by getting them to tell you what to write). Form can be highlighted by using different colour pens, if possible, or/and a substitution table:

You	**should**	wear a suit
		polish your shoes
		comb your hair
		shave
	shouldn't	wear those clothes

You is not unnecessarily repeated, nor is *should*. This way the form is further highlighted. Notice the same subject is used – the sentence *He should wear a suit*, for example, is not giving advice and therefore has a different meaning.

You could, if you feel it appropriate, label the respective parts of the sentence (subject/*should(n't)*/base verb). You could include your concept questions and answers. You could also mark sentence stress and intonation. Whatever degree of detail you decide is necessary, students need to have some 'take-away' record of the form that is being covered in the lesson – the above illustration is probably the minimum requirement.

We will come back to form and pronunciation elsewhere, but the initial 'presentation' via pictures is an approach which is very commonly used, especially at lower levels.

Because it is primarily 'teacher-led' it is a method more suitable for students whose level of English is not high rather than for more advanced students, who need a greater learning challenge (see next two units).

✳ *Tips*

1 If you are using pictures, make sure they are clear, simple and appropriate.
2 If necessary, check essential vocabulary at the start of the lesson.
3 Build the context slowly, carefully and clearly – *guide* the students and keep them involved throughout. Don't tell them what they can tell you. By continually asking them questions you are also checking they understand.
4 Have an obvious 'target' – a model sentence which will be a logical conclusion to your context build. In my example, *You should wear a suit* was a logical 'conclusion' to the storyline.
5 Try to elicit the sentence if you can, otherwise just tell them. If they have never come across this language before, it just tells you the extent of your task, it does not mean your context wasn't clear!
6 If possible, show your context to a colleague or friend beforehand to see if they think it is clear and appropriate.

You should also research the language yourself so you know everything you need to know about it. You need to anticipate possible problems your students might have with this language. These could include:

- confusion about the time reference
- a desire to refer to it as a tense (e.g. present simple – which it isn't!)
- incorrectly adding a *to* – You should to wear a suit   ✗
- thinking that by changing the person (e.g. from *you* to *I*) the function remains the same – it doesn't.

See Unit 19 for more on anticipating language problems.

Task

For each of the sentences below:

a Try to work out the **form**. In simple terms, what is the fixed structure that never changes? Refer to the *should* example earlier. Another example is *I wish I hadn't done it'* = subject + *wish* + subject + *had/hadn't* + past participle. It sounds a bit complicated and 'grammatical' but it helps you, and students, know how a sentence is built.

b Think of a possible **context** and establish the **function** (e.g. 'giving advice') and **meaning**, including the time reference.

c Think about, or even prepare, a possible picture(s) to illustrate your meaning.

Do not worry about the level of the class these sentences might be aimed at, but none would be for a very low-level class, so assume that your students have a reasonable level of English.

1 I shouldn't have done it.

2 I'm going to France next week.

3 Would you mind helping me?

4 Let's go to a concert.

5 I've worked here for ten years.

Do this now, and check it later!

Key to Task

1 I shouldn't have done it.

Form: subject + *should(n't) have* + past participle (verb 3)
Function: maybe 'expressing regret about past action'
Context: student copies from another student in an exam, tells a friend about it the next day and expresses regret.

2 I'm going to France next week.

Form: *to be* (present form) + *going* + base verb (NOT the present continuous/progressive!)
Function: maybe 'giving information about future schedule'
Context: two busy friends with diaries trying to arrange to go out, one explains that the next week is impossible.

3 Would you mind helping me?

Form: *would you mind* + *-ing* form of verb
Function: maybe 'polite request for help' – asking now, but for help in near future.
Context: person at airport struggling with heavy cases, asks a passer-by for help.

4 Let's go to a concert.

Form   *let's* + base verb (NOT 'let us' as this is unnatural in spoken English).
Function: maybe 'making a suggestion' – suggesting now what you could do later.
Context: two friends wanting to go out later, not sure where to go. One looks in the entertainment guide to see what's on, notices a concert and suggests that.

5 I've worked here for ten years.

Form: subject + *to have* (present) + past participle (present perfect simple)
Function: maybe 'giving reason for wanting to find a new job' Time reference is past to present (from ten years ago to now) and maybe continuing into the future, at least for a while.
Context: employee to employer explaining why they have decided to leave the company – wanting a new challenge.

For each of the above **contexts**, meaning can be illustrated either with visuals or some other way (e.g. dialogue on tape). But in each case there is a situation which illustrates the **meaning**. How much detail you give the students about the **form** will depend on their level. Higher-level students are normally comfortable and confident using terminology, lower-level ones much less so – it is pointless referring to 'subject' if they do not know what a 'subject' is!

Review

- The meaning, form and pronunciation of new language needs to be covered, and communicative practice should be given, too.
- Situational presentations are controlled and effective ways to introduce language at certain levels.
- They can be done via visuals as long as these are clear and appropriate.
- Teachers need to be clear in their own minds of the meaning of the target language, how it is formed, and how it is spoken.
- This type of presentation is usually teacher-led, but with student involvement throughout. Let them do as much of the 'storytelling' as possible.
- Have a model sentence to work towards – elicit it if you can, give it if you can't.
- Check they really have understood the meaning (Unit 8).

This approach makes language more real than just a series of rules and explanations. The students see the language in a situation that is real and that provides the meaning. The teacher, therefore, starts with **meaning** and then introduces the language. In a sense, it is grammar in disguise, useful if you have students who are demotivated by thinking they are about to do a grammar lesson. If it works well, the presentation is interesting, even enjoyable. Above all, it is much better than giving an 'explanation' of a language point.

Some other ways to approach language teaching will be covered in the next three units.

5 TEACHING GRAMMAR VIA A TEXT OR RECORDING

The 'presentation' approach featured in the previous unit involved the teacher supplying a situation, possibly using visuals, in order to build a context with her students and eventually leading to a 'model sentence' containing the target language. It was very teacher controlled, though with full student participation, and rather contrived. It is a method which is effective, especially with lower levels, but no one method can, or should, be used exclusively or else it becomes predictable and boring. In addition, the method described is not especially challenging for higher-level groups. What we will look at in this unit, then, is another approach which provides variety from the previous one, a greater challenge, more learner autonomy, and a more realistic context.

 Task 1

Look at the reading texts below and answer the following questions:

a Does the language seem relatively simple, or high level for an intermediate class?

b Does the text seem authentic ('real') or 'made-up'?

c What specific language point appears to be potentially the most difficult/useful to focus on in class?

---

1 *Bertrand's San Francisco experience*

Bertrand is French and he lives and works in the north of France. His English is very good because he studied it at school and uses it in his job. A few months ago, he went to San Francisco for the first time to visit some friends he met in France a few years ago. He stayed for a week and in that time Bertrand and his friends had a very busy time – they visited Fisherman's Wharf, rode the cable cars, saw the sea lions by Pier 39, ate in a different restaurant every day, walked up the steep hills, and did lots of shopping in the fantastic department stores. And, of course, they took lots of photographs.

At dinner one evening Bertrand and his friends – Marie, Myrianne and Norbert – were having dinner when the fire alarm sounded, but the waiters didn't seem to be worried, they just carried on working. Everyone around them carried on eating. They thought it was so weird, everyone carrying on with their meals even though the fire alarm was sounding. Bertrand and his friends decided to get out quickly, but just as they went out of the door they saw a sign by the entrance warning customers that there was going to be a fire alarm test that evening and they should ignore it. Bertrand and his friends quietly sat back down again to continue their meal feeling a bit embarrassed.

The friends continued to tour San Francisco and sample the different food at all the many restaurants in the city, and not listening to any more fire alarms! They had a wonderful time and had a lot of fun. They didn't know when they might meet again but this had been a great experience for them all. Bertrand was sad to leave his friends, especially as he knew he had to go back to work three days later.

2 A special offer

Abibus have been producing top-class cars with unbeatable value for 12 years. We have been talking to you our customers a lot recently to find out if there was any way of making our cars even better than before and you told us that our product is as good as it gets. The only way to make *Abibus* cars even more attractive would be to reduce the price, but that can't be done ... or can it? Well, yes it can! From the beginning of next year, for a period of six months, we are cutting the cost of all our *Abibus* cars by 10% – **yes, 10%**! So from January 1 next year, *Abibus* cars will be 10% *cheaper*, but still top class. We have been doing business with you for 12 years. Now for the next 12!

3 Living in France

When I left my country a few months ago to work in France, I was very excited – I still am! But it hasn't been easy! For a start, there's the language. I used to study French at school, but that was a long time ago, so when I moved to Paris I knew it would be difficult because I had forgotten most of what I'd learned. And I knew that driving would be a problem because I'm used to driving on the left, but in France they drive on the right! So, the first few times I drove here, it was really strange and difficult. It's still not easy, but I'm getting used to it – slowly! I really like the life here, though, – it's fast and exciting. The food is very different – it's generally great, but I'll never get used to eating frogs' legs. I was told that it tastes like chicken, but I don't agree! But the wine is great, and there are so many different types of cheese, I love it!

There are so many places to visit, and things to do, but a lot of people who visit France only come to Paris. Paris is great, but France is a fantastic country – the south of France is beautiful, and with great weather. And the north of France has some great places to visit – Lille is really pretty and is a friendly area. Yes, I'm really glad I live in this country now!

Key to Task 1

1 Bertrand's San Francisco Experience

Level of language

Although this is the longest of the texts, the language is quite simple for intermediate students. *Steep* hills, *sample*, *embarrassed*, and maybe *department stores* are the only items of vocabulary that could cause problems, and the grammatical forms should mostly be familiar.

Authenticity

Definitely made up! It has the feel of having been specially written and graded (in terms of language). And the 'fire alarm' situation is somewhat contrived. This contrasts with *authentic* or *near authentic* texts. An authentic text is a 'real' one, e.g. a newspaper article. A near authentic text is one which is not real, but appears real – maybe something adapted from its original form, or an authentic-sounding made up text. The 'Bertrand' text is neither authentic nor near authentic. This does not mean that it is not valid to use, as we will see. What is important about this text, which makes it okay to use, is that the use of the target language is authentic. There is only one example of the target language, and it is surrounded by other language. So the text, while sounding a little contrived, is valid for classroom use.

Language to focus on

They were having dinner when the fire alarm sounded

In this example from the text, we are using two verb forms – the past progressive (continuous), and the past simple. The progressive is *were having* and the simple is *sounded*. In the sentence, they started dinner **before** the fire alarm sounded, and may or may not have continued after. Thus, the simple **interrupts** the progressive.

This language point is a very useful one to focus on, and appropriate for the level.

How this can be done, we will see later.

2 A Special Offer

Level of language

Higher than in the previous text, though not significantly (*top class*, *unbeatable value*, *cutting the cost*, *to do business*) – certainly not so high that meaning is unclear.

Authenticity

Again, this is not an authentic text. However, it is near authentic because it was adapted from a real text and it sounds more real.

Language to focus on

Present perfect progressive, e.g. *Abibus* have been producing cars for 12 years. This describes an activity which started at a definite point in the past and continues to the present, with the likelihood of it continuing into the future. The present perfect tense, both simple and progressive, is a particularly difficult one for students, and there are differences between British and American use.

3 *Living in France*

Level of language

The vocabulary would be reasonably straightforward for intermediate students, but there is a variety of verb forms used with some complex time references (see below).

Authenticity

Not authentic, but adapted from an article about people working in different countries, so near authentic.

Language to focus on

Different uses of *used to*: *I used to study French at school*
I'm used to driving on the left
I'm getting used to it (driving on the right)
I'll never get used to eating frogs' legs

You may not focus on all four uses at the same time, but with a higher-level class you could. Consider for yourselves how these differ in meaning and time reference!

Using texts to illustrate meaning

Before we look at the advantages of using texts to illustrate the meaning of language that you want to focus on, we need to consider possible methodology. If we refer to the 'Bertrand' text above and assume we are teaching an intermediate group of multilingual adults, we could follow the stages below:

1 Ask the class if anyone has been to / would like to go to San Francisco. If anyone has, they can tell the others about it. If no-one has then they can tell you what they imagine San Francisco. to be like. Alternatively, show pictures of San Francicso (steep hills, cable cars, Fisherman's Wharf etc) to get the class talking about it.

2 Tell students they are going to read about Bertrand, who visited San Francisco. They will have to read the text quickly and answer the questions:

Why did Bertrand visit San Francisco?
How many people was he with?
Why did they stop eating?
Did they finish their meal later?

Students read the text, and then work together to answer the questions. You then check the answers.

3 Highlight the sentence from the text *They were having dinner when the alarm sounded* and focus on the language (MPF).

From this point on, you could follow the procedure described in the previous unit (concept questions, pronunciation work, highlighting the form on the board, eliciting further examples). This is a crucial part of the lesson as it is where you need to focus on the target language and check understanding.

4 Provide students with varied activities such as practice of the progressive / simple 'interrupted' action. See Unit 10 for more on practice activities.

Stage 3 above might appear a 'quick check' which almost gets in the way of everything else, but it is, in fact, the most important part of the lesson because this is where you are checking meaning, form and pronunciation. This is not a 'reading' but a 'grammar' lesson, so Stage 3 must not be a 'rush job' to get on to some fun practice activity that you have planned; students cannot have practice of something which they do not first understand.

So what is the purpose of the Reading part of the lesson? The text does what the picture of Jason in the previous unit did (*you should wear a suit*) – it provides a context, and the context illustrates meaning. However, it is important that the Reading text/tasks are not so demanding that students end up spending much more time trying to understand the text than working on the grammar. The tasks that you provide, therefore, need to be relatively straightforward (as above in Stage 2). They should serve to check basic understanding of the context, as well as directing students towards the specific language that you will focus on later. The lesson effectively starts as a Reading, but the text is there primarily to provide a later language focus (compare this with a Reading comprehension lesson – see Unit 11).

There are, of course, many variations to the stages listed above. But what we have looked at here is one way of dealing with grammar via a text, keeping a basic approach similar to that of the previous unit, but adapting the way meaning is illustrated.

Task 2

Look at Texts 2 and 3 above and the language focus suggested. How might you stage each lesson (following a similar pattern to the first text)? There is no key for this task, but it will help you to come up with possible ideas for yourself.

The three texts above are all 'made-up', though two are adapted from authentic sources. Sometimes it is easier and more convenient to devise your own texts to suit your requirements, but if you do, ensure that the level is appropriate for the students you are teaching, and that there is enough 'other language' in the text, not just a very unnatural over-use of the target language. Even with the 'Bertrand' example above, un-authentic as it is, there is plenty of 'other language' – with only one example of the target language.

Ideally, however, it is better to use authentic texts if possible. Course books (and you are likely to be using one) will usually 'present' language via authentic / near authentic material, but sometimes you may find, for example, **a newspaper article** which contains a piece of language that you want to focus on in class. Similarly, you might introduce language via **a recording**. Have a look at this conversation from a cassette. What piece of language do you think the teacher wanted to focus on later?

Jill	Hello.
Guy	Hello, er, Jill?
Jill	Yes, speaking.
Guy	It's Guy here.
Jill	Oh, hello Guy. How are you?
Guy	Fine, thanks. Er, I was wondering if you could, er, babysit for us tonight?
Jill	Yeah, sure, what time?
Guy	Er, we're going out at about 7.30.
Jill	7.30? Yeah, fine, what time do you want me to come round?
Guy	About seven?
Jill	Fine.
Guy	We'll leave some food for you.
Jill	Great. That would be lovely.

Guy	And some wine, if you like.
Jill	Cool.
Guy	See you later, then … and thanks again.
Jill	No problem, see you later. Bye.

Despite the 'script', this is an authentic recording – what you have above is merely the transcript. The two teachers were given a situation ('Guy – you want Jill to babysit tonight / Jill – you are available to babysit tonight / Guy – call Jill') and from that they improvised a telephone call. This was recorded, and the transcript written up afterwards for the teacher's benefit. So on the recording it sounded like a natural conversation. Most of the time, however, teachers don't need to go to all this trouble because the course book you will be using will have many recordings to introduce language, just as they have many texts. But sometimes you might prefer to prepare your own. If you do, don't script it, and make sure the recording is clear to use in class. And, as with the texts, make sure it is, at the very least, near authentic in sound and containing 'other language'.

The piece of language that the teacher wanted to focus on from the recording was the very hesitant and polite request *I was wondering if you could* + base verb, as a means of introducing other forms of request. The possible procedure could be the same as with the text. For more on Listening comprehension lessons, see Unit 12.

It is important to stress, though, that you will rarely want to make up your own texts or recordings when course books use this approach so much, and when there is so much authentic material available to use.

Advantages of this approach

The previously described situational presentation using pictures is a largely teacher-centred approach, albeit with student involvement. It is, as we saw, an approach particularly effective with lower levels, primarily because there is very little 'other language' for them to deal with. With the approach described in this unit, however, there is a great deal of 'other language'. How much, and at what level, will be influenced by who you are teaching (you can, for example, edit a newspaper article if you feel it is too long

and the reading part of the lesson would take up too much time).
Also you shouldn't allow students to get concerned by unknown
vocabulary in the text/recording if they don't actually need to know
it for this lesson. As we saw before, the principal aim in such a
lesson is language-related, not skills-related.

With the right level, therefore, this approach has a number of
advantages, if executed properly:

- It is more challenging.
- Various skills are required.
- There is generally greater variety/stimulation.
- It is less teacher-centred.
- Students are exposed to the target language in an authentic/near
  authentic setting.
- Consequently, they see/hear the target language before having
  to focus on it.
- The target language is surrounded by 'other language'.
- Students should be encouraged to read/listen in English in their
  everyday lives, to be exposed to 'real English' – using this
  approach in the classroom encourages them to do so.
- The whole approach becomes more natural, less contrived.
- Course books use this approach a lot, making preparation easier
  for the teacher!

Possible dangers

- The text/recording part of the lesson may go on too long, leaving
  inadequate time for the language focus/practice (unintentional
  language neglect)
- It is sometimes tempting for new teachers to devote too little
  time to the language focus part of the lesson (intentional
  language neglect).
- The text/recording may be inappropriate in terms of level.
- The tasks provided may be inappropriate in terms of level, or
  just too time consuming.
- If the text/recording is made up, it may contain too unnatural a
  use/over-use of the target language.

Overall, the approach is very realistic and stimulating.

Review

- Texts and recordings can be a very effective way of illustrating meaning of particular language.
- The approach involves a greater challenge for higher-level students.
- Skills work and language focus are integrated.
- Target language is surrounded by other language, which is more 'real'.
- Students are exposed to the target language before having to focus on it.
- There is greater variety and interest.
- The teacher should ensure that the language focus part of the lesson is given adequate time.
- Texts/recordings can come from course books, authentic sources, or be made up.
- Made-up texts/recordings need to be 'near authentic' (e.g. recordings should not be scripted).

6 | TEACHING GRAMMAR VIA 'TEST TEACH TEST'

We have looked at two different ways of focusing on grammar in the classroom; via a situational presentation, and via a text or recording – the latter usually from a set course book. Another approach is what is often referred to as **Test Teach Test**. In very simple terms, this approach begins with the teacher giving the students some kind of task relating to a particular piece of language with the basic aim of seeing how much the students already know, or do not know, about this language. So let's do it! Have a go at this:

✓ Task

Complete these sentences using *the/a/an*:

 1 I really like ____ actor who plays Johnny Redway.
 2 I'm just calling to say how much I enjoyed ____ concert.
 3 I hope you'll give her _____ gift on her birthday.
 4 ____ unemployed receive money from the government.
 5 Last week I went to buy _____ book for my trip, but when I got to the end of ____ book I discovered some pages were missing!

Why did you choose these answers? Can you think of any rules?

Now complete these sentences with *the,* or add nothing:

 6 The bus can carry _____ 45 passengers.
 7 I wasn't born in _____ England. I was born in _____ United States.
 8 _____ politicians are not generally respected by society.
 9 I spent the evening in _____ bar drinking _____ beer.
10 On _____ last evening of his vacation, Zsigmond and his girlfriend went on _____ beach. They looked at _____ moon and talked about _____ love! He thinks he's so romantic!

Why did you choose these answers?

The object of this exercise, given to the students at the start of the lesson, is to discover how much they know or don't know about articles (*a/an/the*), and to allow the students to discover the same facts for themselves! It gives the students some immediate responsibility for their own learning, especially if they can work on this task together in pairs or groups sharing their knowledge. Assuming you are giving this task to, say, an intermediate class, then your students are likely to get most of the answers correct. What they might have more trouble with is the 'why?'.

This part of the lesson is the first **Test** in 'Test Teach Test'. But 'test' (or 'quiz' in American English) is a somewhat misleading and intimidating word as it gives the impression that students are under pressure and need to provide correct answers. In fact, it is a *discovery* task – but somehow 'Discovery Teach Test' doesn't have the same ring to it!

By having the students work together on the task, the **Test** element is reduced. If they are told to work in silence and not to compare answers afterwards, then they will feel pressure. Of course, it is perfectly valid for them to work individually at first, as long as they then get a chance to compare their answers before the teacher goes through it with them.

The teacher, having set up the task, should now be monitoring and evaluating as she observes how much / how little the students seem to know, and what the particular problems are that will need to be clarified later. Also, she may be able to deal with some individual problems while the class is working.

It is important, too, not to think that this first **Test** must be 'grammatical' in its presentation, such as the above one on articles. It can also be communicative and unstructured, supposing some prior exposure to the language you are to focus on.

For example, for a lesson on prepositions of place and position (*in the corner, on the shelf, under the table* etc) you might begin by asking students to describe to each other a room in their accommodation, or the classroom you are teaching in. Or you might provide a picture of a room and give them the same task. As they work, you listen to see how accurately or inaccurately they use prepositions in their descriptions.

Another example: for a lesson on the second conditional sentence (e.g. *If I won the lottery, I'd buy a house*), you might ask your students to tell each other what they would do if they won a huge amount of money. Better still, give them a greater variety of 'What if?' type questions to discuss.

For experienced teachers and high-level classes, you can also throw open the lesson and ask the students to tell each other everything they know, or think they know, about the language in question. But this really should not be tried by newer teachers, and certainly not by those on training courses!

Having done this first **Test** the teacher now has to do the **Teach** part of the lesson. To be honest, this is a tricky stage in this kind of lesson for a new or training teacher, for the simple reason that you do not know, until the lesson is underway, exactly what you are going to have to spend most time on. If your students have completed the first task very well (going back to our task on articles); or if they have used the target language reasonably accurately in communication (e.g. the prepositions), then your own task at this stage will be less demanding. However, if they have been very inaccurate and confused in their use of the language, then you are going to have to do a lot of clarification. But if they have done a conventional grammar exercise, your first job in the **Teach** is to check the correct answers. So let's do that by going back to the task on articles.

Key to Task 2

1 *The* actor, because there's only one and you know who is being referred to – it is definite.
2 *The* concert, for the same reason (compare this with *I'd love to go to a concert some time* – one concert, any concert).
3 *a gift*, because you don't know what gift, so it is *any* gift, but just one.
4 *The unemployed* (The jobless, The disadvantaged...) – referring to particular groups in society.
5 *a* book / *the* book – going from any book to the specific book that I bought.
6 No article (*the bus can carry 45 passengers*) – there is no article before a number except when referring to a particular

group (e.g. *the 45 passengers on the bus had a lucky escape when...*)

7 *In England, in the United States*. No articles before most countries, except when the country includes a plural, e.g. *The Netherlands*.

8 No article before *politicians* because you are talking about a general group in society (compare with question 4).

9 *a bar* (unless there's only one where you live, or you are referring to your 'usual' bar). No article before names of drinks when you are talking generally. Compare this with *I'd like a beer* (meaning a *glass* of beer), and *the beer in this bar is particularly good*.

10 *the last evening* (referring to a specific evening); *the beach* – only one; *the moon* – only one. No article before *love* – referring to a general and uncountable noun.

As you go through the answers and reasons it is absolutely essential that you do not assume that because one student has given you the correct answer and reason, everybody else in the class agrees or has understood. So you need to ask concept questions and elicit further examples as you go along. This is partly why this approach is harder for new teachers. Unlike previous approaches, and especially the situational presentation, where the teacher has more control over how the lesson flows, this approach may require you to deal with a lot more student questions!

But if you have planned and researched your lesson, as you should have done, this should not be too much of an ordeal! Remember, students ask questions not to trap you, but because they are inquisitive about the language. You may not always have a grammatical answer to give, but more often than not you can answer based on what you instinctively know is correct.

New teachers might feel safer giving students a more communicative first task (such as the two examples above) because this way they won't have to cope with explaining why certain answers are right or wrong. But, of course, they will still have to clarify meaning, form and pronunciation, as in any other kind of language lesson. In this **Teach** stage you are doing the essential tasks we mentioned earlier – you are clarifying meaning (concept questions, more examples etc), form, and pronunciation. You may wonder about pronunciation of articles, but the weak pronunciation of 'a' (/ ə /) is important. Having said that, there will be much less pronunciation work on articles than on a grammatical structure.

You might possibly think that articles are quite straightforward – they are not, especially when you consider their different usage in other languages. The task on page 47 did not even get into areas such as those nouns which sometimes have an article and sometimes don't (*I'll see you at school* vs. *I'll see you at the school*)! So, as with any piece of language – research it, anticipate problems, be prepared.

The final stage in this approach is the second **Test**. If the first Test is really *discovery*, then the second test is really *practice*. And the example task on page 47 could also be used as 'practice'. You may, then, ask 'What's the difference?' The difference is the *aim*. With Test 1, the aim is to find out how much the students know about the language before the teacher focuses on it. The aim of Test 2 is to have practice of the language *after* the teacher has focused on it. So the nature of the activity may be very similar, but the aim – the reason for doing it – is quite different. Test 2 could involve one, or more than one, practice activity, perhaps one that is an 'exercise' and one that is communicative. See Unit 10 for more on practice activities.

It should be noted that all the approaches to teaching grammar that we have looked at can be used with teaching vocabulary as well, and you need to bear this in mind when you read the next unit (Teaching vocabulary).

✳ *Tips*

1 If the first Test is an accuracy-based exercise (like the one on articles here), make sure it only deals with those aspects of the target language that you want to focus on in your lesson.
2 Monitor carefully this first Test in order to assess how much or how little the students know, and what specific problems you might have to deal with later.
3 Help with individual problems as they come up at this stage.

4 Make sure that the task you choose for the first Test is appropriate for the level.

5 Have varied tasks for the respective Test stages of the lesson.

6 Make sure you know what the correct answers are and the reasons for them.

If you are unsure, refer to the previous two units.

Advantages of this approach

As with the text/recording method described in the previous unit, **Test Teach Test** is more student-centred in its approach. Indeed, of the three approaches, this is the one which immediately has the students working together and not responding to the teacher. Even with the text/recording approach, the teacher still has to establish interest and maybe teach some vocabulary. This method, however, presents the task (Test 1) straight away and only requires instructions. For this reason it is particularly effective with higher levels, but it can be used with lower levels (for example, having students categorize nouns represented in pictures into 'countable' and 'uncountable'). Another positive point is that you are assuming some prior exposure to the language and are allowing the students the opportunity to show you, and remind themselves of, what they know. By getting students to work things out for themselves, you are encouraging learner autonomy and reducing over-reliance on the teacher. What's more, this can be a challenging approach – just how challenging will depend on the level and the class. But, unlike the text/recording approach, the focus is very much on the language. Consequently it is clearer to the students what the point of the lesson is, and what the teacher's aims are. As mentioned before, another advantage is that the teacher can discover just how well the students know the language before focusing on it – thinking time!

Just as course books frequently introduce a language focus via a text or recording, so too they frequently introduce language via an activity ('Test'). This is a good way of having students think for themselves. It also helps the busy teacher as it is far easier to use an activity provided in the course book than to have to create one for yourself.

Possible dangers

Picking up on the last advantage, if you are relying on an activity in the book then you will be asking for trouble if you don't do the activity yourself beforehand. You need to check that it covers what you want to cover in the lesson and no more, and that you know what the answers are and why. It can be embarrassing if you confirm the answer given in the Teacher's Book but cannot give a reason for it if asked – or worse, if it is not the answer you would have given!

As mentioned before, the unpredictability of the approach can un-nerve new and training teachers – you just can't be sure how well the students will do that first activity! But if you have researched and planned properly, you'll be okay; and if you have monitored the students during the task, you will have an idea of what problems you will have to sort out in the **Teach.**

With high-level classes there is the possibility that they will complete the first activity successfully and with no difficulties, and they might wonder why they are doing something which is 'easy'. In truth, this might suggest that the teacher has simply chosen something at too low a level. Or maybe the students don't know as much as they think they do, so don't be afraid to challenge them with lots of questions. However, even if they do know it all, there's no harm in reviewing the language and giving them more practice of it. In fact, reviewing familiar language is a very important thing for them to do, so tell them so!

Review

We have looked at three different approaches to the teaching of language – a situational presentation, via a text or recording, and now **Test Teach Test**. Let's again review the main points.

- The approach used should be varied and appropriate for the level, students and language being taught.
- The essential ingredients of language learning are **Meaning**, **Pronunciation**, and **Form** with plenty of varied practice, and whatever the approach, **MPF** should be covered.

- Text/recording or **Test Teach Test** approaches are more initially student-centred and challenging – therefore more appropriate for higher levels.
- Teachers should always research carefully the language they are teaching and anticipate possible student problems.
- Students should always be kept fully involved in the learning process and be encouraged to tell you what they already know.
- Students should be allowed and encouraged to ask questions and raise issues for discussion in relation to what is being taught.
- 'Grammar' should include communicative practice, but the more conventional exercise is good for focusing on accuracy; such exercises are often popular with students who perceive this as 'real' learning!
- Language learning should be varied, interesting, and challenging.

All three of these approaches can be adapted for the teaching of vocabulary, and the next unit looks at other approaches used specifically for the teaching of new words or review of previously taught ones.

7 TEACHING VOCABULARY

We saw in earlier units that **context** is essential if language is to have real meaning that students can relate to. Students learning vocabulary sometimes make the mistake of relying too heavily on dictionaries which give no more than a definition – and this often contains words unknown to the student.

Teachers, too, sometimes resort to 'lazy' ways of clarifying.

Task 1

What is wrong with the following approaches to dealing with meaning in the classroom? The students are pre-intermediate.

1 Teacher asks class "What does *scared* mean?"
2 Teacher asks class "Does anyone know what *scared* means?"
3 Teacher asks student "Tell the rest of the class what *scared* means".
4 Teacher asks class "Do you all know what *scared* means?"
5 Teacher tells class "*Scared* means *frightened*".
6 Teacher tells class "Check the meaning of *scared* in your dictionary".
7 Teacher tells class "*Scared* is when you are in a situation and you feel really worried and think there might be some danger".
8 Teacher tells class "For example, you get *scared* before an exam".

New teachers, especially, often use the above approaches because they are easy and safe and don't require much thought or planning. But in terms of clarifying meaning, they are practically worthless. Here's why.

Key to Task 1

1/2/3   Even if someone does know, they are unlikely to have enough language to clearly tell the rest of the class. And how do you know if the other students can understand?

4   They might think they do, but don't. Maybe they'll say "Yes" just to escape the teacher's attention! And if someone does say "Yes", will the teacher then assume that everybody else knows?

5   If they don't know what *scared* means, they are unlikely to know what *frightened* means.

6   Very lazy! And just how helpful will the definition be? One dictionary definition is '*to be filled with fear or alarm*'. Whereupon the students look up *fear* and *alarm*!

7   A wordy explanation – and misleading, as *worried* is not the same as *scared*.

8   *Scared* before an exam? Or *worried*? Or *nervous*? If you say that *scared* means *worried* then students will think a person is *scared* about a friend's health rather than *worried*.

We will look later at a variety of more effective ways of clarifying the meaning of vocabulary. But first, consider your own experiences of learning new words in a new language.

Problems when learning new words

Meaning

- A word may have more than one meaning (e.g. *odd*, *cut*, *patient*).
- First language (L1) interference such as 'false friends' – words with a similar appearance or sound to words in their L1 but with different meanings. For example, compare *sympathetic* with the French *sympathique*, meaning a 'nice' person. In German *bekommen* means 'to receive', sometimes resulting in a German speaker saying what sounds like 'I'll become a glass of beer' when they mean to say *I'll have a glass of beer*).
- Words may have different connotations, i.e. the meaning can be interpreted in different ways (e.g. *slim/thin/skinny*).
- A student may understand the meaning of a word, but not the

appropriate context in which to use it. This is particularly true of language which is especially formal or informal.

Form

- The spelling may be very different from the sound (e.g. *cough*).
- Students may be competent speakers of the language but poor writers.
- A word may be more than one part of speech (e.g. it may be a verb and a noun).
- If you teach a word like *to rely*, you also need to teach that it is followed by *on*.
- Some words are irregular (e.g. the plural of *person* is *people*, the plural of *sheep* is *sheep)*.
- Different varieties of English have different spellings e.g. *color/ colour*.
- Spelling patterns are not obvious, e.g. *happy/happier* ... *hit/hitting*).

Pronunciation

- The sound may not correspond to the spelling, as **Form** above (another example being *sign*).
- Students may have difficulty knowing how many syllables the word is broken up into (a typical mistake being *clothes* pronounced with two syllables).
- It is hard to know which syllable is stressed.
- Some words have shifting stress, e.g. his *conduct*/to *conduct*.
- A group of consonants together – a consonant cluster – can be difficult, e.g. *crisps*.
- A speaker of a particular language might have difficulty with certain sounds, e.g. the Japanese with / l / and the French with / h /.
- Some words with different spellings and meanings are pronounced the same, e.g. *bear/bare*.

Issues of pronunciation are looked at in Unit 9 but in this unit we will concentrate on how teachers can clarify meaning to a class. However, as with any language, and as with any lesson, teachers need to anticipate possible problems students might have and why they might have them. This way the teacher will be better equipped to help the students. With vocabulary, it is all too easy – as we saw at the start of this unit – to give a quick dictionary-type definition of a word rather than properly to illustrate meaning.

The pros and cons of students using dictionaries

We need first to distinguish between different types of dictionary. A Spanish student, for example, may want to use his English/Spanish dictionary. Or a standard English/English dictionary. Or an English/English dictionary produced for English students (and available at different levels). Of course, this student will need his Spanish/English dictionary out of class. However, in my view, teachers should greatly limit the use of the bilingual dictionary in the classroom, especially at higher levels, because literal translation is not always helpful or accurate. But I believe it is unreasonable for students to be prevented from referring to dictionaries in the classroom. Why buy a dictionary and bring it to class only for a teacher to say "No dictionaries"? Dictionaries, after all, have many advantages. They provide:

- meaning (but see disavantages below!)
- spelling
- a pronunciation guide (syllables, stress, sounds)
- information about the part of speech (e.g. *v* for *verb*)
- and sometimes, example sentences.

However, for students to benefit from all this, they have to know how to use a dictionary, what the symbols mean, how to locate an entry, etc. As teachers, we have to help them with this, maybe even spending a lesson or part of a lesson on dictionary training.

We also need to make students aware of the potential disadvantages of using dictionaries, including:

- All meanings will be given, not just the one your students need – so they can be confusing.
- In an ungraded dictionary, the definitions may be too complex (remember the example earlier with *scared*).
- Students may not know how to find the information they need (thus the need for dictionary training).

- If they are using a bilingual dictionary, the definition given may not be 'correct' (false friend).
- Students may become too reliant on dictionaries.

So the teacher needs to encourage use of English/English dictionaries in the classroom. She needs to provide some training on how to use them. She needs to allow use of the dictionaries, but not at any time. She needs to be realistic in allowing students occasional use of their own dictionary – sometimes it saves a bit of time! But she has to make sure that the student doesn't rely on the dictionary as a substitute teacher, or as a constant 'check' on what the teacher has said!

When do we teach vocabulary?

- You may be doing a vocabulary-based lesson (e.g. words associated with weather).
- You may need to teach words required for the reading or listening coming up later – but only the essential words that the students will need to complete the task(s). Unit 11 deals with the 'selection of vocabulary' for a skills lesson.
- You may be asked a question about what a particular word means.

We saw in Unit 4 that language teaching should cover **MPF** (**Meaning, Pronunciation and Form**), as well as providing **Use** (practice). This applies to vocabulary as much as it does to grammar. However, if you are teaching a word that is only going to be read or heard and not spoken (i.e. vocabulary for passive use – recognition only), pronunciation is less important. But if you are teaching it for active use (it will be spoken), then pronunciation will be very important.

How can we illustrate meaning effectively?

If explanations are to be kept to a minimum, especially at lower levels, then teachers need to employ other more effective methods which serve to *illustrate* and *clarify* meaning.

Task 2

Try to think of a way to illustrate the meanings of the following words, avoiding any wordy explanations. As with the grammar, we want to try and see if any students already know the word, so we try to elicit the word if we can. Therefore, do not write up or say the word and then ask "What does this mean?"! Start with meaning.

If no-one knows the word, tell them, but by this time the meaning should be clear.

> *elementary*
> 1 to jump (e.g. *He was so happy he jumped up and down*)
> 2 tired (e.g. *She was very tired because she had worked all day*)
> 3 credit card
> 4 rich (e.g. *Bill Gates is very rich*)
> 5 heavy (e.g. *The bag is very heavy*)
>
> *intermediate*
> 6 greedy (e.g. *Don't be greedy, you've eaten enough*)
> 7 patient (e.g. *The hospital has 75 patients at the moment*)
> 8 similar (e.g. *These houses are very similar*)
>
> *advanced*
> 9 to take after (e.g. *She takes after her mother, they look so similar*)
> 10 to be hard up (e.g. *Teachers are usually hard up*)

❇ *Tips*

> 1 Use pictures or drawings. For example, rather than explaining what *lorry/truck* is, show them a picture of one.
> 2 Use realia (the actual object). For example, rather than explaining what a *credit card* is, show them one.
> 3 Use mime. For example, rather than explaining what *tired* or *jump* is, mime it. Also, with *heavy*…try to pick up something in the class and exaggerate the fact that is is heavy to lift.
> 4 Use contrasts. For example, with *heavy* you could also pick up something which is not heavy to emphasize the point.
> 5 Use synonyms of a lower level. For example, for *exhausted* you could say 'a word meaning *very tired*'. If you are teaching them *exhausted*, then they should know *tired*. But if the synonym is not of a lower level, you are wasting your time. The same approach can be used with antonyms.
> 6 Use spoken gap-fill sentences. For example: *I ate four pieces of*

cake...I was very _____. If the students know *greedy*, they will complete the sentence. If they don't, you tell them – they now know the meaning. Much better than "What does *greedy* mean?"! You could treat *patient* in a similar way – *Sick people in a hospital are called* _____. This combines the gap-fill with a simplified definition. With *similar* you could combine the gap-fill approach with two drawings on the board of, say, two houses which are similar, but not the same.

7 With higher levels, try to use more student-centred approaches to increase learner independence. This respects the fact that your students already have a good level of English. Higher levels need to be challenged more, and they quickly get bored if there is too much coming directly from the teacher. One way is to provide contextualised sentences – a sentence containing the word and making the meaning clear. Rather like the gap-fill sentence, but with no gap! Students work together, and if they don't know the word, they can deduce it from the context. For example, 'I am so *hard up* I can't even afford a newspaper.' You could use a similar approach with *take after*, maybe combined with pictures. Obviously, you have to check afterwards that the students really have understood the meanings (refer again to Unit 8).

8 Matching exercises can combine some of the above approaches. For example, match words with pictures, words with synonyms, words with simplified definitions, words in sentences with simplified definitions.

9 Following a reading task, students can look at given words in the text and match them with, or deduce, their meaning.

10 Course books usually have an excellent and varied selection of vocabulary tasks including all of the above. In general, teachers should avoid too much of a teacher-centred approach and instead encourage students to work more independently.

This idea of *illustrating* meaning rather than explaining it, and of starting with meaning rather than the written or spoken word, makes the learning of meaning more effective. But the point made in Unit 4 remains valid here. If you successfully elicit a word or correct answer from one student, it does not mean that everyone else has understood. You need to *check* understanding (e.g. with concept questions).

Task 3

How could you check understanding of the words in the previous task? When you're finished, check the suggestions in the **Key** at the end of the unit.

How will the students remember what they have learned?

Students have different ways of remembering. Some prefer to write up their own personalized dictionary with translations into their language, for example. Some try to learn and remember a certain number of words a day/week. Others keep a record of what they have learned, but in a disorganized, random way. So what can the teacher do to help the longer-term learning process?

 ## *Tips*

1 Show students how they might keep an organized record of their work (not just vocabulary) – there is no 'right way', but there are plenty of 'wrong ways'! One possibility is according to themes or topics.
2 The record should include a guide to meaning, form and pronunciation. The word should appear in a contextualized sentence.
3 Students should be encouraged to be selective in choosing the words they really feel are useful to them longer-term. They should disregard the others. They also need to know the difference between 'active' and 'passive' vocabulary.
4 Students need to be encouraged to personally review what they have learned in whatever way they find helpful.
5 Teachers should do review activities in class from time to time – you can't just teach something in one lesson and never return to it!

Key to Task 2 and 3

1 *Jump*   You could do a mime contrast, e.g. hop and ask "Is this the same as *jump*?"
2 *Tired*   "I work all day with no break. Am I tired?"
              "I slept for twelve hours. Am I tired?"

3 *Credit card*	Show examples of credit cards/travel cards/I.D. cards and each time ask "Is this a credit card?" "Can I buy something with a credit card? Do I need real money?"
4 *Rich*	"I have a million pounds/dollars – Am I rich?" "I have ten dollars/pounds in the bank – Am I rich?"
5 *Heavy*	A series of contrasts with objects in the room.
6 *Greedy*	"I eat eight pieces of bread – Am I greedy?" "I share my food with you – Am I greedy?" "Is *greedy* positive or negative?"
7 *Patient*	"Is a patient usually sick?" "Is everyone in a hospital a patient?" "Do patients need to see a doctor?"
8 *Similar*	Use of pictures and realia would be effective (e.g. the clothes different students are wearing).
9 *Take after*	Again, pictures would be best. "My brother and I look completely different. Do I take after him?"
10 *Hard up*	"I'm rich. Am I hard up?" "I have very little money. Am I hard up?"

Notice with the last item that I did NOT say *If I'm rich, am I hard up?*. This is a conditional sentence and may be complex to understand. Keep your check questions simple!

Review

- You should avoid the "What does this mean?" approach.
- It is usually more effective to start with meaning and then elicit or give the word.
- Avoid giving non-simplified definitions or explanations of words.
- Train students how to use their dictionaries effectively. Limit, but do not put a ban on their use in the classroom.
- Be aware of some of the problems students have learning vocabulary as this will help you in your planning.

- Use a variety of ways to illustrate meaning depending upon the level of the class.
- With higher levels especially, but not exclusively, encourage learner independence – student-centred learning.
- Check that meaning really has been understood.
- If appropriate for the lesson, check pronunciation (Unit 9).
- Check that students are clear about the form of the word and sentence-fit.
- Make sure the words are being taught for a reason – that they will be used in the lesson!
- Help students to keep effective records of what they have learned.
- Review what has been learned from time to time.

8 | CHECKING UNDERSTANDING OF MEANING

We saw in earlier units the importance of *illustrating/conveying* meaning of language, rather than *explaining* it. By providing meaning in **context** teachers are providing their students with real language rather than abstract rules. But this is not enough. We have to *make sure* they have understood. If, as in the example in Unit 4, the teacher has used a situational presentation to illustrate the meaning of *should* for giving advice (*You should wear a suit*), the meaning may have been clearly *illustrated*. However, this does not guarantee that it has been *understood*. Teachers, therefore, must employ effective and efficient methods for checking the students' understanding of the meaning which has been illustrated. So how can this be done? And how should it **not** be done?

✗ **a** "Do you understand?", "Okay?", "All right?" etc

✗ **b** "So, Gabriel, what does *should* mean?", "Do you know what this means?"

It is easy to ask these obvious questions. But when, as is usually the case, the students give the answer we are hoping for ("Yes"), it does not mean that we can safely move on to the next part of the lesson. "Do you understand?" – "Yes" proves absolutely nothing. Quite possibly, the student just wants to get the attention of the teacher placed elsewhere and saying "Yes" is a quick way of doing this. Maybe the student thinks she does understand, but doesn't really – "Yes" doesn't tell the teacher this. Maybe the student does indeed understand, but still the teacher has no evidence – and besides, what of the other students?

Asking a student to repeat or give the meaning is flawed for different reasons. Principally, does the student – even if he has understood – possess a sufficient range and level of language to express himself? Would other students understand him? Would the 'explanation' be correct? This approach can only really succeed with very high-level students, and when the teacher knows her class very well. In such circumstances, it is more effective with vocabulary than with grammar.

Concept questions

It is necessary, then, to use other methods of checking understanding, of which the most effective is **concept questions**. Let us consider again the model sentence featured in Unit 4 – *You should wear a suit* (advice to casually dressed friend about to go to a job interview). In order for the teacher to first illustrate the meaning of *should* in this context, and then check it has been understood, she needs to have recognised the **essential meaning** of the language in this context:

- we are talking about a possible future action (wearing the suit)
- the person we are talking to does not have to wear a suit (no obligation)
- it would be a good idea if he did – I want him to (friendly advice)

We can now turn these statements into simple questions:

1 Are we talking about the past, the present, or the future? **future**
2 Does this person *have to* wear a suit? / *Must* he wear a suit? **no**
3 Do I think it would be good for him to wear a suit? **yes**

If students give any answers different from the above, the assumption must be that they have not understood and some further clarification will be required.

✗ *Should he wear a suit?*	Avoid using the grammatical form being tested in the question. If they haven't understood it before, they won't understand it now!
✗ *Will he wear a suit?*	Irrelevant, and impossible to answer, anyway! Avoid questions which focus on the **context** rather than the **concept**.

Let's look at another example – *She's been to Hungary* – and consider the **essential information** of this present perfect simple statement:

- it happened in the past
- the listener doesn't know when, according to the sentence alone
- the listener doesn't know how many times the person went
- she's not there now (compare with *She's gone to Hungary*)

Again we can turn these into **concept questions** as follows:

1 Are we talking about the past, present, or future?   **past**
2 Do we know exactly when she went?   **no**
3 Do we know how many times she went?   **no**
4 Is she there now?   **no**

✗ *Has she been to Hungary?/Have **you** been to Hungary?* etc.

The second question is asking for information, not checking meaning. Both questions use the tense that is being tested – compare with question 3 above, which uses a simple tense instead.

We can use the same method with vocabulary. The **essential meaning** of *library* is:

- we can borrow books from a library (we assume here that our students understand *borrow*)
- we cannot buy books from a library
- we can sit and read in a library

Increasingly, libraries also contain other facilities such as video and music loans, but for the sake of this example we will focus on books!

The **concept questions**, therefore, become as follows:

1 Can you borrow books from a library?   **yes**
2 Can you buy books from a library?   **no**
3 Can you sit and read in a library?   **yes**

For *engaged* (as in *Marc and Laura have got engaged*) the **essential meaning** is:

- they are not married now
- they plan to marry
- they have promised each other / it is a definite plan

1 Are they married now?	**no**
2 Will they marry in the future?	**yes**
3 Is this definite / sure?	**yes**
4 Have they made a promise to marry?	

✳ *Tips to formulate concept questions*

1 First analyse the language and its meaning within the given context.
2 Define the essential meaning in simple statements.
3 Turn these statements into questions.
4 Keep the questions simple in terms of both language and length.
5 Avoid questions which are not relevant to the *meaning* of the language.
6 Avoid using the same grammatical forms in the questions that you are testing.
7 Ask questions which do not require a lot of language in the answer.
8 Make sure the answers are clear and unambiguous.
9 Plan them in advance – until you have more experience and confidence, they will not be easy to think of on the spot.
10 Avoid the 'absurd' (e.g. *honeymoon* – "Can you find honey on the moon"?)

✓ Task 1

Now you try. For each of the following language items (given in bold), think of a context, define the meaning, and devise concept questions and expected answers.

1 **I wish I had** a car. (subject + *wish* + subject + past simple form)
2 **You shouldn't have taken** that book. (subject + *should(n't) have* + past participle)
3 **I'm looking forward to** my holiday. (subject + *to be* + *looking forward to…*)
4 **I had** my suit **cleaned**. (subject + *had* + object + past participle)
5 They're going on their **honeymoon**.
6 That watch is really **expensive**.

When you have finished, look in the **Key** on p. 71 and see if you're on the right lines.

It is also good, when appropriate, to use examples, realia, mime, pictures, etc to check understanding.

Examples   For *difficult* you could contrast 3+3 with 37 x 37 and for each sum ask if it is difficult.

Realia   For *mug* (mug of tea) you could show a plastic cup, and a cup and saucer and ask if they are *mugs*.

Mime   For *slowly* you could mime walking quickly and then slowly and check that students have understood which is which.

Pictures   For *lorry/truck* show pictures of other vehicles and ask for each "Is this a lorry/truck?"

As we saw in Unit 7, any of these methods can also be used for the initial *illustration* of meaning. In other words, you could convey the meaning of *difficult* by using the sums above, then check understanding via a similar example.

Timelines

Time can be represented using many different verb forms. This can be confusing for a student when, for example, we use the present simple tense to refer to future time (e.g. *The train goes at 3.30 this afternoon.*) Concept questions can be used to check understanding of tenses and time, but there is another way which is particularly effective and that is using **timelines**.

Looking at the sentence *I was watching TV at 10pm last night* we can identify the **form** as past continuous (progressive) – subject + *to be / was* or *were* + *-ing*. The **time** reference is an action which started **before** 10pm and may have continued **after** 10pm, but the action was already in progress **at** 10pm. This can be represented on a timeline thus:

past   now   future

10pm yesterday

The **x** here represents the moment of 10pm. The wavy line represents the ongoing action / state of watching TV. The wavy line begins before 10pm and goes right up to the moment of 10pm. The subsequent question mark beyond 10pm represents the fact that we do not know if the person continued to watch TV after 10pm, though they may have done. The wavy line is clearly set in past time only, consistent with this use of the past progressive.

Another example. *I had been waiting for the bus for 10 minutes before it came.* The first clause is in the past perfect progressive (subject + *had been* + *-ing*) and the second is in the past simple (*came* being the past simple form of *come*). This sentence is also set entirely in the past with one action going on for a period of time (waiting) before being stopped by an event (the arrival of the bus). In the timeline below, the wavy line again represents an ongoing action/state whilst the cross represents an event which effectively interrupts or ends the ongoing action.

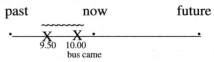

Timelines can be used to help illustrate meaning, and they can be used to check students have understood what has gone before. And here's an important point. If the teacher is using a timeline to check her students have understood the concept, then she must *involve* the students in the process of drawing the timeline. Students, then, need to know what a timeline is, and what the symbols the teacher uses actually mean (e.g. the wavy line). The teacher needs to be consistent in the way she draws timelines. The students can then effectively tell the teacher how to construct the timeline, or even draw it themselves. Timelines can be used in conjunction with concept questions. For example, a teacher may ask "Where do I put the cross? Here? Or here?"

�֎ *Tips for using timelines*

1 Plan how to do them.
2 Practise drawing them.
3 Make sure the line can be clearly seen on the board.
4 Use different colours for effect and clarity.

5 Make sure students know what everything on the line represents.
6 Involve the students.
7 Use them in conjunction with concept questions.
8 Be consistent in how you draw them.
9 Make sure they accurately reflect the essential meaning.
10 Give the students the opportunity to copy the line, and check to see if they have copied it correctly.

Task 2

Now you try. Draw timelines for the following, then check the **Key**.

Start by analysing the language and defining the essential meaning, then draw the timelines.

1 I've lived here for 10 years.
2 This time next week I'll be on the plane.
3 I'll have finished by 3pm tomorrow.
4 I've been to Ghana.
5 I saw *Titanic* yesterday, but I had seen it before.

Key to Task 1 and 2

Task 1: Concept questions

1 Do I have a car now?   **no**
   Do I want a car?   **yes**
   Is it likely I will get one soon?   **no**
2 Did you take the book?   **yes**
   Do I think you were right to take it?   **no**
3 Are we talking about the past, the present, or the future?   **future**
   Will I have a holiday?   **yes**
   Am I excited about my holiday?   **yes**
4 Was my suit dirty?   **yes**
   Is it clean now?   **yes**
   Did I clean it?   **no**
   Did another person clean it?   **yes**
5 Are we talking about the past, the present, or the future?   **future**
   Is a honeymoon a kind of holiday?   **yes**
   When you go on holiday with your friends, is that a honeymoon?   **no**
   Do you go on honeymoon straight after getting married?   **yes**

6 In my opinion does the watch cost a lot of money? **yes**
This watch (pointing) cost me £10 ($15). Was it expensive? **no**
That watch (pointing) cost my friend £300 ($450). Was it expensive? **yes**

Note: These are only sample questions. You may have written different ones which are still relevant.

Task 2: Timelines

Review

In this unit we have established the following main points:

- Teachers have to **check** that students have understood what they have learned.
- Questions such as "Do you understand" are ineffective.
- Asking a student to explain meaning is, with few exceptions, unrealistic and can do more harm than good.
- Concept questions are an effective and efficient way to genuinely check understanding. They should:
  - focus on the essential meaning of the language in the context being provided
  - be short and simple in nature
  - avoid ambiguity
  - require very short answers
  - avoid the use of the grammatical form being tested
- Timelines provide a visual representation of time.
- Students need to be involved in the drawing of a timeline.
- Timelines need to be clear, simple, and accurate.

The process of checking understanding may take up a tiny percentage of your lesson, but it is a crucial stage. If the students remain unsure of meaning now, then they will struggle for the rest of the lesson, and will be unable to use the language naturally or correctly after the lesson. Checking understanding effectively is a difficult teaching skill, but it becomes easier with practice. After a while, you will find that you can do this 'on the spot' because your own confidence with language and its meaning and use will have grown with experience. To begin with, though, analyse the language carefully and plan how you will check understanding.

9 | PRONUNCIATION

What is **phonology**? In general terms we can say that it includes:

sounds the individual sounds of the language

stress both *word* stress and *sentence* stress

intonation the rising and falling of our voice to convey different meaning

connected speech including how sounds and words relate to each other in natural speech; how sounds alter or disappear, how words link etc.

Phonology is the subject of many books. This unit is more of a guide to how pronunciation can be incorporated into lessons rather than a unit about phonology.

Whether you are a Californian or a Texan, a Londoner or a Glaswegian, the Queen of England or a bus driver, there is no *right* and *wrong* way to speak. Just *different* ways. People have different accents and maybe dialects. There are numerous varieties of English spoken around the world – and they are all English. Of course there are differences. Between American and British English, for example, there are many differences, not least in pronunciation. Sometimes it really does seem as if they are different languages, and one is sometimes referred to as 'correct' English, and the other not. But I disagree with this. Indeed, I believe strongly that students should be exposed to as many different accents, dialects and varieties of English as possible and that we are doing students a disservice if we expose them only to what in Britain is sometimes referred to as 'the Queen's English', or 'BBC English'. 'BBC English' refers to the days when all announcers and newsreaders seemed to possess standard south of

England pronunciation. These days it is different, but not everybody is happy. In 1999, a popular news programme in England was presented for the first time by a Welshman. Amazingly enough, there were some complaints that a Welshman was allowed to read the BBC news! As for the 'Queen's English' – well, no comment is required!

As a teacher, I try to play recordings that represent different British accents, and different varieties of English. At first, students find these difficult to understand. And that's the point. A student studying in London who visits Glasgow, or a student studying in San Francisco who visits Georgia, is going to find it difficult at first to understand what is the same language. I also strongly emphasize to my students that there is no right and wrong way to speak English (in terms of varieties of pronunciation), and that it is important for them not to be restricted to 'standard' English, whatever that is. It is sad, in my opinion, when TV programmes in either Britain or the United States feature a character from the other side of the Atlantic who speaks in an extremely unrealistic way, unrelated to any variety of English. Anyone who saw the American comedy 'Friends', when it was based in London, will know what I mean. The British are equally at fault in their portrayal of American accents.

Spelling and pronunciation

There is no clear relationship between spelling and pronunciation in English. Look at the following words:

cough – tough – though – through – they all end *-ough* but the pronunciation is different each time.

How can a student seeing the word *laugh* for the first time have any chance of knowing that *ugh* is pronounced /f/?

heard and *beard* both end *-eard*, but the sound is quite different!

How about *both* and *brother*? Both contain the letters *oth*, but they are pronounced differently.

Wednesday – so is that 'wed – nes – day'? No, it's /wen/ – a silent *d*.

Convinced now? This is a very strange language.

The sounds of English

What students need, then, is some representation of the sounds of English so that when they look up a word in a dictionary and see its phonetic transcription, they can work out the standard pronunciation of that word. This is not to say that we are trying to make our students speak in a certain way, just that they need a guide. Such guides will have alternative versions depending on the variety of English, so please do not consider the following example to be representative of the only 'correct' way to say these sounds; many readers will have additional sounds in their own speech, and may not recognize some of these examples!

The underlined part of each word represents the given sound (phoneme).

Vowels

b<u>ea</u>d	/ i: /
b<u>i</u>d	/ ɪ /
g<u>oo</u>d	/ ʊ /
f<u>oo</u>d	/ u: /
b<u>e</u>d	/ e /
th<u>e</u> *	/ ə /
b<u>i</u>rd	/ 3: /
d<u>oo</u>r	/ ɔ: /
c<u>a</u>t	/ æ /
c<u>u</u>t	/ ʌ /
c<u>a</u>rd	/ ɑ: /
c<u>o</u>d	/ ɒ /

* known as the 'schwa' – the most common sound in English. In a word like *potato*, the first *o* is pronounced with a schwa (weak form); also the *e* in *terrific*.

Consonants

<u>p</u>i<u>p</u>	/ p /
<u>b</u>ul<u>b</u>	/ b /
<u>t</u>ac<u>t</u>	/ t /
<u>d</u>are<u>d</u>	/ d /
<u>ch</u>ur<u>ch</u>	/ tʃ /

judge	/ dʒ /
cake	/ k /
gargle	/ g /
fearful	/ f /
vivid	/ v /
thirtieth	/ θ /
this	/ ð /
cease	/ s /
zones	/ z /
sheepish	/ ʃ /
treasure	/ ʒ /
murmur	/ m /
nanny	/ n /
thing	/ ŋ /
hope	/ h /
lull	/ l /
red	/ r /
went	/ w /
yacht	/ j /

Diphthongs

rear	/ ɪə /
may	/ eɪ /
pure	/ ʊə /
boy	/ ɔɪ /
go	/ əʊ /
rare	/ eə /
buy	/ aɪ /
now	/ aʊ /

Why use such a transcription? Well, it helps give students a record of pronunciation, it helps them to use a dictionary more effectively, and it can aid the teacher doing correction work. Students need to get the chance to hear a sound within a word before being expected to try and repeat it – in other words, teachers should raise students' awareness. How to do this we will look at later.

Task 1

In Unit 19 you will see how important it is for teachers to anticipate problems students might have with the meaning, form, and pronunciation of any language being taught. With this in mind, look at the following, and consider what peculiarities of pronunciation exist in terms of sounds and their relationship with one another.

Example:

> *a pint of beer* the *a* is a schwa, 'of' is pronounced / əv / and *a pint of* sounds like one word.

1 asked
2 "Can you help me?" "Yes, I can"
3 Do you want some?
4 Do you have to go?
5 I used to drive
6 Fish and chips
7 Peas
8 Goodbye
9 Shall we go?
10 I was working at home

Before checking your answers, let's consider one way of raising students' awareness and improving their own pronunciation skills, not just regarding sounds, but all aspects of pronunciation: **drilling**.

Repetition drills involve students repeating what they have heard. Consider the sentence *She's worked here for two years.* The teacher is covering the present perfect simple in her lesson for unfinished state and has illustrated meaning using the context of a teacher who... *has worked here for two years.* The given sentence has been elicited by the teacher and it contains the target language. Students, however, need to keep the language in context – it's no good their repeating *has worked* in isolation, they need to say the whole sentence. A good approach to drilling is as follows:

Having elicited the sentence *She's worked here for two years* the teacher repeats the sentence naturally, two or three times so the students can hear it. Then, the teacher should 'break up' the sentence using her fingers as representations of each word.

"She – has – worked – here – for – two – years"

Then, she repeats "She – has" and joins her first and second fingers and say "She's", and repeats the remainder of the sentence.

Finally, she repeats the whole sentence again naturally, perhaps demonstrating rhythm. This first stage is *giving a model pronunciation.*

Then, giving a clear direction to the students (spoken and gesture), the teacher tells the class to repeat what she has said. At this stage she says "Repeat", and avoids vague instructions such as "would you like to say that?". Students repeat the sentence together first (*choral repetition)* to give them confidence. Then, the teacher gestures to students in turn to say the sentence individually, eliciting correction as required.

With this sentence, the teacher's model needs to include the /iːz/ in *she's*; the /t/ sound of *-ed*; and the weak form with schwa of *for.*

Drilling applies equally to vocabulary, and the aim is for students to hear and give the correct and natural pronunciation of a piece of language. There are teachers who don't like drilling, or don't believe in it. But in my experience, students who are not confident about pronunciation welcome the controlled opportunity to have practice saying a new piece of language. They need to have confidence in their pronunciation, and this is one way to give it to them. Of course, you shouldn't overdo it at any stage, but done properly and appropriately, drilling is valuable and effective.

Key to Task 1

1 **a** is pronounced /aː/ and **d** is pronounced /t/.
2 The first *can* is pronounced weakly /kən/, but the second is not – /kæn/.
3 *do you* is normally pronounced / djə / or /djʊ/.
4 'do you' as above. *v* in *have* is pronounced /f/; *to* has a schwa.
5 *d* is pronounced /t/ and the *to* is weak. The words *used* and *to* are linked as if they were one word.
6 *and* is reduced to /n/ and the whole phrase sounds like a single word.
7 /piːz/ – the /z/ sound is especially strange for students who see the spelling.

8 Say the word quickly and you'll notice *d* disappears and the word sounds like /gʊbaɪ/.

9 Similar to above, the *ll* disappears and *shall we* sounds like /ʃwɪ/.

10 *s* sounds like /z/, and the word is pronounced weakly /wəz/.

So why is all this important? For a start, teachers need to be aware of how English sounds in order to help students produce the natural pronunciation themselves. It's not really a case of being 'correct', but of sounding 'natural'. However, students can't sound natural unless they have clear models to follow.

Teaching monolingual classes, you can focus on specific problems encountered by that nationality. For example, students who have difficulties distinguishing between two different but similar sounds (minimal pairs) such as /i:/ and /ɪ/ can be given tasks where they have to hear and recognize the two sounds in various words; or/and think of examples themselves. You do not need to 'teach' all the sounds, but focus on those which are relevant to your students and which they need help with.

Word stress

Word stress is another area which can cause difficulties. In the word *advertisement* the stress falls on the third syllable in American English (ad-ver-TISE-ment), but on the second in British English (ad-VER-tise-ment).

 Task 2

Where does the main stress fall on the following words? Be careful, there are some tricks included!

1 telephone
2 cassette
3 surprising
4 conduct
5 note-book
6 address
7 worked
8 examination
9 photograph
10 photographer

Before going through this task, why is it important to focus on word stress in the classroom? It is probably less important than other aspects of pronunciation because meaning is rarely affected by mis-stressing a word. But if the pronunciation is so incorrect that it becomes almost unintelligible, then meaning would become affected.

Stress can be marked in different ways.

newspaper něwspaper 'newspaper Ooo NEWSpaper <u>news</u>paper

The third of these is how dictionaries normally mark stress, but most teachers, in my experience, prefer the second (with a small circle or square). The fourth example (stress bubbles) are a very visual way to represent stress, good with very low-level classes. Whatever mark you adopt, be consistent, and make sure your students know what it means. Make sure, too, that students know how to read a dictionary for word stress – this, accompanied by the phonetic transcription, gives the students a good guide to pronunciation.

Key to Task 2

1 Ooo
2 oO
3 oOo
4 Oo (noun) oO (verb)
5 Oo
6 oO (British) Oo (American)
7 There is no stress because there is only one syllable, though many students will mistakenly say work-ed.
8 oooOo
9 Ooo
10 oOoo

In the classroom there are many activities you can do with word stress.

If you have been teaching vocabulary, you could have words on one side of a sheet and stress patterns on the other, and students match them (or any variation thereof). Students can listen to words and decide for themselves what the stress pattern is.

You can give two columns (e.g. oO and Oo) and a series of words which students have to categorize. As mentioned before, drill new words which will be needed to be used actively.

Sentence stress

Sentence stress can certainly affect meaning. Consider this simple sentence:

I gave Gaston a book.

So, which word is stressed most?

 Task 3

How does the meaning of this sentence change when the words are stressed as follows?

1 **I** gave Gaston a book.
2 I **gave** Gaston a book.
3 I gave **Gaston** a book.
4 I gave Gaston **a** book.
5 I gave Gaston a **book.**

Consider, too, the implied difference between *I'm going now* and *I am going now.* And what about *Today's Wednesday – No, it's Thursday*?

*I **am** going now* is emphatic, almost a show of determination to leave, despite efforts to make you stay. This could be correct in the right context, but the more natural pronunciation will be the contracted form, and this is what you should drill. With the 'Thursday' example, the second speaker is correcting information given by the first speaker, the stress falling on the correction – a very useful aspect of sentence stress to focus on.

Key to Task 3

1 It was I who gave him the book, not someone else.
2 I didn't lend him the book, I gave it to him.
3 I gave the book to Gaston, not another person.
4 I gave Gaston only one book, not more.
5 It was a book I gave him, not something else.

Such a task raises students' awareness of the importance of sentence stress and how it affects meaning. What is stressed depends in part on the important information a student wishes to convey.

We can signify sentence stress by saying the sentence ourselves, 'beating' the sentence (e.g. knocking on a table), humming it (e.g. "da da da DA da"), clapping it etc. But context is essential if the task is to be meaningful.

Intonation

Sentence stress goes hand in hand with intonation. Very often, someone we are speaking to may appear rude because of their intonation patterns. It is quite easy to make "Thank you very much for your help" sound rude or sarcastic. But intonation is so hard to teach because students invariably end up exaggerating intonation patterns until they sound completely unnatural. A case in point is *question tags* (e.g. *Your name's Luc, isn't it?*). A rising intonation on the tag (*isn't it?*) implies a real question requiring an answer. But a falling intonation implies that nothing more than confirmation is required of a fact already known (*it's really cold today, isn't it?* – there is no disputing the temperature). But just try making students use tags naturally! They usually end up leaving what sounds like a huge space between the statement and the tag, and then really exaggerating the intonation of the tag.

What students *can* do is to listen and identify polite versus not so polite intonation patterns. The "Thank you very much for your help" example might be said once, naturally and politely, and again, sarcastically; students must identify the difference.

A variation of this is to do the same task with a single word (e.g. *Hello*) and students must interpret the speaker's feelings (pleased to see me, not pleased to see me, surprised to see me etc).

Students could also mark intonation patterns on sentences (which have context) with rising, falling, and level arrows.

Pronunciation is sometimes focused on as a lesson, or part of a lesson, but most of the time, the most valuable pronunciation work takes place spontaneously as problems arise. Drilling, too, is a common feature of many teachers' classes, especially with lower-level students. Pronunciation work is generally more effective with monolingual classes because you can deal with common problems.

You will find that some course books integrate pronunciation work into their units while others keep it quite separate. Some books feature pronunciation work a lot, others much less so, just as some teachers do more pronunciation work than others.

There are lots of good books available which contain various pronunciation tasks for different levels, often with a 'fun' approach (stress bingo, phonetic crosswords, picture 'spot the difference' containing minimal pairs, etc), and you can provide whatever pronunciation work you feel is appropriate and useful for your students. But do not neglect this part of language learning.

Review

- Meaning can be affected by unclear pronunciation, whether sounds, stress, intonation, or any other feature of it.
- As teachers, we should expose our students to the different varieties of pronunciation and not portray one as 'better' than another.
- We should not ignore pronunciation work – we can have specific parts of a lesson devoted to it or deal with issues as they arise.
- Drilling provides students with safe, controlled practice of new language but the activity needs to be executed efficiently.
- Course books provide ideas for pronunciation work, as do other specialized books, but the pronunciation areas to focus on should be determined by the needs of your particular students.
- With monolingual classes, focus on the common problems experienced by that nationality.
- With all classes, focus on any problems which are affecting meaning and intelligibility.

10 | **PRACTICE ACTIVITIES**

In Units 4–9, we looked at various ways of focusing on language. We established that a teacher needs to cover the meaning, form, and pronunciation of new language and to do so in a way which involves the students using appropriate methodology for the level. But the missing element in this learning process is getting students to *use* the language they have learned – to practise it. Without this, students are restricted to learning rules and patterns, albeit in context, without having the opportunity to operate the language they have learned.

We saw that with **Test Teach Test** a task which could be described as 'practice' takes place right at the start, but with the purpose of being diagnostic. Normally, of course, practice tasks take place after the teacher has focused on the language itself.

We will look in this unit at some examples of practice tasks for both vocabulary and grammar, as well as looking at the teacher's role before, during and after the task.

Controlled and less controlled practice

First, let's go back to Unit 4 when we had a context of an unemployed man, Jason, going for a job interview at a bank. Jason was very casual in his appearance, *too* casual for a bank employing a new person. This context led us towards the sentence *You should wear a suit* which contains the language *you should* + base verb with the function here of giving strong friendly advice. We saw in this unit how we could cover the meaning, form and pronunciation. Having done all this, the teacher now needs to provide one, maybe two, practice tasks so the students can gain confidence using this language.

Task 1

Here are two possible activities, but very different in nature. Look at how each is described and consider what the main difference is in the nature of the tasks.

Activity 1

The teacher elicits from the students ideas of what a good café is like. She asks for examples of local ones and what students think of them. She also elicits examples of what would be considered unacceptable in a café. Throughout, the teacher is eliciting and personalizing, and at the same time establishing interest in the topic.

At this point, the teacher says she has a friend called Erik who owns a café. She shows them a picture of Erik in his café. The picture shows a café that is dirty, with three cats inside, and an owner who is smoking and unclean in appearance. The teacher elicits opinions from the students, as well as other ideas of what this café might be like.

All of this will take no more than three or four minutes. The teacher reminds the class that Erik is a friend of hers whose business is doing badly and she wants to advise him on how to improve his position. She elicits an example from the class in which they have to use either *should* or *shouldn't* – for example, *You shouldn't smoke*. Students then work together to say/write more examples using *should/shouldn't* based on the examples of problems given by the students earlier. They might come up with something like this:

> you shouldn't have cats in the café
>
> you should wear clean clothes
>
> you should cook good food
>
> you should clean the café
>
> you shouldn't make your food and drink so expensive

The teacher then gets back from the class what they have come up with and writes sentences on the board as confirmation of correct usage.

Activity 2

The teacher says she has another friend, Javier, who is Spanish. Javier is studying English in Miami but he has a problem. There is a large Spanish-speaking community in Miami and Javier finds himself speaking Spanish most of the time he is not in class. Consequently he is making very slow progress with his English. He calls you and asks for advice. The students imagine Javier is their friend and have to suggest ways in which he can improve his English and give reasons why. The teacher elicits an example – *Try to meet some American friends because they will speak English with you*. The students work together and think of more examples and then the teacher asks the class for their suggestions.

So what's the difference in the nature of these practice activities? The first *requires* the use of the target language because the teacher has set it up that way and told the class to use *you should/you shouldn't*. The second does not *require* this language to be used, but the opportunity is there for the students to use it because strong friendly advice is being given. But this time the teacher did not say "Use *should*", she just said "Give suggestions" *and reasons*. By asking for reasons, the teacher is getting the students to use other language, and by not insisting on *should* and *shouldn't* the teacher is allowing students to use other appropriate language in a more natural way. So, this task is much less controlled in terms of the language being used by the students.

We have, then, a task where the language being used is controlled, and a task where the language being used is guided, but not controlled. By doing the controlled task first, the teacher is giving students confidence in using the language, before asking them to be a little more varied in what they say. Remember that this class is around elementary level, perhaps pre-intermediate.

What the activities have in common, apart from the function and the related language being used, is the teacher's role. In each case she set the context, established interest, gave instructions with an example, and checked answers afterwards. More of this later.

Forms of practice activities

Practice can be spoken or written, controlled or less controlled (to varying degrees). It can also be in the form of an 'exercise' (e.g. gap-fill), or something more communicative.

 Task 2

Prepositions

If we take another piece of language – prepositions of place – we can look at various types of practice. Prepositions of place include *in* the corner, *on* the table, and *under* the bed.

Try to think of five different ways in which students at pre-intermediate level could practise this language.

Prepositions are difficult for students because there are no nice, easy rules to follow, or even much logic to them. You can have prepositions of time (*at* 3 o'clock), prepositions of movement (go *down* the street), and prepositions of place (*under* the bed)...and that's just for starters! Students of all levels generally hate prepositions.

Here are five possible activities:

1 Students see a picture of a room and have to complete five sentences with the appropriate preposition from a given list.
  Example: *The newspaper is ___ the table* – students look at picture and select *on* to complete the sentence.

2 Students are given two very similar pictures of a room and in pairs they have to find differences. They cannot show each other their pictures, but they can ask questions, and describe their own pictures.
  Example: *How many people are sitting on the sofa?*
  I have a cat sleeping under the bed

3 Students have a picture of a room with completed sentences. But each sentence has the wrong preposition which students must correct. A variation of this is: students listen to a recording of someone describing their room and look at the accompanying picture to discover which sentences on the recording fail to match the details in the picture.

4 Students tell each other about a room in their house/apartment and what is there. Afterwards, the teacher asks one student to tell the class about her partner's room.

5 The teacher provides various objects and students tell each other where in the room to place each object.

Activities **1** and **3** are what one might describe as 'traditional exercises'. Students usually like these more than teachers, and they definitely have value, as long as students are doing other types of activity as well.

Activities **2** and **4** are more communicative. **4**, in particular, is much less controlled in terms of the language being used.

3rd conditional

Let us now look at the 3rd conditional (e.g. *If I had forgotten your birthday, you would have been angry with me*) and at typical types of practice activity.

1 Complete the sentence with the words (in brackets) in the correct form:
If I (win) the competition, I (be happy).
If I had won the competition, I would have been happy.

2 Find the mistake in this sentence:
If she hadn't of seen me, she would never have known.
✗ 'of' *If she hadn't seen me…*

3 Rewrite this sentence in the 3rd conditional:
I didn't study hard and I failed my exam.
If I had studied (more/harder), I would have passed/wouldn't have failed…

4 Chain sentences … one student gives the first half of the sentence for another student to complete.

5 The teacher provides the ending of a story (e.g. a man gets dismissed from his job). Students have to write a sentence which explains this ending and which climaxes with a 3rd-conditional.
Example: *If he hadn't slept late, he wouldn't have lost his job.*

6 Students have to think of a conversation between two people. They decide on the context and the relationship between the characters. Somewhere in the conversation a 3rd conditional sentence must appear in a natural and appropriate way.

7 'Noughts and Crosses' ('Tick-Tack-Toe') – half the class are the 'noughts' and the other half are the 'crosses'. Each team selects two verbs from a given list and thinks of a 3rd conditional sentence that uses these verbs. If they produce a correct sentence, they choose in which box to place their 0 or X. A team wins when they form three noughts or crosses in a row, in the usual way. This is an example of using a classic game for language purposes. Most such games can be adapted in some way.

8 Matching sentence halves. In column **a** you have the if-clauses, and in column **b** the main clauses. Students have to match correct halves. A more interesting variation of this is to have the sentence halves on separate pieces of paper, face down on the table. One student picks up an if-clause, the other student picks up a main clause. If they match, put them to one side; if they don't, put them back face down in the same position. This activity is sometimes called 'Pairs', or as 'Pelmanism' and can be used to practise many different language items.

9 Jumbled sentences.
Example: wouldn't – I – eaten – have – if – sick – less – had – I – felt
If I had eaten less, I wouldn't have felt sick.

10 Students talk about their lives so far and include information about the consequences of what they did or didn't do.
Example: *If I had worked harder, I would have gone to university.*

✓ Task 3

Think of one practice activity of any type for the following which is different from the ones given below.

1 Giving directions (elementary)
2 Comparatives (intermediate)
3 Question forms (intermediate)

For each of the above, there are numerous possible activities, and of various types. Normally, you would want your students to do a couple of quite different activities, as explained above (*should*). Of the many possibilities for these three language items, here is one example for each.

Directions

Students work in pairs and each has a diagram of a street, but each diagram has different places shown (e.g. post office). Each student has to ask directions to find the places not listed on her diagram. They listen to the directions and write in the location of the place. Afterwards, students look at each other's pictures to check.

Comparatives

Students compare their country with the country they are now in, if applicable. Alternatively, use pictures to make comparisons of people and places.

Question Forms

Each student in a pair is given an incomplete text about a person. The missing information is different in each text. One student has to ask relevant questions of the other in order to complete their text.

Example: *My name is Marco and I am _____ years old.*
'How old is Marco?'

Many of the activites described so far can be used with vocabulary as well. Let's look at some possibilities.

Imagine you have 'taught' your students words related to the following topics:

weather (elementary)
sport (intermediate)
music (upper intermediate)

Certain types of activity could be used with all of these. Examples of these are **matching** (e.g. word to definition/picture) and **gap-fill** (putting the correct word into the sentence gap). But more communicative activities are also needed; some ideas for Speaking in Unit 13 are appropriate for language practice, too. A good example is **ranking**. This is when students put items in a particular order. With 'sport', this could be according to which they like watching most, or which are considered healthiest. With 'music', it could be according to which is their favourite type of music, or which instrument is the hardest to learn. With 'weather', it could be according to which is the most common type of weather in respective seasons, or favourite weather types and why. The point about ranking is that there are no right and wrong answers, it is designed to promote speaking which allows use of the vocabulary items.

For the weather vocabulary, students could have pictures of different weather types, and then write in the description. For the sport vocabulary, students could mime a sport for the others to guess. For the music vocabulary, students could discuss examples of musicians who are associated with different kinds of music, and whether they like them or not.

For review purposes, students could be given **anagrams** (words with letters jumbled up) of words previously learned and from which they can make sentences. Or, they could be asked to find words in a **word puzzle**.

Example: frtg*football*kdrwqy
               reefgvsawryhfhtedc   Your actual puzzle would
               gnixobjuytedfghytg   be bigger than this!

So where do you get ideas for practice activities from?

- Course books and accompanying workbooks are usually excellent sources. You should adapt and supplement in the way described in Unit 21.
- Other resource books which give various ideas of practice activities for different language points. Make sure the activity is suitable for the level and students you are teaching.
- Your own ideas and creativity! If you come up with a successful activity, keep it for future use.

The role of the teacher

I have already mentioned the teacher's role in the practice stages of a lesson; that role comes in four parts.

1 Before the lesson

The teacher needs to find an activity (activities) appropriate in terms of the class level, age group, composition and nature of the class, time available, and language focus (does it really provide practice of the specific language point you will have focused on earlier in the lesson?). Creating your own activity sounds a good thing to do, but don't spend two hours preparing an activity that will last five minutes!

2 Before the activity

Prepare students linguistically – are there any words you need to check they know before they begin the practice? Set the context and establish interest. Give/elicit an example of what they have to do. Arrange the class in the way you want them to work and with whom.

3 During the activity

Do not participate. Monitor to check they have understood, to give help, to keep things moving, to respond to difficulties or to quick finishers. If they are doing a speaking activity, listen and note down mistakes with the target language for correction later (see Unit 16).

4 After the activity

Go through correct answers with reasons, if applicable. Or get feedback on what has been discussed. Do correction of mistakes noted, if applicable.

If you fulfil your role for these four stages, you will surely achieve your aims for this part of the lesson.

Tips for practice activities

1 Be varied in your choice of activities from lesson to lesson.
2 Do not consider 'traditional exercises' to be boring for the student – they will be only if you do nothing else; otherwise, they tend to be very popular.
3 Do not spend too much of your time preparing original material, especially if you are only going to use it one time.
4 If you are using an activity from a book, check carefully for yourself how it is going to work, what instructions you will need to give etc.
5 If it is an activity with right and wrong answers, make sure you know what the answers are and the reasons for them.
6 Leave sufficient time for the activity to be done properly.
7 Leave sufficient time to go through answers / get feedback.
8 Do further practice another day as review – you must recycle your input and not treat pieces of language as something you 'do' in a single lesson.

Review

- Covering meaning, form and pronunciation is not enough if students do not get any opportunity to use the language and practise it.
- Practice activities need to be carefully selected, and properly set up with instructions and examples.
- Ideally, practice will involve more than one type and include something which is communicative.
- Teachers should familiarize themselves with the activity first, and with why the answers are what they are.
- They should monitor the practice carefully and leave time to get feedback.
- They should not participate in the activity.
- Correction of mistakes made with the target language is essential.
- The language being used in the practice may be controlled, or less controlled, allowing for other language to be used.
- Practice activities should be as relevant and interesting as possible.

11 | RECEPTIVE SKILLS 1: READING

The receptive skills of Reading and Listening (compare with the productive skills of Speaking and Writing) require an understanding of the text (*text* here referring to both reading text and recording) but also some related language work or other productive tasks before or after the text. Therefore, there will be a great deal of overlap when discussing the four skills for the very reason that they are very rarely focused on in isolation in the classroom. Even isolating them in the Unit titles is misleading! However, each of these four units will have a primary focus, and this one is Reading.

Task

Here are some questions for you to think about before you read any further. Make some notes.

Why do I ask you to do this? You may think it is easier just to skip the questions and read the rest of the Unit. Well, the analogy with the classroom is that as teachers we don't just 'give a lecture' but we try to involve our students in the learning process; to make them work! In any learning process, a student is more likely to remember something in the future if they have had an active part in the learning process. That is why this book includes some tasks for you to do, just as, when you teach, you will ask your students to do a task before you go through it with them!

So, look at these ten questions:

1 Think about your own schooldays – How much did you enjoy or not enjoy Reading lessons (in any subject)? Why? What was the role of the teacher in these lessons?
2 Is reading aloud a Reading lesson?

3 Is a Reading lesson going to be boring? What can prevent potential boredom?

4 What different types of reading skill can you think of? What should we focus on in the classroom?

5 In an ESL class, what do you think a Reading lesson might consist of?

6 Should students be given tasks before or after they read the text? Why?

7 Do you think it is necessary or desirable for your students to understand all the vocabulary in a text? If not, how do you decide what they need to know?

8 Is a Reading lesson always going to have a grammar focus (refer to Unit 5)?

9 How will you find texts to use in your lessons? What factors affect the selection of texts?

10 Could one text be used at two quite different levels? If you think 'Yes', how?

One other task! Please go and find a reading text – maybe something from a newspaper or magazine, or a brochure, or a TV guide … anything at all as long as it is not too lengthy. And I will use the text below.

Schooldays

They say that schooldays are the happiest days of your life. Well, they weren't for me! I enjoyed very little of my time at school and was glad to leave. My college days, however, were much more fun and the subjects I was studying were the ones I wanted to study!

At school, I used to have a one-hour bus ride each day there and back. We started school at 9 o'clock and finished at 3.45. There were some lessons that I really hated – mostly the 'practical' classes like Woodwork because I am hopeless at that sort of thing. I was useless at Art, too. Other kids used to have fun trying to guess what it was I had drawn. No, I was good at sitting behind my desk and listening to the teacher and making notes – English, History, Geography etc. Days when I had Woodwork, or Art, or Physical Education – those were the days I hated, the days that most other kids loved!

My teachers were a mixed bunch as well. Those that were good were very good, and those that were bad were very bad. Then there were the student teachers! The kids always gave them a really rough ride and it made you wonder why anyone would want to go into a class of 40 loud and misbehaving kids and try and teach them!

Thinking back, I'm not sure I learned too much that was to be useful in my life. Some of it was interesting for sure, but I wish I had been taught more about the practical things in life – how to deal with interviews, filling in application forms, finance management – rather than trying to figure out what 'x' equalled in some fancy equation.

The strange thing is, I always wanted to be a teacher myself. I had the ambition of starting off in the classroom and ending up as a school principal – of being a popular, successful, respected teacher. Well, it never happened and I don't regret it. Teachers nowadays have too much paper work to do, too many discipline issues to deal with. That's not for me. These days my ambitions are different and more realistic.

Back to the task …

1 The role of the teacher

I can only remember one approach to Reading lessons, whether in French or in English. The teacher would ask us to read a set number of pages either in silence, or taking it in turns to read aloud. This would be followed by questions, done individually, and checking of answers. The teacher's role was minimal, and more often than not he would use the time to mark homework etc! I did not enjoy these lessons. In particular I remember one English Literature teacher whose sole approach throughout the school year was to set us pages to read in the lesson – there weren't even any tasks to focus on, and certainly no communication between students or between students and teacher. It was thoroughly demoralizing. Talking to many trainees over the years, I discovered that a large majority had had similar experiences. Those few that had had positive experiences invariably referred to the role of the teacher in making the lesson interesting, even fun.

2 Is Reading aloud a lesson?

It may be in part, but the actual reading aloud focuses on pronunciation and not on comprehension. Unless your students are advanced, they are going to struggle with such a task, and other students may have trouble understanding. There is the possibility of some students feeling pressured – either as they are reading or as they await 'their turn'. So please don't ask students to do this if your aim concerns reading comprehension. They may (if you are very lucky) read well, but are they 'listening' to what they are saying?

3 Is a Reading lesson boring?

A loaded question! Of course, there is real potential for boredom. Maybe the topic is dull, irrelevant, or too difficult. Maybe there seems no point to the lesson. Maybe there is no communication, no interest. So, yes, such a lesson could well be very boring. But it shouldn't be, not if the teacher stages the lesson appropriately. It is the teacher's responsibility to keep interest high.

4 Different types of Reading skill

You may be rushing out of your home one day and there's a newspaper on the table and you see a headline which attracts your interest ... but you must hurry to catch your train, you have just 30 seconds. So you quickly look at the story to see what it's about, to get the **gist.** You are not concerned at this stage with detail, and you are not looking for particular information. You just want to know what the story is about. This is **skim** reading, or reading for **gist**. It implies reading quickly. It is something we do a lot.

Maybe you are looking at a train timetable. You need to know what time your train departs and where from. Or maybe you are searching for ticket prices and journey length. You have a timetable and information booklet, but it contains a huge amount of information which is of no use to you whatsoever. So, when you are reading you can – and should – disregard this information and look for what you need as quickly as possible. This is **scan** reading, or reading for *particular information.* This also is something we do a lot.

As you settle down to read a good book, or to study a contract you must sign, or maybe an article about a movie star you like, you will not be in a hurry. Far from it, you need, or want to, read carefully ... detail is important. You need to understand what you are reading. This is **intensive** reading, or reading for *detail*. And yes, we do this a lot, too!

We should, therefore, provide tasks which require a combination of all these different skills, not necessarily in a single lesson, but over a period. How this can be done, we shall see next.

5 An approach to a Reading lesson

There are many different approaches to a Reading lesson, or part of a lesson. Here we will look at just one – a fairly standard approach, commonly used by teachers and in course books. Have the text you selected earlier handy for this, and also refer to the one above ('Schooldays' p. 96). What we will concentrate on here are **stages** and **aims of stages**, and this will further prepare us for Unit 16.

Here, then, is a possible approach:

Stage 1

Check understanding of 'essential' vocabulary (see question **7** for more on this). Don't spend too long on this!

Aim: for students to understand the meaning of words essential to the completion of set tasks.

Stage 2

Establish interest in the topic. In any lesson, but especially a skills-based lesson, this is an absolutely crucial stage. This is where the skills of the teacher are contributing to stopping this from becoming a boring "read this" lesson. What you need to do here is to get the students interested in the topic, to make them *want to read* the text. This shouldn't take up too much of your lesson, but nor should it be a 10-second effort along the lines of "Did you like school?" "Yes/No" and on to the text! Looking at the text above on Schooldays, and a lesson for intermediate students, you might have your students discussing their likes and dislikes from their time at school; or describing their typical day; or remembering different subjects studied; or teachers they remember and why.

With other text-types this stage might involve *prediction* (e.g. from a newspaper headline, or from words in the text, or from a picture accompanying the text); *contextualization* (e.g. using pictures to set the scene of the topic); or *building interest* (e.g. describing an almost impossible scenario and then say "Well, it really happened and we're going to read about it"; or sound effects from a story – this can be combined with prediction).

More often than not, this stage involves either some personalized discussion based on the topic, or prediction, but it depends on the text and the level of the class you are teaching. Using a basic text on vacations for a low-level class, you might just ask them to list as many names of countries as they can think of. But whatever you do, you are working to create interest in the text that is to follow.

Look at the text you chose. How would you create interest for whatever level you think you might use it with?

Aim: to generate students' interest in the topic of the text.

The two stages so far have merely prepared your students for the Reading task. What follows will depend on a number of factors, but usually a combination of two (maybe three) different tasks based on the text will provide the main focus of the lesson.

Stage 3

If you set two different reading tasks then you should give the easier task first, to build confidence. This would be a task which requires either skim or scan reading. A task to encourage skim reading would be any that focuses on the gist, the overall theme, of the text. A task to encourage scan reading would be one that requires the students to find specific information in the text (names, numbers, places, prices etc) without having to read carefully or understand in detail. So, with 'Schooldays' a scan reading task might include questions such as:

Did he start school before or after 8.30am?
Did he love or hate Woodwork, Art, and Physical Education?
When he was younger, what job did he hope to have?

None of these questions requires anything more than 'pulling out information' from the text, just as you have to 'pull out' the time of your train from a timetable. Consequently, the teacher should impose a fairly strict time limit for the completion of the task to encourage 'quick reading' (which is what scan reading involves). If the teacher allows five minutes to find the answers to those questions, then the students will read slowly and carefully, and be concerned about understanding everything they read.

Scan reading tasks could include multiple choice / true or false / matching pictures with sentences. Revising the first question above, you could ask "He started school at (a) 8.00 (b) 8.30 (c) 9.00 (d) 9.30". Or "He started school at 9.00 – True or False?". Or, use four clocks showing different times – a visual multiple choice.

What scan reading tasks could you use for your selected text? Try to think of three or four different questions, and at least two different question types.

Aim: for students to have practice in scan reading skills.

Stage 4

Now you can move on to providing a task(s) that require more detailed reading. Any of the question types mentioned already could also be used here, but this time requiring *understanding* – and this could be linked to the vocabulary covered in Stage 1. To clarify this, look at these questions for 'Schooldays'.

Was he good at Art? (No, he was *useless* at it).
Were all his teachers of the same standard? (No, they were a *mixed bunch*)
Did student teachers have an easy time in class? (No, they were given a *rough ride*)
Does he still want to be a teacher? (No, interpretation of final paragraph)

These are just some possible questions. The first three demand understanding of essential vocabulary: *useless*, *mixed bunch*, *to be given a rough ride*. To answer the questions the students must understand the meanings – so these would be words to check in Stage 1. The final question requires more of a general understanding of the final paragraph. 'Why?' questions are also good for this kind of reading. And another possible *interpretation* question could be "Do you think he liked *Metalwork* at school?" No, because he didn't like practical subjects – the answer is implied but not stated.

This kind of task, then, needs careful reading and greater understanding – but not of everything (see question 7).

It is important, though, not to just give a series of questions every time – you should vary the types of questions that you ask. The nature of the text may suggest a certain type of task. For example, matching headlines from newspapers with stories would be intensive reading. One possibility with much lower levels is matching meaning with pictures.

This stage obviously needs more time, and students should be allowed to work together and help each other – the task should be communicative, not a test.

In terms of true/false or multiple-choice type questions, however, it is very important that the teacher does not just give/elicit the answers, but asks for reasons why. Otherwise, an answer could be a guess with no understanding.

What intensive reading task could you set for your chosen text?

Aim: for students to have practice in intensive reading.

Stage 5

This is very much an optional stage, but it is nice to finish such a lesson with something communicative. So, any of the suggestions made for Stage 2 could be expanded and used here as a class speaking activity. Anything relating to education, schooling, teaching, or even ambitions would be an appropriate topic for a Speaking activity (see Unit 13 for more on Speaking). If you were continuing with the same class in the next lesson you might use this Reading text as a springboard for some language work (e.g. *I wish* + past perfect from the text), or for further skills work (e.g. some Listening on the same topic, or Writing). What you do will depend on time, but try to have *something* as a follow-up to your Reading activity.

What Speaking activity could you provide for your students based on the text you have selected?

Aim (for example): for students to describe and discuss their own school-days in a free-speaking activity

You now have a 'sandwich effect' – the main part of the lesson (Reading skills) is sandwiched between other activities (generating interest / vocabulary and Speaking). This makes the 'heavy' part of the lesson more digestable and far less boring than the "read this" approach which so many of us experienced at school.

Main lesson aim: for students to have practice in scan and intensive reading skills.

Secondary aim: for students to have practice describing and discussing school experiences in a free-speaking activity.

So, what would you do if you only had this class for one lesson and the Reading was taking longer than anticipated. Would you (a) skip the last part of the Reading and go on to the Speaking, or (b) continue and finish the Reading and drop the Speaking?

Faced with that choice, go for 'b'. The Reading is the main part of the lesson and teachers should try to achieve their main aims. If there is no time for the Speaking in this lesson, it doesn't matter. Failure to complete all you planned is not a sign of a failed lesson if you have still achieved your main aims. More of this in Unit 16.

Once again it needs to be stressed that this is only one of many ways to approach a Reading lesson – it is an effective and popular approach, but do not feel this is the only way!

6 Give task before or after text?

Give the tasks **before** handing out the texts. This gives the students a reason for reading – an aim. If you let them read it, before handing out the task, they will just have to read it all over again. And the first time, they will read carefully and try (and want) to understand every single word! This would not be skim or scan reading, and it would be pointless anyway with no task.

7 Which vocabulary?

We have already established the need to deal only with those words which will be needed for the completion of the tasks. In the 'Schooldays' text, three words were selected, but words such as *misbehaving*, *fancy* (equation), and *ambition* did not have to be understood in order to complete the tasks. If you check too many words, you won't have enough time for the Reading part of your lesson, and your main aims will be affected. Having said that, your

Stage 5 could be Vocabulary based, maybe guessing meaning from context, possibly using multiple choice. It is good to focus on Vocabulary **after** as well as **before** a Reading / Listening task.

8 Reading and grammar

Reading texts may or may not lead in to 'grammar', as we saw in Unit 5. But there is nothing wrong in focusing primarily on skills work along with vocabulary.

9/10 Selection of texts

If you are using a course book, this will have a rich supply of texts and tasks, though you may need or want to adapt the tasks to make them appropriate for who you are teaching. Otherwise, use authentic texts (see Unit 21) and create appropriate tasks. One text could be used for different levels – the degree of difficulty will depend on what you ask your students to do with the text. Of course, this is not to say that any text can be used with any level! But there is an element of adaptability possible to allow one text to be used with, say, two different levels. Apart from the level, your choice should be influenced by suitability (choice of topic, consideration of cultural issues etc), and interest (is this something which will be useful? or interesting? or maybe both?).

✳ *Tips*

1 Select texts which will be interesting, relevant and suitable.
2 Provide tasks which will be appropriate for the level and the students.
3 Stage the lesson so that you establish interest and give a variety of tasks.
4 If possible, conclude with a follow-up activity based on the text.
5 Remember that the main aims relate to the Reading.
6 Give the tasks before the text.
7 Give an approximate time limit which is appropriate for the task / reading skill.
8 Check answers / reasons for answers.
9 Go on to some language work from the text (optional).

Task 2

1 How might you stage a Listening skills lesson?
2 What different Listening skills can you think of?

Why Listening is difficult

First, let us go back to the question of why Listening is difficult for most students.

Within the classroom environment, students become used to the voice, accent, or dialect of their regular teacher(s). Furthermore, the teacher grades her language and perhaps speaks just a little slower than is usual. The teacher constantly uses gestures, and will check understanding. A student can interrupt and ask for repetition or clarification. But with a recording, none of these factors are present.

A natural-sounding recording will usually consist of two or more speakers using ungraded language. Their voices, accents, and maybe dialects will be new to your students. These 'new' voices often come as a shock first time round. Often, too, the conversation is, or seems, fast. There is no body language and there are no gestures to help understanding. Students cannot ask the voices to slow down, or clarify... the cassette starts and continues. There is no control – at least, not on the students' part. Compare this with a Reading. Students have some control over how fast or slowly they read (though they may have been given a time limit). They have to contend with new language, as they do on a cassette, but they don't have the problem of accents and speed. They can quickly re-read a line that they have not understood, but they cannot rewind the cassette. With a Reading, then, students have some control. With a Listening they have none. Add to this the fact that they will be trying to answer questions as they listen and the obvious result is *pressure*.

Staging a Listening skills lesson

As teachers, of course, we must be aware of these difficulties and try to help our students in every way we can. Just as with a Reading, we shouldn't just go in and say "Read pages 20–25 then I'll give you some questions", so we shouldn't just go in, play a cassette, and hand out questions. We have to *stage* the lesson. The first part of Task 2 asked you to consider how this might be done. Well, the simple answer is that a Listening lesson *could* be (but doesn't have to be) staged in a similar way to the Reading stages described in Unit 11. If we take an example of a recording about two people discussing a movie they saw on TV, the lesson could go as follows:

Stages 1–5

1 Pre-teach vocabulary *essential for the completion of tasks* (as per Reading).

2 Establish interest in the topic (as per Reading) – maybe talking about movies recently seen, stars they particularly like, identifying actors in photos etc. Another possibility is *prediction.* You might, for example, show some pictures in the book from the Listening and ask students to predict what they think the recording will be about.

3 A Listening for **gist** task (e.g. "What are they talking about?" or "Did they like the movie they saw?");
or
Listening for **specific information** (e.g. "How many people are talking?" / "Where did they see the movie?" / "Which actors were in it?")
Either of these tasks would provide a gentler introduction to the recording, with relatively easy questions to build confidence and get students used to the voices and speed etc. Of course, the precise type of questions will depend on the recording and the level of your class.

4 Listening for **detail** (as Reading for detail!) – maybe asking how you know they liked or didn't like the movie (requiring identification and understanding of adjectives).

5 A follow-up activity based on the topic. This could include listing and talking about the types of movie that they like, favourite actors, writing a review of a recently seen film etc. With a low-level class this could be adapted so that students give a mark out of ten for different types of film (e.g. *Western*) identified in photos;
or

using the cassette to focus on a specific language point in the recording (as for a Reading text).

By staging the lesson in this way we are 'sandwiching' the Listening as we did with the Reading. In other words, we prepare the students for what is to come, and we try and create interest so that they end up wanting to listen.

Problems of 'real-life' listening

In order to genuinely improve our students' listening ability we need to focus on specific skills. However, if you are teaching in a country where English is not the first language, there is limited value in Listening lessons, given that students will only be listening to English in the classroom. That does not mean that you shouldn't do Listening, just that it is of much less value than it would be as an activity while in an English-speaking country.

So let's consider some examples of 'real-life' listening in a country where English is the first language:

- listening to announcements at a bus station / airport etc
- face-to-face conversation
- speaking on the telephone
- watching TV

For someone standing at a station wanting to catch a train to Edinburgh, there are different ways to seek the required information (time of train, platform number etc.) One is to read the monitors that display the information. Another is to **listen for announcements**. Now, you don't have to be a visitor from another country to find it very difficult to understand announcements at public places! The sound is often distorted, the accent or dialect may be difficult to understand, the delivery may be fast, there is a lot of background noise, and you can't stop and ask for the person to repeat any information. On the other hand, you do usually have the visual back-up, announcements are often repeated, and you can always ask someone for assistance.

In order to help our students improve their listening skills, we need to be aware of, and to acknowledge, the specific difficulties they might have with a certain type of listening; but we also need to direct students to the type of help they can get.

With this kind of listening the most important skill is 'listening for and extracting specific information' (times, numbers etc) – the equivalent of scan reading. Therefore, if we can provide an authentic-sounding recording of a station announcement, then we can provide a task which replicates real life, even if we can't replicate the exact conditions.

Face-to-face conversation has many difficulties in common with the announcements – fast delivery, different accents or dialects, background noise, having to participate in the conversation rather than 'just' listening. However, the other person's gestures and body language will help; you can ask for repetition, clarification and for the person to speak more slowly; and the context of the conversation will be established (whereas on a recording you may not have any idea what the conversation will be about until the cassette is played!).

The skills are likely to be different from those needed for announcements, though. Normally, in face-to-face conversation, you are focusing on gist, or detail. Maybe the person is giving directions, or telling you what you could do over the weekend. Or maybe they are telling you their impressions of a movie they saw. You *can* have 'face-to-face' Listening in the classroom – roleplays. And you can provide appropriate tasks.

Speaking on the telephone is really difficult and stressful for most students. For a start there are no gestures to help. There is the possibility of distorted sound as well as different accents and dialects. There is specific telephone language ("Please hold the line" etc) to deal with. The problem of 'turn taking' can be difficult (when is it your 'turn' to speak?). Dealing with 'menus' can frustrate anyone ("if you want X press 4, if you want Y press 5" etc!). Also, a student may have difficulty making himself understood. If the student is making the call, he might be able to prepare questions or check what information he needs to find, in advance. In any case, he can ask for repetition – unless he's listening to a machine. If he is, there is nothing he can do but repeat the process and listen all over again (something I have had to do more than once!). As teachers, we can help by teaching students telephone language and procedure, and by giving them a variety of tasks – including listening for gist, specific information, and detailed understanding; taking a message; making notes; and leaving a message on an answering machine. Best of all, maybe, is to give them a real task to do which involves them calling someone, maybe to get information.

Watching TV is much less pressurized, but the language tends to be fast and often colloquial. It is obviously not graded in any way, and as such can be very difficult to understand. But the visuals are a huge help. If the programme has been recorded, then students can rewind and listen again. All kinds of listening skills are required, but detail is particularly important. Thus, getting students to watch and listen to a news broadcast is very valuable, and you can provide varied tasks to help improve the students' listening ability.

These examples show the importance of recognizing the factors that help and hinder students in different kinds of listening, and some of the various skills that they need to improve. Therefore, we need to select types of recordings and types of tasks which will be relevant and useful as well as interesting. These recordings will normally come from the course book you will be using, but you can add authentic ones as well. As we have already seen, texts and recordings can also be used as a springboard for language focus work.

As for the issue of different dialects (e.g. south of England vs. north of England) or different varieties (e.g. American vs. British), it is a good thing for students to be exposed to as many types as possible – though it is often very difficult for them!

By breaking the lesson into stages, choosing appropriate material and tasks, you will be decreasing the stress levels of the students. There are other things you can do as well.

✳ *Tips (to help the students)*

1 Try to place the machine as close as you can to your students (or have them move close to it).
2 Maybe play a few seconds first, to check they can hear the cassette.
3 Give out the task before they listen and give them time to read it.
4 Make the first task relatively straightforward to give them confidence.
5 Reassure the students – tell them not to worry if they find it difficult at first, that you'll play it again; that they don't need to understand everything, etc.

6 Let them discuss their answers after the first play. While they do that, you rewind the cassette. Communication is important – it's not a test.

7 Nearly always you should play the cassette a second time (occasionally three times if it is not too long).

8 Sometimes, it is appropriate to play the cassette and 'pause' at crucial moments, isolating the answers. This usually helps a lot.

9 Go through the reasons for the answers.

10 Have a copy of the tapescript to hand so you can clarify important dialogue in case students are unclear about something.

It is not only students that can find Listening lessons stressful, so can teachers! As with any kind of lesson using technology, things can go wrong – and they do!

✳ *Tips (to help the teacher)*

1 Check there is a cassette player in your room. If not, find out where you can get one.

2 Check you know how to use it!

3 Get the cassette you need well in advance of the lesson – if you leave it to the last moment and you can't find it, you will have a problem!

4 Check the correct cassette is in the box – sometimes teachers put the wrong one back!

5 Listen to the recording before the lesson – check it is there and that it is audible.

6 Cue your cassette up to the start – I am always amazed at trainee teachers who wait until the lesson has started before they do this! It adds to your stress, and it gives a poor impression to the students.

7 If the machine has a counter number button, press it to '000' – this way, when you want to rewind the cassette you know how far to rewind it!

8 Have a copy of the tapescript just in case you need it.

9 Know what you will do if something goes horribly wrong (e.g. the tape gets chewed up by the machine) – be prepared! Don't panic.

10 Never trust any technology!

If you have access to a video recorder and monitor then you have the possibility of extending your Listening into Listening and Watching. There are many specially made videos on the market which you can use. This does not mean a 'watching TV' lesson, it means staging the lesson as you would any other, having clear aims, appropriate tasks etc. But with video you have something different – *variety*. See Unit 24 for more on using video.

Integrating the skills

So far, we have looked at Reading and Listening as almost separate entities. But we have seen that there are considerable similarities in terms of types of task, types of skill, and typical lesson staging. But if you look at typical course books, you will find that more often than not, Reading and Listening feature in a single lesson presentation. A Reading activity about someone shopping for food might be followed by a Listening activity on the same subject (or vice versa). A Listening activity about a tourist visiting New York could be followed by a Reading passage about life in this city. And so on.

We can go further than that. In the next two units we will look at the productive skills of Speaking and Writing. A truly integrated skills lesson (or, more likely, back-to-back lessons) might feature all four skills. Let's consider an elementary class doing a fully integrated skills lesson on the theme of holidays/vacations. A possible lesson frame could be:

1 Check essential vocabulary.
2 Establish interest in the topic (e.g. postcards from around the world).
3 Reading about a holiday in Australia (with a couple of tasks, maybe).
4 Listening with two people discussing a trip to Australia (again, a couple of tasks).
5 Writing – students write a postcard from one of the characters in the Listening based on the information on the tape.
6 A Speaking activity – maybe students rank favourite holiday destinations from 1 to 10.

Obviously, each of these stages needs detail – and for stages **1** to **4** you can refer back to the relevant units. For stages **5** and **6** you should refer to the next two units. The principal point is that you shouldn't always treat these skills in isolation – in fact, you should rarely do so, for the simple reason that we don't in real life! We might watch something on TV and then talk about it. We might listen to a news story and then read about it in the next day's newspaper. We might write a letter or e-mail to a friend and then read their reply. So, *integrating* skills work is important.

In addition, as mentioned in Unit 5, skills work can lead on to (or follow on from) language focus work (other than the checking of essential vocabulary). To focus on a particular piece of language from a text or recording is useful and relevant.

What must be stressed, then, is that there are no prescriptive lesson types. Of course, trainers, and books such as this one, and also all course books, will suggest approaches based on experience and research. New teachers usually appreciate such guidance early on, to have lesson 'patterns' to follow. But as long as you have achievable and appropriate aims for your lesson, any lesson frame is possible. We have already seen in an earlier unit how a 'grammar' lesson might begin with free speaking. It may not fit the traditional image of a grammar lesson, but that does not matter. So, as your experience increases, have the confidence to vary your lesson patterns and approaches, to integrate more. Don't do it a particular way because 'that's how the book does it' if you do not think the 'book' way is appropriate for your class! Don't follow a suggested approach given in this book if you can think of a more suitable approach for your class. You know your students, I don't!

Review

- Listening is a difficult skill for students, more so than Reading.
- Teachers need to stage Listening lessons in such a way as to reduce the difficulty.
- We should be aware of the specific problems that students might have with different types of Listening – but also of the factors that help understanding.
- We should provide tasks which are relevant and interesting (though their relevance will be less in a non English-speaking country).
- Listening can, and should, be integrated with other skills and language work.
- Teachers should check the machine and cassette well in advance.
- It is important for students to work together on tasks in order to make the activity less pressured. Don't play the cassette and then immediately ask for answers to the task!
- Play the cassette more than once and pause it, if appropriate.
- Use of the video recorder, if available, also provides good practice of Listening skills.

13 | PRODUCTIVE SKILLS 1: SPEAKING

A school I worked at used to ask students on their first day to put in order what they most wanted to 'do' on their course, from a list including Grammar, Vocabulary, Reading, Listening, Speaking, Writing, and Pronunciation. I would estimate that 90% (at least) put Speaking as their number one choice. But what is Speaking? Is it talking to the teacher? Is it talking to each other about a grammar task? Is it reading aloud? Is it spoken practice of language learned? Or is it chatting? I wonder how many of the students completing that first day questionnaire knew what was meant by Speaking. And is a Speaking lesson / stage of a lesson one of the easiest for a teacher to set up? Or one of the hardest? In short, Speaking as an activity in the classroom is much more complex than one might at first think!

 Task

Think about the following questions and make some notes before reading on.

1 What's the difference between **fluency** and **accuracy** and which is more important?

2 What do you understand by Speaking in a lesson?

3 Do you think it is relatively easy for a teacher to set up a Speaking activity? Why? Why not?

4 What's the point of doing Speaking? It's not 'learning', so isn't it a waste of time?

5 What problems might occur in the execution of a student-speaking activity?

6 What do you think is the teacher's role before, during and after a Speaking activity?

7 If the students are speaking, should the teacher correct their mistakes sometimes? never? always?

8 What kinds of Speaking activities could you do in a lesson?

1 Fluency vs. accuracy

In simple terms, **fluency** is the ability to talk fairly freely, without too much stopping or hesitating. To 'keep going'. More than that, it also requires that the listener understands what is being said, so there must be intelligibility and meaning. With **accuracy** the emphasis is on 'correct English' – the right grammar, the right vocabulary.

Can you have one without the other? Yes, definitely. I have taught many students who were very fluent but not very accurate. They could talk a lot. I could understand them. But their speech was littered with errors. On the other hand, some students focus their minds so much on being accurate that they lose their fluency. Just a few weeks before writing this, I taught a student who was obsessed by accuracy. He was of an intermediate level and had a pretty good range of language, but he was simply unable – or maybe unwilling – to just talk. He would begin a sentence and ask me or himself grammatical questions about how to continue the sentence. Time and time again I said that I could not answer him because I did not know what he was trying to say. For example, he might say "Yesterday I…what's the word?" Well, how could I know? He was so determined to construct his sentences bit by bit that the listener could rarely understand what he was trying to say. He also took an eternity to say anything, as you might imagine!

So, is one more important than the other? The easy answer is that they are equally important. But to be honest, I don't think they are. It also depends on who you are teaching. If you are teaching a high-level student who is about to take an exam to enter college, accuracy is very important indeed. However, in most ESL exams the accuracy is tested most in a 'grammar' part of the test with gap-fills, multiple choice etc. In the speaking part of the exam the emphasis is as much on fluency as it is on accuracy. But accuracy *is* important.

If you are teaching beginners, or other very low levels, you cannot possibly expect fluency. Here, you really do have to help them build their language accuracy bit by bit until they reach the stage when they can speak more fluently.

With a reasonably high-level student who does not need to prepare for an exam, and especially if they are living in or are going to visit an English-speaking country, then I would argue that fluency is much more important. Does it really matter if they make the odd mistake when asking someone in the street for directions? Does it really matter if they use the past simple instead of the present perfect when telling an English-speaking friend about their last vacation? No, it does not – though some students think it does, as I said earlier! What is far more important is that they can make themselves understood, and can talk relatively fluently. But even if their level is not high, the ability to be understood is greater than the need for accuracy in the situations just described. For this to be possible, we as teachers need to give our students confidence when speaking and not to be obsessed with constant correction (see Unit 15 for more on correction). We need to make it clear to our students exactly why we want them to just speak for 20 minutes in a lesson. The accuracy-fixated student I described earlier thought Speaking was "boring" – "I want to do grammar". Unfortunately, as accurate as his English *sometimes* was, he had no fluency and he drove everyone crazy as we waited for him to finish a sentence! Fluency is very important!

2 What is Speaking in a lesson?

First, let me say what Speaking is **not**. It is not reading aloud (pronunciation). It is not reading the answer to a grammar question (accuracy). It is not reading the answer to a Reading/Listening question (comprehension). In each of these cases the aims are not Speaking-related.

In a lesson, Speaking is one of two things. It could be a Speaking activity designed to give practice of language just learned or reviewed. Here the emphasis is as much on accuracy as it is on fluency and correction at some stage will be important. This type of Speaking was covered in Unit 10. In this Unit we are looking at the other type of activity – Speaking designed to promote fluency. That said, if you are teaching a low-level class then fluency is not a realistic aim, so we will look later at how you can still give low-level classes Speaking activities.

3 / 4 / 5 Practical difficulties

It sounds straightforward – just give them something to talk about. Unfortunately, at least with some classes, Speaking stages can be the most difficult to execute successfully. There are many reasons for this:

- teachers might get complacent and not plan this stage properly
- students may not see it as 'real learning' (as with the student I described before)
- the class may be very small
- the class may be very big (making 'control' a problem)
- the class may (and probably will be) all from the same country – so they will end up speaking in their own language
- they might perceive it as an 'exercise' that has to be finished as quickly as possible
- they might not have sufficient language to do what you have asked them to do

Let's look at each of these problems in turn:

Need for planning

We all do it. It's been a long day, we're tired, we don't have anything planned for the final 30 minutes, so we say "Okay, I want you to talk about food for the rest of the lesson". If you'll pardon the pun, that's a recipe for disaster. Speaking must be planned properly and with clear aims. See question 6 for much more on this.

'Speaking is not real learning'

No, they might not see it as 'real learning'. And, in fact, it's **not** *learning*; this is an important point. Learning a new language must include a high proportion of practice, of using the language. Students should not always be *learning* and teachers should not always be *teaching*. We saw in Unit 2 that the teacher has many roles to perform, and the teaching of language is only one of these. Another is to set up opportunities for students to use the language they are learning, and fluency activities is one example. New and training students often complain after such an activity "But I didn't teach them anything", to which I would reply "No, well done". You must not feel that you are not doing your job if you are not constantly teaching them. If you do, there will be severe overload.

As for the reluctant students, the school questionnaire I referred to at the start of this Unit (and supported by my own experience over the last seventeen years) confimed that Speaking for fluency is a very high priority for most students, so convincing them of its worth is rarely a problem – unless you do too much of it. But when you do have a student or class that needs convincing, tell them why you're asking them to do it. Tell them about accuracy vs. fluency. Tell them about the importance of using the language confidently. In other words, *involve* them in the teaching rationale.

Problems of large and small classes

If you teach in the U.K. in winter time you could well find yourself with a class of two, three or four students. This does not mean that you cannot do Speaking. It does mean that it will be harder work for you setting it up and keeping it going, and harder work for them. It may also mean that the activity may be over quicker. But this is all okay. You must not deny them the opportunity to speak because you don't think you have enough students. Similarly, if you have a very quiet class, don't give up.

In many contexts you might have classes of 20, 30 or 40 students, possibly mono-lingual. This is a much greater problem than the class of three. It is unrealistic to expect such a class not to use their own language for at least part of the time, and if it is a big class, you cannot properly monitor. You may even have a situation where the class becomes excessively noisy. So be realistic. If you have 30 Spanish-speaking students, they are certain to use as much Spanish as English. But remind them of the importance of using only English and encourage them to do so. Set up an activity with a clear purpose and a time limit. However, if you have your 30 Spanish students, and you are in Barcelona and when class is over they have no need to speak English, the value of Speaking activities is lessened. Motivation is lower compared with students who need to use their English outside school. You may, though, find that some of your 30 Spanish students (to continue with that example) need to speak English in their jobs, or they are going to do an exam. So do not abandon the idea of doing Speaking activities, because if you do, you will not be providing the complete learning package.

Students' attitudes

Students who do not understand, or who are not convinced by, doing Speaking for fluency in the classroom, sometimes view such activities as they do a 'grammar exercise' – as a task to be finished as soon as possible. This can be infuriating for the teacher. You set up an interesting speaking activity designed for 20 minutes of conversation, and after two minutes one group proudly says "We've finished". It's this kind of student who will treat a "Where did you last go on vacation?" question, with a one word answer "Brazil", and not follow it up as anyone would in 'real' conversation with questions about Brazil. This is in sharp contrast with the class who will happily chat for half an hour and very reluctantly stop when the teacher says time is up. This is one of the big problems with this kind of lesson, it is so unpredictable. You just don't know how a class is going to respond to your Speaking activity. If you want it to last twenty minutes, and it lasts two, you have a problem. For solutions – see question 6!

If the teacher sets up a Speaking activity which is beyond the students' language ability, then the teacher has either misjudged the level of the class, or she has failed to ensure that the students have the necessary language to complete the activity. It is a planning issue. This, too, relates to question 6.

6 The teacher's role

Speaking is no different from any other kind of classroom activity – it needs planning and it needs proper execution. The role of the teacher is as important with Speaking as it is with teaching grammar, but the role is very different. Remember what we said before – this type of activity is not 'teaching' or 'learning'. We are focusing here on Speaking for fluency, rather than practice of specific language. With this in mind, we can consider the role of the teacher in three stages: before, during and after.

Stage 1: Before ...

In the previous two units we looked at Listening and Reading skills, and described typical lesson 'patterns'. In each, it was suggested that such a lesson could finish up with what was described as a 'follow-up activity based on the topic'. For example, a lesson where the main focus was a Reading about Culture could end with a Speaking activity on the same topic. A Listening activity about Education could finish with students talking about the educational system in their countries (or, for a much lower level, ranking favourite subjects at school from 'favourite' to 'least favourite'), and so on. Typically, then, your Speaking activity will fit in to a lesson structure naturally rather than just be 'thrown in' for no apparent reason – it will be part of the lesson package. Sometimes, however, teachers will give a Speaking lesson (the entire lesson will be Speaking). Such lessons might have a less obvious aim, and be harder to sustain, but all teachers do them from time to time. Sometimes, for example, you have a class that is tired and is in the mood for just 'chatting', or it is your perception that they would benefit from a 'relaxed' lesson with no pressures of language learning or practice. This is not uncommon, but you shouldn't over-do such lessons.

Before the lesson, then, know why you want your students to speak, – what your aim is. In all likelihood, it will be part of a lesson containing other skills and/or language work and so the 'lesson fit' will be obvious not only to you but also to your students. Choose an activity (possibly from the course book) that is suitable for their level and interests, and that is appropriate in terms of the lesson fit and aims. Plan your instructions and class management. Anticipate any possible problems you might have, and that the students might have. If, for example, you would like pair work, what will you do if you get an odd number (usually two students can work as 'one')? How do you want the students sitting (if you have a choice)? Opposite each other? In a circle? Semi-circle? In other words, **plan** how you will set up the activity

That is before the lesson, but you also have a role before the activity itself. You should give very clear instructions, with an example of the activity if possible. The task should be clear (the students need to know exactly what you want them to talk about and why). You need to arrange the class how you want. If the activity involves role playing, the roles should be clear as should the purpose for speaking. And give an approximate time limit. If you emphasise that you want the students talking for ten minutes ("So don't tell me after two minutes that you've finished!"), then they are more likely to speak at greater length. With much lower-level classes, give a specific task – in other words, avoid "Talk about..." instructions which require a lot of language. Instead, give them an achievable goal (e.g. ten jobs to rank according to which they would most like to do, and with a very brief reason why). If you say "talk about jobs you would like to do" they won't have enough language to do so.

Stage 2: During ...

If you are teaching a class of under twenty, say, then you should remove yourself from the central position of the room. As long as you are there, the students will feel you are part of the activity and will try to keep you involved in it. By getting out of the way you are effectively saying "get on with it". With classes which are reasonably small (say 1–20), turning on background music (instrumental) can have a similar effect. With very small classes, the music can make it less intimidating for the students to speak. That's up to you, some teachers like using music, others don't!

Once the activity is under way, monitor it as best you can (but it won't be easy if you have sixty in the class!) to ensure everyone has understood, is speaking, is using mostly English, and doesn't have any problems. Also, you are there to ensure one or two students in a group don't dominate the activity, and that a group doesn't finish far too quickly. You are **not** monitoring in order to join in, and you are **not** monitoring to interrupt with error correction. If you do there will not be any fluency and you will not have achieved your aim. If you have planned and set up the activity properly, it should work well, but you always face the danger of accuracy-obsessed students who don't want to speak without constant correction, and of the very shy students who are not used to such activities in the classroom. But these are problems for you to anticipate and be ready to deal with if they occur. We have discussed already the accuracy-obsessed student, but for the very shy one you could gently interrupt the group briefly and gesture for that student to say something (e.g. "Sonoko, what do you think?"), and then walk away so as not to intimidate her. You can't, and shouldn't, do any more than this. It's all down to confidence. Very recently, I had three Korean female students who were out of their country for the

first time ever. I was their teacher in a communication class –
which was quite a challenge because they simply would not speak!
They were shy and lacking in confidence. When learning English
in Korea they just listened to the teacher giving explanations, and
then did exercises. As a result, their accuracy was very good, but
their fluency was non-existent. Three months later, after much
gentle encouragement from the several teachers they had, you
couldn't keep them quiet! Their confidence had grown, they had
observed the communicative approach, and slowly had come out of
their shells. So, never be discouraged by quiet students – help them,
guide them, but don't put pressure on them.

Stage 3: After ...

Try to allow enough time to get some feedback from the class. For
example, one group could tell another group what they decided/
discovered/discussed, and vice versa. Or you could select certain
students to give some feedback. It is nice to get some kind of
feedback so that you can show your interest, as well as for the
students to share some of what they have talked about. This needs
only to be a few minutes.

7 When to correct?

Don't interrupt the Speaking activity with correction because you
will stop the flow and the students will not be able to improve their
fluency. But some correction could (should?) take place *after* the
activity has finished. See Unit 16 for details on this.

8 Types of Speaking activities

Of course, it depends on the level and who is in your class. Avoid
culturally-sensitive topics, and avoid politics and religion at all
times! As you will see more, when you read Unit 21, you should not
assume an activity is 'okay' just because it is in your course book.
You have the responsibility to decide what is appropriate in every
sense. Having said that, course books do provide a multitude of
ideas for Speaking, nearly always as part of a lesson 'package'.
What you must do, as we saw earlier, is to set it up and make it work.

Beyond your course book, there are many Speaking resource books
on the market. Again, be confident to say "No way" if that's how

you feel. Be prepared to adapt ideas to make them more suitable. But do explore whatever resources are available in your place of work. Quite often, I come up with my own, very simple, ideas. That's the beauty of Speaking stages, you don't actually need materials every time, you can sometimes verbalize the task.

There are certain types of activity that tend to be very popular. **Ranking** has been mentioned before (putting items in a particular order). If *Sport* is your topic, students could rank sports from those they like most to those they like least; or from the most beneficial for the body to the least beneficial. If *Food* is your topic, you could again do favourite to least favourite; or most healthy to least healthy. If *Travel* is your topic, it could be most comfortable to least comfortable, and so on. The point of ranking is that there are no 'correct' answers, because it is designed to promote discussion. Higher levels can add reasons for their choices.

Sometimes, such activities may appear unrealistic or contrived, but the speaking skill involved is very real. Take the old favourite, the **Balloon Debate**. This is where students take on the identities of famous people travelling together on a hot air balloon which is overloaded. One person has to go over the side to lighten the load. The students, in their roles, argue who it should be and why. Hardly a realistic activity, but the skill of giving opinions and justifying them is very real indeed. Incidentally, of all the many times I have done this activity with students, it has never yet failed to produce a near-riot!

Debates can be organized, though only for high-level classes. Make sure the topic has obvious points on both sides and is not too controversial.

Students can prepare and carry out **surveys/questionnaires.** This involves finding out information and reporting back. If possible, let your students go beyond your own classroom, maybe to interview other students or staff or people on the street!

Describing visuals is another good area to explore, and can be a good way to allow students with imagination and sufficient language to extend their normal classroom speaking.

Role plays can work with any levels. With lower levels, give them a realistic context (especially if in an English-speaking country) like shopping, and a realistic role (i.e. the customer, not the shop assistant).

Student talks are good for more advanced levels. Maybe, each day, one student could give a talk for about five to ten minutes on a subject of interest to them, and the other students can then ask questions. Make sure the students have had notice of this so they can prepare some notes.

Discussion on topics of interest, or maybe in the news, is a relaxed way to promote fluency, but make sure the students have something specific to talk about, and that you don't end up doing more speaking than them.

These are just a few ideas. Remember, use your course book for initial ideas, then supplementary materials. Don't expect fluency from very low-level classes – give them something much more controlled.

✳ *Tips*

1 Plan the activity properly
2 Know what you'll do if not everyone turns up and your hoped-for even numbers don't materialise.
3 Think in advance about issues of classroom management, especially if you have a very large class, rows of tables etc.
4 You may have to 'justify' the activity to the class if you are teaching a mono-lingual class in a non English-speaking country.
5 Plan your instructions carefully so it is clear what the activity is, what students' 'roles' are etc.
6 Give an idea of time so as to increase the likelihood of the activity lasting as long as you want it to! (e.g. "You have ten minutes, so keep talking for that time!").
7 Get out of the way! If you remain the focal point in the class, they are likely to be more inhibited. With small classes, light background music can help, too. With very large classes, however, you might need to remain at the front in order to keep everything under control!
8 Make sure the activity is appropriate in terms of language required and topic choice.

9 With much lower-level classes, do not expect fluency – provide a clear task.
10 Monitor the activity as best you can, but do not interrupt. Encourage quieter individuals or groups, though.
11 Try to allow enough time afterwards for some feedback to the activity.
12 Maybe do some delayed correction work (see Unit 16).

Remember that whereas some students want to do as much speaking as possible, others can be resistant to the idea of 'just chatting' and may need convincing of the worth of doing such an activity. Don't be afraid to do this. If you are teaching adults, as this book pre-supposes, then they should be treated as adults, and that includes explaining the rationale of what you are asking them to do if they seem sceptical!

14 PRODUCTIVE SKILLS 2: WRITING

Maybe this unit should be subtitled 'The Forgotten Skill' since less time appears to be spent on this lesson type than any other, except in exam or literacy classes. In truth, few students seem to mind. In my experience, very few ever complain about not having done enough writing. However, it is not a skill that should be ignored. The problem for the teacher seems to be how to 'teach' it.

With the receptive skills of Reading and Listening it was apparent that there were more similarities than differences. Most things we said about Reading also applied to Listening, and vice versa. But the same is not true of the productive skills.

 ## Task 1

Answer these questions before reading on.

1 How does Writing differ from Speaking as a productive skill?
2 What makes Writing difficult for students? And what makes it easier than Speaking?
3 Why teach Writing?

Motivation for Writing

For a start, many students need the Writing skill much less than they need any of the others, and if they are studying General English in their own country then their interest in Writing lessons may be minimal. Even if they are studying in an English-speaking country there is often limited value for them in doing Writing. If they write a postcard home, they will not be using English, for example. On the other hand, they might have to complete an application form in English. If they are working in England then they might need to do some kind of writing in their work.

If students are preparing for an exam, then Writing is much more important because they will be tested in some way. Exams usually require students to write at least two different types of answer, e.g. discursive, descriptive, analytical, letters etc.

You might be teaching students who are now living permanently in an English-speaking country and who have no writing skills at all, not even an awareness of the script and how to form it.

So with Writing lessons more than most, the motivation is very varied. Some students will need, or will want, to do it, but for others, Writing will be seen as a waste of time.

But Writing should never be totally neglected, and if it is an important part of the course you are teaching, it should be focused on extensively. It can be difficult to get students to write in class (or at home for homework), but as with any other kind of activity, the teacher can 'sell' it. The lesson staging can be such that the students are more inclined to write with at least some enthusiasm.

Written language is often more formal than spoken language. Of course, this is not always the case. E-mail style tends to be very informal, even more so than conversation at times!

A student studying in an English-speaking country might have to do many different types of Writing including forms, phone messages, formal letters, and C.Vs or resumés. In addition, students preparing for an exam will need to write compositions, including formal and informal letters, descriptions, discursive essays, graph interpretation, narratives etc. These different types vary in their degree of formality, and each requires specific skills and language. It is this variety which makes Writing much more difficult than Speaking.

Differences between Writing and Speaking

With Writing, the emphasis on *accuracy* is much greater than with Speaking. For a start, the written form is visible and mistakes are seen. With Speaking, students often make 'slips of the tongue' – they have said something wrong, but if they could hear a recording of what they said, they could correct the 'mistake' themselves. As we saw in the last unit, 'mistakes' when speaking are often not that important. Written tasks on the other hand often require accuracy (e.g. completing an application form) and formal language. Because they recognize this, many students feel under pressure when writing. However, with Writing students can proof-read and self correct. They can go more slowly and carefully than when they are speaking. It is an important skill we must teach our students – read what you have written!

Punctuation is another factor absent from Speaking. Increasingly these days, one might question the importance of correct punctuation, but whereas one can accept that the correct use of colons or semi-colons is not really so important, surely the correct use of capitalization and question marks, for example, does matter.

Spelling may also cause problems, something which mother-tongue speakers have difficulty with, never mind learners from another country! Again, people differ in their views of the importance of correct spelling, but the fact remains that, as teachers, we have to recognize what is 'correct' writing, and what is 'incorrect'. If we cannot recognize a mistake, then we cannot correct it.

With Writing students do not have to concern themselves with aspects of pronunciation, or being fluent. Those students I mentioned in the last unit who are much more interested in accuracy than fluency, are often very good when writing. It is very common to find students who have had accuracy-based language learning, writing extremely well and accurately, but struggling to speak more than a sentence or two!

Writing tends to be more *economical* in its use of the language. There are no 'hesitators' ("mmm", "er", "well" etc) that litter our conversation. Written language is direct and efficient.

Another difference between Writing and Speaking is that when writing students cannot express themselves visually. There is no 'body language' or gestures to aid understanding. With Writing, students rely exclusively on the language itself.

Speaking is naturally acquired and students in an English-speaking country are immersed in the language all the time. But Writing is something which has to be taught. What's more, Writing is a lonely skill – we speak with other people but we write individually. All of this makes Writing harder than Speaking.

Writing skills

What students need is to have specific writing skills developed. It's not enough for the teacher to say "Okay, write about a country you like". This is writing for the sake of writing, but the teacher isn't 'doing anything'. For the students' writing to improve, they need to have help in particular areas. So let's consider three different types of Writing that a student might do in the classroom – a postcard, a letter, a discursive essay. We have already said that it is rare for a student to write a postcard in English if they are sending it to their own country, but once they return home they might send postcards back to friends met when they were learning English. We will take a letter of application for a job as an example of letter writing. As for the discursive essay, (one which argues for and against a particular topic) we will assume this is for an exam class who will need to write this kind of essay.

Writing a postcard (elementary level) requires the writer to use informal style writing, use simple linkers (e.g. *but* rather than *on the other hand*), know how to begin and end a postcard, and, arguably, to spell correctly. The writer also needs to know the type of subjects covered in a postcard (e.g. weather) and the layout including which side the name and address are written.

Task 2

What different skills are required to successfully complete a letter of application (intermediate)?

Before looking at this Task, let's consider some of the general skills required when writing. We have already mentioned punctuation, spelling, and style of language. To these we can add **correct structure**, **correct layout** (on a letter), **linking expressions** (*on the other hand*, *nevertheless firstly*, etc), **paragraphing**, **organizing**, **appropriate vocabulary**, **language of comparison and contrast** (e.g. *similarly, conversely, in conclusion*), **precis writing** (taking messages on the phone we need to summarize the main points), and **presentation** (e.g. presenting a neatly and correctly filled in form). Obviously, the skills required will vary according to the task being set and the level being taught. We saw with the postcard-writing at elementary level that the skills required are fairly basic. This is not the case with the other types of Writing mentioned.

Key to Task 2

For the **letter of application**, students will need to know how to correctly lay out the letter – where to put their address, their name, the recipient's address etc. Layout conventions vary and change from time to time, but there are basic layouts which can be taught. Students also need to know about punctuation and capitalization. They need to know how to organize the letter, and that each paragraph should have a specific purpose. They need to know how to be economical with their language so as to keep the letter to a reasonable length. They need to know about the style of language to use (formal), and about how to use linking words and deconstructed forms (not *I'm* but *I am*).

One way of helping students to recognize an appropriate style of language for letters is to present a selection of language from two different letters – one formal and one informal. So, phrases like "Dear Sir", "Dear Marc", "I'd like to have this job", "I would like to apply for the position..." are all mixed together, and students have to separate the formal from the informal. Maybe the different examples could be from one complete letter so students can then put the letter together and have a model of a complete letter.

For the **discursive essay**, students will need to know spelling, punctuation, language of contrast and comparison, language of summing up, vocabulary related to the topic, how to organize, how to develop and support ideas, and how to be concise.

If we take, for example, the language of contrast and comparison (including *while, whereas, despite* etc), the teacher could present these expressions in sentences that are incorrect in some way, and the students try to correct them.

To practise the skill of organization, the teacher could present a complete, but jumbled up, letter and ask the students to put the paragraphs in the correct order.

By knowing which skills different writing tasks require, the teacher will be better able to help her students with their writing.

How to teach Writing

As with Speaking, we can distinguish between a lesson that *includes* Writing, and one which is *exclusively* Writing. At the time of writing this, I am teaching an exam class that will have to do a 60-minute Writing exam (one task of 150 words over 20 minutes, and one of 250 words over 40 minutes). With this class I frequently have lessons devoted to Writing skills which focus on what they will have to do in their exam. I am also teaching a General English class where Writing figures far less prominently. When it does, it is always part of a lesson containing other activities.

We saw in previous units that skills work is not usually dealt with in isolation, and when this is the case it is the role of the teacher to 'package' the lesson in a way which is appropriate for the level, the students, and the aims of the lesson. As an example of this, we shall go back to the earlier examples of writing a postcard (elementary class), and writing a letter of application for a job (intermediate level). We shall assume a lesson of about 60 minutes and look at possible lesson packages to include these writing activities.

For the 'postcard lesson' we could follow this procedure – one of many variations possible.

1 Show real postcards from different countries and students try to guess the countries.
   Aim: for students to become interested in theme of lesson.
   5 mins
2 Stick six completed postcards around the room (writing side on display) and give students **Reading** task (e.g. they have a form to fill in showing the countries each writer is in, who they are writing to, and if they seem happy or unhappy in this place) Teacher then gets feedback.
   Aim: for students to have practice in scan and gist reading skills.
   10 mins
3 Students are given extracts of language from a letter and from a postcard (formal vs. informal) and they have to distinguish which is which). Teacher then checks.
   Aim: for students to recognize appropriate style of language.
   10 mins

4 Teacher elicits what types of topic are usually written about in postcards.
   Aim: for students to be able to identify appropriate postcard topics.   5 mins

5 Students are each given one of the completed postcards from Stage 2 as a model of postcard-writing. They also get a blank postcard. They have to write a real postcard to someone they know who speaks English (it could be you!) and to whom they will send this postcard. This adds authenticity to the task (compare this with them writing an imaginary postcard on a piece of paper!).
   Aim: for students to have practice in postcard-writing skills. 15 mins

6 Once the postcards have been written, and later sent, you could have a follow-up **Speaking** activity – maybe ranking the six countries featured in the model postcards in order of which place they would most like to visit.
   Aim: for students to have practice in Speaking (putting items in order and giving reasons).   10 mins

In a lesson package such as this, Writing is just part of the lesson, but it fits in with all the other parts. The teacher has established interest, provided a model of the writing task, prepared students with the style and type of language required, established what needs to be included, and also included Reading and Speaking. All of this is suitable for an elementary class, but it remains a full lesson of language and skills. The Writing task is real-life.

 ## Task 3

Now try and think of a lesson package for an intermediate class doing a letter of application.

Before going on, let's consider further examples of the types of Writing that can be done at these three broad levels.

At elementary level, in addition to postcards, students could do gap-fills, dialogues, informal letters, simple forms, and guided writing. By 'guided' writing I mean that students are, for example, given part of a written text already completed, and students complete the rest. 'Parallel texts' can also provide lower-level guidance – as with the example above students see a complete model of the type of task that they will have to do. At this level, though, the amount of writing required will be much less than at higher levels.

At intermediate level, in addition to formal letters, students could do simple report writing, simple discursive essays, guided narratives, and complex form filling.

At higher levels, in addition to discursive writing, students could do unguided narratives, note-taking, precis writing, statistical interpretation, report writing, and speculative writing.

Key to Task 3

Going back to the last Task, one possible lesson package for a lesson including writing a letter of application could be as follows:

1 Class discussion on types of jobs.
   Aim: for students to start thinking about basic theme of lesson.
   2 mins
2 **Reading 1** – students sequence the process of getting a job (sentences on seeing a job ad., completing an application etc)
   Aim: for students to have practice in reading for detail.   8 mins
3 Teacher shows an ad. for a job for which the students will apply. Discussion as to why they want to apply for the job and why they would be suitable.
   Aim: for students to start thinking about reasons for applying for the job.   5 mins
4 **Reading 2** – students read a completed letter of application and look at a list of paragraph functions. They match paragraphs with functions.
   Aim: for students to identify the purpose of paragraphs in such a letter.   5/10 mins
   or
   Students are given a completed letter of application which contains some inappropriately informal language. They have to identify and improve the inappropriate language.
   Aim: for students to focus on appropriateness of language. 5/10 mins (if time allows, you might include both of these stages)

5 **Writing** – students write a letter of application for the job featured in Stage 3. This could be a letter already partly written (perhaps with letter layout completed).
   Aim: for students to have practice in writing a letter of application.   15/20 mins
6 Students read (or are read) all the letters and choose the best applicant.
   Aim: for students to have practice in evaluating qualities of job applicants.   10 mins

As with the 'postcard lesson', the above staging is only an example; there are many possible variations. But what the two examples given show is how a Writing activity can be integrated into a lesson which focuses on other areas, too. Of course, and especially with an exam class, you might instead have a lesson which is exclusively concentrated on Writing. In such a case you will almost certainly be using a course book designed for the preparation of this exam, and it will guide you as to how to approach such lessons. When teaching General English, as we said before, you will probably be using a course book as well, and this book, too, will give you ideas of how to do Writing in the classroom. Most of the time it will suggest an integrated approach, as I have in this unit. Sometimes, you might find it more appropriate to set the actual Writing task for homework – that depends on your context with your students.

One very important aspect of the teacher's role has not been mentioned yet. Given the emphasis on accuracy in Writing, the need for 'correction' is important. We will be looking at correction of both oral and written work, in the next unit. For now, it is enough to say that correction of written work does need to occur, and that there are various ways of approaching it.

✳ *Tips*

1 Do not ignore the Writing skills.
2 Integrate Writing with other activities, linking and staging the different parts of the lesson.
3 Be prepared to have lessons which focus entirely on Writing when appropriate.

4 Be aware of which writing skills are required for a particular task and help develop your students' ability to use these skills.

5 Be aware of which style of writing is required for a particular activity and help develop your students' ability to write in this way.

6 With low levels especially, provide models of the activity.

7 With low levels especially, provide guided writing activities.

8 In all cases, provide writing activities which are relevant to your students' needs both inside and outside the classroom.

9 Use ideas given in course books and specialist resource books.

10 When appropriate, let students work together in the preparation of a Writing activity. This will reduce the pressure and feeling of isolation.

11 Work with the students when they are writing, help and support them, for the same reason as '10'.

12 Use varied techniques for correction (see Unit 16).

Review

We have seen in this unit that Writing is a skill that is often neglected, partly because its relevance is not always clear, and partly because teachers don't always know the best way to approach it. The potential for boredom is great! But we have also seen that for many students, Writing is very important, either for their own lives if in an English-speaking country, or in their jobs. It can also be very important for those students preparing for an exam of which Writing will be one part. So teachers need to work on developing their students' writing skills.

- Teachers have to develop the specific skills required for a particular task, and make sure the students know what type of language to use.

- Teachers should also know how to integrate Writing into a lesson.

- Correction is very important because the focus on accuracy is much greater than it is with Speaking.

- Teachers should work on ways to reduce the pressure that students feel when doing a Writing activity:

by properly staging the lesson
by supporting students during the activity
by allowing students to help each other.

So Writing should not be ignored. However, the amount of time you spend on it compared with the other skills will be less, and will vary a lot according to who you are teaching and in which country you are teaching.

15 | SPOKEN AND WRITTEN MISTAKES

If you are trying to use a language which you are learning, you will make mistakes. It is inevitable. Some of these mistakes are 'slips of the tongue or pen' which you can correct yourself without prompting; others are what one might call 'real' mistakes, where you do not know what the problem is and need someone to tell you. Mistakes can be spoken, or written. This unit will cover both types, and will not differentiate between *mistakes* and *errors*. As for the role of the teacher, when it comes to correction, this is by no means as straightforward as you might think, or as the student who says "I want all my mistakes corrected" might think. **Correction**, in short, is a complex part of the teacher's role.

We will look at the following areas in this unit and the next one:

- Why students make mistakes
- Types of mistake
- When the correction should take place
- Different ways of correction
- Techniques of correction

Why students make mistakes

First, consider this. A child of five utters the sentence *Yesterday I went to school*. Two years later that same child says *Yesterday I goed to school*. What is your conclusion regarding this child's learning progress?

The obvious, easy answer is that the child has regressed. He has gone from correct grammar to incorrect grammar. But, in fact, the opposite is true. The initial correct utterance was probably imitation of a parent using *went* with *yesterday*. However, the second utterance suggests that the child has learned the rule that verbs in the past end with *-ed*. What that child hasn't yet learned is that English has many irregular verbs where this rule does not apply. But the child is trying to apply a rule he has learned, albeit incorrectly, and this is a good thing. This is the first point that needs to be made – 'mistakes' are part of the learning process and as such should be treated positively, assuming the student can learn from his mistakes. Students need to realize that they will make mistakes because it is impossible not to if you are learning a language. What they must do is know *why* it is a mistake so as not to repeat it. Teachers have the responsibility of making the whole process of mistakes and correction a positive one that all students can benefit from.

 Task 1

1 Why do students make mistakes (written and spoken)?
2 Think of as many examples of types of mistake as you can.

Stop now and do the task

As I have said before, some students are obsessed with accuracy and want every mistake corrected. When a teacher does correction work, out come their pens and they write everything down as if this is the part of the lesson when they are truly learning. You get the students who want to know which mistakes were theirs. And you get the students who insist on correcting other students' mistakes whenever they hear them! In other words, it is easy for some students to start seeing language learning as essentially learning rules and correcting errors. Fortunately, such students are in the minority because neither students nor teachers should view constant correction as desirable; indeed, it can be counter productive and become a negative experience when, as I said at the start, mistakes should be seen as a positive part of the learning process.

Key to Task 1

Students, as you might expect, make mistakes for many reasons. Some of these reasons are common to speaking and writing, others are not.

  1 **First language (L1) interference**. An example of this is 'false friends' (see Unit 7).

But it is any case where a student's first language may 'get in the way'. Another example is different nationalities experiencing particular problems. Japanese learners, for example, have difficulties with using articles.

2 **Forgetfulness**. Students learn. Students forget. We cannot expect a language learner to remember everything we teach them!

3 **Lack of understanding.** Maybe the student simply hasn't understood what you have taught them. They are still trying to correctly apply a rule or form.

4 **Slip of the tongue.** As mentioned before, in an attempt to achieve fluency a student will 'trip over' and make mistakes which are not genuine ones.

5 **Unclear teaching!** Yes, for whatever reason, maybe the teacher has not taught a language point or checked it sufficiently clearly.

6 **Incorrect application of rules.** A student may try very hard to use what they have learned, but is as yet unaware of 'exceptions' or development of this language point. For example, you cannot cover all ways of talking about the future at once, so students will only use what they have learned. Refer, too, to the example earlier (*went* and 'goed').

7 **Incorrect model.** Students will hear 'incorrect' English all the time – from each other and from speakers outside the school. They will copy what they hear.

8 **Habitual**. Some students continue to make the same mistakes time after time however often they are corrected. I'm not sure there's much you can do about this!

9 **Unawareness**. If students don't know (haven't learned) something, they cannot be expected to get it right! Most of the above can apply, too, to writing (with 'slip of the tongue' becoming 'slip of the pen'). In addition we can add the following:

10 **Alternative script.** Consider, for example, the Japanese or Arabic scripts – they are completely different from the English script and this makes writing in English especially difficult.

11 **Pressure**. As we saw in the last unit, writing is a pressurized activity and there is a strong focus on accuracy. Such pressure can induce mistakes.

12 **Style of language.** The specific styles of written language, the discourse etc, provide specific problems and the likelihood of mistakes.

13 **Lack of practice.** Students generally get much less practice, and even less 'teaching' of Writing than practically any other classroom activity.

Types of mistake

By recognizing *why* students make mistakes with the language, teachers are in a better position to anticipate problems and to help. It also helps to be able to recognize different types of mistake. These can be very broadly categorized as follows:

■ **Grammar.** Any mistake of structure or form.

■ **Vocabulary.** Any vocabulary mistake, which may or may not affect meaning.

■ **Pronunciation.** We looked at this area in Unit 9.

■ **Appropriateness.** The utterance is technically correct, but not appropriate in the context.

Apart from pronunciation, all of these can apply equally to Writing. To this list we can add:

■ **Layout**

■ **Spelling**

■ **Punctuation**

■ **Organization and general cohesion**

 Task 2

Look at the following examples of students' **spoken** mistakes. What do you think was the cause, and what type of mistake is it?

1 He writed me a letter.
2 I am seeing the television.
3 When I will see you again?
4 We need airballs for the party.
5 Will you borrow me your dictionary?
6 Teacher: "Ask her what her name is."
   Student: "What your name is?"

  7 He drive very slow.
  8 I've watched television last night.
  9 (Student in pub:) "Give me a beer."
10 She go home later.

Key to Task 2

1 This is a further example of a student trying to apply a rule she has learned (add -*ed* to a verb to put it in the past) without knowing that many verbs have irregular past forms. So it is a grammar mistake, but a very understandable one.

2 The student has learned to use the present progressive to describe an action in progress at the time of speaking (e.g. *It is raining*). What she does not yet know is that some verbs are not formed with -*ing* in this way, and *see* is one such verb. In this case *watching* is the verb we would use. Another grammar mistake.

3 The student has learned the typical sentence structure in English (subject + verb + object – *I will see you again*) but has not yet learned the correct question form, and is unaware that we invert the auxiliary verb and subject ('When *will I* see you again?'). It is another example of a student making a mistake because she does not know the rule.

4 A vocabulary mistake. This student wanted to use the word *balloon* but just didn't know the word. So he invented the word '*airball*' a much more descriptive word than 'balloon'!

5 Some languages do not make the distinction between *borrow* and *lend*, but even after students learn the difference, they invariably have great difficulty remembering the distinction. It is a vocabulary mistake, and one I have been asked about more times than just about any other vocabulary item.

6 Here's a mistake for which the teacher must take some responsibility, and it relates to the example in **3** above as well. *What* is a question word, and the teacher has said "Ask her *what*…" and the student, has listened very carefully to the rest of the sentence – "…*what* her name is" – and has repeated this as a direct question, which it isn't!

I once observed a lesson given by a trainee teacher. He was setting up a Speaking activity based on questions. He gave examples of the 'questions to ask' by saying something like "Ask your partner where they come from", and as he spoke he wrote on the board what he was saying. So on the board we had 'Where they come from', 'Where they are living now', 'What food they like' etc. As the Speaking activity began, sure enough, the teacher to his horror heard student A asking student B "Where you come from?" etc. An easy mistake for the teacher to have made, but with serious consequences!

7 The student either has not learned the distinction between an adjective and an adverb (*slowly*), or has simply forgotton to use it, as often happens. Another problem students sometimes have with adverbs is they think that they all end -*ly*, which they don't (e.g. *She drove fast, He worked hard*).

8 The present perfect tense is a complex one, and is used differently in British and American English. But one aspect is the same, you do not use a definite time reference (*last night*) with the present perfect. In this case you would use the past simple – I *watched television last night*.

9 "Give me a beer" is correct English, but if you go into a bar and use the words it might be considered rude (though this depends partly on your intonation). At the very least it would be considered unusually direct. So this is inappropriate language.

10 I go, you go, he/she *goes* – the verb adds -*s*/-*es* in the third person singular. However, whatever their level, students continually forget this! Forgetfulness is another consideration for the teacher. As well as treating mistakes as a positive part of the learning process, teachers should remember that students are trying to absorb so much new language and so many rules that they cannot possibly be expected to remember everything. If you are currently doing a teacher-training course, you will appreciate how hard it is to remember and put into practice everything your tutors have been telling you! And if you do anything 'wrong', it's all part of *your* learning experience as a new teacher!

'Positive' mistakes

Let's look now at an authentic piece of written work done together by a group of elementary students. They wrote a supposed letter of appreciation to a departing teacher. This is what they wrote. What is your overall impression?

> Dear Matt,
>
> we want to say good buy and hand over you a little present.
>
> We are most grateful to you for your help. You are a very good teacher.
>
> You've got empathy and you shaped the lessons very interesting. We think, do you like teaching.
>
> We hope we'll meet again on the occasion of the term of autumn.
>
> Today, we wish you many beautiful journey trought the German country.
>
> We wish you good luck and all the best!

This is a fascinating piece of work because it illustrates so well the idea of 'positive' mistakes. At first glance you might think there are lots of mistakes and it is a poor piece of work. But consider two points. First, the meaning is clear throughout. Second, how well the students are trying to use the language. For example, "you've got empathy and shaped the lessons very well". "Empathy" may not be the right word here, but it is a word they have acquired and are trying to use, and the meaning behind the sentiment is clear. As for "you shaped the lessons very well", it may be a clumsy phrase, but it is a tremendous effort and clearly means the teacher staged the lessons well. "On the occasion of the term of autumn" may be technically correct, but is inappropriate and unnatural. But arguably it *is* 'correct'!

Of course, there are examples of good and appropriate language. "We are most grateful to you for your help" may be a little formal, but no sensitive teacher is going to do anything other than praise the sentence.

Overall, if I had been the teacher, I would have been delighted with this attempt from elementary-level students!

Review

- Whether we are talking about spoken or written mistakes, teachers must recognize why mistakes might have happened (this, as we will see, can help with correction), and should treat them as a positive part of the learning process.
- If we perceive mistakes as a sign of poor teaching or lazy learning, we will not be helping ourselves or our students.
- If students are hard on themselves because their English is less than perfect – we must help them to understand they are being unrealistic and ultimately hindering their own progress. This applies equally to trainee teachers who say their lesson was a 'disaster' because a couple of parts did not go quite according to plan.
- We should focus as much on the 'correct' as the 'incorrect'.
- As teachers we must adjust our focus on accurate English according to who we are teaching.

Mistakes are far more of an issue for students preparing for an exam than for those who are seeking only to improve their English. For the latter, it is more important to speak fluently and make themselves understood.

16 CORRECTION

In the last unit we looked at some examples of spoken and written mistakes, and some reasons for them. We also discussed what the teacher's – and student's – attitude to spoken and written mistakes should be. In this unit we will look at when correction should take place, different ways of correction, and correction techniques.

 Task 1

What are your views on the following questions about correction of **spoken** mistakes?

1 Should all mistakes be corrected? Should *any* mistakes be corrected?
2 If you answered 'no' to the first question, which mistakes should be corrected?
3 Should correction take place immediately, or later?
4 Who should do the correction – the teacher, the same student or another student in the class?
5 How should the correction take place?

Stop now and do this task!

The role of correction in the classroom is far from clear-cut. However, all teachers do need to know the skills of correction work, and trainee teachers on courses are assessed on their ability to recognize and correct both spoken and written mistakes.

1 Should all mistakes be corrected?

The first question is easy to answer – 'No', all mistakes should not be corrected. What's more, even if you thought they should be (as some students do), it would be impossible. Imagine a class of twenty students, all talking at different points, all making mistakes, and you trying to correct every single mistake. Impossible. Should *any* mistakes be corrected? Again, an easy question – 'Yes', of course. How can any person learn any new skill without having important mistakes corrected? How can a person learn to be a teacher without knowing or being told when something has gone wrong?

As we have already seen, some students will say "I want all my mistakes corrected". Such a student is being unrealistic, but as a teacher you might have to tell that student, or the class, why it is unrealistic as well as undesirable. I have done so many times. With a class of the right level it is no bad thing to have a discussion about learning stategies, and the rationale of what the teacher does or does not do in lessons.

2 Which mistakes should be corrected?

Now it gets more difficult! Having established that some, but not all, mistakes should be corrected, we are faced with the question of *which* mistakes should be corrected. Ultimately, of course, this is for the individual teacher to decide, based on the context of the lesson and the nature of the class. However, I think we can make some general points about what mistakes should be corrected. If students are doing a practice activity, using specific language previously 'presented', as well as other language, then mistakes involving the 'target language' should be focused on for correction (*when* and *how* will be discussed later). In this type of activity, the focus is more on accuracy, and therefore, at least some of the inaccuracies with the target language need to be corrected. Common mistakes are also suitable for correction. If you are teaching a monolingual class, then you are more likely to get similar mistakes being repeated, and everyone will benefit from the correction. What a teacher might determine to be a 'significant' mistake, for whatever reason, can be corrected. For example, a mistake which results in unclear meaning, or a mistake which results in the wrong meaning. If a student, for example, says "Can I lend your dictionary", then this is a significant mistake.

If some mistakes *should* be corrected, then others should not – or do not have to be. For example, if your students have done a Reading comprehension and, in answering a question about the text, a student gives the right answer, but uses incorrect grammar, then the teacher can ignore the mistake because the aim of the task was comprehension, not accuracy. If you are chatting with students before or after the lesson, and they make mistakes when talking, ignore the mistakes as long as meaning is clear. If the students have done a practice activity and made mistakes with the target language, focus on those mistakes and worry less about any others, unless they are significant in some other way, or are very common. To correct or not, then, depends on the aim of the lesson or activity. A mistake that occurs in a controlled-practice activity might not be corrected if it occurs in a fluency activity, or a comprehension exercise.

In other words, be selective in deciding which mistakes need to be corrected, and which don't. Some teachers, myself included, might argue that from time to time higher-level students should be allowed to do a Speaking activity where the sole focus is fluency and the teacher should ignore all mistakes.

3 When to correct mistakes?

Sometimes, you might feel that a mistake should be corrected immediately. For example, pronunciation mistakes would normally be corrected straight away. If a student was answering a question concerning target language and a mistake with the target language was made in the answer, again the mistake would be corrected immediately. Sometimes, in the normal flow of a lesson, if a student is talking and a mistake or mistakes in his speech impede your understanding, you might feel that you need to correct.

Besides the many mistakes that would go uncorrected, there are also those that you would not want to correct straight away. If your students are doing any kind of Speaking activity, and especially if you want them to achieve some fluency, then to interrupt with corrections would stop the flow and possibly intimidate them.

One advantage of delayed correction is that the correction becomes less 'personal' – students will not necessarily remember whose mistake it was (though many insist on guessing!). You may wonder, though, if you do not do correction straight away, how you will remember what you hear – see Question 5.

In conclusion, some mistakes will be corrected straight away, some later, and some not at all.

4 Who should do the correction?

There are three possibilities – **teacher** correction, **self**-correction, and **student-to-student** correction.

 Task 2

What do you think are the advantages and disadvantages of each type of correction?

Stop now and do this task!

In any class, there will be a combination of all three types. In the next question we will consider techniques of correction, and also how correction can become a lesson in itself. But now let us go back to the Task.

A) Teacher correction

Advantages

- The correction will, or should be, quicker, more efficient, and 'accurate'.
- Student confidence – 'It is the teacher's job', 'The teacher is always right'!
- The teacher can ensure correction is handled sensitively and fairly.
- The teacher can use proper and varied techniques of correction.

Disadvantages

- It fails to encourage learner-independence.
- It can be intimidating for students to have 'the teacher' correcting their mistakes.
- Students might feel embarrassed, however sensitive the teacher might be.
- Too much teacher correction might be demoralising for students.
- Other students might feel 'left out' of the lesson while a mistake is being corrected.

Despite the disadvantages expressed, the teacher should do some correction work, but must ensure it is done properly. As to how much of the correction should come from the teacher, there is no answer to that. It depends on a multitude of factors and opinions.

B) Self-correction

Advantages

- Students are involved in the process – this renews confidence if they can correct themselves.
- Self-corrected mistakes are more memorable and less likely to be repeated.
- It encourages learner independence.
- It gives the teacher feedback on the student's knowledge, ability and awareness.

Disadvantages

- Students may not be able to self-correct and consequently become demotivated.
- Students may feel pressured or embarrassed.
- Students may 'correct' what was already correct and do more harm than good.
- It is potentially more time consuming.

On balance, self-correction is definitely more positive than negative. The first three advantages listed above are really important points.

C) Student-to-student correction

Advantages

- Some students might find it easier being corrected by a peer (but many won't!).
- It involves other students ... gives them self confidence.
- It encourages other students to stay involved in the lesson.
- It encourages an atmosphere of cooperation.
- It helps the teacher 'assess' awareness of other students.

Disadvantages

- The student doing the correction may get it wrong!
- It can be slower and less effective (if student does not have correction skills).
- The student being corrected may not understand the other student (pronunciation).
- The student being corrected may not know if the correction is accurate.
- Some students may feel it is the job of the teacher to correct.

Some student-to-student correction takes place naturally, and within reason, it should be encouraged. But beware the student who constantly tries to correct other students as if they were the teacher. This can be counter-productive and irritate almost everyone in the class!

Overall, the ideal is for there to be a mixture of all three 'correctors', but the precise balance will vary according to many factors. But avoid using teacher correction because that produces a total reliance on the teacher and no encouragement of learner independence.

5 How to correct?

Obviously, you want to avoid "No, that's wrong", or anything like it. That becomes negative and critical and goes against the idea of correction being part of the learning process. How you correct depends partly on who is doing the correction, the type of mistake, the context of the lesson, and the nature of the mistake. But we can make some very general points.

✳ *Tips for immediate correction*

1 Indicate that a mistake has been made. How? Maybe facial gesture, verbal remark ("Again"/"Sorry?") or just a gesture for the student to stop their sentence, thus indicating a problem.

2 Locate *where* the mistake is in the sentence. How? A verbal gap-fill, perhaps. For example, if the student has said "Everyday she go to the shops", the teacher might say "everyday she ____ the shops". Use your fingers, too. With each finger representing one word, when you get to word four in this sentence indicate with your finger that this is the problem word. It might sound an odd thing to do, but it works! By locating the mistake the student will be able to focus on the correct part of the sentence, and not end up 'correcting' what doesn't need to be corrected!

3 Categorize the *type* of mistake. How? In this case you might say "3rd person singular?", if the students know the rule of adding -*s* or -*es* in these cases. This suggests the mistake was no more than a slip of the tongue and it will be possible to elicit self-correction, or student-to-student correction.

By indicating that a mistake has been made, where, and what type, the teacher is using her correction skills to make the correction efficient and successful, and is involving the students.

4 At this point, either the student or another student is likely to correct. If they can't, you do it.

5 Maybe get the student to repeat the whole sentence for reinforcement.

6 You could repeat the entire sentence, with mistake, with rising (questioning) intonation to make it obvious that the sentence is incorrect. Or just repeat the incorrect part in the same way.

7 Similar to point **2**, repeat the sentence only up to where the mistake occurred.

8 As mentioned before, use your fingers (clearly!) to indicate, for example, that two words are the wrong way round in a sentence (e.g. "Where you will go?"). You can use your fingers to indicate a 'swap over' of words.

9 Use word prompts to categorize. e.g. "Tense?" "Preposition?" etc.

10 If a student has incorrectly repeated what a previous student said, go back to the other student and ask her to repeat what she said before.

All of this might sound like a lengthy process. Normally it takes a matter of seconds for the simple reason that the student, or another, can correct easily. The process only slows down if the mistake was a real one and no-one in the class knows the problem.

Delayed correction

We saw in Unit 13 that if students are doing a fluency activity, or a Speaking activity where they are getting practice of language previously presented, then to interrupt with corrections would spoil the fluency and potentially intimidate the students. In addition, assuming they are working in groups, other students would not benefit from the correction. So how can a teacher do correction later in the lesson?

The usefulness of the 'Tips' below depends on the size of the class and the ability of the teacher to monitor. They are most valuable with a class of no more than, say, twenty, in a room where the teacher can walk around fairly freely.

✳ *Tips for delayed correction*

1 Arrange the class so that it is possible for you to walk around the room, e.g. they work in groups of four in a circular-type formation.

2 Get a pen and a piece of paper.

3 Keep a 'low profile', listen to each group in turn and try to pick out mistakes.

4 Concentrate on mistakes in the target language, if appropriate, or common mistakes, or mistakes you view as important or useful to focus on.

5 Make a note, too, of particularly good sentences.

6 If you have real trouble hearing (it can be difficult!), note down mistakes which could easily be made in such an activity – a correction phase does not have to consist of genuine mistakes made by students in your class!

7 At the end of the activity, or just before, write up all your sentences on the board. Tell the students that some sentences are correct and others are not. They have to decide and correct as necessary. If time, let them work on this in groups. If not, go through it together. This way, everyone benefits from the correction and it is 'depersonalized'. You will almost always have someone who wants to know "Was that my mistake?". Tell them it does not matter!

8 When going through the mistakes, locate and categorize as indicated above, and write in the correction on the board.

9 Another possibility, especially if time is short and you have lots of sentences, is to do the correction phase in the next lesson.

10 Linked to the previous point, you might 'collect' mistakes from the week and have a correction lesson as a kind of review at the end of the week. At least this way you won't have students trying to work out whose mistakes they were, because they won't remember!

The idea of including correct sentences is to emphasize that a lot of what they said was very good, but that is up to you. I sometimes do it, but not always.

My experience of working on teacher-training courses, where the new teacher's skills of correction are part of their assessment, is that there is often considerable reluctance to correct. There seems to be a number of reasons for this. One is the teacher's own lack of confidence in grammar and fear of being asked questions. Trust your instincts. You know what is right and what is wrong. Others feel that it is too negative to correct. To which I ask, how can someone learn any new skill (including teaching), if you are not made aware of what you are doing wrong? To the teacher who avoids correction I say "you are not doing your job". What you have to do is be sensitive and use your skills. In all my years of teaching, the one activity that students have always reacted to positively is correction, and especially 'delayed correction', as described above. I have heard students saying "Too much grammar", "Too much speaking" etc. I have *never* heard students say "Too much correction". Too *little* correction, yes, but never too much.

Correcting written work

So far we have looked primarily at spoken mistakes. Much of what has already been said can apply equally to the correction of **written** mistakes, but there are other factors, too, that need to be considered.

Most people's memories of written work at school being returned by the teacher probably include a long wait, the sight of red ink, lots of corrections and little praise. Or, they may remember pages of written work being returned with a single comment (e.g. "Good"), making you wonder just how carefully the teacher read your work!

1 When do we correct written work?
2 How do we correct written work?

As with spoken mistakes, so with written mistakes, there is an argument for leaving some mistakes uncorrected. Look again at Unit 14 and the example of written work given and the points made about 'praiseworthy mistakes'. Consider, too, the demoralizing effect of a student getting work back which is filled with corrections.

Yes, you might argue that to leave a mistake uncorrected implies that it is not a mistake – and you would be right! But it is a question of balance and personal judgment in the context of the teaching.

If students are writing in class you might go round during the activity giving one-to-one help and corrections. Or, you might collect in their work, correct at home, and return it – preferably – the next lesson. In other words, you again have the choice between 'immediate' or 'delayed' correction.

Symbols

Just as with spoken mistakes, teachers should try to involve the students in the correction process of written work. One way of doing this is to employ correction symbols. The teacher might underline or circle the mistake (locate it), then indicate the type of mistake (categorize) by putting in the margin the appropriate symbol (e.g. 'G' for grammar). The students have a copy of all the symbols and what they mean. They then go away and try to correct the work for themselves – though this still requires the teacher to check again afterwards, so it is a lengthier process. Symbols might include 'V' (vocabulary), 'WO' (word order), 'P' (punctuation) etc. You must be consistent in your use of symbols, and still include praise for good work.

If you use the more traditional approach to correction (and, as ever, you should aim for variety), then do not litter the work with corrections. Praise as you can, encourage, be supportive.

Class correction

Another principle of correction of spoken mistakes which can apply to written correction is the one of class correction. Take the work away and make a note of common or significant mistakes. Make a copy of this list (or write it directly on the board) for the next lesson, then have time where everybody looks at all the mistakes from the different pieces of work. This depersonalizes the correction and everybody benefits from it.

With higher levels, students can exchange their work with a partner and look for mistakes in each other's work (student-to-student correction).

Grading

As for 'grading' the work, this is up to you. Personally, I do not, unless the students are preparing for an exam, in which case I would grade according to exam standards for students to have an idea of their level in relation to the exam. Otherwise, I would always write a summary comment, as positive as possible, but no 'grade'. However, that's a personal policy!

Review

- Correction should be part of the learning process and treated positively.
- It should include a variety of teacher correction, self-correction, and student-to-student correction (spoken and written).
- The teacher should indicate where the mistake is and categorize it.
- The teacher should employ facial gesture, finger work, verbal prompts etc to help with correction.
- With written work, the teacher should use a variety of approaches as well.
- Use of symbols for written correction should be clear and consistent.
- Not all mistakes should be corrected – focus on target language mistakes, common mistakes, or mistakes which you feel are significant for some other reason.
- If the aim is comprehension (the correct answer to a question), focus on the correct answer, not the correct grammar.
- Don't interrupt fluency work – make a note of mistakes.
- Be positive, praise good work.
- **Do** correct – don't avoid it!

17 | LESSON PLANNING 1: AIMS

If you are or are about to be a trainee teacher on a training course you will be required to submit a detailed lesson plan each time you teach. If you are a relatively new teacher, you may still be preparing less-detailed plans for your lessons. Once you have more experience, you will rarely produce an actual plan (unless you are being observed, maybe), but you will still plan what you are going to do. So in the following units we are looking at both planning in general, and producing actual lesson plans. This unit focuses on **aims.**

I have heard some experienced people in the profession argue against the notion of planning, suggesting that teaching and learning should be spontaneous according to the wishes and reactions of the students on a particular day. For some teachers this might work and be successful. However, I would argue that there are very few teachers, however experienced, who would be comfortable just walking into a class with no idea of what will be done in the lesson. Further, I do not believe that most students in my experience would be satisfied with such an approach to learning. But whatever the arguments, such an approach is definitely not acceptable for new or training teachers! What is true, however, is that teaching should be flexible; students must have an influence on what you teach and how you teach it, and this has an important influence on the teacher's planning

A good lesson plan will be one which is a clear working document. In other words, another teacher would be able to teach your lesson for you by using your plan as a guide. In my experience as a trainer, I can normally predict the quality of the lesson I am about to observe from the quality of the plan which has been handed in.

Obviously, the length and precise contents of your plan will depend on the length of your lesson, the type of lesson, and also the stage of the course. But all plans are likely to include at least the following:

> level of class, length of lesson, aims(s) of lesson, target language (if appropriate), personal aims, anticipated problems (for the students) and possible solutions, stages, aim of each stage, estimated timings, interaction, procedure, board layout.

We will look at most of these points in the next unit, but for now we will concentrate on aims.

Lesson aims

The **aim** of your lesson, or the aim of the stage of the lesson, is what you hope to achieve, or what you hope your students will achieve. It answers the question 'Why am I doing this?'. Clear and realistic aims are essential if a lesson is going to be successful.

Some teachers phrase the aims from the teacher's perspective, others from the students' perspective. Compare these examples:

'to provide practice in scan reading skills'

'for students to have practice in scan reading skills'

Personally, I much prefer the latter. As teachers, it is more beneficial if we think of what we are doing from the students' perspective whenever possible. New teachers can get too involved in the teaching process (quite understandably!) and neglect to give sufficient importance to the students' learning. As a teacher, of any length of experience, you should think about the students as much as, and probably more than, your own teaching. When observing classes, it is good to focus on one or two students, or the class as a whole, and observe their involvement and progress in the lesson – how has their learning been affected by the teaching?

Formulating clear aims is not easy for the new teacher and the result is often inappropriate or aims that are far too general. You don't always need the rest of a plan to recognize poor aims, as the following task will show.

Task 1

Which of the following lesson aims seem inadequate? Can you improve them?

1 'to teach the present perfect'
2 'for students to have practice in listening for gist'
3 'for students to learn new vocabulary'
4 'for students to have reading practice'
5 'to teach the positive and negative forms of the present progressive and to give written and spoken practice'
6 'for students to improve their pronunciation'
7 'to set up a speaking activity clearly and efficiently'
8 'Reading and Speaking'
9 'for students to learn the correct usage of selected conjunctions in written English and to have practice in a letter-writing task'
10 'to teach Writing'

Before looking at this task, I must emphasize the importance of a teacher having clear aims. I said before, I can usually predict the quality of the lesson I am about to see according to the quality of the plan. The first section I always look at is the one marked 'Aims'. If the stated aims are clear, appropriate, and achievable (according to the rest of the plan), then I would expect to see an effective and successful lesson. But a plan with vague aims, or inappropriate aims (for whatever reason) is almost certainly going to result in an ineffective lesson.

Let's look now at the aims above.

Key to Task 1

1 It is not uncommon in the early stages of training courses to see an aim such as this one. Personally, I am not keen on the focus on the teacher (the aim says what the teacher will do rather than what the students will achieve), but the real problem is that the aim is vague and unachievable. The present perfect tense is a complex one in any case, but all tenses (verb-forms) have a variety of forms. You have the positive (affirmative), negative, question (interrogative), and answer forms. Is this teacher going to 'teach' all four forms in one lesson? Then you have to consider the different uses of this

tense (bearing in mind that the tense is used differently in British and American English). Thus, *I've been to France* refers to a past experience, whereas *I've worked here for ten years* refers to something started in the past and continuing to now. Again, it would be unrealistic to include more than one usage with, say, a pre-intermediate class. The importance, therefore, of pre-lesson **language analysis** by the teacher is essential. Equally imortant is the teacher's own **language awareness** – actually knowing and understanding the language that she is to teach. We will look at this in Unit 19.

So, this teacher probably has little understanding of what the present perfect is and what is involved in teaching it. I would expect to see a lesson which covers far too much, and done via explanations and lots of exercises.

A better aim would be 'for students to learn and have spoken practice in the positive and negative forms of the present perfect simple (indefinite past event)'. Seeing such an aim would lead me to expect a focused lesson.

2 This is a good aim. It sees the lesson from the point of view of the students, but more than that, it provides a clear and specific objective. It's not 'to have practice in listening' (too vague), but listening 'for gist'. So the teacher is aware of different types of listening skills, and has prepared a lesson which focuses on one particular skill.

3 This aim is written from the students' perspective, which is good. However, like number **1**, it is far too vague. What vocabulary? How much? For what purpose? Will there be any practice? If so, what type? There's nothing wrong in having a vocabulary lesson (rather than teaching vocabulary which will be needed, say, for a later Reading task), but it needs to have focus.

An improvement would be 'for students to learn the meaning, form and pronunciation of ten words about *television* and to have practice using them in a written gap-fill and a free-speaking activity'. Better than '...to learn vocabulary'!

4 Again, it is good that it is written from the students' perspective. However, unlike number two, it is too general. Which reading skills will be used?

Compare this with 'for students to have practice in gist reading (scanning), and reading for detail'. A teacher who plans according to the original aim is one who will probably have given little thought to the tasks being given, and is unaware of what type of reading is required of the students. This teacher might, for example, give three gist questions and allow ten minutes for the task! As we saw in an earlier unit, gist reading should be relatively quick. If it takes ten minutes it is no longer gist reading!

5   A good, clear aim – albeit written from the teacher's perspective – and a big improvement on number **1**. If this aim came from the lesson plan of a trainee at the end of a course, and the first aim in the task had come from the same trainee early in the course, it would demonstrate real progress.

6   I would be pleased to see a teacher doing a lesson focusing on pronunciation, and I would be very glad not to be reading 'to teach pronunciation'! However, the aim remains far too vague. Which aspect(s) of pronunciation?

An improvement would be 'for students to have practice in recognizing and using rising and falling intonation in question tags and to recognize the different meanings implied'.

7   This aim is what the teacher personally wants to achieve, it has nothing to do with the reason for doing the activity. It is a *personal* aim, not a lesson aim. There is nothing wrong in having personal aims (in fact, they are very important), but they should not be confused with teaching and learning aims.

8   There is a very common confusion for new teachers on training courses – the difference between the 'What?' and the 'Why'? **Aims** are the 'Why?' ('Why am I asking my students to do this?'), **stages** are the 'What?' ('What will the students do?'). Both Reading and Speaking are 'What?'s. Refer back to number four for a possible improvement.

9   This is fine.

10   This is not fine. Which aspects of writing? The aim as written suggests that the teacher is going to teach the class *how* to write – and all in one lesson!

An improvement would be 'for the students to have practice in writing a letter of application, using appropriate layout, language and paragraphing'.

Sometimes a lesson may have 'main' and 'secondary' aims. For example:

Main (primary) aim: 'for students to have practice in reading for gist and for detail.'
Secondary aim: 'for students to have practice in free-speaking on topic of text.'

We saw in Unit 11 that a teacher needs to be flexible according to various factors, including her aims. We concluded that a teacher should not 'rush through' the Reading part of this lesson (or skip part of it) in order to do the Speaking. It is important to achieve the main aim properly even if the secondary aim is not achieved. Teachers, then, need to be aware themselves of what 'must be done and done properly' in a lesson, and what can be adapted or even dropped (or kept for the next lesson). We will look at this 'flexibility' issue in the next unit.

If you are observing a lesson, try to see if you can write down what you thought the the aims of the lesson were and then compare your aims with the actual ones written by the teacher. If the lesson was effective and successful, the task should be easy!

Stage aims

Everything a teacher does, and everything a teacher says, should have a purpose. Well, almost everything! Aims, therefore, are not only for the lesson itself but also for each part of it. A trainee on a teaching course, submitting lesson plans, should always indicate the aim of every stage. An experienced teacher knows instinctively why they are doing or saying what they are. The word 'stage' refers to what the students will be doing – it's a kind of 'title' for the part of the lesson. Examples would be 'Reading', 'Practice', 'Boardwork'. The aim would then clarify the reason for this.

 Task 2

For each of the following stages, try to give a possible aim.
Example:
Reading: 'for students to have practice in reading for gist'.

1 Writing
2 Boardwork
3 Practice
4 Context
5 Language focus
6 Drilling
7 Checking answers
8 Speaking
9 Feedback
10 Correction

Key to Task 2

Of course, there are no single 'correct' answers because there are many possibilities according to the lesson being taught, but here are some suggested answers. Yours may be different but still 'correct'.

1 'for students to have practice in using narrative tenses'
2 'for students to have a record of the target language'
3 'for students to have controlled practice of the 2nd Conditional via a gap-fill'
4 'to establish student interest in the topic of the lesson (food)'
5 'to clarify the meaning, form and pronunciation of '*wish* + past perfect' (NB: each of the three points here would have separate sub-stages and sub-aims)
6 'for students to have controlled practice in repeating the new words correctly and naturally'
7 'for students to see if they have the right answers and to understand the reasons for the answers'
8 'for students to have practice in speaking fluently on the lesson topic (food), and ranking food from favourite to least favourite.'
9 'for students to tell the rest of the class what they learned from the pair work questionnaire'
10 'for students' significant mistakes with the target language to be corrected in groups and then clarified by teacher'.

✳ *Tips*

1 Think of the lesson and its stages from the *students'* perspective – what they will do or achieve, rather than what you will do.
2 Keep the aim as precise and clear as possible.
3 Make sure the aim is realistic (in terms of the level, the students, the time).
4 If appropriate, distinguish between main/primary and secondary aims.
5 Give priority to the achievement of the main aim.
6 Ensure that every stage of your lesson also has an aim – these aims will usually also be from the students' perspective.
7 Make sure you are clear about what the aims are as well!

Having looked at aims of the lesson and aims of each stage of the lesson, we can go one step further. An experienced teacher always knows why she is doing or saying something. What might appear to an observer to be an aimless chat could in reality be the teacher guiding the students towards thinking about a particular question. A teacher showing a class a picture of a famous star, or a hot sunny country, might be going on to a lesson on 'I wish'. A teacher briefly telling the class what she likes and dislikes about a place might just be giving an example of what he wants her students to talk about together. Such a teacher won't have to write out a plan in advance to know why she is doing what she's doing, or saying what she's saying. But until you have that experience, almost everything has to be planned in one form or another. Trainee teachers on courses will teach in one month what some teachers do in just one day, so they have the time to write a proper plan, and of course, this is expected of them. We will look in the next unit at what else, apart from aims, needs to be included in a plan, and how to approach it.

Review

- Aims need to be clear, precise, achievable, and preferably stated from the point of view of students learning rather than of the teacher teaching.
- Teachers with vague aims in language lessons have probably not understood, or given sufficient thought to, the language being taught.
- The main aim of a lesson needs to be achieved, to some extent at least, even if this means being less effective in achieving the secondary aim.
- Every action, every word of the teacher, should have a purpose – whether this has been planned, or is instinctive.
- You should review the success of your lesson and to what extent you have achieved your aims.
- Resist the temptation to dismiss a lesson as a 'disaster'.
- You should think more about the students' learning than about the teaching process in isolation.
- Ask yourself "what did the students do/learn/practise in my lesson?" – you might be pleasantly surprised!

However, such a review might be misleading in one respect. It might give the impression that a teacher should stick, inflexibly, to her plan. Not only is this not true, but also it is a dangerous conclusion. As we will see in the next unit, a teacher must always teach with the students' needs in mind, even if this means departing from a carefully prepared plan!

18 LESSON PLANNING 2: WRITING A LESSON PLAN

This unit is specifically for those who are on training courses and who have to submit formal lesson plans for each lesson that they teach. We looked in the last unit at **aims** and how to formulate them, and aims are one of the items on any lesson plan. Indeed, they are the *key* item, since the aims say why the lesson is happening and what the students will get out of it. The aims of stages state the reason for each part of the lesson.

As a trainee on a course you will find lesson plans something of a chore in the sense that they take time and have to be done properly. However, they are an essential part of any training course. So, apart from the fact that you have to submit plans on such courses, what are the benefits of writing plans for teachers and students?

Most trainees on courses have never taught before and everything they are doing and hearing on the course is new. There is a lot of information, a lot of jargon, a lot of theories and techniques to digest. On many such courses the trainees have to start teaching practice right away .... a kind of 'learning by doing'. It is easy, then, to get confused and frustrated and to lose sight of the main focus of a lesson and how to teach it. Preparing a lesson plan will greatly help the teacher to think more clearly about *what* she is going to do, *why* she is doing it, and *how* she will do it. The result should be a more confident teacher, a more focused and effective lesson, and happy students. Students benefit most because they will be getting a well-thought-out lesson with clear learning objectives.

A good lesson plan ...

- is a clear working document for the teacher. It should be a document that a teacher can work from in the lesson, so it should be neat, clear, well-organized.

- has clear and appropriate and achievable aims.
- should include assumptions of what will help the students achieve the aims – their existing skills and knowledge as known or assumed by the teacher.
- should anticipate problems of any kind in the lesson **and** possible solutions. (see next unit).
- should have clear and logical stages which link naturally. Ideally, the students as well as the teacher should see where the lesson is going and how the stages link.
- should include a variety of activities and interaction.
- should reflect how that lesson fits in with others being taught before or/and after.
- should be realistic in terms of what can be achieved in the time available.
- should not be too long and detailed (see point **1**)
- should be flexible – a really important point which we'll come back to later.

Lesson plan contents (1)

Typical features of lesson plans were listed in the last unit. Here we will look at each of these features and see an example of an authentic plan written by a trainee on a training course. The example is not presented as a perfect model but as a typical plan produced by a successful candidate on a course. Aspects of the plan are unclear, but overall it is a satisfactory plan from a Pass-standard trainee.

The central feature of any plan will be the **lesson aim(s)**. This was looked at in detail in the previous unit. In our example plan the aims were stated as:

for students to review and have practice using *like* and *don't like*
for students to learn the use of *would like* and have practice using it

(this was a 40-minute lesson given to an elementary group of learners).

This teacher had covered *like* and *don't like* in the previous lesson. This, then, is indicated in the section **timetable fit and presumed student knowledge**.

follows on from previous lesson (*like / don't like*)
will continue tommorrow

This emphasizes the point that lessons should be part of an overall syllabus or structure, with links not just within a single lesson, but between different lessons. In other words, aims should apply to a series of lessons / a course, as well as to a single lesson.

Plans normally also indicate the **materials and resources** to be used:

Headway Elementary (Soars – O.U.P.) Pages 62/63
My own visuals and word/sentence cards

Note here, in reference to Unit 20, how this trainee did not rely exclusively on the course book being used, but supplemented this with her own materials and ideas.

Good teachers – experienced as well as new – should always be working on improving their own skills and focusing on areas in which they have a weakness. On training courses, especially, there is an emphasis in assessment of *progress*; how trainees improve in areas where they have previously had problems. You should limit yourself to two or three areas in an individual lesson. The trainee in this lesson wrote...

to encourage students more to use the new/revised
language in a communicative way.

A major part of planning is **anticipating problems** with accompanying possible solutions. This is something the next unit will focus on in more detail, but in the example plan the trainee wrote:

students might be confused by use of 'like' and 'would like'
solution: ask concept questions

It should be said here that this part of the plan needed to be much more detailed!

Boardwork needs to be neat, organized, and a clear 'take-away' record for students who copy from it. Particularly if a teacher is going to use the board to write up target language covered in a lesson, she should plan how to lay out her boardwork. The teacher in the example here planned two phases of boardwork. The first part was planned as follows:

---

like

subject	like	*verb + -ing*	
I	like	flying	
She/He	likes	flying	= *enjoyment*
We	like	flying	

Negative:

subject	do not like	*verb + -ing*
I	don't like	flying
She/He	doesn't like	flying
We	don't like	flying

For the second part of the lesson the boardwork plan was as follows:

---

would like

subject	would like	infinitive	
I	would like	to play tennis	= *polite way of*
short form:	I'd like	to play tennis	*saying "I want"*

Negative:

I would not (wouldn't) like to play tennis

Note that there is no reference here to question forms and answers. It would have been unrealistic to cover that as well at this level in just 40 minutes, so it was covered in a lesson the next day. The boardwork is not perfect – there is no sentence stress marked, for example, but it does provide clear examples and highlighting of the form.

The remainder of the plan refers to the actual stages and procedure employed to achieve the stated aims. The five sections of the lesson itself are **stage/aim/procedure/interaction/time**.

The **stage** is effectively a title for a part of a lesson – it states *what* you will do. The **aim** refers to the aim of that specific part (stage) of the lesson; virtually everything that a teacher says or does in a lesson is said or done for a reason. It states *why* you are doing something, or why the students are doing something. The

procedure is a description of *how* something will be done. The **interaction** concerns who is working with whom – individual work / pair work / group work / whole class. The **time** is the very approximate estimate of how long something is likely to take. As we shall see later, this is very flexible according to various factors.

Going back to the example plan, this is how the remainder of the plan looked:

Stage: **introduction**
Aim: to get students interested in the subject
Procedure: introduce imaginary friend, Eric, and ask students what they think Eric does and doesn't like
Interaction: teacher to class
Time: 3 mins

Stage: **focus on form (1)**
Aim: clarify form of *like* (revision)
Procedure: elicit when we use *I like* meaning something we enjoy doing. Give examples and elicit more examples, including the negative.
Interaction: teacher to class
Time: 5 mins

Stage: **practice**
Aim: students have practice of *like* (questions and answers)
Procedure: divide class into two groups, give each group a set of questions and answers, students match them up / I monitor
Interaction: groups
Time: 5 mins

Stage: **check answers**
Aim: clarification of answers
Procedure: get students to give correct answers / I will confirm or correct
Interaction: teacher to class
Time: 2 mins

Stage: **practice 2**
Aim: students have spoken practice of *like*
Procedure: divide students into pairs, give each student a picture and they have to tell their partner whether they like or

don't like the items. Then get each student to tell the
class about their partner's likes and dislikes
Interaction: pairs
Time: 7 mins

Stage: **focus on language**
Aim: clarify meaning / form / pronunciation of *would like*
Procedure: elicit when we use *would like* as a polite way of saying
I want. Show form on board, including negatives.
Check meaning via concept questions, and
pronunciation via drilling
Interaction: teacher to class
Time: 5 mins

Stage: **practice 3**
Aim: students have practice using *would like*
Procedure: in pairs students get worksheet with sentences
containing *like* and *would like* and students have to
choose which to use
Interaction: pairs
Time: 8 mins

Stage: **check answers**
Aim: clarification of answers
Procedure: elicit answers from students, get others to say if they
are right or wrong
Interaction: teacher to class
Time: 5 mins

The resulting lesson, though not strong, did provide a clear focus on
the target language, and practice – although this was not very varied.

Need for flexibility of timing

However important a lesson plan is for trainee teachers, and
however important lesson planning is for any teacher, the key word
must always be *flexibility*. A successful lesson is not defined as
'finishing the plan'; teachers have to make changes to their plan as
they teach, in response to how the lesson is going and how the
students are reacting to, and dealing with, what you are giving them
to do. Consequently, the timings for each stage are estimated. They
are very approximate guidelines for the teacher of how long each

stage might take. If one stage is down as five minutes, and it takes six, seven, eight ... okay. But if it takes fifteen, then the teacher either has misjudged the amount of time likely to be needed, or has allowed the activity to continue for too long. On a training course, teachers usually have very fixed time slots for their lessons, e.g. 40 minutes; they do not have the luxury of being able to carry on with their plan in another lesson, as working teachers do. So time becomes an issue, and rightly so because it forces the trainee to think about achievement of aims and the allocation of time to different stages in order to achieve the aims.

Look at this example of time allocation for stages in a 40-minute Reading lesson:

		minutes
1	Pre-teach essential vocabulary	10
2	Establish interest in topic	5
3	Reading task 1	5
4	Reading task 2	10
5	Speaking follow-up	10

Stages **1** and **2** prepare the students to complete the Reading tasks. Stages **3** and **4** are the actual reading parts of the lesson. Stage **5** is a Speaking activity related to the topic of the text. Let's imagine the main aim of the lesson is 'for students to have practice in scan reading and intensive reading skills'. As it stands, 15 minutes of the lesson is devoted to Reading, but with the necessary vocabulary input this is effectively 25 minutes. But, what if the vocabulary takes 20 minutes, and establishing interest takes 10? Suddenly, most of the lesson has gone without any Reading. It has become more of a vocabulary lesson. On a training course this is important because assessment is based in large part on achievement of aims. In a 'real' teaching situation it would not matter if the teacher was able to continue the same lesson after the break, for example. But trainee teachers need to show they can achieve aims in a given space of time. In the example above, this means making sure the students get their 15 (or 12, or 18...) minutes of reading.

So, what if the teacher has taken longer than planned to get to the Reading? The students are nearing the end of Reading 1, but there is only ten minutes left. What do you do – continue with Reading 2 and forget the Speaking, or go on to Speaking and leave out

Reading 2? Or do you do both and finish your lesson ten minutes late? Well, if it's a training course you definitely don't do both and go over time because you are expected to finish your lesson more or less on time. And in a 'real' teaching situation you probably wouldn't be able to over-run by so much either.

The answer is, scrap the Speaking and do Reading 2 because if you don't, then half of your reading aims will not have been achieved. The Speaking is a 'bonus', but does not relate to your main aims.

Need for on-the-spot decisions

So, the teacher has to make on-the-spot decisions all the time if, as usually happens, activities take longer, or less time, than anticipated. Sometimes you will need to 'stretch' an activity because something before took less time than expected. Sometimes, you will need to add something to your lesson for the same reason. But whatever you do, the main requirement for you on a training course is to achieve your aims in the time given. I recently observed a lesson where the teacher spent 20 minutes 'establishing interest' but only two minutes on the main focus of the lesson – the aims were not achieved. This is why trainees need to estimate the timings of each stage, and know what part(s) of the lesson need the main focus. Experienced teachers can do this instinctively, but if you are on a training course, or if you are a very new teacher, you must estimate times according to your aims.

It is very important in any lesson to respond to what is happening in your class. If you are teaching new language and some students are finding it very difficult, you should not move on just because you have more to 'get through'! You must spend more time on this piece of language and clarify it further. You must know when to 'slow down' or 'speed up', when to quieten things down or liven them up, when to stop students working on a task even if they haven't all finished, when to abandon an activity because it is far too difficult. These are very difficult decisions for new teachers who feel safer following a plan. But as your experience and confidence grow, you will make all these on-the-spot decisions instinctively. You will 'read' your students' faces and know how to change your lesson as you teach it. Sometimes, the time of day, or week will affect how a lesson goes. Or the weather. Or what the students did the previous night. Or something in the news. So many factors can affect the mood and performance of a class, and the teacher has to know how to respond to these factors. If you don't, you are not responding to your students and their needs. All of this relates to being flexible with your plan!

✳ *Tips*

NB: These tips are mostly for trainee teachers doing courses where lesson plans have to be submitted

1 Don't leave the writing up of the plan to the last minute – this will result in greater stress and less confidence. The earlier you complete it, the better for you!

2 When writing it, make sure it is neat, organized and easy to follow.

3 Always think of your lesson and your aims from the point of view of your students.

4 Plan the lesson not just according to the level of the class, but also the composition of the class.

5 When teaching language, analyse that language beforehand and consider the most appropriate way to teach it.

6 Anticipate any problems students might have with the language you are teaching and with any other aspect of the lesson – and think of possible solutions.

7 Be flexible with your plan according to the reactions of your students.

8 Highlight stages that could be shortened if necessary, or dropped, and stages that must have a significant focus in relation to the achievement of stated aims.

9 In relation to point **8**, be flexible with your estimated timings.

10 After the lesson, ask yourself to what extent you achieved your aims; what worked and what didn't work (and why). In other words, review your plan and the lesson in order to continue developing your skills and awareness as a teacher.

We will look now at another authentic lesson plan written by a trainee teacher.

Lesson plan contents (2)

This was a 60-minute lesson given to intermediate students.

Aim(s):	for students to learn and practise modal verbs of deduction for students to have practice in gist and intensive reading skills
Target Language:	3rd person singular affirmative and negative (*must, could, might, may*)
	e.g. *He must be rich / He can't be rich* for making deductions
Anticipated Problems:	students will be aware of *must* for obligation and might find this use of the word confusing
	opposite of *must* for deduction is *can't*, not *mustn't*
	students might produce the *-ing* form (not focus of my lesson)
	must is pronounced weakly with a silent *t* (*I must go*)
Personal Aims:	not to rush instructions
	clear concept checking
Boardwork:	

Modal verbs of deduction

expressing degree of certainty about the present	expressing degrees of certainty about the past
e.g. 'He must be in love'	e.g. 'He could have been in love'

certainty	possibility	certainty	possibility
must/can't	*might/may/could*	*must have*	*might*
		can't have	*may have*
			could

(NB: boardwork fails to indicate what kind of verb follows the modal in each case)

Stage:	**context**
Aim:	to generate student interest in topic
Procedure:	introduce topic of family relationships, elicit types of relationship; individual students read and think about questions (1min), then discuss in pairs and report back three pieces of information
Interaction:	teacher to class / individual / students to class
Time:	5 mins
Stage:	**reading**
Aim:	for students to practise gist and intensive reading skills and locate target language
Procedure:	tell students topic of text / give out worksheet / 30 secs. for gist task / 5 mins for intensive task / students compare answers / teacher checks answers / ask concept questions relating to the modal verbs in the answers
Interaction:	individual, pairs, whole class
Time:	15 mins
Stage:	**board stage**
Aim:	to provide a written record of the target language (as above)
Procedure:	elicit sentences and structures from students / further concept checking as required.
Interaction:	teacher to students
Time:	10 mins
Stage:	**practice**
Aim:	grammar reinforcement / check understanding
Procedure:	worksheet – "What can you deduce from the following situations?" give example / students work in pairs
Interaction:	pairs
Time:	10 mins
Stage:	**follow-up activity (speaking)**
Aim:	authentic practice of modals of deduction
Procedure:	set the scene / students read task and I check they understand / give out clue cards / give instructions and check / students do task and guess the murderer
Interaction:	teacher to class / then groups
Time:	20 mins

What the procedure does not refer to is any reference to pronunciation work, even though the teacher referred to potential problems of pronunciation earlier in the plan. However, the resulting lesson *did* have a pronunciation focus, and it was a very successful lesson!

Review

- Having established aims which are both achievable in the time available and appropriate for the particular group of students being taught, teachers must prepare stages that link naturally and lead towards these aims.
- Teachers need to be flexible with all aspects of their plan and should at all times teach according to the needs and reactions of the students.
- Plans focus the teacher's mind on what needs to be achieved and the best way of achieving it – students benefit from this planned approach.
- Experienced teachers can be more 'spontaneous' in their teaching, but until you reach that stage always plan carefully, not just a particular lesson, but a series of lessons – you must consider the 'timetable fit' and so avoid producing a series of 'one-off' lessons.

19 LESSON PLANNING 3: ANTICIPATING PROBLEMS/LANGUAGE ANALYSIS

In the previous units we have looked at writing lesson plans, and formulating clear and appropriate aims. But teachers must never assume that a lesson will always, or even sometimes, go exactly as planned. They should anticipate problems, and be ready with solutions. This is an essential aspect of planning. This unit will focus on language analysis, and anticipating problems with language; the next will look at more general problems that can occur when we are teaching. Before reading further, then, do the following task.

Language analysis

Task 1

For each of the following sentences, decide what the **form** of the underlined part of the sentence is; and what the **meaning** is in the sentence given. You need to be clear about this before we anticipate problems. Bear in mind that for these sentences there is no context, so more than one meaning might be possible.

Example: I**'ve been** work**ing** here for two years. (intermediate)

Form: subject + *has/have* + verb-*ing* (= present perfect continuous/progressive)

Meaning: something you started doing in the past, are still doing now, and might continue doing in the future.

1 She**'s gone** to France. (pre-intermediate)
2 I**'ll have finished** by lunchtime. (intermediate)
3 It **must have been** a mistake. (intermediate)
4 You **have to pay** taxes. (elementary)
5 Can you **put me up** for the night? (pre-intermediate)

  6 I**'m** meet**ing** Zsigi for lunch tomorrow. (elementary)

  7 Could you help me with my bags, please? (low elementary)

  8 She **goes** to the gym every day. (low elementary)

  9 This time next week I**'ll be** ly**ing** on the beach. (high intermediate)

10 I **used to believe** in Santa Claus. (pre-intermediate)

The **form**, as referred to in Unit 4, is the fixed part of the sentence, the part that does not change. Look at this simple example:

> I'm going to leave
> > work
> > eat
> She's going to drive
> > sing

Here we see that the fixed part is 'subject (*I*, *She*, etc) + *going to* + base verb' If we said I*'m going to leave tomorrow*, then *tomorrow* is not part of the 'grammar' of the sentence.

If we have the verb t*o depend*, the form includes the spelling, the fact that *to depend* is followed by '*on* + object', and that it is a regular verb.

The **meaning** is what this particular bit of language means in the context given. The sentence *You've lost weigh*t could be an expression of concern, or congratulating someone who has been on a diet and successfully lost weight.

With these points in mind, let's go back to the earlier task.

Key to Task 1

  1 This is the present perfect simple. *Sh*e is the subject and is followed by *has* or *have*, and the past participle. The past participle is the third form of the verb – *go, went, **gone**; eat, ate, **eaten**; work, work, **worked**. Work* is an example of a regular verb because its past endings are both *-ed*. The other verbs are irregular verbs because they have different and varying past endings.

     The meaning here is that the person went to France at some unspecified past point and is still there now; for how long, we don't know.

2 This is the future perfect simple. The subject is followed by '*will have* + past participle'. You are saying that whatever it is you are doing, you started at an unspecified past time, have not completed it, but definitely will at some point before lunchtime.

3 '*must have* + past participle' follows the subject. *Must* is a modal auxiliary verb, and *must have* places the time in the past. It is being used for *deduction* (informed guess) about a past event, so you think it was a mistake because there is no other likely explanation.

4 'Subject + *have to* + base verb' is used to state something which is an obligation, usually an external obligation such as a rule or law.

5 'to *put someone up* (*put* + object + up)' for a period of time means to give someone a place to sleep for a short period of time.

6 'Subject + *is/am/are* + verb-*ing*' (= present continuous/progressive) here means that you and Zsigi have made an arrangement for lunch – it is a definite future arrangement.

7 '*Could you* + base verb + object' is used in this example to make a polite request.

8 The present simple in the 3rd person singular (*I go, you go, she goes*) is used here to talk about daily routine.

9 Following the time reference, we have 'subject + *will be* + verb-*ing*'. It means here that at a specified future time (*this time next week*) an action (*lying on the beach*) will be in progress, but will have begun at an unspecified time before, and will continue for a period after this time as well.

10 'Subject + *used to* + base verb' talks about a past habit or state, no longer true.

We have ten pieces of language with likely levels specified and we have established the form and the likely meaning (concept). As new teachers, you may be unsure yourselves of some of this language, but it is important to think of the problems from the students' point of view.

Anticipating problems

 Task 2

For each of the ten examples, consider what specific problems students might have with the **form**, **meaning** and **pronunciation** of the language (NB: aspects of pronunciation were also dealt with in Unit 9).

Example: I've been working here for two years.

- Students might have trouble with the concept of past, present and future time.
- The name of the tense – *present* perfect progressive – may add to this problem.
- Students might confuse this with the present progressive.
- Students might be unfamilar with some past participle verb forms.
- Possible confusion with contracted form (*She's*) … *is* or *has*?
- Confusion with other use of same tense (e.g. *I've been playing tennis* – but not now).
- *been* is pronounced /bɪn/ in natural speech

This is not a complete list of possible problems, but it gives you an idea of the types of problems that students might encounter. In many cases we could add 'first language interference', and this certainly applies to this example. There is also a different usage of the present perfect in British and American English, though this applies more to the present perfect simple. So, go ahead now and think of problems for the ten examples. Refer to a grammar book for help!

Before looking at this task, you may wonder if it is really so important to analyse language in this way before teaching it. It is. Ninety-nine per cent of all trainee teachers that I meet tell me that they are worried about 'grammar'. That is because so few people, in my experience, studied English grammar at school, and this results in the new teacher having to teach something about which they appear to know very little. This contrasts with the typical mid- to high-level student who appears to know much more than the new teacher. They know what a conditional sentence is. What a past participle is. What a preposition is. But I said that the teacher 'appears' to know very little, and the student 'appears' to know a lot. In one respect it is true – the 'terminology', what things are, lots of rules maybe. However, the nervous new teacher, if her first language is English, has an *instinctive* awareness of what is right and what is wrong; what is natural and what is not; and what is appropriate and what is not. What this teacher does not have – yet – is a ready understanding of 'Why?'. It is this 'Why?' question that teachers fear being asked, as well as the identification of terms.

Of course, many teachers do not have English as their first language. Many trainees that I have worked with have come from many different countries. What these teachers have, usually, is an excellent knowledge and thus greater confidence. On the other hand, their choice of language is occasionally 'artificial' – for example, they might use idiomatic language in the wrong context.

For these reasons, it **is** essential for the new teacher to analyse the language she is going to teach. Any teacher needs to know the form of the language, the meaning in the context being used in the lesson, the phonological aspects of the language, and the most appropriate way for students to have practice using it. This will give the teacher confidence, and in turn the students will benefit.

When planning the lesson, the teacher needs to anticipate the problems that her students might have, at their level, with the language being covered in that particular lesson. By doing this, the teacher is better prepared to deal with those problems, and will be much better prepared to cope with all aspects of what is to be taught. And the biggest fear of new teachers will be greatly reduced – "What if someone asks me a question I can't answer?"! Again, the main beneficiaries of the teacher analysing the language and anticipating problems will be the students.

In doing the above task there was no specific group of students, no specific lesson plan. Therefore, the points we make are general ones that would not necessarily all apply to any one class; they are not exhaustive. So let's go through the task now.

Key to Task 2

1 She's gone to France.

- The tense is called *present* perfect, but it refers to past time.
- Confusion with past simple (*She went to France*).
- Students may incorrectly add a time reference ('...*last week*').
- Students may confuse it with other uses of the same tense.
- British and American use of the present perfect is different.
- Students may not know the correct past participle verb forms.
- Students might confuse the contracted form of *has* with *is*.
- In the word *she's* the sound is /iːz/.
- In this sentence we have the preposition *to* and this is pronounced weakly.
- With this particular sentence, students may confuse *gone* with *been*.

The teacher, then, needs to provide a clear context which illustrates the time reference and to check understanding with concept questions / a timeline. Pronunciation needs to be worked on with the teacher providing a clear model and getting students to repeat it. Any task comparing this tense with the past simple would be valuable. Boardwork could provide a clear record of form.

These points apply to all these examples (but timelines are appropriate only for numbers **1, 2, 6, 9, 10**).

2 I'll have finished by lunchtime.

- It's a complex sentence with a double contraction likely in the spoken form.
- The pronunciation of *have* will be weak.
- Students are likely to confuse this with other future forms.
- The concept of the sentence is difficult (something which will already be finished at a given time in the future).
- Students may think of this sentence as only ever applying to future time, when it may refer to something already begun.
- The preposition *by* may be confused with *at*.
- Students may have problems with past participle forms.
- With *she* and *he* students may say *has* rather than *have*.

3 **It must have been a mistake.**

- Students might find the concept of 'deduction' difficult.

- Students might confuse this use of *must* with its meaning of obligation (e.g. *You must do your homework*).

- Students are likely to have problems with the form (modal + *have* + past participle).

- Students will probably think this is present perfect because of the '*have* + past participle', but it's not! If it was present perfect, for example, we would say *he has*, but here it is *he must **have** gone*.

- Students may be unfamiliar with some past participle forms.

- Students may be unclear about the time being referred to in this sentence.

- The degree of probability (compare with *it might have been*) may be unclear.

- Students will think the negative of this is *mustn't* (it's actually *can't*, as in *it can't have been a mistake*).

- Pronunciation of *must have* with the weak form of *have* in the contraction. With the negative, if being taught, there is the natural double contraction (*can't've*).

4 **You have to pay taxes.**

- Students may want to use *must* and be unclear of the fine difference between *must* and *have to*.

- The concept of 'external obligation' may be difficult.

- Students might confuse this use of *have* with the possessive use (*I have a pen*).

- The time reference may not be understood (general time in this case, but it isn't always – it can refer to future time, as in *I have to be at work early tomorrow*).

- The pronunciation of *have* is with an /f/ sound, not /v/, whereas the possessive use of *have* the sound is /v/ not /f/! There is the weak form of *to* with the schwa.

- Students may use incorrect form of *have* (for example, they might say *She have to…*).

5 **Can you put me up for the night?**

- Students may try to work out the meaning literally – not possible with phrasal verbs.
- The time reference may not be clear.
- The idea of providing someone with temporary accommodation could be difficult.
- The form is difficult. In this case the two parts of the verb are separated and the object goes in the middle. However, you can also say *Can you put up Marek for the night?*, but you cannot say *Can you put up him for the night?*. The form of phrasal verbs is complex!
- Students can be overwhelmed by the number and different types of phrasal verbs, and in some cases by different meanings of the same verb!
- In this sentence, the weak form of *can* would be pronounced (kən).

6 **I'm meeting Zsigi for lunch tomorrow.**

- The name of the tense (*present* progressive) conflicts with the time reference (future).
- Students may want to use the present simple (which would be incorrect here, but correct if you meet Zsigi for lunch every, say, Tuesday).
- Students might want to use *will*, which they think of as *the* way to talk about the future.
- The present progressive has various uses which the students might confuse. The first they are likely to have learned is for an action in progress at the time of speaking (e.g. *I'm watching TV right now*).
- The idea of a definite future arrangement, compared with intention, or other possible future function, is likely to confuse students.
- Spelling forms of the verb could be a problem (e.g. *meet* + *ing*, but **not** '*shine* + *ing*').
- Some verbs rarely if ever take the -*ing* form (e.g. *know*, *hear*, *like*).

- If students don't use the contracted form *I'm* and say *I am*, it changes the meaning somewhat and becomes more of an emphatic statement.
- If saying *he's* students might think they hear *his*.
- Similarly, *you're* may be heard as *your*, and *we're* as *were*.
- With *I'm meeting* the /m/ of *I'm* effectively disappears when spoken naturally.
- Students might be confused by the alternative negative forms (e.g. *you aren't* / *you're not*).
- Some languages have a single tense covering uses of both simple and progressive in English (first language interference might be an issue with most of these language points).

There are various other possible problems with this sentence/tense, but the above points highlight the complexity of it!

7 **Could you help me with my bags, please?**

- Students will ask about the alternative modal *can*.
- Idea of a polite request contrasting with other types of requests may be difficult.
- Weak form of *you* in pronunciation.
- Appropriate intonation important to convey politeness.
- Some students might use *to*-infinitive (*Could you to help me...?*)
- Students might confuse this use of *could* with others.

8 **She goes to the gym every day.**

- Concept of habitual action / routine might be difficult.
- Possible confusion with other uses of present simple that students might have been exposed to.
- Students may try to use present progressive instead, thinking of 'continuity' matching the idea of repeated action.
- Pronunciation of *goes* /z/ – the *s* can be pronounced /s/, /z/ or /ɪz/.
- Students may forget to add *-s* for 3rd person singular (*I go, You go, She goes*).

- Students may incorrectly add *-s* to plural form (*They comes*).
- Spelling problems (*play* + *-s* / *go* + *-es* etc).
- Students may try to add an auxiliary (*She does go*).
- Some languages have one tense covering the English simple and progressive tenses.

9 This time next week I'll be lying on the beach.

- The form is long and difficult.
- The concept is a very hard one for students (action in progress at a specified future time).
- Linked to the last point, the fact that the action will have started some time before and may continue beyond the specified time, is a difficult one to understand.
- Students will confuse this with other future forms, especially present and future progressive tenses.
- As mentioned in an earlier example, some verbs cannot usually be put into the progressive form.
- The contracted *'ll* is difficult for students to hear, and the contraction is difficult to produce.
- Students may omit the contraction and say *I will* which makes the meaning more emphatic and different from the usual contracted form.
- Spelling patterns, as mentioned before, will cause problems.

10 I used to believe in Santa Claus.

- Students might confuse this with the past simple (*I believed in Santa Claus*).
- The concept of past habit (*I used to smoke 20 cigarettes a day*) might be confused with past state (*I used to live in San Francisco*).
- Either or both of these concepts may be difficult for students to understand.
- Students may be further confused with other forms of *used* (e.g. *I'm used to* + *-ing* / *I'm getting used to* + *-ing*).
- The pronunciation is difficult – the *d* is pronounced /t/, and the *to* is weak and linked to the /t/.

- Students may pronounce the *s* as /z/ as in the verb (*I use a computer*).
- Students might pronounce *used* with two syllables, when there is only one.
- If the negative form is being taught, students may well incorrectly add *-d*. The correct form is *I didn't use to smoke* and **not** *I didn't used to smoke*.

Review

- Analyse the language point you are going to teach. Ask yourself: What is the **meaning** in the context of the lesson (supplied by a text, cassette, or some other means)?
  What is the **form** of the language?
  What is the correct and natural **pronunciation**?
- Anticipate the **problems** that the students might have with this language. Which problems and how many will depend on the level and nature of your class, and on whether it is a monolingual or multilingual class.
- If you don't analyse the language and anticipate problems, you **will** be 'caught out'! And you will deserve to be!
- The longer you teach, the more you will instinctively know, and the less unsure you will be. Until you reach that point, deal with each language point as it comes up (as you have to teach it) and your language awareness will steadily increase.
- Don't be afraid to ask more experienced colleagues for help as well as consulting reference books.

20 | LESSON PLANNING 4: WHAT WOULD YOU DO IF...?

So, you know your aims; you have planned your lesson; you have analysed any language being taught and anticipated problems students might have with it. However, there is still one more aspect involved when planning – you need to anticipate what might go wrong in your lesson, or any other lesson, and be prepared so you know what you would do. It is sometimes these unexpected problems that panic new or training teachers, especially the latter because they are being observed and assessed! If, however, you can think in advance of virtually any unexpected development in your lesson then nothing can go wrong...well, almost nothing.

Here, then, are twenty such situations to consider...and they have all happened to me at some time!

Unexpected problems

 Task

What would you do if...

1 you planned for an activity where students would work in pairs, but one student was absent and you had an odd number of students in your class?
2 a student arrived very late as you were teaching and had missed an important part of the lesson?
3 you had a student who said your lesson was "boring"?
4 you had a student who said your lesson was "too easy"?
5 you were doing a Listening activity and the machine chewed up the tape at the start?
6 you had a student who dominated the class and would never let anyone else speak?
7 you had left your lesson plan / materials / book at home?

8 you were asked a grammar question you could not answer?

9 you were teaching a multilingual class which included two students of the same nationality who continually spoke to each other in their language?

10 your class was very big (at least twenty students)?

11 a student(s) always finished tasks well before anyone else?

12 the students found what you are doing very difficult?

13 you had a truly mixed-ability class?

14 your planned lesson finished very early and you had to 'fill time'?

15 the lesson time was up but you still had lots of your planned lesson to do?

16 you had two students in the class who clearly did not like each other?

17 you gave out a task and a student said "I did this with another teacher"?

18 a student constantly used their dictionary in class?

19 a student's mobile phone went off in class and they took the call?

20 you explained a rule but a doubting student said "But my other teacher said …"?

There does come a point in your teaching life when you have encountered just about every possible situation (I say 'just about' because you can never be sure!), have experienced every little or not so little problem, and feel completely confident as you walk in to teach. It is good to experience situations like the ones above, because only then will you learn what to do, and your confidence will grow so that you do not fear the same things happening again. Well, not as much!

Let's look at these twenty situations and consider some possible solutions – though the precise solutions will depend upon the context and exact circumstances.

Key to Task

1 Pair work with an odd number of students

I have seen this situation panic trainee teachers far more than it should, given the relative smallness of the problem. You should not become the odd student's partner because if you do, you effectively become a student in the class and your role as a teacher is lost. You need to monitor the class, help the students as required, generally keep an eye on things – and you can't do that if you are participating in the task. So, depending on the task, two students could act as one. In your instructions you might say "okay, you're A, you're B, you're A, you're B…Javier and Ella, you're A together, and Bruno is B".

Another approach definitely to avoid, and I have seen it happen more than once, is to say to the odd student "Okay, you just watch"! This is leaving the student out of the task completely and is the worst thing you could do.

2 Late arrivals

You should acknowledge this student's arrival ("Hello"), and gesture for them to sit down quickly and quietly. At this stage, do not stop your teaching to tell the latecomer what they have missed – this would annoy those students who were there on time, and effectively send the message that it is okay to arrive late. Carry on, but when the rest of the class begin a task individually or together, go to the latecomer and fill them in on what you have done so far.

If a student is persistently late, have a quiet word to find out why. It could be a case of transport problems, but make it clear that they are losing out by missing part of your lessons on a regular basis. Above all, don't let latecomers disrupt your lesson!

3 'Your lesson is boring'

It will happen! If you have been monitoring your students and 'reading their faces' then you may have detected this student's lack of interest. It is usually easy to notice when a student appears bored, then you can do something about it before the dreaded complaint. You also need to think why this student might be bored. In all likelihood, it is because they think the lesson is too easy (see **4**), or maybe they have covered this material before (see **17**). In each of these situations, you can deal with the problem. But maybe they really are bored – perhaps they are reluctant students, sent by their parents. Maybe they have a low interest threshold. Or maybe, just maybe, your lesson *was* boring! Did you establish interest? Did you adapt the material to suit the needs and interests of your class?

Were you displaying some enthusiasm and energy when teaching? Or were you just following the book and not making the lesson come alive?

Whatever the reason for the student's comment, you still need to react properly to it. Don't get angry or defensive or make an issue of the comment in front of the other students! Have a word privately with the student after class and establish the reason for the comment, then try and respond so this student will not feel bored again. However, it needs to be said that sometimes you get a student who will be bored regardless of what you do because they just don't want to be there! Try to include in your lesson topics that you know this student is interested in, maybe involve them actively in some way. Keep them busy!

4 'Your lesson is too easy'

Any of the points made in **3** could apply here. But in addition, it could be that the lessons *are* too easy. This could be because you are not pitching your lessons to the general and intended level of the class, in which case other students would probably be voicing the same complaint. Or it could be that this student should be in another class. Discuss this student with your Director of Studies and maybe it will be possible to move her to a different class. If it isn't, you will need to give this student extra and more challenging tasks to do.

5 Chewed up tape

This is one of any number of things that can go wrong when you are doing a Listening activity and such problems were discussed more fully in Unit 12. If the worst happens and you cannot rescue the cassette, and you do not have the tapescript in the book for students to refer to (thus making it a Reading stage!), then you will have to lose that part of your lesson and expand on whatever else you have planned. Most potential problems in a Listening lesson can be avoided, but if the machine decides to eat the tape, there is nothing you can do! And it will happen to you at some point, believe me!

6 Dominant students

A student may dominate the class for different reasons. They might be of a good level and know a lot and are keen to show you what they know. Or they might be naturally talkative and outgoing. And it is good to have a student in the class who participates well because a class that never speaks is a nightmare! However, you should not let a student 'take over' so that even when you ask another student to speak, the dominating student answers for them! Sometimes, such students like to play the role of the teacher and start correcting other students when you don't want them to! You must, therefore, curb the enthusiasm of this student in a way which is subtle and which does not result in him or her withdrawing from that point on. Using students' names can help (ask an open question and the dominating student (d.s.) will always answer; ask another student by name to speak and the d.s. may keep quiet!). A polite gesture to the d.s., to allow someone else to speak, can work. Or, saying something like "Okay, thank you, but someone else". So you are not telling the d.s. to be quiet, but encouraging others to participate more.

7 Missing materials

A nightmare scenario! But you still have to go and teach the class! There is no simple solution to this, but the longer you teach, the more materials and 'standby lessons' you will build up, and you can fall back on these in an emergency. You will also become more familiar with resource books and know where you can find a lesson at short notice. Failing that, and if the context allows it, give your class a challenging Speaking lesson to keep them busy while you think of what to do later. If the class is very low level, give them a task which revises what they have done before (e.g. write the days of the week) which again will give you a bit of time to think. Ask a colleague, or your Director of Studies for help. Best of all, check before you leave home that you have what you need!

8 Difficult questions

The moment dreaded by all new or training teachers! The last unit gave the best advice – if you have analysed the language, and anticipated problems students might have with the language, then you are much less likely to be asked a question you cannot answer.

But if you still get asked a tricky question, there are various possible reactions. If you know the answer but can't think of it right then, throw the question open to the class to see if someone else can answer. Or even throw it back to the student herself – "Well, what do **you** think?!" While the class/student is trying to answer, you get a valuable minute to collect your thoughts. If this is not going to be enough time, you could set it as a homework task and then make sure you go back to it the next day. If the question is about a point irrelevant to that lesson, say you'll come back to it later but you need to move on. Or, just be honest and say you're not sure but will check. However, don't do this too often or the students will lose confidence in you!

9 Students chatting

Teachers have different views on how best to deal with this situation. Some will make a point at the start of the class to make sure – if possible – that no-one is sitting next to another student who speaks the same language (e.g. place names on seats!). Others will separate students during a class (but do it sensitively – move several students so it is not obvious that you are targeting particular people). One way to do this, if the class is not too big and it is easy for people to move around, is to say to students in turn "you're **1**, you're **2**, you're **3**, you're **1**, you're **2**, you're **3**…okay, all the **1**'s sit here, all the **2**'s sit here" etc). But if students are sitting in rows behind desks, you can't do that!

Other teachers will let students sit where they want, but strongly remind and encourage them to speak only English. Don't get too strict, make a joke out of it ("That doesn't sound like English!"). Separate them as a last resort. But don't forget, as has been said elsewhere in this book, if your class is multilingual you really must try and keep it to English only or else many students will get annoyed. If it is monolingual, all you can do is to minimize the use of first language to an acceptable level.

10 Large classes

It's all very well planning pair and group work, and wanting to monitor students' work etc when you have a class of ten students and plenty of space. But if you have 20, 30 or 40 students, sitting in rows behind fixed desks, then what? Clearly you will need to re-think and adjust your expectations, particularly with regards to class management. You will find it difficult to see, or be seen by, students at the back, to move students, to monitor and evaluate progress, and maybe to keep the class in order. There, are, however, some tips to follow. You can still do pair work, it's just that each student will be restricted to working with either of the students they are sitting next to. Maybe, however, they can turn around and pair up with a student in the row behind. If this is possible, then so is group work. A resulting problem is how to get the students' attention back when you want them to stop. Some teachers might clap their hands, or raise one hand, or have some other agreed signal so the students know when they should stop working.

Certain students can help you (e.g. handing out worksheets, acting as a spokesperson for a group etc).

You need to project your voice well, and make sure that anything you are showing to the class (including boardwork) is big enough to be seen by everyone.

Finally, give clear ground-rules about what is and is not permitted; you have to stamp your authority on big classes so that the class knows what to do, when to do it, and how to do it.

11 Quick students

When students are working individually, in pairs, or in groups, you should monitor their work. One of the things you should look out for is a student who finishes a task very quickly, too quickly for you to be prepared to stop the others. There are various options to deal with this. You could ask this student to help another. You could extend the task ("Think of five more examples"). You could have another task ready for quick finishers. You could, if appropriate to the task, ask the student to review what they have done and look for mistakes. But whatever you do, don't leave this student idle because then she will get bored and frustrated. Keep your students busy and challenged.

12 Materials too difficult

Maybe the student is in the wrong class and they would benefit from a lower-level class, if one is available. Talk to your Director of Studies and see what the possibilities are. But don't rush to quick judgements if that student has only just joined your class. Students need time to settle into a class and find their level within it. They often start by thinking the class is too difficult, but are happy enough a few days later.

It could be that the class as a whole is finding the lessons too difficult, so maybe you are pitching the lessons at the wrong level. Sometimes teachers need to lower or raise the level of the lessons to adjust to the students in the class. As you will see in Unit 27, terms such as 'elementary' are very broad and some elementary classes are stronger than others. The teacher has to respond and adapt her teaching accordingly.

13 Mixed ability

To some extent this point is discussed elsewhere in the book. In Unit 27, for example, I suggest that you should avoid ending up teaching groups separately within a single class – it is difficult, and can lead to frustration for all concerned. However, sometimes you might feel this is the only option if you have a student of a very different level, and there is no other class for him to go to. In this case, prepare separate material, tasks for this student to work on, and spend time going over the tasks with that particular student. But I am talking about a student who is genuinely at a different level, not just a little weaker or stronger than others in the class, because this happens all the time.

Another alternative – again time-consuming for the teacher when planning – is to have the same material for the whole class, but different tasks. This is another last resort, in my opinion, simply because of the extra workload for the teacher. You might have the option of letting this student use the school's self-access materials, if they exist. For example, language laboratory work, or reading/listening to graded readers (story books graded for particular levels).

Sometimes a student will gradually adjust, maybe with the help of another student who can work with them. So pairing the student off with a sensitive stronger student might help as well.

Whatever you do don't just let this student struggle alone.

14 Filling time

However well you know your class, and however well you have
planned your lesson, you can never be sure that your lesson won't
finish early. You will have to fill that time because you cannot just
finish early – your students have paid for a certain amount of
teaching time and they must get it. My own favourite experience
was when, despite having had over ten years' teaching experience,
I found myself in a 60-minute lesson having 30 minutes to fill,
instead of the 5 or 10 minutes you might expect! I had completely
misjudged my students' capabilities with the lesson I had planned,
which just goes to show that however experienced you might be,
you can still get it very wrong sometimes!

Ideally, you should always have something 'up your sleeve' – an
extension to what you have already done, or something quite
different. This could be a quick language game, a discussion topic,
a review of earlier work, or maybe a quick information exchange
("Find out what everyone had for breakfast today"!). Whatever it
is, it should be something that does not require any preparation
time because you are unlikely to do it anyway! But always be
prepared ... just in case.

15 Not enough time

This is the opposite problem, but actually not really a problem at
all. For a multitude of possible reasons, there are times when your
lesson time is up and you have covered barely half of what you
planned to do. This may actually reflect well on your teaching,
rather than badly, if you have responded to your students in the
lesson and realized that they were having more difficulty than you
had anticipated and 'slowed down' as a result. Much better to do
that than finish what you had planned, regardless of your students'
needs. Of course, it may be that you have misjudged the level, but
as long as you are aware of this and compensate for it in future
lessons, no harm is done. Normally, you would be teaching the
class again and can continue the 'unfinished lesson' – this has the
added bonus of less preparation being needed for the next lesson.

16 Student antipathy

Sometimes a teacher can remain unaware of two students not liking each other, or one not liking the other, because they remain outwardly polite and friendly, and there is no problem. But there are tell-tale signs to watch out for in case the problem is more serious. The most obvious is when particular students do not work together when in pairs, when you have seen them working happily with others in the class. Or when a student goes out of her way to avoid sitting next to a particular student. Sometimes it is more open and obvious; sometimes – as sensitive as you might be to such situations – you have no idea until it is too late. This happened to me when I was still a fairly new teacher. I taught a class for three months and remained quite unaware that two young men in the class hated each other – until, that is, they started fighting after a class one day!

All you can do is to keep your eyes and ears open – back again to 'reading faces' – and if you sense a problem, do some tactful class management to keep the pair apart. Maybe talk to them individually if you feel this might help. However, this is a rare problem.

17 'We've already done this'

I hate this! But when students are being taught by different teachers, and there are certain materials in the school, available to any teacher, it is bound to happen to you. And at times, **you** will be that 'other teacher'! Usually, it is impossible to change your lesson there and then – and if this only concerns one or two students, you shouldn't anyway. You could give them different tasks to do, get them to work with students who haven't done it before, or – if possible – give them something quite different to do. If, however, it's a language point ("Oh, I've done the 2nd conditional") be honest with them and tell them that they still won't remember or know everything about it, or tell them that you will be using different material from the other teacher, and that it is important for them to review what they have done before and not think that they have ever 'done' a piece of language.

A more difficult case is when you are doing a particular skills lesson and the whole class did the same text or recording with another teacher. If you can think on your feet and come up with a different lesson on the spot, great. If not, you'll just have to repeat the lesson but make it as individual in style and approach as possible. I'm afraid this will happen to you on many an occasion! Sometimes it is good to ask other teachers beforehand if they have used a particular lesson recently with that class.

18 Over-use of dictionaries

How much you allow your students to use dictionaries, and what kind of dictionaries, is ultimately your decision. My own view is that they are a valuable learning tool as long as students know how to use them properly to get the most from them. However, the dictionary should not become a substitute for the teacher. For example, you write up a new word and immediately a student goes for her dictionary rather than listening to you. You have lost the attention of that student, she is not listening to you, and the dictionary has become a crutch for the student. Or a student checks everything you say, not quite believing you until the dictionary confirms what you have said! In these situations you need to be quite firm – and even more so with noisy electronic dictionaries, especially those that have 'voices'! Tell the students when they can and cannot use their dictionaries, and teach them how to use them. Encourage the use of English/English dictionaries, especially those produced specially for language learners. But when the situation demands it, say "No dictionaries".

19 Mobile phones in class

This is a particular pet hate of mine! It is just polite in any public situation to turn off the phone – whether it's in a concert hall, theatre, or classroom. But when a student not only fails to turn off the machine, but actually takes the call while your class is in progress, well I am afraid this is pure rudeness! Gesture to that student to leave the classroom to continue the conversation, but also have a sign to tell the class that all mobile phones should be switched off before the start of the lesson.

20 'But my other teacher said ...'

More often than not the student has misheard or misunderstood what that teacher actually said. It happens a lot, and when you check with the teacher concerned, you nearly always find that the teacher said nothing of the sort! Students worry when teachers appear to have contradicted each other, so always check with the other teacher so you can resolve the issue with the student. Usually this just means clarifying to them what the other teacher really did say! There will be times when **you** will be that 'other teacher'!

Review

In the last four units we have seen that a great deal is involved in planning a lesson. In this unit we have looked at aspects of lessons which can cause problems at any time, mostly to do with the smooth running of classes. Most, if not all, of the problems mentioned in this unit will be experienced by you at some time in your career; and as you deal with each one in the best way according to your circumstances, so you will 'grow' as a teacher. The last four units should have left you with one obvious conclusion...a successful lesson depends on proper planning. Be prepared!

21 | USING COURSE BOOKS AND OTHER MATERIALS

When I trained to be a teacher, I remember having to 'create' all my lesson materials (admittedly for only a total of six hours' teaching in four weeks!) because we were not using a course book. But no teacher could teach on a daily basis without access to resources, either a course book or 'in-house' materials. Yet, there seems to remain some resistance to the whole idea of using a book – usually, I find, among the least expected group – training teachers. This is understandable because we probably all had some experience at school of a teacher just setting work from the book while she proceeded to mark homework, or whatever. And this is the point. Using a course book is only the right option if it is used properly. In other words, teachers need to know *how* to teach from a book – and this is not as easy as it might sound. But if it is done properly, it is an enormous time-saver for the teacher and it benefits the students in a number of ways.

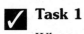 **Task 1**

When using a course book teachers need to **Use**, **Reject**, **Adapt**, and **Supplement**. Later in this unit we will examine what is meant by this, but for now spend a few minutes trying to understand what these words mean, and consider possible examples.

Why use a course book?

As suggested above there are advantages for the teacher and the students. From the teacher's point of view the book provides material, a syllabus structure, and consequently much less time is needed to prepare lessons. Personally, I don't see how any full-time teacher can function without a book! From the student's point of view there is a visual record of progress, easier access to material, the possibility to review work at home, and a feel of greater professionalism (compare this with lots of photocopied worksheets – environmentally unfriendly as well!). In theory, then, using a course book would seem to be highly desirable. There is certainly no shortage of books on the market, but there are numerous potential problems with their use, as we shall see.

Essentials of a 'good' course book

It is very hard to find consensus on what is a 'good' book – what one teacher thinks is good, another thinks is bad. The truth is that most books are usable. How 'good' one is depends on how the teacher uses it. However, there are some basic requirements:

It should have been written by experienced teachers and be roughly appropriate for the level intended. You can assume both. In addition, though, the book should ideally be appropriate for the country you are teaching in, and this can be a problem. Unfortunately, a great many books are British in content and culture, and if not solely British then European. It's all very well reading an article about British food when you are in London, not when you are in Tokyo or Jakarta or NewYork. Photographs of 'famous people' are rarely recognized by students in Britain, let alone anywhere else. However, because of the increasing awareness of this problem, more and more 'non-British' books are appearing. I was reminded of this problem just a week before writing this when a fellow teacher at a school in London, an Australian, remarked "It's so much easier using (name of book) in London than Sydney…at least here it means something".

Your book, too, should be appropriate not just for the level you are teaching, but for the students you are teaching. We will develop this point later, but enough now to say that just because something is in the book doesn't mean you should use it with your students.

Ideally, the book should be visually attractive and well laid out – clear, and easy to follow. Preferably, the pages won't be too cluttered. It's nice to have a clear Contents page, and the language covered should have a logical progression (starting the book at one level, but finishing it at a higher level).

As for topics, I have to say that most experienced teachers, and a fair number of students, get a bit tired of talking about 'weather', 'food', 'holidays', 'countries' and various other standard topics that appear in most books. However, in defence of these topics, I should say that for lower levels they are 'easy' and 'safe'. Higher levels can deal with more advanced topics such as 'technology' because they have more language to talk about them. What I find difficult to accept, speaking very personally, are topics which include gruesome pictures of suffering people, or contentious issues like war and religion. Teachers, too, sometimes wonder why we are teaching students the vocabulary of, say, crime. Just how useful and relevant is this? I would argue that weather, food, holidays etc *are* in fact, more useful topics, but that much higher levels can, and should, be challenged with more thought-provoking issues.

An ideal course book should have variety and balance – language work, skills work, pronunciation, review units or sections and grammar summaries. In other words, it should be a complete package for the student.

Using a course book

Use, reject, adapt, supplement

The teacher's dream scenario: You look at the next unit in the book. The topic is interesting, the material is good and appropriate, the 'staging' is logical, the tasks are just right – in short, it's perfect. So, **use** it as it is. This doesn't happen often, but when it does, make the most of it! There is no need to re-invent the wheel, and new teachers, especially, sometimes feel that they have to create their own materials and do lots of cutting up and pasting. If it's not necessary – don't do it.

The teacher's nightmare scenario: You look at the next unit in the book and the topic is quite inappropriate for your students. Or the material is very dry or difficult or easy or unclear. So, **reject** it. However, having said that, if you are working in a school where you have to cover everything in the book, you will not be allowed to reject material (i.e. not use it). But if you do have flexibility, don't be afraid to say "I can't use this with my students". This is a nightmare because you then have to find alternative material to cover the language you need to focus on, and that takes time. Some teachers are more 'sensitive' than others, for better or for worse. I tend to **reject** quite often, but some material which I have found 'untouchable' another teacher has had no problems with, so maybe I am too topic-sensitive!

To be honest, **use** and **reject** are the least common courses of action when teaching from a book. Far more likely – indeed, most of the time – you will have material which is okay to an extent, but it needs to be changed a bit. Maybe the Listening task is too simple and you want to extend it. Maybe you don't need to use one part of a page. Maybe you would prefer to use your own practice activity rather than the one in the book. In other words, **adapt** (change certain parts as necessary), and **supplement** (use other material – your own, or from other sources). So you are still using much of what is in the book, but you are not slavishly doing everything simply because it is there! If you simply go through the book from start to finish, setting work from it, you are not demonstrating any teaching skills, you are not responding to your students' individual needs and backgrounds (go back to Units 1 and 2!) and you are not doing your job properly! Even if it's a great book, it still needs exploiting. This takes us back to the comment in the introduction to this unit – teachers, especially new or training ones, sometimes feel it is 'wrong' to use a course book for the simple reason that they had the experience of a teacher at school who simply 'followed the book'. If you do 'follow the book' then your students will indeed be bored and demotivated – and so will you!

There are other considerations, too. A course book will normally come with accompanying cassettes/CDs, a Teacher's Book, and maybe a Workbook with extra practice tasks. The Teacher's Book is invaluable because it provides answers, and sometimes explanations for answers. Some will also have suggested procedure. Either this book or the Student's Book will have the tapescripts. But there is a danger in relying too much on the Teacher's Book. For a start, any book is liable to have the odd misprint or two – so although a book says the answer to a question is 'A', it might be incorrect! And even if the answer is correctly printed, as it is most of the time, that doesn't mean that it's the answer *you* would give! So, check the answers beforehand, and make sure you agree with them and know why the answer is what it is.

Make the course book suitable

The authors of your book have written it knowing the approximate level it is intended for. But only you know your students. You have to make the book suitable for the people you are teaching.

There are many books available these days, but the chances are that you will be assigned a book to use. You may be lucky and be given some choice; you may be very unlucky and not have a book at all. But whatever book you hopefully have, be critical in a positive sense. Don't, as many new or training teachers do, have a quick look at it and say it's no good. I have heard many excellent books dismissed in this way on training courses, partly because of the feeling that 'it's not fun'. Similarly, new teachers are sometimes tempted to describe almost any practice activity as a 'game'. This should most definitely be avoided, assuming you are teaching adults. For many students 'game' = 'not learning'. But even if your students do like the occasional game, (and there's nothing wrong with games in the classroom from time to time) it doesn't mean that you have to keep using the word because you don't like what's in the book! You'll be surprised at what students like. In my experience, the so-called 'conventional grammar exercise' is actually very popular, as long as it is balanced with more communicative work. And if it is done in pairs or groups, or if students have the opportunity to discuss and compare their answers afterwards, so much the better. In other words, don't be too quick to assume that something is 'boring'. You need to set the activity up in an interesting way, prepare the students for it etc, but do not ignore lots of good material in the book because it's not 'fun'! The important thing is for learning to be interesting and challenging.

The other general point I would make is about the level. The levels we use ('elementary', 'intermediate' etc) are very broad and you will always have a mixed-level class to some extent. A common complaint among training teachers is "They're not all the same level". Well, what do you expect? How is every student in the class going to be of the same level in all aspects of the language? Of course, there are occasions when, for various reasons, you will have someone of the wrong level in the class, but that is different from having a range of intermediate students in one class. The book, therefore, is for this very general level. Consequently, you will need to adapt, supplement and reject material in order to make it right for the precise level of your class.

✳ *Tips*

1 Critically evaluate the material in the book in relation to who you are teaching.
2 Reject the unsuitable and use the ideal. Otherwise, adapt and supplement the material.
3 Use the book if there is one (you may have to anyway) – the students expect it. Besides, you'd be crazy not to!
4 Check answers to exercises, and make sure you can justify them if necessary.
5 Make real use of the material, don't just say "Okay, do exercise 3 on page 40". You must use your skills – establish interest, prepare them, clear instuctions, etc.
6 Don't skip about in the book (Unit 3, then 6, then 14, then 2…) – the units are in an order for a reason. You can't just pick the parts *you* like the best!!
7 Make sure that the way you use the material in the book is appropriate in every way to the students in your class.
8 Teach your students how to get the most out of the book in their free time (assuming they have their own copy). For example, books normally have separate grammar summaries, vocabulary summaries, pronunciation guides, and maybe tapescripts.
9 Make it clear to them why you are not using everything in the book (assuming you are allowed not to cover everything), and why you are using other material. They are adults and you can involve them in the teaching rationale. Work not done in class can always be done at home.
10 Make full use of the material – new teachers tend to cover in a week what an experienced teacher would cover in a month!

Using other materials

The point has already been made several times that teachers need to supplement the material in the course book with other material. Fortunately, there is a very good range and variety of resource books on the market – so look at what your school has. You also have the 'home-grown' material for those who are creative and have some time to do it! But it doesn't have to be anything special – maybe just some questions of your own to accompany a Listening task. Or, very commonly, an idea of your own to get the students speaking together.

Also, teachers should make good use of **authentic** material.

What is authentic material?

This is real-life material not specially published or adapted for classroom use. Examples include brochures, posters, newspapers, magazines, leaflets, signs, notices, songs, radio, TV. It is best to use something reasonably up-to-date, at least as far as reading matter is concerned (so, today's or yesterday's newspaper, not last month's). It is good to use authentic material simply because it is *real* and something your students might be exposed to on any day. However, the value of such material is more limited when teaching in a country where English is not the first language because the students *won't* find this material so easily in English. Newspapers, however, are an exception because in most countries you can access English-language newspapers, and students can be encouraged to read them from time to time. In fact, it is important in non English-speaking countries to direct your students to material of all sorts which is available in English (magazines, books, English-speaking radio and TV).

Exploiting it

Units 11–14 cover approaches to skills work so you should read those units for ways to use material for purely skills focus. However, you may want to use the material, at least in part, to focus on language. What we covered in Unit 5 is particularly appropriate for this.

In general, having chosen appropriate material, devise tasks which are graded for the level being taught (and the students in your class), making these as practical, useful and interesting as possible. Decide, too, which aspects of the material you want to focus on. This will be partly determined by the level of the class – what you want an elementary class to do with a newspaper article will be different from what an upper-intermediate class will do with the same article. The level will dictate the nature of the tasks.

Newspaper articles are an especially good source of authentic material. Here are eight ways to use newspapers in the classroom.

1 You can focus on one article/story for Reading, Speaking or Vocabulary.

2 You can focus on one article for specific language.

3 You can compare how different papers cover the same story.

4 You can focus on the use of headlines (e.g. prediction of the story to follow).

5 You can focus on the different parts of a paper.

6 You can use the articles as a springboard for debate with higher-level classes.

7 You can give students a headline and they write an article; then they compare their article with the original.

8 If you have the facilities, you can compare the TV news report of a story with the newspaper one.

Task 2

Think of six different ways to use a holiday brochure in the classroom (say, a brochure advertising vacations in Mexico). I'll give you some suggestions below.

Key to Task 2

Here are just a few possibilities.

As with the newspaper articles, you can use brochures to focus on specific language, or as a Reading comprehension. You could also focus on the specific vocabulary featured in brochures (e.g. *discount*, *package tour* etc), or on the style of language used – for example, what 'not far from the beach', and 'cosy room' really mean!

Speaking possibilities are many: students could choose their favourite trip from a selection of brochures and role-play a conversation asking for more information or booking the trip. You might also get the students to imagine that they *did* go on the trip and to discuss what the reality was. As a Writing activity, students could prepare a brochure advertising their home town or a place that they have visited. Maybe you could use your brochure on, for example, Mexico as an introduction to a text or recording you have on the country, or as a follow-up to the text or recording.

Review

- Course books are there to be used and they should be used critically by the teacher. You should make full use of the material, and follow the guidelines of **Use, Reject, Adapt, Supplement**. You need to make the material suitable for the students in your class.
- But don't be too dismissive of a book or some of its contents. And don't just 'pick out' the parts you like!
- Books save preparation time, provide a structure, give ideas and material, language notes and guidance, a visual record, and a daily reference.
- British-based books are less appropriate in other countries, but seek out any books which might have been published in the country you are teaching in.
- Check the contents of a book, including the Teacher's Book, to make sure you know how something is supposed to work, that you agree with the answers, and can justify the answers.
- Yes, a book may be boring, especially if you just go from one page to the next. So make it interesting and stimulating.
- Authentic material provides exposure to real English and it can be used in many different ways. It also provides variety.
- There are also many other resource books around for alternative ideas. Again, by using other material you add variety to your lessons.

22 | TESTING STUDENTS

Your first thoughts on seeing the title of this unit will probably be 'exams', or 'tests'. These are indeed, part of what is referred to here as 'testing students', but only a part. Teachers test students all the time because testing means to find out what students know and remember. Five examples of classroom testing are:

- concept questions; test if students understand meaning
- eliciting; tests what students know and remember
- language exercises; test what students know
- comprehension tasks; test what students understand
- drilling; tests students' pronunciation

However, beyond normal classroom practice, students may take the following types of tests:

- placement
- diagnostic
- progress
- achievement

We will look at these types of testing and at the difference between subjective and objective assessment.

First, though, a test for you!

Task 1

Complete the following:

1 Tests are motivating for students.

Students _____ (rewrite first sentence)

1 mark

2 The Cambridge FCE, CAE and CPE exams are: (choose one)
   (a) placement tests (b) diagnostic tests   *1 mark*
   (c) progress tests (d) achievement tests
3 We test students informally in class all the time. True or False?
                                                        2 marks
4 The marking system should be _____ (rely)   *1 mark*
5 The reliability of the marking system is affected by the
   _____ given to each section.   *5 marks*
6 When I was at school I _____ tests   *1 mark*
7 be-interesting-do-to-tests-format-and-in-varied-should
   (rewrite using all words)   *8 marks*
8 One of the biggest disadvantages of testing is _____
                                                        10 marks
9 Objective assessment is _____
   Subjective assessment is _____ *1 mark*
10 Which word is spelt correctly? (no dictionaries)
   accomodation – benefitted – neccesary – reciept – businness
                                                        1 mark

Progress tests are what teachers will give their students from time
to time. At some schools they are a requirement; at others, they are
strongly encouraged. Typically, they take place once a week / every
two weeks and their aim is usually seen as testing what students can
remember of what they have recently learned. This undoubtedly is
one aim of such tests, but I also like to present them to my students
as an important opportunity for them to review what they have
done in class, and to clarify what they are still having difficulty
with. Sometimes, I will even let students work together on such
tests, but with one important proviso – they cannot refer back to
notes or books. Why? Because if they can make themselves
remember without such help, they are more likely to remember
more easily in the future.

Such tests do not have to be in test format, like the above example
(which we go back to shortly). They can be done as quizzes, for
example, with students in teams answering questions. You could
also get your students to make up the questions for a quiz, telling
each team which language points they need to think of questions
about. Or, students could prepare posters which display important
points learned in the recent lessons.

In other words, be *varied* in how you conduct progress tests. That said, there is nothing wrong from time to time in giving students an authentic progress test. Many students welcome it, seeing it as a 'proper' part of learning. But you need to give them notice that they are going to be tested soon (tell them when), and what they are going to be tested on. Some course books will have ready-made 'tests' after every few units, or you can prepare your own.

Let's go back to the task above to make some more points about testing.

Key to Task 1

1 The correct answer is *Students are motivated by tests.*

Yes, students **are** usually motivated by tests, but the great danger, of course, is that they might do badly and be discouraged. This is why I personally prefer students to help each other when doing such tests, because for me the aim is for them to review what they have learned, rather than to be examined on what they can remember.

As for this question, you were required to write a passive sentence, and using a different tense from the first. Not an easy task…but only one mark awarded!

2 The correct answer is *achievement.*

An **achievement test** can take two forms. It can be an *internal* test given by you at the end of the course, usually done in exam format. And it can be an *external* exam written and marked by an outside body. The examples given in the question are all UCLES exams – First Certificate in English, Certificate of Advanced English, and Certificate of Proficiency in English. You are unlikely to be asked to teach an exam class until you have a lot of experience, so this unit will not focus on the exams themselves. However, you may need to administer an internal achievement test. If you do find yourself having to teach an exam class, without much experience (it happened to me!), you will work from an exam course book which will tell you most of what you need to know. Beyond this, seek help from your Director of Studies. There is a huge range of public (external) examinations that students can take around the world at all levels. Some are a passport to universities, jobs, promotion etc. The TOEFL exam, now available to be done entirely on computer with an immediate result, is primarily for entry into universities in the United States. It is, as you would expect, all in American English.

Internal achievement tests, like progress tests, are generally welcomed by students because it gives them what they perceive to be a measure of their progress. If I sound less than convinced by my own words it is because I think 'exams' are limited in their ability to prove to students what they know and understand. Students have to know how to do an exam, how to prepare for it, how to tackle it. Exams have an important place in learning, but maybe not *as* important as some students believe. How many times have students and teachers been very surprised at someone's result?! But, of course, it is a wonderful feeling for both parties when a student does very well in an achievement test.

3 *True* – a point discussed at the start. Note that two marks were on offer here for a true/false question, compared to only one for the first question which required grammatical knowledge.

4 *Reliable* (and logical). The marking system for this test is hardly that! If you are setting a test, award marks (if you are going to award any at all – you don't have to) that reflect the difficulty of the question, and the type of answer required.

5 *Weight.* If a test is divided up into different sections (e.g. Reading, Vocabulary, Grammar), does each section carry equal weight? Should it? Do your students know how they are being marked? They should. Five marks on offer for this! (see **4**).

6 This asks for an opinion, it does not test anything and you cannot assess opinions! As far as the question itself is concerned, most students have negative memories of tests – rows and rows of desks, silence, conditions, invigilators, stress. Should it be like this? Maybe exams have to be, but not internal tests.

7 This should read *tests should be varied in format and interesting to do* (and hardly deserves eight marks!).

If you are devising a test, vary the types of question because this will make it a little more interesting. For example, you could include multiple choice, true/false, correction of mistakes, writing sentences, sentence transformation etc. Make the test as varied and unpredictable as you can.

8 Well, there are several! The point is that you cannot mark an answer to such a question when there is no obvious answer. How will you award the ten marks? Only to students whose opinion is the same as yours? This type of question has no place in a test.

9 Objective tests are those which have right and wrong answers, with no ambiguity. Example: The capital city of Scotland is (a) Glasgow (b) Aberdeen (c) Edinburgh. The TOEFL test is a classic example of objectivity as it is entirely multiple choice, as above. Marking is unaffected by the marker, consistent, accuracy based, and easy to assess.

Subjective tests are those which require the marker to assign marks according to a marking scheme. Obvious examples are composition (essay) writing, and exam 'interviews' (Speaking exams). With such testing, it is imperative that students know what the marking criteria are. Marking is affected by the interpretation of the marker and is therefore less consistent and harder to assess.

Many tests are a combination of these two types, and this is probably ideal.

In terms of establishing proper and fair criteria for marking, and conditions for testing, you should consider the following points:

- Does it test what you want it to test?
- Is the test reliable/long enough to allow students to complete the test successfully?
- Is the marking system reliable and fair, and known to the students?
- Is it culturally fair?
- Is the format varied and interesting?
- Is the 'test' going to be carried out in an atmosphere which is not unnecessarily stressful?

Make sure you go over all the answers and reasons with the class.

10 The answer is that *none* of them is spelt correctly! It is a trick question – hardly fair. For the record, the correct spellings are: accommodation – benefited – necessary – receipt – business.

We have looked at **progress** and **achievement** tests so far – now for two other types.

Placement tests are carried out by a school when students first arrive. They are designed to find out the approximate level of the student in order to place her in the right class. These tests tend to involve grammar (often multiple-choice questions), and speaking (a brief chat with a teacher). Sometimes they are more involved, but when a school has thirty students to test and place in two hours, it is not possible to have a full and proper test. Sometimes, as a result, students need to be moved to a more appropriate class after a day or two. Most placement tests are standard and are used each time there are new arrivals.

Diagnostic tests include placement tests, but also include other types. Teachers give diagnostic tests or tasks on a frequent basis when teaching (as in the 'Test Teach Test' method described in Unit 6). In fact, any time the teacher gives the students something to do with the aim of finding out how much they already know, this is diagnostic testing.

 Task 2

Already we have looked at various types of testing that take place in the classroom when we are teaching. For each of the below, think of two examples of tasks we might give our students, and whether the 'assessment' is subjective or objective.

1 Grammar **2** Vocabulary **3** Reading **4** Listening
5 Speaking **6** Writing

Suggested answers will come later!

Every school I have ever worked at has used placement tests and progress tests. The progress tests are usually left to the teacher to administer because the teacher is in the best position to know what kind of test to give, and what needs to be included.

Some schools I have worked at have had an internal achievement test – a standard one based on the syllabus followed for that particular level. Such a test is sometimes also used to allow a student to go up to the next level (I'm not keen on this because in my experience a student is allowed to move up regardless of their result, so what's the point?). It is also common for students to be issued with an internal certificate and teacher's report on their progress when they leave. This rounds off the course of study nicely and does give the student a sense of achievement with or without an actual test.

Continuous assessment is a type of testing not normally found in EFL/ESL except in the form described above, where teachers write a report for the student on the basis of their work and progress on the course, and also when teachers consider whether a student could move up (or down) a level based on their work in class. External assessment, however, is nearly always done on the basis of exams.

Back to the task…

Key to Task 2

1 Grammar

transformation exercises (see Q1 in the 'test' at the start of this unit) – objective

open-ended sentence completion (more than one answer possible) – subjective and objective since one answer might be considered to be 'better' than another.

2 Vocabulary

matching – objective   use of vocabulary in essays
                                   – subjective

3 Reading

true/false – objective   questions requiring interpretation
                                   – subjective

4 Listening

multiple choice – objective   note taking (e.g. main points)
                                   – subjective

5 Speaking

repetition – objective   description of pictures
                                   – subjective

6 Writing

gap-filling – objective   essays (compositions)
                                   – subjective

Review

■ Testing is an integral part of teaching – teachers do it all the time in their lessons. Most tasks we give, most eliciting we do, most questions we ask have the aim of 'testing' our students in some way.

■ Beyond normal classroom practice, students take placement tests and progress tests, and sometimes achievement tests (internal and/or external).

■ Teachers most have to concern themselves with progress tests. They should know how to administer tests which are fair, varied, and interesting and which have clear testing objectives and criteria, and which are familiar to the students.

■ Question types may require either objective or subjective assessment, or a combination of the two.

■ Teachers should vary the types of test they give, and allow interaction between students if they feel this is appropriate.

■ Testing should always be seen as part of the learning process and should be done in as pleasant an atmosphere as is possible in the circumstances.

23 | TEACHING ONE-TO-ONE

This book is almost entirely a general introduction to classroom teaching. This unit, however, does look at one specific class type – teaching one-to-one. The reason for this is that I have yet to meet an experienced teacher who has not done this kind of teaching, but training courses rarely prepare teachers for it. As you might imagine, it requires a very different role for the teacher – and for the student. I have done a great deal of one-to-one teaching and I enjoy it; but I know other teachers who avoid it if at all possible. But it seems to me that more and more students want to learn this way, and more and more schools are offering specialized one-to-one courses, so it is a type of teaching that cannot be ignored here.

There are different one-to-one contexts. For most teachers, it is a way to supplement their salary; their one-to-one classes are out of school hours and privately arranged; or offered directly by the school the teacher is working at.

For others, one-to-one classes are offered by the school in normal school hours and the teacher is required to teach them.

A minority specialize in one-to-one as their main teaching income, usually supplemented by income from elsewhere. Increasingly common are 'Homestay' courses where the student lives in the teacher's home. The teacher not only gives lessons, but also goes out with the student, cooks for him etc. It is 'total immersion' on both sides!

If you are a teacher working in a country which is not your own, it is essential that you find out first if you are permitted to teach one-to-one classes. You should check your contract and, if in doubt, ask your employer, because if you are not supposed to teach one-to-one classes, and you do, you may find yourself on a plane home. Even in your own country, there may be something in your contract to this effect, the reason being that you may take potential business away from the school. Schools sometimes prefer to arrange private classes themselves on behalf of teachers, but the problem with this arrangement is that a substantial part of the student's fees goes to the school and not to the teacher.

For my own part, I have taught one-to-one classes arranged by my school, arranged privately, and for a company which sent teachers out to companies to teach members of its staff. From a financially selfish point of view, I prefer privately-arranged classes because then the whole fee goes to me. It is very frustrating to learn how much schools often charge for this arrangement, and how little of it goes to the teacher.

 Task

There are many issues to consider, especially with privately-arranged one-to-one lessons, and not only regarding the actual teaching. But let's begin with the teaching and learning considerations. Consider the following questions.

1 Why do some students choose to learn this way rather than going into a regular class?
2 What are the learning advantages and disadvantages for the student?
3 How can a teacher prepare for a one-to-one course with a student when that student has approached that teacher and asked for, say, a month's worth of lessons, four hours a week?
4 What are the main differences involved for the teacher when teaching one-to-one rather than a regular class?
5 What do you think are the rewards and headaches of this kind of teaching?

Before looking at these points, it is important to approach a series of one-to-one lessons in a way which is both professional, and which protects your and your student's rights. If you are making a purely private arrangement with a student, follow these points:

✳ *Tips*

1 Agree on a price either for each lesson, or for the series of lessons. Make sure the price is neither significantly higher nor significantly lower than the 'going rate'.

2 If you have to travel, travel costs should be included.

3 Agree on procedure if lessons are cancelled by either side. This might mean that if the student cancels less than 24 hours before the lesson, they still pay you all or most of the fees. Or if you cancel, the lesson is offered at a later date. Whatever is decided, make sure both sides discuss and agree the terms.

4 Agree, also, on when and how you will be paid.

5 If the student later asks if a friend can also participate, it is no longer private tuition, so you cannot charge double the price!

6 Make sure that the lessons will take place in an appropriate setting.

7 If you are being paid for, say, 60 minutes' tuition, that means 60 minutes, not 55. Lessons should be uninterrupted (e.g. don't stop to take a phone call).

8 If you are teaching a student who is also a friend, make it clear when the lesson itself begins and when it ends.

9 Do not arrange very long lessons because one-to-one is tiring for both parties. For most teachers and students, one hour is long enough.

10 Plan as thoroughly as you would for any other class; do not treat one-to-one classes casually.

Back to the task …

1 Reasons for students wanting one-to-one

From my own experience, I would say that there are many reasons for students doing one-to-one classes.

■ Schools usually give total beginners 1-1 teaching when they first arrive until their English is good enough to go into a class.

■ Students preparing for an exam sometimes feel they need more intensive practice.

■ Students sometimes feel that they want to have as much tuition as possible to improve their English quickly.

- Students often need to improve their English for their jobs.
- Students wanting to work in an English-speaking environment will need to bring their level of English up before applying.
- Some students just prefer one-to-one tuition over classes.

Whatever the student's reason for wanting one-to-one, you need to know what it is if you are to provide for that student's needs.

2 Advantages and disadvantages

The advantages are obvious enough – the student does not have to 'share' you with other students. Their learning is more intensive and more geared to their own needs, they can ask as many questions as they want, ask you to teach them skills that they need or want, and have lessons pitched at their precise level and desired pace. They will normally cover much more material and language than they would in a regular class. As such, progress is normally faster, and more visible to both parties.

But the disadvantages are equally obvious. Of course, the student cannot work with other students in the class; so there is no pair or group work, or even just someone to talk to. This restricts teacher/students interaction which makes learning hard work for the student, with considerable concentration and attention required at all times. Some students feel the need for interaction and company, while others much prefer the intensive nature of one-to-one learning. For those students with limited time available, and a definite goal, one-to-one can be advantageous. But for students on holiday, only wanting to improve their skills, one-to-one makes little sense.

3 How can a teacher prepare?

If a student approaches you and asks you to give them private tuition, the temptation can be to think it is a soft option which will bring in more money. But the correct way to react to such a request, if interested, is to discuss with the student exactly why they want one-to-one and what their needs are. If it is a student you already know, it is easier. But if it is not, then there is much you need to find out before commencing. The best way to approach this is via a *needs analysis*. Students pay for one-to-one lessons because they want a teacher who will meet their needs and accommodate their learning style – and they usually want some control over what goes on in the lessons. Consequently, it is crucial to evaluate the needs, level and expectations of that student.

You need to find out three things:

- The student's current *level*. You could use a grammar review test from a course book of the approximate level you think your student is. But you need to supplement this with an informal 'interview' to gauge the spoken ability of that student in terms of both accuracy and fluency. If possible, give them some reading, listening and writing, too.

- The student's *needs*. How much input does she want? Does she want to follow a set syllabus? Which skills are important for her to study? In what context(s) will she be needing to use her English? Is she preparing for an exam? Does she require English for her job? Find out as much as you can about what your student wants from you (this could be done in discussion and the 'interview' could form part of your level analysis). You must also be fair to yourself and the student if they ask you for a specialized course which you are not equipped to teach. For example, if you are asked to teach medical English and you have no knowledge of this, nor access to relevant materials, tell the student that you cannot teach him/her.

- The student's *previous language learning experience*. This will help you teach the student in the way which is best for him/her.

Having done the above, together you can agree on a specific study course with clear aims. Once the course is under way, be very flexible and adjust the content and approach of lessons as you see fit, or as the student requires.

Beware of students who only want "speaking". It sounds an easy option, but it is a vague request and it is hard work for the teacher to maintain nothing but conversation for a whole hour, maybe several times a week!

There is one other aspect which you need to think about before agreeing to teach the student. Do you think the two of you will 'get on'? It is important to have a good rapport with the student you are teaching, and this is another reason for meeting them beforehand.

4 Differences between one-to-one and regular classes

You obviously can't do pair or group work, and doing Reading lessons seems inappropriate because they can do reading at home. The student is paying for tuition, not to be given work to do individually which they could do outside the lesson.

But virtually everything else mentioned in this book applies equally to one-to-one teaching; it's just that it will be very teacher centred.

- You will need to prepare more because material is covered quicker. The material itself can be selected, at least in part, by the student. Have at least 50% more material than you think you need because it is much harder to 'fill time' with one-to-one. Any unused material can be used next time.
- You will need to be very flexible.
- You will have to concentrate hard throughout.
- You should fit in with the student's needs and go at their desired pace.
- You can be less concerned with time because there are no other students to teach.
- It might be appropriate to use more of the student's own language if you know it.
- It is easy to monitor progress.
- It is potentially more rewarding because of the last point.

5 Rewards and headaches

Most of the rewards and headaches have been mentioned. For me, the toughest aspect of this kind of teaching is the demanding nature of it; you get no opportunity to 'relax' (e.g. when you ask students to work in groups in a regular class). The most rewarding aspect for me is building a unique relationship with the student and giving them what they need, and then seeing their progress if all goes well.

I have had many rewarding one-to-one classes, but let me describe one such class. I was teaching in a school where one of my new students was a near beginner called Zsigmond Sipos, from Hungary. Zsigmond could only come to classes twice a week for three hours a time but he wanted to improve his English quickly in order to take the Cambridge First Certificate exam ten months later. He asked me if I would teach him privately for the remainder of the year and prepare him for the exam. To have gone from his then level to the level required to pass the exam in such a short space of time would have been beyond most students. But Zsigmond knew exactly what he wanted and needed, and he played his part fully in influencing the direction and nature of the lessons. He was the ideal one-to-one student because he had very clear goals and a determination to succeed which was admirable. He worked incredibly hard, asked many questions, did lots of homework, and used English outside our lessons whenever possible. The lessons were intense, but at the same time they were very easy for me because the requirements were so clear, and the student so motivated. By December he was ready to take the exam, and he passed it with an 'A' grade. The credit for this lay almost entirely with Zsigmond, not with me, because he approached one-to-one learning in exactly the right way. But when he called me to tell me of his result, I was very pleased and proud. And **that** is the principal reward of one-to-one teaching!

However, as I mentioned before, some teachers do not like the intensity of this type of teaching, preferring the interaction of a regular class. But sometimes you will have to do one-to-one teaching even if you would prefer not to, if only to get some extra money!

Tips

1 Keep a record of all work done; particularly important given the flexible nature of the lessons.
2 Maybe provide the student with the record, or encourage her to keep her own, to make review of work done easier.
3 Keep notes on points that come up as you teach to come back to later. It is too easy to forget otherwise.
4 When doing speaking practice, you could record the conversation and then replay the tape for correction work.
5 Focus more on pronunciation as it is easier to do so one-to-one.

6 Still elicit and encourage learner independence. If you don't, it will be too easy for the student to rely on you entirely and take no responsibility for her own learning.

7 Spend time training the student how best to use reference books, such as dictionaries and grammar books.

8 Review work previously covered. This is always important, but with one-to-one you cover so much more that it is especially important to review progress.

9 If you are using a course book, you will need to adapt a lot of the tasks as they are mostly designed for pair and group work.

10 The student won't mind you finishing late, but you should never finish early.

Review

The most important points to take from this unit refer to 'being professional'.

■ The demand for one-to-one learning is growing, but for the student it is expensive. In return for paying high hourly fees, students rightly expect value for money.

■ Students usually have clear and specific requirements, for which the teacher needs to prepare by doing a needs analysis.

■ The teacher should not commit herself to a course she is not equipped to teach.

■ The teacher must also plan throughly for each lesson, and be extra flexible in her teaching.

■ There should be an agreement between parties regarding time, location and fees for lessons as well as travel and photocopying costs, and arrangements if lessons have to be cancelled by either side.

■ Lessons should begin punctually and not finish early. Nor should they be interrupted by the teacher because of phone calls etc.

24 | USING VIDEO

When the possibility exists to move your lesson away from the traditional classroom format, you have the chance to add variety to your course; to do something different. Such possibilities might include:

■ **Computers**

'C.A.L.L.' is 'Computer Assisted Language Learning and is obviously a hugely popular option in these times. Students have many options for both independent and guided learning and practice, and schools realize the importance of C.A.L.L. Complete learning courses on the internet with no face-to-face teacher access will surely become popular with students who are unable for any reason to attend a school. The TOEFL examination can now be done exclusively on computer, and as this is probably the most-taken EFL exam in the world, the need for computer access in schools becomes even more necessary.

■ **Language laboratory**

Perhaps considered old-fashioned by some, the language lab. remains popular with many students. Many types of activity are possible, ranging from 'Readers' (graded stories in print and on tape), to exam practice. Students can work at their own pace, and choose the type of task they wish to do. It is a good way to improve listening skills.

■ **Excursions**

Going out with your class to places of interest, and linking the trip to set tasks (before, during or after) is an excellent way to make learning authentic and interesting. You need to ensure that you are allowed to make such trips if they are to be in normal lesson hours.

- **Self access**

    Schools often have some kind of self-access centre where students can work independently outside lesson times. Typically, self-access material includes reading and listening materials at different levels. Computer facilities may be part of the self-access centre or may be housed separately. Teachers should not only encourage students to use the self-access centre, but also make sure that the students know how best to use it effectively.

Access to video

Access to video brings with it many advantages and potential problems for the teacher, but the range of activities you can do is considerable. There are many **specialized video courses** available, some linked to a course book. These courses cover all levels from beginner to advanced and give the possibility of regular video use featuring the same characters. Ready-made material comes as part of the package (which, as with a course book, you will need to adapt and supplement for your particular students). The tasks may focus on skills work, grammar, functional language, or a combination.

You can also do **'one-off' video lessons** with recorded broadcasts (but you must check copyright requirements … all suggestions made here assume that this has been done). TV news or other programmes can be an excellent source of material for lessons which may be independent of others you have taught on the course, or linked in some way. For example, if you have been doing a lesson(s) on newspapers and news coverage, it is logical to focus on TV presentation of news. If you have been doing 'humour' as a topic, then extracts from comedy shows (with tasks) would tie in well.

The third way of using the video is the one that students would probably prefer, but which is usually the least appropriate, and that is **'just watching'** something, with no tasks. It is a pleasantly easy option for the teacher (and I, in common with most teachers I'm sure, have chosen this easy option from time to time!) but one which has no aims. It is an option which is probably best suited for an end-of-course 'reward', but otherwise some students may wonder what the point is.

As with any lesson, the teacher must plan and stage what she is to do. The lesson should have clear aims, whether language or skills-based. And the students need to be prepared for the lesson (necessary vocabulary, interest established). For students who question the validity of watching a video, you will need to stress what the point of the lesson is.

Ways to use the video

You can do all of the activities below with the video, but they will not all be appropriate. Your choice of activity will depend on the usual factors – level of class, composition of class, nature of class, aims of lesson, age group, and time available.

1 Silent viewing

The students watch the picture with the sound turned down and predict what they think is being said. They then re-view with sound to check. The extract may be just half a minute or so, perhaps, longer. This can be done at any level – with beginners, for example, it can be as simple as "Good morning, how are you?".

2 Sound only

The reverse of **1**, the students turn away from the screen and only listen. They can predict the context ("where are they?" "who are they?" etc).

3 What happens next?

This prediction exercise is perfect for story-videos. One way of doing this, given the right video, is to have students predict how they think the story will end. It is also good when something unusual or unexpected is about to happen. It doesn't have to be restricted to action – students can also predict what someone is going to say. I remember using one video where two people met, one wearing a hat. As they greeted each other one person raised his hat and the students had to predict what the other person was saying. They all came up with variations of "Pleased to see you" only to discover the actual words "You've lost all your hair"!

4 Description

Students work in pairs – one watching the next part of the video with sound off, and describing the action to their partner, who is looking away. This gives practice of descriptive language, interpretation of action, gestures and context, and – for the 'listener' – listening for detail. Only for higher levels!

5 Picture freeze

Students see a still from the video which is 'paused'. They can do any kind of prediction task described above. A variation would be to use adjectives to describe how the person might be feeling, what they might be thinking, and what they look like. Beyond this, students could also predict the context, maybe even the time of year etc.

6 Comprehension

Students view video or part of video and complete a comprehension task. This is a 'visual listening' comprehension lesson following all the guidelines given in Unit 12.

7 Making a review

As a follow-up, students can be critics and write a review of what they have seen. Alternatively, this can be done as a speaking activity. When watching the video, therefore, students make notes which will be used for the review. You would need to elicit possible 'headings' for this review before viewing.

8 Quiz questions

Students watch an extract from the video and, working in teams, they have to write, say, two questions to ask the other teams about the extract. They could both be comprehension, or one might be an observation question (e.g. "how many chairs were there in the room?"). At the end, each team asks their questions to the others. Finally, you replay the tape to confirm all the answers.

9 Observation task

Play part of the video and ask your question(s) **after** viewing. This could be an observation question (as in **8**). Make sure your students know what type of question they are going to be asked, to make sure they concentrate hard on the action.

10 What happened before?

You show a later part of a video (maybe the ending) and students have to predict what happened before to lead to this.

Whatever 'follow up' you might do after the video itself (and a couple of suggestions were given above) might take place in the next lesson, or the next day, since video lessons tend to be quite long and demanding for all concerned!

Task 1

What are the advantages and disadvantages of video use for both students and teacher?

Think about this before reading on

Tips for using the equipment

Let's deal straight away with the disadvantage which often scares teachers off from using the video at all – the machine! If you are very lucky, there will be a video machine and monitor in your classroom. Otherwise, there may be a video 'room' which you have to reserve in advance, or a video machine /monitor which you will have to book and transport to your classroom. If reserving is required, make sure you double-check your time and day so that you don't find at the appointed time that you have made a mistake and cannot use the video.

Having made sure of this, check beforehand that you know how to operate the machine – especially the fast forward and rewind buttons, the volume and pause buttons. If there is a remote control, this is much better to use; but again, you must be familiar with its controls. You may also need to 'install' the machine and know which cables plug in where. This is my personal nightmare!

Finally, you have to 'know your videotape'. Watch it beforehand and become familiar with it. If you can, use the time counter so you know at what point in the video key points occur. Know whereabouts you need to pause the tape. Know whereabouts answers to tasks are featured. And know how long the tape lasts!

This last point is important because you could get a 60-minute lesson out of a 5-minute tape. So if the tape lasts 30 minutes and your lesson lasts 40, considerable adapting will be be needed; or you could continue on another day if this is possible.

In short, know your equipment. Don't avoid using the video because of fears of what might go wrong. That said, be prepared just in case there is a catastrophe and you cannot use the video!

Key to Task 1

Advantages

1 Using the video gives your lessons *variety*. This is very important for maintaining interest and motivation, especially when you are teaching a group for a long period. Do not, therefore, overuse this type of lesson. Once every few weeks is probably sufficient.
2 Partly because of the above, students' *interest* is usually high. But this is also because of the fact that 'television' is worldwide, and something students want to understand and enjoy outside class.
3 The visuals and sound make the language more *memorable*; context is clear.
4 Videos can increase awareness of *culture*; especially useful if the students are not studying in their own country.
5 Video can be used in a number of ways. It can be used to focus on specific language and skills, as well as providing a link with earlier or later lessons.

Disadvantages

1 The concerns about using the equipment, as discussed earlier.
2 Students may not see it as 'real learning' – so tell them what the purpose is.
3 You may have to prepare your own material and tasks.
4 Students sometimes get frustrated by the 'stop start' nature of the lesson.
5 With very big classes not everyone may be able to see the screen properly, and some of the techniques described above become very difficult.

Task 2

Imagine you are going to use a 5-minute TV news broadcast with a group of upper intermediate students in a 50-minute lesson. What might you do, and with what aims? And what might you have done/ might do after this video lesson?

Using a camcorder

Before we look at a possible response to this task, there is one other option available to the teacher with the video. If your school has one, you can use a video camera. This adds another list of technical concerns because you have to know how to properly operate the camera, as well as how to replay the recording afterwards! But if you have this facility, and are not discouraged by the things that could go wrong, it adds another dimension to your teaching – with the right class.

There are a number of ways to use this piece of equipment. Here are five possibilities:

- role play (e.g. in a shop)
- debates
- mini talks (a student prepares and gives a talk on a subject of interest)
- surveys (e.g. interviewing staff, students, or outsiders on a given topic)
- plays (with each student taking on a role)

When viewing what you have recorded, supply a task for students to do while they watch. You can note down significant mistakes for later correction. Be aware, though, that some students do not like being filmed. My view is, even if only one student is uncomfortable, don't do it.

Key to Task 2

An earlier lesson could have featured:
- style of language featured in news broadcasts
- a discussion about news coverage in students' country/countries
- vocabulary for news topics

- a look at how newspapers cover news, and contrast this with TV
- a discussion of recent news items

For the lesson itself, interest could be consolidated by getting students to predict from headlines what news stories might be about. They could also predict which would be the main item. This could also be done from the attached pictures shown to accompany the headlines.

When watching the broadcast, gist listening could be focused on ("What are the stories about?"); and more detailed listening ("Make a note of the most important points of each story"). Alternatively, comprehension questions could be provided.

You might focus on all the stories, a selection, or only one. You could also extract specific language to focus on. Maybe, for example, students have to listen out for selected words and deduce the meaning from context.

As a follow-up, any of the speaking topics mentioned in the pre-viewing stage could be used. Or, students could write a newspaper report of a story in the typical style of a newspaper.

Review

Video is one of a number of ways in which the teacher can do a lesson away from the normal classroom setting. Using the video brings special challenges for the teacher in terms of using equipment, but this concern should not stop the teacher from making use of this resource.

- You can use the video for any level, and use a whole range of techniques to make the most of it. But you should not over-use the video.
- Video lessons, like any other, must be carefully planned with clear aims and stages.
- If the facility exists, you can also record your students doing work and view later with tasks and correction.
- Teachers should also encourage students to use other out-of-class resources such as self-access, computers, and language laboratories.

25 | PROFESSIONAL AWARENESS AND DEVELOPMENT

There are a great many people who 'train' to be teachers on short courses, often of about four weeks full time. Some, of course, undertake much longer training courses; others have no training – they learn 'on the job' (not recommended!). When I trained to be a teacher, back in 1984, I enrolled on a recognized four-week training course. This involved a combination of teaching practice (a total of six hours, teaching two different levels), written work, input from trainers, and observations of classes given by experienced and qualified teachers. I have worked on such courses as a trainer for many years, and many thousands of prospective teachers enrol on such courses each year. Often, trainees on such courses think, hope, even expect, that at the end of the four weeks they will be 'teachers'. Employers sometimes wonder how someone can become a teacher in such a short space of time. The answer is, you can't. There is sometimes a misunderstanding about what such courses sets out to do, and what can realistically be achieved in such a time. **Professional development** continues, and has to continue, throughout a teacher's career.

Sometimes, too, it can be forgotten that a good teacher doesn't necessarily make a good employee, and vice versa. The expectations placed on a teacher are more than only knowing how to teach. **Professional awareness** in any workplace is an essential expectation that any employer will have of an employee.

This unit looks at these two key areas of a teacher's profession.

Professional Awareness

 Task

Consider what expectations different groups have of a teacher. Think about **students, staff, employers**. It might help to think about your own experiences of 'good' teachers (not just the ability to teach well); and of your experiences in the workplace, either as a fellow employee or as an employer.

Stop reading now and make a list of points.

Expectations of students

Students are fee-paying clients of a school or other educational establishment. Yes, they want a good teacher, but they want much more besides from the school, and also from the teacher.

1 **They expect a teacher to know the subject she is teaching and how to teach it.**

   This was covered in the first two units and is an obvious expectation.

2 **They expect a teacher to be reliable and punctual.**

   If a lesson is scheduled to begin at 9am then the teacher should be in the classroom at that time – ideally, a few minutes earlier to prepare the classroom and welcome the arriving students. The teacher shouldn't wait for everyone to arrive before starting the lesson; then late-comers know that they can continue to arrive late without missing out on anything, and the punctual arrivers become understandably upset at the 'dead time'. There is no excuse for teachers sitting in the staffroom chatting when they should be in the classroom.

3 **They expect the teacher to be more than a teacher to them.**

   A teacher should not see her relationship with the students as only going in, teaching, and leaving again. The teacher should show an interest in the students, chat to them outside class, be available – all points made in the first units.

4 **They expect the teacher to be understanding of their particular set of circumstances.**

   So, find out something about the students you are teaching .

5 **They expect the teacher to be professional, fair, and polite –**
   and consistent in their behaviour both to the group and to the
   individuals within it.
6 **In some circumstances, they expect the teacher to be dressed
   in a way which is appropriate to the culture of the country
   they are in.**

I have witnessed situations where a good teacher was unpopular
with the students and commanded little respect simply because
they were not friendly or approachable outside lesson time, or even
during lesson time. I have heard students describe someone as an
'average' teacher, but as well-liked because of the way they acted
towards students. It is professional to give more than your teaching
skills to those that you teach; it is the same that you would expect
of someone teaching you.

Expectations of staff

Teaching is a stressful occupation and one which involves far more
than 'just' teaching 25 hours a week, or whatever. It involves
planning, marking, attending staff meetings, going to workshops or
conferences, maybe participating in social activities, definitely
being available to students outside class time. It requires you to
have a lesson to teach at the appointed time – knowing what you
are going to do and what you need for the lesson, as well as always
having following lessons at the back of your mind. To the students
you need to be unfailingly polite and friendly, whatever you may
think of them personally.

All of this can result, at times, in high degrees of stress and
frustration, and sometimes the staff-room is the place where this
frustration is expressed. What you shouldn't do is add to this stress
level.

1 **Respect other people's space.**
   For example, if you and a colleague share a desk, don't put
   your books etc in the space of your colleague!
2 **Offer support and encouragement.**
   The staff-room tends to be the place where teachers express
   frustrations, so offer a sympathetic ear.

3 Choose the right moment to seek help and encouragement, or to ask questions.

If a colleague is hurrying to a lesson, or appears stressed themselves for some reason, don't pester them with something that can wait until later. You have to respect their mood!

4 Put things back in the right place!

Few things irritate teachers more than being unable to find a tape or book etc because it has not been returned to the right place, or has simply not been returned.

5 Don't hog the photocopier or other resources.

If you have access to a photocopier, don't spend too long using it if someone else wants to copy. So, don't start looking for the right page while you're at the machine; do it before. If you're not in a hurry and there are others waiting behind you, give way. If the paper runs out while you are using the photocopier, or if the machine jams, don't just leave it for the next person to deal with.

6 Share ideas and experiences.

It is good on both a personal and a professional level to share your successes and 'failures' in the classroom. If a teacher, for example, is searching for a practice activity for the present perfect, and you know where there is a good one, tell them.

7 Maintain the records required.

If a teacher has to take your class at short notice but you haven't filled in your record of work done, you won't be very popular! (see the next unit for more on this)

8 Clean your board before leaving the classroom.

You should leave the room how you would hope to find it if you were about to teach there. So, a clean board and a reasonably neat and organized room. And if another class is coming into your room, vacate the room on time – don't over-run.

9 Don't take something out of another teacher's room without asking.

What has happened to me many times is a teacher 'borrowing' my room's cassette recorder without asking me first, because they don't have one! If you have asked first and borrowed

something, return it – don't expect the other teacher to have to
come and collect it.

10 Don't complain too much.
Don't continually moan and complain as if you are the only
person working there with niggles. Don't bring such levels of
stress into work that it rubs off on everybody else.

I have experienced many types of staff-room – most, very friendly
and mutually supportive, and 'student-friendly'. I have been in
others, where staff barely spoke to each other and students knocking
on the door were sent away because the teachers were on a break.
Fortunately, the majority of experiences have been excellent and
teachers are a very mutually supportive group. But I do remember
certain individuals – one who hogged the photocopier all the time,
another who was a full-time complainer, another who just wouldn't
cooperate with anyone. As in any workplace, you meet a variety of
different characters. But the overall point is that teaching is a
stressful (but very enjoyable) job, and you need to respect the
people you work with – their characters, backgrounds, interests etc
– just as you need to respect the students. If you do not, your
employment is unlikely to be long term!

Expectations of employers

Some time ago, I gave a questionnaire to three employers of
teachers about teacher behaviour, and their answers are included
here.

1 Punctuality and reliability.
Teaching is not one of those jobs where you can arrive ten
minutes late and just catch up on the work. The students will be
there expecting to be taught. Of course, sometimes it can't be
helped, but if you make a habit of it then you will earn yourself
a reputation for being unreliable and this may reduce your
chances of further work at that school.

2 Record keeping.
As already mentioned.

3 Neat dress.
Some schools will have a dress code, others may be relaxed. In

any case, find out what is appropriate and acceptable to wear, especially with regard to the local culture.

4 Professional behaviour.
This is an overall heading for everything mentioned in this unit, but in particular your behaviour towards students and fellow staff.

5 Don't criticize the school or staff when talking to students.
Whatever your feelings, it is not professional to express them to the students.

6 Read your contract.
Find out what you are and are not permitted to do, and what you are obliged to do.

7 Attend staff meetings and in-service training sessions.
This may be compulsory, anyway!

8 Be willing to be observed and to observe others.
See the section on **Professional development** (below).

The directors I spoke to mentioned their common complaints about teachers:

"teachers losing registers (rolls)"
"teachers not returning materials"
"teachers being constantly late"
"teachers who don't have a positive attitude towards students"

Professional development

At the beginning of this unit, the point was made that some training teachers, and some employers, maybe expect too much from a short training course. Also, some employers can be very doubtful about the value of such courses, wondering how anyone can 'become' a teacher so quickly. Well, you can't. Four-week training courses, validated by different bodies, are popular worldwide – and there are courses which last longer but involve the same amount of work. But participants and employers need to be clear about the typical content of such courses (mentioned at the beginning of this unit), and the aims.

These courses typically involve participants teaching for a total of about six hours at two different levels. That's it. Six hours. Many teachers do that in a single day. Participants, then, should be realistic. Don't get upset or disappointed if your observing tutor still gives you 'points to work on' in your final lesson! Don't imagine or expect that after six hours of carefully controlled teaching that you are ready to deal with all types of class and student. What such courses do is to provide you with essential teaching skills – the methodology – as well as some theory behind it. But even the methodology is limited in its scope, as it is in this book, because you can only 'learn' so much in the time available. What you have to do to pass the teaching component of these courses is show convincingly and consistently that you can do what is asked of you (the assessment criteria for teaching practice). If you pass, then you are equipped to begin teaching, with support and supervision – you have shown that you can teach reasonably successfully and have demonstrated an appropriate degree of self-awareness which will allow you to continue to develop your skills. In a sense, the real training starts once you begin your first teaching assignment. It is then that you have to begin teaching different levels, age groups, types of student, types of class; and then that you have the sole responsibility for your own teaching. It is at this point that your **development** begins.

Most experienced teachers, myself included, would say that it took them a good two years of teaching before they felt reasonably confident and able in most teaching contexts. Also, it takes this long to gain full confidence about the language itself that you are teaching. If new teachers accept these facts, and do not expect the impossible from the short training courses, then they will have more realistic expectations of themselves. The short training courses described are excellent and invaluable, but only as a starting point.

Employers need to recognize the limits of these courses, too. They need to offer support and guidance to new teachers, whilst at the same time recognizing that no training course providers are suggesting that they are providing you with the 'finished article'. As we all know, you can be teaching for many many years but still learning and developing – or you should be!

If we can accept these points, then we can think about how this 'post-training' development can happen; or, in the case of teachers teaching with no prior training, how 'on-the-spot' training and development might occur.

✳ *Tips*

1 Seek the opportunity from time to time to observe other teachers' lessons.

2 Ask your Director of Studies or an experienced teacher to observe you teaching at regular intervals and to give you feedback.

3 Continue to self-assess your own teaching (as you would have done on any training course you did /are doing) and to develop your strengths and work on your weaknesses.

4 If available, watch videos your school might have of lessons.

5 Attend in-service sessions at your school or nearby (staff development).

6 Attend workshops and conferences if you get the chance. In particular, the annual conferences held by IATEFL and TESOL *

7 Read new publications and periodicals about the profession, usually available at your place of work.

8 Talk to more experienced teachers and learn from them.

9 Try out new ideas you have read or heard about – don't just continue churning out the same old lessons in the same old way!

10 When you're ready, do further teaching qualifications (e.g. an MA related to the profession, such as Applied Linguistics; or a Diploma course). Normally this would require an absolute minimum two years' experience first. Your place of work can advise you of the options available.

11 Show an interest and desire to be involved in other aspects of school life, perhaps helping with administration, or social activities, or materials design.

12 Join professional organisations such as TESOL or IATEFL if you are working freelance. If you are not, your school is likely to have institutional membership.

13 With more experience, write articles for ESL publications and give talks or run sessions at workshops.

14 Eventually move into other areas as well as teaching; such as writing, training, examining.
15 Recognize that the profession itself develops and moves in new directions. Keep in touch and move with it, maybe even influence its direction in some small way!
16 Keep in touch on the web! There are many useful sites – just do an ESL or EFL search and you'll be given many addresses to explore.

*TESOL – Teachers of English to Speakers of Other Languages. Has over 20,000 members worldwide. e-mail: tesol@tesol.edu.

TESOL Central Office
700 South Washington Street
Suite 200
Alexandria
Virginia 22314
U.S.A.

*IATEFL – International Association of Teachers of English as a Foreign Language. Has over thirty branches worldwide and over 1000 delegates attend its annual conference. e-mail: iatefl@compuserve.com

IATEFL
3 Kingsdown Chambers
Whitstable
Kent
CT5 2DJ
England

Review

- Teachers and employers need to be realistic about what can be achieved on a short training course.
- These courses provide the essential basics of teaching and theory.
- New teachers need to always maintain their development beyond any training.
- Teachers with no training need to find ways to learn the basics on the job.
- This development can take place in a variety of ways, as described.
- Teachers, like all employees anywhere, need to have professional awareness.
- Students, colleagues and employers all have expectations of how you should 'behave'.
- In short, good teachers need to be more than good teachers, they need to be good colleagues and employees, and they need to remember that they are working in a very responsible profession which aims to maintain high standards.

26 | CAREER PROSPECTS AND JOB HUNTING

People go into this profession for different reasons. Many only want to spend a few years travelling and see teaching as the ideal way to earn some money along the way. Others have expectations of making it their long-term career. Sometimes, experienced teachers from the state sector decide that they want to try a different kind of teaching. Many want a career change, to leave their current employment in order to do something completely different. Then there are people like me, who fall into none of the above categories! Having finished university I still wasn't sure what I wanted to do, until one day I saw an advertisement for an EFL training course and thought I would send away for the information. That eventually resulted in my doing the course and starting my first job the Monday after it finished. The rest, as they say ...

This unit looks at two main areas – finding work, and developing career prospects. It is not intended to 'sell' the profession and make it seem more glamorous than it actually is. My honest opinion, shared by virtually every colleague I have worked with, is that it is a mixture of good and bad, like most jobs. The 'good' includes the opportunity to travel; meeting people from all over the world; working with many different colleagues; having a good degree of independence once you're inside your classroom; not being in a traditional '9-5' job; variety; usually longer holidays than other jobs; and working with young people – a fact appreciated more as you get older!

On the negative side – and I am speaking personally here, not everyone would agree with me – the money is generally not good. In fact, given that most teachers have graduate status, the salaries paid by some schools is scandalous, especially when you compare salaries with other graduate professions. Actual career prospects are also comparatively limited. In addition, too many teachers in certain countries have little security, having work only as long as there are enough students to teach.

I am not here to make personal recommendations, but the one institute I will mention is now closed, so I feel I can talk about it without being accused of trying to influence readers. But I once had a truly rewarding time working at a teacher-training centre which also had ESL classes, where going to work every day was a pleasure. English International in San Francisco, owned and run by Jeff Mohamed and Deanne Manwaring, was a model of how this profession should be. Conditions were very comfortable, resources were excellent and frequently upgraded, staff were very well looked after, students were extremely happy, and everyone was treated with respect. It was the only occasion when I woke up in the morning happy to go to work.* I would like to think that there are many other institutes like this one in the world – and I have experienced other good schools, too – but I have worked in other places where conditions and resources were poor, prospects were limited, and staff were treated with little respect. The profession is improving, but still has some way to go before it can compare favourably with many other graduate professions, and I make no apologies for saying so. Still, the positive points are considerable, which is why I am still in the profession, seventeen years after starting my first job.

*There is now a website for English International which provides an excellent source of information for those in the profession, or entering it (see Unit 30).

The UCLES CELTA course alone sees around 7,000 new teachers trained every year. Add to that figure those training on other courses, those without qualifications looking for work, and experienced teachers looking for work, and you will see that there is a lot of competition for jobs. It is extremely important, therefore, to make any job application as professional as possible, and to know what to expect at any interview.

Finding a job

Increasingly, people look for work on the internet (see Unit 30). Others will go through Recruitment Agencies (Unit 30), and others will see jobs advertised in newspapers – in the U.K. the Tuesday *Guardian* and Friday's *Times Educational Supplement* remain the best bets.

Many people will find their first work with chain schools – those with 'branches' in many different countries. Perhaps the most famous of these is International House. There are definite advantages to trying to find work at such a chain because any central application can lead you to one of many different 'branches', unless you are seeking work in one particular area. Normally, you apply to the central office. Some of these chains also run training courses (International House, for example, runs CELTA), and training with the chain can be advantageous when applying for work with them! Recruitment agencies will allow you to be considered for different schools in different countries, rather than one particular organization. Again, there is a 'central' office where you go through the application procedure, and many employers will seek staff through such agencies. Requirements and conditions will vary widely, but the agency can advise you on this.

The British Council is one of the largest employers in the profession, with schools in over fifty countries. They will usually, however, ask for some experience, but you can contact them directly and find out their up-to-date requirements (Unit 30 again).

Some teachers start out working for voluntary organizations. The Peace Corps in the United States and Voluntary Services Overseas in the U.K. are two examples, and their addresses are also in Unit 30. VSO places nearly 2,000 volunteers a year in more than fifty different countries. Volunteers receive a local salary, as well as flights, accommodation and insurance. The Peace Corps has placements in around seventy countries and they have a toll-free number in the United States to contact.

There are some restricting factors, however, which you need to be aware of before applying. If you are a European Union citizen applying for work outside the E.U. you may, depending on the country, require a university degree (usually in any subject), and need to gain a work permit and visa – not necessarily a formality. The same applies to United States citizens applying for work outside the United States. In addition, some schools have a policy of, or at least a preference for, employing teachers whose first language is English.

Applying for a job

Whichever of the above paths you choose, you will need to submit an application. This will involve having a resumé (C.V.), a covering letter if you are applying by mail, and probably your educational and training certificates. You will then need to have an interview.

Your **resumé** needs to be professional in appearance, not handwritten, not too long, and containing only relevant information.

✳ *Tips*

- Be honest!
- Don't over-sell yourself.
- Make the resumé easy to handle and to refer to.
- Include full contact details.
- Make sure all time-periods are covered.
- List your time in chronological order.
- Include all relevant qualifications and experience.
- Highlight any teaching experience you have.
- Specify which training course you did or are doing.
- Do not include irrelevant details.
- Mention any skills you possess, especially knowledge of foreign languages.
- Include names of referees – at least one of whom should be a professional referee.

Rightly or wrongly, many employers will insist on a photograph! Others will also want to know your age. If you do have to submit a photo, make sure it is a recent one and you look casually smart. I was once shown an application from a young man who had submitted a photo of himself wearing only a pair of shorts! That is not the way to present a professional image!

A **covering letter** should accompany your resumé. Schools and agencies receive many applications and I'm afraid that some applicants do not appear to put much effort into them judging by some of the ones I have seen.

✳ *Tips*

- Make sure you **do** include a covering letter.
- Keep it brief.
- State clearly what job you are applying for and where you saw it advertised.
- If you have to hand-write, write neatly.
- Use proper paper, not something torn out of a notebook.
- Say why you want the job and why you should be considered for it, but don't over-do it!
- Proof-read your letter for spellings etc.

The interview

All being well, you will be invited for an interview, either at the institute itself, or at a central office which handles all applications. I have interviewed many people over the years, and it has been interesting to see the different approaches adopted by interviewees – from the very formal, well-prepared candidate to the very casual, take-it-as-it-comes candidate. Neither is necessarily better than the other, but it is advisable to 'play safe' and follow certain guidelines.

✳ *Tips*

- Arrive on time! It creates a bad first impression if you don't.
- Dress appropriately! At least look as if you have made an effort.

- Don't make it seem as if you are doing the employer a favour by coming to the interview – over confidence, or even arrogance, do not go down well.
- Answer questions fully and honestly.
- Come ready with questions of your own (see later).
- Know something about the institute you are applying to.
- Be relaxed and friendly in your manner; try to let your personality show through.
- Talk about your previous experience when asked; if you do not have any, talk about your training course and what it included and what your teaching experience was like.
- Show enthusiasm; make it seem as if you really *do* want the job!

Another important part of any interview is knowing what sorts of questions you might be asked, and what information you need to gain from the interviewer. Yes, you are the one who is looking for a job and being interviewed, but that doesn't mean that you should accept a job offer without knowing sufficient details about the job and the context.

Here are some typical questions that you should expect to be asked:

What types of class have you taught? – including levels, sizes, age groups. If you have no experience, talk about the classes you taught on your training course, if you have done one.

What materials did you use? – including course books and supplementary materials.

What did you think of the course book(s) you used? – give a fair evaluation and avoid "it was fantastic" or "it was terrible" because all books have their pros and cons.

What level(s) do you prefer teaching and why? – it is perfectly okay to have preferred levels (most teachers do), as long as you make it clear that you would be prepared to teach any level.

What different nationalities have you taught? – be prepared to discuss the particular problems they encountered and how you helped them to deal with these problems. Avoid making any sweeping general statements about a particular nationality.

Why did you become a teacher? – it's the obvious question to expect if you are applying for your first job, so be prepared with a good answer!

Why have you applied for this job? – another obvious question to expect, so know something about the institute you have applied to and show an interest in the organization (a few questions of your own about its history would go down well!).

What are your future plans? – the only answer to avoid here is something like "Well, I only want this job for a few months because I want to get a better job once I have more experience"!

Why do you want to work in this particular country?/Do you speak the language? / What languages do you speak? – if you do not speak the language of the country you want to work in, show an interest in learning it.

What do you know about this school? – as mentioned earlier … do your homework!

Do you have any questions? – have some ... but not too many!

So what questions should you ask? Well, a good interviewer should cover all the essential points, if they haven't been dealt with in any literature sent out to you before the interview. However, you shouldn't leave the interview without knowing the answers to the following questions – but if you need to ask all of these, then it suggests the interviewer didn't do a very good job!

- How many hours a week will I have to teach (25 is a standard 'maximum')?
- How will these hours be split (maybe a 'block' or a 'split shift')?
- Will I have to work evenings or weekends?
- Will all my teaching be in one building or will I have to travel?
- If travelling is involved, how much?
- What levels will I have to teach?
- What age group(s) will I have to teach (this might mean teaching children)?
- Do I have to be in the school even when I am not teaching?
- Am I paid for this 'non-contact' time?

- How much will I be paid?
- Will I be paid weekly or monthly?
- How will I be paid?
- Are national holidays paid?
- What holiday or vacation allowance do I get?
- When can I take my holiday/vacation?
- Is it paid or unpaid?
- Do I get sick pay?
- What's the cost of living (what seems like a high or low salary might in real terms be different depending on the cost of living – in some countries a little money goes a long way, but in others a lot of money goes not very far at all)?
- What taxes do I have to pay?
- What length of contract do I get and what notice is required on either side?
- Are my flights paid for?
- Does the school provide or give help finding accommodation?
- If the school provides it, who will I share with, and how far is the accommodation from the school?
- Does the school arrange or give help getting work permits etc?
- How well equipped is the school in terms of facilities and resources?
- How many other teachers are there?
- What in-service professional development does the school provide?

Do not sign any contract without reading it carefully and checking anything you are not sure about. Make sure you have a copy of that contract. If you do not feel comfortable with any aspect of the job or conditions, don't take it.

But this is a two-way process, and many employers or agencies will tell you of teachers who have agreed to everything, then called at the last minute to say they have taken another job (or not called at all). This is highly unprofessional.

Once in a job, seek help and guidance from your Director of Studies as well as more experienced colleagues.

Prospects

For those who decide that they do want to make this their career, if only for a few years, there are ways to improve your pay and conditions and to do things other than only teaching. But there is more to it than that. You need to develop your own skills and professional awareness (see Unit 25) so that you keep in touch with developments in the profession, and new teaching ideas and theories. As well as following the tips mentioned in Unit 25, you need also to have an idea of possible career development opportunities.

The majority of teachers start out by doing a 'Certificate' course (e.g. Cambridge/RSA CELTA, or a Trinity College Certficate), but others will do a similar, less-recognized course. Some will find a teaching job despite having had no training! After a minimum of two years (maybe less with Trinity) of actual teaching experience, there is the option in most countries of doing a 'Diploma' course. This includes practical and written exams as well as observed and assessed teaching practice. Many centres will expect you to have the Certificate as well. In the United States the common 'next stage' is an MA in TESOL. The Diploma has little recognition in the States. MAs are becoming increasingly popular in Europe, too, as the demand for permanent posts increases.

By doing a 'post-Certificate' course and gaining an extra qualification, you are improving your prospects of work outside the classroom, as well as of a permanent contract in your own country. In the U.K., for example, it is virtually impossible to get a long-term contract without the Diploma.

So what are the possibilities? Some teachers, like myself, move into teacher-training. To become an UCLES trainer, for example, you need several years of varied teaching experience, and then complete a trainer-training course. The training courses are thorough and carefully monitored but if you are successful you become an approved trainer.

Another possibility is writing! This does not necessarily mean writing books, but maybe articles for specialized ESL publications.

Presenting papers at conferences/conventions (e.g. the annual TESOL or IATEFL conferences/conventions) is a good experience to have, if you don't mind addressing an audience of a hundred people!

Teaching Business or other English for Special Purposes (ESP) classes can be more financially rewarding, as well as broadening your experience.

Administrative positions within a school are what many people eventually seek. Acting as an 'ADOS' (Assistant Director of Studies) leaves you equipped to become a 'DOS' (Director of Studies), in charge of the academic direction of the school and its academic staff.

Others, who start teaching adults (16+), choose to move into the field of teaching Young Learners; some go in the opposite direction.

A good move after you have some experience is to become an examiner. This will not earn you a lot of extra money, but it will give you good experience.

For the very ambitious (and brave!) your target might be to one day set up your own school...good luck!

Review

In this unit we have seen some of the ways to find work; how best to apply for work; the essentials of an interview, including questions you will be asked and should ask yourself; the considerations before agreeing to take a job; and what prospects exist after you have some experience. Prospects and conditions are not fantastic, but the job itself often is – the actual experience of teaching students from all over the world, and the possibilities of travelling all over the world. As was emphasized in the last unit as well, there is a great need for professional behaviour from you, and a recognition of the need for professional development. There is also the hope that you yourself will be treated with professional respect, and rewarded in a way which your qualifications and background merit.

27 | A GUIDE TO LEVELS

A common question asked by new teachers is "What does 'Intermediate' mean?" (or any other such term). And a common complaint is "I have a mixed-level class".

Labels to describe levels of students, such as 'intermediate' and 'elementary' are very broad and a class described as, say, 'elementary' could well include students who seem to be of different levels. It is important, then, to remember that such labels are more guides to the general and approximate level of students in a class, and that students may be better at one aspect of the language (e.g. speaking) than another (e.g. 'grammar'). You do need to pitch your classes to the general level of your class, adapting material as required, bearing in mind that one intermediate group of students may be 'stronger' than another.

A class which is in some way mixed-ability – and the bigger the class the more likely this is – will always be more difficult to teach. What you should avoid, if possible, is effectively teaching separate groups within the class, with different tasks etc. If you do this, you will no longer be teaching and managing 'a class' but separate groups. This is both difficult and ultimately counter-productive. You should give extra support and guidance to the weaker students in the class (for example, when others are working together), but do not pitch the lessons to the needs of the lowest-level students within the class, because if you do that you will frustrate and handicap those whose level is generally higher and who are in the majority in the class.

Do not hesitate, for example, to stop group work even if some people have not finished a task. Students who have already finished something do not want to have to wait every time for all the others to finish. You have to allow a reasonable amount of time for a task, but to be prepared to say "Okay, stop there" and then go through answers together as a class.

Be aware, too, that in some schools a system of 'continuous enrolment' operates. This means that new students can join an established class at any time (often this means any Monday). So the composition of your class will fluctuate as students leave and others arrive even though you are teaching a fixed course. You will find, as a result, that the overall level of a class can change to varying extents on a fairly regular basis. It is not ideal, but as teachers we have to adjust our lessons and plans accordingly. Have a look at Unit 20 for more on this.

The following guide should give you some idea of what a 'typical' student at a given level should be able to do at the end of his or her course. Six labels are defined here, but some schools may have further levels.

(NB: This is an adapted guide courtesy of SKOLA Schools, London.)

Beginner

A student finishing a Beginner's course should be able to do or know the following:

- the alphabet
- tell the time
- give simple information about himself/herself
- ask other people for simple information about himself/herself
- communicate in simple terms in everyday places
  e.g. ask for directions
  >order a meal
  >make suggestions
  >buy items in a shop/store
- listen to and understand simple, natural speech and conversation, and complete simple tasks
- read some authentic texts such as timetables or brochures and find the information they need
- do simple writing tasks such as taking or leaving a message

Elementary

A student finishing an Elementary course should be able to do or know the following:

- talk about past time
- make future plans or arrangements
- describe people using simple adjectives
- describe places using simple adjectives
- express his / her hopes or intentions
- ask people to do something for them
- cope in simple social situations
  e.g. talk about the weather
       describe simple physical problems
       give advice
- listen to natural speech and conversation, and understand the gist of what is being said
- use authentic newspapers and find specific information required
- write simple letters and be able to link ideas together

Pre-Intermediate

A student finishing a Pre-Intermediate course should be able to do or know the following:

- talk about more abstract ideas
- participate in discussions and be able to politely interrupt, and take turns in communicating
- give a logical argument to support a view
- converse with relative ease in everyday social situations
- use formal and informal English in appropriate social situations
- help people with problems by making suggestions and giving advice
- listen to natural speech and conversation and begin to understand nuances and subtleties
- read most authentic texts and books for pleasure
- write letters, both formal and informal

Intermediate

A student finishing an Intermediate course should be able to do or know the following:

- express her or his opinion
- agree or disagree with other people's opinions
- express how he or she is feeling and their general emotions
- use conventional social language in greeting and welcoming people
- talk about hypothetical situations in the past and present
- talk about past habits
- speculate about the present and future
- listen to natural speech and conversation, and understand detail
- read authentic texts such as newspapers and magazines and understand the general meaning
- write letters to friends

Upper Intermediate

A student finishing an Upper Intermediate course should be able to do or know the following:

- use all the main structures and tenses in English with a high degree of accuracy
- begin to express sarcasm and irony
- use appropriate speech in a wide range of social situations
- understand and express humour
- argue an opinion with confidence
- follow lectures and other monologues and be able to take notes
- read most authentic texts and be able to understand them with relative ease
- write letters, compositions, reports, summaries

Advanced

A student completing an Advanced course should be able to do or know the following:

- express himself/herself easily

- integrate well with people whose first language is English
- vary stress and intonation to affect meaning
- follow and understand most forms of entertainment
- use a vocabulary of about 3,000 words
- study for a high-level qualification in English

This unit is designed to give you as simple as possible a guide to just some language terminology that you will need to recognize and use as you start out teaching. But do treat it as no more than a guide ... for more details, consult your grammar book.

Verb forms and tenses

Here are five examples of verbs, each with four forms.

	a	b	c	d	e
1	to go	to eat	to read	to work	to listen
2	went	ate	read	worked	listened
3	gone	eaten	read	worked	listened
4	going	eating	reading	working	listening

Form 1 = the infinitive
Form 2 = the past simple
Form 3 = the past participle
Form 4 = the present participle

Look first at examples **d** and **e**. You will see that forms 2 and 3 are identical; the verb ends -*ed* both times. Such verbs are called **regular verbs**, and they are very straightforward, except for some irregular spellings. For example, a verb such as *cry* in form 2 and 3 would becomed *cried*.

Examples **a**, **b** and **c** are different because the 2nd and 3rd forms do not follow this pattern, or even have a separate consistent pattern. These are **irregular verbs** and students hate them! There are many irregular verbs and each has its own pattern. Some have no variation at all (e.g. *to hit, hit, hit*), while the example above in **c** has the same form for 2 and 3, but different pronunciation!

You may wonder what the difference is between the past simple and past participle. The past participle (form 3, or 'verb 3') is used in conjunction with the auxiliary verb *has/have/had*. For example *I have* **eaten**, *She has* **gone**, whereas the past simple form would be *I* **ate** (*a sandwich earlier*), and *She* **went** (*home yesterday*).

With regular verbs there is no apparent difference – *I* **worked** (*yesterday*) and *I have* **worked** a lot recently. But the first case is form 2, and the second is form 3!

The infinitive presents its own problems to students. Consider these two sentences:

I like **to watch** *TV.*   and   *I may* **watch** *TV later.*

The first sentence has the full *to*-infinitive form, the second does not (in this case because it follows a modal verb – *may*). Teachers and grammar books vary in how they label this second example. Some call it the 'infinitive without *to*'; others call it the 'bare infinitive'; while others call it the 'base verb'. Just be consistent in whatever terms you use.

Finally, students (and some teachers!) confuse the present participle with a **gerund**. Both have the *-ing* form, but the present participle (form 4 above) is a verb, while a gerund is a noun. Look at these two sentences:

I'm eating my lunch.   *I like swimming.*

In the first case, the *-ing* form is a verb; it says what you are doing. In the second, it is a noun and *like* is the verb, just as in *I like books*, *I like music* etc.

Higher-level students are usually very familiar with all these terms, but lower-level ones may not be, and you may not want to burden them with such terminology until their level of English is much better. You will find that some students will refer to verb 3, for example, maybe asking you "What's verb 3 of *sleep*?". Now you know what they mean!

You need to know such terms when identifying **tenses**, and you also need to be aware of the difference between **tense** and **time**.

Task 1

Look at the following sentences and use a grammar book to help you. For each one, identify the **tense** and the **time** it refers to. Unit 19 will also help!

Example: *I left* the country last week – *past simple, referring to finished past action.*

1 I work every Saturday.
2 She leaves on Saturday.
3 I'm writing this sentence.
4 I'm meeting Matt later.
5 I was watching TV last night.
6 There was a good movie on TV last night but **I had seen it before**.
7 I've been to Canada.
8 I've worked here for a year.
9 I'm sorry I'm so dirty, **I've been working on the car**.
10 That's the phone ... **I'll get it**.
11 I'll have finished by lunchtime.
12 I'll be leaving tomorrow.

Doing this task you should quickly become aware that very often in English the name of a tense does not relate to the time it refers to. This causes many problems as, for example, when a teacher asks "What's the tense?" and a student replies "Future" when she means the sentence is talking about the future. As far as future time is concerned, low-level students will often use *will* for any future reference believing that this is the future tense. Here, too, you must be careful. We do not have a single 'future tense' even though we refer, for example, to the 'future simple' tense. There are many different ways to express future time (and present time, and past time) in English and they do not all require the use of tenses. Consider *I used to smoke* and *I may leave later*. One refers to past time and one to future time, but neither is a tense.

That said, tenses are very important, and teachers who refuse to use terms such as 'future simple' are not really helping their students when most course books and grammar books recognize such labels. But you must always be clear when teaching to convey the time reference of any piece of language.

Back to the task…

Key to Task 1

1 present simple, referring to habitual action – so 'general time'
2 present simple, referring to future time
3 present progressive (also known as 'continuous'), referring to action in progress at the time of speaking – so present time.
4 present progressive, referring to future time
5 past progressive, referring to past time
6 past perfect simple, referring to earlier of two past events – so past time
7 present perfect simple, referring to (indefinite) past time
8 present perfect simple, referring to unfinished past state – so past and present time
9 present perfect progressive, referring to recent past time
10 future simple (or 'will future'), referring to spontaneous decision (present time) about something you will immediately do (future time)
11 future perfect simple, referring to future time
12 future progressive, referring to future time

These are merely examples to demonstrate how, sometimes, the name of the tense does not necessarily refer to the time being described.

Tense forms

Present simple is the 1st form of the verb
Past simple is the second form of the verb
Past progressive is the verb *to be* in the past (*was/were*) + verb -*ing*
Past perfect simple is *had* + past participle
Past perfect progressive is *had been* + verb -*ing*
Present progressive is the verb *to be* in the present (*is/are/am*) + verb -*ing*
Present perfect simple is *have/has* + past participle
Present perfect progressive is *have/has* + been + verb -*ing*
Future simple is *will* + base verb
Future progressive is *will be* + verb -*ing*
Future perfect simple is *will have* + past participle
Future perfect progressive is *will have been* + verb -*ing*

In each case, of course, there is a subject at the start!

1st/2nd/3rd person

You also need to be aware of the rule regarding verbs in the 3rd person singular.

1st person   the person/people who is/are speaking (*I/We*)
2nd person   the person/people you are speaking to (*You/You*)
3rd person   the person/people you are speaking about
                 (*He, She, It, They*)

With 3rd person singular (*He/She/It*) the verb takes *-s* or *-es*:

   I work / You work / She work**s**   I go / You go / He go**es**

Students at all levels have trouble remembering this rule.

Language terminology

Now let's look at a few other selected examples of language terminology.

Articles – *a/an* (indefinite), *the* (definite)
Adjectives – usually modify a noun: *big, expensive, blue, loud*
Adverbs – usually modify a verb: *quickly, efficiently, hard, well, successfully* (but *hard* can be an adverb or an adjective)
Auxiliary verbs – *be, do, have*
Clause – part of a sentence containing a subject and a verb
Common nouns – those that do not begin with a capital letter: *table, chair, pen*
Compound noun – a noun made up of two words: *address book, ashtray, door-handle*
Conditional sentences – (see below)
Conjunctions – joining words: *although, but, however*
Contraction – *I am = I'm   They are = They're*
Countable noun – a noun that can be counted: *pen(s), book(s), car(s)*
Functions – the 'reason' for speaking: *invitation, suggestion, apology*
Modal auxiliary verbs ('Modals') – *may, can, should, might, would* (see below)
Phrasal verb – verbs with more than one part which do not have a literal meaning: *to look up* (*a word in the dictionary*)

Possessive adjectives – *my, your, their* etc
Prepositions – e.g. *by, on, under, across, through, beside, in, out* etc
Pronouns – *I, we, You* are examples of subject pronouns
            me, us, you are examples of object pronouns
            mine, yours, hers are examples of possessive pronouns
Proper nouns – 'names': *Spain, Thomas, Everest*
Question Tags – e.g. *It's cold, **isn't it**?   She's late, **isn't she?***

Conditional sentences

Conditional Sentences feature in all course books from around elementary level. There are, in fact, a great many types of conditional sentence and not just the three or four presented in most books. But given that you will have to teach the basic types at some stage, here is a simplistic guide to **conditionals 1,2,3** ... but bear in mind that this merely scratches the surface of this bit of language!

"Look at that sky. If it rains, we'll go back inside."

The sky threatens rain, so the possibility of this is quite high. We are referring to a possible future happening. The definite consequence of rain would be the speakers going back inside. The first part of the sentence is '*If* + subject + present simple', and the second is 'subject + *will* + base verb'. This grammar with this use is the standard and simple type of **1st conditional sentence.**

"If I won the lottery this week, I'd quit my job."

We are again referring to future time, but on this occasion, the possibility of the *if*-clause being met is remote, though still possible. Compared with the chance of rain in the first example, the contrast is clear. We have past simple in the first clause, and 'the modal verb (*would*) + base verb, in the second. This 'low possibility'/hypothetical future is one type of **2nd conditional sentence.** Again, this is a simplistic and incomplete definition, but it's a start!

"If I hadn't eaten that undercooked meat, I wouldn't have become sick."

...but I **did** eat the meat and I **did** become sick. So this sentence is set in the past and you cannot change what has happened. The *if* clause has the past perfect, and the main clause has 'a modal (*would*) + *have* + past participle'.This is a standard **3rd condtional sentence.**

You need to know this because students often ask about conditionals and, as I said, course books are very fond of these three conditionals, so you will teach them on a regular basis. Once you start teaching higher levels, you and your students will be exposed to the numerous variations of conditional which we use in our everyday spoken language.

Modal verbs

Finally, a further look at modal verbs. There are so many functions attached to modals, which again underlines the importance of illustrating meaning properly, and checking understanding. When students ask "what does *must* mean?" there is no answer, as the following task should demonstrate.

 Task 2

Look at the modals in these sentences and say what you think the function might be.

Example: You **must** do your homework (teacher to student) –
          obligation

 1 "Who's that?" "It **must** be the delivery boy."
 2 "**Can** I go now?"
 3 "I **can** speak five languages."
 4 "You **can't** smoke in this school."
 5 "**Could** you help me?"
 6 "I **could** play squash for an hour when I was younger."
 7 "You **may** leave if you want."
 8 "It **may** rain later."
 8 "You **may** not go out tonight." (parent to child)
10 "**May** it always be so!"

This task serves as a taster of the complexity of modals in terms of their various meanings and different shades of meaning. Which functions you teach will depend in part on the level of the class, but virtually all levels of course books feature uses of modal verbs.

Back to the task...

Key to Task 2

1 This is making a **deduction** – you ordered a pizza, there's someone at the door, so in all likelihood it is the pizza boy.
2 **Asking permission** to leave.
3 Talking about present **ability**.
4 **Prohibition**.
5 Making a **request**.
6 Talking about **past ability**.
7 **Giving permission**.
8 Expressing **possibility**.
9 **Prohibition / refusing a request**.
10 Expressing a **wish** (a rather specialized use of this modal).

Review

In this unit we have defined just a few common language terms. Obviously, you should refer to a grammar book and the grammar guides in any course book for more detailed and contextualized information. But you do need to know terminology, especially when teaching higher-level students, as they are familiar with such terms. In other words, you need to have the *knowledge* before you can begin the planning of how to *teach* language. This is a gradual learning process. It is often this 'terminology' that most worries new teachers – all you can do is to learn it yourself as you go along. It's almost like learning a second language!

29 | A GLOSSARY OF TERMS

Here is a list of selected terms that you will encounter in the profession, most of which have featured elsewhere in this book. Please note, however, that I could have listed four hundred if I had wanted to, so this list is far from exhaustive!

aims
Top alphabetically, and top in importance. Your aims are your goals, what you want to achieve and what you want your students to achieve, both in the lesson as a whole and in each part (**stage**) of a lesson. You can have main aims, and secondary aims. Try to think of your aims from the point of view of what students will do/learn, rather than what the teacher will do/teach.

appropriateness
This covers a number of points. In terms of *language*, what we say may be technically correct, but not appropriate to the context. It may be too formal, for example. In terms of *material*, what we ask our students to do / read / listen to etc must be appropriate according to the culture and background of the group, as well as their age, sex, general interests, the course and the lesson aims.

authentic
Authentic material is real-life material, such as a newspaper or brochure. It has not been altered for classroom use, but the accompanying tasks will make it appropriate for the level and needs of the class. Authentic language is real, unscripted and ungraded language, such as a news broadcast from the radio.

clarification
This is when you want to check that students have understood meaning (see **concept questions** and **timelines**), or when you need to go over something which has not been understood. It is also the

checking of instructions and tasks. Avoid questions like "Do you understand?" because this is not an effective way to clarify.

closed pairs
This is when students work in pairs with someone sitting directly next to them (c/f **open pairs**)

concept questions
One of the best ways to clarify/check that students have understood an item of language is to ask concept questions. These are short questions or prompts, usually requiring very short and simple answers. They check understanding of the concept (meaning) of language. When checking understanding of structures within a sentence, avoid using the structure in your concept questions.

contractions
When the subject and auxiliary verb are 'joined', e.g. *She's* (*She is*).

controlled practice
When students do a practice activity where the nature of the activity means there is a tight control over the language that they use. The teacher restricts the students' language so that the target language is used a lot. Written gap-fills are a good example.

context
The context is the situation, and the situation provides meaning. Context could be in a text, a recording, pictures, spoken scenario from the teacher etc. Language must have context because language must have meaning.

diagnostic
A diagnostic test is to find out the level of a student; how much she knows. Within a lesson, a diagnostic task (**test teach test**) is to see how much students already know about a piece of language being focused on in your lesson.

drilling
This is when students repeat a piece of language with the aim of pronouncing it correctly and naturally. They do this having first heard a model to copy. *Choral* drilling is when they all say it together (confidence-building), and *individual* drilling is when they say it by themselves. There are different types of drill, but the basic repetition drill is probably the most common type used by teachers. (See also **front chaining**.)

eliciting
Try to get the students to tell you as much as possible, rather than you telling them everything. Ask them questions, find out what they already know, involve them. It relates to keeping the students actively involved in the lesson and participating fully.

feedback (plenary)
If students have done a task together, get some reaction afterwards as a class. For example, if they have been discussing preferred holiday destinations, find out afterwards what some of them decided. If students have been working on an exercise together, you have to go through the answers with the class afterwards and clarify as required. 'Plenary' is 'whole class feedback'

front chaining / back chaining
When drilling a long or otherwise difficult sentence, it can help students if you 'break it down' from the *front* or the *back*. For example: "I'm getting used to driving on the left" becomes "on the left"... "to driving on the left"... "getting used to driving on the left" ... "I'm getting used to driving on the left" (back chaining). Each part is drilled in turn with the whole sentence being repeated at the end.

function
There is always a reason for saying something – all speech has a function. Examples of functions are invitation, suggestion, apology, refusal, criticism, praise, complaint, deduction, request, offer, giving permission and order. Functions can be illustrated via **context**.

gist
When reading a text or listening to a recording, students can be asked to listen for the gist – what the text or recording is about; the general sense or meaning of it.

illustrating meaning
Rather than *explaining* what something means, teachers should illustrate meaning in some way – via context, visuals, mime, realia etc. In other words, the teacher should *show* meaning, not explain it. This is also known as **conveying** meaning.

L1
The student's first language or 'mother tongue'.

language grading
The language teachers use in the classroom must be of a level which will be understood by the students, without it becoming overly simple. Obviously, the lower the level of the students, the more graded the teacher's language has to be. Speaking more slowly is not sufficient if the language being used is too difficult to understand.

lexis
Another word for vocabulary.

mingling
When students get up and move around (as with a "Find someone who..." exercise). With very large classes, and especially when they are sitting behind desks, mingling is not a realistic option.

model sentence
This is a sentence in context which includes a stucture which the teacher wishes to focus on. It allows the structure to be looked at within a meaningful sentence rather than in isolation. For example: *I'm leaving tomorrow morning on the first flight* contains 'subject + *to be* + verb-*ing*' (present progressive). It is sometimes also known as a **marker sentence**.

monitoring
When students are working together, the teacher should go round as best she can and observe their work and provide help and support. Monitoring also involves 'keeping an eye' on the class, looking to see if someone has finished a task, for example. Even when the class and teacher are working together, the teacher should maintain eye contact with the students and observe reactions and problems. Watching and reacting is a key teaching skill.

monolingual classes
Classes where all the students are of the same nationality. Unless you are teaching in an English-speaking country, most classes you teach will be of this type.

MPF
The meaning, pronunciation, and form of a piece of language.

multilingual classes
Classes with more than one nationality.

open pairs
If one student interacts with another student in the room, but not one sitting next to them, this is **open pairs**. This might occur, for example, when one student is asked by the teacher to ask a question of another student in the room.

productive skills
Speaking and Writing – where the students must *produce*.

receptive skills
Listening and Reading – where students *receive* and comprehend.

rapport
Teachers need to build a good relationship with their students, but also foster good relations between the students. This is **rapport**. A good classroom atmosphere and good relations between all in the class is essential.

schwa
The only sound in the language with its own name, the schwa / ə / is the most common sound in English.

situational presentation
Typically, this is when the teacher provides a situation (**context**) in order to focus on an item of language contained within a model sentence. It is a relatively teacher-centred means of introducing language, but students are involved through the teacher eliciting from them in order to build the context.

scan reading
Reading in order to find specific information without needing to focus on or understand other parts of the text.

skim reading
Reading for overall gist.

stage
A stage in a lesson is a separate part of it. Each stage should have its own aim (objective).

STT
Student Talking Time.

structure
Some sentences will have a fixed grammatical structure (invariable) which may be a tense (*I'll have gone by 3pm* contains

'subject + *will have* + past participle' = future perfect simple); or maybe not (*I used to smoke* contains 'subject + *used to* + base verb'. When asking concept questions to check understanding, avoid using the structure in the question.

target language
This is the language being focused on in a lesson.

test teach test
This is a way of focusing on language in a lesson which first involves some diagnostic task; then the teacher clarifies meaning; then the students do practice tasks.

timelines
The visual representation of time; very useful with tenses.

TTT
Teacher Talk Time – which shouldn't be unnecessarily high!

warmers
Short activities used at the start of a lesson which are not necessarily related to what is to follow. They are useful to wake a class up, give latecomers more time to arrive etc. Such activities could also be used as *Fillers* when you have time on your hands at the end of your lesson, or during a lesson to vary the pace. Activities can include simple tasks like "Think of ten adjectives beginning with c"; or "Find out what everyone in the room had for dinner last night."

weak forms
Typically these are vowel sounds in unstressed syllables. In '*Can* you help me?', the word *can* is pronounced /kən/.

30 SOME USEFUL ADDRESSES

These are just a few selected addresses you might find useful now, or later.

United Kingdom

UCLES (University of Cambridge Local Examinations Syndicate)
CILTS Unit
1 Hills Road
Cambridge
CB1 2EU
England
tel: 01223 553311 extension CILTS Unit

IATEFL (International Association of Teachers of English as a
Foreign Language)
3 Kingsdown Park
Tankerton
Whitstable
Kent
CT5 2DJ
England
tel: 01227 276528

KELTIC Bookshop
25 Chepstow Corner
Chepstow Place
London W2 4XE
England
tel: 020 7229 8560

Trinity College
London External Exam Board
16 Park Crescent
London
W1N 4AH
England
tel: 020 7323 2328

Saxoncourt Recruitment (Shane Schools)
59 South Moulton Street
London
W1Y 1HH
England
tel: 020 7499 8533

English Worldwide
The Italian Building
Dockhead
London
SE1 2BS
England
tel: 020 7252 1402

VSO (Voluntary Services Overseas)
317 Putney Bridge Road
London
SW15 2PN
England
tel: 020 8780 7527

ARELS (Association of Recognised English Language Services)
2 Pontypool Place
Valentine Place
London
SE1 8QF
England
tel: 020 7242 3136

British Council
10 Spring Gardens
London
SW1A 2BN
England
e-mail: www.britishcouncil.org/jobs

EL Gazette / EL Prospects
Dilke House
1 Malet Street
London
WC1E 7JA
England
tel: 020 7255 1969

United States

JET (Japan Exchange and Teaching) Program
Japanese Embassy
2520 Massachusetts Ave NW
Washington, DC 20008
USA
tel: 202 939 6722

Peace Corps
Room 8500
1990 K St NW
PO Box 941
Washington, DC 20526
USA
tel: 202 606 3780

TESOL (Teaching of English to Speakers of Other Languages)
700 South Washington Street
Suite 200
Alexandria
Virginia 22314
USA
tel: 703 836 0774

American Language Academy
1401 Rockville
MD 20852 A080
USA

TESOL Placement Agency
address as above

ECIS (European Council of International Schools)
– North America
105 Tuxford Terrace
Basking Ridge
New Jersey 07920
USA

tel: 1 908 903 0552

AASA (Association of American Schools in South America)
14750 NW 77 Court
Suite 210
Miami Lakes
Florida 33016
USA

Internet Sites

Please note that I have been unable to check each of these addresses personally, but each has come recommended! Otherwise, do a search for 'efl' / 'esl' / 'tefl' / 'tesol' and you are sure to get a huge selection of sites!

www.english-international.com (especially recommended to check out)
www.eslcafe.com
www.aaaefl.co.uk
www.jobs.edunet.com/
www.teflnetnation.com/
www.globalesl.net
www.britishcouncil.org/jobs
www.ihworld.com
www.tefl.net
www.tefl.com
www.jobs.tes.co.uk
www.educationunlimited.co.uk/tefl
www.saxoncourt.com
www.edunet.com/english/grammar/index.cfm
www.dictionary.com

INDEX